Unified Framework of Control and Detection

Steven X. Ding

Unified Framework of Control and Detection

Applications to Fault Diagnosis, Tolerant Control, and Secure CPSs

Springer

Steven X. Ding
Institute for Automatic Control
and Complex Systems
University of Duisburg-Essen
Duisburg, Germany

ISBN 978-3-662-72416-3 ISBN 978-3-662-72417-0 (eBook)
https://doi.org/10.1007/978-3-662-72417-0

This Springer imprint is published by the registered company Springer-Verlag GmbH, DE, part of Springer Nature.
The registered company address is: Heidelberger Platz 3, 14197 Berlin, Germany

To My Parents and Eve Limin

Preface

This book introduces a unified framework for control and detection that brings together concepts from the unified design of control and detection systems, an active research area in the 1990s, and extends them into a comprehensive methodology for analysing and synthesising control, fault diagnosis, and fault-tolerant control systems. The framework departs from conventional input-output modeling by representing system dynamics as subspaces in the space of process input-output signals, where nominal and uncertain system behaviours are modelled by distinct subspaces. A one-to-one mapping from process input-output signals to residual and latent variables forms the foundation of this approach, enabling consistent analysis of control and detection behaviour across a wide spectrum of dynamic systems and supporting the uniform and optimal design of advanced control and detection schemes. This signal subspace based formulation provides an effective means of addressing uncertainties and faults within the residual subspace while simultaneously creating a common platform that integrates traditionally independent research paradigms: model-based, data-driven, and machine learning-aided methods for control and fault diagnosis. By bridging these domains, the framework fosters new possibilities for innovative and cohesive design methodologies. A particularly impactful application is in secure cyber-physical systems (CPSs), where the effects of cyber-attacks and technical faults are modelled in distinct subspaces, thereby enabling simultaneous fault and attack detection as well as unified strategies for fault-tolerant and resilient control.

The book is structured in three parts. Part I lays the conceptual and computational groundwork, beginning with essential control-theoretic and mathematical preliminaries, followed by the introduction of key definitions and principles of the framework, and concluding with computational methods and algorithms that support its implementation. The part closes with an outlook on future applications and emerging research directions that build upon the framework's unifying capabilities. Part II develops advanced methods to address uncertainties in control systems. After a concise review of established fault diagnosis and control techniques, it presents analysis and synthesis methods for diagnosis, control, and estimation, with Chaps. 9–11 focusing on projection-based detection and estimation as powerful tools within the framework for handling uncertainties. Part III applies the framework to the domain of cyber-security, proposing systematic approaches to analysing industrial cyber-physical systems under cyber-attacks, detecting cyber-attacks in CPSs, and enabling simultaneous detection of faults and attacks alongside unified approaches to fault-tolerant and resilient control.

This book would not have been possible without the support of many individuals. I am deeply indebted to Prof. Dr.-Ing. L. Li from the University of Science and Technology Beijing for long-term collaboration and extensive editorial contributions, and I sincerely thank Ms. S. Ding and Ms. X. Liu for their meticulous proofreading of the chapters.

Finally, I would like to express my appreciation to the editorial team at Springer-Verlag for their valuable support and professional management.

Duisburg, Germany Steven X. Ding
July 2025

Competing Interests The author has no competing interests to declare that are relevant to the content of this manuscript.

Contents

Notations

$\forall$	For all
$\in$	Belong to
$\subset$	Subset
$\cup$	Union
$\cap$	Intersection
$\oplus$	Direct sum
$\approx$	Approximately equal
$A := B,\ B =: A$	A is defined as B
$\Rightarrow$	Implies
$\Leftrightarrow$	Equivalent to
$>>\ (<<)$	Much greater (less) than
max (min)	Maximum (minimum)
sup (inf)	Supremum (infimum)
$\lceil x \rceil$	Ceiling function
$\mathbb{R}$ and $\mathbb{C}$	Field of real and complex numbers
$\mathbb{R}^n$ and $\mathbb{C}^n$	Space of real and complex n-dimensional vectors
$\mathbb{R}^{n\times m}$ and $\mathbb{C}^{n\times m}$	Space of n by m real and complex matrices
$\mathcal{L}_2(-\infty, \infty)(\mathcal{L}_2)$	Signal space of all signals with bounded energy
$\mathcal{L}_2[0, \infty)(\mathcal{H}_2) \subset \mathcal{L}_2$	Signal space of all signals defined in $[0, \infty)$
$\mathcal{L}_2(-\infty, 0] \subset \mathcal{L}_2$	Signal space of all signals defined in $(-\infty, 0]$
$\mathcal{H}_\infty(\mathcal{H}_\infty^{n\times m})$	Function space of all transfer functions of stable systems (n by m-dimensional transfer function matrices of stable systems)
$\mathcal{RH}_\infty(\mathcal{RH}_\infty^{n\times m})$	Space of all rational transfer functions of stable systems (n by m-dimensional rational transfer function matrices)
$\mathcal{L}_\infty$	Lebesgue space of all functions essentially bounded on the imaginary axis
X^T	Transpose of matrix X
$X^\sim$	Conjugate transpose of (complex) matrix X
$X^\perp$	Orthogonal complement of matrix X
X^{-1}	Inverse of matrix X

X^{+}	Pseudo-inverse of matrix X
X^{-}	Left or right inverse of matrix X
$diag(X_{1,} \cdots, X_n)$	Block diagonal matrix formed with $X_{1,} \cdots, X_n$
$X(i:j,\ p:q)$	Submatrix consisting of the i-th to the j-th rows and the p-th to the q-th columns of matrix X
$rank(X)$	Rank of matrix X
$tr(X)$	Trace of matrix X
$\det(X)$	Determinant of matrix X
$\lambda(X)$	Eigenvalue of matrix X
$\bar{\sigma}(X)$ $(\sigma_{\max}(X))$	Largest (maximum) singular value of matrix X
$\underline{\sigma}(X)$ $(\sigma_{\min}(X))$	Least (minimum) singular value of matrix X
$\sigma_i(X)$	The i-th singular value of matrix X
$\Vert \cdot \Vert$	Euclidean norm of a vector
$\Vert \cdot \Vert_F$	Frobenius norm of a matrix
$\mathcal{L}_2$-bounded	Set of signals with bounded energy
$G(z)$	Transfer matrix
$G^{\sim}(z)$, $G^{\sim}(e^{j\theta})$	Conjugate of $G(z)$, $G(e^{j\theta})$
$\mathcal{P}_{\mathcal{V}}$	Operator of projection onto $\mathcal{V}$
$\mathcal{P}_{\mathcal{V}}^{*}$	Adjoint of operator $\mathcal{P}_{\mathcal{V}}$
$(A,\ B,\ C,\ D)$	Shorthand for the state space representation
$rank(G(z))$	Normal rank of $G(z)$
$\Vert G \Vert_2$	$\mathcal{H}_2$ norm of (stable) transfer function matrix G
$\Vert G \Vert_\infty$	$\mathcal{H}_\infty$ norm of (stable) transfer function matrix G
$Pr(a < b)$	Probability that $a < b$
$\mathcal{N}(a, \Sigma)$	Gaussian distribution with mean vector a and covariance matrix Σ
$x \sim \mathcal{N}(a, \Sigma)$	x is distributed as $\mathcal{N}(a, \Sigma)$
$\chi^2(m)$	Chi-squared distribution with m degrees of freedom
$\mathbb{E}(x)$	Expectation of x

Chapter 1
Introduction

1.1 Background

Fault diagnosis and tolerant control are essential parts of today's automation control systems. Handling of uncertainties remains a challenging problem in dealing with fault diagnosis and control, particularly in the context of modern industrial automation, where systems operate under feedback control loops, dynamic setpoints, and process uncertainties. Let us illustrate some examples without rigorous control-theoretic handlings.

Example 1.1 Consider a feedback control loop under the influence of unknown inputs and faults,

$$\begin{cases} y = G_u u + G_d d + G_f f \\ u = K y + v, \end{cases} \tag{1.1}$$

where $\left(G_u, G_d, G_f\right)$ are stable transfer functions, K is the transfer function of the controller, (u, y) are the plant input and output vectors, v is the reference signal, and (d, f) are disturbance and fault vectors, respectively. The closed-loop dynamic is governed by

$$y = (I - G_u K)^{-1} \left(G_u v + G_d d + G_f f\right). \tag{1.2}$$

It is an understandable view that a robust controller that rejects the influence of the unknown input vector d reduces simultaneously the system sensitivity to the fault vector f. On the other hand, generating a residual vector

$$r = R\left(y - G_u u\right)$$

yields

$$r = R\left(G_d d + G_f f\right). \tag{1.3}$$

S. X. Ding, *Unified Framework of Control and Detection*,
https://doi.org/10.1007/978-3-662-72417-0_1

Hereby, R, called post-filter, is a dynamic system that is to be set to detect the fault vector f under the existence of d. This is a standard fault detection problem. Comparing the system dynamics (1.2) and (1.3) illustrates that the (feedback) controller and fault detector design can be well performed separately. It is noteworthy that

- both systems (1.2) and (1.3) are driven by the same signal

$$r_0 = y - G_u u = G_d d + G_f f,$$

  which is in fact a residual signal carrying information about uncertainties in the system dynamics. In our case, they are d and f,
- although both systems are driven by the same signal r_0, the design of a feedback controller and fault detector can be addressed separately, thanks to the well-known separation principle.

Example 1.2 Now, the system model (1.1) is extended to

$$\begin{cases} y = G_{u,\Delta} u + G_d d + G_f f, G_{u,\Delta} = G_u + \Delta G_u \\ u = K y + v, \end{cases} \tag{1.4}$$

in order to take into account of uncertainties caused by possible parameter variations ΔG_u. It is interesting to observe that the residual signal $r_0 = y - G_u u$ is given by

$$r_0 = \Delta G_u K y + \Delta G_u v + G_d d + G_f f, \tag{1.5}$$

and the system output can be expressed by

$$y = G_u u + r_0. \tag{1.6}$$

Note that the system dynamics of (1.4) and (1.6) are equivalent. In our subsequent work, the model form (1.6) is called kernel-based input-output model. It follows from (1.6) that the closed-loop dynamic is subject to

$$y = (I - G_u K)^{-1} (r_0 + G_u v), \tag{1.7}$$

which can be further written into

$$r_0 = (I - G_u K)(I - G_u K - \Delta G_u K)^{-1} \begin{pmatrix} \Delta G_u K (I - G_u K)^{-1} v \\ +\Delta G_u v + G_d d + G_f f \end{pmatrix}, \tag{1.8}$$

$$y = (I - G_u K - \Delta G_u K)^{-1} \begin{pmatrix} \Delta G_u K (I - G_u K)^{-1} v \\ +\Delta G_u v + G_d d + G_f f \end{pmatrix} + (I - G_u K)^{-1} G_u v. \tag{1.9}$$

Again, it can be observed that both the closed-loop (1.7) and the fault detector

$$r = R\left(y - G_u u\right)$$

are driven by the residual signal r_0, which comprises full information about the uncertainties in the system. On the other hand, the mutual interference of the uncertainties, reference signal, unknown input and fault considerably complicates the system analysis and design.

From the above two examples, it becomes plain that the existence of process uncertainties makes analysis and synthesis of control and diagnosis of automatic control systems considerably difficult and challenging. The crucial point is the inherent coupling of the nominal dynamics and process uncertainties. On the other hand, the residual signal r_0 seems capable of representing the process uncertainties and useful for system analysis and synthesis.

In engineering applications, uncertainties are a crucial concern. They may be present in various forms. For instance, aging of components in an automatic control system, or in a machine and an asset causes variations in system parameters. It is inevitable, as machines and assets operate over years. Uncertainties like aging are generally a longtime process. They begin with incipient variations that do not affect the system dynamics and performance significantly. In such transitional phase, the variations are interpreted as uncertainties, namely, the tolerable variations from the nominal set-up and setting parameters, which result in incipient degradation of the system performance. As the variations become larger and the performance degradation approaches the tolerable margin, they should be timely detected and identified as faults, in order to prevent critical and even potentially dangerous or unsafe system operations. From the modelling viewpoint, parameter variations caused by such aging processes are typically modelled as parameter uncertainties for incipient variations or multiplicative faults when they endanger system operations. It should be emphasised that they are well characterised by their longtime and incipient process, significantly distinguishing them from those actuator or sensor faults typically handled in the literature as abrupt changes modelled in the form of additive faults. Therefore, it is reasonable to model such parameter variations uniformly as multiplicative uncertainties. On the other hand, it is of practical importance to distinguish incipient variations, i.e. uncertainties that slightly affect the system dynamics and performance, and variations leading to performance degradation approaching the tolerable margin, namely faults.

Roughly speaking, control, detection and estimation schemes are based on the system input and output dynamic, including its state space representation. When uncertainties, particular in the multiplicative form, are present in the system under consideration, the signal space of the output variable y is fully spanned by the uncertain signals, as showcased in the above two examples. It is hard to distinguish the nominal dynamics from the influence of uncertainties, and thus makes system classification, analysis and synthesis considerably complicated.

In recent years, machine learning (ML) based methods have drawn remarkably increasing attention in both academic and industrial fields. One of the most popular application areas of ML methods is classification, to which fault detection as a typical one-class classification problem belongs. Towards this end, intensive research efforts have been made in representation learning. Roughly speaking, the basic idea behind ML-based fault detection methods lies in the reconstruction of process variables in the nominal process operation state. We observe that in the ML-based classification research domain, uncertainties are often handled in an obviously different manner. It is supposed that the nominal system behaviour lies in a lower-dimensional subspace or manifold. Uncertainties (noise, disturbances, parameter variations) cause data to spread into a higher-dimensional space. In order to recover the intrinsic behaviour of the system, techniques like dimensionality reduction methods help to remove redundancy in the data. Consequently, the lower-dimensional subspace modelling the nominal system behaviour is reconstructed. Inspired by this practice, we now re-visit our examples.

Example 1.3 We firstly consider the system model (1.1) given in Example 1.1. The nominal system dynamic $y = G_u u$ forms a subspace in the signal space spanned by (u, y). To be specific, the relation

$$\begin{bmatrix} -G_u & I \end{bmatrix} \begin{bmatrix} u \\ y \end{bmatrix} = 0$$

defines a kernel subspace

$$\mathcal{K}_{G_u} = \left\{ \begin{bmatrix} u \\ y \end{bmatrix}, \begin{bmatrix} -G_u & I \end{bmatrix} \begin{bmatrix} u \\ y \end{bmatrix} = 0 \right\} \subset \mathcal{S}$$

in the system signal space $\mathcal{S}$,

$$\mathcal{S} = \left\{ \begin{bmatrix} u \\ y \end{bmatrix}, (u, y) \text{ are generated by the system (1.1)} \right\}.$$

Note that

$$\dim \left(\mathcal{K}_{G_u}\right) = \dim(y) =: m < \dim\left(\mathcal{S}\right) = m + p, \, p := \dim(u).$$

Let

$$\mathcal{I}_{G_u} = \left\{ \begin{bmatrix} u \\ y \end{bmatrix}, \begin{bmatrix} u \\ y \end{bmatrix} = \begin{bmatrix} (I - KG_u)^{-1} \\ G_u (I - KG_u)^{-1} \end{bmatrix} v \right\} \subset \mathcal{S}$$

be an image subspace in $\mathcal{S}$. Because

$$\begin{bmatrix} -G_u & I \end{bmatrix} \begin{bmatrix} (I - KG_u)^{-1} \\ G_u (I - KG_u)^{-1} \end{bmatrix} = 0,$$

the kernel subspace $\mathcal{K}_{G_u}$ is equivalent to the image subspace $\mathcal{I}_{G_u}$. $\mathcal{I}_{G_u}$ is an alternative model of the nominal system dynamic. We now recast the closed-loop system dynamic (1.2) in the signal space $\mathcal{S}$ as follows,

$$\begin{bmatrix} u \\ y \end{bmatrix} = \begin{bmatrix} (I - KG_u)^{-1} \\ G_u (I - KG_u)^{-1} \end{bmatrix} v + \begin{bmatrix} K (I - G_u K)^{-1} \\ (I - G_u K)^{-1} \end{bmatrix} r_0. \tag{1.10}$$

Hereby,

$$\begin{bmatrix} (I - KG_u)^{-1} \\ G_u (I - KG_u)^{-1} \end{bmatrix} v \in \mathcal{I}_{G_u}$$

and represents the nominal system dynamic, while

$$\begin{bmatrix} K \\ I \end{bmatrix} (I - G_u K)^{-1} r_0 \in \mathcal{I}_K,$$

$$\mathcal{I}_K := \left\{ \begin{bmatrix} u \\ y \end{bmatrix}, \begin{bmatrix} u \\ y \end{bmatrix} = \begin{bmatrix} K \\ I \end{bmatrix} (I - G_u K)^{-1} r_0 \right\} \subset \mathcal{S}$$

describes the influence of the uncertainty on the closed-loop dynamic. It is apparent that

$$\mathcal{I}_K \neq \mathcal{I}_{G_u} \text{ and } \mathcal{S} = \mathcal{I}_{G_u} \oplus \mathcal{I}_K.$$

Next, consider the system model (1.4) given in Example 1.2. It follows from (1.4) and (1.6) immediately that

$$\begin{bmatrix} u \\ y \end{bmatrix} = \begin{bmatrix} (I - KG_u)^{-1} \\ G_u (I - KG_u)^{-1} \end{bmatrix} v + \begin{bmatrix} K \\ I \end{bmatrix} (I - G_u K)^{-1} r_0, \tag{1.11}$$

where r_0 is given by (1.8) and contains the impact of the uncertainties on the closed-loop system dynamic.

Equations (1.10) and (1.11) impressively demonstrate that the system dynamics of nominal and uncertain operations, including faulty operations, can be modelled by means of two different subspaces in the signal space (u, y), where the residual signal r_0 plays an essential role. Moreover, it is of interest to observe that

$$\begin{bmatrix} -G_u & I \end{bmatrix} \begin{bmatrix} K \\ I \end{bmatrix} (I - G_u K)^{-1} = I, \begin{bmatrix} I & -K \end{bmatrix} \begin{bmatrix} (I - KG_u)^{-1} \\ G_u (I - KG_u)^{-1} \end{bmatrix} = I$$
$$\Longrightarrow \begin{bmatrix} v \\ r_0 \end{bmatrix} = \begin{bmatrix} I & -K \\ -G_u & I \end{bmatrix} \begin{bmatrix} u \\ y \end{bmatrix},$$
$$\begin{bmatrix} I & -K \\ -G_u & I \end{bmatrix} \begin{bmatrix} (I - KG_u)^{-1} & -K (I - G_u K)^{-1} \\ G_u (I - KG_u)^{-1} & (I - G_u K)^{-1} \end{bmatrix} = I.$$

These relations, in particular the one-to-one mapping between the process data (u, y) and the signals (v, r_0), are fundamental in our study. In the context of the image subspaces $\mathcal{I}_{G_u}$ and $\mathcal{I}_K$, (v, r_0) are called latent variables as well.

Modelling of the system nominal and uncertain dynamics as different subspaces or manifolds in the signal space of the system input and output (u, y) is the foundation of the unified framework of control and detection. On this basis, analysis and synthesis issues of control and diagnosis systems are addressed, and alternative and even more capable methods are developed.

In the ML research domain, projection methods are widely applied for dimensionality reduction. They are a standard technique in statistic learning and data-driven classification. The most popular algorithms are principal component analysis (PCA) and partial least squares (PLS) algorithms. Thanks to their uniform formulation and clear geometric interpretation of projection onto subspaces, PCA and PLS are not only widely accepted in industry, but also recognised as a conceptional basis for the development of advanced fault diagnosis methods. It is worth mentioning that, in order to deal with fault detection issues in nonlinear processes, some known manifold learning methods have been applied to fault detection, the so-called orthogonal locality preserving projection is a representative one among them. Such manifold learning methods mainly focus on dimensionality reduction based fault diagnosis.

To address and handle analysis and synthesis of control and diagnosis systems in different signal subspaces in the (u, y) space, projection methods promise a capable tool. To this end, endeavours are made to explore projection methods and develop projection-based diagnosis and control schemes for linear and nonlinear systems. Projection-based methods will serve as the major tool in the unified framework of control and detection. In this regard, we notice a further class of projection methods, the so-called autoencoder (AE) technique, an efficient ML-based tool for dimensionality reduction and feature learning. As showcased in Fig. 1.1, an AE consists of an encoder that compresses the process data (u, y) into the latent (hidden) variable v, and a decoder driven by the latent variable reconstructs the process variables $(\hat{u}, \hat{y})$ reflecting nominal system operations. Using neural networks (NNs), the encoder (NN_{en}) and decoder (NN_{de}) are learnt in such a way that the latent variable contains exclusive information (features) of the nominal systems operations, from which the process variables (u, y) in the nominal state are then fully reconstructed. How to learn such an ideal latent variable is the challenging issue in this learning process. Inspired by this idea, our endeavours to leverage projection methods to deal with control and detection issues will focus on optimising the projection system that is config-

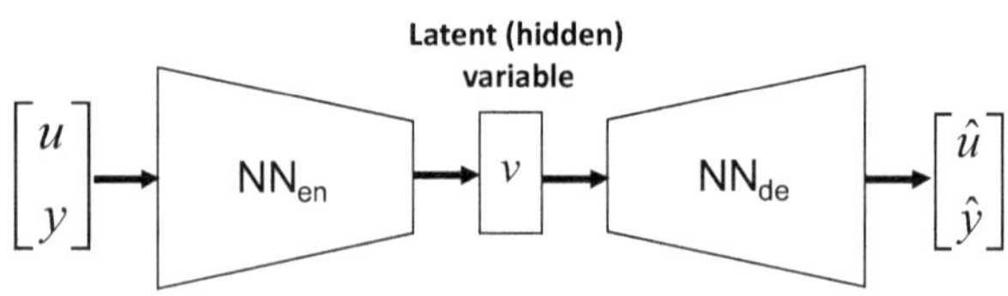

Fig. 1.1 Schematic description of an autoencoder

ured like an AE so that the latent variable contains exclusive information needed for reconstructing the target variables to be estimated. In the literature on representation learning, the concept of information bottleneck has been introduced, which suggests that all the information from the input variables required in the reconstructing (estimating) target should be contained in the learned latent representation in the NNs. In the information-theoretic framework, it has been proposed to generate latent variable as the so-called minimal sufficient statistic. For our purpose, methods of the control-theoretic counterpart of such a variable and its design will be investigated.

We now attend to the role of the residual signal, an essential variable in the unified framework. In the model-based fault diagnosis framework, residual generation is the centrepiece of any fault detection systems. Explicitly embedding residual generation in a control system was rarely researched. The development of the unified framework of control and detection has been motivated and driven by the endeavours to achieve unified design of control and detection systems and implementation of feedback control systems with an embedded residual generation. The basic idea of this work is the cognition that feedback control and fault detection share the common information basis, the residual signal, as demonstrated in the aforementioned examples. Below, we present a further example to reveal the interconnection of feedback control and residual generation, and showcase the implementation of feedback control systems with an embedded residual generation.

Example 1.4 We consider a trivial feedback control loop

$$\begin{cases} y = G_u u + G_d d + G_f f \\ u = K(v - y), \end{cases} \tag{1.12}$$

where the plant dynamic is modelled by (1.1). The closed-loop dynamic is governed by

$$y = G_u \left(I + KG_u\right)^{-1} Kv + \left(I + G_u K\right)^{-1} \left(G_d d + G_f f\right). \tag{1.13}$$

Assume that the closed-loop system is stable, i.e.

$$\left(I + KG_u\right)^{-1} K, \left(I + G_u K\right)^{-1}$$

are stable. Let

$$Q = \left(I + KG_u\right)^{-1} K,$$

which yields

$$\begin{aligned} K &= Q \left(I - G_u Q\right)\right)^{-1} \Longrightarrow u = \left(I - QG_u\right)\right)^{-1} Q(v - y) \\ &\Longrightarrow u = Qv + Q \left(G_u u - y\right) = Qv - Qr_0. \end{aligned} \tag{1.14}$$

The control law (1.14) is the well-known internal model control (IMC). It is of interest to notice that although the IMC system,

$$\begin{cases} y = G_u u + G_d d + G_f f \\ u = Qv + Q\left(G_u u - y\right) = Qv - Qr_0, \end{cases} \tag{1.15}$$

is derived from the control system (1.12), the IMC scheme is distinct from the control scheme (1.12) in the configuration, realisability and, in particular cases, also the performance. Specifically, the IMC scheme comprises a residual generator, $r_0 = y - G_u u$. Furthermore, all sub-systems in the IMC system, $\left(Q, G_u, G_d, G_f\right)$ are stable, while the controller K in the control scheme (1.12) can be an unstable transfer function. Concerning the control performance, comparing the loop dynamic of the control system (1.12) with that of IMC system given by

$$\begin{aligned} y &= G_u Q\left(v - r_0\right) + G_d d + G_f f \\ &= G_u Q v + \left(I - G_u Q\right)\left(G_d d + G_f f\right) \end{aligned}$$

demonstrates that in the case that G_u is invertible with a stable inverse, the IMC system delivers a perfect control performance

$$Q = G_u^{-1} \Longrightarrow y = G_u Q v = v.$$

In the literature, successful applications of the unified design of control and detection systems to fault diagnosis in feedback control systems and fault-tolerant control have been reported. Most of these investigations have been conducted on the basis of process input-output representations. In the unified framework of control and detection, study on analysis and design of control and detection systems as a whole and issues of fault-tolerant control will be continued as a major thematic field. Analysis of the system behaviour in various system signal subspaces and projection-based methods will serve as the main research instrument.

With the increasing complexity of automatic control systems, from single-loop feedback control systems to networked control systems and nowadays cyber-physical systems (CPSs), maintaining reliable and safe system operations is of elemental importance for optimally managing industrial complex industrial systems over the whole operation life-cycle. In recent years, automatic control systems as the centrepiece of CPSs become a main target of cyber-attacks. In this context, management of cyber-attacks imposes, besides functionality maintenance, cyber security demands in the framework of monitoring, diagnosis and tolerant control. To this end, remarkable research endeavours have been made to extend the well-established observer-based detection schemes, for instance, integrating a moving target or an auxiliary system or injecting watermark signals. On the other hand, it can be observed that little attention has been paid to technical specifications and configurations of the controllers and control systems embedded in the CPS under consideration. Our recent work on the application of the unified framework of control and detection to analysis of CPSs under cyber-attacks and detection of cyber-attacks has highlighted new insightful aspects that are useful for fault and attack detection as well as tolerant control. Below is an example for a better understanding of the related issues.

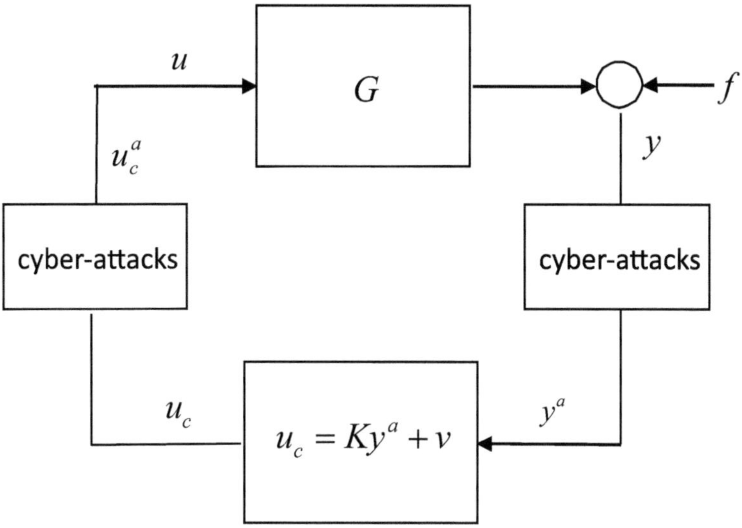

Fig. 1.2 A control system embedded in a CPS under cyber-attacks

Example 1.5 Consider a control system of a part of an industrial CPS control system shown in Fig. 1.2 with the plant model

$$y = Gu + f,$$

where f represents the fault vector. The plant is connected with the controller

$$u_c = Ky^a + v$$

over a communication network. Here, y^a denotes the plant output y under cyber-attacks. The plant input under cyber-attacks is denoted by u_c^a, and it holds

$$u = \begin{cases} u_c, & \text{attack-free} \\ u_c^a, & \text{under cyber-attacks.} \end{cases}$$

We firstly consider the closed-loop dynamic under cyber-attacks modelled by

$$u_c^a = u_c + a_{u_c}, \; y^a = y + a_y$$

with $\left(a_{u_c}, a_y\right)$ denoting the additive type of cyber-attacks. Similar to Example 1.3, it turns out by some routine computations,

$$\begin{bmatrix} u \\ y \end{bmatrix} = \begin{bmatrix} (I - KG)^{-1} \\ G(I - KG)^{-1} \end{bmatrix} (v + a) + \begin{bmatrix} K \\ I \end{bmatrix} (I - GK)^{-1} r_0, \tag{1.16}$$

$$a = Ka_y + a_{u_c}, r_0 = f. \tag{1.17}$$

Now, suppose that cyber-attacks are modelled by

$$u_c^a = \Delta_u u_c, \; y^a = \Delta_y y,$$

where $\left(\Delta_u, \Delta_y\right)$ represent some unknown and stable dynamic systems that manipulate the system input and output signals. They are called multiplicative cyber-attacks. Hence, the system dynamic under cyber-attacks is described by

$$\begin{cases} y = G\Delta_u u_c + f \\ u_c = K\Delta_y y + v. \end{cases}$$

The closed-loop dynamic is governed by

$$\begin{bmatrix} u \\ y \end{bmatrix} = \begin{bmatrix} (I - K_\Delta G)^{-1} \\ (I - GK_\Delta)^{-1} G \end{bmatrix} \Delta_u v + \begin{bmatrix} K_\Delta \\ I \end{bmatrix} (I - GK_\Delta)^{-1} \Delta_y f, \tag{1.18}$$

$$K_\Delta = \Delta_u K \Delta_y. \tag{1.19}$$

Example 1.5 showcases that

- the additive cyber-attacks affect the system image subspace $\mathcal{I}_G$, while the fault acts on the image subspace $\mathcal{I}_K$, as described by (1.16),
- the multiplicative cyber-attacks can be equivalently modelled as uncertainties in the feedback controller, namely $K_\Delta = \Delta_u K \Delta_y$.

Moreover, notice that for residual generation, in the case of additive cyber-attacks,

$$r_0 = \begin{bmatrix} -G_u & I \end{bmatrix} \begin{bmatrix} u \\ y \end{bmatrix} = f,$$

$$\begin{bmatrix} I & -K \end{bmatrix} \begin{bmatrix} u \\ y \end{bmatrix} = \begin{bmatrix} I & -K \end{bmatrix} \begin{bmatrix} (I - KG)^{-1} \\ G(I - KG)^{-1} \end{bmatrix} (v + a) = v + a.$$

That means, the output residual signal r_0 is decoupled from the cyber-attacks, while the input residual signal defined by

$$r_u := \begin{bmatrix} I & -K \end{bmatrix} \begin{bmatrix} u \\ y \end{bmatrix} \Longrightarrow a = Ka_y + a_{u_c} = r_u - v,$$

is capable of a reliable attack detection.

The following conclusions can be drawn from the above observation.

- The traditional methods of output residual-based detection and robust control method are inefficiently in dealing with attack detection and resilient control.
- As against, the input residual-based detection and control promise more efficient alternative solutions.
- The impact of cyber-attacks on feedback control systems can be modelled in a dual form of unknown inputs and plant uncertainties.

In the context of the unified framework of control and detection, the aforementioned issues can be generally formulated as

- modelling of system dynamics under cyber-attacks as anomalies in the system signal space (u, y),
- analysis of such type of anomalies and their influence on the system image subspace based on the duality, and
- detection, estimation and control of faults and cyber-attacks in different system signal subspaces.

Putting it in a nutshell, the unified framework of control and detection is capable of detecting and mitigating faults and cyber-attacks simultaneously and promises to enhance safety and security of CPSs. Exploration of analysis, detection and control of CPSs under cyber-attacks is a further key task of the unified framework.

1.2 Outline of the Contents

This book is composed of three parts. Part I serves as a common basis for the subsequent studies, Parts II - III are dedicated to various application thematic areas. They are self-contained and so structured that they can also be used for self-study on the concerned topics.

1.2.1 Part I: Unified Framework of Control and Detection: Concepts, Methods, and Computation Tools

The objectives of this part are to (i) summarise and review necessary control-theoretic and mathematical preliminaries, (ii) introduce basic definitions and concepts in the unified framework, (iii) review the related analysis and design schemes of control, detection and estimation, and explore alternative and new schemes specially dedicated to the control, detection and estimation issues in the unified framework, (iv) present computation methods and algorithms needed for the subsequent work, and (v) showcase perspective applications of the unified framework and related research topics. The main attention is paid to the analysis of nominal and uncertain system dynamics in various system signal subspaces.

Factorisations of dynamic systems are well-established mathematical and control-theoretic methods for system representations, analysis and synthesis. In Chap. 2, coprime factorisations of linear time-invariant (LTI) systems are introduced. In the foreground are the control engineering interpretations rather than rigorous mathematical definitions and concepts. Specifically, the interpretations of (i) system left and right coprime factorisations (LCF and RCF) as an observer-based residual generator and a state feedback controller, and (ii) the LCF and RCF of a feedback controller as an observer-based state feedback controller are detailedly delineated. In particular, the primal form of the so-called Bezout identity, the core of the unified framework, is deduced from these interpretations. It is instrumental for the reader to keep them in mind in the subsequent study. On the basis of LCF and RCF, the alternative system representation forms, the stable kernel and image representations (SKR and SIR) are introduced, followed by the presentation of image, kernel and residual subspaces as $\mathcal{H}_2$ and $\mathcal{L}_2$ subspaces in Hilbert space of (u, y). As special classes of coprime factorisations, the normalised LCF (SKR) and RCF (SIR) and their computation algorithms are described.

Chapter 3 is devoted to the parameterisations of stabilising feedback controllers and observers. While the so-called Youla parameterisation is well-known in robust control theory, the observer parameterisation and its applications are rarely reported. Related to the Youla parameterisation, the general form of the Bezout identity is introduced. In the unified framework of control and detection, the observer-based realisation form rather than the original form of the Youla parameterisation is preferred. The former clearly reveals the central role of the residual signal in a feedback control system and gives a unified form of feedback control and detection. In a comparison study, the advantages of the observer-based realisation form are showcased. Finally, in terms of the observer parameterisation, it is delineated that any output feedback controller is an observer for the estimation of a state feedback controller. A remarkable result is that a feedback of the residual signal is at the core of both parameterisations.

There exist numerous variations and extensions of the Youla parameterisation. In Chap. 4, some of them are presented, which are applied to addressing different control and detection issues in the unified framework. The plug-and-play (PnP) realisation of the Youla parameterisation is a fault-tolerant control method aiming at recovering system performance suffered from degradation. By feeding back residual signal, it enables controller tuning in a plug-and-play manner without any modification on the default controller and without affecting the system stability. Furthermore, two dual forms of the Youla parameterisation are introduced. Particularly, the first one dealing with parameterisation of system uncertainties is widely adopted in the subsequent work. Finally, two issues as extensions of the Youla parameterisation are investigated. The first one is the so-called observer-based unified system model, which unifiedly describes the plant, controller and closed-loop dynamics in a simple state space model. Parameterisation of the closed-loop dynamics, in particular the response to the coprime factor uncertainties, is the second issue. In fact, the parameterisation of the closed-loop dynamics rather than the Youla parameterisation of stabilising feedback controllers is the centrepiece of the unified framework.

The unified framework of control and detection is established for solving control and detection problems met in dynamic control systems, including linear time-varying (LTV) and nonlinear systems. In Chap. 5, a systematic summary of (i) the LCF and RCF, in particular the normalised LCF and RCF, (ii) the concepts of SIR and SKR and image, kernel and residual subspaces, (iii) the observer-based realisations of stabilising controllers, (iv) the Bezout identity and, based on it, the relations between the process input and output signals and the residual signal pair, and (v) the parameterisation of LTV observers is presented. The study in this chapter on system representations, analysis and synthesis of linear discrete time-varying (LDTV) systems is not a straightforward extension of the results presented in the previous chapters on LTI systems. It is highly motivated by developments in research and application domains of control and detection in the recent decades.

The main objective of Chap. 6 is to present the unified framework for nonlinear dynamic systems. The endeavours are made to introduce and illustrate the major mathematical and control-theoretic tools, which are known in the literature and needed for modelling, analysis and synthesis of nonlinear systems. These include (i) system factorisation technique, (ii) operator theory-based analysis and synthesis of nonlinear systems, and (iii) Hamiltonian systems. In particular the last topic plays a crucial role in the subsequent work. The Hamiltonian systems theory not only serves as a capable instrument to deal with control and detection issues of nonlinear systems, but also exposes a new aspect to deal with detection and tolerant control, performance-based fault detection and tolerant control.

Serving as a bridge from the theoretical foundation of the unified framework to its applications, Chap. 7 highlights some perspective explorations of the unified framework. To begin with, dynamic systems are explored from their input-output data structures. Specifically, the input-output data structure is defined in terms of the image/kernel and residual subspaces. It is the inherent nature of the system data and independent of the system configuration. The idea behind this aspect is the one-to-one mapping from the system input-output data to the input and output residual signals, defined by the Bezout identity. This study not only depicts inherent relations between controller optimisation and the system data structure, but also establishes a bridge between the model-based and data-driven methods for control and detection. On the basis of the observer-based uniform system model, the so-called kernel-based system model representation is introduced. The kernel-based model is driven by the output residual and is of the same input-output dynamic like the original system. Thanks to its major advantage to be able to run real-time and fully model the system dynamics under all operation conditions, this model form is widely applied in the subsequent investigations. An interesting and special application of the unified framework is modelling of system dynamics by means of finite samples long input and output signals. Emerging from behavioural systems theory as the system-theoretic basis, data-driven system analysis and control become a new research area in recent years. Inspired by the recent work in this thematic area, (i) the image and kernel representations of plants and controllers of finite samples long signals, (ii) image, kernel and residual subspaces of finite samples long (input-output) signals, and (iii) some related properties, are presented. This exploration provides us with a data-driven version of the unified framework of control and detection. Furthermore, alternative proofs of the so-called fundamental lemma are delineated on the basis of the data-driven image and kernel representations. Finally, three potential application domains are explored. The focus is on those possible new aspects of leveraging control and detection methods in the unified framework to existing control and detection topics. Concerning detection issues, two substantial messages are conveyed, (i) any uncertainties, including faults, in the plant are captured by the output residual signal, and (ii) the input residual signal comprises uncertainties caused by, for instance, cyber-attacks. A further application deals with tracking control and MPC. They can be alternatively and elegantly handled in the system image subspace and by means of optimising the corresponding latent variable. Moreover, an alternative prediction model is developed that enables a compensation of a part of the system uncertainties using control and detection methods developed in the unified framework. The last application is dedicated to multi-agent systems (MASs). The main intention of this work is to explore the capability of the unified framework. As an example, the trivial consensus problem among agents are alternatively represented in the SIR and SKR forms. It is showcased that the local feedback control in the agents are decoupled from the consensus controller. As a remarkable result, the local controllers in the agents can be designed independent of the consensus algorithm to approach control and detection problems like robust control, fault detection and fault-tolerant control. On account of this cognition, alternative and more capable detection methods are proposed, which can be applied to heterogeneous MASs as well.

1.2.2 Part II: Application of the Unified Framework to Fault Diagnosis and Tolerant Control in Uncertain Dynamic Systems

After a brief review of the major fault diagnosis and control methods, the main focus of this part is on developing advanced analysis and synthesis methods to solve diagnosis, control and estimation problems in control systems with uncertainties.

In Chap. 8, basic observer-based methods for residual generation, which are applied both for control and fault detection purposes, are firstly reviewed. The main attention is paid to the existing optimal fault detection schemes when additive disturbances (unknown input vectors) and multiplicative model uncertainties are present in the process under consideration. Although modifications and new observer-based schemes are presented, efficient and optimal fault detection in dynamic systems with model uncertainties remains an open issue.

The fact of the lack of systematic and efficient design schemes for optimal fault detection in uncertain dynamic systems is one of main motivations for developing alternative fault detection techniques. The projection-based detection and estimation methods are an answer to the open problem, which are described in Chaps. 9–11

After a short introduction to the mathematical preliminaries for function (signal) spaces, operator theory and the related concepts and definitions of orthogonal projections in Hilbert space and gap metric, basic ideas and schemes of fault detection using the method of orthogonal projection onto a system subspace in Hilbert space, like image and residual subspaces, are described in Chap. 9. The design of the corresponding fault detection system follows a uniform procedure with three major steps: (i) definition and determination of an orthogonal projection operator, (ii) construction of the projection-based residual generator and realisation of the online implementation algorithm, and (iii) gap metric-based threshold setting. This scheme provides an alternative framework for fault detection in dynamic systems to the well-established observer-based methods. A significant feature of this framework is that the parameter settings are well determined by the orthogonal projection and the associated gap (metric) computation, and can be calculated using the existing algorithms. This distinguishes the projection-based framework from the observer-based one essentially. While the projection-based method formulates fault detection as a one-class classification problem, the observer-based one addresses it from the control-theoretic viewpoint and formulate it as a trade-off between robustness (against uncertainties) and sensitivity (to the faults). Consequently, in the projection-based framework, (i) identifying how far the process (measurement) data is from the nominal process operation modelled by the system image subspace, and (ii) determining whether the identified deviation lies in the tolerant interval, build the two major tasks. In this chapter, two projection-based fault detection schemes are presented, respectively based on orthogonal projections onto $\mathcal{L}_2$ and $\mathcal{H}_2$ image subspaces. In particular, the scheme based on the projection onto $\mathcal{L}_2$ subspace is useful viewed from two aspects: (i) the needed mathematical handling is less demanding and thus helpful to understand the underlying ideas towards optimal fault detection, and (ii) the equivalence to the observer- (SKR-) based residual generation is a useful conclusion for practical applications, which allows to realise an orthogonal projection-based optimal fault detection using an observer-based fault detection system. At the end of this chapter, an extension of the projection-based fault detection methods is explored, which highlights a uniform formulation of projection-based fault detection of dynamic systems as a one-class classification problem. Based on it, a perspective solution in this framework is proposed.

Roughly speaking, the projection-based methods proposed in Chap. 9 solve the related fault detection problems by determining to which subspace, the image or the residual subspace, the collected data belong. Thereby, the available information about the system dynamics is the system image representation and the corresponding image subspace that describe the nominal system operation. This is a typical one-class classification task. Chapter 10 continues the work on projection-based fault diagnosis and addresses binary classification issues, which serves as the basis for the extension to multi-class classification towards fault isolation, and generalisation of projection-based fault detection methods. For binary classification, two process models are given with the corresponding SIRs, representing nominal and faulty system dynamic respectively. Understandably, two projection systems are to be constructed and, based on them, classification decision is then made. The crucial scenario is the similarity between the projections respectively generated by both projection-based residual generators. This may remarkably hinder a correct classification. In the first part of this chapter, properties of the similarity of two image subspaces are analysed with the aid of operator theory. It reveals that the binary fault classification problem is hard to solve as the system operates in a transitional phase from fault-free to faulty operations. To solve this problem, modification schemes are investigated and two approaches are proposed. The first one is to leverage available knowledge of the latent variable, for instance, presented in the form of a convex set, to embed a further projection in the projection-based detection system. Another approach deals with the issue by modelling the faulty dynamic using the dual Youla parameterisation, a result presented in Chap. 4. On account of such additional information, a classification approach is elaborated. The work is completed by extensions of the aforementioned binary classification-based fault detection to multi-classification towards fault isolation. Finally, noticing that the projection-based classification methods might become less efficient, when the gap metric-based threshold setting is too conservative, an integrated method is developed, which comprises a combination of the projection-based fault classification schemes with data-driven algorithms. As a result of this endeavour, considerably tighter threshold setting can be achieved, which remarkably improves the detection performance.

The theoretical basis for a successful application of the orthogonal projection technique to fault detection in linear systems is the Hilbert Projection Theorem. Concretely, the normalised SIR-based orthogonal projection of the process data onto the image subspace in Hilbert space, describing the nominal system dynamics, allows an orthogonal decomposition of the process data into an optimal estimate (projection) and the residual. An orthogonal projection operator is a linear operator and cannot be applied to dealing with nonlinear systems. In order to deal with projection-based one-class fault detection for nonlinear systems, projections onto system image and residual manifolds are studied in Chap. 11. Inspired by the manifold learning methods, the construction of the projection systems is based on the normalised SIR of the nonlinear system under consideration. Note that an essential assumption in manifold learning is that the data are generated from a lower-dimensional manifold that is embedded inside of a higher-dimensional space composed of redundant and irrelevant information (uncertainties). The basic idea of manifold learning methods is to create a data mapping (projection) from the higher-dimensional manifold to the lower-dimensional one and preserving local neighbourhoods at every point of the underlying manifold. The projection system is idempotent and thus preserves the underlying manifold in the local neighbourhoods. Specifically, based on Hamiltonian systems theory the projection system is constructed by connecting the normalised SIR and its adjoint system. As one of the major results in this work, Bregman divergence, as a measure of difference between two points in a space, is proposed to solve the problem of distance measure in non-Euclidean space. It is a distinguishing contribution that the normalised SIR and its adjoint as a Hamiltonian system and the Bregman divergence induced by the corresponding Hamiltonian functions are integrated applied for the projection-based detection. This scheme not only enables the realisation of the objective of performance-oriented fault detection, but also uncovers the information geometric aspect of the projection system. It is known that the input and output of a Hamiltonian system build a Legendre transform and there exists a Hamiltonian system as the Legendre dual system. As a result, the computation of the corresponding Bregman divergence can be expressed in terms of the dual Hamiltonian functions. In particular, a Bregman divergence together with a Legendre transform induces a dually flat Riemannian manifold in the data space that is regarded as a dualistic extension of the Euclidean space. In this context, the Hamiltonian system based projection can be interpreted as a geodesic projection onto the system image manifold. So far, this part of work is an integration of control theory and information geometry. Concerning uncertainties in data and their estimation, a dual projection system to the aforementioned SIR-based projection system is proposed. To be specific, a Hamiltonian system is constructed by means of the normalised SKR and its adjoint system. Since the SKR is a residual generator and the residual contains uncertainty information, the SKR-based Hamiltonian system represents a geodesic projection of the process data onto the residual manifold. An immediate application of this result is fault estimation. That means, variations in the process data caused by faults, in particular those process faults, are modelled like uncertainties by means of the uncertainty model and the residual manifold represents the corresponding set in the data space. The SKR-based Hamiltonian system serves as an optimal estimator for such variations in the process data.

The objective of Chap. 12 is twofold. On the one hand, the existing unknown input estimation and disturbance control methods are re-visited in the unified framework of control and detection. On the other hand, the projection-based and co-inner-based least squares (LS) estimation methods developed in the unified framework are leveraged to explore advanced estimation and control methods and algorithms. At first, the unknown input observer (UIO) and disturbance observer-based control (DOBC) schemes are reviewed. Concerning the UIO schemes, the role of UIOs as estimators of the unknown input vector is investigated. In the unified framework of control and detection, the well-established DOBC schemes are explored. It is showcased that the compensation of the unknown input vector, a key step in the DOBC schemes, is indeed a special case of the output residual generation embedded in the unified framework of control and detection. Furthermore, general existence conditions to achieve a decoupling of closed-loop dynamics from the disturbance are provided. Concerning DOBC design for systems with uncertainties, the stability concern is closely examined. LS estimation schemes of unknown inputs (together with state variables) and uncertainties corrupted in the process data build another focus of this chapter. In this regard, the definition of the LS estimation problem is formulated. The LS algorithm delivers the estimated unknown input vector and the (unknown) initial state vector with the minimal norms and, importantly, subject to the system equations. This feature distinguishes the LS estimation schemes from those H_∞-optimal ones. The further work addresses estimating uncertainties corrupted in process data. To this end, the projection technique is leveraged for uncertainties estimation. It is remarkable that both aforementioned estimation schemes have one point in common that backward computations are an essential step of the estimation algorithms. The successful exploration of the projection technique for uncertainty estimation has inspired the endeavour to learn fault-tolerant observers. The basic idea is to address variations in the system dynamics caused by faults by means of estimation and compensation in a learning iteration. Herewith, the term learning is understood in the regard of the so-called reinforcement learning (RL). It is of considerable interest to notice that the work on optimal fault-tolerant observers is dual to the RL-aided LQ optimal control. To be specific, the optimisation problem is formulated as an LS estimation, dual to the well-known LQ control, and it is solved with respect to the co-state variable and the conjugate of the system SKR, dual to the state feedback and the system SIR. The learning algorithm is performed sequentially backwards in time, dual to feedforward in time as usually done in RL technique. As an example of applying LS estimation algorithms to fault-tolerant control and performance recovery, an observer-based fault-tolerant LQ control scheme is presented at the end of this chapter.

As a data-driven system model, finite samples system representations are widely applied to data-driven control of dynamic systems and build the mainstream in the research domain of data-driven control. In the first part of Chap. 13, applications of model-based finite samples system representations to fault detection and predictive control are explored. The first results exemplify numerous benefits of such applications, amongst others, (i) significant reduction of computation efforts, particularly, for online implementation, (ii) engineering-equivalent realisation, and (iii) release of existence conditions for optimal solutions. Concerning the application of the finite samples system representations to data-driven detection and control, the data-driven version of the unified framework is firstly established, including, (i) the normalised image and kernel representations, (ii) the data-driven Bezout identity, (iii) the relation between the process input-output data and the residual signals, and (iv) description of uncertainties in the data set. Based on it, two data-driven fault detection schemes are presented, (i) data-driven fault detection in systems with unknown inputs, and (ii) data-driven projection-based fault detection. At the end of this chapter, data-driven control issues are addressed. Noticing that existing data-driven control schemes are mainly implemented in the moving window pattern, an alternative data-driven image representation is developed, which models the system dynamics in the moving window pattern. On this basis, a data-driven feedback control scheme is presented. The resulting control system is fully residual-driven and fits the moving window operating pattern. Moreover, the data-driven predictive control with embedded feedback control is considered.

1.2.3 Part III: Secure Industrial Cyber-Physical Control Systems

The objectives of this part are (i) analysis of industrial cyber-physical control systems under cyber-attacks, (ii) detection of cyber-attacks in CPSs, (iii) simultaneous detection of faults and cyber-attacks, and (iv) fault-tolerant and resilient control. The first two topics are key issues in the context of secure industrial cyber-physical systems, while the latter two deal with safety and security (S&S) of CPSs, which are drawing considerable attention both in engineering application and research domains. The goal of this work is to deal with those relevant cyber-security issues in the unified framework of control and detection towards systematic analysis and unified, capable design methods.

To begin with, a comprehensive analysis of CPS control systems under cyber-attacks is conducted in Chap. 14. To this end, the configuration and models of industrial CPSs under consideration are specified. It is followed by an analysis of the CPS dynamics under the so-called integrity attacks, in particular the aspect of possible impairment of system dynamics caused by such type of cyber-attacks. On this basis, the potential of the existing control and detection methods to manage the system performance impairment due to the cyber-attacks is assessed, which leads to the formulation of cyber-attack detection and resilient control issues to be solved. Specifically, the responses of CPS control systems to the cyber-attacks are described. Among a number of achieved results, the most important and distinguishing result is the conclusion that the cyber-attacks exclusively affect the plant image subspace. In other words, the impacts of the cyber-attacks on the closed-loop dynamics are equivalently modelled (i) as an unknown input added to the reference signal, when the additive type of attacks act on the CPS, and (ii) as uncertainty in the feedback controller in the case of the multiplicative attacks. This result is so far of considerable importance, because the immediate consequences/claims follow, namely, (i) detection and control of CPSs under cyber-attacks and dynamic processes suffering uncertainties or faults are two different but dual problems, and (ii) the real challenges of detection and control of cyber-attacks are the development of novel detection and (feedback) control schemes that are able to detect and control the uncertainties in the plant image subspace. Based on the aforementioned results, some elemental system properties of CPSs under cyber-attacks are analysed with a focus on the role of the input residual signal. All these results motivate the problem formulation for the subsequent work.

Chapter 15 is devoted to the issues of detecting cyber-attacks, in particular the so-called stealthy attacks on CPS control systems. At first, a uniform model form of cyber-attacks is introduced, which models attacks that are injected into the system either as unknown inputs or as feedback of the system variables. The latter type of attacks are particularly critical, because they may cause remarkable changes in the system eigen-dynamic. Differing from typical stealthy attacks addressed in the literature, the concept of kernel attacks is introduced, which are defined in terms of system input and output signals in CPSs that lie in the system kernel subspace.

On this basis, relations of kernel attacks and typical stealthy attacks addressed in the literature are explored, and existence conditions of kernel attacks are examined with respect to various configurations of CPSs. In the last part of this chapter, capable detection schemes of cyber-attacks, in particular, kernel attacks, are proposed. Distinctly differing from the existing detection schemes, the core of the proposed schemes is the generation of the input residual and fusion of the input and output residuals. Proceeding from these detection schemes, issues of simultaneous detection of cyber-attacks and faults and configuration of resilient CPSs are addressed at the end of this chapter.

Resilient and fault-tolerant control of CPSs under cyber-attacks, including simultaneous detection of faults and cyber-attacks, is the main objective of Chap. 16. It comprises two major parts. The first part is devoted to potential control schemes that enable suppression of control performance degradation caused by cyber-attacks using the input residual signal. The proposed control scheme with a feedback of the input residual signal is then integrated in the design of a resilient and fault-tolerant control system for CPSs. In the second part of this chapter, a comprehensive design scheme towards a resilient and fault-tolerant control system of CPSs is proposed, which consists of four sub-systems, (i) fault detection, (ii) cyber-attack detection, (iii) performance optimisation and recovery, and (iv) resilient control of CPS under cyber-attacks, whereby issues (ii) and (iv) are the main study focus.

Putting it in a nutshell, the work on secure CPSs in the above three chapters is conducted in the unified framework of control and detection. It apparently differs from state of the art research in this thematic area in the following four aspects, (i) system integrity, (ii) system configuration, (iii) the transmitted signals, and (iv) system analysis and synthesis.

The last chapter, Chap. 17, serves as a summary of the unified framework of control and detection as a whole. The Bezout identity is the mathematical basis for the parameterisation of all stabilising controllers and stability analysis of feedback control systems and thus crucial for controller synthesis and robustness analysis. More importantly, it models a mapping from the system input-output signals to the residual signals. The Bezout identity and its variations are summarised as the core of the unified framework. On this basis, the SIR and SKR systems representations are introduced and, correspondingly, the image, kernel and residual subspaces or manifolds are established. These concepts form the cornerstones of the unified framework. As the basic theoretic methods, normalised coprime factorisations, projection methods and Hamiltonian systems are reviewed.

1.3 Notes and References

The unified design of control and detection systems for fault diagnosis in feedback control system and fault-tolerant control was an active research area in the 90 s. A comprehensive review about the main activities and publications in this area has been reported in [35]. Although the unified design of control and detection systems has initiated the establishment of the unified framework of control and detection, the latter is not only limited to system design. The unified framework of control and detection comprises alternative and often more efficient methods that enables us to analyse and design the control and detection system as a whole and in the system signal subspaces. It opens new perspectives to deal with complex control and detection problems met in modern industrial automation control systems. Some results have been, for instance, reported in [40, 42, 45]. More importantly, it lays the foundation to bridge model-based, data-driven and ML-based methods.

In the last three decades, a great number of monographs on process monitoring, diagnosis and fault-tolerant control have been published. Among them, [8, 13, 20, 36, 39, 71, 77, 94, 130, 135, 158, 184, 236] are dedicated to the model-based fault diagnosis techniques, [23, 37, 66, 130, 174, 201] are focused on the data-driven process monitoring and diagnosis methods, and [13, 130, 133, 150] deal with fault-tolerant control issues. Concerning the surveys on these topics, we recommend the reader the following representative survey papers, [58, 60, 64, 93, 136, 197, 224] on the model-based fault diagnosis techniques, [38, 65, 67, 85, 164, 165, 196, 219] on the data-driven process monitoring and diagnosis, and [93, 220, 232] on fault-tolerant control methods.

Although ML-based fault diagnosis and tolerant control methods are not explored in this book, the rapid development in the ML research domain has enormous impact on our work. In particular, many brilliant ideas from the representation learning [11] have inspired our endeavours to develop advanced methods within the unified framework. For instance, the autoencoder technique for dimensionality reduction and feature learning, and the concepts of information bottleneck [192] and minimal sufficient statistic [68, 192] for learning the latent variable.

Concerning the needed control-theoretic and mathematical preliminaries, we would like to recommend the following monographs, [17, 57, 194, 199, 243].

Part I
Unified Framework of Control and Detection: Concepts, Methods, and Computation Tools

Chapter 2
System Factorisations, Kernel and Image Representations and Subspaces

For linear time invariant systems, transfer function and state space representation are the two model forms mostly used in control engineering and theory. Apart from their capability of modelling system input and output transfer dynamics, the state space representation is the basis for the description of system internal dynamics as well as the associated structural properties like controllability and observability. In this chapter, we will introduce two other system representation forms, the so-called left and right coprime factorisations, and associated with them, the kernel and image representations. These system representation forms are closely related to the transfer function and state space representations, but address dynamic relations between the system input and output, treated as process measurement variables, and two latent variables, the so-called residual signals. These relations build the basis of the unified framework of control and detection, and exemplify the information aspect of feedback control as well as the intimate relations between feedback control and detection.

Although coprime factorisation methods are a mathematical and control theoretical tool, our introduction focuses not only on computational formulas and algorithms but also on control engineering interpretations and useful relations for our subsequent work. The reader is referred to the literature cited at the end of this chapter for more detailed descriptions with rigorous mathematical definitions and concepts.

2.1 LCF and RCF, Residual, and Observer-Based State Feedback Control

Consider discrete-time LTI systems modelled by

$$y(z) = G(z)u(z),\ y \in \mathbb{C}^m,\ u \in \mathbb{C}^p \tag{2.1}$$

S. X. Ding, *Unified Framework of Control and Detection*,
https://doi.org/10.1007/978-3-662-72417-0_2

with u and y as the plant input and output vectors. It is assumed that $G(z)$ is a proper real-rational matrix and its minimal state space realisation is given by

$$x(k+1) = Ax(k) + Bu(k), x(0) = x_0, \tag{2.2}$$
$$y(k) = Cx(k) + Du(k), \tag{2.3}$$

where $x \in \mathbb{R}^n$ is the state vector and x_0 is the initial condition of the system. Matrices A, B, C, D are appropriately dimensioned real constant matrices. A factorisation of a transfer function matrix over $\mathcal{RH}_\infty$ gives a further system representation form and factorises the transfer matrix into two stable transfer matrices. The left and right factorisations of $G(z)$ are given by

$$G(z) = \hat{M}^{-1}(z)\hat{N}(z) = N(z)M^{-1}(z), \tag{2.4}$$

whose state space representations are

$$\hat{M}(z) = (A - LC, -L, WC, W), \hat{N}(z) = (A - LC, B - LD, WC, WD), \tag{2.5}$$
$$M(z) = (A + BF, BV, F, V), N(z) = (A + BF, BV, C + DF, DV), \tag{2.6}$$

where (real) matrices F and L are selected such that $A + BF$ and $A - LC$ are Schur matrices, and W and V are invertible. It is apparent that

$$\hat{M}^{-1}(z)\hat{N}(z) = N(z)M^{-1}(z) \Longleftrightarrow$$
$$\hat{M}(z)N(z) - \hat{N}(z)M(z) = \begin{bmatrix} -\hat{N}(z) & \hat{M}(z) \end{bmatrix} \begin{bmatrix} M(z) \\ N(z) \end{bmatrix} = 0. \tag{2.7}$$

Before introducing the next concept, we would like to illustrate the state space presentations (2.5)–(2.6). By introducing a state vector $\hat{x} \in \mathbb{R}^n$ and output vector $r_0 \in \mathbb{R}^m$, (2.5) yields

$$\hat{x}(k+1) = A\hat{x}(k) + Bu(k) + L\left(y(k) - \hat{y}(k)\right), \tag{2.8}$$
$$r_y(k) = Wr_0(k), r_0(k) = y(k) - \hat{y}(k), \hat{y}(k) = C\hat{x}(k) + Du(k), \tag{2.9}$$

which can be, in terms of $\left(\hat{M}(z), \hat{N}(z)\right)$, written as

$$r_y(z) = \hat{M}(z)y(z) - \hat{N}(z)u(z). \tag{2.10}$$

Notice that for $\hat{x}(0) = x(0), \forall k, r_0(k) = 0$, which implies, according to (2.10),

$$\hat{M}(z)y(z) - \hat{N}(z)u(z) = 0 \Longrightarrow y(z) = \hat{M}^{-1}(z)\hat{N}(z)u(z).$$

System (2.8) is a (full) state observer, and together with the output Eq. (2.9), it builds a so-called observer-based residual generator. The vector r_0 is called primary residual vector, which plays a central role in our subsequent study. It is worth noting that r_0 is generally deviated from zero, for instance, due to an unknown initial value $x(0)$ or uncertainties in the system parameters.

Remark 2.1 Since matrix W, also called (algebraic) post-filter, is invertible, r_y and r_0 as residual vectors contain the same amount of information. On the other hand, r_0 as the estimation error of the output vector is of clear system technical meaning. Therefore, in our subsequent work, the residual vector is often referred to r_0, or W is set equal to a unity matrix, when this does not leads to confusion. In fact, it is state of the art in the relevant literature that W is assumed to be a unity matrix in LCF (2.5).

Concerning the state space representation given in (2.6), we define the input vector as

$$u(k) = Fx(k) + Vv(k),$$

where $Fx(k)$ can be interpreted as a state feedback and, in this context, $v \in \mathbb{R}^p$ is understood as the reference vector of the control system. It yields,

$$x(k+1) = (A + BF)x(k) + BVv(k) \tag{2.11}$$

$$\begin{bmatrix} u(k) \\ y(k) \end{bmatrix} = \begin{bmatrix} F \\ C + DF \end{bmatrix} x(k) + \begin{bmatrix} V \\ DV \end{bmatrix} v(k). \tag{2.12}$$

This is exactly the state space representation given in (2.6), which can be further written into

$$\begin{bmatrix} u(z) \\ y(z) \end{bmatrix} = \begin{bmatrix} M(z) \\ N(z) \end{bmatrix} v(z) \tag{2.13}$$

$$\Longrightarrow y(z) = N(z)v(z) = N(z)M^{-1}(z)u(z).$$

Remark 2.2 Analogous to W, V is called (algebraic) pre-filter, and often assumed to be a unity matrix in RCF (2.6).

As a summary, the left and right factorisations of $G(z)$, $\left(\hat{M}(z), \hat{N}(z)\right)$ and $(M(z), N(z))$, can be realised by means of observer-based residual generator and state feedback controller design for the given plant model (2.1) or (2.2)–(2.3). It is of interest to notice that

- the observer design and state feedback controller design are two dual (design) problems and essential in control theory,
- in the regard of these factorisations, two additional variables, r_y and v, are introduced; r_y acts as the output of the observer-based residual generator, while v serves as the input of the state feedback control loop, and moreover,
- the input and output relation of the plant, $y(z) = G(z)u(z)$, is implicitly described in terms of the dynamics of these two systems.

We now introduce the concept of coprime factorisations. The left and right factorisations, $\left(\hat{M}(z), \hat{N}(z)\right)$ and $(M(z), N(z))$, are called left and right coprime factorisation, respectively, if there exist $\left(\hat{X}(z), \hat{Y}(z)\right)$ and $(X(z), Y(z))$ over $\mathcal{RH}_\infty$ so that

$$\begin{bmatrix} \hat{N}(z) & \hat{M}(z) \end{bmatrix} \begin{bmatrix} \hat{Y}(z) \\ \hat{X}(z) \end{bmatrix} = \begin{bmatrix} -\hat{N}(z) & \hat{M}(z) \end{bmatrix} \begin{bmatrix} -\hat{Y}(z) \\ \hat{X}(z) \end{bmatrix} = I, \tag{2.14}$$

$$\begin{bmatrix} X(z) & Y(z) \end{bmatrix} \begin{bmatrix} M(z) \\ N(z) \end{bmatrix} = I. \tag{2.15}$$

Sequentially, $\left(\hat{M}(z), \hat{N}(z)\right)$ and $(M(z), N(z))$ build left and right coprime pairs (LCP and RCP). Next, we explore the state space representations of $\left(\hat{X}(z), \hat{Y}(z)\right)$ and $(X(z), Y(z))$ that satisfy (2.14)–(2.15) for given LCP and RCP $\left(\hat{M}(z), \hat{N}(z)\right)$ and $(M(z), N(z))$.

Departing from the state space realisation of $\left(\hat{M}(z), \hat{N}(z)\right)$, (2.8)–(2.9), defining

$$u(k) = F\hat{x}(k)$$

yields

$$\hat{x}(k+1) = (A + BF)\hat{x}(k) + Lr_0(k), \tag{2.16}$$

$$\begin{bmatrix} u(k) \\ y(k) \end{bmatrix} = \begin{bmatrix} F \\ C + DF \end{bmatrix} \hat{x}(k) + \begin{bmatrix} 0 \\ I \end{bmatrix} r_0(k). \tag{2.17}$$

Now, let the system $\begin{bmatrix} u \\ y \end{bmatrix}$ as the input of the observer-based residual generator (2.10). It turns out

$$\begin{aligned} \forall r_y, r_y(z) &= \begin{bmatrix} -\hat{N}(z) & \hat{M}(z) \end{bmatrix} \begin{bmatrix} u(z) \\ y(z) \end{bmatrix} \\ &= \begin{bmatrix} -\hat{N}(z) & \hat{M}(z) \end{bmatrix} \begin{bmatrix} F(zI - A_F)^{-1} LW^{-1} \\ C_F(zI - A_F)^{-1} LW^{-1} + W^{-1} \end{bmatrix} r_y(z), \\ A_F &= A + BF, C_F = C + DF, \end{aligned}$$

which implies

$$\begin{bmatrix} -\hat{N}(z) & \hat{M}(z) \end{bmatrix} \begin{bmatrix} -\left(-F(zI - A_F)^{-1} LW^{-1}\right) \\ C_F(zI - A_F)^{-1} LW^{-1} + W^{-1} \end{bmatrix} = I. \tag{2.18}$$

We now rewrite the observer system (2.8)–(2.9) as

$$\hat{x}(k+1) = (A - LC)\hat{x}(k) + (B - LD)u(k) + Ly(k), \tag{2.19}$$

and set

$$u(k) = F\hat{x}(k) + V\bar{v}(k),$$

which leads to

$$\bar{v}(k) = V^{-1}\left(u(k) - F\hat{x}(k)\right) \Longrightarrow \tag{2.20}$$

$$\bar{v}(z) = V^{-1}\left(I - F(zI - A + LC)^{-1}(B - LD)\right)u(z) \tag{2.21}$$

$$-V^{-1}F(zI - A + LC)^{-1}Ly(z). \tag{2.22}$$

Recalling the RCP model (2.13), we have,

$$\forall\bar{v}(z),\ \bar{v}(z) =$$

$$\begin{bmatrix} V^{-1}\left(I - F(zI - A_L)^{-1}B_L\right) & -V^{-1}F(zI - A_L)^{-1}L \end{bmatrix}\begin{bmatrix} M(z) \\ N(z) \end{bmatrix}\bar{v}(z),$$

$$A_L = A - LC,\ B_L = B - LD,$$

which gives

$$\begin{bmatrix} V^{-1}\left(I - F(zI - A_L)^{-1}B_L\right) & -V^{-1}F(zI - A_L)^{-1}L \end{bmatrix}\begin{bmatrix} M(z) \\ N(z) \end{bmatrix} = I. \tag{2.23}$$

Now, set $\bar{v}(z) = 0$ and let u, y be the output of the system (2.16)–(2.17). It turns out

$$\begin{bmatrix} V^{-1}\left(I - F(zI - A_L)^{-1}B_L\right) & -V^{-1}F(zI - A_L)^{-1}L \end{bmatrix}\cdot \begin{bmatrix} -\left(-F(zI - A_F)^{-1}LW^{-1}\right) \\ C_F(zI - A_F)^{-1}LW^{-1} + W^{-1} \end{bmatrix} = 0. \tag{2.24}$$

By (2.18), (2.23) and (2.24), it is demonstrated that

$$\hat{X}(z) = \left(A + BF, LW^{-1}, C + DF, W^{-1}\right), \hat{Y}(z) = \left(A + BF, LW^{-1}, -F, 0\right), \tag{2.25}$$

$$X(z) = \left(A - LC, B - LD, -V^{-1}F, V^{-1}\right), Y(z) = \left(A - LC, L, -V^{-1}F, 0\right), \tag{2.26}$$

$$\begin{bmatrix} X(z) & Y(z) \end{bmatrix}\begin{bmatrix} -\hat{Y}(z) \\ \hat{X}(z) \end{bmatrix} = 0. \tag{2.27}$$

It follows from the state space representations of $\left(\hat{X}(z), \hat{Y}(z)\right)$, $(X(z), Y(z))$, (2.16)–(2.17) and (2.19)–(2.20), respectively, that

- $\left(\hat{X}(z), \hat{Y}(z)\right)$ describes an observer-based control system with r_y (r_0) and $\begin{bmatrix} u \\ y \end{bmatrix}$ as its input and output variables, respectively, that is,

$$\begin{bmatrix} u(z) \\ y(z) \end{bmatrix} = \begin{bmatrix} -\hat{Y}(z) \\ \hat{X}(z) \end{bmatrix} r_y(z), \tag{2.28}$$

- $(X(z), Y(z))$ can be interpreted as a closed-loop "residual generator" with $\begin{bmatrix} u \\ y \end{bmatrix}$ and $\bar{v}$ as its input and output variables satisfying

$$\bar{v}(z) = \begin{bmatrix} X(z) & Y(z) \end{bmatrix} \begin{bmatrix} u(z) \\ y(z) \end{bmatrix} \Longrightarrow$$
$$\bar{v}(z) = 0, u(z) = -X^{-1}(z)Y(z)y(z).$$

Moreover, writing (2.27) into

$$X^{-1}(z)Y(z) = \hat{Y}(z)\hat{X}^{-1}(z) \tag{2.29}$$

gives an output feedback controller of the form

$$u(z) = K(z)y(z), K(z) = -X^{-1}(z)Y(z) = -\hat{Y}(z)\hat{X}^{-1}(z). \tag{2.30}$$

Equations (2.7), (2.14), (2.15) and (2.27) can be written into the compact form

$$\begin{bmatrix} X(z) & Y(z) \\ -\hat{N}(z) & \hat{M}(z) \end{bmatrix} \begin{bmatrix} M(z) & -\hat{Y}(z) \\ N(z) & \hat{X}(z) \end{bmatrix} = \begin{bmatrix} I & 0 \\ 0 & I \end{bmatrix}. \tag{2.31}$$

It is the well-known (double) Bezout identity. Note that, according to (2.7), (2.14)–(2.27), $(X(z), Y(z))$ and $\left(\hat{X}(z), \hat{Y}(z)\right)$ are LCP and RCP of the following transfer function matrix

$$X^{-1}(z)Y(z) = \hat{Y}(z)\hat{X}^{-1}(z),$$

whose control engineering interpretation is an output feedback controller. The control engineering interpretation of the factorisations is schematically sketched in Fig. 2.1.

Remark 2.3 Hereafter, we may drop out of the domain variable z or k when there is no risk of confusion.

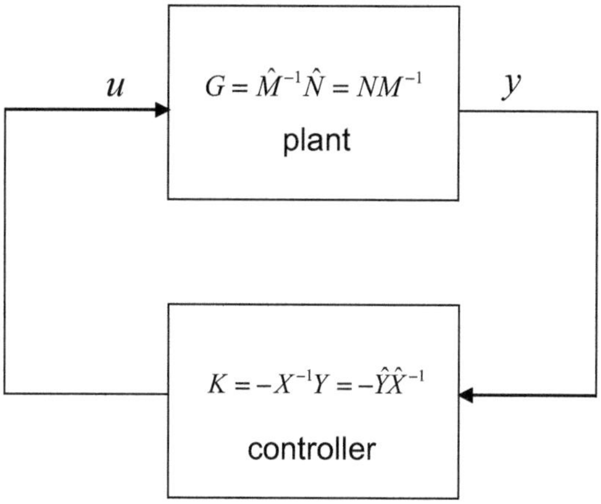

Fig. 2.1 Interpretation of system factorisations as a feedback control loop

For our subsequent study, we now introduce a special type of LCF and RCF of the plant dynamics. They are called normalised coprime factorisations, denoted by $\left(\hat{M}_0, \hat{N}_0\right)$ and (M_0, N_0), respectively, and satisfy

$$\left[\hat{N}_0(z)\ \hat{M}_0(z)\right]\begin{bmatrix}\hat{N}_0^{\sim}(z)\\ \hat{M}_0^{\sim}(z)\end{bmatrix} = \hat{N}_0(z)\hat{N}_0^{\sim}(z) + \hat{M}_0(z)\hat{M}_0^{\sim}(z) = I, \tag{2.32}$$

$$\left[M_0^{\sim}(z)\ N_0^{\sim}(z)\right]\begin{bmatrix}M_0(z)\\ N_0(z)\end{bmatrix} = M_0^{\sim}(z)M_0(z) + N_0^{\sim}(z)N_0(z) = I. \tag{2.33}$$

Here, $T^{\sim}(z) = T^T(z^{-1})$ denotes the conjugate of $T(z)$. The state space realisations of $\left(\hat{M}_0, \hat{N}_0\right)$ and (M_0, N_0) are summarised in the following theorem.

Theorem 2.1 *Given the system model (2.1)–(2.3),* $\left(\hat{M}_0, \hat{N}_0\right)$ *and* (M_0, N_0) *are LCP and RCP with* $L = L_0$, $W = W_0$, $F = F_0$ *and* $V = V_0$, *as given below:*

$$L = L_0 = \left(BD^T + APC^T\right)\left(I + DD^T + CPC^T\right)^{-1}, \tag{2.34}$$

$$W = W_0 = U\left(I + DD^T + CPC^T\right)^{-1/2}, \tag{2.35}$$

$$UU^T = I \Longleftrightarrow W_0\left(I + DD^T + CPC^T\right)W_0^T = I, \tag{2.36}$$

$$F = F_0 = -\left(I + D^TD + B^TSB\right)^{-1}\left(D^TC + B^TSA\right), \tag{2.37}$$

$$V = V_0 = \left(I + D^TD + B^TSB\right)^{-1/2}J, \tag{2.38}$$

$$J^TJ = I \Longleftrightarrow V_0^T\left(I + D^TD + B^TSB\right)V_0 = I. \tag{2.39}$$

Here, $P > 0$, $S > 0$ *solve the following Riccati equations respectively,*

$$P = APA^T + BB^T - L_0\left(I + DD^T + CPC^T\right)L_0^T, \tag{2.40}$$

$$S = A^T SA + C^T C - F_0^T \left(I + D^T D + B^T SB\right) F_0. \tag{2.41}$$

The proof of this theorem is known in the literature, see the reference given at the end of this chapter.

The following property of normalised LCF and RCF is useful for our subsequent work.

Theorem 2.2 *Given the normalised LCF* $\left(\hat{M}_0, \hat{N}_0\right)$ *and RCF* (M_0, N_0) *of the system model (2.1)–(2.3), it holds*

$$\begin{bmatrix} -\hat{N}_0^{\sim}(z) \\ \hat{M}_0^{\sim}(z) \end{bmatrix} \begin{bmatrix} -\hat{N}_0(z) & \hat{M}_0(z) \end{bmatrix} + \begin{bmatrix} M_0(z) \\ N_0(z) \end{bmatrix} \begin{bmatrix} M_0^{\sim}(z) & N_0^{\sim}(z) \end{bmatrix} = I. \tag{2.42}$$

Proof It follows from Bezout identity (2.31) and (2.32)–(2.33) that

$$\begin{bmatrix} M_0^{\sim} & N_0^{\sim} \\ -\hat{N}_0 & \hat{M}_0 \end{bmatrix} \begin{bmatrix} M_0 & -\hat{N}_0^{\sim} \\ N_0 & \hat{M}_0^{\sim} \end{bmatrix} = \begin{bmatrix} I & 0 \\ 0 & I \end{bmatrix}.$$

Since $\begin{bmatrix} M_0 & -\hat{N}_0^{\sim} \\ N_0 & \hat{M}_0^{\sim} \end{bmatrix}$ is the inverse of $\begin{bmatrix} M_0^{\sim} & N_0^{\sim} \\ -\hat{N}_0 & \hat{M}_0 \end{bmatrix}$, it holds, as well,

$$\begin{bmatrix} M_0 & -\hat{N}_0^{\sim} \\ N_0 & \hat{M}_0^{\sim} \end{bmatrix} \begin{bmatrix} M_0^{\sim} & N_0^{\sim} \\ -\hat{N}_0 & \hat{M}_0 \end{bmatrix} = \begin{bmatrix} I & 0 \\ 0 & I \end{bmatrix}.$$

We would like to conclude this section with the following remark. The aforementioned discussions on the control engineering interpretations of LCF and RCF showcase their close relations to the controller and observer-based residual generator design. They will serve as the major tool for the introduction of the unified framework of control and detection.

2.2 Inner-Outer and Co-Inner-Outer Factorisations

Inner-outer factorisation (IOF) and co-inner-outer factorisation (CIOF) are a class of factorisations of transfer functions. They are closely related to the coprime factorisations introduced in the previous section and provide a system aspect of optimal design of control and detection systems. For example, IOF technique is a powerful tool for solving robustness related control problems. For the optimal fault detection purpose, CIOF plays an important role. In this section, the basics of standard inner-outer and co-inner-outer factorisations and some of their exemplary applications to control and detection are briefly presented. Corresponding to our applications, only stable systems modelled by (2.1) and (2.2)–(2.3) are under consideration.

2.2.1 Essentials

To begin with, definitions of inner and outer systems are introduced.

Definition 2.1 Given $G(z) \in \mathcal{RH}_\infty^{m\times p}$, it is called inner when

$$G^{\sim}(z)G(z) = I,$$

and it is called co-inner if

$$G(z)G^{\sim}(z) = I.$$

Definition 2.2 Given $G(z) \in \mathcal{RH}_\infty^{m\times p}$, it is called outer when it has a right inverse $G^{-}(z)$,

$$G(z)G^{-}(z) = I,$$

which is analytic for $|z| > 1$, including at infinity. It is called co-outer if it has a left inverse $G^{-}(z)$,

$$G^{-}(z)G(z) = I,$$

which is analytic for $|z| > 1$, including at infinity.

Remark 2.4 The above definitions of inner and co-inner are standard in the literature. There exists however an enhanced version that gives inner/co-inner a control-theoretic explanation and is popular in the study on nonlinear systems. The relevant topics will be addressed in Sect. 6.3.2.

A rational matrix G is called surjective (injective) if it has full row (column) rank for almost all z. It is obvious that an inner (co-inner) should be injective (surjective), and an outer (co-outer) is surjective (injective).

Theorem 2.3 *Given $G(z) \in \mathcal{RH}_\infty^{m\times p}$ with the minimal state space representation (A, B, C, D), then (i) $G(z)$ is an inner if and only if*

$$B^T XA + D^T C = 0, D^T D + B^T XB = I,$$

(ii) $G(z)$ is a co-inner if and only if

$$CYA^T + DB^T = 0, DD^T + CYC^T = I,$$

where $X > 0, Y > 0$ are the solutions of Lyapunov equations

$$A^T XA - X + C^T C = 0,$$
$$AYA^T - Y + BB^T = 0,$$

respectively.

Definition 2.3 Given $G(z) \in \mathcal{RH}_\infty$, an IOF of $G(z)$ is a factorisation

$$G(z) = G^i(z)G^o(z), \tag{2.43}$$

where G^i is inner and G^o is the outer.

CIOF is a dual form of IOF and can be expressed in terms of IOF as

$$G^T(z) = \left(G^i(z)G^o(z)\right)^T = \left(G^o(z)\right)^T \left(G^i(z)\right)^T.$$

Definition 2.4 Given $G(z) \in \mathcal{RH}_\infty$, a CIOF of $G(z)$ is a factorisation

$$G(z) = G^{co}(z)G^{ci}(z), \tag{2.44}$$

where G^{ci} is co-inner and G^{co} is the co-outer.

Next, we introduce the existence conditions for IOF and CIOF and the associated computation algorithms. The following results are well-known in robust control theory.

Theorem 2.4 *Given $G(z) \in \mathcal{RH}_\infty^{m\times p}$ that is injective and satisfies $\forall\theta \in [0, 2\pi]$,*

$$rank\left(G(e^{j\theta})\right) = p \Longleftrightarrow G^\sim(e^{j\theta})G(e^{j\theta}) > 0,$$

then there exists an IOF. As the dual result, for $G(z) \in \mathcal{RH}_\infty^{m\times p}$, being surjective and satisfying $\forall\theta \in [0, 2\pi]$,

$$rank\left(G(e^{j\theta})\right) = m \Longleftrightarrow G(e^{j\theta})G^\sim(e^{j\theta}) > 0,$$

then there exists a CIOF.

Theorem 2.5 *Given $G(z) \in \mathcal{RH}_\infty^{m\times p}$ that is injective and satisfies $\forall\theta \in [0, 2\pi]$,*

$$rank\left(G(e^{j\theta})\right) = p \Longleftrightarrow G^\sim(e^{j\theta})G(e^{j\theta}) > 0,$$

then there exists an RCF $G(z)$, which also yields an IOF,

$$G(z) = N(z)M^{-1}(z) = G^i(z)G^o(z)$$

with $G^o(z) = M^{-1}(z) \in \mathcal{RH}_\infty$ as outer and $G^i(z) = N(z)$ as inner. This factorisation is unique up to a constant unitary matrix. Furthermore, corresponding to the state space realisation (2.2)–(2.3), when $\forall\theta \in [0, 2\pi]$

$$rank\begin{bmatrix} A - e^{j\theta}I & B \\ C & D \end{bmatrix} = n + p,$$

then the above RCP (M, N) is given by

$$\begin{aligned} &M(z) = (A + BF, BV, F, V), N(z) = (A + BF, BV, C + DF, DV), \\ &V = \left(D^T D + BXB^T\right)^{-1/2}, F = -\left(D^T D + BXB^T\right)^{-1}(B^T XA + D^T C), \end{aligned} \tag{2.45}$$

where $X > 0$ $(X \geq 0)$ is the stabilising solution of the Riccati equation

$$A^T XA - X + C^T C - F^T \left(D^T D + BXB^T\right) F = 0.$$

As the dual result, for $G(z) \in \mathcal{RH}_\infty^{m\times p}$, being surjective and satisfying $\forall\theta \in [0, 2\pi]$,

$$G(e^{j\theta})G^\sim(e^{j\theta}) > 0,$$

then there exists an LCF $G(z) = \hat{M}^{-1}(z)\hat{N}(z)$ that also gives a CIOF

$$G(z) = \hat{M}^{-1}(z)\hat{N}(z) = G^{co}(z)G^{ci}(z)$$

with $G^{co}(z) = \hat{M}^{-1}(z)$ as co-outer and $G^{ci}(z) = \hat{N}(z)$ as co-inner. This factorisation is unique up to a constant unitary matrix. Moreover, if $\forall\theta \in [0, 2\pi]$

$$\mathit{rank} \begin{bmatrix} A - e^{j\theta} I & B \\ C & D \end{bmatrix} = n + m,$$

then the above LCF is given by

$$\begin{aligned} &\hat{M}(z) = (A - LC, -L, WC, W), \hat{N}(z) = (A - LC, B - LD, WC, WD), \\ &W = \left(DD^T + CYC^T\right)^{-1/2}, L = (A^T YC^T + BD^T)(DD^T + CYC^T)^{-1}, \end{aligned} \tag{2.46}$$

where $Y > 0$ $(Y \geq 0)$ is the stabilising solution of the Riccati equation

$$AYA^T - Y + BB^T - L(DD^T + CYC^T)L^T = 0.$$

It is noteworthy that Theorem 2.5 establishes a connection between IOF (CIOF) and RCF (LCF), which is useful for our latter study.

2.2.2 Application Examples

As mentioned, as a powerful analysis and design tool, IOF and CIOF are widely applied in control and detection areas. Below are some examples.

Normalised RCF and LCF: It is evident that the normalised RCF and LCF, written as

$$\begin{bmatrix} M_0 \\ N_0 \end{bmatrix} = \left(A + BF_0, BV_0, \begin{bmatrix} F_0 \\ C + DF_0 \end{bmatrix}, \begin{bmatrix} V_0 \\ DV_0 \end{bmatrix} \right) \text{ and}$$
$$\begin{bmatrix} \hat{N}_0 & \hat{M}_0 \end{bmatrix} = \left(A - L_0 C, \begin{bmatrix} B - L_0 D & -L_0 \end{bmatrix}, W_0 C, \begin{bmatrix} W_0 D & W_0 \end{bmatrix} \right),$$

are inner and co-inner, respectively. According to Theorem 2.5, any RCF and LCF,

$$\begin{bmatrix} M \\ N \end{bmatrix} = \left(A + BF, BV, \begin{bmatrix} F \\ C + DF \end{bmatrix}, \begin{bmatrix} V \\ DV \end{bmatrix} \right) \text{ and}$$
$$\begin{bmatrix} \hat{N} & \hat{M} \end{bmatrix} = \left(A - LC, \begin{bmatrix} B - LD & -L \end{bmatrix}, WC, \begin{bmatrix} WD & W \end{bmatrix} \right),$$

can be factorised as

$$\begin{bmatrix} M(z) \\ N(z) \end{bmatrix} = \begin{bmatrix} M_0(z) \\ N_0(z) \end{bmatrix} V(z),$$
$$\begin{bmatrix} \hat{N}(z) & \hat{M}(z) \end{bmatrix} = R(z) \begin{bmatrix} \hat{N}_0(z) & \hat{M}_0(z) \end{bmatrix}$$

with outer $V(z)$ and co-outer $R(z)$ respectively given by

$$V(z) = V_0^{-1} \left(I + (F - F_0)(zI - A - BF)^{-1} B \right) V,$$
$$R(z) = W \left(I - C(zI - A + LC)^{-1} (L - L_0) \right) W_0^{-1}.$$

See also Lemma 3.1 for the derivation.

Linear quadratic optimal control: Linear quadratic (LQ) control is a popular optimal (state) feedback control scheme and often serves as a benchmark. For our purpose, we firstly introduce LQ optimal control scheme shortly and then address its relations to system inner-outer factorisation.

Consider a system described by state Eq. (2.2) and suppose that (i) (A, B) is controllable (stabilisable), and (ii) the state vector x is measurable. An LQ controller is a state feedback,

$$u(k) = Fx(k),$$

that minimises the cost function

$$J = \sum_{k=0}^{\infty} \begin{bmatrix} x^T(k) & u^T(k) \end{bmatrix} W \begin{bmatrix} x(k) \\ u(k) \end{bmatrix}, \tag{2.47}$$

$W \geq 0 \in \mathbb{R}^{(n+p)\times(n+p)}$ is a weighting matrix. Without loss of generality, W is written into

$$W = \begin{bmatrix} C^T \\ D^T \end{bmatrix} \begin{bmatrix} C & D \end{bmatrix}, C \in \mathbb{R}^{(n+p)\times n}, D \in \mathbb{R}^{(n+p)\times p},$$

and it is assumed that (C, A) is observable (reconstructable) and $D^T D > 0$. Hence,

$$J = \sum_{k=0}^{\infty} \left(Cx(k) + Du(k) \right)^T \left(Cx(k) + Du(k) \right). \tag{2.48}$$

Here, we would like to place a greater emphasis on the abuse of notations C and D. They are adopted only for the reason to ensure notationally compatible with the notations used in the inner-outer factorsation, and thus should not be interpreted as the ones given in the output Eq. (2.3).

Remark 2.5 Often, the cost function is of the form,

$$J = \sum_{k=0}^{\infty} \left(x^T(k) Q x(k) + u^T(k) R u(k) \right), \, Q \geq 0, \, R > 0,$$

a special case of (2.47). It is also common practice that $\mathcal{L}_2$-norm of a control output vector, not identical with the measurement output as given in (2.3), is defined as the cost function. Let us denote by

$$y_c(k) = Cx(k) + Du(k)$$

the control output. The cost function (2.48) is then written as

$$J = \sum_{k=0}^{\infty} y_c^T(k) y_c(k) = \sum_{k=0}^{\infty} \| y_c(k) \|^2 . \tag{2.49}$$

In the context of linear quadratic gaussian (LQG) or $\mathcal{H}_2$ control, the cost function could be $\mathcal{L}_2$-norm of the system output (measurement) vector, namely

$$J = \sum_{k=0}^{\infty} \| y(k) \|^2$$

with $y(k)$ given by the output Eq. (2.3).

It is a known result that, on the assumption of cost function (2.48), the LQ controller is given by

$$u(k) = Fx(k),$$
$$F = -\left(D^T D + B X B^T \right)^{-1} (B^T X A + D^T C),$$

that is, the feedback gain matrix F is identical with the one for the inner-outer factorisation in (2.45). Extending u to

$$u(k) = Fx(k) + Vv(k),$$
$$F = -\left(D^T D + B X B^T \right)^{-1} (B^T X A + D^T C),$$

with $v \in \mathbb{R}^p$ as the reference signal, and $V = \left(D^T D + BXB^T\right)^{-1/2}$ as a pre-filter, the response of the control output y_c to the reference signal v,

$$y_c(z) = N(z)\,v(z) = \left((C + DF)(zI - A - BF)^{-1} B + D\right) V v(z),$$

is an inner. As a result,

$$J = \sum_{k=0}^{\infty} y_c^T(k) y_c(k) = x_0^T X x_0 + \|y_c(k)\|_2^2 = x_0^T X x_0 + \|v(k)\|_2^2 .$$

It is of interest to observe that

$$u(z) = \left(F(zI - A - BF)^{-1} B + I\right) V v(z) = M(z)v(z).$$

In other words, on the assumption of the plant dynamic,

$$y_c(z) = G_c(z)u(z),$$

which is equivalently factorised into

$$G_c(z) = N(z)M^{-1}(z) = G_c^i(z)G_c^o(z) \Longrightarrow u(z) = \left(G_c^o(z)\right)^{-1} v(z),$$

the LQ controller can be interpreted as a compensation of the invertible part of the plant dynamic expressed in terms of the outer.

Optimal fault detection filter: A fault detection filter (FDF) is an observer-based residual generator with a full-order observer, as described by (2.8)–(2.9). Taking an unknown input as disturbance into account, the model (2.2)–(2.3) is extended to

$$\begin{aligned} x(k+1) &= Ax(k) + Bu(k) + E_d d(k), \\ y(k) &= Cx(k) + Du(k) + F_d d(k), \end{aligned}$$

where $d(k) \in \mathbb{R}^{k_d}$ represents an $\mathcal{L}_2$-bounded unknown input vector, and E_d, F_d are known matrices. Correspondingly, the dynamic of the FDF is governed by

$$\begin{aligned} r_y(z) &= \hat{M}(z)y(z) - \hat{N}(z)u(z) = \hat{N}_d(z)d(z), \\ \hat{N}_d(z) &= (A - LC, E_d - LF_d, WC, WF_d) . \end{aligned}$$

The so-called unified solution gives an optimal design of the observer gain matrix L and post-filter W, given by

$$\begin{aligned} L &= (A^T Y C^T + BD^T)(DD^T + CYC^T)^{-1}, \\ W &= \left(DD^T + CY(k)C^T\right)^{-1/2}, \end{aligned}$$

which results in a maximal fault detectability. Observe that (L, W) are exactly the ones adopted in the co-inner-outer factorisation given in (2.46). Therefore, $\hat{N}_d(z)$ is a co-inner, leading to

$$\left\| r_y \right\|_2 = \|d\|_2$$

In fact, given the transfer function from d to y,

$$y(z) = G_d(z)d(z), G_d(z) = F_d + C\left(zI - A\right)^{-1} E_d,$$

by an LCF and an equivalent co-inner-outer factorisation,

$$G_d(z) = \hat{M}_d^{-1}(z)\hat{N}_d(z) = G_d^{co}(z)G_d^{ci}(z)$$
$$\Longrightarrow \hat{M}_d(z) = \left(G_d^{co}(z)\right)^{-1}, \hat{N}_d(z) = G_d^{ci}(z),$$

we have

$$r_y(z) = \hat{M}(z)y(z) - \hat{N}(z)u(z) = \left(G_d^{co}(z)\right)^{-1} G_d^{co}(z)G_d^{ci}(z)d(z)$$
$$= G_d^{ci}(z)d(z) = \hat{N}_d(z)d(z).$$

We can draw an interesting conclusion from this example that the unified solution implies a co-inner-outer factorisation of G_d.

2.3 SKR and SIR, Kernel and Image Subspaces

By means of the LCF and RCF of the plant dynamics, we have introduced in the last section two stable and dual systems, (2.10) and (2.13), in which two system relations between the process data (u, y) and two (hidden) variables, $\left(r_y, v\right)$, are given. Named according to the (mathematical) role of (u, y), these two systems are called stable kernel representation (SKR) and stable image representation (SIR) of the plant, respectively. That is,

$$\text{SKR} : r_y(z) = K_G(z)\begin{bmatrix} u(z) \\ y(z) \end{bmatrix}, K_G(z) = \begin{bmatrix} -\hat{N}(z) & \hat{M}(z) \end{bmatrix}, \tag{2.50}$$

$$\text{SIR} : \begin{bmatrix} u(z) \\ y(z) \end{bmatrix} = I_G(z)v(z), I_G(z) = \begin{bmatrix} M(z) \\ N(z) \end{bmatrix}. \tag{2.51}$$

Corresponding to them, we introduce the definition of two subspaces in Hilbert space $\mathcal{H}_2$.

Definition 2.5 Given system model (2.1)–(2.3) and the corresponding LCP and RCP $\left(\hat{M}, \hat{N}\right)$ and (M, N), the $\mathcal{H}_2$ subspace $\mathcal{K}_G$ defined by

$$\mathcal{K}_G = \left\{ \begin{bmatrix} u \\ y \end{bmatrix} \in \mathcal{H}_2 : \begin{bmatrix} -\hat{N} & \hat{M} \end{bmatrix} \begin{bmatrix} u \\ y \end{bmatrix} = 0 \right\} \tag{2.52}$$

is called kernel subspace of G, and the $\mathcal{H}_2$ subspace $\mathcal{I}_G$ defined by

$$\mathcal{I}_G = \left\{ \begin{bmatrix} u \\ y \end{bmatrix} \in \mathcal{H}_2 : \begin{bmatrix} u \\ y \end{bmatrix} = \begin{bmatrix} M \\ N \end{bmatrix} v, v \in \mathcal{H}_2 \right\} \tag{2.53}$$

is image subspace.

It is evident that

- the kernel and image subspaces $\mathcal{K}_G$ and $\mathcal{I}_G$ consist of all (bounded) input and output pairs (u, y) generated by v,
- they are closed subspaces in $\mathcal{H}_2$, and
- they describe the same signal set, (u, y) generated by v, and thus

$$\mathcal{K}_G = \mathcal{I}_G.$$

When $v \in \mathcal{L}_2, \mathcal{L}_2 = \mathcal{H}_2 \oplus \mathcal{H}_2^{\perp}$, the output of the SIR builds a subspace in $\mathcal{L}_2$,

$$\begin{bmatrix} u \\ y \end{bmatrix} = I_G v \in \mathcal{L}_2.$$

Accordingly, there are image and kernel subspaces in Hilbert space $\mathcal{L}_2$,

$$\mathcal{I}_G^{\mathcal{L}_2} = \left\{ \begin{bmatrix} u \\ y \end{bmatrix} \in \mathcal{L}_2 : \begin{bmatrix} u \\ y \end{bmatrix} = \begin{bmatrix} M \\ N \end{bmatrix} v, v \in \mathcal{L}_2 \right\}, \tag{2.54}$$

$$\mathcal{K}_G^{\mathcal{L}_2} = \left\{ \begin{bmatrix} u \\ y \end{bmatrix} \in \mathcal{L}_2 : \begin{bmatrix} -\hat{N} & \hat{M} \end{bmatrix} \begin{bmatrix} u \\ y \end{bmatrix} = 0 \right\}. \tag{2.55}$$

The image and kernel subspaces $\mathcal{I}_G^{\mathcal{L}_2}$ and $\mathcal{K}_G^{\mathcal{L}_2}$ will be considered in our investigation on projection-based fault detection schemes.

Remark 2.6 In the sequel, image and kernel subspaces are generally adopted for $\mathcal{I}_G$ and $\mathcal{K}_G$ in $\mathcal{H}_2$, when they are not specified.

Analogously to $\mathcal{I}_G$ and $\mathcal{K}_G$, the image and kernel subspaces of the controller, in the context of the feedback control configuration shown in Fig. 2.1, are defined as follows.

Definition 2.6 Given system model (2.30) and the corresponding LCP and RCP (X, Y) and $\left(\hat{X}, \hat{Y}\right)$, the $\mathcal{H}_2$ subspaces $\mathcal{K}_K$ and $\mathcal{I}_K$ defined by

$$\mathcal{K}_K = \left\{ \begin{bmatrix} u \\ y \end{bmatrix} \in \mathcal{H}_2 : \begin{bmatrix} X & Y \end{bmatrix} \begin{bmatrix} u \\ y \end{bmatrix} = 0 \right\}, \tag{2.56}$$

$$\mathcal{I}_K = \left\{ \begin{bmatrix} u \\ y \end{bmatrix} \in \mathcal{H}_2 : \begin{bmatrix} u \\ y \end{bmatrix} = \begin{bmatrix} -\hat{Y} \\ \hat{X} \end{bmatrix} r_y, r_y \in \mathcal{H}_2 \right\} \tag{2.57}$$

are called kernel and image subspaces of K.

Remark 2.7 Given that the image and kernel subspaces of the plant are under consideration in most of our works, the terms "image and kernel subspaces" are generally adopted for the ones of the plant.

We would like to emphasise that image and kernel subspaces are signal (data) spaces. Handling of detection and control issues in this regard enables the realisation of proposed design schemes and algorithms in the data-driven fashion.

2.4 Uncertainties and Residuals

Roughly speaking, given a process model, for instance, the system model (2.1)–(2.3), uncertainties refer to any changes in the process that cause deviations of the dynamics from the system model. In real applications, uncertainties are inevitable. The major concern in the context of system performance, reliability and safety is those uncertainties, which are unmeasurable and thus cannot be compensated. Uncertainties could be caused by variations in system model parameters (model uncertainties), environment disturbances (unknown inputs) or technical faults and even cyber attacks in networked control systems. At this stage of our discussions, we will not distinguish uncertainties caused by faults and model uncertainties, and address them in a uniform manner. Our attention is to the description of various types of uncertainties and, above all, their relations to the residual vector introduced in the previous section.

2.4.1 *Description of Process Uncertainties Under Consideration*

With the aid of the coprime factorisation technique, process uncertainties under consideration are shortly summarised in this subsection. We denote the nominal (uncertainty-free) plant model by $G_0(z)$ whose minimal realisation is given by (2.2)–(2.3). Without loss of generality, it is assumed that the LCF and RCF of $G_0(z)$ are normalised,

$$G_0(z) = \hat{M}_0^{-1}(z)\hat{N}_0(z) = N_0(z)M_0^{-1}(z).$$

Unknown inputs are a typical type of uncertainties and described by $G_d(z)d(z)$. The system model with unknown input vector d is given by

$$y(z) = G_0(z)u(z) + G_d(z)d(z), d \in \mathbb{C}^{k_d},$$

where $G_d(z)$ is a known transfer function. For simplicity, in the subsequent study it is assumed that the state space representation of the above model is

$$x(k+1) = Ax(k) + Bu(k) + E_d d(k), \tag{2.58}$$
$$y(k) = Cx(k) + Du(k) + F_d d(k) \tag{2.59}$$

with known matrices E_d and F_d. Note that the unknown input vector d does not affect the system stability. Differently, variations of parameters in the system matrices (A, B, C, D) result in system eigen-dynamic related changes in the open- or closed-loop system configuration and therefore have to be addressed with care. In the factorisation framework, the so-called coprime factor uncertainties are introduced as follows. Let

$$G(z) = \hat{M}^{-1}(z)\hat{N}(z) = N(z)M^{-1}(z)$$

be the process under consideration. The LCP and RCP, $\left(\hat{M}, \hat{N}\right)$ and (M, N), are corrupted with uncertainties modelled by

$$\hat{M}(z) = \hat{M}_0(z) + \Delta_{\hat{M}}(z), \hat{N}(z) = \hat{N}_0(z) + \Delta_{\hat{N}}(z), \tag{2.60}$$
$$M(z) = M_0(z) + \Delta_M(z), N(z) = N_0(z) + \Delta_N(z), \tag{2.61}$$

where $\Delta_{\hat{M}}, \Delta_{\hat{N}}, \Delta_M, \Delta_N \in \mathcal{H}_\infty$ denote some unknown stable systems and are called coprime factor uncertainties.There are numerous ways to model these terms in more details, depending on available *a prior* knowledge. For instance,

$$\Delta_{\hat{M}}(z) = \hat{M}_0(z)\delta_{\hat{M}}(z), \Delta_{\hat{N}}(z) = \hat{N}_0(z)\delta_{\hat{N}}(z), \delta_{\hat{M}}(z), \delta_{\hat{N}}(z) \in \mathcal{H}_\infty$$

represent those uncertainties or faults with known and unknown parts. It is also often the case that only the bounds of uncertainties are known, for instance, expressed by

$$\left\| \begin{bmatrix} -\Delta_{\hat{N}} & \Delta_{\hat{M}} \end{bmatrix} \right\|_\infty \leq \delta_{\Delta_K}, \left\| \begin{bmatrix} \Delta_M \\ \Delta_N \end{bmatrix} \right\|_\infty \leq \delta_{\Delta_I}.$$

Remark 2.8 Without loss of generality, it is assumed, throughout the book, that,

$$\delta_{\Delta_K} < 1, \delta_{\Delta_I} < 1.$$

Remember that SKR and SIR are alternative system representations. Let

$$K_{G_0}(z) = \begin{bmatrix} -\hat{N}_0(z) & \hat{M}_0(z) \end{bmatrix}, I_{G_0}(z) = \begin{bmatrix} M_0(z) \\ N_0(z) \end{bmatrix}$$

be the normalised SKR and SIR of the nominal system under consideration. In our subsequent work, we also consider uncertain SKR and SIR described by

$$K_G(z) = \left[-\hat{N}_0(z) \; \hat{M}_0(z) \right] (I + \Delta_K(z)), \tag{2.62}$$

$$I_G(z) = (I + \Delta_I(z)) \begin{bmatrix} M_0(z) \\ N_0(z) \end{bmatrix}, \tag{2.63}$$

where $\Delta_K, \Delta_I \in \mathcal{RH}_\infty$ are unknown and assumed to be bounded by

$$\|\Delta_K\|_\infty < 1, \, \|\Delta_I\|_\infty < 1.$$

Below are some useful relations between the uncertainties modelled by (2.60), (2.61) and (2.62)–(2.63).

Theorem 2.6 *Given an uncertain system $G = \hat{M}^{-1}\hat{N} = NM^{-1}$ with the nominal model $G_0 = \hat{M}_0^{-1}\hat{N}_0 = N_0 M_0^{-1}$, then the SKR and SIR of G,*

$$K_G(z) = \left[-\hat{N}(z) \; \hat{M}(z) \right], \, I_G(z) = \begin{bmatrix} M(z) \\ N(z) \end{bmatrix},$$

can be equivalently written as

$$\left[-\hat{N}(z) \; \hat{M}(z) \right] = \left[-\hat{N}_0(z) \; \hat{M}_0(z) \right] (I + \Delta_K(z)),$$

$$\Delta_K(z) = \begin{bmatrix} -\hat{Y}_0(z) \\ \hat{X}_0(z) \end{bmatrix} \left(\left[-\hat{N}(z) \; \hat{M}(z) \right] - \left[-\hat{N}_0(z) \; \hat{M}_0(z) \right] \right) \in \mathcal{RH}_\infty,$$

$$\begin{bmatrix} M(z) \\ N(z) \end{bmatrix} = (I + \Delta_I(z)) \begin{bmatrix} M_0(z) \\ N_0(z) \end{bmatrix},$$

$$\Delta_I(z) = \left(\begin{bmatrix} M(z) \\ N(z) \end{bmatrix} - \begin{bmatrix} M_0(z) \\ N_0(z) \end{bmatrix} \right) \left[X_0(z) \; Y_0(z) \right] \in \mathcal{RH}_\infty,$$

where (X_0, Y_0) and $\left(\hat{X}_0, \hat{Y}_0\right)$ are the LCP and RCP corresponding to (M_0, N_0) and $\left(\hat{M}_0, \hat{N}_0\right)$. Furthermore, for some $\Delta_I(z) \in \mathcal{RH}_\infty$ and

$$\begin{bmatrix} M(z) \\ N(z) \end{bmatrix} = (I + \Delta_I(z)) \begin{bmatrix} M_0(z) \\ N_0(z) \end{bmatrix},$$

it holds

$$\left[\hat{M} \; \hat{N} \right] = \left[\hat{M}_o \; \hat{N}_o \right] (I + \Delta_K(z))^{-1}, \, \Delta_K(z) = \Delta_I(z).$$

The proof of this theorem can be found in the references given at the end of this chapter.

Example 2.1 Inspired by the relations given in Theorem 2.6, we present two parameterisation forms of SKR and SIR of G with uncertainties. Given $G = \hat{M}^{-1}\hat{N} = NM^{-1}$, an SKR of G can be modelled by

$$\begin{bmatrix} -\hat{N} & \hat{M} \end{bmatrix} = \begin{bmatrix} -\hat{N}_0 & \hat{M}_0 \end{bmatrix} \left(I + \begin{bmatrix} -\hat{Y}_0 \\ \hat{X}_0 \end{bmatrix} \Delta_{K_G} \right), \tag{2.64}$$

where

$$\Delta_{K_G} = \begin{bmatrix} -\hat{N} & \hat{M} \end{bmatrix} - \begin{bmatrix} -\hat{N}_0 & \hat{M}_0 \end{bmatrix} \in \mathcal{RH}_\infty$$

parametrises uncertainty. On the assumption that

$$\left\| \begin{bmatrix} -\hat{Y}_0 \\ \hat{X}_0 \end{bmatrix} \Delta_{K_G} \right\|_\infty < 1,$$

$I + \begin{bmatrix} -\hat{Y}_0 \\ \hat{X}_0 \end{bmatrix} \Delta_{K_G}$ is invertible over $\mathcal{RH}_\infty$ with

$$\left(I + \begin{bmatrix} -\hat{Y}_0 \\ \hat{X}_0 \end{bmatrix} \Delta_{K_G} \right)^{-1} = I - \begin{bmatrix} -\hat{Y}_0 \\ \hat{X}_0 \end{bmatrix} \left(I + \Delta_{K_G} \begin{bmatrix} -\hat{Y}_0 \\ \hat{X}_0 \end{bmatrix} \right)^{-1} \Delta_{K_G} \in \mathcal{RH}_\infty.$$

Therefore, (M, N) satisfying

$$\begin{bmatrix} M \\ N \end{bmatrix} = \left(I + \begin{bmatrix} -\hat{Y}_0 \\ \hat{X}_0 \end{bmatrix} \Delta_{K_G} \right)^{-1} \begin{bmatrix} M_0 \\ N_0 \end{bmatrix} \tag{2.65}$$

$$= \left(I - \begin{bmatrix} -\hat{Y}_0 \\ \hat{X}_0 \end{bmatrix} \left(I + \Delta_{K_G} \begin{bmatrix} -\hat{Y}_0 \\ \hat{X}_0 \end{bmatrix} \right)^{-1} \Delta_{K_G} \right) \begin{bmatrix} M_0 \\ N_0 \end{bmatrix} \tag{2.66}$$

is right-coprime and

$$\begin{bmatrix} -\hat{N} & \hat{M} \end{bmatrix} \begin{bmatrix} M \\ N \end{bmatrix} = 0.$$

Analogous to it, for

$$\begin{bmatrix} M \\ N \end{bmatrix} = \left(I + \Delta_{I_G} \begin{bmatrix} X_0 & Y_0 \end{bmatrix} \right) \begin{bmatrix} M_0 \\ N_0 \end{bmatrix},$$

$$\Delta_{I_G} = \begin{bmatrix} M \\ N \end{bmatrix} - \begin{bmatrix} M_0 \\ N_0 \end{bmatrix} \in \mathcal{RH}_\infty,$$

$$\left\| \Delta_{I_G} \begin{bmatrix} X_0 & Y_0 \end{bmatrix} \right\|_\infty < 1,$$

it holds

$$\begin{bmatrix} -\hat{N} & \hat{M} \end{bmatrix} = \begin{bmatrix} -\hat{N}_0 & \hat{M}_0 \end{bmatrix} \left(I + \Delta_{I_G} \begin{bmatrix} X_0 & Y_0 \end{bmatrix} \right)^{-1}$$

$$= \begin{bmatrix} -\hat{N}_0 & \hat{M}_0 \end{bmatrix} \left(I - \Delta_{I_G} \left(I + \begin{bmatrix} X_0 & Y_0 \end{bmatrix} \Delta_{I_G} \right)^{-1} \begin{bmatrix} X_0 & Y_0 \end{bmatrix} \right). \tag{2.67}$$

Equations (2.66)–(2.67) provide us with the uniform parameterisation of uncertain SKR and SIR, either by Δ_{K_G} or by Δ_{I_G}. Observe that (2.64) can be written as

$$\begin{bmatrix} -\hat{N} & \hat{M} \end{bmatrix} = \begin{bmatrix} -\hat{N}_0 & \hat{M}_0 \end{bmatrix} + \Delta_{K_G}.$$

Corresponding to it, the SIR is given by

$$\begin{bmatrix} M \\ N \end{bmatrix} = \begin{bmatrix} M_0 \\ N_0 \end{bmatrix} + \Delta_{I,K_G},$$

$$\Delta_{I,K_G} = -\begin{bmatrix} -\hat{Y}_0 \\ \hat{X}_0 \end{bmatrix} \left(I + \Delta_{K_G} \begin{bmatrix} -\hat{Y}_0 \\ \hat{X}_0 \end{bmatrix} \right)^{-1} \Delta_{K_G} \begin{bmatrix} M_0 \\ N_0 \end{bmatrix}. \tag{2.68}$$

Δ_{I,K_G} is interpreted as the corresponding variation in the SIR. Similarly, for

$$\begin{bmatrix} M \\ N \end{bmatrix} = \begin{bmatrix} M_0 \\ N_0 \end{bmatrix} + \Delta_{I_G},$$

it holds

$$\Delta_{K,I_G} = -\begin{bmatrix} -\hat{N}_0 & \hat{M}_0 \end{bmatrix} \Delta_{I_G} \left(I + \begin{bmatrix} X_0 & Y_0 \end{bmatrix} \Delta_{I_G} \right)^{-1} \begin{bmatrix} X_0 & Y_0 \end{bmatrix} \tag{2.69}$$

$$= -\begin{bmatrix} -\hat{N}_0 & \hat{M}_0 \end{bmatrix} \left(I + \Delta_{I_G} \begin{bmatrix} X_0 & Y_0 \end{bmatrix} \right)^{-1} \Delta_{I_G} \begin{bmatrix} X_0 & Y_0 \end{bmatrix}. \tag{2.70}$$

A possible application of the results in this example is the modelling of multiplicative faults. For instance, let

$$\Delta_{K_G}(z) = E_f(z)\delta_f(z) F_f(z)$$

be the fault model with known $E_f(z), F_f(z) \in \mathcal{RH}_\infty$ representing the "directions of the fault" and unknown $\delta_f(z) \in \mathcal{RH}_\infty$ denoting the "size of the fault". The corresponding faulty effect in SIR is, according to (2.68),

$$\Delta_{I,K_G} = -\begin{bmatrix} -\hat{Y}_0 \\ \hat{X}_0 \end{bmatrix} \left(I + E_f \delta_f F_f \begin{bmatrix} -\hat{Y}_0 \\ \hat{X}_0 \end{bmatrix} \right)^{-1} E_f \delta_f F_f \begin{bmatrix} M_0 \\ N_0 \end{bmatrix}.$$

A further class of uncertainties is the so-called linear fractional transformation (LFT) model and described by

$$x(k+1) = Ax(k) + Bu(k) + Ew(k), \tag{2.71}$$

$$y(k) = Cx(k) + Du(k) + Fw(k), \tag{2.72}$$

$$d(k) = C_d x(k) + D_d u(k) + F_d w(k) \in \mathbb{R}^{k_d}, \tag{2.73}$$

$$w(z) = \Delta d(z), \Delta \in \mathcal{H}_\infty, \tag{2.74}$$

where $w \in \mathbb{R}^{k_w}$ is an unknown input vector, and Δ is an unknown stable system. They represent the system uncertainty. E, F, C_d, D_d, F_d are known matrices of appropriate dimensions. It is straightforward that the system input-output dynamic, $y(z) = G(z)u(z)$, is described by

$$\begin{aligned} G(z) &= G_0(z) + G_{12}(z)\Delta \left(I - G_{22}(z)\Delta\right)^{-1} G_{21}(z), \\ G_0 &= (A, B, C, D), G_{12} = (A, E, C, F), \\ G_{21} &= (A, B, C_d, D_d), G_{22} = (A, E, C_d, F_d). \end{aligned} \tag{2.75}$$

Example 2.2 As an illustrating example, we would like to derive the SKR and SIR of G given in (2.75). According to the state space realisations of LCF and RCF, (2.5) and (2.6), G_{12}, G_{21} and G_{22} can be respectively factorised as

$$\begin{aligned} G_{12}(z) &= N_{12}(z)M_{12}^{-1}(z), G_{21}(z) = \hat{M}_{21}^{-1}(z)\hat{N}_{21}(z), \\ G_{22}(z) &= \hat{M}_{22}^{-1}(z)\hat{N}_{22}(z) = N_{22}(z)M_{22}^{-1}(z), \\ M_{12}(z) &= M_{22}(z), \hat{M}_{21}(z) = \hat{M}_{22}(z), \end{aligned}$$

which yields

$$\begin{aligned} G_{12}(z)\Delta \left(I - G_{22}(z)\Delta\right)^{-1} G_{21}(z) &= G_{12}(z) \left(I - \Delta G_{22}(z)\right)^{-1} \Delta G_{21}(z) \\ &= G_{12}(z)\Delta \left(\hat{M}_{21}(z) - \hat{N}_{22}(z)\Delta\right)^{-1} \hat{N}_{21}(z) \\ &= N_{12}(z) \left(M_{12}(z) - \Delta N_{22}(z)\right)^{-1} \Delta G_{21}(z). \end{aligned}$$

Below, it is assumed that

$$\Delta_1 := \Delta \left(\hat{M}_{21}(z) - \hat{N}_{22}(z)\Delta\right)^{-1} \hat{N}_{21}(z) \in \mathcal{RH}_\infty, \tag{2.76}$$

$$\Delta_2 := N_{12}(z) \left(M_{12}(z) - \Delta N_{22}(z)\right)^{-1} \Delta \in \mathcal{RH}_\infty. \tag{2.77}$$

Let $\left(\hat{M}_0, \hat{N}_0\right)$, (M_0, N_0) be the normalised LCP and RCP of G_o. The SKR and SIR of G are respectively given by

$$\begin{aligned} G &= \hat{M}_0^{-1}\left(\hat{N}_0 + \Delta_{\hat{N}}\right) = \left(N_0 + \Delta_N\right) M_0^{-1} \Longrightarrow \\ K_G &= \begin{bmatrix} -\hat{N}_0 - \Delta_{\hat{N}} & \hat{M}_0 \end{bmatrix} = \begin{bmatrix} -\hat{N}_0 & \hat{M}_0 \end{bmatrix} \left(I + \Delta_{K_G}\right), \\ \Delta_{\hat{N}} &= W_0 \left(F + C\left(zI - A + L_0 C\right)^{-1}\left(E - L_0 F\right)\right) \Delta_1 \in \mathcal{RH}_\infty, \\ \Delta_K &= \begin{bmatrix} -\hat{Y}_0 \\ \hat{X}_0 \end{bmatrix} \begin{bmatrix} -\Delta_{\hat{N}} & 0 \end{bmatrix}, \\ I_G &= \begin{bmatrix} M_0 \\ N_0 + \Delta_N \end{bmatrix} = \left(I + \Delta_I\right) \begin{bmatrix} M_0 \\ N_0 \end{bmatrix}, \end{aligned}$$

$$\Delta_N = \Delta_2 \left(D_d + (C_d + D_d F_0)(zI - A - BF_0)^{-1} B\right) V_0 \in \mathcal{RH}_\infty,$$
$$\Delta_I = \begin{bmatrix} 0 \\ \Delta_N \end{bmatrix} \begin{bmatrix} X_0 & Y_0 \end{bmatrix}.$$

Here, (L_0, W_0) and (F_0, V_0) are the observer gain and post-filter as well as the feedback gain and pre-filter matrices of the normalised LCF and RCF $\left(\hat{M}_0, \hat{N}_0\right)$ and (M_0, N_0), respectively, (X_0, Y_0) and $\left(\hat{X}_0, \hat{Y}_0\right)$ are the LCP and RCP corresponding to (M_0, N_0) and $\left(\hat{M}_0, \hat{N}_0\right)$.

2.4.2 Residual Dynamics with Respect to Uncertainties

Given the nominal (fault- and uncertainty-free) system model (2.2)–(2.3), system (2.8)–(2.10) builds an observer-based residual generator. We have learnt that the residual vector r_0 contains informations about the uncertainties in the process under consideration. Now, by means of models of uncertainties introduced in the previous subsection, we would like to specify this claim. Without loss of generality, it is supposed in the sequel that the normalised SKR is applied for the residual generation purpose. The dynamic of the residual vector r_y with respect to unknown input vector d in the model (2.58)–(2.59) is governed by

$$e(k+1) = (A - L_0 C)\, e(k) + (E_d - L_0 F_d)\, d(k),$$
$$r_y(k) = W_0 \left(Ce(k) + F_d d(k)\right), e(k) = x(k) - \hat{x}(k).$$

In the case of coprime factor uncertainties, (2.60)–(2.61) or (2.62)–(2.63), the dynamics of the residual vector r_y are described respectively by

$$r_y = -\begin{bmatrix} -\Delta_{\hat{N}} & \Delta_{\hat{M}} \end{bmatrix} \begin{bmatrix} u \\ y \end{bmatrix} = -\Delta_{K_G} \begin{bmatrix} u \\ y \end{bmatrix} = -K_{G_0} \Delta_K \begin{bmatrix} u \\ y \end{bmatrix},$$
$$r_y = \begin{bmatrix} -\hat{N}_0 & \hat{M}_0 \end{bmatrix} \begin{bmatrix} u \\ y \end{bmatrix} = K_{G_0} \Delta_{I_G} \begin{bmatrix} M_0 \\ N_0 \end{bmatrix} v, \Delta_{I_G} = \begin{bmatrix} \Delta_M \\ \Delta_N \end{bmatrix}. \tag{2.78}$$

Note that the hidden variable v in (2.78) is often not available during online operations. To solve this problem, we are able to use the relation (2.69) in Example 2.1, and rewrite (2.78) as

$$r_y = -\Delta_{K,I_G} \begin{bmatrix} u \\ y \end{bmatrix}$$
$$= \begin{bmatrix} -\hat{N}_0 & \hat{M}_0 \end{bmatrix} \Delta_{I_G} \left(I + \begin{bmatrix} X_0 & Y_0 \end{bmatrix} \Delta_{I_G}\right)^{-1} \begin{bmatrix} X_0 & Y_0 \end{bmatrix} \begin{bmatrix} u \\ y \end{bmatrix}.$$

Now, consider the LFT uncertainty (2.71)–(2.74). Using the result in Example 2.2, it holds

$$r_y = -\begin{bmatrix} -\Delta_{\hat{N}} & \Delta_{\hat{M}} \end{bmatrix} \begin{bmatrix} u \\ y \end{bmatrix} = W_0 \left(F + C (zI - A + L_0 C)^{-1} (E - L_0 F) \right) \Delta_1 u,$$

where Δ_1 is given by (2.76).

To put it in a nutshell, residual vector r_y (r_0) is driven by all possible uncertainties in the process under consideration. In other words, it is an information provider and indicator for any anomaly in the system dynamics caused by the uncertainties. It is remarkable that all aforementioned types of uncertainties are caused by variations of the (plant) model parameters or unknown inputs acted on the plant and hence affect the plant dynamics. Managing such uncertainties is classical issues in robust control, fault diagnosis and fault-tolerant control. In industrial CPSs, cyber-attacks can lead to uncertainties in the embedded controllers. As will be addressed in this book, such uncertainties can be effectively managed by means of the residual signals generated on the controller side. The formalism of this handling setting will be shortly introduced in the next section.

2.5 Signal and Residual Subspaces

In traditional control engineering and theory, the major focus is on process input and output signals, (u, y). On the other hand, in the previous sections, the so-called residual signals, $\left(r_y, \bar{v}\right)$, have been introduced. They are generated by the process signals (u, y) and serve as information signals about anomalies in the system. In this section, we will analyse relations between the process signal pair (u, y) and residual/information signals $\left(r_y, \bar{v}\right)$ and study the role of the residual/information signals $\left(r_y, \bar{v}\right)$ in control and detection, in comparison with the process signals (u, y). For our purpose, the control system configuration shown in Fig. 2.2 is under consideration. The system input pair, (r, v), will be explained in the context of our subsequent investigation.

To begin with, we consider the closed-loop dynamics of the control system shown in Fig. 2.2. It is straightforward that

$$\begin{bmatrix} u(z) \\ y(z) \end{bmatrix} = \begin{bmatrix} I & -K(z) \\ -G(z) & I \end{bmatrix}^{-1} \begin{bmatrix} v(z) \\ r(z) \end{bmatrix}.$$

By means of the LCF of K and G, $K = -X^{-1}Y$, $G = \hat{M}^{-1}\hat{N}$, it can be further written as

$$\begin{bmatrix} u(z) \\ y(z) \end{bmatrix} = \begin{bmatrix} X(z) & Y(z) \\ -\hat{N}(z) & \hat{M}(z) \end{bmatrix}^{-1} \begin{bmatrix} X(z)v(z) \\ \hat{M}(z)r(z) \end{bmatrix}$$

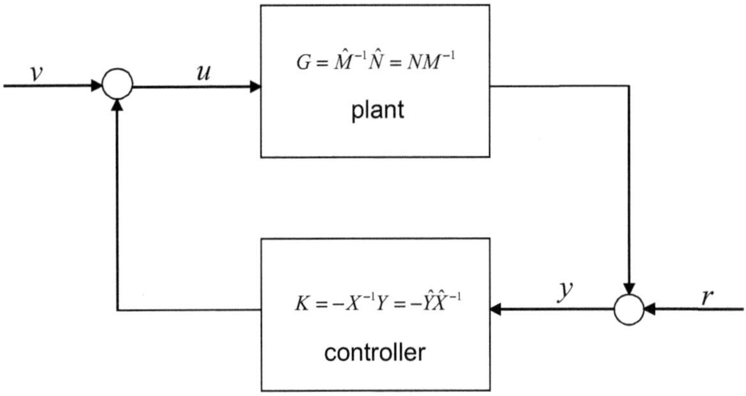

Fig. 2.2 Feedback control loop considered in Sect. 2.4

$$\Longleftrightarrow \begin{bmatrix} X(z)v(z) \\ \hat{M}(z)r(z) \end{bmatrix} = \begin{bmatrix} X(z) & Y(z) \\ -\hat{N}(z) & \hat{M}(z) \end{bmatrix} \begin{bmatrix} u(z) \\ y(z) \end{bmatrix}.$$

Recalling (2.10) and (2.22),

$$\hat{M}(z)r(z) = \begin{bmatrix} -\hat{N}(z) & \hat{M}(z) \end{bmatrix} \begin{bmatrix} u(z) \\ y(z) \end{bmatrix} = r_y(z), \tag{2.79}$$

$$X(z)v(z) = \begin{bmatrix} X(z) & Y(z) \end{bmatrix} \begin{bmatrix} u(z) \\ y(z) \end{bmatrix} = \bar{v}(z) \tag{2.80}$$

are the (process) residual vector and (control side) residual vector, respectively. On the other hand, according to Bezout identity (2.31), we have

$$\begin{aligned} \begin{bmatrix} u(z) \\ y(z) \end{bmatrix} &= \begin{bmatrix} X(z) & Y(z) \\ -\hat{N}(z) & \hat{M}(z) \end{bmatrix}^{-1} \begin{bmatrix} \bar{v}(z) \\ r_y(z) \end{bmatrix} \\ &= \begin{bmatrix} M(z) & -\hat{Y}(z) \\ N(z) & \hat{X}(z) \end{bmatrix} \begin{bmatrix} \bar{v}(z) \\ r_y(z) \end{bmatrix}. \end{aligned} \tag{2.81}$$

From the mathematical viewpoint, (2.79)–(2.81) imply that there exists a one-to-one mapping between the process signal pair (u, y) and the residual/information signal pair $(\bar{v}, r_y)$. Observe that

$$\begin{bmatrix} u(z) \\ y(z) \end{bmatrix} = \begin{bmatrix} M(z) \\ N(z) \end{bmatrix} \bar{v}(z) + \begin{bmatrix} -\hat{Y}(z) \\ \hat{X}(z) \end{bmatrix} r_y(z). \tag{2.82}$$

It follows from Definitions 2.5–2.6 that $\begin{bmatrix} M \\ N \end{bmatrix} \bar{v}$ and $\begin{bmatrix} -\hat{Y} \\ \hat{X} \end{bmatrix} r_y$ build the system and controller image subspaces $\mathcal{I}_G$ and $\mathcal{I}_K$, respectively. More importantly, the Bezout identity implies that $\mathcal{I}_G$ and $\mathcal{I}_K$ are complementary. Consequently, the process signal space $\mathcal{S}$ is constructed by

$$\mathcal{S} = \left\{ \begin{bmatrix} u \\ y \end{bmatrix} \in \mathcal{H}_2^{m+p} \right\} = \mathcal{I}_G \oplus \mathcal{I}_K. \tag{2.83}$$

We summarise this statement in form of a theorem.

Theorem 2.7 *The image subspaces of the plant and the controller defined in Definitions 2.5–2.6, $\mathcal{I}_G$ and $\mathcal{I}_K$, are complementary, namely*

$$\mathcal{I}_G \cap \mathcal{I}_K = 0.$$

Proof The fact that $\mathcal{I}_G$ and $\mathcal{I}_K$ are complementary is apparent. Otherwise, there would exist $\alpha \in \mathcal{I}_G, \alpha \neq 0$, so that α belongs to $\mathcal{I}_K$. Since for any $\beta \in \mathcal{I}_K, \gamma \in \mathcal{I}_G, \beta \neq 0, \gamma \neq 0$,

$$\begin{bmatrix} -\hat{N} & \hat{M} \end{bmatrix} \beta \neq 0, \begin{bmatrix} -\hat{N} & \hat{M} \end{bmatrix} \gamma = 0.$$

it is clearly impossible that $\alpha \in \mathcal{I}_K \cap \mathcal{I}_G$. As a result, $\mathcal{I}_G \cap \mathcal{I}_K = 0$.

Similarly, it holds

$$\mathcal{K}_G \cap \mathcal{K}_K = 0, \tag{2.84}$$

which implies that the subspace consisting of $(\bar{v}, r_y)$ and denoted by $\mathcal{R}$ can be expressed as

$$\begin{aligned} \mathcal{R} &= \left\{ \begin{bmatrix} \bar{v} \\ r_y \end{bmatrix} \in \mathcal{H}_2^{m+p}, \text{generated by} \begin{bmatrix} u \\ y \end{bmatrix} \in \mathcal{H}_2^{m+p} \text{ according to (2.79)–(2.80)} \right\} \\ &= \mathcal{R}_{\bar{v}} \oplus \mathcal{R}_{r_y}, \\ \mathcal{R}_{\bar{v}} &= \left\{ \begin{bmatrix} \bar{v} \\ 0 \end{bmatrix}, \bar{v} \text{ is generated by} \begin{bmatrix} u \\ y \end{bmatrix} \in \mathcal{H}_2^{m+p} \text{ according to (2.80)} \right\}, \\ \mathcal{R}_{r_y} &= \left\{ \begin{bmatrix} 0 \\ r_y \end{bmatrix}, r_y \text{ is generated by} \begin{bmatrix} u \\ y \end{bmatrix} \in \mathcal{H}_2^{m+p} \text{ according to (2.79)} \right\}. \end{aligned}$$

Remember that the signal vector $\bar{v}$ has two interpretations, depending on its role in the plant image subspace $\mathcal{I}_G$ and in subspace $\mathcal{R}_{\bar{v}}$. That is, $\bar{v}$ can be interpreted as a reference signal in $\mathcal{I}_G$, while in $\mathcal{R}_{\bar{v}}$ $\bar{v}$ is the residual vector of the control input side. Aiming at gaining deeper insights into the process signal spaces, and their role in the unified framework of control and detection, in the sequel, we take a close look at these two cases.

Remark 2.9 In subsequent chapters, the so-called residual subspace will be defined as the orthogonal complement of the system image subspace $\mathcal{I}_G$ and denoted by $\mathcal{R}_G$.

2.5.1 Feedback Control with $\bar{v}$ as Reference

It is known from (2.82) that the closed-loop variables (u, y) consist of two components,

$$\begin{bmatrix} u \\ y \end{bmatrix} = \begin{bmatrix} u_v \\ y_v \end{bmatrix} + \begin{bmatrix} u_r \\ y_r \end{bmatrix},$$

which are complementary to each other, and

- (u_v, y_v) is driven by the reference signal $\bar{v}$,

$$\begin{bmatrix} u_v(z) \\ y_v(z) \end{bmatrix} = \begin{bmatrix} M(z) \\ N(z) \end{bmatrix} \bar{v}(z),$$

- while (u_r, y_r) by the residual signal r_y,

$$\begin{bmatrix} u_r(z) \\ y_r(z) \end{bmatrix} = \begin{bmatrix} -\hat{Y}(z) \\ \hat{X}(z) \end{bmatrix} r_y(z). \tag{2.85}$$

It is obvious that (u_v, y_v) corresponds the feedforward controller u_v and the response of the plant variables to it. The dynamics of (u_v, y_v) have no influence on the closed-loop eigen-dynamics like stability. System description (2.85) reveals that the feedback dynamic of the closed-loop system shown in Fig. 2.2 is indeed an observer-based state feedback control loop,

$$\hat{x}(k+1) = (A + BF)\,\hat{x}(k) + LW^{-1}r_y(k), \tag{2.86}$$

$$\begin{bmatrix} u_r(k) \\ y_r(k) \end{bmatrix} = \begin{bmatrix} F \\ (C + DF) \end{bmatrix} \hat{x}(k) + \begin{bmatrix} 0 \\ W^{-1} \end{bmatrix} r_y(k), \tag{2.87}$$

and driven by the residual vector r_y. Consider that the residual vector r_y is an indicator of uncertainties in the process (plant). Hence, the spirit of feedback control is the feedback of the uncertainties in the plant dynamics. It is worth to emphasise that the feedforward and feedback dynamics, $\begin{bmatrix} u_v \\ y_v \end{bmatrix}$ and $\begin{bmatrix} u_r \\ y_r \end{bmatrix}$, are vectors from two complementary subspaces, $\mathcal{I}_G$ and $\mathcal{I}_K$. Thus, for the analysis and synthesis of feedback control loops, it is sufficient to consider $\begin{bmatrix} u_r \\ y_r \end{bmatrix}$ exclusively.

Notice that (u_r, y_r) is driven by r_y, therefore it can be used for the fault detection purpose. To this end, (u_r, y_r) is computed either based on (2.82),

$$\begin{bmatrix} u_r(z) \\ y_r(z) \end{bmatrix} = \begin{bmatrix} u(z) \\ y(z) \end{bmatrix} - \begin{bmatrix} M(z) \\ N(z) \end{bmatrix} \bar{v}(z),$$

with known $\begin{bmatrix} M \\ N \end{bmatrix} \bar{v}$, or alternatively, by an orthogonal projection of $\begin{bmatrix} u \\ y \end{bmatrix}$ onto the image subspace of the controller, $\mathcal{I}_K$, which is spanned by $\begin{bmatrix} -\hat{Y} \\ \hat{X} \end{bmatrix}$. The latter method is of considerable practical interest, since the only information needed for the realisation of the orthogonal projection is the SIR of the controller, which is known and independent of variations in the plant dynamics. This enables us a (process) model-free fault detection. In Part II of this book, we will describe this approach in more detail.

Remark 2.10 In Part II, the orthogonal projection-based fault detection framework will be introduced.

2.5.2 Feedback Control with $\bar{v}$ as Residual

We now study the case that $\bar{v}$ is an unknown residual vector and generated by (u, y) according to (2.80). In a traditional feedback control system, the controller,

$$K(z) = -X^{-1}(z)Y(z),$$

is independent of any variation is the plant and hence

$$\bar{v}(z) = \begin{bmatrix} X(z) & Y(z) \end{bmatrix} \begin{bmatrix} u(z) \\ y(z) \end{bmatrix} \equiv 0.$$

Unfortunately, as a part of an industrial CPS, today's feedback control systems may be under cyber-attacks. A consequence of disturbed communications or software manipulations is possible changes in the controller structure or dynamics. Denoting such changes by (Δ_X, Δ_Y), it turns out

$$\bar{v}(z) = \begin{bmatrix} X(z) + \Delta_X & Y(z) + \Delta_Y \end{bmatrix} \begin{bmatrix} u(z) \\ y(z) \end{bmatrix} = \begin{bmatrix} \Delta_X & \Delta_Y \end{bmatrix} \begin{bmatrix} u(z) \\ y(z) \end{bmatrix} \neq 0.$$

In other words, the residual vector $\bar{v}$ enables us to detect any manipulations on the controller K. In our subsequent work, as a residual vector (of the controller side), $\bar{v}$ is also denoted by r_u.

We would like to call the reader's attention to the relations (2.82) and (2.83). That is, in the signal space $\mathcal{S}$, the influences of plant uncertainties and manipulations of the controller on system signals (u, y) can be easily decoupled, for instance, by means of (orthogonal) projections onto $\mathcal{I}_G$ and $\mathcal{I}_K$ subspaces, respectively.

Example 2.3 In this example, an interesting application of the relation (2.82) and complementary subspaces $\mathcal{I}_G$ and $\mathcal{I}_K$ to fault and cyber-attack detection in an interconnected control system with a structure sketched in Fig. 2.3 is introduced. When

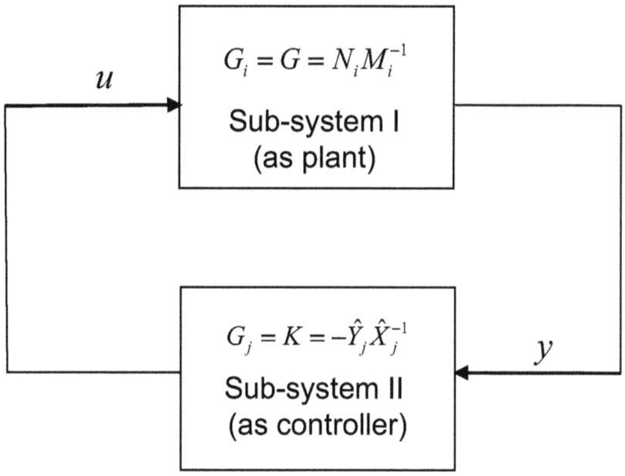

Fig. 2.3 An interconnected system as a feedback control loop

such an interconnected system is stable and the Bezout identity holds, it can be equivalently modelled as a feedback control loop shown in Fig. 2.1 with one of the sub-systems as the plant and another one as the feedback controller.

Let the transfer functions of the two sub-systems be $G_i, i = 1, 2$, and suppose that G_i is the plant G and $G_j, j = 1, 2, j \neq i$, is the controller K. The SIR of G_i is expressed by $\begin{bmatrix} M_i \\ N_i \end{bmatrix}$ and accordingly the SIR of G_j by $\begin{bmatrix} -\hat{Y}_j \\ \hat{X}_j \end{bmatrix}$. Denote the interconnection signals by u and y that are available at both sub-systems as measurement data. Viewed from the i-th sub-system, the system (loop) dynamic is described by

$$\begin{bmatrix} u(z) \\ y(z) \end{bmatrix} = \begin{bmatrix} M_i(z) \\ N_i(z) \end{bmatrix} r_j(z) + \begin{bmatrix} -\hat{Y}_j(z) \\ \hat{X}_j(z) \end{bmatrix} r_i(z), i, j = 1, 2, j \neq i,$$

where r_i, r_j represent the variations in G_i and G_j, respectively. As a result, the i-th sub-system can detect possible variations in the j-th sub-system from its nominal dynamic by projecting $\begin{bmatrix} u \\ y \end{bmatrix}$ onto $\mathcal{I}_G$ spanned by $\begin{bmatrix} M_i \\ N_i \end{bmatrix}$ without model knowledge of the j-th sub-system described by $\begin{bmatrix} -\hat{Y}_j \\ \hat{X}_j \end{bmatrix}$.

Remark 2.11 An extension to interconnected systems with N, $N > 2$, sub-systems is straightforward by different combinations of the N sub-systems. A potential application of such a detection scheme is distributed performance recovery, in which, for instance, the i-th sub-system can update its controller to compensate performance degradations caused by variations in other sub-systems. It should be mentioned that the above example is only a conceptual study on certain assumptions.

2.6 Notes and References

Coprime factorisations of dynamic systems are a standard mathematical and control-theoretic tool for system representations and widely adopted in robust control theory. In [57, 198, 199, 243], the interested reader can find well-structured and detailed descriptions about this useful tool. As a special coprime factorisation, the normalised LCF and RCF play an important role in our subsequent investigation, for instance, in the framework of projection-based fault diagnosis. The results given in Theorem 2.1 are presented and proved by Hoffmann [86].

We have consciously refrained from rigorous mathematical definitions and concepts. Instead, the focus is on the control engineering interpretations. From the very beginning, the interpretations of system LCF and RCF as an observer-based residual generator and a state feedback controller, the centerpieces of the unified control and detection framework, are introduced. And associated with them, the Bezout identity and another two relevant RCP and LCP are presented in the context of a standard feedback control loop with the configuration shown in Fig. 2.1. In this style, we are going to continue the presentation of the unified framework in the subsequent chapters.

Inner-outer and co-inner-outer factorisations are another type of system factorisations and closely related to the coprime factorisations. We refer the reader to [243] for a textbook styled description of this topic and [25, 121, 151, 193] for a comprehensive study. The proof of Theorems 2.3–2.5 can be found, for instance, in [243]. As demonstrated in the subsequent application examples, IOF and CIOF are widely applied in dealing with optimal control and detection issues. Concerning the LQ optimal control, the reader is referred to [19, 105]. The unified solution of optimal FDF design has been proposed by Ding et al. [41]. The achieved optimal FDF is in fact an $\mathcal{H}_2$ observer and of the maximum fault detectability, as proved in [36, 39].

On the basis of system LCF and RCF, two new system representation forms are introduced. This marks a decisive step from the widely used system input-output dynamics to the dynamic models

- SKR with process variables (u, y) as input and the residual vector r_y as output, and
- SIR with a latent variable v as input and the process variables (u, y) as output.

The signal pair $(r_y, \bar{v})$ are two hidden system variables. While r_y is the output of the observer-based residual generator and hence an indicator for uncertainties in the process under consideration, $\bar{v}$ serves as the reference signal of the state feedback controller. With regard to the feedback control loop and the Bezout identity, r_y as the input drives the SIR of an observer-based state feedback controller, and $\bar{v}$ can be interpreted as a residual vector generated by means of the SKR of the controller. These results reveal two important facts.

- a feedback controller is driven by the uncertainties in the process, and
- a controller can also act as a residual generator.

The former fact is helpful to gain deep insight into the feedback control mechanism and to develop control performance recovering systems, and the latter is useful for cyber-attack detection to be addressed in this book as well.

By means of SKR and SIR, the so-called kernel and image subspaces are established. They are the subspaces of the process signal pair (u, y), and build subspaces in Hilbert space. The introduction of kernel and image subspaces enables handling of many control and detection issues directly in signal spaces and in the data-driven fashion. This is an essential aspect of the unified framework of control and detection. The reader is referred to [194, 199] for detailed descriptions about system SKR, SIR, and kernel and image subspaces.

Handling of uncertainties in control systems is a major concern in robust control theory. Besides of unknown input modelling, representations of coprime factor uncertainties and LFT model are widely adopted in the research domain of robust control [199, 243]. In the fault detection framework, the major attention on studying uncertainty issues is how to receive sufficient information about the existence and type of the uncertainties under consideration. Against this background, in parallel with uncertainty modelling, attention has been paid to the observer-based residual generators, serving as indicators for uncertainties, and their dynamics with respect to the system uncertainties. The results described in Theorem 2.6 are a summary of the relevant results in [75] and Chap. 9 in [39] with some slight modifications. We would like to call the reader's attention to the result in Example 2.1 about the equivalent descriptions of uncertainties in the SKR and SIR models. They are useful in our fault diagnosis work.

The final effort in this chapter has been devoted to the relations between the process signals (u, y) and residual pairs $(r_y, \bar{v})$ as well as corresponding subspaces. Thanks to the Bezout identity, there exists a one-to-one mapping over $\mathcal{RH}_\infty$ between (u, y) and $(r_y, \bar{v})$. More importantly, the responses of (u, y) to r_y and $\bar{v}$ are analysed, and it is demonstrated that the image subspaces of the plant and controller, $\mathcal{I}_G$ and $\mathcal{I}_K$, are complementary. Some potential applications of this result are briefly summarised. It is noteworthy that the study in the signal spaces, instead of the model framework like transfer functions or state space representations, is a decisive step for data-driven realisations of model-based schemes and concepts. It builds a basis for an explainable application of ML-methods to dealing with control engineering issues as well.

Chapter 3
Parameterisations of Feedback Controllers and Observers

Parameterisations of stabilising feedback controllers and observers are the theoretical core of the unified framework of control and detection. While the so-called Youla parameterisation is well-known as the parameterisation form of stabilising feedback controllers and intensively applied in various control engineering relevant thematic fields, applications of observer parameterisation, apart from observer-based fault detection, are rarely reported. In this chapter, by means of Youla parameterisation and observer parameterisation, it will be delineated that all output feedback controllers are observer-based state feedback controllers, and the separation principle holds in a more generalised sense. This claim is a cornerstone of the unified framework.

In a similar style to the previous chapter, we will introduce the main results of parameterisations of stabilising feedback controllers and observers with focuses on control engineering interpretations, control and detection relations, information aspects and with a deliberate avoidance of rigorous mathematical treatment, when there is no risk of confusion.

3.1 Youla Parameterisation and Its Observer-Based Realisation Form

Since stabilisation is the main objective of controllers, the feedback control loop configuration sketched in Fig. 3.1 is the object of our consideration. Here, $\left(\hat{M}, \hat{N}\right)$ and (M, N) are LCP and RCP of the plant G, $\left(\hat{V}, \hat{U}\right)$ and (V, U) are RCP and LCP of the feedback controller K. Without loss of generality, it is supposed that v and d represent the reference and (unknown) disturbance vectors, respectively.

Now, we consider the closed-loop dynamics,

$$\begin{bmatrix} u \\ y \end{bmatrix} = \begin{bmatrix} I & -K \\ -G & I \end{bmatrix}^{-1} \begin{bmatrix} v \\ d \end{bmatrix}, \tag{3.1}$$

S. X. Ding, *Unified Framework of Control and Detection*,
https://doi.org/10.1007/978-3-662-72417-0_3

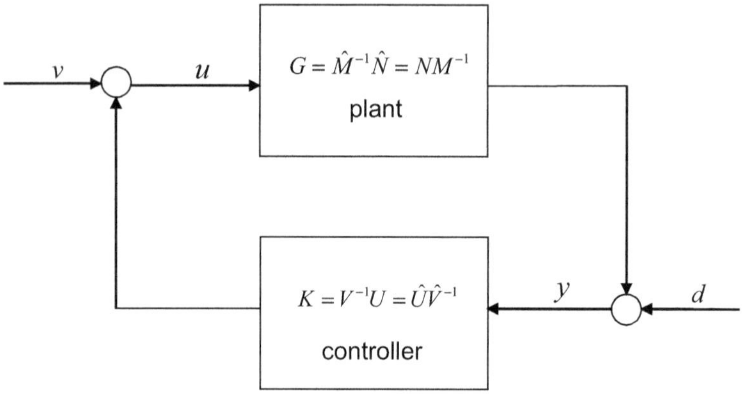

Fig. 3.1 Feedback control loop under consideration

which can be further written into

$$\begin{bmatrix} u \\ y \end{bmatrix} = \begin{bmatrix} V & -U \\ -\hat{N} & \hat{M} \end{bmatrix}^{-1} \begin{bmatrix} Vv \\ \hat{M}d \end{bmatrix}$$
$$= \begin{bmatrix} M & 0 \\ 0 & -\hat{V} \end{bmatrix} \begin{bmatrix} M & \hat{U} \\ N & \hat{V} \end{bmatrix}^{-1} \begin{bmatrix} v \\ -d \end{bmatrix}.$$

Hence, the closed-loop system is stable if and only if

$$\begin{bmatrix} V & -U \\ -\hat{N} & \hat{M} \end{bmatrix}^{-1} \in \mathcal{RH}_\infty, \begin{bmatrix} M & \hat{U} \\ N & \hat{V} \end{bmatrix}^{-1} \in \mathcal{RH}_\infty. \tag{3.2}$$

Remark 3.1 Throughout the book, the commonly used term "stability of a closed-loop" implies the internal stability of a feedback control loop. For its definition, the reader is referred to the given references.

Observe that due to coprimeness of the plant and controller, (3.2) is true if and only if

$$(VM - UN)^{-1} \in \mathcal{RH}_\infty, \left(\hat{M}\hat{V} - \hat{N}\hat{U}\right)^{-1} \in \mathcal{RH}_\infty, \tag{3.3}$$

since

$$\begin{bmatrix} V & -U \\ -\hat{N} & \hat{M} \end{bmatrix} \begin{bmatrix} M & \hat{U} \\ N & \hat{V} \end{bmatrix} = \begin{bmatrix} VM - UN & 0 \\ 0 & \hat{M}\hat{V} - \hat{N}\hat{U} \end{bmatrix}.$$

Recall that a special case for $\left(\hat{V}, \hat{U}\right)$ and (V, U) has been studied in Chap. 2,

$$\left(\hat{V}, \hat{U}\right) = \left(\hat{X}, -\hat{Y}\right), (V, U) = (X, -Y),$$

and the closed-loop stability is guaranteed thanks to the Bezout identity (2.31). Motivated by this observation, we extend the Bezout identity to

$$\begin{bmatrix} M(z) & -\hat{Y}(z) \\ N(z) & \hat{X}(z) \end{bmatrix} \begin{bmatrix} I & Q(z) \\ 0 & I \end{bmatrix} \begin{bmatrix} I & -Q(z) \\ 0 & I \end{bmatrix} \begin{bmatrix} X(z) & Y(z) \\ -\hat{N}(z) & \hat{M}(z) \end{bmatrix} = \begin{bmatrix} I & 0 \\ 0 & I \end{bmatrix},$$

where $Q(z) \in \mathcal{RH}_\infty$ is an arbitrary transfer function (matrix) with an appropriate dimension. It yields

$$\begin{bmatrix} M & -\hat{Y} + MQ \\ N & \hat{X} + NQ \end{bmatrix} \begin{bmatrix} X + Q\hat{N} & Y - Q\hat{M} \\ -\hat{N} & \hat{M} \end{bmatrix} = \begin{bmatrix} I & 0 \\ 0 & I \end{bmatrix}. \tag{3.4}$$

That means, $\forall Q(z) \in \mathcal{RH}_\infty$,

$$\begin{aligned} K &= \hat{U}\hat{V}^{-1} = -\left(\hat{Y} - MQ\right)\left(\hat{X} + NQ\right)^{-1} \\ &= V^{-1}U = -\left(X + Q\hat{N}\right)^{-1}\left(Y - Q\hat{M}\right) \end{aligned}$$

stabilises the closed-loop. Along this line and considering $\begin{bmatrix} X & Y \\ -\hat{N} & \hat{M} \end{bmatrix}$ and $\begin{bmatrix} M & -\hat{Y} \\ N & \hat{X} \end{bmatrix}$ being invertible over $\mathcal{RH}_\infty$, any (V, U) and $\left(\hat{V}, \hat{U}\right)$ can be written as

$$\text{for some } \begin{bmatrix} \bar{U} & \bar{V} \end{bmatrix} \in \mathcal{RH}_\infty, \begin{bmatrix} V & -U \end{bmatrix} = \begin{bmatrix} \bar{U} & \bar{V} \end{bmatrix} \begin{bmatrix} X & Y \\ -\hat{N} & \hat{M} \end{bmatrix},$$

$$\text{for some } \begin{bmatrix} \tilde{U} \\ \tilde{V} \end{bmatrix} \in \mathcal{RH}_\infty, \begin{bmatrix} \hat{U} \\ \hat{V} \end{bmatrix} = \begin{bmatrix} M & -\hat{Y} \\ N & \hat{X} \end{bmatrix} \begin{bmatrix} \tilde{U} \\ \tilde{V} \end{bmatrix}.$$

It is evident, by the Bezout identity, that

$$\begin{bmatrix} \bar{U} & \bar{V} \end{bmatrix} = \begin{bmatrix} V & -U \end{bmatrix} \begin{bmatrix} M & -\hat{Y} \\ N & \hat{X} \end{bmatrix} = \begin{bmatrix} VM - UN & -V\hat{Y} - U\hat{X} \end{bmatrix},$$

$$\begin{bmatrix} \tilde{U} \\ \tilde{V} \end{bmatrix} = \begin{bmatrix} X & Y \\ -\hat{N} & \hat{M} \end{bmatrix} \begin{bmatrix} \hat{U} \\ \hat{V} \end{bmatrix} = \begin{bmatrix} X\hat{U} + Y\hat{V} \\ \hat{M}\hat{V} - \hat{N}\hat{U} \end{bmatrix}.$$

Remember that (3.3) is true, when K with the LCP and RCP (V, U) and $\left(\hat{V}, \hat{U}\right)$ stabilises the closed-loop. As a result, (V, U) and $\left(\hat{V}, \hat{U}\right)$ are respectively,

$$\begin{aligned} \begin{bmatrix} V & -U \end{bmatrix} &= (VM - UN) \begin{bmatrix} I & Q_1 \end{bmatrix} \begin{bmatrix} X & Y \\ -\hat{N} & \hat{M} \end{bmatrix}, \\ Q_1 &= (VM - UN)^{-1}\left(-V\hat{Y} - U\hat{X}\right) \in \mathcal{RH}_\infty^{p\times m}, \end{aligned}$$

$$\begin{bmatrix} \hat{U} \\ \hat{V} \end{bmatrix} = \begin{bmatrix} M & -\hat{Y} \\ N & \hat{X} \end{bmatrix} \begin{bmatrix} Q_2 \\ I \end{bmatrix} \left(\hat{M}\hat{V} - \hat{N}\hat{U} \right),$$
$$Q_2 = \left(X\hat{U} + Y\hat{V} \right) \left(\hat{M}\hat{V} - \hat{N}\hat{U} \right)^{-1} \in \mathcal{RH}_\infty^{p \times m}.$$

Moreover, it follows from

$$\begin{bmatrix} V & -U \\ -\hat{N} & \hat{M} \end{bmatrix} \begin{bmatrix} M & \hat{U} \\ N & \hat{V} \end{bmatrix} = \begin{bmatrix} VM - UN & 0 \\ 0 & \hat{M}\hat{V} - \hat{N}\hat{U} \end{bmatrix} \Longleftrightarrow$$
$$\begin{bmatrix} (VM - UN)^{-1} & 0 \\ 0 & I \end{bmatrix} \begin{bmatrix} V & -U \\ -\hat{N} & \hat{M} \end{bmatrix} \begin{bmatrix} M & \hat{U} \\ N & \hat{V} \end{bmatrix} \begin{bmatrix} I & 0 \\ 0 & \left(\hat{M}\hat{V} - \hat{N}\hat{U} \right)^{-1} \end{bmatrix} = \begin{bmatrix} I & 0 \\ 0 & I \end{bmatrix}$$

that

$$\begin{bmatrix} M & \hat{U} \\ N & \hat{V} \end{bmatrix} \begin{bmatrix} (VM - UN)^{-1} & 0 \\ 0 & \left(\hat{M}\hat{V} - \hat{N}\hat{U} \right)^{-1} \end{bmatrix} \begin{bmatrix} V & -U \\ -\hat{N} & \hat{M} \end{bmatrix}$$
$$= \begin{bmatrix} M & -\hat{Y} \\ N & \hat{X} \end{bmatrix} \begin{bmatrix} I & Q_2 \\ 0 & I \end{bmatrix} \begin{bmatrix} I & Q_1 \\ 0 & I \end{bmatrix} \begin{bmatrix} X & Y \\ -\hat{N} & \hat{M} \end{bmatrix} = \begin{bmatrix} I & 0 \\ 0 & I \end{bmatrix},$$

which implies

$$-Q_1 = Q_2 =: Q \Longrightarrow$$
$$K = -\left(\hat{Y} - MQ \right) \left(\hat{X} + NQ \right)^{-1} = -\left(X + Q\hat{N} \right)^{-1} \left(Y - Q\hat{M} \right). \quad (3.5)$$

Summarising the above discussion demonstrates that a feedback controller K stabilises the closed-loop system if and only if it is parametrised by (3.5). This is the well-known Youla parameterisation of all stabilising controllers. Here, $Q(z) \in \mathcal{RH}_\infty$ is the parameterisation transfer function.

Remark 3.2 At the chapter end, references are given for the proof of Youla parameterisation. It is worth mentioning that due to the possibly different notations adopted for LCF and RCF as well as the Bezout identity, the presentation formulas of Youla parameterisation in the literature could slightly vary.

In Chap. 2, it has been delineated that a controller $K = -\hat{Y}\hat{X}^{-1} = -X^{-1}Y$ is equivalent to an observer-based state feedback controller. Inspired by this result, we examine

$$\begin{aligned} u &= Ky + v \\ &= -\left(\hat{Y} - MQ \right) \left(\hat{X} + NQ \right)^{-1} y + v \\ &= -\left(X + Q\hat{N} \right)^{-1} \left(Y - Q\hat{M} \right) y + v \end{aligned} \quad (3.6)$$

in the context of observer-based feedback control. Recalling the Bezout identity (3.4), $\left(\left(\hat{X}+NQ\right),-\left(\hat{Y}-MQ\right)\right)$ and $\left(\left(X+Q\hat{N}\right),-\left(Y-Q\hat{M}\right)\right)$ are RCP and LCP, respectively. This implies,

$$\begin{bmatrix} u-v \\ y \end{bmatrix} = \begin{bmatrix} -\left(\hat{Y}-MQ\right) \\ \hat{X}+NQ \end{bmatrix} h, \tag{3.7}$$

$$\left(X+Q\hat{N}\right)(u-v)+\left(Y-Q\hat{M}\right)y=0, \tag{3.8}$$

where h is a hidden (latent) variable to be determined, and equals to r_y for $Q=0,\ v=0$, as proved in Chap. 2. We start with (3.8). According to (2.6) and (2.26), it holds

$$\tilde{x}(k+1)=(A-LC)\tilde{x}(k)+(B-LD)(u(k)-v(k))+Ly(k), \tag{3.9}$$

$$\tilde{r}(k)=W\left(y(k)-\tilde{y}(k)\right),\ \tilde{y}(k)=C\tilde{x}(k)+D\left(u(k)-v(k)\right), \tag{3.10}$$

$$V^{-1}\left(u(k)-v(k)-F\tilde{x}(k)-r_Q(k)\right)=0, \tag{3.11}$$

$$r_Q(z)=Q(z)\,\tilde{r}(z). \tag{3.12}$$

It is interesting to note that (3.10) is the time-domain realisation of (3.8), and means

$$u(k)=F\tilde{x}(k)+r_Q(k)+v(k).$$

That is, it is an observer-based state feedback controller together with a feedback of a residual signal and control reference. In the sequel, we will have a close look at such a controller. Concerning with the RCP (3.7), we have, following (2.5) and (2.25),

$$\tilde{x}(k+1)=(A+BF)\tilde{x}(k)+BVr_Q(k)+LW^{-1}\tilde{r}(k), \tag{3.13}$$

$$\begin{bmatrix} u(k)-v(k) \\ y(k) \end{bmatrix} = \begin{bmatrix} F \\ C+DF \end{bmatrix}\tilde{x}(k)+\begin{bmatrix} 0 \\ W^{-1} \end{bmatrix}\tilde{r}(k)+\begin{bmatrix} I \\ D \end{bmatrix}Vr_Q(k). \tag{3.14}$$

It becomes apparent that the hidden variable h is the residual vector $\tilde{r}$.

Note that systems (3.9) and (3.13) are of the same dynamics. It is the state observer for the system

$$x(k+1)=Ax(k)+B\left(u(k)-v(k)\right)=Ax(k)+B\left(F\tilde{x}(k)+r_Q(k)\right), \tag{3.15}$$

$$u(k)=F\tilde{x}(k)+r_Q(k)+v(k), \tag{3.16}$$

$$y(k)=Cx(k)+D\left(u(k)-v(k)\right), \tag{3.17}$$

which describes the dynamics of the system (2.2)–(2.3) with an observer-based state feedback controller exclusive of the influence of the reference signal v. Remember the relation (2.82), the discussion in Subsection 2.5.1 and the claim that feedback control is the feedback of the residual signal. The dynamics described by (3.15)–(3.17) exactly reflect this fact. In this regard, (3.7) and (3.8) can be equivalently written as

$$\hat{x}(k+1) = (A - LC)\hat{x}(k) + (B - LD)u(k) + Ly(k) \tag{3.18}$$

$$= (A + BF)\hat{x}(k) + BV\left(r_Q(k) + \bar{v}(k)\right) + LW^{-1}r_y(k), \tag{3.19}$$

$$r_y(k) = W\left(y(k) - \hat{y}(k)\right), \hat{y}(k) = C\hat{x}(k) + Du(k), \tag{3.20}$$

$$V^{-1}\left(u(k) - \bar{v}(k) - F\hat{x}(k) - r_Q(k)\right) = 0, \tag{3.21}$$

$$\begin{bmatrix} u(k) \\ y(k) \end{bmatrix} = \begin{bmatrix} F \\ C + DF \end{bmatrix}\hat{x}(k) + \begin{bmatrix} 0 \\ W^{-1} \end{bmatrix} r_y(k) + \begin{bmatrix} I \\ D \end{bmatrix} V\left(r_Q(k) + \bar{v}(k)\right), \tag{3.22}$$

$$r_Q(z) = Q(z)r_y(z), \bar{v}(z) = \left(X(z) + Q(z)\hat{N}(z)\right)v(z). \tag{3.23}$$

The expression (3.21) implies that

- the controller $u = Ky + v$ can be equivalently expressed by

$$u(k) = F\hat{x}(k) + r_Q(k) + \bar{v}(k), \tag{3.24}$$

- a "residual vector" at the control input side can be constructed by

$$\bar{r}_u(k) := V^{-1}\left(u(k) - F\hat{x}(k)\right) = V^{-1}\left(r_Q(k) + \bar{v}(k)\right),$$

  which is capable, for instance, for detecting cyber-attacks on control systems.

A further interesting result is that (3.19) and (3.22) can also be written into

$$\begin{bmatrix} u \\ y \end{bmatrix} = \begin{bmatrix} -\hat{Y} + MQ \\ \hat{X} + NQ \end{bmatrix} r_y + \begin{bmatrix} M \\ N \end{bmatrix}\bar{v}, \tag{3.25}$$

which is the generalised form of (2.82) and implies that the closed-loop variables (u, y) consist of the feedback and feedforward dynamics, respectively expressed by

$$\begin{bmatrix} -\hat{Y} + MQ \\ \hat{X} + NQ \end{bmatrix} r_y \text{ and } \begin{bmatrix} M \\ N \end{bmatrix}\bar{v},$$

and driven by the (output) residual r_y and reference $\bar{v}$, respectively.

Example 3.1 In this example, an alternative form of the RCF of the controller, (3.7), is derived. Starting from (3.25) and noting the following identity due to the (extended) Bezout identity (3.4)

$$\begin{bmatrix} M \\ N \end{bmatrix}\left(X + Q\hat{N}\right) = \begin{bmatrix} I - \left(\hat{Y} - MQ\right)\hat{N} \\ \left(\hat{X} + NQ\right)\hat{N} \end{bmatrix}, \tag{3.26}$$

it turns out

$$\begin{bmatrix} u - v \\ y \end{bmatrix} = \begin{bmatrix} -\hat{Y} + MQ \\ \hat{X} + NQ \end{bmatrix} \left(r_y + \hat{N} v \right). \tag{3.27}$$

That is, the hidden (latent) variable of the SIR of the controller,

$$u - v = Ky = -\left(\hat{Y} - MQ\right)\left(\hat{X} + NQ\right)^{-1} y,$$

is $r_y + \hat{N} v$.

At the end of this section, the major results are summarised in the following theorem.

Theorem 3.1 *Given the control loop (3.1) with the plant model (2.1)–(2.3), then all stabilising controllers are parametrised by*

$$u(z) = F\hat{x}(z) + Q(z) r_y(z) + \bar{v}(z), \; Q(z) \in \mathcal{RH}_\infty, \tag{3.28}$$

where $\hat{x}$ and r_y are the state estimate and residual vector delivered by the observer-based residual generator (2.8)–(2.9), and Q is the parameterisation transfer function (matrix), F is the state feedback gain adopted in the RCF of G, $\bar{v}$ represents the control reference defined in (3.23). Moreover, the SKR and SIR of the controller are, respectively,

$$r_u = \begin{bmatrix} X + Q\hat{N} & Y - Q\hat{M} \end{bmatrix} \begin{bmatrix} u - v \\ y \end{bmatrix},$$

$$\begin{bmatrix} u - v \\ y \end{bmatrix} = \begin{bmatrix} -\hat{Y} + MQ \\ \hat{X} + NQ \end{bmatrix} \left(r_y + \hat{N} v \right),$$

and the dynamic of the loop variables (u, y) is governed by

$$\begin{bmatrix} u \\ y \end{bmatrix} = \begin{bmatrix} M \\ N \end{bmatrix} \bar{v} + \begin{bmatrix} -\hat{Y} + MQ \\ \hat{X} + NQ \end{bmatrix} r_y. \tag{3.29}$$

Here, $\left(\hat{M}, \hat{N}\right)$, (M, N) and $\left(X + Q\hat{N}, Y - Q\hat{M}\right)$, $\left(\hat{X} + NQ, -\hat{Y} + MQ\right)$ are LCP and RCP of the plant and controller, respectively, and satisfy the Bezout identity (3.4).

Definition 3.1 Hereafter, controller (3.28) is called observer-based realisation of Youla parameterisation, while

$$u_{fb}(z) := F\hat{x}(z) + Q(z) r_y(z), \; u_{ff}(z) := \bar{v}(z),$$

are called feedback and feedforward controllers, respectively.

Expression (3.29) is the extended form of (2.82) and describes the one-to-one mapping between the process signal pair (u, y) and the residual/information signal pair $(\bar{v}, r_y)$ in the closed-loop. Here, $\begin{bmatrix} -\hat{Y} + MQ \\ \hat{X} + NQ \end{bmatrix} r_y$ builds the (feedback) controller image subspace

$$\mathcal{I}_K = \left\{ \begin{bmatrix} u \\ y \end{bmatrix} : \begin{bmatrix} u \\ y \end{bmatrix} = \begin{bmatrix} -\hat{Y} + MQ \\ \hat{X} + NQ \end{bmatrix} r_y, r_y \in \mathcal{H}_2 \right\},$$

and the relation

$$\mathcal{S} = \left\{ \begin{bmatrix} u \\ y \end{bmatrix} \in \mathcal{H}_2^{m+p} \right\} = \mathcal{I}_G \oplus \mathcal{I}_K \tag{3.30}$$

remains valid.

Remark 3.3 It is noteworthy that the feedforward controller $\bar{v}$ is the equivalent expression of the reference signal v of the controller (3.6) in form of the observer-based realisation (3.28). In case that an observer-based controller is directly implemented as

$$u(z) = F\hat{x}(z) + Q(z)r_y(z) + Vv(z),$$

then

$$\bar{v} = Vv.$$

3.2 Youla Parameterisation Versus Its Observer-Based Realisation

Although the observer-based realisation (3.28) is equivalently derived from the Youla parameterisation (3.5), the realisations of both control schemes are evidently different. As shown in Fig. 3.2, the Youla parameterisation controller,

$$\begin{aligned} u &= Ky + v, \\ K &= -\left(\hat{Y} - MQ\right)\left(\hat{X} + NQ\right)^{-1} = -\left(X + Q\hat{N}\right)^{-1}\left(Y - Q\hat{M}\right), \end{aligned}$$

is realised in an open-loop configuration, while its observer-based one is closed-loop configured. More importantly, any observer-based realisation is a stable system, and, differently, a Youla parameterisation controller may be unstable. This difference is momentous for the real-time realisation, for instance implementation on a micro-controller. It is of considerable practical interest to investigate structural properties of these two different forms of stabilising feedback controllers and to get deeper insights into feedback control strategies.

Below is a motivation example.

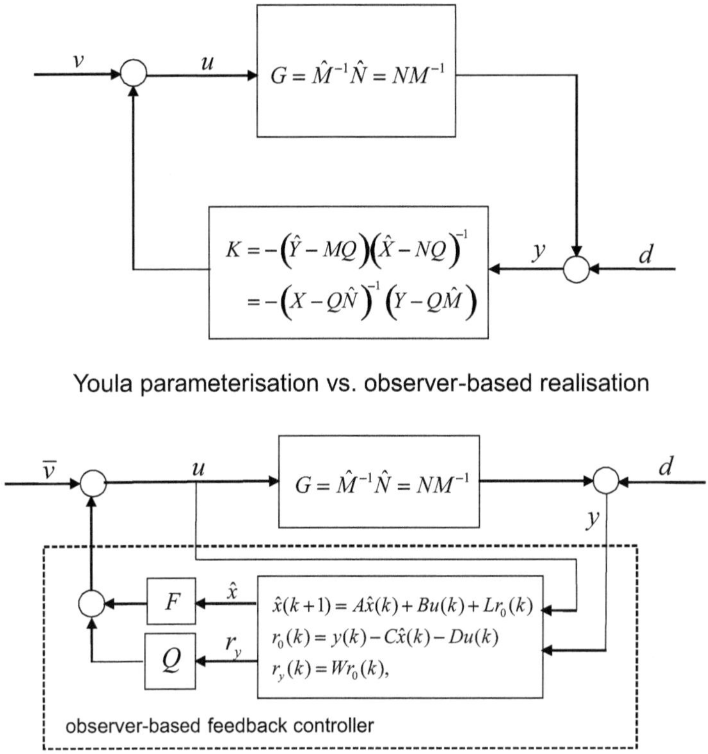

Fig. 3.2 Realisations of Youla parameterisation and its observer-based form

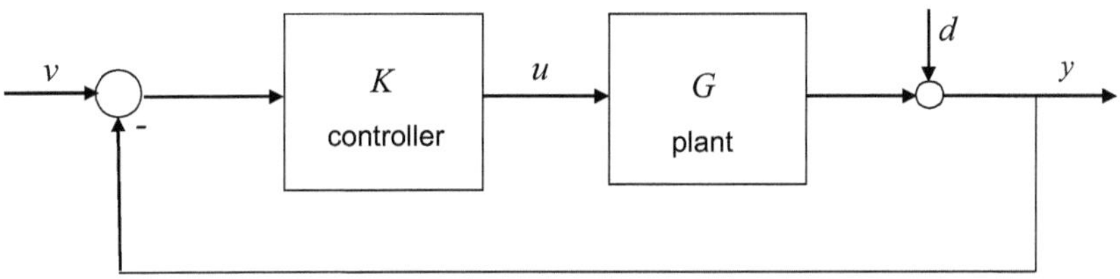

Fig. 3.3 A standard feedback control loop

Example 3.2 Consider a control loop sketched in Fig. 3.3 with a stable plant G and controller K that stabilises the loop. It is well-known in the literature that K can be equivalently parametrised by

$$\begin{aligned} K &= (I - (-Q)G)^{-1}(-Q) = -Q(I - G(-Q))^{-1}, \\ -Q &= K(I + GK)^{-1} = (I + KG)^{-1}K \in \mathcal{RH}_\infty. \end{aligned} \tag{3.31}$$

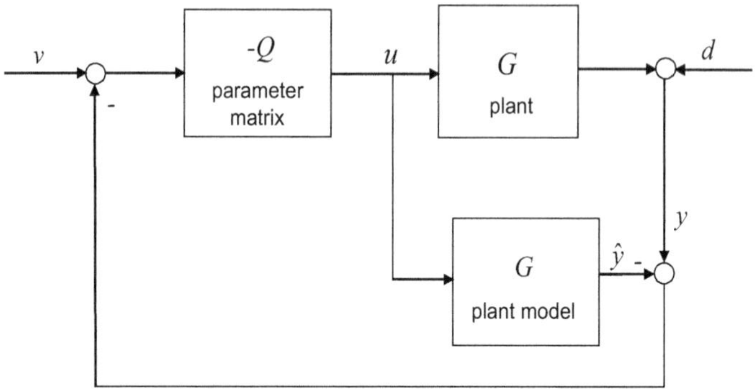

Fig. 3.4 IMC loop configuration

The parameterisation form (3.31) leads to the well-known internal model control (IMC) configuration, as delineated in Fig. 3.4. Note that controller (3.31) is a special case of the observer-based realisation (3.28) for

$$
\begin{aligned}
& L = 0,\, F = 0,\, W = I,\, V = I \Longrightarrow \\
& \hat{M} = I,\, M = I,\, \hat{N} = N = G,\, \hat{X} = I,\, \hat{Y} = 0,\, X = I,\, Y = 0 \Longrightarrow \\
& u = F\hat{x} + Qr_y + \bar{v},\, \bar{v} = -Qv \Longleftrightarrow \\
& u = -Qv + Q\,(y - Gu) = -Qv + Qd,
\end{aligned}
$$

since the plant is stable. It is straightforward that using IMC,

$$v - y = v + GQv - GQd - d.$$

Now, assume that G is invertible over $\mathcal{RH}_\infty$. Setting

$$-Q = G^{-1} \in \mathcal{RH}_\infty$$

yields

$$v - y = 0.$$

For a comparison, let us examine the control performance, when the control loop in Fig. 3.3 is realised. It holds

$$v - y = (I + GK)^{-1}\, v - (I + GK)^{-1}\, d.$$

It is clear that in this case, for any realisable controller K and independent of G being invertible, the control error, $v - y$, exists. The reason for this different result is obvious, since in the control law (3.31), $-Q$ should be set equal to G^{-1}. This results in, however, an unrealisable controller K. This example demonstrates that the observer-based realisation of stabilising controllers may deliver better control performance.

3.2.1 On Well-Posedness

Roughly speaking, well-posedness is a basic requirement for controllers and control systems to be realisable in the context of system causality. A dynamic system with rational transfer function S is well-posed, if S is proper. For the feedback control loop shown in Fig. 3.1, it is well-posed, when the controller K and the closed-loop dynamics, $\begin{bmatrix} I & -K \\ -G & I \end{bmatrix}^{-1}$, are well-posed. It is well-known that the invertibility of $I - K(\infty) G(\infty)$ is a necessary and sufficient condition for the well-posedness of the closed-loop (3.1), and the reader is referred to the references given in the last section. To motivate our study, we first consider the following example.

Example 3.3 Given the plant model

$$G(z) = \frac{z+b}{z+a}, |a| < 1, b \neq 0,$$

which is a stable system, and thus for $L = 0, F = 0, W = I, V = I$,

$$\hat{M} = I, M = I, \hat{N} = N = G, \hat{X} = I, \hat{Y} = 0, X = I, Y = 0.$$

As a special form of the observer-based realisation (3.28), an IMC is designed, as given in Fig. 3.4,

$$-Q(z) = \frac{z+a}{z}.$$

The closed-loop dynamic is governed by

$$\begin{bmatrix} u \\ y \end{bmatrix} = \begin{bmatrix} -Q(z)\left(v(z) - r_y(z)\right) \\ G(z)u(z) + d(z) \end{bmatrix} = \begin{bmatrix} -Q(z) \\ -G(z)Q(z) \end{bmatrix} v(z) + \begin{bmatrix} Q(z) \\ I + G(z)Q(z) \end{bmatrix} d(z)$$
$$= \begin{bmatrix} \frac{z+a}{z} \\ \frac{z+b}{z} \end{bmatrix} v(z) - \begin{bmatrix} \frac{z+a}{z} \\ \frac{b}{z} \end{bmatrix} d(z),$$

and hence it is stable and well-posed. Now, we realise the same controller in the standard feedback loop configuration in Fig. 3.3. By Youla parameterisation, it turns out

$$K = -(I + QG)^{-1} Q = -\frac{z+a}{b}.$$

That is, K is not well-posed and cannot be realised.

Motivated by the above example, we examine the well-posedness of the observer-based realisation (3.28) and have the following result.

Theorem 3.2 *Consider a feedback control loop with (i) the plant model*

$$y(z) = G(z)u(z) + d(z),$$

where G is modelled by (2.1)–(2.3) and d is the unknown input vector, (ii) the observer-based stabilising controller

$$u(z) = F\hat{x}(z) + Q(z)r_y(z) + \left(X(z) + Q(z)\hat{N}(z)\right) v(z),\ Q(z) \in \mathcal{RH}_\infty, \quad (3.32)$$

where $\hat{x}$ and r_y are the state estimate and residual vector delivered by the observer-based residual generator (2.8)–(2.9), and Q is the parameterisation transfer function (matrix), F is the state feedback gain adopted in the RCF of G, v represents the control reference. Then, the controller (3.32) and the closed-loop dynamics are well-posed.

Proof That the controller (3.32) is well-posed is obvious, since the observer and observer-based residual generator, parameterisation matrix Q and $\left(X + Q\hat{N}\right)$ are proper. Concerning with the closed-loop dynamics, it follows from Theorem 3.1 that

$$\begin{aligned}\begin{bmatrix} u \\ y \end{bmatrix} &= \begin{bmatrix} -\hat{Y} + MQ \\ \hat{X} + NQ \end{bmatrix} r_y + \begin{bmatrix} M \\ N \end{bmatrix} \bar{v} \\ &= \begin{bmatrix} -\hat{Y} + MQ \\ \hat{X} + NQ \end{bmatrix} \hat{M} d + \begin{bmatrix} M \\ N \end{bmatrix} \left(X + Q\hat{N}\right) v.\end{aligned}$$

Here, $\left(\hat{M}, \hat{N}\right)$, (M, N) and $\left(X + Q\hat{N}, Y + Q\hat{M}\right)$, $\left(\hat{X} + NQ, -\hat{Y} + MQ\right)$ are LCP and RCP of the plant and controller, respectively, satisfy the Bezout identity (3.4) and well-posed. As a result, the closed-loop dynamics is well-posed.

Theorems 3.1, 3.2 and Example 3.3 assert that, given the plant model (2.1)–(2.3),

- the well-posedness of the closed-loop dynamics is guaranteed, when the feedback controller is designed by means of the Youla parameterisation (3.5), but
- the controller with the open-loop configuration,

$$K = -\left(\hat{Y} - MQ\right)\left(\hat{X} + NQ\right)^{-1} = -\left(X + Q\hat{N}\right)^{-1}\left(Y - Q\hat{M}\right),$$

  may be ill-posed.

Note that the first fact is independent of the implementation form of the controller, while the second one suggests a careful examination of the controller K when it is realised in the open-loop form. To simplify such an examination, we present the following condition.

Theorem 3.3 *Given plant model (2.1)–(2.3), a feedback controller parametrised by*

$$K = -\left(\hat{Y} - MQ\right)\left(\hat{X} + NQ\right)^{-1} = -\left(X + Q\hat{N}\right)^{-1}\left(Y - Q\hat{M}\right),$$

is well-posed, if

$$I + N(\infty)\, Q(\infty)\, W, I + VQ(\infty)\, \hat{N}(\infty) \text{ are invertible,} \tag{3.33}$$

where $\left(\hat{M}, \hat{N}\right), (M, N)$ *and* $(X, Y), \left(\hat{X}, \hat{Y}\right)$ *as well as* W, V *are given in (2.5)–(2.6) and (2.25)–(2.26), respectively.*

Proof It is apparent that K is well-posed if $\left(\hat{X} + NQ\right)^{-1}, \left(X + Q\hat{N}\right)^{-1}$ are well-posed. According to (2.25)–(2.26),

$$\hat{X}(\infty) = W^{-1}, X(\infty) = V^{-1}.$$

Hence, $\left(\hat{X} + NQ\right)^{-1}$ *and* $\left(X + Q\hat{N}\right)^{-1}$ *are well-posed, if*

$$\hat{X}(\infty) + N(\infty)Q(\infty) = (I + N(\infty)Q(\infty)W)\, W^{-1} \Longleftrightarrow I + N(\infty)Q(\infty)W,$$
$$X(\infty) + Q(\infty)\hat{N}(\infty) = V^{-1}\left(I + VQ(\infty)\hat{N}(\infty)\right) \Longleftrightarrow I + VQ(\infty)\hat{N}(\infty)$$

are invertible.

Remark 3.4 Often, W and V are set to be unity matrix. In that case, the condition becomes simply $I + N(\infty)Q(\infty)$ and $I - Q(\infty)\hat{N}(\infty)$ being invertible. The result in Example 3.3 can be immediately achieved by examining the condition (3.33).

3.2.2 On the Influences of F and L

Matrices F and L in the RCF and LCF (2.5)–(2.6) are two parameters that are, as state feedback and observer gain matrices, of important control engineering significance. The results presented in Theorem 3.5 and Corollary 3.4 enhance their important role in controller design and synthesis. On the other hand, it is known that their selection has no influence on the transfer function G. That is, G is invariant to F and L. Against this background, we would like to investigate the influences of F and L on

- Youla parameterisation controllers (3.5),
- their observer-based realisation form (3.28), as well as
- the closed-loop dynamics.

Remark 3.5 For the sake of simplicity and without loss of generality, it is assumed, in the subsequent study, that $W = I, V = I$.

To begin with, a useful and known result is introduced.

Lemma 3.1 *Given G modelled by (2.1)–(2.3) with two LCFs and RCFs,*

$$\hat{M}_i = \left(A_{L_i}, -L_i, C, I\right), \hat{N}_i = \left(A_{L_i}, B - L_i D, C, D\right),$$
$$M_i = \left(A_{F_i}, B, F_i, I\right), N_i = \left(A_{F_i}, B, C + DF_i, D\right),$$
$$A_{L_i} = A - L_i C, A_{F_i} = A + BF_i, i = 1, 2,$$

then it holds

$$\left[\, \hat{M}_2(z)\ \hat{N}_2(z) \,\right] = R_{21}(z) \left[\, \hat{M}_1(z)\ \hat{N}_1(z) \,\right], \tag{3.34}$$
$$R_{21}(z) = I - C\left(zI - A_{L_2}\right)^{-1}(L_2 - L_1) \in \mathcal{RH}_\infty,$$
$$\begin{bmatrix} M_2(z) \\ N_2(z) \end{bmatrix} = \begin{bmatrix} M_1(z) \\ N_1(z) \end{bmatrix} V_{21}(z), \tag{3.35}$$
$$V_{21}(z) = I + (F_2 - F_1)\left(zI - A_{F_2}\right)^{-1} B \in \mathcal{RH}_\infty.$$

Moreover, $R_{21}(z)$ and $V_{21}(z)$ are invertible over $\mathcal{RH}_\infty$, and $R_{21}^{-1}(z)$ and $V_{21}^{-1}(z)$ satisfy

$$R_{21}^{-1}(z) = I - C\left(zI - A_{L_1}\right)^{-1}(L_1 - L_2) =: R_{12}(z), \tag{3.36}$$
$$V_{21}^{-1}(z) = I + (F_1 - F_2)\left(zI - A_{F_1}\right)^{-1} B =: V_{12}(z). \tag{3.37}$$

Proof Due to the dual form of (3.34) and (3.35), we only give the proof of (3.34) and (3.36). It is straightforward that

$$\begin{aligned} R_{21}^{-1} &= \left(I + C\left(zI - A_{L_2}\right)^{-1}(L_1 - L_2)\right)^{-1} \\ &= I - C\left(\left(zI - A_{L_2}\right) + (L_1 - L_2)C\right)^{-1}(L_1 - L_2) \\ &= I - C\left(zI - A_{L_1}\right)^{-1}(L_1 - L_2), \end{aligned}$$

and furthermore,

$$R_{21}\hat{M}_1 = \left(I - C\left(zI - A_{L_2}\right)^{-1}(L_2 - L_1)\right)\hat{M}_1$$
$$= I - C\left(zI - A_{L_2}\right)^{-1} L_2 + C\left(\begin{array}{c} \left(zI - A_{L_2}\right)^{-1} - \left(zI - A_{L_1}\right)^{-1} + \\ \left(zI - A_{L_2}\right)^{-1}(L_2 - L_1)C\left(zI - A_{L_1}\right)^{-1} \end{array}\right) L_1$$
$$= I - C\left(zI - A_{L_2}\right)^{-1} L_2 = \hat{M}_2,$$
$$R_{21}\hat{N}_1 = \left(I - C\left(zI - A_{L_2}\right)^{-1}(L_2 - L_1)\right)\left(D + C\left(zI - A_{L_1}\right)^{-1}(B - L_1 D)\right)$$

$$= D + C\left(\begin{array}{c}\left(zI - A_{L_1}\right)^{-1}(B - L_1 D) - \\ \left(zI - A_{L_2}\right)^{-1}(L_2 - L_1)\left(D + C\left(zI - A_{L_1}\right)^{-1}(B - L_1 D)\right)\end{array}\right)$$

$$= D + C\left(zI - A_{L_2}\right)^{-1}(B - L_1 D) - C\left(zI - A_{L_2}\right)^{-1}(L_2 - L_1) D$$

$$= D + C\left(zI - A_{L_2}\right)^{-1}(B - L_2 D) = \hat{N}_2.$$

It follows from this lemma immediately that

- the LCF and RCF of G are invariant with regard to L and F, and
- the image and kernel subspaces of G are invariant with regard to L and F, that is,

$$\begin{aligned}
\mathcal{I}_G &= \left\{\begin{bmatrix} u \\ y \end{bmatrix} \in \mathcal{P}_2 : \begin{bmatrix} u \\ y \end{bmatrix} = \begin{bmatrix} M_i \\ N_i \end{bmatrix} v_i, v_i \in \mathcal{P}_2\right\} \\
&= \left\{\begin{bmatrix} u \\ y \end{bmatrix} \in \mathcal{P}_2 : \begin{bmatrix} u \\ y \end{bmatrix} = \begin{bmatrix} M_j \\ N_j \end{bmatrix} v_j, v_j \in \mathcal{P}_2\right\}, \\
\mathcal{K}_G &= \left\{\begin{bmatrix} u \\ y \end{bmatrix} \in \mathcal{P}_2 : \begin{bmatrix} -\hat{N}_i & \hat{M}_i \end{bmatrix}\begin{bmatrix} u \\ y \end{bmatrix} = 0\right\} \\
&= \left\{\begin{bmatrix} u \\ y \end{bmatrix} \in \mathcal{P}_2 : \begin{bmatrix} -\hat{N}_j & \hat{M}_j \end{bmatrix}\begin{bmatrix} u \\ y \end{bmatrix} = 0\right\}, \\
& i \neq j, \mathcal{P}_2 = \{\mathcal{L}_2, \mathcal{H}_2\},
\end{aligned}$$

since $R_{ij}(z)$, $V_{ij}(z)$ are invertible over $\mathcal{RH}_\infty$.

Next, parameterisation of possible changes in the LCF and RCF of controller,

$$K(z) = -X^{-1}(z)Y(z) = -\hat{Y}(z)\hat{X}^{-1}(z),$$

as a result of variations in L and F is under consideration. To be specific, the relations between two LCFs and RFSs corresponding to two different settings of the state feedback gain and observer gain matrices, $F_i, L_i, i = 1, 2,$ are investigated. Following notations (X_i, Y_i), $\left(\hat{M}_i, \hat{N}_i\right)$, $\left(\hat{X}_i, \hat{Y}_i\right)$, (M_i, N_i), $i = 1, 2$, are adopted for our study,

$$\begin{aligned}
& X_i = \left(A_{L_i}, -B_{L_i}, F_i, I\right), Y_i = \left(A_{L_i}, -L_i, F_i, 0\right), \\
& \hat{M}_i = \left(A_{L_i}, -L_i, C, I\right), \hat{N}_i = \left(A_{L_i}, B_{L_i}, C, D\right), \\
& \hat{X}_i = \left(A_{F_i}, L_i, C_{F_i}, I\right), \hat{Y}_i = \left(A_{F_i}, -L_i, F_i, 0\right), \\
& M_i = \left(A_{F_i}, B, F_i, I\right), N_i = \left(A_{F_i}, B, C_{F_i}, D\right), \\
& A_{L_i} = A - L_i C, A_{F_i} = A + BF_i, B_{L_i} = B - L_i D, C_{F_i} = C + DF_i.
\end{aligned}$$

Theorem 3.4 *Given* (X_i, Y_i), $\left(\hat{M}_i, \hat{N}_i\right)$, $\left(\hat{X}_i, \hat{Y}_i\right)$, (M_i, N_i), $i = 1, 2$, *it holds*

$$\begin{bmatrix} X_1(z) & Y_1(z) \end{bmatrix} = V_{21}(z) \begin{bmatrix} X_2(z) & Y_2(z) \end{bmatrix} + \bar{V}_{21}(z) \begin{bmatrix} -\hat{N}_2(z) & \hat{M}_2(z) \end{bmatrix}, \tag{3.38}$$

$$\begin{bmatrix} -\hat{Y}_1(z) \\ \hat{X}_1(z) \end{bmatrix} = \begin{bmatrix} -\hat{Y}_2(z) \\ \hat{X}_2(z) \end{bmatrix} R_{21}(z) + \begin{bmatrix} M_2(z) \\ N_2(z) \end{bmatrix} \bar{R}_{21}(z), \tag{3.39}$$

$$V_{21}(z) = V_{12}^{-1}(z) = I + (F_2 - F_1) \left(zI - A_{F_2}\right)^{-1} B \in \mathcal{RH}_\infty,$$

$$R_{21}(z) = R_{12}^{-1}(z) = I - C \left(zI - A_{L_2}\right)^{-1} (L_2 - L_1) \in \mathcal{RH}_\infty,$$

$$\bar{V}_{21}(z) = \begin{bmatrix} X_1(z) & Y_1(z) \end{bmatrix} \begin{bmatrix} -\hat{Y}_2(z) \\ \hat{X}_2(z) \end{bmatrix} \in \mathcal{RH}_\infty,$$

$$\bar{R}_{21}(z) = \begin{bmatrix} X_2(z) & Y_2(z) \end{bmatrix} \begin{bmatrix} -\hat{Y}_1(z) \\ \hat{X}_1(z) \end{bmatrix} \in \mathcal{RH}_\infty.$$

Proof By reason of the Bezout identity (2.31), $\begin{bmatrix} -\hat{Y}_1 \\ \hat{X}_1 \end{bmatrix}$ and $\begin{bmatrix} X_1 & Y_1 \end{bmatrix}$ can be parametrised respectively by

$$\begin{bmatrix} -\hat{Y}_1(z) \\ \hat{X}_1(z) \end{bmatrix} = \begin{bmatrix} M_2(z) & -\hat{Y}_2(z) \\ N_2(z) & \hat{X}_2(z) \end{bmatrix} \begin{bmatrix} Q_1(z) \\ Q_2(z) \end{bmatrix},$$

$$\begin{bmatrix} X_1(z) & Y_1(z) \end{bmatrix} = \begin{bmatrix} \bar{Q}_1(z) & \bar{Q}_2(z) \end{bmatrix} \begin{bmatrix} X_2(z) & Y_2(z) \\ -\hat{N}_2(z) & \hat{M}_2(z) \end{bmatrix},$$

for some $Q_1, Q_2, \bar{Q}_1, \bar{Q}_2 \in \mathcal{RH}_\infty$. It follows from the Bezout identity (2.31) and Lemma 3.1 that

$$\begin{bmatrix} -\hat{N}_1(z) & \hat{M}_1(z) \end{bmatrix} \begin{bmatrix} -\hat{Y}_1(z) \\ \hat{X}_1(z) \end{bmatrix} = I \Longleftrightarrow$$

$$R_{12}(z) \begin{bmatrix} -\hat{N}_2(z) & \hat{M}_2(z) \end{bmatrix} \begin{bmatrix} M_2(z) & -\hat{Y}_2(z) \\ N_2(z) & \hat{X}_2(z) \end{bmatrix} \begin{bmatrix} Q_1(z) \\ Q_2(z) \end{bmatrix} = R_{12}(z) Q_2(z) = I$$

$$\Longrightarrow Q_2(z) = R_{12}^{-1}(z) = R_{21}(z),$$

$$\begin{bmatrix} X_1(z) & Y_1(z) \end{bmatrix} \begin{bmatrix} M_1(z) \\ N_1(z) \end{bmatrix} = I \Longleftrightarrow$$

$$\begin{bmatrix} \bar{Q}_1(z) & \bar{Q}_2(z) \end{bmatrix} \begin{bmatrix} X_2(z) & Y_2(z) \\ -\hat{N}_2(z) & \hat{M}_2(z) \end{bmatrix} \begin{bmatrix} M_2(z) \\ N_2(z) \end{bmatrix} V_{12}(z) = \bar{Q}_1(z) V_{12}(z) = I$$

$$\Longrightarrow \bar{Q}_1(z) = V_{12}^{-1}(z) = V_{21}(z).$$

Moreover,

$$\begin{bmatrix} X_1(z) & Y_1(z) \end{bmatrix} \begin{bmatrix} -\hat{Y}_2(z) \\ \hat{X}_2(z) \end{bmatrix} = \begin{bmatrix} \bar{Q}_1(z) & \bar{Q}_2(z) \end{bmatrix} \begin{bmatrix} X_2(z) & Y_2(z) \\ -\hat{N}_2(z) & \hat{M}_2(z) \end{bmatrix} \begin{bmatrix} -\hat{Y}_2(z) \\ \hat{X}_2(z) \end{bmatrix}$$

$$= \bar{Q}_2(z) = \bar{V}_{21}(z),$$

$$\left[X_2(z)\ Y_2(z) \right] \begin{bmatrix} -\hat{Y}_1(z) \\ \hat{X}_1(z) \end{bmatrix} = \left[X_2(z)\ Y_2(z) \right] \begin{bmatrix} M_2(z) & -\hat{Y}_2(z) \\ N_2(z) & \hat{X}_2(z) \end{bmatrix} \begin{bmatrix} Q_1(z) \\ Q_2(z) \end{bmatrix}$$

$$= Q_1(z) = \bar{R}_{21}(z).$$

The theorem is proved.

An immediate result of Theorem 3.4 is the parameterisation of feedback controller K with respect to various state feedback gain and observer gain matrices.

Corollary 3.1 *Given* $(X_i, Y_i), \left(\hat{M}_i, \hat{N}_i\right), \left(\hat{X}_i, \hat{Y}_i\right), (M_i, N_i), i = 1, 2$, *a stabilising controller* K *can be equivalently expressed by*

$$\begin{aligned} K(z) &= -X_1^{-1}(z)Y_1(z) = -\hat{Y}_1(z)\hat{X}_1^{-1}(z) \\ &= -\left(X_2(z) + Q_{21}(z)\hat{N}_2(z)\right)^{-1}\left(Y_2(z) - Q_{21}(z)\hat{M}_2(z)\right) && (3.40) \\ &= -\left(\hat{Y}_2(z) - M_2(z)Q_{21}(z)\right)\left(\hat{X}_2(z) + N_2(z)Q_{21}(z)\right)^{-1}, && (3.41) \\ Q_{21}(z) &= -V_{21}^{-1}(z)\bar{V}_{21}(z) = \bar{R}_{21}(z)R_{21}^{-1}(z). && (3.42) \end{aligned}$$

Moreover, the Youla parameterisation of K *is invariant to the LCF and RCF of the plant model* G.

Proof On account of Theorem 3.4, we have

$$\begin{aligned} -X_1^{-1}Y_1 &= -\left(V_{21}X_2 - \bar{V}_{21}\hat{N}_2\right)^{-1}\left(V_{21}Y_2 + \bar{V}_{21}\hat{M}_2\right), \\ -\hat{Y}_1\hat{X}_1^{-1} &= -\left(\hat{Y}_2R_{21} - M_2\bar{R}_{21}\right)\left(\hat{X}_2R_{21} + N_2\bar{R}_{21}\right). \end{aligned}$$

Recall V_{21} and R_{21} are invertible over $\mathcal{RH}_\infty$, which yields

$$\begin{aligned} -X_1^{-1}Y_1 &= -\left(X_2 - V_{21}^{-1}\bar{V}_{21}\hat{N}_2\right)^{-1}\left(Y_2 + V_{21}^{-1}\bar{V}_{21}\hat{M}_2\right), \\ -\hat{Y}_1\hat{X}_1^{-1} &= -\left(\hat{Y}_2 - M_2\bar{R}_{21}R_{21}^{-1}\right)\left(\hat{X}_2 + N_2\bar{R}_{21}R_{21}^{-1}\right). \end{aligned}$$

Furthermore,

$$\left[X_1\ Y_1(z) \right] \begin{bmatrix} -\hat{Y}_1 \\ \hat{X}_1 \end{bmatrix} = 0 \Longleftrightarrow$$

$$\left[V_{21}\ \bar{V}_{12} \right] \begin{bmatrix} X_2 & Y_2 \\ -\hat{N}_2 & \hat{M}_2 \end{bmatrix} \begin{bmatrix} M_2 & -\hat{Y}_2 \\ N_2 & \hat{X}_2 \end{bmatrix} \begin{bmatrix} \bar{R}_{21} \\ R_{21} \end{bmatrix} = 0 \Longleftrightarrow$$

$$V_{21}\bar{R}_{21} + \bar{V}_{12}R_{21} = 0.$$

Consequently, (3.40)–(3.42) are proved. Observe that (3.40) and (3.41) are identical with Youla parameterisation form given in (3.5). As a result, independent of an (arbitrary) LCF and RCF of G, $\left(\hat{M}, \hat{N}\right)$ and (M, N), *and the corresponding RCP and LCP* $\left(\hat{X}, \hat{Y}\right)$ *and* (X, Y), all stabilising controllers can be parametrised by

$$K = -\left(\hat{Y} - MQ\right)\left(\hat{X} + NQ\right)^{-1} = -\left(X + Q\hat{N}\right)^{-1}\left(Y - Q\hat{M}\right).$$

Example 3.4 In this example, parameterisation matrices $\bar{V}_{21}$ and $\bar{R}_{21}$,

$$\bar{V}_{21}(z) = \begin{bmatrix} X_1(z) & Y_1(z) \end{bmatrix} \begin{bmatrix} -\hat{Y}_2(z) \\ \hat{X}_2(z) \end{bmatrix},$$
$$\bar{R}_{21}(z) = \begin{bmatrix} X_2(z) & Y_2(z) \end{bmatrix} \begin{bmatrix} -\hat{Y}_1(z) \\ \hat{X}_1(z) \end{bmatrix},$$

are specified. By the state space representations of (X_i, Y_i) and $\left(\hat{X}_i, \hat{Y}_i\right)$, we have

$$Y_1\hat{X}_2 - X_1\hat{Y}_2 = F_2\left(zI - A_{F_2}\right)^{-1} L_2 - F_1\left(zI - A_{L_1}\right)^{-1} L_1$$
$$-F_1\left(zI - A_{L_1}\right)^{-1}\left(B_{L_1}F_2 + L_1 C_{F_2}\right)\left(zI - A_{F_2}\right)^{-1} L_2.$$

Note that

$$\begin{aligned} & F_1\left(zI - A_{L_1}\right)^{-1}\left(B_{L_1}F_2 + L_1 C_{F_2}\right)\left(zI - A_{F_2}\right)^{-1} L_2 \\ &= F_1\left(zI - A_{L_1}\right)^{-1}\left(BF_2 + L_1 C\right)\left(zI - A_{F_2}\right)^{-1} L_2 \\ &= F_1\left(zI - A_{L_1}\right)^{-1}\left(-\left(zI - A_{F_2}\right) + \left(zI - A_{L_1}\right)\right)\left(zI - A_{F_2}\right)^{-1} L_2 \\ &= -F_1\left(zI - A_{L_1}\right)^{-1} L_2 + F_1\left(zI - A_{F_2}\right)^{-1} L_2. \end{aligned}$$

Therefore, it holds

$$\begin{aligned} \bar{V}_{21} &= Y_1\hat{X}_2 - X_1\hat{Y}_2 \\ &= \left(F_2 - F_1\right)\left(zI - A_{F_2}\right)^{-1} L_2 - F_1\left(zI - A_{L_1}\right)^{-1}\left(L_1 - L_2\right). \end{aligned} \tag{3.43}$$

Due to the dual form of $\bar{V}_{21}$ and $\bar{R}_{21}$, it is straightforward that

$$\begin{aligned} \bar{R}_{21} &= Y_2\hat{X}_1 - X_2\hat{Y}_1 \\ &= \left(F_1 - F_2\right)\left(zI - A_{F_1}\right)^{-1} L_1 - F_2\left(zI - A_{L_2}\right)^{-1}\left(L_2 - L_1\right). \end{aligned} \tag{3.44}$$

Based on the above results, we now examine the influences of F and L on the observer-based realisation form (3.28) as well as the closed-loop dynamics. The results are summarised in the following two corollaries.

Corollary 3.2 *Given* $(X_i, Y_i), \left(\hat{M}_i, \hat{N}_i\right), \left(\hat{X}_i, \hat{Y}_i\right), (M_i, N_i), i = 1, 2,$ *the observer-based realisation of a stabilising controller*

$$\hat{x}(k+1) = (A + BF_1)\hat{x}(k) + B\bar{v}_1(k) + L_1 r_1(k), \tag{3.45}$$

$$r_1(k) = y(k) - C\hat{x}(k) - D\left(F_1\hat{x}(k) + \bar{v}_1(k)\right), \tag{3.46}$$

$$u(k) = F_1\hat{x}(k) + \bar{v}_1(k), \tag{3.47}$$

$$\bar{v}_1(z) = X_1(z)v(z), \tag{3.48}$$

can be equivalently realised by

$$\hat{x}(k+1) = (A + BF_2)\hat{x}(k) + B\left(r_Q(k) + \bar{v}_2(k)\right) + L_2 r_2(k), \tag{3.49}$$

$$r_2(k) = y(k) - C\hat{x}(k) - D\left(F_2\hat{x}(k) + r_Q(k) + \bar{v}_2(k)\right), \tag{3.50}$$

$$u(k) = F_2\hat{x}(k) + r_Q(k) + \bar{v}_2(k), \tag{3.51}$$

$$\bar{v}_2(z) = \left(X_2(z) + Q(z)\hat{N}_2(z)\right) v(z), \tag{3.52}$$

$$r_Q(z) = M_2(z)Q(z)r_2(z), \tag{3.53}$$

$$Q(z) = \bar{R}_{21}(z)R_{12}(z) = -V_{12}(z)\bar{V}_{21}(z). \tag{3.54}$$

Proof It is known from the observer-based Youla parameterisation described in Theorem 3.1 that

$$u = F_1\hat{x} + \bar{v}_1 = -\hat{Y}_1 r_1 + M_1\bar{v}_1.$$

By Lemma 3.1, Theorem 3.4 and Corollary 3.1, u can be equivalently written into

$$\begin{aligned} u &= -\hat{Y}_2 R_{21} r_1 + M_2\bar{R}_{21} r_1 + M_2 V_{12}\left(V_{21}X_2 - \bar{V}_{21}\hat{N}_2\right) v \\ &= -\hat{Y}_2 r_2 + M_2\bar{R}_{21}R_{12} r_2 + M_2\left(X_2 - V_{12}\bar{V}_{21}\hat{N}_2\right) v \\ &= -\hat{Y}_2 r_2 + M_2 Q r_2 + M_2\left(X_2 + Q\hat{N}_2\right) v = F_2\hat{x} + r_Q + \bar{v}_2. \end{aligned}$$

Corollary 3.3 *Given* $(X_i, Y_i), \left(\hat{M}_i, \hat{N}_i\right), \left(\hat{X}_i, \hat{Y}_i\right), (M_i, N_i), i = 1, 2,$ *the dynamic of the closed-loop*

$$\begin{bmatrix} u(z) \\ y(z) \end{bmatrix} = \begin{bmatrix} -\hat{Y}_1(z) \\ \hat{X}_1(z) \end{bmatrix} r_1(z) + \begin{bmatrix} M_1(z) \\ N_1(z) \end{bmatrix} \bar{v}_1(z)$$

with r_1 *and* $\bar{v}_1$ *being defined in Corollary 3.2 is equivalent with*

$$\begin{bmatrix} u(z) \\ y(z) \end{bmatrix} = \begin{bmatrix} -\hat{Y}_2(z) + M_2(z)Q(z) \\ \hat{X}_2(z) + N_2(z)Q(z) \end{bmatrix} r_2(z) + \begin{bmatrix} M_2(z) \\ N_2(z) \end{bmatrix} \bar{v}_2(z),$$
$$\bar{v}_2(z) = \left(X_2(z) + Q(z)\hat{N}_2(z) \right) v(z),$$
$$Q(z) = \bar{R}_{21}(z)R_{12}(z) = -V_{12}(z)\bar{V}_{21}(z).$$

The proof is analogous to the one of Corollary 3.2 by means of Lemma 3.1, Theorem 3.4 and Corollary 3.1 and is thus omitted.

Corollaries 3.1–3.3 describe the influences of variations of F and L on (i) Youla parameterisation controllers (3.5), (ii) their observer-based realisation form (3.28), and (iii) the closed-loop dynamics. The most important message from these results consists in the fact that these influences can be exactly modelled by residual signals. In other words, given a stable control loop, by integrating residual signals any stabilising feedback controller and the associated closed-loop dynamics can be achieved. We illustrate this claim with the aid of the following example.

Example 3.5 Given a stable control loop (3.1) with the plant model (2.1)–(2.3) and feedback controller

$$u_0(z) = K_0(z)y(z),$$

the task is to approach a feedback controller

$$u(k) = K(z)y(z),$$

by adding a suitable signal Δu to u_0, that is

$$u(k) = u_0(k) + \Delta u(k).$$

This is a trivial tuning problem and the popular solution is to tune the feedback of y so that it approaches to

$$\Delta u(z) = (K(z) - K_0(z))\, y(z).$$

Below, we present an alternative solution, in which a feedback of residual signal instead of y is used for the tuning purpose. To begin with, u_0 and u are rewritten, based on Theorem 3.1, into their observer-based implementation form,

$$\begin{aligned} \hat{x}_0(k+1) &= A\hat{x}_0(k) + Bu_0(k) + L_0 r_0(k), \\ r_0(k) &= y(k) - C\hat{x}_0(k) - Du_0(k), \\ u_0(z) &= F_0\hat{x}_0(z) + Q_0(z)r_0(z), \\ \hat{x}(k+1) &= A\hat{x}(k) + Bu(k) + L_0 r(k), \\ r(k) &= y(k) - C\hat{x}(k) - Du(k), \\ u(z) &= F\hat{x}(z) + Q_1(z)r(z). \end{aligned}$$

Note that, during nominal operations, $r_0 = r$. It follows from Theorem 3.4 and Corollary 3.2 that

$$\begin{aligned}
F\hat{x}(k) &= F_0\hat{x}_0(k) + r_Q(k), \\
r_Q(z) &= M_0(z)Q(z)r_0(z), \, M_0 = \left(A_{F_0}, B, F_0, I\right), \\
Q(z) &= \left(I - (F_0 - F)(zI - A_F)^{-1} B\right)\left((F_0 - F)\left(zI - A_{F_0}\right)^{-1} L_0\right) \\
&= (F_0 - F)(zI - A_F)^{-1} L_0.
\end{aligned}$$

As a result, we have

$$\Delta u(z) = u(z) - u_0(z) = (Q_1(z) + M_0(z)Q(z) - Q_0(z))\, r_0(z).$$

This implies that tuning the feedback matrix of r_0 leads to approaching Ky. A potential application of this solution is online controller optimisation, for instance, aiming at recovering the system performance suffered from degradation. In that case, the optimal controller K is generally unknown and is to be approached by a tuning algorithm. The major advantage of feedback of r_0 instead of y is to enhance the so-called "control data privacy" in the context of cyber security, as will be discussed in the last part of this book.

3.3 Parameterisation of Observers

While Youla parameterisation is widely used in design and synthesis of feedback control systems, observer parameterisation of LTI systems is less popular and its use in dealing with control engineering and theoretical issues is rarely reported. For this reason, but also motivated by our intention to demonstrate that the feedback controller in observer-based realisation of Youla parameterisation (3.28), $u = F\hat{x} + Qr_y$, is in fact an observer for Fx, we will introduce, in this section, the observer parameterisation with detailed descriptions.

Definition 3.2 Consider the LTI plant model (2.2)–(2.3). A dynamic system

$$x_p(z) = O(z)\begin{bmatrix} u(z) \\ y(z) \end{bmatrix}, \, O(z) \in \mathcal{RH}_\infty, \tag{3.55}$$

is called an observer for $Px(k)$ if

$$\forall x(0), u(k), \lim_{k\to\infty}\left(Px(k) - x_p(k)\right) = 0. \tag{3.56}$$

Here, $P \in \mathbb{R}^{q\times n}$ is a given matrix with an arbitrary integer $q \in [1, n]$, and the initial state of the dynamic system $O(z)$ is zero.

Remark 3.6 In the sequel, it is assumed that

$$rank\,(P) = q.$$

To begin with, we consider the case $x(0) = 0$. Hence, Px can be written as

$$Px(z) = P\,(zI - A)^{-1}\,Bu(z).$$

By means of the system SIR, condition (3.56) is equivalently written as

$$\begin{aligned} Px(z) = P\,(zI - A)^{-1}\,BM(z)v(z) = O(z)\begin{bmatrix} M(z) \\ N(z) \end{bmatrix} v(z) \\ \Longrightarrow P\,(zI - A - BF)^{-1}\,B = O(z)\begin{bmatrix} M(z) \\ N(z) \end{bmatrix}. \end{aligned}$$

The last step is based on the relation

$$P\,(zI - A)^{-1}\,B = P\,(zI - A - BF)^{-1}\,BM^{-1}(z).$$

Write $O(z)$ as

$$O(z) = \begin{bmatrix} O_1(z) \; O_2(z) \end{bmatrix}\begin{bmatrix} X(z) & Y(z) \\ -\hat{N}(z) & \hat{M}(z) \end{bmatrix}$$

with $O_1(z)$, $O_2(z)$ to be determined. It turns out

$$\begin{aligned} P\,(zI - A - BF)^{-1}\,B &= \begin{bmatrix} O_1(z) \; O_2(z) \end{bmatrix}\begin{bmatrix} X(z) & Y(z) \\ -\hat{N}(z) & \hat{M}(z) \end{bmatrix}\begin{bmatrix} M(z) \\ N(z) \end{bmatrix} \\ &= \begin{bmatrix} O_1(z) \; O_2(z) \end{bmatrix}\begin{bmatrix} I \\ 0 \end{bmatrix}, \end{aligned}$$

which results in

$$O_1(z) = P\,(zI - A - BF)^{-1}\,B,$$

and $O_2(z)$ being any (arbitrary) $\mathcal{RH}_\infty$ transfer function matrix of the appropriate dimension. We denote it by $O_2(z) = R(z)$. Consequently, we have

$$\begin{aligned} x_p(z) &= O(z)\begin{bmatrix} u(z) \\ y(z) \end{bmatrix} \\ &= \begin{bmatrix} P\,(zI - A - BF)^{-1}\,B \;\; R(z) \end{bmatrix}\begin{bmatrix} X(z) & Y(z) \\ -\hat{N}(z) & \hat{M}(z) \end{bmatrix}\begin{bmatrix} u(z) \\ y(z) \end{bmatrix} \\ &= P\,(zI - A - BF)^{-1}\,B\,(X(z)u(z) + Y(z)y(z)) + r(z), \qquad (3.57) \\ r(z) &= R(z)\left(\hat{M}(z)y(z) - \hat{N}(z)u(z)\right). \end{aligned}$$

Dynamic system (3.57) is a parameterisation form of all LTI observers with R as the parameterisation transfer function matrix and on the assumption $x(0) = 0$.

Next, we derive a state-space realisation of (3.57). Note that r is indeed a residual vector. Hence, we concentrate on the first term in (3.57). For our purpose, BF and P are rewritten as

$$BF = B^e F^e, B^e = \begin{bmatrix} B & 0 \end{bmatrix}, F^e = \begin{bmatrix} F \\ F_1 \end{bmatrix}, rank\left(F^e\right) = n, P = P_0 F^e,$$

for some matrix F_1, and, based on it, the first term in (3.57) is further written as

$$\begin{aligned} &P_0 F^e \left(zI - A_F\right)^{-1} B^e \left(X^e(z)u^e(z) + Y^e(z)y(z)\right), \\ A_F &= A + BF = A + B^e F^e, u^e = \begin{bmatrix} u \\ 0 \end{bmatrix}, \\ X^e &= \left(A - LC, -(B^e - LD^e), F^e, I\right), Y^e = \left(A - LC, -L, F^e, 0\right), \\ D^e &= \begin{bmatrix} D & 0 \end{bmatrix}. \end{aligned}$$

Since

$$F^e \left(zI - A_F\right)^{-1} B^e = M^e(z) - I, M^e = \left(A_F, B^e, F^e, I\right),$$

it holds

$$\begin{aligned} &P_0 F^e \left(zI - A_F\right)^{-1} B^e \left(X^e(z)u^e(z) + Y^e(z)y(z)\right) \\ &= P_0 \left(M^e(z)X^e(z) - X^e(z)\right) u^e(z) + P_0 \left(M^e(z)Y^e(z) - Y^e(z)\right) y(z). \end{aligned}$$

It follows from the Bezout identity,

$$\begin{gathered} M^e(z)X^e(z) - I = -\hat{Y}^e(z)\hat{N}^e(z), M^e(z)Y^e(z) = \hat{Y}^e(z)\hat{M}^e(z), \\ \hat{Y}^e = \left(A_F, -LW^{-1}, F^e, 0\right), \hat{N}^e = \left(A - LC, B^e - LD^e, WC, WD^e\right), \\ \hat{M}^e = \hat{M} = (A - LC, -L, WC, W), \end{gathered}$$

that

$$\begin{aligned} &P_0 F^e \left(zI - A_{^eF}\right)^{-1} B^e \left(X^e(z)u^e(z) + Y^e(z)y(z)\right) \\ &\quad = P_0 \hat{Y}^e(z) \left(\hat{M}^e(z)y(z) - \hat{N}^e(z)u^e(z)\right) - P_0 Y^e(z)y(z) \\ &\qquad + P_0 F^e \left(zI - A + LC\right)^{-1} (B^e - LD^e)u^e(z). \end{aligned}$$

Bearing in mind that

$$P_0 \hat{Y}^e(z) = \left(A_F, -LW^{-1}, P_0 F^e, 0\right) = \left(A_F, -LW^{-1}, P, 0\right) =: \hat{Y}_p(z) \in \mathcal{RH}_\infty,$$

$$\hat{M}^e(z)y(z) - \hat{N}^e(z)u^e(z) = \hat{M}(z)y(z) - \hat{N}(z)u(z) = r_y(z),$$
$$\alpha(z) := P_0 \left(F^e (zI - A + LC)^{-1} (B^e - LD^e)u^e(z) - Y^e(z)y(z)\right) \Longleftrightarrow$$
$$\begin{cases} \hat{x}(k+1) = (A - LC)\,\hat{x}(k) + (B - LD)\,u(k) + Ly(k), \\ \alpha(k) = P_0 F^e \hat{x}(k) = P\hat{x}(k), \end{cases}$$

we have

$$x_p(z) = O(z)\begin{bmatrix} u(z) \\ y(z) \end{bmatrix} = P\hat{x}(z) + \left(R(z) + \hat{Y}_p(z)\right)\left(\hat{M}(z)y(z) - \hat{N}(z)u(z)\right).$$

Since R is any $\mathcal{RH}_\infty$ transfer function matrix, defining

$$Q(z) = R(z) + \hat{Y}_p(z) \in \mathcal{RH}_\infty$$

leads to

$$x_p(k) = P\hat{x}(k) + r_Q(k),$$
$$r_Q(z) = Q(z)r_y(z),$$

which guarantees that $\forall u$, condition (3.56) holds.

In order to prove that $\forall x(0)$, condition (3.56) keeps true as well, let us examine the estimation error $Px(k) - x_p(k)$,

$$Px(k) - x_p(k) = Pe(k) - r_Q(k), e(k) = x(k) - \hat{x}(k).$$

It is apparent that

$$e(k+1) = (A - LC)\,e(k), r_y(k) = WCe(k).$$

As a result, we finally have

$$\forall x(0), \lim_{k\to\infty} \left(Px(k) - x_p(k)\right) = \lim_{k\to\infty} \left(Pe(k) - r_Q(k)\right) = 0.$$

Summarising the aforementioned results gives the proof of the following theorem.

Theorem 3.5 *Given the plant model (2.2)–(2.3), all LTI observers for Px in sense of Definition 3.2 are parametrised by*

$$\hat{x}(k+1) = (A - LC)\,\hat{x}(k) + (B - LD)\,u(k) + Ly(k),$$
$$x_p(k) = P\hat{x}(k) + r_Q(k), r_y(k) = W\left(y(k) - \hat{y}(k)\right),$$
$$r_Q(z) = Q(z)r_y(z), Q(z) \in \mathcal{RH}_\infty$$

with Q as the parameterisation transfer function matrix.

As a result of Theorems 3.1–3.5, the following conclusion is obvious.

Corollary 3.4 *Given the control loop (3.1) with the plant model (2.1)–(2.3), any feedback controller of the control law $u = Ky + v$, that stabilises the loop and is parametrised by*

$$u_{fb}(z) = F\hat{x}(z) + Q(z)r_y(z),$$

is an observer for Fx.

Corollary 3.4 reveals an information aspect of output feedback control. It is a reconstruction of the state feedback controller Fx. In this sense, it can be interpreted as a general form of separation principle for designing feedback controller:

$$\text{Controller design} \;=\;$$
$$\text{Design of state feedback controller } Fx + \text{ design of an observer for } Fx.$$

3.4 Notes and References

Internal stability is an essential concept of feedback control systems. The reader is referred, for instance, to [243] for the definition and detailed explanation. As a basic requirement, internal stability of a feedback control loop is of fundamental importance. A feedback controller that guarantees the closed-loop stability is called stabilising controller. The original parameterisation form (3.5) of all stabilising feedback controllers was developed by Youla [221, 222] and Kucera [106] in the 70 s independently, and hence is called Youla or Youla-Kucera parameterisation, after the name of the authors. Youla parameterisation is a fundamental result in control theory and widely applied in design and synthesis of feedback control systems. For a comprehensive review of Youla parameterisation and its main applications, the reader is referred to the review papers [4, 134]. Although the observer-based form (3.28) of Youla parameterisation has not received appreciable attention and is rarely applied in control system analysis and controller design, it can be found in [4, 243].

Besides the presentation of the original form of Youla parameterisation based on the Bezout identity (2.31) and its extended form (3.4), our introduction has been focused on the application of the control engineering interpretations of LCF and RCF of the plant and feedback controller to the deduction of the observer-based implementation form (3.28). In this regard, we have generalised the results presented in Chap. 2 that

- a feedback controller is driven by residual signal r_y,
- the closed-loop dynamics can be expressed by the response of (u, y) to r_y and $\bar{v}$, which is modelled by the image subspaces of the plant and controller, $\mathcal{I}_G$ and $\mathcal{I}_K$, spanned by $\begin{bmatrix} M \\ N \end{bmatrix}$ and $\begin{bmatrix} -\hat{Y} + MQ \\ \hat{X} + NQ \end{bmatrix}$, respectively, and

- the SKR and SIR of the controller K can be interpreted as a residual generator r_u as well as an image subspace with a latent variable $r_y + \hat{N}v$.

These results are summarised in Theorem 3.1.

It is the common opinion that the original parameterisation form (3.5) and the observer-based implementation form (3.28) are equivalent. Noticing the different configurations of their implementation forms, open-loop for the parameterisation form (3.5) and closed-loop for the observer-based one (3.28), realisations of the Youla parameterisation and its observer-based form have been examined. Theorem 3.2 reveals that the observer-based (closed-loop) realisation (3.32) guarantees the well-posedness of the controller and the closed-loop dynamics, while Theorem 3.3 gives an existence condition for the Youla parameterisation controller (3.5) being well-posed.

A further investigation on the influences of varying state feedback and observer gain matrices on the controller and the closed-loop dynamics demonstrates the relations given in Theorem 3.4 and Corollary 3.1. It is worth to mention that some of the results in Theorem 3.4 and Corollary 3.1 have been proved in [39, 45]. Differently, the proof of Theorem 3.4 and Corollary 3.1 has been achieved with the aid of the double Bezout identity. In Corollary 3.2, an important result is presented. It is delineated that, given a stable control loop, any favoured stabilising feedback controller and the associated closed-loop dynamics can be achieved by feeding back residual signals. This result is useful for our subsequent work on control performance recovery.

The results on observer parameterisation presented in Theorem 3.4 are known since a couple of decades, for instance, reported in [51]. Our proof of Theorem 3.5 is based on the double Bezout identity. In addition, some proof steps in [51] without rigorous mathematical handling have been corrected. A by-product of Theorem 3.5 is the proof of the fact that any stabilising controller is an observer for the estimation of $Fx(k)$. This claim is a cornerstone of the unified control and detection framework.

Chapter 4
Variations of Stabilising Controller Parameterisations

In the previous chapter, we introduced the original form of the Youla parameterisation of all stabilising controllers and its observer-based realisation form. During the past decades, numerous variations of the Youla parameterisation have been reported. They have been developed for various control engineering purposes, mainly in the field of robust and adaptive control, but also for fault-tolerant control. This chapter is dedicated to the introduction of some existing variations of the Youla parameterisation and the development of new ones. They are the main ingredients of the unified framework of control and detection and will be applied to dealing with various issues addressed in the subsequent parts.

4.1 Modular Functionalisation of Controllers

To begin with, we analyse the functionality of controller parameters in the context of control performance and modular configuration of controllers, which also motivates applications of different Youla parameterisation forms.

Consider the two configurations of feedback control loops shown in Fig. 3.2. The observer-based configuration is modularly structured, which enables tuning of each controller parameter, F, L and Q, individually and in a targeted manner, while in the configuration with the controller $u = Ky + v$, the feedback controller K is generally tuned or optimised as a whole, which may unintentionally lead to a complex optimisation problem. Against this background, we start our analysis from the observer-based Youla parameterisation form (3.28). An observer-based stabilising controller consists of several functional modules:

S. X. Ding, *Unified Framework of Control and Detection*,
https://doi.org/10.1007/978-3-662-72417-0_4

- an observer and an observer-based residual generator,

$$\hat{x}(k+1) = A\hat{x}(k) + Bu(k) + Lr_0(k),$$
$$r_0(k) = y(k) - \hat{y}(k),\ \hat{y}(k) = C\hat{x}(k) + Du(k),$$
$$r_y(k) = Wr_0(k),$$

  which serve as an information provider for the controller and diagnostic system and deliver a state estimation, $\hat{x}$, and the primary residual, $r_0 = y - \hat{y}$, as well as r_y,
- controllers

$$u(z) = F\hat{x}(z) + Q(z)r_y(z) + \bar{v}(z),$$

  including

  - the feedback controller: $u_{fb} = F\hat{x} + Qr_y$ and
  - the feedforward controller: $u_{ff} = \bar{v} = \left(X + Q\hat{N}\right)v,$

  and in addition, it can be extended for the detection purpose, by connecting a stable post-filter R to r_0 to build
- a detector

$$r(z) = R(z)r_0(z),\ R(z) \in \mathcal{RH}_\infty.$$

This modular structure provides us with clear parameterisations of the functional modules:

- the state observer and the residual generator are parametrised by L and W,
- the feedback controller by F, Q,
- the feedforward controller by $V = X + Q\hat{N}$, and
- the detector by R.

Although all five parameters listed above are available for the design and online optimisation objectives, they have evidently different functionalities, as summarised below:

- F, L determine the stability and eigen-dynamics of the closed-loop,
- R, V have no influence on the system stability, and V serves for the tracking behaviour, while R acts as an optimiser for the fault detectability,
- Q is capable to perform multi-functionality, including
  - enhancement of the system robustness and control performance,
  - improvement of the estimation performance, when u_{fb} is interpreted as an estimator for Fx, as claimed in Corollary 3.4, and
  - enhancement of the system stability, as described by Theorem 3.4 and Corollary 3.2.

On the other hand, it should be mentioned that in case of existence of uncertainties in the loop, tuning Q may lead to instability as well, as will be demonstrated in the sequel.

It is evident that, due to their different functionalities, the above five parameters have to be treated with different priorities. Recall that the system stability and eigendynamics are the fundamental requirements on an automatic control system. This requires that tuning of F, L is of the highest priority to guarantee the stability. Differently, R, V are used to optimise control or detection performance. In case that a temporary system performance degradation is tolerable, the real-time demand and the priority for an online optimisation of R and V are relatively low. Due to its multi-functionality, the priority of the parameter Q depends on the targeted tuning purpose. In fact, most of the variations of the Youla parameterisation are related to the selection of Q.

When an automatic control system is integrated into a complex and networked system like a CPS, cyber security becomes a critical issue. In this context, the functionalisation of controllers offers a useful design tool towards a cyber security-conscious system configuration. To delineate potential applications, consider a controller in the form (3.5) and suppose that the plant is networked with a control station, as sketched in Fig. 4.1. It is clear that, during system operations, the process data (u, y) should be real-time transmitted over the network. Consequently, an adversary is able to gain system knowledge, for instance, the plant dynamics, privacy in the process data (u, y) or the control strategy, by means of eavesdropping attacks. The situation could become more critical, when the controller K as a whole is to be updated online and real-time to meet system optimisation requirements or to recover performance, since the adversary could manipulate the updating procedure and disturb the system operation or even damage the system. In order to meet control and system performance requirements on the one hand, and to enhance cyber security on

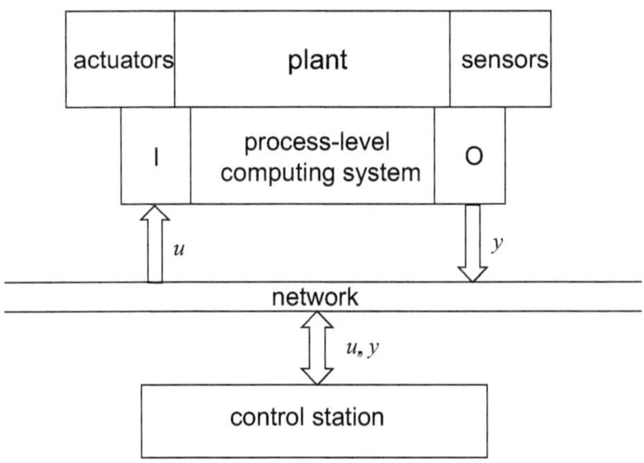

Fig. 4.1 A networked control system

the other hand, the observer-based controller form (3.28) provides us with an ideal configurable implementation platform with modular functional units. Based on the functionalisation of the controller, the observer-based controller (3.28) can be so configured that the performance and cyber security requirements are simultaneously satisfied. For instance,

- an observer and an observer-based residual generator are implemented on the plant side making use of the available computing capacity, and deliver the functionality $u_0 = F\hat{x}$,
- the remaining functional modules of the controller, $Qr_y + \bar{v}$, run at the control station, and
- the needed data transformations between the plant and the control station are, from the plant to the control station, the residual r_y, and from the control station to the plant, $u_c = Qr_y + \bar{v}$.

This configuration is of several benefits:

- the controller at the plant side, $u_0 = F\hat{x}$, operates independent of the network and data transmissions, which guarantees the basic fail-safe requirement, namely system stability,
- the transmission of the signal pair $\left(u_c, r_y\right)$ instead of (u, y) prevents adversary from gaining system knowledge by means of eavesdropping attacks,
- when performance optimisation or recovery actions are needed, real-time tuning Q can be realised in the control station, where sufficient computation power is generally available.

In summary, the functionally modular and configurable observer-based controller form (3.28) enables "control functionally sharing" among the control functional units, which is similar to the secret sharing scheme well-known in cryptography, and can be utilised for efficient detection of cyber-attacks and enhancement of cyber security of automatic control systems.

4.2 PnP Configured Youla Parameterisation

An immediate application of the observer-based realisation of (3.28) is the establishment of the so-called plug-and-play (PnP) configuration sketched in Fig. 4.2, which can be viewed as a variation of the Youla parameterisation. Given the plant with model (2.1)–(2.3), and suppose that the feedback controller K_0 stabilises the feedback loop. Without loss of generality, it is further supposed that a normalised SKR of G,

$$r_y = K_{G_0} \begin{bmatrix} u \\ y \end{bmatrix} = \begin{bmatrix} -\hat{N}_0 & \hat{M}_0 \end{bmatrix} \begin{bmatrix} u \\ y \end{bmatrix}, \tag{4.1}$$

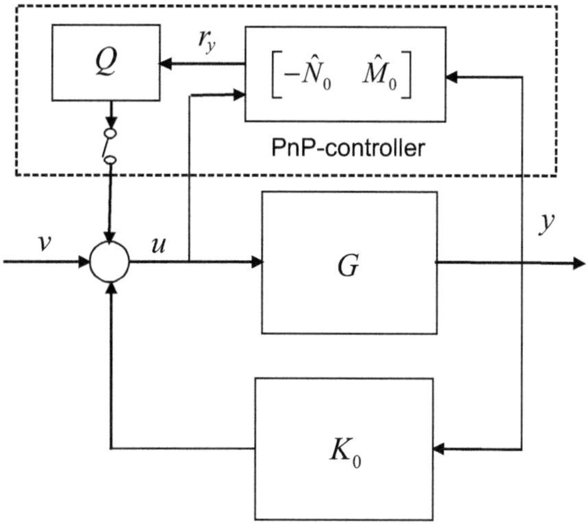

Fig. 4.2 PnP configured Youla parameterisation

exists with observer (2.8)–(2.9). A PnP stabilising controller is described by

$$u(z) = \begin{cases} K_0(z)y(z) + v(z), & \text{nominal operation,} \\ K_0(z)y(z) + v(z) + Q(z)r_y(z), & \text{switching of } Qr_y \text{ is triggered.} \end{cases} \tag{4.2}$$

Here, $Q(z) \in \mathcal{RH}_\infty$, and switching of Qr_y is triggered by certain designed control actions over a defined time interval. As described in the following theorem, the closed-loop stability with PnP controller (4.2) is guaranteed.

Theorem 4.1 *Given the plant with model (2.1)–(2.3) with the normalised SKR K_{G_0} and a stabilising controller K_0, then all stabilising controllers can be parametrised by*

$$u(z) = K_0(z)y(z) + Q(z)r_y(z), \; Q(z) \in \mathcal{RH}_\infty, \tag{4.3}$$

with parameterisation matrix Q, where r_y is the residual vector generated by residual generator (4.1).

Proof The proof is straight-forward. It follows from Theorem 3.1 that $K_0 y$ can be written as

$$K_0(z)y(z) = F\hat{x}(z) + Q_0(z)r_y(z)$$

for some $Q_0(z) \in \mathcal{RH}_\infty$. It is obvious, according to Theorem 3.1, that

$$u(z) = F\hat{x}(z) + Q_0(z)r_y(z) + Q(z)r_y(z)$$

parametrises all stabilising controllers.

We call the parameterisation form (4.3) PnP configured Youla parameterisation, since its major application is the PnP stabilising controller (4.2).

Remark 4.1 It should be mentioned that the feedback of Qr_y will cause change in the response to the reference signal v. Since our major concern is the feedback dynamics, we will not pay special attention to the tracking behaviour (the response to the reference signal), when it is not addressed.

The PnP controller (4.2) is of considerable practical interest in industrial applications. Different from the well-established adaptive control technique, the update of the controller is realised without any modification on the existing (default) controller K_0, which is often an essential safety requirement. Moreover, thanks to Theorem 4.1, the feedback of Qr_y does not, in general, imperil the system stability and can be thus reliably and safely utilised for control performance recovery. In our subsequent work, the PnP configuration will be frequently applied to recovering system performance and enhancing cyber security.

4.3 State Observer-Based Youla Parameterisation

On the basis of the claim in Corollary 3.4 that the feedback controller u_{fb} is an estimate for Fx, a further variant of the Youla parameterisation can be realised. Construct a state estimator as follows,

$$\hat{x}(k+1) = A_L\hat{x}(k) + B_L u(k) + Ly(k), A_L = A - LC, B_L = B - LD, \tag{4.4}$$

$$r_0(k) = y(k) - C\hat{x}(k) - Du(k) = y(k) - \hat{y}(k), \tag{4.5}$$

$$\hat{x}^*(z) = \hat{x}(z) + R(z)r_0(z), r_0(z) = \hat{M}(z)y(z) - \hat{N}(z)u(z), \tag{4.6}$$

where $R(z) \in \mathcal{RH}_\infty$,and Rr_0 is added to increase the degree of design freedom with parameterisation matrix R. It is known from Theorem 3.5 that $\hat{x}^*$ is an estimate for x, and so $F\hat{x}^*$ for Fx. Now, setting

$$u(k) = F\hat{x}^*(k) \tag{4.7}$$

results in, according to (2.26),

$$X(z)u(z) = -Y(z)y(z) + Q(z)r_0(z), Q(z) = FR(z) \Longrightarrow$$
$$u(z) = K(z)y(z),$$
$$K(z) = -\left(X(z) + Q(z)\hat{N}(z)\right)^{-1}\left(Y(z) - Q(z)\hat{M}(z)\right),$$

which is a Youla parameterisation feedback controller.

Without loss of generality, it is assumed that

$$n \geq p, rank(F) = p.$$

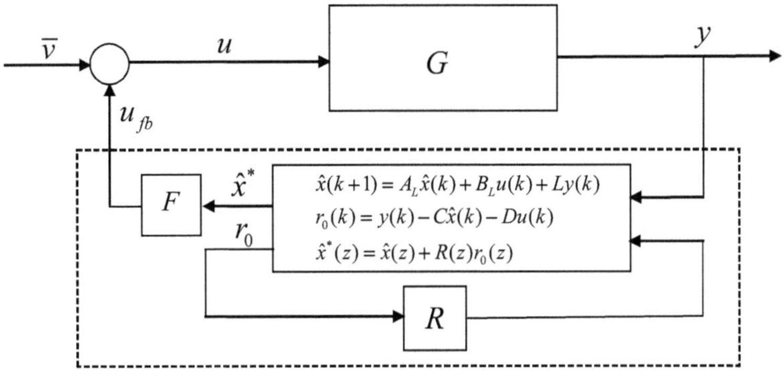

Fig. 4.3 State observer-based Youla parameterisation

We have the following state observer-based Youla parameterisation.

Theorem 4.2 *Given the plant with model (2.1)–(2.3), then dynamic system (4.4)–(4.7) parametrises all stabilising controllers, where R is the parameterisation matrix.*

Proof The feedback controller (4.7) can be rewritten as

$$u(z) = F\hat{x}(z) + FR(z)r_0(z) = F\hat{x}(z) + Q_0(z)r_0(z), \tag{4.8}$$
$$F^-Q_0(z) = R(z), FF^- = I.$$

It becomes, according to Theorem 3.1, clear that (4.8), and thus the system (4.4)–(4.7) parametrises all stabilising controllers.

We would like to call the reader's attention to the state estimator (4.4)–(4.6). The feedback of the residual signal Rr_0 provides us with higher degree of design freedom and thus could be utilised to better estimation and detection performance. A realisation of the system configuration is sketched in Fig. 4.3.

4.4 Dual Youla Parameterisation I

As showcased in Fig. 4.4, without considering the reference signal, a feedback control system is of an absolutely symmetric configuration. So far, it is plain and logical that dual forms of the Youla parameterisation would exist. In this section, we briefly present the first dual form on the assumption that

- the nominal plant model is given with the LCP and RCP, $\left(\hat{M}, \hat{N}\right)$ and (M, N), and
- a stabilising feedback controller K of the form (3.5) is applied to the closed-loop in Fig. 4.4, where $\left(\hat{M}, \hat{N}\right)$, (M, N), $\left(\hat{X}, \hat{Y}\right)$ and (X, Y) are given by (2.5)–(2.6) and (2.25)–(2.26), respectively, and satisfy the Bezout identity (2.31).

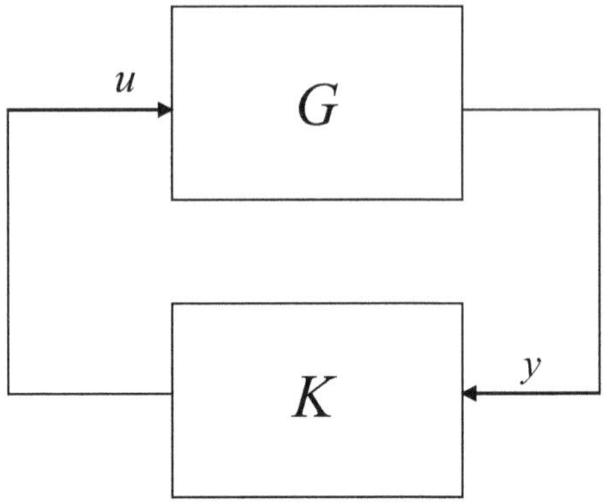

Fig. 4.4 A feedback control system

4.4.1 Case: $Q = 0$

At first, we assume that in (3.5) $Q = 0$. Let $S(z) \in \mathcal{RH}_\infty$ be of appropriate dimension. By the following operation, the Bezout identity is extended to

$$\begin{bmatrix} I & 0 \\ -S & I \end{bmatrix}\begin{bmatrix} X & Y \\ -\hat{N} & \hat{M} \end{bmatrix}\begin{bmatrix} M & -\hat{Y} \\ N & \hat{X} \end{bmatrix}\begin{bmatrix} I & 0 \\ S & I \end{bmatrix}$$
$$= \begin{bmatrix} X & Y \\ -\hat{N} - SX & \hat{M} - SY \end{bmatrix}\begin{bmatrix} M - \hat{Y}S & -\hat{Y} \\ N + \hat{X}S & \hat{X} \end{bmatrix} = \begin{bmatrix} I & 0 \\ 0 & I \end{bmatrix}. \tag{4.9}$$

Analogue to the discussion in Sect. 3.1 about the Youla parameterisation and thus as a dual result of Youla parameterisation (3.5), the following theorem, known as dual Youla parameterisation in the literature, can be obviously proved.

Theorem 4.3 *Given the control loop shown in Fig. 4.4, the LCPs* $\left(\hat{M}, \hat{N}\right)$ *and* (X, Y), *RCPs* (M, N) *and* $\left(\hat{X}, \hat{Y}\right)$, *and the controller*

$$K = -\hat{Y}\hat{X}^{-1} = -X^{-1}Y, \tag{4.10}$$

then all transfer function matrices parametrised by $S \in \mathcal{RH}_\infty$ *as follows,*

$$G = \left(\hat{M} - SY\right)^{-1}\left(\hat{N} + SX\right) = \left(N + \hat{X}S\right)\left(M - \hat{Y}S\right)^{-1}, \tag{4.11}$$

represent the plant dynamics in the control loop, which can be stabilised by the controller K. *Here,* $\left(\hat{M} - SY, \hat{N} + SX\right)$ *and* $\left(M - \hat{Y}S, N + \hat{X}S\right)$ *are LCP and RCP, respectively.*

Against the background that $\left(\hat{M}, \hat{N}\right)$ and (M, N) are the LCP and RCP of the nominal plant model, the dynamics modelled by (SY, SX) and $\left(\hat{Y}S, \hat{X}S\right)$ are generally interpreted as (plant) model uncertainties. In this context, let

$$K_{G_0} = \begin{bmatrix} -\hat{N} & \hat{M} \end{bmatrix}, I_{G_0} = \begin{bmatrix} M \\ N \end{bmatrix}$$

be the SKR and SIR of the nominal system. The SKR and SIR of the plant dynamics (4.11) can be modelled by

$$K_G = \begin{bmatrix} -\hat{N} & \hat{M} \end{bmatrix} + \begin{bmatrix} -\Delta_{\hat{N}} & \Delta_{\hat{M}} \end{bmatrix}, \begin{bmatrix} -\Delta_{\hat{N}} & \Delta_{\hat{M}} \end{bmatrix} = S \begin{bmatrix} -X & -Y \end{bmatrix}, \tag{4.12}$$

$$I_G = \begin{bmatrix} M \\ N \end{bmatrix} + \begin{bmatrix} \Delta_M \\ \Delta_N \end{bmatrix}, \begin{bmatrix} \Delta_M \\ \Delta_N \end{bmatrix} = \begin{bmatrix} -\hat{Y} \\ \hat{X} \end{bmatrix} S. \tag{4.13}$$

Example 4.1 In Example 2.1, we have studied the relations of uncertainties in the plant SKR and SIR. We now demonstrate that the above relations, (4.12)–(4.13), are a special case of the relations (2.68)–(2.69) in Example 2.1. It follows from (2.68) that for

$$\Delta_{K_G} = \begin{bmatrix} -\Delta_{\hat{N}} & \Delta_{\hat{M}} \end{bmatrix} = S \begin{bmatrix} -X & -Y \end{bmatrix},$$

it holds

$$\begin{aligned} \Delta_{I,K_G} &= -\begin{bmatrix} -\hat{Y} \\ \hat{X} \end{bmatrix} \left(I + \Delta_{K_G} \begin{bmatrix} -\hat{Y} \\ \hat{X} \end{bmatrix} \right)^{-1} \Delta_{K_G} \begin{bmatrix} M \\ N \end{bmatrix} \\ &= -\begin{bmatrix} -\hat{Y} \\ \hat{X} \end{bmatrix} S \begin{bmatrix} -X & -Y \end{bmatrix} \begin{bmatrix} M \\ N \end{bmatrix} = \begin{bmatrix} -\hat{Y} \\ \hat{X} \end{bmatrix} S = \begin{bmatrix} \Delta_M \\ \Delta_N \end{bmatrix}. \end{aligned}$$

Similarly, according to (2.69), for

$$\Delta_{I_G} = \begin{bmatrix} \Delta_M \\ \Delta_N \end{bmatrix} = \begin{bmatrix} -\hat{Y} \\ \hat{X} \end{bmatrix} S,$$

we have,

$$\begin{aligned} \Delta_{K,I_G} &= -\begin{bmatrix} -\hat{N} & \hat{M} \end{bmatrix} \Delta_{I_G} \left(I + \begin{bmatrix} X & Y \end{bmatrix} \Delta_{I_G} \right)^{-1} \begin{bmatrix} X & Y \end{bmatrix} \\ &= -\begin{bmatrix} -\hat{N} & \hat{M} \end{bmatrix} \begin{bmatrix} -\hat{Y} \\ \hat{X} \end{bmatrix} S \begin{bmatrix} X & Y \end{bmatrix} = S \begin{bmatrix} -X & -Y \end{bmatrix} = \Delta_{K_G}. \end{aligned}$$

We now analyse the closed-loop dynamics. The uncertainty in the SKR results in

$$r_y = \begin{bmatrix} -\hat{N} & \hat{M} \end{bmatrix} \begin{bmatrix} u \\ y \end{bmatrix} = -\Delta_{K_G} \begin{bmatrix} u \\ y \end{bmatrix} = S \begin{bmatrix} X & Y \end{bmatrix} \begin{bmatrix} u \\ y \end{bmatrix}.$$

On the other hand, the controller setting (4.10) yields

$$u = -X^{-1}Yy \Longleftrightarrow \begin{bmatrix} X & Y \end{bmatrix} \begin{bmatrix} u \\ y \end{bmatrix} = 0 \Longrightarrow$$
$$r_y = \begin{bmatrix} -\hat{N} & \hat{M} \end{bmatrix} \begin{bmatrix} u \\ y \end{bmatrix} = S \begin{bmatrix} X & Y \end{bmatrix} \begin{bmatrix} u \\ y \end{bmatrix} = 0. \tag{4.14}$$

That is, in the system setting shown in Fig. 4.4, the variation in the plant dynamics cannot be detected using an observer-based residual generator. We now consider a more practical case with

$$u = Ky + v = -\hat{Y}\hat{X}^{-1}y + v = -X^{-1}Yy + v,$$

where $v \neq 0$ is the reference vector. It turns out

$$Xu + Yy = Xv := \bar{v} \Longrightarrow$$
$$r_y = \begin{bmatrix} -\hat{N} & \hat{M} \end{bmatrix} \begin{bmatrix} u \\ y \end{bmatrix} = S \begin{bmatrix} X & Y \end{bmatrix} \begin{bmatrix} u \\ y \end{bmatrix} = S\bar{v}. \tag{4.15}$$

Consequently, we have, following Theorem 3.1, the closed-loop dynamics

$$\begin{bmatrix} u \\ y \end{bmatrix} = \begin{bmatrix} -\hat{Y} \\ \hat{X} \end{bmatrix} r_y + \begin{bmatrix} M \\ N \end{bmatrix} \bar{v} = \left(\begin{bmatrix} M \\ N \end{bmatrix} + \begin{bmatrix} -\hat{Y} \\ \hat{X} \end{bmatrix} S \right) \bar{v}. \tag{4.16}$$

Residual dynamic (4.15) implies that the variation in the plant dynamics can be well detected as far as $v \neq 0$. In fact, by sufficient excitations in v, model (4.15) can be used for identifying the uncertainty dynamic S.

4.4.2 Case: $Q \neq 0$

We now extend our study to the case $Q \neq 0$. It is apparent that

$$\begin{bmatrix} X + Q\hat{N} & Y - Q\hat{M} \\ -\hat{N} - SX & \hat{M} - SY \end{bmatrix} \begin{bmatrix} M - \hat{Y}S & -\hat{Y} + MQ \\ N + \hat{X}S & \hat{X} + NQ \end{bmatrix}$$
$$= \left(\begin{bmatrix} X & Y \\ -\hat{N} & \hat{M} \end{bmatrix} + \begin{bmatrix} Q\hat{N} & -Q\hat{M} \\ -SX & -SY \end{bmatrix} \right) \left(\begin{bmatrix} M & -\hat{Y} \\ N & \hat{X} \end{bmatrix} + \begin{bmatrix} -\hat{Y}S & MQ \\ \hat{X}S & NQ \end{bmatrix} \right)$$
$$= \begin{bmatrix} I & 0 \\ 0 & I \end{bmatrix} + \begin{bmatrix} 0 & -Q \\ -S & 0 \end{bmatrix} + \begin{bmatrix} 0 & Q \\ S & 0 \end{bmatrix} + \begin{bmatrix} -QS & 0 \\ 0 & -SQ \end{bmatrix} = \begin{bmatrix} I - QS & 0 \\ 0 & I - SQ \end{bmatrix}$$

such that the Bezout identity as the mathematical essential for the closed-loop stabilisation does not hold. In order to find the existence conditions for the stabilisation, we examine the dynamics of the closed-loop

$$\begin{cases} y = Gu + d, \\ u = Ky + v, \end{cases} \tag{4.17}$$

where (d, v) represent external input vectors, and

$$G = \left(\hat{M} - SY\right)^{-1}\left(\hat{N} + SX\right) = \left(N + \hat{X}S\right)\left(M - \hat{Y}S\right)^{-1}, \tag{4.18}$$

$$K = -\left(\hat{Y} - MQ\right)\left(\hat{X} + NQ\right)^{-1} = -\left(X + Q\hat{N}\right)^{-1}\left(Y - Q\hat{M}\right). \tag{4.19}$$

By the following routine computations,

$$\begin{aligned} \begin{bmatrix} u \\ y \end{bmatrix} &= \begin{bmatrix} I & -K \\ -G & I \end{bmatrix}^{-1} \begin{bmatrix} v \\ d \end{bmatrix} \\ &= \begin{bmatrix} X + Q\hat{N} & Y - Q\hat{M} \\ -\left(\hat{N} + SX\right) & \hat{M} - SY \end{bmatrix}^{-1} \begin{bmatrix} \bar{v} \\ \bar{d} \end{bmatrix} \\ &= \begin{bmatrix} M - \hat{Y}S & -\hat{Y} + MQ \\ N + \hat{X}S & \hat{X} + NQ \end{bmatrix} \begin{bmatrix} I - QS & 0 \\ 0 & I - SQ \end{bmatrix}^{-1} \begin{bmatrix} \bar{v} \\ \bar{d} \end{bmatrix}, \\ \bar{v} &= \left(X + Q\hat{N}\right)v, \bar{d} = \left(\hat{M} - SY\right)d, \end{aligned}$$

it becomes clear that

$$\begin{bmatrix} I - QS & 0 \\ 0 & I - SQ \end{bmatrix}^{-1} \in \mathcal{RH}_\infty \tag{4.20}$$

is a necessary and sufficient condition for the closed-loop stability. Since

$$\begin{aligned} y = Gu + d &\Longrightarrow \left(\hat{M} - SY\right)y = \left(\hat{N} + SX\right)u + \bar{d} \Longrightarrow \\ r_y &= \begin{bmatrix} -\hat{N} & \hat{M} \end{bmatrix} \begin{bmatrix} u \\ y \end{bmatrix} = S \begin{bmatrix} X & Y \end{bmatrix} \begin{bmatrix} u \\ y \end{bmatrix} + \bar{d}, \\ u = Ky + v &\Longrightarrow \left(X + Q\hat{N}\right)u = -\left(Y - Q\hat{M}\right)y + \bar{v} \Longrightarrow \\ r_u &:= \begin{bmatrix} X & Y \end{bmatrix} \begin{bmatrix} u \\ y \end{bmatrix} = Q \begin{bmatrix} -\hat{N} & \hat{M} \end{bmatrix} \begin{bmatrix} u \\ y \end{bmatrix} + \bar{v}, \end{aligned}$$

we finally have the following known theorem.

Theorem 4.4 *Given the control loop (4.17) with the plant model (4.18) and feedback controller (4.19), the loop is stable if and only if the feedback control system*

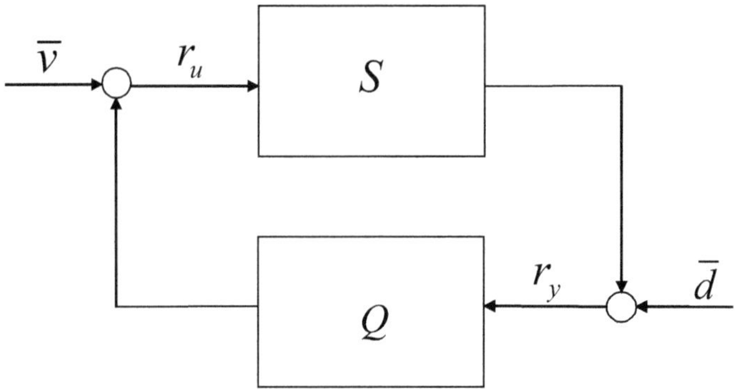

Fig. 4.5 The equivalent feedback control system (4.21)–(4.22)

$$r_y = Sr_u + \bar{d}, \tag{4.21}$$

$$r_u = Qr_y + \bar{v} \tag{4.22}$$

is stable and well-posed.

Proof The proof is straightforward, since the system (4.21)–(4.22) is stable if and only if

$$\begin{bmatrix} r_y \\ r_u \end{bmatrix} = \begin{bmatrix} 0 & S \\ Q & 0 \end{bmatrix} \begin{bmatrix} r_y \\ r_u \end{bmatrix} + \begin{bmatrix} \bar{d} \\ \bar{v} \end{bmatrix} \Longrightarrow$$

$$\begin{bmatrix} r_y \\ r_u \end{bmatrix} = \begin{bmatrix} I & -S \\ -Q & I \end{bmatrix}^{-1} \begin{bmatrix} \bar{d} \\ \bar{v} \end{bmatrix} = \begin{bmatrix} (I - SQ)^{-1} & S(I - SQ)^{-1} \\ -(I - QS)^{-1} Q & (I - QS)^{-1} \end{bmatrix} \begin{bmatrix} \bar{d} \\ \bar{v} \end{bmatrix}$$

is stable, which is equivalent to condition (4.20).

Figure 4.5 shows the configuration of the control loop (4.21)–(4.22). It is of considerable interest to notice the following facts:

- the closed-loop stability of the original system is equivalent with the stability of the loop shown in Fig. 4.5;
- this alternative control loop consists of two stable systems, S and Q, in which S, as uncertainty in the plant, is unknown and acts as "the plant" in the loop, Q as a part of the controller serves as "the feedback controller" in the loop and can thus be tuned, and moreover,
- in the loop, residual signal pair (r_y, r_u), instead of the process signal pair (u, y), is the output variable and the information carrier.

Example 4.2 By this example, we would like to call the reader's attention to the relation between the process data (u, y) and the residual data (r_u, r_y). Recall that

$$\begin{bmatrix} r_u \\ r_y \end{bmatrix} = \begin{bmatrix} X & Y \\ -\hat{N} & \hat{M} \end{bmatrix} \begin{bmatrix} u \\ y \end{bmatrix}$$

and the LCPs $\left(\hat{M}, \hat{N}\right)$, (X, Y) are determined by the nominal system model and independent of the model uncertainty. On the other hand, by means of the Bezout identity (2.31), (u, y) can be fully re-constructed by $\left(r_u, r_y\right)$ as

$$\begin{bmatrix} u \\ y \end{bmatrix} = \begin{bmatrix} M & -\hat{Y} \\ N & \hat{X} \end{bmatrix} \begin{bmatrix} r_u \\ r_y \end{bmatrix}.$$

It is interesting to note that if (M, N) and $\left(\hat{M}, \hat{N}\right)$ are the normalised RCF and LCF, as given in Theorem 2.1, then it holds, according to Theorem 2.2,

$$\begin{bmatrix} -\hat{N}_0^{\sim} \\ \hat{M}_0^{\sim} \end{bmatrix} \begin{bmatrix} -\hat{N}_0 & \hat{M}_0 \end{bmatrix} + \begin{bmatrix} M_0 \\ N_0 \end{bmatrix} \begin{bmatrix} M_0^{\sim} & N_0^{\sim} \end{bmatrix} = I.$$

It yields

$$\begin{aligned} \begin{bmatrix} u \\ y \end{bmatrix} &= \left(\begin{bmatrix} -\hat{N}_0^{\sim} \\ \hat{M}_0^{\sim} \end{bmatrix} \begin{bmatrix} -\hat{N}_0 & \hat{M}_0 \end{bmatrix} + \begin{bmatrix} M_0 \\ N_0 \end{bmatrix} \begin{bmatrix} M_0^{\sim} & N_0^{\sim} \end{bmatrix} \right) \begin{bmatrix} M_0 & -\hat{Y}_0 \\ N_0 & \hat{X}_0 \end{bmatrix} \begin{bmatrix} r_u \\ r_y \end{bmatrix} \\ &= \begin{bmatrix} M_0 \\ N_0 \end{bmatrix} r_u + \left(\begin{bmatrix} -\hat{N}_0^{\sim} \\ \hat{M}_0^{\sim} \end{bmatrix} + \begin{bmatrix} M_0 \\ N_0 \end{bmatrix} \begin{bmatrix} M_0^{\sim} & N_0^{\sim} \end{bmatrix} \begin{bmatrix} -\hat{Y}_0 \\ \hat{X}_0 \end{bmatrix} \right) r_y. \end{aligned} \tag{4.23}$$

Since

$$\begin{bmatrix} M_0 & -\hat{N}_0^{\sim} \\ N_0 & \hat{M}_0^{\sim} \end{bmatrix}^{\sim} = \begin{bmatrix} M_0^{\sim} & N_0^{\sim} \\ -\hat{N}_0 & \hat{M}_0 \end{bmatrix},$$

$$\begin{bmatrix} M_0 & -\hat{N}_0^{\sim} \\ N_0 & \hat{M}_0^{\sim} \end{bmatrix}^{\sim} \begin{bmatrix} M_0 & -\hat{N}_0^{\sim} \\ N_0 & \hat{M}_0^{\sim} \end{bmatrix} = \begin{bmatrix} M_0^{\sim} & N_0^{\sim} \\ -\hat{N}_0 & \hat{M}_0 \end{bmatrix} \begin{bmatrix} M_0 & -\hat{N}_0^{\sim} \\ N_0 & \hat{M}_0^{\sim} \end{bmatrix} = I,$$

it turns out

$$\left\| \begin{bmatrix} u \\ y \end{bmatrix} \right\|_2 = \left\| \begin{bmatrix} I & -M_0^{\sim} \hat{Y}_0 + N_0^{\sim} \hat{X}_0 \\ 0 & I \end{bmatrix} \begin{bmatrix} r_u \\ r_y \end{bmatrix} \right\|_2. \tag{4.24}$$

Equation (4.24) implies that the energy in the process signal (u, y) is bounded by the information amount in the residual signal $\left(r_u, r_y\right)$, which allows us to express the often adopted quadratic cost function of (u, y) in terms of $\left(r_u, r_y\right)$. Moreover, it follows from Theorem 4.4 that

$$\left\| \begin{bmatrix} u \\ y \end{bmatrix} \right\|_2 \leq \left\| \begin{bmatrix} r_u \\ r_y \end{bmatrix} \right\|_2 = \left\| \begin{bmatrix} (I - SQ)^{-1} & S (I - SQ)^{-1} \\ (I - QS)^{-1} Q & (I - QS)^{-1} \end{bmatrix} \begin{bmatrix} \bar{d} \\ \bar{v} \end{bmatrix} \right\|_2 \tag{4.25}$$

$$\leq \left\| \begin{bmatrix} (I - SQ)^{-1} & S(I - SQ)^{-1} \\ (I - QS)^{-1} Q & (I - QS)^{-1} \end{bmatrix} \right\|_{\infty} \left\| \begin{bmatrix} \bar{d} \\ \bar{v} \end{bmatrix} \right\|_{2}. \tag{4.26}$$

These relations can be utilised for our subsequent investigations on performance degradation monitoring and recovering strategies. It is of considerable interests to observe that the relations in (4.25)–(4.26) are independent of the (nominal) process model and process data (u, y).

4.5 Dual Youla Parameterisation II

We consider the control loop in Fig. 4.4 again. Due to the absolute symmetry of the loop configuration, the roles of the plant and controllers could be exchanged without changing the loop dynamics. This property could be useful for detection, monitoring and identification applications. In this regard, we introduce an additional dual form of the Youla parameterisation.

Given a feedback controller

$$u(z) = K(z)y(z).$$

Denote by (A_K, B_K, C_K, D_K) the minimal state space realisation of K, i.e.

$$x_K(k+1) = A_K x_K(k) + B_K y(k), \tag{4.27}$$
$$u(k) = C_K x_K(k) + D_K y(k). \tag{4.28}$$

Without loss of generality, it is supposed that the state vector $x_K \in \mathbb{R}^n$ and its initial value $x_K(0) = 0$. Below, we will present a parameterisation of all plants dynamics G, which, together with the controller K, build a (internally) stable feedback control loop shown in Fig. 4.4.

Remark 4.2 For the sake of simplicity, we will below adopt the notations used in our previous study with a sub-index "K" standing for "controller".

Let $\left(\hat{M}_K, \hat{N}_K\right)$ and (M_K, N_K) be the LCF and RCF of K satisfying (2.5)–(2.6) with (A, B, C, D) and (L, F, W, V) being substituted by (A_K, B_K, C_K, D_K) and (L_K, F_K, W_K, V_K), respectively. Together with the RCP $\left(\hat{X}_K, \hat{Y}_K\right)$ and LCP (X_K, Y_K) satisfying (2.25)–(2.26) by the above-mentioned (matrix) substitutions, they build the following Bezout identity

$$\begin{bmatrix} X_K & Y_K \\ -\hat{N}_K & \hat{M}_K \end{bmatrix} \begin{bmatrix} M_K & -\hat{Y}_K \\ N_K & \hat{X}_K \end{bmatrix} = \begin{bmatrix} I & 0 \\ 0 & I \end{bmatrix}. \tag{4.29}$$

On the basis of the Bezout identity (4.29), a parameterisation of the plant dynamics is described in the following theorem.

Theorem 4.5 *Consider the control loop shown in Fig. 4.4. Suppose that the LCF and RCF of the controller K are* $\left(\hat{M}_K, \hat{N}_K\right)$ *and* (M_K, N_K) *and the closed-loop is stable. Then, the transfer function matrices of the plant can be parametrised by*

$$G = -\left(\hat{Y}_K + M_K S\right)\left(\hat{X}_K - N_K S\right)^{-1} = -\left(X_K - S\hat{N}_K\right)^{-1}\left(Y_K + S\hat{M}_K\right), \tag{4.30}$$

where $S \in \mathcal{RH}_\infty$ *is the parameterisation matrix, and* $\left(\hat{X}_K, \hat{Y}_K\right)$ *and* (X_K, Y_K) *are RCP and LCP satisfying the Bezout identity (4.29). Moreover, if a reference signal* v *and an unknown input* d *are added to the control signal* u *and the plant output* y*, respectively,*

$$\begin{aligned} u &= Ky + v = \hat{M}_K^{-1}\hat{N}_K y + v = N_K M_K^{-1} y + v, \\ y &= Gu + d = -\left(\hat{Y}_K + M_K S\right)\left(\hat{X}_K - N_K S\right)^{-1} u + d \\ &= -\left(X_K - S\hat{N}_K\right)^{-1}\left(Y_K + S\hat{M}_K\right) u + d, \end{aligned}$$

it holds

$$r_{u,K} = \hat{M}_K u - \hat{N}_K y = \hat{M}_K v =: \bar{v}, \tag{4.31}$$

$$r_{y,K} = X_K y + Y_K u = -S r_{u,K} + \bar{d}, \bar{d} =: \left(X_K - S\hat{N}_K\right) d, \tag{4.32}$$

$$\begin{bmatrix} u \\ y \end{bmatrix} = \begin{bmatrix} -\hat{Y}_K - M_K S \\ \hat{X}_K - N_K S \end{bmatrix} \bar{v} + \begin{bmatrix} M_K \\ N_K \end{bmatrix} \bar{d} \tag{4.33}$$

$$= \begin{bmatrix} -\hat{Y}_K \\ \hat{X}_K \end{bmatrix} r_{u,K} + \begin{bmatrix} M_K \\ N_K \end{bmatrix} r_{y,K}. \tag{4.34}$$

Proof Equation (4.30) is the dual form of the Youla parameterisation of all stabilising controllers given in (3.5) and $r_{u,K}, r_{y,K}$ in (4.31)–(4.32) are the residual signals, while (4.33) represents the plant dynamics in the loop and is the dual result of Theorem 3.1. The proof of all these results is obvious by means of the duality and thus omitted.

Example 4.3 In this example, the state space representations of the residual generators (4.31)–(4.32) and the system dynamics (4.33) are given. It follows from (2.25)–(2.26) and (2.5)–(2.6) that

$$\begin{aligned} \hat{x}_K(k+1) &= (A_K - L_K C_K)\hat{x}_K(k) + (B_K - L_K D_K) y(k) + L_K u(k) \\ &= (A_K + B_K F_K)\hat{x}_K(k) - L_K W_K^{-1} r_{u,K}(k) + B_K V_K r_{y,K}(k), \\ r_{u,K}(k) &= W_K\left(u(k) - \hat{u}(k)\right), \hat{u}(k) = C_K \hat{x}_K(k) + D_K y(k), \\ r_{y,K}(k) &= V_K^{-1}\left(y(k) - \hat{y}(k)\right), \hat{y}(k) = F_K \hat{x}_K(k), \end{aligned}$$

$$\begin{bmatrix} u(k) \\ y(k) \end{bmatrix} = \begin{bmatrix} F_K \\ C_K + D_K F_K \end{bmatrix} \hat{x}_K(k) + \begin{bmatrix} 0 \\ W_K^{-1} \end{bmatrix} r_{u,K}(k) + \begin{bmatrix} I \\ D_K \end{bmatrix} V_K r_{y,K}(k),$$

$$r_S(z) := S(z) r_{u,K}(z) = S(z)\bar{v}(z).$$

From this example it can be clearly seen as well that the output feedback controller is an observer-based state feedback controller.

4.6 An Observer-Based Unified System Model

This section is in fact a summary of the major results in the previous sections and chapters, including the RCFs and LCFs of the plant and controller, Youla parameterisation and the closed-loop dynamics, in a state space model form, the so-called observer-based unified system model. The intention here is twofold:

- summarising these essential concepts and schemes in a unified abstract form, and
- providing an alternative system model for online implementations.

Suppose that the nominal plant model (2.1)–(2.3) is under consideration. Let

$$\hat{x}(k+1) = A\hat{x}(k) + Bu(k) + L\left(y(k) - \hat{y}(k)\right), \tag{4.35}$$

$$r_0(k) = y(k) - \hat{y}(k), \hat{y}(k) = C\hat{x}(k) + Du(k) \tag{4.36}$$

be a state observer as well as an observer-based residual generator. It is remarkable that rewriting this system as

$$\hat{x}(k+1) = A\hat{x}(k) + Bu(k) + Lr_0(k), \tag{4.37}$$

$$y(k) = r_0(k) + C\hat{x}(k) + Du(k), \tag{4.38}$$

the input-output open-loop dynamics can be fully recovered. According to (4.37)–(4.38), the RCFs and LCFs of the plant and controller are given, respectively, by

- RCF of the plant: for $u(k) = Fx(k) + Vv(k), \hat{x}(0) = x(0)$,

$$\begin{gathered}\hat{x}(k+1) = A\hat{x}(k) + Bu(k) + Lr_0(k) \Longrightarrow \\ x(k+1) = (A+BF)x(k) + BVv(k), \\ \begin{bmatrix} u(k) \\ y(k) \end{bmatrix} = \begin{bmatrix} F \\ C+DF \end{bmatrix} x(k) + \begin{bmatrix} I \\ D \end{bmatrix} Vv(k) \Longleftrightarrow \\ \begin{bmatrix} u \\ y \end{bmatrix} = \begin{bmatrix} M \\ N \end{bmatrix} v;\end{gathered}$$

- LCF of the plant: for $\hat{x}(0) = x(0)$,

$$\begin{aligned}\hat{x}(k+1) &= A\hat{x}(k) + Bu(k) + Lr_0(k) \\ &= (A-LC)\hat{x}(k) + (B-LD)u(k) + Ly(k),\end{aligned}$$

$$r_y(k) = Wr_0(k) = W\left(-C\hat{x}(k) + \begin{bmatrix} -D & I \end{bmatrix}\begin{bmatrix} u(k) \\ y(k) \end{bmatrix}\right) = 0 \Longleftrightarrow$$
$$r_y = \begin{bmatrix} -\hat{N} & \hat{M} \end{bmatrix}\begin{bmatrix} u \\ y \end{bmatrix};$$

- RCF of the controller with $Q = 0$, $v(k) = 0$,

$$\begin{aligned}
\hat{x}(k+1) &= A\hat{x}(k) + Bu(k) + Lr_0(k) \\
&= (A + BF)\hat{x}(k) + LW^{-1}r_y(k), \\
\begin{bmatrix} u(k) \\ y(k) \end{bmatrix} &= \begin{bmatrix} F \\ C + DF \end{bmatrix}\hat{x}(k) + \begin{bmatrix} 0 \\ W^{-1} \end{bmatrix} r_y(k) \Longleftrightarrow \\
\begin{bmatrix} u \\ y \end{bmatrix} &= \begin{bmatrix} -\hat{Y} \\ \hat{X} \end{bmatrix} r_y;
\end{aligned}$$

- LCF of the controller with $Q = 0$, $v(k) = 0$,

$$\begin{aligned}
\hat{x}(k+1) &= A\hat{x}(k) + Bu(k) + Lr_0(k) \\
&= (A - LC)\hat{x}(k) + (B - LD)u(k) + Ly(k), \\
r_u(k) &= V^{-1}\left(-F\hat{x}(k) + \begin{bmatrix} I & 0 \end{bmatrix}\begin{bmatrix} u(k) \\ y(k) \end{bmatrix}\right) = 0 \Longleftrightarrow \\
r_u &= \begin{bmatrix} X & Y \end{bmatrix}\begin{bmatrix} u \\ y \end{bmatrix}.
\end{aligned}$$

In the case $Q \neq 0$, $v \neq 0$,

- RCF of the controller

$$\begin{aligned}
\hat{x}(k+1) &= A\hat{x}(k) + Bu(k) + Lr_0(k) \\
&= (A + BF)\hat{x}(k) + B\left(r_Q(k) + \bar{v}(k)\right) + LW^{-1}r_y(k), \qquad (4.39) \\
\begin{bmatrix} u(k) \\ y(k) \end{bmatrix} &= \begin{bmatrix} F \\ C + DF \end{bmatrix}\hat{x}(k) + \begin{bmatrix} 0 \\ W^{-1} \end{bmatrix} r_y(k) + \begin{bmatrix} I \\ D \end{bmatrix}\left(\bar{v}(k) + r_Q(k)\right), \\
r_Q(z) &= Q(z)r_y(z), \ \bar{v}(z) = \left(X(z) + Q(z)\hat{N}(z)\right)v(z) \Longleftrightarrow \\
\begin{bmatrix} u \\ y \end{bmatrix} &= \begin{bmatrix} M \\ N \end{bmatrix}\bar{v} + \begin{bmatrix} -\hat{Y} + MQ \\ \hat{X} + NQ \end{bmatrix} r_y; \qquad (4.40)
\end{aligned}$$

- LCF of the controller

$$\begin{aligned}
\hat{x}(k+1) &= A\hat{x}(k) + Bu(k) + Lr_0(k) \\
&= (A - LC)\hat{x}(k) + (B - LD)u(k) + Ly(k),
\end{aligned}$$

$$r_u(k) = V^{-1}\left(-F\hat{x}(k) + \begin{bmatrix} I & 0 \end{bmatrix}\begin{bmatrix} u(k) \\ y(k) \end{bmatrix}\right) = r_Q(k) + \bar{v}(k) \Longleftrightarrow$$
$$r_u = \begin{bmatrix} X & Y \end{bmatrix}\begin{bmatrix} u \\ y \end{bmatrix} = Qr_y + \bar{v}.$$

Remark 4.3 In the RCF realisation (4.39)–(4.40), according to the result in Example 3.1, the system dynamics can be equivalently written as

$$\begin{bmatrix} u - v \\ y \end{bmatrix} = \begin{bmatrix} -\hat{Y} + MQ \\ \hat{X} + NQ \end{bmatrix}\left(r_y + \hat{N}v\right)$$

as well.

Definition 4.1 The dynamic system (4.37)–(4.38) is called observer-based unified (input-output) system model.

Concerning the observer-based input-output system model, as a unified model, the following two remarkable points should be emphasised:

- as given by (4.40), this model describes the closed-loop dynamics as well, and
- the basic model (4.37)–(4.38) can be, independent of the existence of any model uncertainties, implemented on online. Moreover, it is a stable system and thus reliably operates online.

As showcased in Sect. 2.4, all types of uncertainties affect the model parameters or/and the dynamics of the system under consideration directly. In general, they are not measurable and accessible. On the contrary, the existence of the system uncertainties will not cause any change in the observer-based input-output system model (4.37)–(4.38) explicitly. Information about the uncertainties is fully embedded in the residual vector, as demonstrated in the above study, which is available and accessible in the model (4.37)–(4.38). These two different model forms for the same system under consideration are schematically demonstrated in Fig. 4.6, in which Δ is used to denote system uncertainties schematically.

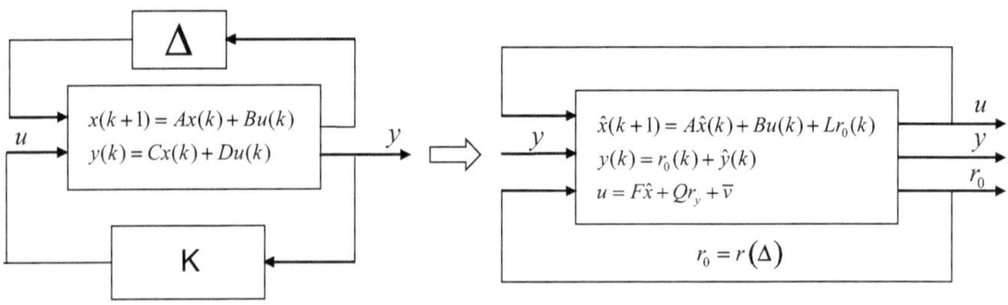

Fig. 4.6 From the standard model to the observer-based I/O system model: a schematic description

4.7 Parameterisation of System Responses with Respect to Uncertainties

Analysis of system responses with respect to model uncertainties is the objective of this section. To be specific, a parameterisation of impacts of model uncertainties on loop responses will be presented.

4.7.1 On Model Uncertainties

We confine our study to coprime factor uncertainties described by

$$\Delta_{I_G} = \begin{bmatrix} \Delta_M \\ \Delta_N \end{bmatrix} \in \mathcal{RH}_\infty, \Delta_{K_G} = \begin{bmatrix} -\Delta_{\hat{N}} & \Delta_{\hat{M}} \end{bmatrix} \in \mathcal{RH}_\infty,$$

on the assumption of RCF and LCF of the plant being modelled by

$$G = (N_0 + \Delta_N)(M_0 + \Delta_M)^{-1} = \left(\hat{M}_0 + \Delta_{\hat{M}}\right)^{-1} \left(\hat{N}_0 + \Delta_{\hat{N}}\right). \tag{4.41}$$

Due to the Bezout identity (2.31), Δ_{I_G} and Δ_{K_G} can be written into

$$\Delta_{I_G} = \begin{bmatrix} M_0 \\ N_0 \end{bmatrix} \delta_{I_G} + \begin{bmatrix} -\hat{Y}_0 \\ \hat{X}_0 \end{bmatrix} \delta_{I_K}, \tag{4.42}$$

$$\Delta_{K_G} = \delta_{K_G} \begin{bmatrix} -\hat{N}_0 & \hat{M}_0 \end{bmatrix} + \delta_{K_K} \begin{bmatrix} X_0 & Y_0 \end{bmatrix}, \tag{4.43}$$

where $\delta_{I_G}, \delta_{I_K}, \delta_{K_G}, \delta_{K_K} \in \mathcal{RH}_\infty$ are unknown, and the sub-indices (I_G, I_K) and (K_G, K_K) stand for image and kernel subspaces of (G, K) (plant and controller), respectively. The following lemma provides us with an interesting relation between Δ_{I_G} and Δ_{K_G}.

Lemma 4.1 *Given the plant model (4.41) with the coprime factor uncertainties described by (4.42)–(4.43) and satisfying*

$$\left(I + \begin{bmatrix} -\hat{Y}_0 \\ \hat{X}_0 \end{bmatrix} \Delta_{K_G}\right)^{-1} \in \mathcal{RH}_\infty, \left(I + \Delta_{I_G} \begin{bmatrix} X_0 & Y_0 \end{bmatrix}\right)^{-1} \in \mathcal{RH}_\infty,$$

then it holds

$$\delta_{I_K} = -\left(I + \delta_{K_G}\right)^{-1} \delta_{K_K} \left(I + \delta_{I_G}\right), \delta_{K_K} = -\left(I + \delta_{K_G}\right) \delta_{I_K} \left(I + \delta_{I_G}\right)^{-1}.$$

Proof It follows from Theorem 2.6 and Example 2.1 that there exist, given Δ_{K_G}, Δ_{I_G},

$$\begin{bmatrix} \bar{M} \\ \bar{N} \end{bmatrix} := \begin{bmatrix} M_0 \\ N_0 \end{bmatrix} + \Delta_{I,K_G},$$

$$\begin{aligned} \Delta_{I,K_G} &= -\begin{bmatrix} -\hat{Y}_0 \\ \hat{X}_0 \end{bmatrix} \left(I + \Delta_{K_G} \begin{bmatrix} -\hat{Y}_0 \\ \hat{X}_0 \end{bmatrix} \right)^{-1} \Delta_{K_G} \begin{bmatrix} M_0 \\ N_0 \end{bmatrix} \\ &= -\begin{bmatrix} -\hat{Y}_0 \\ \hat{X}_0 \end{bmatrix} \left(I + \delta_{K_G} \right)^{-1} \delta_{K_K}, \end{aligned}$$

$$\begin{aligned} \begin{bmatrix} -\hat{\bar{N}} & \hat{\bar{M}} \end{bmatrix} &= \begin{bmatrix} -\hat{N}_0 & \hat{M}_0 \end{bmatrix} + \Delta_{K,I_G}, \\ \Delta_{K,I_G} &= -\begin{bmatrix} -\hat{N}_0 & \hat{M}_0 \end{bmatrix} \Delta_{I_G} \left(I + \begin{bmatrix} X_0 & Y_0 \end{bmatrix} \Delta_{I_G} \right)^{-1} \begin{bmatrix} X_0 & Y_0 \end{bmatrix} \\ &= -\delta_{I_K} \left(I + \delta_{I_G} \right)^{-1} \begin{bmatrix} X_0 & Y_0 \end{bmatrix}, \end{aligned}$$

respectively. Note that

$$\begin{bmatrix} \bar{M} \\ \bar{N} \end{bmatrix} \left(I + \delta_{I_G} \right), \left(I + \delta_{K_G} \right) \begin{bmatrix} -\hat{\bar{N}} & \hat{\bar{M}} \end{bmatrix}$$

are respectively the RCF and LCF of G. Furthermore, the parameterisations given in (4.42)–(4.43) are unique. As a result, it holds

$$\begin{aligned} \begin{bmatrix} M \\ N \end{bmatrix} &= \begin{bmatrix} \bar{M} \\ \bar{N} \end{bmatrix} \left(I + \delta_{I_G} \right) = \begin{bmatrix} M_0 \\ N_0 \end{bmatrix} \left(I + \delta_{I_G} \right) + \Delta_{I,K_G} \left(I + \delta_{I_G} \right) \\ &\Longrightarrow \Delta_{I_G} = \begin{bmatrix} M_0 \\ N_0 \end{bmatrix} \delta_{I_G} + \begin{bmatrix} -\hat{Y}_0 \\ \hat{X}_0 \end{bmatrix} \delta_{I_K}, \\ \delta_{I_K} &= \Delta_{I,K_G} \left(I + \delta_{I_G} \right) = -\left(I + \delta_{K_G} \right)^{-1} \delta_{K_K} \left(I + \delta_{I_G} \right), \\ \begin{bmatrix} -\hat{N} & \hat{M} \end{bmatrix} &= \left(I + \delta_{K_G} \right) \begin{bmatrix} -\hat{\bar{N}} & \hat{\bar{M}} \end{bmatrix} = \left(I + \delta_{K_G} \right) \left(\begin{bmatrix} -\hat{N}_0 & \hat{M}_0 \end{bmatrix} + \Delta_{K,I_G} \right) \\ &\Longrightarrow \Delta_{K_G} = \delta_{K_G} \begin{bmatrix} -\hat{N}_0 & \hat{M}_0 \end{bmatrix} + \delta_{K_K} \begin{bmatrix} X_0 & Y_0 \end{bmatrix}, \\ \delta_{K_K} &= \left(I + \delta_{K_G} \right) \Delta_{K,I_G} = -\left(I + \delta_{K_G} \right) \delta_{I_K} \left(I + \delta_{I_G} \right)^{-1}. \end{aligned}$$

Remark 4.4 It is apparent from the above proof that the system (4.41) with the coprime factor uncertainties described by (4.42)–(4.43) can be equivalently factorised into

$$\begin{aligned} G &= \left(N_0 + \hat{X}_0 \delta_{I_K} \left(I + \delta_{I_G} \right)^{-1} \right) \left(M_0 - \hat{Y}_0 \delta_{I_K} \left(I + \delta_{I_G} \right)^{-1} \right)^{-1} \\ &= \left(\hat{M}_0 + \left(I + \delta_{K_G} \right)^{-1} \delta_{K_K} Y_0 \right)^{-1} \left(\hat{N}_0 + \left(I + \delta_{K_G} \right)^{-1} \delta_{K_K} X_0 \right). \end{aligned}$$

That is, Δ_{I_G} and Δ_{K_G} are respectively

$$\Delta_{I_G} = \begin{bmatrix} -\hat{Y}_0 \\ \hat{X}_0 \end{bmatrix} \delta_{I_K} \left(I + \delta_{I_G}\right)^{-1},$$
$$\Delta_{K_G} = \left(I + \delta_{K_G}\right)^{-1} \delta_{K_K} \begin{bmatrix} X_0 & Y_0 \end{bmatrix}.$$

This also explains, from the factorisation viewpoint, why there is no variation in the closed-loop response if $\delta_{I_K} = 0$ (equivalently $\delta_{K_K} = 0$), as will be demonstrated in the sequel.

4.7.2 Parameterisation of System Responses

We now study the dynamics of the closed-loop depicted in Fig. 4.7. The closed-loop dynamic is governed by

$$\begin{bmatrix} u \\ y \end{bmatrix} = \begin{bmatrix} I & -K \\ -G & I \end{bmatrix}^{-1} \begin{bmatrix} v \\ 0 \end{bmatrix}, \tag{4.44}$$

$$K = V^{-1}U = \hat{U}\hat{V}^{-1}, \tag{4.45}$$

$$V = X_0 + Q\hat{N}_0, U = -Y_0 + Q\hat{M}_0, \hat{V} = \hat{X}_0 + N_0 Q, \hat{U} = -\hat{Y}_0 + M_0 Q, \tag{4.46}$$

where (V, U) and $\left(\hat{V}, \hat{U}\right)$ are the LCF and RCF of the controller and given by the Youla parameterisation. The following lemma is a known result.

Lemma 4.2 *Given control-loop depicted in Fig. 4.7 and described by (4.44) with*

$$G = NM^{-1} = (N_0 + \Delta_N)(M_0 + \Delta_M)^{-1},$$

then

$$\begin{bmatrix} u \\ y \end{bmatrix} = \left(\begin{bmatrix} M_0 \\ N_0 \end{bmatrix} + \begin{bmatrix} \Delta_M \\ \Delta_N \end{bmatrix}\right) \left(I + \begin{bmatrix} V & -U \end{bmatrix} \begin{bmatrix} \Delta_M \\ \Delta_N \end{bmatrix}\right)^{-1} \bar{v},$$
$$\bar{v} = Vv.$$

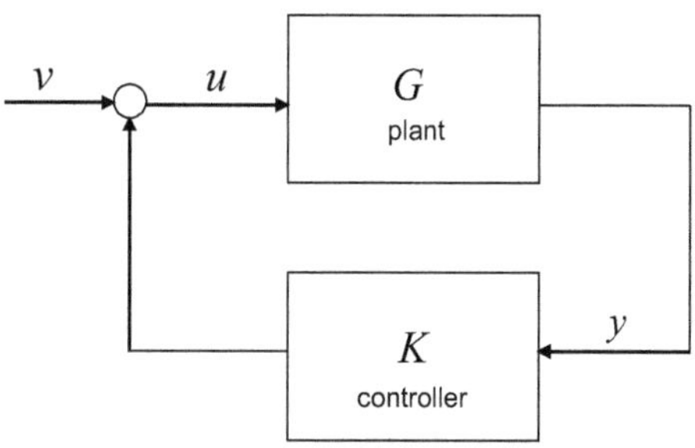

Fig. 4.7 The feedback loop under consideration

Proof It is a straightforward computation,

$$\begin{bmatrix} u \\ y \end{bmatrix} = \begin{bmatrix} I & -K \\ -G & I \end{bmatrix}^{-1} \begin{bmatrix} v \\ 0 \end{bmatrix} = \begin{bmatrix} M & 0 \\ 0 & V \end{bmatrix} \begin{bmatrix} M & -\hat{U} \\ -N & \hat{V} \end{bmatrix}^{-1} \begin{bmatrix} I \\ 0 \end{bmatrix} v$$
$$= \begin{bmatrix} M & 0 \\ 0 & -\hat{V} \end{bmatrix} \begin{bmatrix} M & \hat{U} \\ N & \hat{V} \end{bmatrix}^{-1} \begin{bmatrix} I \\ 0 \end{bmatrix} v.$$

By the Bezout identity (3.4), it turns out

$$\begin{bmatrix} M & \hat{U} \\ N & \hat{V} \end{bmatrix}^{-1} = \left(\begin{bmatrix} M_0 & \hat{U} \\ N_0 & \hat{V} \end{bmatrix} + \begin{bmatrix} \Delta_M & 0 \\ \Delta_N & 0 \end{bmatrix} \right)^{-1}$$
$$= \begin{bmatrix} I + \begin{bmatrix} V & -U \end{bmatrix} \begin{bmatrix} \Delta_M \\ \Delta_N \end{bmatrix} & 0 \\ \begin{bmatrix} -\hat{N}_0 & \hat{M}_0 \end{bmatrix} \begin{bmatrix} \Delta_M \\ \Delta_N \end{bmatrix} & I \end{bmatrix}^{-1} \begin{bmatrix} V & -U \\ -\hat{N}_0 & \hat{M}_0 \end{bmatrix}.$$

which yields

$$\begin{bmatrix} u \\ y \end{bmatrix} = \begin{bmatrix} M & 0 \\ 0 & -\hat{V} \end{bmatrix} \begin{bmatrix} (I+\Delta_1)^{-1} & 0 \\ -\Delta_2 (I+\Delta_1)^{-1} & I \end{bmatrix} \begin{bmatrix} V \\ -\hat{N}_0 \end{bmatrix} v,$$
$$= \begin{bmatrix} M (I+\Delta_1)^{-1} V \\ \hat{V} \left(\Delta_2 (I+\Delta_1)^{-1} V + \hat{N}_0 \right) \end{bmatrix} v,$$
$$\Delta_1 = \begin{bmatrix} V & -U \end{bmatrix} \begin{bmatrix} \Delta_M \\ \Delta_N \end{bmatrix}, \Delta_2 = \begin{bmatrix} -\hat{N}_0 & \hat{M}_0 \end{bmatrix} \begin{bmatrix} \Delta_M \\ \Delta_N \end{bmatrix}.$$

The following equations hold due to the Bezout identity (3.4),

$$\hat{V} \hat{N}_0 = N_0 V, \hat{V} V = N_0 \hat{N}_0, -\hat{V} U + N_0 \hat{M}_0 = I,$$

by which we have

$$\hat{V} \left(\Delta_2 (I+\Delta_1)^{-1} V + \hat{N}_0 \right) = \left(\hat{V} \Delta_2 + N_0 (I+\Delta_1) \right) (I+\Delta_1)^{-1} V$$
$$= \left(N_0 + \begin{bmatrix} \hat{V} & N_0 \end{bmatrix} \begin{bmatrix} V & -U \\ -\hat{N}_0 & \hat{M}_0 \end{bmatrix} \begin{bmatrix} \Delta_M \\ \Delta_N \end{bmatrix} \right) (I+\Delta_1)^{-1} V$$
$$= (N_0 + \Delta_N) (I+\Delta_1)^{-1} V = N (I+\Delta_1)^{-1} V.$$

As a result,

$$\begin{bmatrix} u \\ y \end{bmatrix} = \begin{bmatrix} M \\ N \end{bmatrix} (I+\Delta_1)^{-1} \bar{v}.$$

The lemma is proved.

Example 4.4 In this example, the analogue result of Lemma 4.2 is examined. Consider a control-loop depicted in Fig. 4.7 and described by (4.44) with

$$G = \hat{M}^{-1}\hat{N} = \left(\hat{M}_0 + \Delta_{\hat{M}}\right)^{-1}\left(\hat{N}_0 + \Delta_{\hat{N}}\right).$$

It is straightforward that

$$\begin{aligned} \begin{bmatrix} u \\ y \end{bmatrix} &= \begin{bmatrix} I & -K \\ -G & I \end{bmatrix}^{-1} \begin{bmatrix} v \\ 0 \end{bmatrix} = \begin{bmatrix} V & -U \\ -\hat{N} & \hat{M} \end{bmatrix}^{-1} \begin{bmatrix} \bar{v} \\ 0 \end{bmatrix} \\ &= \left(I + \begin{bmatrix} \hat{U} \\ \hat{V} \end{bmatrix} \begin{bmatrix} -\Delta_{\hat{N}} & \Delta_{\hat{M}} \end{bmatrix}\right)^{-1} \begin{bmatrix} M_0 \\ N_0 \end{bmatrix} \bar{v} \end{aligned}$$

$$= \left(I - \begin{bmatrix} \hat{U} \\ \hat{V} \end{bmatrix} \left(I + \begin{bmatrix} -\Delta_{\hat{N}} & \Delta_{\hat{M}} \end{bmatrix} \begin{bmatrix} \hat{U} \\ \hat{V} \end{bmatrix}\right)^{-1} \begin{bmatrix} -\Delta_{\hat{N}} & \Delta_{\hat{M}} \end{bmatrix}\right) \begin{bmatrix} M_0 \\ N_0 \end{bmatrix} \bar{v} \tag{4.47}$$

which is also equivalent to

$$\begin{aligned} \begin{bmatrix} u \\ y \end{bmatrix} &= \begin{bmatrix} M_0 & \hat{U} \\ N_0 & \hat{V} \end{bmatrix} \left(I + \begin{bmatrix} 0 & 0 \\ -\Delta_{\hat{N}} & \Delta_{\hat{M}} \end{bmatrix} \begin{bmatrix} M_0 & \hat{U} \\ N_0 & \hat{V} \end{bmatrix}\right)^{-1} \begin{bmatrix} \bar{v} \\ 0 \end{bmatrix} \\ &= \begin{bmatrix} M_0 & \hat{U} \\ N_0 & \hat{V} \end{bmatrix} \begin{bmatrix} I & 0 \\ \hat{\Delta}_1 & I + \hat{\Delta}_2 \end{bmatrix}^{-1} \begin{bmatrix} \bar{v} \\ 0 \end{bmatrix}, \\ \hat{\Delta}_1 &= \begin{bmatrix} -\Delta_{\hat{N}} & \Delta_{\hat{M}} \end{bmatrix} \begin{bmatrix} M_0 \\ N_0 \end{bmatrix}, \hat{\Delta}_2 = \begin{bmatrix} -\Delta_{\hat{N}} & \Delta_{\hat{M}} \end{bmatrix} \begin{bmatrix} \hat{U} \\ \hat{V} \end{bmatrix}. \end{aligned}$$

We are now in the position to present the main results in this section.

Theorem 4.6 *Given control-loop depicted in Fig. 4.7 and described by (4.44) with*

$$\begin{aligned} G &= NM^{-1} = (N_0 + \Delta_N)(M_0 + \Delta_M)^{-1} \\ &= \hat{M}^{-1}\hat{N} = \left(\hat{M}_0 + \Delta_{\hat{M}}\right)^{-1}\left(\hat{N}_0 + \Delta_{\hat{N}}\right), \\ K &= V^{-1}U = \hat{U}\hat{V}^{-1}, \end{aligned}$$

(V, U) *and* $\left(\hat{V}, \hat{U}\right)$ *satisfying (4.46), it is supposed that*

$$\Delta_{I_G} = \begin{bmatrix} \Delta_M \\ \Delta_N \end{bmatrix} \in \mathcal{RH}_\infty, \Delta_{K_G} = \begin{bmatrix} -\Delta_{\hat{N}} & \Delta_{\hat{M}} \end{bmatrix} \in \mathcal{RH}_\infty$$

satisfy (4.42)–(4.43), then it holds

$$\begin{bmatrix} u \\ y \end{bmatrix} = \begin{bmatrix} M_0 \\ N_0 \end{bmatrix} \bar{v} + \begin{bmatrix} M_0 \\ N_0 \end{bmatrix} Q\delta_{I_K} \left(I + \delta_{I_G} - Q\delta_{I_K}\right)^{-1} \bar{v} + \begin{bmatrix} -\hat{Y}_0 \\ \hat{X}_0 \end{bmatrix} \delta_{I_K} \left(I + \delta_{I_G} - Q\delta_{I_K}\right)^{-1} \bar{v}, \tag{4.48}$$

$$\begin{bmatrix} u \\ y \end{bmatrix} = \begin{bmatrix} M_0 \\ N_0 \end{bmatrix} \bar{v} - \begin{bmatrix} M_0 \\ N_0 \end{bmatrix} Q \left(I + \delta_{K_G} + Q\delta_{K_K}\right)^{-1} \delta_{K_K} \bar{v} - \begin{bmatrix} -\hat{Y}_0 \\ \hat{X}_0 \end{bmatrix} \left(I + \delta_{K_G} + Q\delta_{K_K}\right)^{-1} \delta_{K_K} \bar{v}. \tag{4.49}$$

Proof We first prove (4.48). It follows from Lemma 4.2 that

$$\begin{aligned} \begin{bmatrix} u \\ y \end{bmatrix} &= \left(\begin{bmatrix} M_0 \\ N_0 \end{bmatrix} (I + \delta_{I_G}) + \begin{bmatrix} -\hat{Y}_0 \\ \hat{X}_0 \end{bmatrix} \delta_{I_K} \right) \left(I + \delta_{I_G} - Q\delta_{I_K}\right)^{-1} \bar{v} \\ &= \begin{bmatrix} M_0 \\ N_0 \end{bmatrix} \left(I + Q\delta_{I_K} \left(I + \delta_{I_G} - Q\delta_{I_K}\right)^{-1} \right) \bar{v} \\ &+ \begin{bmatrix} -\hat{Y}_0 \\ \hat{X}_0 \end{bmatrix} \delta_{I_K} \left(I + \delta_{I_G} - Q\delta_{I_K}\right)^{-1} \bar{v} \\ &= \begin{bmatrix} M_0 \\ N_0 \end{bmatrix} \bar{v} + \begin{bmatrix} M_0 \\ N_0 \end{bmatrix} Q\delta_{I_K} \left(I + \delta_{I_G} - Q\delta_{I_K}\right)^{-1} \bar{v} \\ &+ \begin{bmatrix} -\hat{Y}_0 \\ \hat{X}_0 \end{bmatrix} \delta_{I_K} \left(I + \delta_{I_G} - Q\delta_{I_K}\right)^{-1} \bar{v}. \end{aligned}$$

Equation (4.49) is proved by means of (4.47) in Example 4.4 as follows,

$$\begin{aligned} \begin{bmatrix} u \\ y \end{bmatrix} &= \left(I - \begin{bmatrix} \hat{U} \\ \hat{V} \end{bmatrix} \left(I + \begin{bmatrix} -\Delta_{\hat{N}} & \Delta_{\hat{M}} \end{bmatrix} \begin{bmatrix} \hat{U} \\ \hat{V} \end{bmatrix} \right)^{-1} \begin{bmatrix} -\Delta_{\hat{N}} & \Delta_{\hat{M}} \end{bmatrix} \right) \begin{bmatrix} M_0 \\ N_0 \end{bmatrix} \bar{v} \\ &= \begin{bmatrix} M_0 \\ N_0 \end{bmatrix} \bar{v} - \begin{bmatrix} \hat{U} \\ \hat{V} \end{bmatrix} \left(I + \delta_{K_G} + Q\delta_{K_K}\right)^{-1} \delta_{K_K} \bar{v}. \end{aligned}$$

Thus, the theorem is proved.

Example 4.5 In this example, the equivalence between (4.48) and (4.49) is demonstrated by means of Lemma 4.1. Consider $\delta_{I_K} \left(I + \delta_{I_G} - Q\delta_{I_K}\right)^{-1}$ and write it into

$$\delta_{I_K} \left(I + \delta_{I_G} - Q\delta_{I_K}\right)^{-1} = \delta_{I_K} \left(I + \delta_{I_G}\right)^{-1} \left(I + Q \left(I - \delta_{I_K} \left(I + \delta_{I_G}\right)^{-1} Q \right)^{-1} \right) \delta_{I_K} \left(I + \delta_{I_G}\right)^{-1}.$$

It follows from Lemma 4.1 that

$$\delta_{I_K} \left(I + \delta_{I_G}\right)^{-1} = -\left(I + \delta_{K_G}\right)^{-1} \delta_{K_K},$$

which yields

$$\delta_{I_K}\left(I+\delta_{I_G}-Q\delta_{I_K}\right)^{-1}=$$
$$-\left(I-\left(I+\delta_{K_G}\right)^{-1}Q\left(I+\left(I+\delta_{K_G}\right)^{-1}\delta_{K_K}Q\right)^{-1}\right)\left(I+\delta_{K_G}\right)^{-1}\delta_{K_K}$$
$$=-\left(I+\delta_{K_G}+\delta_{K_K}Q\right)^{-1}\delta_{K_K}.$$

With the aid of this relation, the equivalence between (4.48) and (4.49) can be immediately derived.

The following interesting facts are immediately observed from the above theorem:

- when $\delta_{I_K}=0 \Longleftrightarrow \delta_{K_K}=0$, it holds $\begin{bmatrix} u \\ y \end{bmatrix}=\begin{bmatrix} M_0 \\ N_0 \end{bmatrix}\bar{v}$, that is, if

$$\Delta_{I_G}=\begin{bmatrix} M_0 \\ N_0 \end{bmatrix}\delta_{I_G}\Longrightarrow\begin{bmatrix} M \\ N \end{bmatrix}=\begin{bmatrix} M_0 \\ N_0 \end{bmatrix}\left(I+\delta_{I_G}\right)\Longleftrightarrow$$
$$\Delta_{K_G}=\delta_{K_G}\left[-\hat{N}_0\ \hat{M}_0\right]\Longrightarrow\left[-\hat{N}\ \hat{M}\right]=\left(I+\delta_{K_G}\right)\left[-\hat{N}_0\ \hat{M}_0\right],$$

  there is no deviation from the ideal tracking performance $\begin{bmatrix} M_0 \\ N_0 \end{bmatrix}\bar{v}$. Note that $\left(I+\delta_{I_G}\right)^{-1}$ or equivalently $\left(I+\delta_{K_G}\right)^{-1}$ should be stable;
- for $Q=0$,

$$\begin{bmatrix} u \\ y \end{bmatrix}=\begin{bmatrix} M_0 \\ N_0 \end{bmatrix}\bar{v}+\begin{bmatrix} -\hat{Y}_0 \\ \hat{X}_0 \end{bmatrix}\delta_{I_K}\left(I+\delta_{I_G}\right)^{-1}\bar{v}$$
$$=\begin{bmatrix} M_0 \\ N_0 \end{bmatrix}\bar{v}+\begin{bmatrix} -\hat{Y}_0 \\ \hat{X}_0 \end{bmatrix}\left(I+\delta_{K_G}\right)^{-1}\delta_{K_K}\bar{v}$$

  and the closed-loop stability is determined by $\left(I+\delta_{I_G}\right)^{-1}$ or equivalently by $\left(I+\delta_{K_G}\right)^{-1}$;
- the dynamics of the residual signal,

$$r_u=u-\hat{u},\ \hat{u}=F\hat{x}+Qr_y+\bar{v}\Longrightarrow r_u=Vu-Uy-\bar{v},$$
$$r_y=y-\hat{y}=\hat{M}_0y-\hat{N}_0u,$$

  are governed by

$$\begin{bmatrix} r_u \\ r_y \end{bmatrix}=\begin{bmatrix} V & -U \\ -\hat{N}_0 & \hat{M}_0 \end{bmatrix}\begin{bmatrix} u \\ y \end{bmatrix}+\begin{bmatrix} -\bar{v} \\ 0 \end{bmatrix}$$
$$=\begin{bmatrix} 0 \\ \delta_{I_K}\left(I+\delta_{I_G}-Q\delta_{I_K}\right)^{-1}\bar{v} \end{bmatrix}. \tag{4.50}$$

Equation (4.50) presents an interesting result. It implies that the residual r_u on the controller side is, independent of the uncertainties in the system dynamics, constantly zero.

Below, we would like to delineate the above results in an alternative way that helps us to gain a deeper insight into the feedback mechanism. To this end, consider the observer-based unified system model (4.39)–(4.40), which can be written as

$$\begin{bmatrix} u \\ y \end{bmatrix} = \begin{bmatrix} M_0 \\ N_0 \end{bmatrix} \bar{v} + \begin{bmatrix} -\hat{Y}_0 + M_0 Q \\ \hat{X}_0 + N_0 Q \end{bmatrix} r_y. \tag{4.51}$$

Since

$$\begin{bmatrix} -\hat{N} & \hat{M} \end{bmatrix} \begin{bmatrix} u \\ y \end{bmatrix} = 0 \Longleftrightarrow r_y = -\begin{bmatrix} -\Delta_{\hat{N}} & \Delta_{\hat{M}} \end{bmatrix} \begin{bmatrix} u \\ y \end{bmatrix},$$

the dynamic (4.51) builds a loop,

$$\begin{cases} \begin{bmatrix} u \\ y \end{bmatrix} = \begin{bmatrix} M_0 \\ N_0 \end{bmatrix} \bar{v} + \begin{bmatrix} \hat{U} \\ \hat{V} \end{bmatrix} r_y \\ r_y = -\begin{bmatrix} -\Delta_{\hat{N}} & \Delta_{\hat{M}} \end{bmatrix} \begin{bmatrix} u \\ y \end{bmatrix}, \end{cases}$$

as sketched in Fig. 4.8. It is apparent that

$$r_y = -\begin{bmatrix} -\Delta_{\hat{N}} & \Delta_{\hat{M}} \end{bmatrix} \begin{bmatrix} u \\ y \end{bmatrix} = -\begin{bmatrix} -\Delta_{\hat{N}} & \Delta_{\hat{M}} \end{bmatrix} \left(\begin{bmatrix} M_0 \\ N_0 \end{bmatrix} \bar{v} + \begin{bmatrix} \hat{U} \\ \hat{V} \end{bmatrix} r_y \right)$$

$$\Longrightarrow r_y = -\left(I + \begin{bmatrix} -\Delta_{\hat{N}} & \Delta_{\hat{M}} \end{bmatrix} \begin{bmatrix} \hat{U} \\ \hat{V} \end{bmatrix} \right)^{-1} \begin{bmatrix} -\Delta_{\hat{N}} & \Delta_{\hat{M}} \end{bmatrix} \begin{bmatrix} M_0 \\ N_0 \end{bmatrix} \bar{v},$$

$$\begin{bmatrix} u \\ y \end{bmatrix} = \begin{bmatrix} M_0 \\ N_0 \end{bmatrix} \left(I - \begin{bmatrix} \hat{U} \\ \hat{V} \end{bmatrix} \left(I + \begin{bmatrix} -\Delta_{\hat{N}} & \Delta_{\hat{M}} \end{bmatrix} \begin{bmatrix} \hat{U} \\ \hat{V} \end{bmatrix} \right)^{-1} \begin{bmatrix} -\Delta_{\hat{N}} & \Delta_{\hat{M}} \end{bmatrix} \right) \bar{v}.$$

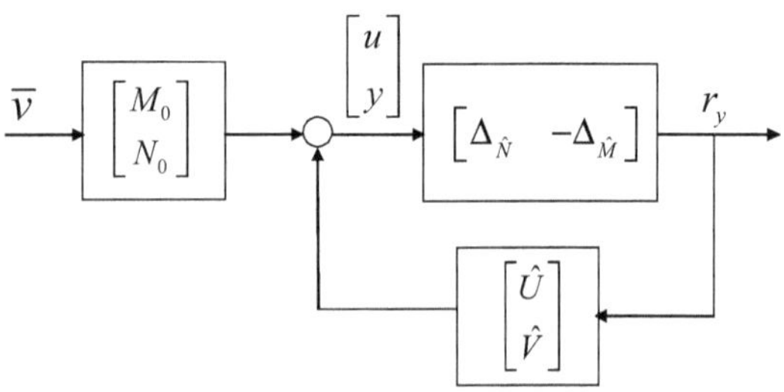

Fig. 4.8 An equivalent closed-loop

4.8 Notes and References

While the previous two chapters provide us with

- the basis of the unified framework of control and detection, and
- the efficient mathematical and control theoretical tools for addressing control and detection issues in the framework,

this chapter presents the structure of the framework. The modular functionalisation of controllers enables a more flexible and target-oriented design and, in particular, online optimisation or the update of controllers. The latter is of considerable importance for performing fault-tolerant control and control performance recovery. The core one among the functional modules of a controller is the observer and observer-based residual generator, which, as the information provider, serves both the feedback control and detection.

The Youla parameterisation realisation in the PnP configuration has been proposed in [129] and successfully applied in practical real-time control systems towards recovering system performance suffered from degradation [131, 212, 213]. Although it is a simple and straightforward use and extension of the Youla parameterisation, the PnP configured Youla parameterisation is of remarkable practical significance in industrial applications. It enables controller tuning

- without any modification on the default controller,
- generally without affecting the system stability, and in particular
- in a plug-and-play manner.

Also the mathematical proof of the state observer-based Youla parameterisation is evident. Nevertheless, its control engineering interpretation is of research interest. It tells us that a parameterisation of all stabilising controllers is equivalent, according to Theorem 3.5, to the parameterisation of all state observers. This verifies, as well, the claim in Corollary 3.4 that any (dynamic) output feedback controller is an estimate for a state feedback controller. This result can also be viewed as a general form of the separation principle that an output feedback controller can be handled as a state feedback controller and an observer. In this regard, the observer design for the detection and control purposes can be efficiently achieved in an integrated manner.

The dual form of the Youla parameterisation summarised in Theorem 4.3 and the associated results given in Theorem 4.4 are well-known [4, 189]. In [190], an elaborate study on the Youla parameterisation and its dual form with comprehensive results have been reported. In a recent survey paper [134], successful applications of this dual form in the past decades have been reviewed. They include

- closed-loop identification (also called Hansen scheme) [4, 81],
- adaptive control [4, 189],
- fault-tolerant control [148, 149], and
- PnP and multi-model adaptive control [10, 186].

Another dual form of the Youla parameterisation presented in Theorem 4.5 has been presented in [190]. Due to the symmetric configuration of the feedback control loop, this form of the Youla parameterisation is obvious and redundant to the original form of the Youla parameterisation. So far, it receives almost no attention in research on synthesis and design of feedback control systems. On the other hand, in dealing with detection, performance degradation monitoring and recovering issues for interconnected systems, this parameterisation form may enable more efficient handlings.

Finally, two issues closely related to the Youla parameterisation are investigated. In Sect. 4.6, the so-called observer-based unified system model has been introduced, which

- unified describes the plant, controller and closed-loop dynamics, and
- summarises the major results on the RCFs and LCFs of the plant and controller, Youla parameterisation and the closed-loop dynamics, in a simple state space model form.

It is remarkable that the observer-based unified system model is stable and can be implemented online reliably and independent of the existence of system uncertainties. Information about the uncertainties is fully embedded in the residual vector, which is one input of the observer-based unified system model.

Section 4.7 has addressed analysis and parameterisation of the loop response to the coprime factor uncertainties. Lemma 4.1 reveals an internal relation between the right and left coprime factor uncertainties, and Theorem 4.6 provides us with the parameterisation of the loop responses (u, y) to the coprime factor uncertainties. An alternative study on the loop dynamics at the end delineates the important role of the residual r_y in the feedback mechanism. It is of considerable significance that the residual r_u is, independent of the uncertainties in the system dynamics, constantly zero. In a nutshell, it is sufficient to make use of the information carried by r_y to regulate the loop dynamics. The results achieved in this section are useful for our subsequent works on networked control systems [218], in which disturbances in communication channels can be modelled as uncertainties in controllers and thus handled as a dual form of the plant model uncertainties.

Chapter 5
The Unified Framework for Linear Time-Varying Systems

In the previous chapters, coprime factorisations as a mathematical tool and the stabilising controller and observer parameterisations as the basic structure of the unified framework of control and detection have been introduced for LTI systems. In the past decades, methods and techniques of coprime factorisations and the stabilising controller parameterisation for linear time-varying (LTV) systems have been well established. In this chapter, we give a systematic summary of the major results with a focus on

- left and right coprime factorisations, in particular the normalised LCF and RCF,
- the concepts of kernel and image representations and subspaces,
- observer-based realisations of stabilising controllers and the relations between the process signal pair (u, y) and the residual signal pair $\left(r_u, r_y\right)$, as well as
- parameterisation of LTV observers.

They build the unified control and detection framework for LTV systems.

5.1 System Models and Basic Concepts

We consider linear discrete-time varying (LDTV) systems, as the nominal process model, described by

$$x(k+1) = A(k)x(k) + B(k)u(k), \tag{5.1}$$

$$y(k) = C(k)x(k) + D(k)u(k), \tag{5.2}$$

where $x(k) \in \mathbb{R}^n$, $u(k) \in \mathbb{R}^p$ and $y(k) \in \mathbb{R}^m$ represent the state, input and output vector, respectively. It is assumed that $u(k)$ is $\mathcal{L}_{2,[0,N]}$ bounded. $A(k)$, $B(k)$, $C(k)$ and $D(k)$ are real matrices, bounded and of appropriate dimensions.

S. X. Ding, *Unified Framework of Control and Detection*,
https://doi.org/10.1007/978-3-662-72417-0_5

For any k_0, $u(k) = 0$ and $k > k_0$,

$$x(k) = \Phi(k, k_0)\, x(k_0),$$
$$\Phi(k, k_0) := \prod_{i=1}^{k-k_0} A(k_0 + i) \in \mathbb{R}^{n \times n},$$

and $\Phi(k, k_0)$ is called transition matrix with $\Phi(k_0, k_0) = I$. The subsequent definitions and concepts are well-known in the literature and essential for introducing the system coprime factorisations.

Definition 5.1 The LDTV system (5.1) is uniformly exponentially stable if there exist $0 < \lambda < 1$, $\gamma > 0$ such that for any k_0, $u(k) = 0$

$$\|x(k)\| \leq \gamma \|x(k_0)\| \lambda^{k-k_0}, k > k_0. \tag{5.3}$$

It is obvious that condition (5.3) is equivalent to

$$\|\Phi(k, k_0)\| \leq \gamma \lambda^{k-k_0}. \tag{5.4}$$

Definition 5.2 The system (5.1)–(5.2) is uniformly stabilisable and detectable if there exist bounded matrix functions $F(k)$ and $L(k)$ such that the systems

$$x(k+1) = (A(k) + B(k)F(k))\, x(k),$$
$$x(k+1) = (A(k) - L(k)C(k))\, x(k)$$

are uniformly exponentially stable.

Hereafter, it is assumed that the nominal system (5.1)–(5.2) is uniformly stabilisable and detectable.

Remember that the coprime factorisations presented in Chap. 2 and applied for system dynamics analysis are related to transfer function matrices of LTI systems, which are apparently not applicable for LTV systems. Alternatively, linear operators are a standard model form for LTV systems. For the LDTV system (5.1)–(5.2), a linear operator $\mathcal{G}$ that maps $(x(0), u)$ to y is defined by

$$\mathcal{G} : \mathbb{R}^n \times \mathcal{L}_{2,[0,N]} \to \mathcal{L}_{2,[0,N]}, \tag{5.5}$$
$$\mathcal{G}(x(0), u(k)) = C(k)\Phi(k, 0)x(0) + C(k)\sum_{j=0}^{k-1} \Phi(k, j+1)B(j)u(j) + D(k)u(k).$$

Remark 5.1 For the sake of simplicity, in the sequel, $\mathcal{G}$ will be frequently used for the LDTV system (5.1)–(5.2). Also the notation,

$$\mathcal{G} = (A(k), B(k), C(k), D(k)),$$

is adopted.

Recall that the normalised factorisation of a transfer function matrix of an LTI system, for instance the RCF of $G(z)$, is defined in terms of $I_G^{\sim}(z) = I_G^T(z^{-1})$, namely

$$I_G^{\sim}(z) I_G(z) = I.$$

Against this background, we now introduce the adjoint of $\mathcal{G}$ for our purpose. Given $x_1, x_2 \in \mathbb{R}^n$ and $\mathcal{L}_{2,[0,N]}$ functions $u_1(k), u_2(k) \in \mathbb{R}^p$, the inner products in $\mathbb{R}^n \times \mathcal{L}_{2,[0,N]}$ and in $\mathcal{L}_{2,[0,N]}$ are defined, respectively, by

$$\langle (x_1, u_1(k)), (x_2, u_2(k)) \rangle_{2,[0,N]} := x_1^T x_2 + \sum_{k=0}^{N} u_1^T(k)\, u_2(k),$$

$$\langle u_1(k), u_2(k) \rangle_{2,[0,N]} := \sum_{k=0}^{N} u_1^T(k)\, u_2(k).$$

According to operator theory, the adjoint of $\mathcal{G}$ is an operator

$$\mathcal{G}^* : \mathcal{L}_{2,[0,N]} \to \mathbb{R}^n \times \mathcal{L}_{2,[0,N]}$$

that is determined by

$$\langle y(k), \mathcal{G}(x(0), u(k)) \rangle_{2,[0,N]} = \left\langle \mathcal{G}^* y(k), (x(0), u(k)) \right\rangle_{2,[0,N]}. \tag{5.6}$$

It turns out

$$\mathcal{G}^*(\bar{y}(k)) = \begin{bmatrix} \sum_{i=0}^{N} \Phi^T(i,0) C^T(i) \bar{y}(i) \\ \sum_{i=k+1}^{N} B^T(k) \Phi^T(i,k+1) C^T(i) \bar{y}(i) + D^T(k) \bar{y}(k) \end{bmatrix}. \tag{5.7}$$

It is of interest to notice that the state space realisation of $\mathcal{G}^*$ can be written as

$$\bar{x}(k) = A^T(k) \bar{x}(k+1) + C^T(k) \bar{u}(k+1), \tag{5.8}$$

$$\bar{y}(k+1) = B^T(k) \bar{x}(k+1) + D^T(k) \bar{u}(k+1). \tag{5.9}$$

Remark 5.2 The identity (5.6) is called duality identity as well. Corresponding to

$$\langle y(k), \mathcal{G}(x(0), u(k)) \rangle_{2,[0,N-1]} = \left\langle \mathcal{G}^* y(k), (x(0), u(k)) \right\rangle_{2,[0,N-1]}$$

and the state space representation (5.8)–(5.9), the duality identity can be expressed as

$$\sum_{k=0}^{N-1} u^T(k)\, \bar{y}(k+1) + x^T(0) \bar{x}(0) = \sum_{k=0}^{N-1} y^T(k)\, \bar{u}(k+1) + x^T(N) \bar{x}(N).$$

Alternatively, with regard to

$$\langle y(k), \mathcal{G}(x(0), u(k))\rangle_{2,[0,N]} = \left\langle \mathcal{G}^* y(k), (x(0), u(k))\right\rangle_{2,[0,N]}$$

and the state space representation

$$\bar{x}(k) = A^T(k)\bar{x}(k+1) + C^T(k)\bar{u}(k), \tag{5.10}$$

$$\bar{y}(k) = B^T(k)\bar{x}(k+1) + D^T(k)\bar{u}(k), \tag{5.11}$$

the duality identity is expressed as

$$\sum_{k=0}^{N} u^T(k)\,\bar{y}(k) + x^T(0)\bar{x}(0) = \sum_{k=0}^{N} y^T(k)\,\bar{u}(k) + x^T(N+1)\bar{x}(N+1).$$

In our subsequent work, these two different forms will be interchangeably applied, depending on which one is more convenient in the context.

5.2 Coprime Factorisations

For our study purpose, we focus on the state space representations of coprime factorisations of LDTV systems and their control theoretical interpretations, as presented in Chap. 2.

Given the LDTV system (5.1)–(5.2), the following stable LDTV system,

$$\hat{x}(k+1) = A_L(k)\hat{x}(k) + B_L(k)u(k) + L(k)y(k), \tag{5.12}$$

$$r_y(k) = W(k)r_0(k),\, r_0(k) = y(k) - \hat{y}(k),\, \hat{y}(k) = C(k)\hat{x}(k) + D(k)u(k), \tag{5.13}$$

$$A_L(k) = A(k) - L(k)C(k),\, B_L(k) = B(k) - L(k)D(k),$$

is an observer-based residual generator. Here, $L(k)$ is the observer gain matrix and selected so that the system is uniformly exponentially stable, and $W(k)$, as the post-filter, is an invertible matrix. Let $\left(\hat{\mathcal{M}}, \hat{\mathcal{N}}\right)$ represent two linear operators defined by

$$\hat{\mathcal{M}} = (A_L(k), -L(k), W(k)C(k), W(k)), \tag{5.14}$$

$$\hat{\mathcal{N}} = (A_L(k), B_L(k), W(k)C(k), W(k)D(k)). \tag{5.15}$$

The above residual generator can then be written as

$$r_y(k) = \hat{\mathcal{M}}(y(k)) - \hat{\mathcal{N}}(u(k)). \tag{5.16}$$

Set

$$u(k) = F(k)\hat{x}(k),$$

where $F(k)$ uniformly stabilises

$$x(k+1) = (A(k) + B(k)F(k))\,x(k),$$

and rewrite (5.1) as

$$\hat{x}(k+1) = A_F(k)\hat{x}(k) + L(k)W^{-1}(k)r_y(k). \tag{5.17}$$

It yields

$$\begin{bmatrix} u(k) \\ y(k) \end{bmatrix} = \begin{bmatrix} F(k) \\ C_F(k) \end{bmatrix} \hat{x}(k) + \begin{bmatrix} 0 \\ W^{-1}(k) \end{bmatrix} r_y(k), \tag{5.18}$$
$$C_F(k) = C(k) + D(k)F(k).$$

With the notations

$$\hat{\mathcal{X}} = \left(A_F(k), L(k)W^{-1}(k), C_F(k), W^{-1}(k)\right), \tag{5.19}$$
$$\hat{\mathcal{Y}} = \left(A_F(k), L(k)W^{-1}(k), -F(k), 0\right), \tag{5.20}$$

system (5.17)–(5.18) can be written into

$$\begin{bmatrix} u(k) \\ y(k) \end{bmatrix} = \begin{bmatrix} -\hat{\mathcal{Y}}\left(r_y(k)\right) \\ \hat{\mathcal{X}}\left(r_y(k)\right) \end{bmatrix}.$$

It is obvious that $\forall x(0), \hat{x}(0)$,

$$r_y(k) = \hat{\mathcal{M}}\left(y(k)\right) - \hat{\mathcal{N}}\left(u(k)\right) = \left(\hat{\mathcal{N}} \circ \hat{\mathcal{Y}} + \hat{\mathcal{M}} \circ \hat{\mathcal{X}}\right)\left(r_y(k)\right)$$
$$\Longrightarrow \hat{\mathcal{N}} \circ \hat{\mathcal{Y}} + \hat{\mathcal{M}} \circ \hat{\mathcal{X}} = \mathcal{I}. \tag{5.21}$$

Identity (5.21) implies that $\left(\hat{\mathcal{M}}, \hat{\mathcal{N}}\right)$ are left coprime and $\left(\hat{\mathcal{X}}, \hat{\mathcal{Y}}\right)$ are right coprime. That is, $\left(\hat{\mathcal{M}}, \hat{\mathcal{N}}\right)$ build an LCP of the plant $\mathcal{G}$, while $\left(\hat{\mathcal{X}}, \hat{\mathcal{Y}}\right)$ are a RCP of the observer-based state feedback system (5.17)–(5.18).

Now, set

$$\hat{x}(0) = x(0) \Longrightarrow \forall k, \hat{x}(k) = x(k),$$
$$u(k) = F(k)\hat{x}(k) + V(k)v(k) \Longrightarrow u(k) = F(k)x(k) + V(k)v(k)$$

with invertible $V(k)$. It holds

$$\begin{aligned} x(k+1) = \hat{x}(k+1) &= A_F(k)\hat{x}(k) + B(k)V(k)v(k) \\ &= A_F(k)x(k) + B(k)V(k)v(k), \end{aligned} \tag{5.22}$$

$$\begin{bmatrix} u(k) \\ y(k) \end{bmatrix} = \begin{bmatrix} F(k) \\ C_F(k) \end{bmatrix} x(k) + \begin{bmatrix} V(k) \\ D(k)V(k) \end{bmatrix} v(k), \tag{5.23}$$

as well as

$$\hat{x}(k+1) = A_L(k)\hat{x}(k) + B_L(k)u(k) + L(k)y(k), \tag{5.24}$$

$$u(k) = F(k)\hat{x}(k) + V(k)v(k) \Longleftrightarrow V^{-1}(k)\left(u(k) - F(k)\hat{x}(k)\right) = v(k). \tag{5.25}$$

Define the following linear operators,

$$\mathcal{M} = (A_F(k), B(k)V(k), F(k), V(k)), \tag{5.26}$$

$$\mathcal{N} = (A_F(k), B(k)V(k), C_F(k), D(k)V(k)), \tag{5.27}$$

$$\mathcal{X} = \left(A_L(k), -B_L(k), V^{-1}(k)F(k), V^{-1}(k)\right), \tag{5.28}$$

$$\mathcal{Y} = \left(A_L(k), -L(k), V^{-1}(k)F(k), 0\right). \tag{5.29}$$

The systems (5.22)–(5.23) and (5.24)–(5.25) can be respectively written as $\forall v(k)$,

$$\begin{aligned} \begin{bmatrix} u(k) \\ y(k) \end{bmatrix} &= \begin{bmatrix} \mathcal{M}(v(k)) \\ \mathcal{N}(v(k)) \end{bmatrix}, \\ \mathcal{X}(u(k)) + \mathcal{Y}(y(k)) &= \mathcal{X} \circ \mathcal{M}(v(k)) + \mathcal{Y} \circ \mathcal{N}(v(k)) = v(k). \end{aligned} \tag{5.30}$$

Consequently, the identity

$$\mathcal{X} \circ \mathcal{M} + \mathcal{Y} \circ \mathcal{M} = \mathcal{I}. \tag{5.31}$$

holds. Hence, $(\mathcal{M}, \mathcal{N})$ build a RCP of the plant $\mathcal{G}$ and $(\mathcal{X}, \mathcal{Y})$ is an LCP of the observer-based state feedback system (5.24)–(5.25).

Observe that for $\hat{x}(0) = x(0)$, $u(k) = F(k)\hat{x}(k) = F(k)x(k)$, and $\forall v(k)$

$$\begin{aligned} r_y(k) &= \hat{\mathcal{M}}(y(k)) - \hat{\mathcal{N}}(u(k)) = 0 \\ &= \hat{\mathcal{M}} \circ \mathcal{N}(v(k)) - \hat{\mathcal{N}} \circ \mathcal{M}(v(k)) \\ &\Longrightarrow \hat{\mathcal{M}} \circ \mathcal{N} = \hat{\mathcal{N}} \circ \mathcal{M}. \end{aligned} \tag{5.32}$$

Similarly, $\forall x(0), \hat{x}(0), v(k) = 0$,

$$\begin{aligned} \mathcal{X}(u(k)) + \mathcal{Y}(y(k)) &= \mathcal{Y} \circ \hat{\mathcal{X}}\left(r_y(k)\right) - \mathcal{X} \circ \hat{\mathcal{Y}}\left(r_y(k)\right) = 0 \\ &\Longrightarrow \mathcal{Y} \circ \hat{\mathcal{X}} = \mathcal{X} \circ \hat{\mathcal{Y}}. \end{aligned} \tag{5.33}$$

Identities (5.21), (5.31), (5.32) and (5.33) build the Bezout identity for the given LDTV system (5.1)–(5.2).

Analogue to the coprime factorisations of LTI systems, we now introduce the following definitions.

Definition 5.3 Given the LCP and RCP of the LDTV system $\mathcal{G}$ and, $\left(\hat{\mathcal{M}}, \hat{\mathcal{N}}\right)$ and $(\mathcal{M}, \mathcal{N})$, which satisfy (5.14)–(5.15) and (5.26)–(5.27), respectively, the following linear operators

$$\mathcal{K}_{\mathcal{G}} : \mathcal{L}_{2,[0,N]} \times \mathcal{L}_{2,[0,N]} \rightarrow \mathcal{L}_{2,[0,N]},$$

$$\mathcal{K}_{\mathcal{G}}\left(u(k), y(k)\right) := \hat{\mathcal{M}}\left(y(k)\right) - \hat{\mathcal{N}}\left(u(k)\right) = r_y(k), \tag{5.34}$$

$$\mathcal{I}_{\mathcal{G}} : \mathcal{L}_{2,[0,N]} \rightarrow \mathcal{L}_{2,[0,N]} \times \mathcal{L}_{2,[0,N]},$$

$$\mathcal{I}_{\mathcal{G}}\left(v(k)\right) := \begin{bmatrix} \mathcal{M}\left(v(k)\right) \\ \mathcal{N}\left(v(k)\right) \end{bmatrix} = \begin{bmatrix} u(k) \\ y(k) \end{bmatrix}, \tag{5.35}$$

are called stable kernel and image representations (SKR and SIR) of the LDTV system $\mathcal{G}$.

Definition 5.4 Given the LCP and RCP $\left(\hat{\mathcal{X}}, \hat{\mathcal{Y}}\right)$ and $(\mathcal{X}, \mathcal{Y})$ satisfying (5.19)–(5.20) and (5.28)–(5.29), respectively, the following linear operators

$$\mathcal{I}_{\mathcal{K}} : \mathcal{L}_{2,[0,N]} \rightarrow \mathcal{L}_{2,[0,N]} \times \mathcal{L}_{2,[0,N]},$$

$$\mathcal{I}_{\mathcal{K}}\left(r_y(k)\right) := \begin{bmatrix} -\hat{\mathcal{Y}}\left(r_y(k)\right) \\ \hat{\mathcal{X}}\left(r_y(k)\right) \end{bmatrix} = \begin{bmatrix} u(k) \\ y(k) \end{bmatrix}, \tag{5.36}$$

$$\mathcal{K}_{\mathcal{K}} : \mathcal{L}_{2,[0,N]} \times \mathcal{L}_{2,[0,N]} \rightarrow \mathcal{L}_{2,[0,N]},$$

$$\mathcal{K}_{\mathcal{K}}\left(u(k), y(k)\right) := \mathcal{X}\left(u(k)\right) + \mathcal{Y}\left(y(k)\right) = v(k), \tag{5.37}$$

are called stable image and kernel representations (SIR and SKR) of the observer-based state feedback control system (5.17)–(5.18).

By means of the defined SKRs and SIRs, the Bezout identity can also be written as

$$\mathcal{K}_{\mathcal{K}} \circ \mathcal{I}_{\mathcal{G}} = \mathcal{I}, \mathcal{K}_{\mathcal{G}} \circ \mathcal{I}_{\mathcal{K}} = \mathcal{I}, \tag{5.38}$$

$$\mathcal{K}_{\mathcal{K}} \circ \mathcal{I}_{\mathcal{K}} = \mathcal{O}, \mathcal{K}_{\mathcal{G}} \circ \mathcal{I}_{\mathcal{G}} = \mathcal{O}. \tag{5.39}$$

Analogue to LTI systems, image and kernel subspaces of LDTV systems are introduced on the basis of the plant SKR and SIR.

Definition 5.5 Given system model (5.1)–(5.2) and the corresponding SIR and SKR, the $\mathcal{L}_2$ subspace $\mathcal{I}_G$ and $\mathcal{K}_G$ defined by

$$\mathcal{I}_G = \left\{ \begin{bmatrix} u \\ y \end{bmatrix} : \begin{bmatrix} u \\ y \end{bmatrix} = \mathcal{I}_{\mathcal{G}}(v) = \begin{bmatrix} \mathcal{M}(v) \\ \mathcal{N}(v) \end{bmatrix}, v \in \mathcal{L}_2 \right\}, \tag{5.40}$$

$$\mathcal{K}_G = \left\{ \begin{bmatrix} u \\ y \end{bmatrix} : \mathcal{K}_{\mathcal{G}}(u, y) = \hat{\mathcal{M}}(y) - \hat{\mathcal{N}}(u) = 0 \right\}, \tag{5.41}$$

are called image and kernel subspaces of the plant $\mathcal{G}$, respectively.

5.3 Normalised Factorisations

The normalised coprime factorisations of LTI systems play a special role in stabilising controller parameterisation, feedback control system analysis and synthesis. As described in Chap. 2, the normalised LCF and RCF of the LTI system G, $\left(\hat{M}_0, \hat{N}_0\right)$ and (M_0, N_0), are defined by

$$\begin{aligned}\hat{N}_0(z)\hat{N}_0^{\sim}(z) + \hat{M}_0(z)\hat{M}_0^{\sim}(z) &= K_{G_0}(z)K_{G_0}^{\sim}(z) = I, \\ M_0^{\sim}(z)M_0(z) + N_0^{\sim}(z)N_0(z) &= I_{G_0}^{\sim}(z)I_{G_0}(z) = I.\end{aligned}$$

Concerning the LDTV system $\mathcal{G}$, new definitions are necessary for the normalised LCF and RCF of $\mathcal{G}$, $\mathcal{K}_{\mathcal{G}}$ and $\mathcal{I}_{\mathcal{G}}$. In this section we will address these and some related issues.

5.3.1 *Isometric and Co-isometric Operators*

For LDTV systems, the normalised LCF and RCF are defined and addressed with the aid of the isometric and co-isometric operators.

Definition 5.6 An operator $\mathcal{P}$ is said to be isometric if

$$\mathcal{P}^* \circ \mathcal{P} = \mathcal{I}.$$

It is called co-isometric when

$$\mathcal{P} \circ \mathcal{P}^* = \mathcal{I}.$$

Definition 5.7 The normalised LCF and RCF of $\mathcal{G}$, denoted by $\mathcal{K}_{\mathcal{G}_0}$ and $\mathcal{I}_{\mathcal{G}_0}$, are co-isometric and isometric operators defined, respectively, by

$$\mathcal{K}_{\mathcal{G}_0} \circ \mathcal{K}_{\mathcal{G}_0}^* = \mathcal{I}, \mathcal{I}_{\mathcal{G}_0}^* \circ \mathcal{I}_{\mathcal{G}_0} = \mathcal{I}.$$

The corresponding LCP and RCP are denoted by $\left(\hat{\mathcal{M}}_0, \hat{\mathcal{N}}_0\right)$ and $(\mathcal{M}_0, \mathcal{N}_0)$, respectively.

The computation algorithms for $\mathcal{K}_{\mathcal{G}_0}$ and $\mathcal{I}_{\mathcal{G}_0}$ are known in the literature and summarised in the following theorem.

Theorem 5.1 *Given the system model (5.1)–(5.2), $\left(\hat{\mathcal{M}}_0, \hat{\mathcal{N}}_0\right)$ and $(\mathcal{M}_0, \mathcal{N}_0)$ are as follows*

$$\hat{\mathcal{M}}_0 = \left(A_{L_0}(k), -L_0(k), W_0(k)C(k), W_0(k)\right), \tag{5.42}$$

$$\hat{\mathcal{N}}_0 = \left(A_{L_0}(k), B_{L_0}(k), W_0(k)C(k), W_0(k)D(k)\right), \tag{5.43}$$

$$L_0(k) = \left(B(k)D^T(k) + A(k)P(k)C^T(k)\right) W_0^2(k), \tag{5.44}$$

$$W_0(k) = \left(I + D(k)D^T(k) + C(k)P(k)C^T(k)\right)^{-1/2}, \tag{5.45}$$

as well as

$$\mathcal{M}_0 = \left(A_F(k), B(k)V_0(k), F_0(k), V_0(k)\right), \tag{5.46}$$

$$\mathcal{N}_0 = \left(A_{F_0}(k), B(k)V_0(k), C_{F_0}(k), D(k)V_0(k)\right), \tag{5.47}$$

$$F_0(k) = -V_0^2(k)\left(D^T(k)C(k) + B^T(k)S(k+1)A(k)\right), \tag{5.48}$$

$$V_0(k) = \left(I + D^T(k)D(k) + B^T(k)S(k+1)B(k)\right)^{-1/2}, \tag{5.49}$$

where $P(k) > 0$ and $S(k) > 0$ are the solutions of the following Riccati recursions, respectively,

$$\begin{aligned} P(k+1) &= A_{L_0}(k)P(k)A_{L_0}^T(k) + L_0(k)L_0^T(k) + B_{L_0}(k)B_{L_0}^T(k) \qquad (5.50)\\ &= A(k)P(k)A^T(k) + B(k)B^T(k) - L_0(k)W_0^{-2}(k)L_0^T(k),\ P(0) = P_0,\\ S(k) &= A_{F_0}^T(k)S(k+1)A_{F_0}(k) + F_0^T(k)F_0(k) + C_{F_0}^T(k)C_{F_0}(k) \qquad (5.51)\\ &= A^T(k)S(k+1)A(k) + C^T(k)C(k) - F_0^T(k)V_0^{-2}(k)F_0(k),\ S(N+1) = S_{N+1}, \end{aligned}$$

where P_0 and S_{N+1} are boundary conditions, whose setting will be explained in Sect. 5.3.4.

As mentioned, the computation algorithms given in the above theorem are known and their proofs can be found in the literature. Some of them are cited at the end of this chapter. Below, we discuss some control theoretically related issues, which also prove the above computation algorithms alternatively. Moreover, these discussions are useful as well to help us to gain a deeper insight into control theoretical aspects.

5.3.2 Inner and Co-inner

The normalised SIR (respectively, SKR) is a stable system, and also called inner (respectively, co-inner). In the literature, the results summarised in Theorem 5.1 have been well investigated in the so-called inner-outer (respectively co-inner-outer) factorisation technique. Below, we briefly outline some key conditions of inner and co-inner (operator) systems.

At first, the conditions for inner are examined. For the sake of notational simplicity, a (stable) LDTV inner system $\mathcal{G}_i = (A(k), B(k), C(k), D(k))$, is considered, for which there exists $S(k) > 0$ so that

$$S(k) = A^T(k)S(k+1)A(k) + C^T(k)C(k). \tag{5.52}$$

Define

$$\bar{x}(k) := S(k)x(k).$$

It holds

$$\begin{aligned}\bar{x}(k) &= \left(A^T(k)S(k+1)A(k) + C^T(k)C(k)\right)x(k)\\ &= A^T(k)\bar{x}(k+1) - A^T(k)S(k+1)B(k)u(k) + C^T(k)\left(y(k) - D(k)u(k)\right)\\ &= A^T(k)\bar{x}(k+1) + C^T(k)y(k) - \left(A^T(k)S(k+1)B(k) + C^T(k)D(k)\right)u(k).\end{aligned}$$

Now, let

$$\bar{y}(k+1) = B^T(k)\bar{x}(k+1) + D^T(k)y(k),$$

which yields

$$\begin{aligned}\bar{y}(k+1) &= B^T(k)S(k+1)x(k+1) + D^T(k)\left(C(k)x(k) + D(k)u(k)\right)\\ &= \left(B^T(k)S(k+1)A(k) + D^T(k)C(k)\right)x(k)\\ &+ \left(B^T(k)S(k+1)B(k) + D^T(k)D(k)\right)u(k).\end{aligned}$$

Thus, conditions

$$A^T(k)S(k+1)B(k) + C^T(k)D(k) = 0, \tag{5.53}$$

$$B^T(k)S(k+1)B(k) + D^T(k)D(k) = I \tag{5.54}$$

yield

$$\bar{x}(k) = A^T(k)\bar{x}(k+1) + C^T(k)\bar{u}(k+1), \bar{u}(k+1) = y(k), \tag{5.55}$$

$$\bar{y}(k+1) = B^T(k)\bar{x}(k+1) + D^T(k)\bar{u}(k+1) = u(k). \tag{5.56}$$

Note that system (5.55)–(5.56) is the dual system of the LDTV system $\mathcal{G}_i$ with $y(k)$ and $u(k)$ as input and output variables, respectively. As a result, we have

$$\begin{aligned}\bar{y}(k+1) = \mathcal{G}_i^*\left(y(k)\right) = \mathcal{G}_i^* \circ \mathcal{G}_i\left(u(k)\right) = u(k)\\ \Longrightarrow \mathcal{G}_i^* \circ \mathcal{G}_i = \mathcal{I}.\end{aligned}$$

That is, under conditions (5.52), (5.53) and (5.54), the LDTV system $\mathcal{G}_i$ is inner. Concerning the analogue results for a co-inner system, we consider

$$x(k+1) = A(k)x(k) + E(k)d(k), \tag{5.57}$$
$$y(k) = C(k)x(k) + F(k)d(k), \tag{5.58}$$

that is co-inner, and the adjoint of (5.57)–(5.58) is denoted by $\mathcal{G}_{ci}$ with the state space realisation

$$\bar{x}(k) = A^T(k)\bar{x}(k+1) + C^T(k)\bar{d}(k+1), \tag{5.59}$$
$$\bar{y}(k+1) = E^T(k)\bar{x}(k+1) + F^T(k)\bar{d}(k+1). \tag{5.60}$$

We delineate that

$$\begin{aligned} y(k) &= \mathcal{G}_{ci}\left(\bar{y}(k+1)\right) = \mathcal{G}_{ci} \circ \mathcal{G}_{ci}^* \left(\bar{d}(k+1)\right) = \bar{d}(k+1) \\ &\Longrightarrow \mathcal{G}_{ci} \circ \mathcal{G}_{ci}^* = \mathcal{I}, \end{aligned}$$

when there exists $P(k) > 0$ so that

$$P(k+1) = A(k)P(k)A^T(k) + E(k)E^T(k), \tag{5.61}$$
$$C(k)P(k)A^T(k) + F(k)E^T(k) = 0, \tag{5.62}$$
$$C(k)P(k)C^T(k) + F(k)F^T(k) = I. \tag{5.63}$$

Moreover, it holds

$$P(k)\bar{x}(k) = x(k). \tag{5.64}$$

We firstly examine (5.64). To this end, observe that when (5.64) holds, it turns out

$$\begin{aligned} x(k+1) &= A(k)x(k) + E(k)E^T(k)\bar{x}(k+1) + E(k)F^T(k)\bar{d}(k+1) \Longrightarrow \\ P(k+1)\bar{x}(k+1) &= A(k)P(k)\bar{x}(k) + E(k)E^T(k)\bar{x}(k+1) + E(k)F^T(k)\bar{d}(k+1). \end{aligned} \tag{5.65}$$

By (5.61)–(5.62), we obtain

$$\begin{aligned} A(k)P(k)A^T(k)\bar{x}(k+1) &= A(k)P(k)\bar{x}(k) + E(k)F^T(k)\bar{d}(k+1) \Longrightarrow \\ A(k)P(k)\left(A^T(k)\bar{x}(k+1) + C^T(k)\bar{d}(k+1)\right) &= A(k)P(k)\bar{x}(k) \Longrightarrow \\ \bar{x}(k) &= A^T(k)\bar{x}(k+1) + C^T(k)\bar{d}(k+1), \end{aligned}$$

which illustrates that $\bar{x}(k)$ given in (5.64) is the state vector of the adjoint system (5.59) and thus (5.64) is true. To show $\mathcal{G}_{ci} \circ \mathcal{G}_{ci}^* = \mathcal{I}$, consider

$$y(k) = C(k)x(k) + F(k)E^T(k)\bar{x}(k+1) + F(k)F^T(k)\bar{d}(k+1). \tag{5.66}$$

According to (5.59), (5.64) and (5.62), it holds

$$
\begin{aligned}
y(k) &= \left(C(k)P(k)A^T(k) + F(k)E^T(k)\right)\bar{x}(k+1) \\
&+ \left(C(k)P(k)C^T(k) + F(k)F^T(k)\right)\bar{d}(k+1) = \bar{d}(k+1),
\end{aligned}
$$

Namely $\mathcal{G}_{ci} \circ \mathcal{G}_{ci}^* = \mathcal{I}$.

The above results are summarised in the subsequent theorem.

Theorem 5.2 *Given the LDTS (5.57)–(5.58) and its adjoint (5.59)–(5.60). It is inner, when conditions (5.52), (5.53) and (5.54) are satisfied. And it is co-inner, if (5.61)–(5.63) hold. Moreover, for the inner system,*

$$
\bar{x}(k) = S(k)x(k),
$$

and for the co-inner,

$$
P(k)\bar{x}(k) = x(k). \tag{5.67}
$$

We now demonstrate that the setting of $F_0(k)$ and $V_0(k)$ given in Theorem 5.1 results in $\mathcal{I}_{\mathcal{G}_0}$ being inner. Examining (5.46)–(5.47) shows that

$$
\begin{aligned}
S(k) &= A_{F_0}^T(k)S(k+1)A_{F_0}(k) + \left[\, F_0^T(k)\ C_{F_0}^T(k) \,\right]\begin{bmatrix} F_0(k) \\ C_{F_0}(k) \end{bmatrix} \\
&= A_{F_0}^T(k)S(k+1)A_{F_0}(k) + F_0^T(k)F_0(k) + C_{F_0}^T(k)C_{F_0}(k), \\
&A_{F_0}^T(k)S(k+1)B(k)V_0(k) + \left[\, F_0^T(k)\ C_{F_0}^T(k) \,\right]\begin{bmatrix} V_0(k) \\ D(k)V_0(k) \end{bmatrix} \\
&= \left(A_{F_0}^T(k)S(k+1)B(k) + F_0^T(k) + C_{F_0}^T(k)D(k)\right) = 0, \\
V_0^T(k)B^T(k)&S(k+1)B(k)V_0(k) + \left[\, V_0^T(k)\ V_0^T(k)D^T(k) \,\right]\begin{bmatrix} V_0(k) \\ D(k)V_0(k) \end{bmatrix} \\
&V_0^T(k)\left(B^T(k)S(k+1)B(k) + I + D^T(k)D(k)\right)V_0(k) = I.
\end{aligned}
$$

Substituting $(A(k), B(k), E(k), F(k))$ by

$$
\left(A_{F_0}(k),\ B(k)V_0(k), \begin{bmatrix} F_0(k) \\ C_{F_0}(k) \end{bmatrix}, \begin{bmatrix} V_0(k) \\ D(k)V_0(k) \end{bmatrix}\right)
$$

leads to (5.52), (5.53) and (5.54). That is, $\mathcal{I}_{\mathcal{G}_0}$ is inner.

In fact, behind $F_0(k)$ and $V_0(k)$ settings is an inner-outer factorisation. To delineate it, consider

$$
\begin{bmatrix} u(k) \\ y(k) \end{bmatrix} = \begin{bmatrix} \mathcal{I}(u(k)) \\ \mathcal{G}(u(k)) \end{bmatrix} =: \hat{\mathcal{G}}(u(k)).
$$

An inner-outer factorisation of $\hat{\mathcal{G}}$, roughly speaking, is to factorise $\hat{\mathcal{G}}$ into

$$\hat{\mathcal{G}} = \mathcal{I}_{\mathcal{G}_0} \circ \mathcal{O}_{\mathcal{G}_0},$$

where $\mathcal{I}_{\mathcal{G}_0}$ is, as mentioned before, inner and $\mathcal{O}_{\mathcal{G}_0}$ is an outer system which is invertible and whose inverse is a stable system as well. Recall a RCF of $\mathcal{G}$,

$$\mathcal{G} = \mathcal{N}_0 \circ \mathcal{M}_0^{-1}$$

with $\mathcal{M}_0$ given (5.46). It is obvious that $\mathcal{M}_0^{-1}$ is invertible and

$$\left(\mathcal{M}_0^{-1}\right)^{-1} = \mathcal{M}_0 = (A_F(k), B(k)V_0(k), F_0(k), V_0(k))$$

is stable. As a result, setting

$$\mathcal{O}_{\mathcal{G}_0} = \mathcal{M}_0^{-1}, \mathcal{I}_{\mathcal{G}_0}(v(k)) = \begin{bmatrix} \mathcal{M}_0 (v(k)) \\ \mathcal{N}_0 (v(k)) \end{bmatrix}, v(k) = \mathcal{M}_0^{-1} (u(k)),$$

we have

$$\begin{bmatrix} u(k) \\ y(k) \end{bmatrix} = \hat{\mathcal{G}} (u(k)) = \mathcal{I}_{\mathcal{G}_0} \circ \mathcal{M}_0^{-1} (u(k)) = \mathcal{I}_{\mathcal{G}_0}(v(k)).$$

As a dual form of the inner operator $\mathcal{I}_{\mathcal{G}_0}$, it is obvious that$\mathcal{K}_{\mathcal{G}_0}$ is a co-inner operator. And the interpretation of $\mathcal{K}_{\mathcal{G}_0}$ as a result of co-inner-outer factorisation can be verified as well. That is, $\mathcal{K}_{\mathcal{G}_0}$ is stable, satisfies

$$\mathcal{K}_{\mathcal{G}_0} \circ \mathcal{K}_{\mathcal{G}_0}^* = \mathcal{I},$$

and can be achieved by the following co-inner-outer factorisation

$$\begin{aligned} \hat{\mathcal{R}} (u(k), y(k)) &:= \mathcal{I} (y(k)) - \mathcal{G} (u(k)) =: \hat{\mathcal{O}}_{\mathcal{G}_0} \circ \mathcal{K}_{\mathcal{G}_0} (u(k), y(k)), \\ \hat{\mathcal{O}}_{\mathcal{G}_0} &= \left(\hat{\mathcal{M}}_0\right)^{-1}, \end{aligned}$$

where $\hat{\mathcal{O}}_{\mathcal{G}_0}$ is co-outer whose inverse is $\hat{\mathcal{M}}_0$ and thus stable.

5.3.3 Losslessness

As known, for LTI systems, the normalised RCF is of the property that for zero initial value,

$$\left\| \begin{bmatrix} u(k) \\ y(k) \end{bmatrix} \right\|_2 = \|v(k)\|_2,$$

which is interpreted as non-dissipative, a property known as lossless. In this subsection, we explore the lossless property of inner and co-inner systems, and demonstrate that the normalised RCF and LCF are lossless.

Definition 5.8 The LDTV system (5.1)–(5.2) is called lossless with respect to the supply rate

$$s(u, y) = u^T(k)u(k) - y^T(k)y(k),$$

when there exists a (storage) function $V(x(k)) \geq 0$, $V(0) = 0$ so that

$$V(x(k+1)) - V(x(k)) = u^T(k)u(k) - y^T(k)y(k).$$

Theorem 5.3 *Given the LDTV system (5.1)–(5.2) that is inner, then it holds*

$$x^T(k+1)S(k+1)x(k+1) - x^T(k)S(k)x(k) = u^T(k)u(k) - y^T(k)y(k), \quad (5.68)$$

$$\|y(k)\|^2_{2,[0,N-1]} + x^T(N)S(N)x(N) = x^T(0)S(0)x(0) + \|u(k)\|^2_{2,[0,N-1]} \quad (5.69)$$
$$\Longrightarrow \|y(k)\|^2_{2,[0,N-1]} \leq x^T(0)S(0)x(0) + \|u(k)\|^2_{2,[0,N-1]},$$

where $S(k)$ is the solution of (5.52).

Proof Since

$$\begin{gathered} x^T(k+1)S(k+1)x(k+1) = \\ x^T(k)A^T(k)S(k+1)A(k)x(k) + u^T(k)B^T(k)S(k+1)B(k)u(k) \\ +2x^T(k)A^T(k)S(k+1)B(k)u(k), \end{gathered}$$

it turns out, according to (5.52), (5.53) and (5.54), that

$$\begin{aligned} & x^T(k+1)S(k+1)x(k+1) - x^T(k)S(k)x(k) \\ &= -x^T(k)C^T(k)C(k)x(k) + 2x^T(k)A^T(k)S(k+1)B(k)u(k) \\ &= u^T(k)u(k) - y^T(k)y(k). \end{aligned}$$

Then, (5.69) follows immediately.

Below, we prove that $\mathcal{I}_{\mathcal{G}_0}$ is such a lossless system. To this end, consider the LDTV system (5.1)–(5.2), its SIR given in (5.35), and define

$$\begin{gathered} V(x(k)) = x^T(k)S(k)x(k), \; S(k) > 0, \\ u(k) = F(k)x(k) + \bar{v}(k), \; \bar{v}(k) = V(k)v(k), \\ H(x, v) = V(x(k+1)) - V(x(k)) + y^T(k)y(k) + u^T(k)u(k) = \\ V(x(k+1)) - V(x(k)) + (C_F(k)x(k) + D(k)\bar{v}(k))^T (C_F(k)x(k) + D(k)\bar{v}(k)) \\ + (F(k)x(k) + \bar{v}(k))^T (F(k)x(k) + \bar{v}(k)). \end{gathered}$$

It turns out

$$H(x, v) = x^T(k)\left(A_F^T(k)S(k+1)A_F(k) - S(k) + F^T(k)F(k) + C_F^T(k)C_F(k)\right)x(k)$$
$$+ \bar{v}^T(k)\left(B^T(k)S(k+1)B(k) + D^T(k)D(k) + I\right)\bar{v}(k)$$
$$+ 2x^T(k)\left(A_F^T(k)S(k+1)B(k) + C_F^T(k)D(k) + F^T(k)\right)\bar{v}(k).$$

It is evident that setting

$$F(k) = -B^T(k)S(k+1)A_F(k) + D^T(k)C_F(k) \Longrightarrow$$
$$\begin{pmatrix} I + B^T(k)S(k+1)B(k) \\ +D^T(k)D(k) \end{pmatrix} F(k) = -\begin{pmatrix} B^T(k)S(k+1)A(k) \\ +D^T(k)C(k) \end{pmatrix}$$
$$\Longrightarrow F(k) = F_0(k) = -V_0^2(k)\left(D^T(k)C(k) + B^T(k)S(k+1)A(k)\right),$$
$$V(k) = V_0(k) = \left(I + D^T(k)D(k) + B^T(k)S(k+1)B(k)\right)^{-1/2}$$

results in

$$H(x, v) = V(x(k+1)) - V(x(k)) + y^T(k)y(k) + u^T(k)u(k) = v^T(k)v(k).$$

That is, $\mathcal{I}_{\mathcal{G}_0}$ is lossless with respect to the supply rate

$$s\left(v, \begin{bmatrix} u \\ y \end{bmatrix}\right) = v^T(k)v(k) - \left(u^T(k)u(k) + y^T(k)y(k)\right).$$

Concerning the dual form of the lossless property of an inner system, there merely exist publications addressing this issue. As cited at the end of this chapter, in a work on optimal fault detection by means of co-inner-outer factorisation, the dual form of the energy lossless behaviour of an inner system has been introduced in the context of lossless information transform. Inspired by this formulation, we present, in the following, the normalised LCF $\mathcal{K}_{\mathcal{G}_0}$ as a system with lossless information transform.

Definition 5.9 Given an LDTV system

$$x(k+1) = A(k)x(k) + E(k)d(k), \tag{5.70}$$
$$r(k) = C(k)x(k) + F(k)d(k), \tag{5.71}$$

and its dual form

$$\bar{x}(k-1) = A^T(k-1)\bar{x}(k) + C^T(k-1)\bar{r}(k), \tag{5.72}$$
$$\bar{d}(k) = E^T(k-1)\bar{x}(k) + F^T(k-1)\bar{r}(k), \tag{5.73}$$

system (5.70)–(5.71) is called lossless with respect to information transform rate

$$s(\bar{r}, \bar{d}) = \bar{r}^T(k)\bar{r}(k) - \bar{d}^T(k)\bar{d}(k),$$

if there exists a function $V(\bar{x}(k)) \geq 0$, $V(0) = 0$ such that

$$V(\bar{x}(k-1)) - V(\bar{x}(k)) = \bar{r}^T(k)\bar{r}(k) - \bar{d}^T(k)\bar{d}(k). \tag{5.74}$$

Theorem 5.4 *Given the LDTV system (5.70)–(5.71) that is co-inner, it holds*

$$\bar{x}^T(k-1)P(k-1)\bar{x}(k-1) - \bar{x}^T(k)P(k)\bar{x}(k) = \bar{r}^T(k)\bar{r}(k) - \bar{d}^T(k)\bar{d}(k), \tag{5.75}$$

$$\begin{gathered} \|x(0)\|^2_{P^{-1}(0)} + \left\|\bar{d}(k)\right\|^2_{2,[1,N]} = \|\bar{r}(k)\|^2_{2,[1,N]} + \|x(N)\|^2_{P^{-1}(N)} \\ \Longrightarrow \|x(0)\|^2_{P^{-1}(0)} + \|d(k)\|^2_{2,[0,N-1]} \geq \|r(k)\|^2_{2,[0,N-1]}, \end{gathered} \tag{5.76}$$

where $P(k)$ is the solution of (5.61).

Proof Because

$$\begin{gathered} \bar{x}^T(k-1)P(k-1)\bar{x}(k-1) = \\ \bar{x}^T(k)A(k-1)P(k-1)A^T(k-1)\bar{x}(k) + \bar{r}^T(k)C(k-1)P(k-1)C^T(k-1)\bar{r}(k) \\ +2\bar{x}^T(k)A(k-1)P(k-1)C^T(k-1)\bar{r}(k), \end{gathered}$$

it turns out, according to (5.61)–(5.63), that

$$\begin{aligned} &\bar{x}^T(k-1)P(k-1)\bar{x}(k-1) - \bar{x}^T(k)P(k)\bar{x}(k) \\ &= -\bar{x}^T(k)E(k-1)E^T(k-1)\bar{x}(k) + 2\bar{x}^T(k)A(k-1)P(k-1)C^T(k-1)\bar{r}(k) \\ &= \bar{r}^T(k)\bar{r}(k) - \bar{d}^T(k)\bar{d}(k). \end{aligned}$$

It follows from Theorem 5.2 that $\bar{x}(k) = P^{-1}(k)x(k)$, leading to

$$x^T(k-1)P^{-1}(k-1)x(k-1) - x^T(k)P^{-1}(k)x(k) = \bar{r}^T(k)\bar{r}(k) - \bar{d}^T(k)\bar{d}(k).$$

Considering further for $\bar{r}(k) = r(k-1)$, $\bar{d}(k) = d(k-1)$, it holds

$$\begin{aligned} &\|x(0)\|^2_{P^{-1}(0)} + \|d(k)\|^2_{2,[0,N-1]} = \|r(k)\|^2_{2,[0,N-1]} + \|x(N)\|^2_{P^{-1}(N)} \Longrightarrow \\ &\|x(0)\|^2_{P^{-1}(0)} + \|d(k)\|^2_{2,[0,N-1]} \geq \|r(k)\|^2_{2,[0,N-1]}. \end{aligned}$$

The theorem is proved.

Before demonstrating $\mathcal{K}_{\mathcal{G}_0}$ being lossless, we would like to discuss the interpretation of $\mathcal{K}_{\mathcal{G}_0}$ in the context of the above definition. As illustrated in the previous chapters, the residual vector as the output of $\mathcal{K}_{\mathcal{G}_0}$ contains information about uncertainties in the process data (u, y) as the input variable of $\mathcal{K}_{\mathcal{G}_0}$ (corresponding to d in system (5.70)–(5.71)). In other words, these uncertainties are the component in

(u, y) that lead to $r(k)$ different from zero (refer to Example 5.1). Condition (5.74) can be interpreted as no loss of information, when the system $\mathcal{K}_{\mathcal{G}_0}$ transfers this information.

Now, we delineate that $\mathcal{K}_{\mathcal{G}_0}$ is lossless in the regard of Definition 5.9. Consider the LDTV system (5.1)–(5.2) and its SKR $\mathcal{K}_{\mathcal{G}}$ given in (5.34), whose adjoint $\mathcal{K}_{\mathcal{G}}^*$ is

$$\bar{x}(k-1) = A_L^T(k-1)\bar{x}(k) + C^T(k-1)W^T(k-1)\bar{r}(k), \tag{5.77}$$

$$\bar{d}(k) = \begin{bmatrix} B_L^T(k-1) \\ -L^T(k-1) \end{bmatrix} \bar{x}(k) + \begin{bmatrix} D^T(k-1)W^T(k-1) \\ W^T(k-1) \end{bmatrix} \bar{r}(k). \tag{5.78}$$

Here,

$$\bar{d}(k) = \begin{bmatrix} \bar{u}(k) \\ \bar{y}(k) \end{bmatrix}.$$

Let

$$\begin{aligned} V(\bar{x}(k)) &= \bar{x}^T(k)P(k)\bar{x}(k),\ P > 0, \\ H\left(\bar{x}, \bar{d}\right) &= V(\bar{x}(k)) - V(\bar{x}(k-1)) + \bar{d}^T(k)\bar{d}(k) \end{aligned}$$

$$= V(\bar{x}(k)) - V(\bar{x}(k-1)) + \begin{pmatrix} \begin{bmatrix} B_L^T(k-1) \\ -L^T(k-1) \end{bmatrix} \bar{x}(k) + \\ \begin{bmatrix} D^T(k-1)W^T(k-1) \\ W^T(k-1) \end{bmatrix} \bar{r}(k) \end{pmatrix}^T$$

$$\cdot \left(\begin{bmatrix} B_L^T(k-1) \\ -L^T(k-1) \end{bmatrix} \bar{x}(k) + \begin{bmatrix} D^T(k-1)W^T(k-1) \\ W^T(k-1) \end{bmatrix} \bar{r}(k) \right).$$

Substituting the relations in (5.77)–(5.78) into $H\left(\bar{x}, \bar{d}\right)$ yields

$$\begin{aligned} H\left(\bar{x}, \bar{d}\right) &= \bar{x}^T(k) \begin{pmatrix} A_L(k-1)P(k-1)A_L^T(k-1) - P(k) \\ +L^T(k-1)L(k-1) + B_L^T(k-1)B_L(k-1) \end{pmatrix} \bar{x}(k) \\ &+ \bar{r}^T(k)W(k-1) \begin{pmatrix} C(k-1)P(k-1)C^T(k-1) \\ +D(k-1)D^T(k-1) + I \end{pmatrix} W^T(k-1)\bar{r}(k) \\ &+ 2\bar{x}^T(k) \begin{pmatrix} A_L(k-1)P(k-1)C^T(k-1) + \\ B_L(k-1)D^T(k-1) - L(k-1) \end{pmatrix} W^T(k-1)\bar{r}(k). \end{aligned}$$

It is apparent that with the setting

$$\begin{aligned} & L(k) = A_L(k)P(k)C^T(k) + B_L(k)D^T(k) \Longrightarrow \\ & \begin{pmatrix} I + C(k)P(k)C^T(k) \\ +D(k)D^T(k) \end{pmatrix} L(k) = \left(A(k)P(k)C^T(k) + B(k)D^T(k)\right) \\ & \Longrightarrow L(k) = L_0(k) = -W_0^2(k)\left(A(k)P(k)C^T(k) + B(k)D^T(k)\right), \\ & W(k) = W_0(k) = \left(I + C(k)P(k)C^T(k) + D(k)D^T(k)\right)^{-1/2}, \end{aligned}$$

we have

$$H\left(\bar{x}, \bar{d}\right) = V(\bar{x}(k-1)) - V(\bar{x}(k)) + \bar{d}^T(k)\bar{d}(k) = \bar{r}^T(k)\bar{r}(k).$$

That is, $\mathcal{K}_{\mathcal{G}_0}$ is lossless with information transform.

Example 5.1 By this example, the information transferring process is explained. Given the LDTV system (5.1)–(5.2) and its SKR $\mathcal{K}_{\mathcal{G}}$,

$$r_y(k) = \mathcal{K}_{\mathcal{G}}(u(k), y(k)) = \hat{\mathcal{M}}(y(k)) - \hat{\mathcal{N}}(u(k)),$$

it is evident that for any $(u, y) \in \mathcal{I}_G$ (refer to Definition 5.5), it holds,

$$r_y(k) = 0.$$

When $(u, y) \in \mathcal{I}_G \oplus \mathcal{I}_G^{\perp}$, where $\mathcal{I}_G^{\perp}$ is the orthogonal complementary subspace of $\mathcal{I}_G$ and $\mathcal{L}_2 = \mathcal{I}_G \oplus \mathcal{I}_G^{\perp}$, it can be decomposed into

$$\begin{bmatrix} u \\ y \end{bmatrix} = \begin{bmatrix} u_0 \\ y_0 \end{bmatrix} + \begin{bmatrix} u^{\perp} \\ y^{\perp} \end{bmatrix}, \begin{bmatrix} u_0 \\ y_0 \end{bmatrix} \in \mathcal{I}_G, \begin{bmatrix} u^{\perp} \\ y^{\perp} \end{bmatrix} \in \mathcal{I}_G^{\perp}.$$

Consequently, we have

$$r_y(k) = \mathcal{K}_{\mathcal{G}}(u(k), y(k)) = \mathcal{K}_{\mathcal{G}}\left(u^{\perp}(k), y^{\perp}(k)\right). \tag{5.79}$$

Equation (5.79) implies that $r_y(k)$ contains exclusively information about deviations from the nominal system operation, and moreover,

$$\bar{d}(k+1) = \begin{bmatrix} u^{\perp}(k) \\ y^{\perp}(k) \end{bmatrix}, \bar{r}(k+1) = r_y(k).$$

Since $\mathcal{K}_{\mathcal{G}_0}$ is co-inner with

$$\|\bar{r}(k+1)\|_{2,[0,N]} = \left\|r_y(k)\right\|_{2,[0,N]} = \left\|\bar{d}(k+1)\right\|_{2,[0,N]} = \left\|\begin{bmatrix} u^{\perp}(k) \\ y^{\perp}(k) \end{bmatrix}\right\|_{2,[0,N]},$$

the losslessness of $\mathcal{K}_{\mathcal{G}_0}$ means that $r_y(k)$ contains the same information amount like $\begin{bmatrix} u^{\perp}(k) \\ y^{\perp}(k) \end{bmatrix}$. That is, there exists no loss of information about the deviation.

5.3.4 LQ Regulation and LS Estimation

A further interesting control-theoretic aspect related to the normalised RCF is the well-known result that the computation of an inner system is equivalent to linear quadratic regulation (LQR) optimisation. An LQR optimisation problem is formulated as: given an LDTV system

$$x(k+1) = A(k)x(k) + B(k)u(k),$$

find $u(k)$ so that the cost function

$$J(N) = x^T(N+1)S_{N+1}x(N+1) + \sum_{k=0}^{N} \left(x^T(k)Q(k)x(k) + u^T(k)R(k)u(k)\right)$$

is minimised. Corresponding to the inner system given above, the weighting matrices $Q(k)$, $R(k)$ are given by

$$Q(k) = C^T(k)C(k) \geq 0, \; R(k) = I + D^T(k)D(k) > 0.$$

This optimisation problem is solved by

$$u(k) = F_0(k)x(k)$$

with $F_0(k)$ given in (5.48).

The dual result that a co-inner system is an LS (least squares) estimator is less known. Below, we briefly introduce this claim. The formulation of an LS estimation problem is as follows: given an LDTV system

$$\begin{aligned} x(k+1) &= A(k)x(k) + E(k)d(k), \\ y(k) &= C(k)x(k) + F(k)d(k), \end{aligned}$$

and measurement data, $y(k), k \in [1, N]$, find an estimate for $d(k), x(0) = x_0$ so that the cost function

$$J = x_0^T P_0^{-1} x_0 + \|d\|_{2,[1,N]}^2, \; \|d\|_{2,[1,N]}^2 = \sum_{k=1}^{N} d^T(k)d(k)$$
$$\text{s.t.} \begin{cases} x(k+1) = A(k)x(k) + E(k)d(k) \\ y(k) - C(k)x(k) = F(k)d(k) \end{cases}$$

is minimised. Recall that the system response is a linear operator $\mathcal{G}$ that maps (x_0, d) to y defined by

$$\mathcal{G} : \mathbb{R}^n \times \mathcal{L}_{2,[0,N]} \to \mathcal{L}_{2,[0,N]},$$
$$\mathcal{G}\left(x_0, d(k)\right) = C(k)\Phi(k,0)x(0) + C(k)\sum_{j=0}^{k-1}\Phi(k, j+1)E(j)d(j) + F(k)d(k).$$

Thus, the above defined estimation problem is indeed to find the solution of

$$y = \mathcal{G}\left(x_0, d\right),$$

which is of the minimum norm (size), i.e. the so-called LS solution. It is well-known that, under the assumption that the operator $\mathcal{G} \circ \mathcal{G}^*$ is invertible, where $\mathcal{G}^*$ is the adjoint of $\mathcal{G}$, the LS estimate is given by

$$\left(x_0, d\right) = \mathcal{G}^* \left(\mathcal{G} \circ \mathcal{G}^*\right)^{-1}(y).$$

When $\mathcal{G}$ is normalised, i.e. $\mathcal{G} = \mathcal{G}_i$ is inner, then

$$\left(x_0, d\right) = \mathcal{G}_i^*\left(y_n\right), \; y_n = \mathcal{G}_i\left(d\right).$$

Here, the state space realisation of $\mathcal{G}_i\left(d\right)$ is given by

$$\begin{aligned}\hat{x}(k+1) &= A_{L_0}(k)\hat{x}(k) + E_{L_0}(k)d(k),\\ y_n(k) &= C(k)\hat{x}(k) + F(k)d(k),\end{aligned}$$

which is equivalent a residual generator, and the LS estimate is given by

$$\bar{x}(k-1) = A_{L_0}^T(k-1)\bar{x}(k) + C^T(k-1)y_n(k), \tag{5.80}$$
$$\hat{d}(k+1) = E_{L_0}^T(k-1)\bar{x}(k) + F^T(k-1)y_n(k). \tag{5.81}$$

The above backward recursive computation begins with

$$\bar{x}(N) = P^{-1}(N)\hat{x}(N).$$

Figure 5.1 summarises the normalised right and left coprime factorisations and the related issues addressed in this section.

5.4 Parameterisation of LDTV Stabilising Controllers

Consider the feedback control loop shown in Fig. 5.2. The parameterisation of LDTV stabilising controllers is known in the literature and can be expressed in an analogue form of the Youla parameterisation for LTI systems. For our study in the unified

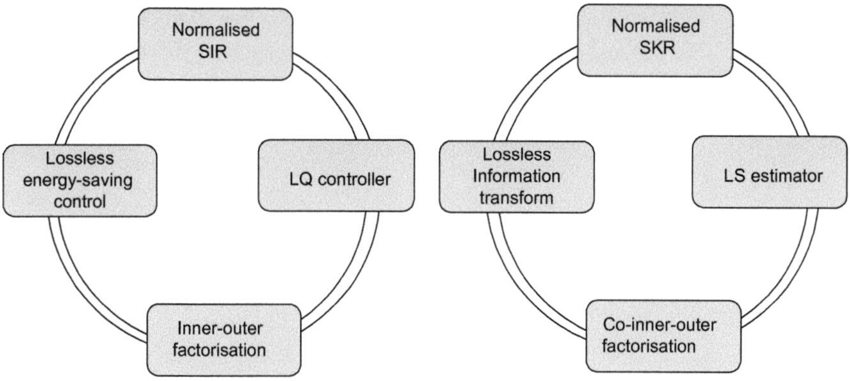

Fig. 5.1 Normalised coprime factorisations and the related issues

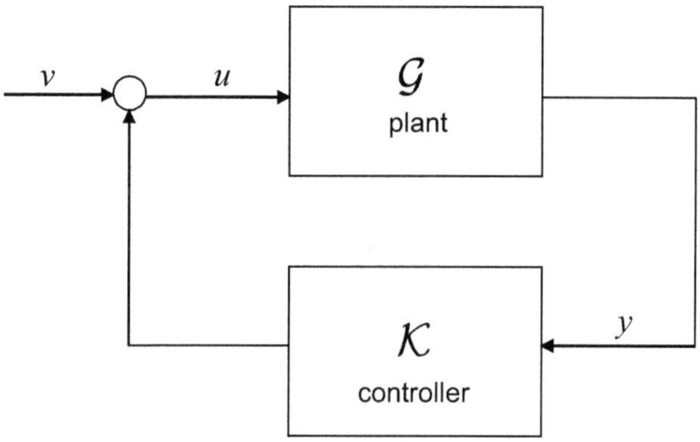

Fig. 5.2 The standard feedback control loop of LDTV systems

framework of control and detection, below the observer-based realisation of the parameterisation and the corresponding closed-loop dynamics are summarised.

Given the system (5.1)–(5.2), the core of the parameterisation form is an observer and the observer-based residual generator given below

$$\begin{aligned} \hat{x}(k+1) &= A(k)\hat{x}(k) + B(k)u(k) + L(k)r_0(k), \\ r_0(k) &= y(k) - \hat{y}(k), r_y(k) = W(k)r_0(k), \\ \hat{y}(k) &= C_F(k)\hat{x}(k) + D(k)u(k). \end{aligned} \tag{5.82}$$

Let $\mathcal{Q}$ be a linear operator defining a stable LDTV system whose state space representation is

$$\mathcal{Q} : \begin{cases} x_Q(k+1) = A_Q(k)x_Q(k) + B_Q(k)r_0(k) \\ r_Q(k) = C_Q(k)x_Q(k) + D_Q(k)r_0(k), \end{cases} \tag{5.83}$$

where $x_Q(k) \in \mathbb{R}^{n_Q}$ represents the state vector of the system, and $A_Q(k)$, $B_Q(k)$, $C_Q(k)$ and $D_Q(k)$ are real matrices, bounded and of appropriate dimensions. Then, all stabilising controllers are parametrised by $\mathcal{Q}$ and expressed by

$$u(k) = F(k)\hat{x}(k) + \bar{v}(k) + r_Q(k). \tag{5.84}$$

Here, $\bar{v}(k)$ is the output of the feedforward controller $\mathcal{X}_Q = \mathcal{X} + \mathcal{Q} \circ \hat{\mathcal{N}}$ given by

$$\bar{v}(k) = \left(\mathcal{X} + \mathcal{Q} \circ \hat{\mathcal{N}}\right)(v(k)). \tag{5.85}$$

Analogue to LTI systems, the closed-loop dynamics with the plant (5.1)–(5.2) and controller (5.82)–(5.85) are given by

$$\hat{x}_v(k+1) = A_F(k)\hat{x}_v(k) + B(k)V(k)\left(\bar{v}(k) + r_Q(k)\right) + L(k)W^{-1}(k)r_y(k), \tag{5.86}$$

$$\begin{bmatrix} u(k) \\ y(k) \end{bmatrix} = \begin{bmatrix} F(k) \\ C_F(k) \end{bmatrix} \hat{x}_v(k) + \begin{bmatrix} V(k) \\ D(k)V(k) \end{bmatrix} \left(\bar{v}(k) + r_Q(k)\right) + \begin{bmatrix} 0 \\ W^{-1}(k) \end{bmatrix} r_y(k). \tag{5.87}$$

It is of interest to note that for $v = 0$, the dynamics (5.86)–(5.87) are the SIR of the controller $u(k) = \mathcal{K}(y(k))$, namely

$$\begin{bmatrix} u(k) \\ y(k) \end{bmatrix} = \begin{bmatrix} \left(-\hat{\mathcal{Y}} + \mathcal{M} \circ \mathcal{Q}\right)(r_y(k)) \\ \left(\hat{\mathcal{X}} + \mathcal{N} \circ \mathcal{Q}\right)(r_y(k)) \end{bmatrix} \tag{5.88}$$

with r_y as the latent variable.

Example 5.2 In this example, we showcase that the controller (5.82)–(5.85) can be equivalently written as the SKR of the controller $u(k) = \mathcal{K}(y(k))$. Observe that (5.82) is equivalent to

$$\begin{aligned} \hat{x}(k+1) &= A_L(k)\hat{x}(k) + B_L(k)u(k) + L(k)W^{-1}(k)y(k), \\ A_L(k) &= A(k) - L(k)V(k),\ B_L(k) = B(k) - L(k)D(k). \end{aligned} \tag{5.89}$$

It follows from the control law $u(k) = F(k)\hat{x}(k) + r_Q(k)$ that

$$V^{-1}(k)\left(u(k) - F(k)\hat{x}(k) - \mathcal{Q}\left(W(k)\left(y(k) - C(k)\hat{x}(k) - D(k)u(k)\right)\right)\right) = 0. \tag{5.90}$$

On account of (5.28)–(5.29) as well as (5.14)–(5.15), (5.90) can be equivalently written as

$$\mathcal{X}(u(k)) + \mathcal{Y}(y(k)) - \mathcal{Q} \circ \hat{\mathcal{M}}(y(k)) + \mathcal{Q} \circ \hat{\mathcal{N}}(u(k)) = 0,$$

which is the SKR of the controller.

5.5 Parameterisation of LDTV Observers

An elemental result in the unified framework for LTI systems is that any output feedback controller is an estimator for Fx. This result is the consequence of parameterisation of LTI observers. Another known application of observer parameterisation is the design of observer-based residual generators. Remember that the proof of the observer parameterisation is attributed to the Bezout identity, which holds for LDTV systems as well, as given by (5.38)–(5.39). So far, it is natural to argue the existence of a parameterisation of LDTV observers in a similar form, and to prove this result along the lines of the study in Sect. 3.3.

To begin with, a general definition of an LDTV observer is introduced.

Definition 5.10 Given the uniformly detectable LDTV system model (5.1)–(5.2), a stable dynamic system represented by linear operator $\mathcal{O} : \mathcal{L}_{2,[0,N]} \times \mathcal{L}_{2,[0,N]} \rightarrow \mathcal{L}_{2,[0,N]}$,

$$x_p(k) = \mathcal{O}\left(\begin{bmatrix} u(k) \\ y(k) \end{bmatrix}\right) \tag{5.91}$$

is called an observer for $P(k)x(k)$ if

$$\forall x(0), u(k), \lim_{k \to \infty} \left(P(k)x(k) - x_p(k)\right) = 0. \tag{5.92}$$

Here, $P(k) \in \mathbb{R}^{q \times n}$ is a given matrix with an arbitrary integer $q \in [1, n]$ and

$$rank\,(P(k)) = q,$$

and the initial state of the dynamic system $\mathcal{O}$ is zero.

In the sequel, a parameterisation of observer (5.91) will be derived. Firstly, for $x(0) = 0$, $P(k)x(k)$ is defined as the output of system

$$\mathcal{G}_x := (A(k), B(k), P(k), 0)\,.$$

By means of RCF of $\mathcal{G}$ and $\mathcal{G}_x$, the system SIR (5.35), condition (5.92) is equivalently expressed as, $\forall\, u(k)$, $\exists v(k)$ so that

$$u(k) = \mathcal{M}(v(k)), \mathcal{M} = (A_F(k), B(k), F(k), I),$$
$$\mathcal{N}_x(v(k)) = \mathcal{O} \circ \mathcal{I}_{\mathcal{G}}(v(k)), \mathcal{N}_x = (A_F(k), B(k), P(k), 0),$$

which yields

$$\mathcal{N}_x = \mathcal{O} \circ \mathcal{I}_{\mathcal{G}}.$$

Let system $\mathcal{O}$ be parametrised by

$$\mathcal{O}\left(\begin{bmatrix} u(k) \\ y(k) \end{bmatrix}\right) = \mathcal{O}_1 \circ \mathcal{K}_{\mathcal{K}}\left(\begin{bmatrix} u(k) \\ y(k) \end{bmatrix}\right) + \mathcal{O}_2 \circ \mathcal{K}_{\mathcal{G}}\left(\begin{bmatrix} u(k) \\ y(k) \end{bmatrix}\right).$$

It follows from the Bezout identity (5.38)–(5.39) that

$$\mathcal{K}_{\mathcal{K}}\left(\begin{bmatrix} u(k) \\ y(k) \end{bmatrix}\right) = v(k), \mathcal{K}_{\mathcal{G}}\left(\begin{bmatrix} u(k) \\ y(k) \end{bmatrix}\right) = 0.$$

Consequently, $\mathcal{O}_2$ can be any stable system and

$$\mathcal{O}_1 = \mathcal{N}_x \Longrightarrow$$
$$\mathcal{O}\left(\begin{bmatrix} u(k) \\ y(k) \end{bmatrix}\right) = \mathcal{N}_x \circ \mathcal{K}_{\mathcal{K}}\left(\begin{bmatrix} u(k) \\ y(k) \end{bmatrix}\right) + \mathcal{R} \circ \mathcal{K}_{\mathcal{G}}\left(\begin{bmatrix} u(k) \\ y(k) \end{bmatrix}\right), \tag{5.93}$$

where $\mathcal{R}$ represents a stable dynamic post-filter connected to the residual generator

$$r_y(k) = \mathcal{K}_{\mathcal{G}}\left(\begin{bmatrix} u(k) \\ y(k) \end{bmatrix}\right),$$

and parametrises the observer $x_p(k) = \mathcal{O}\left(\begin{bmatrix} u(k) \\ y(k) \end{bmatrix}\right)$. As the last step, the state space representation of the term

$$\mathcal{N}_x \circ \mathcal{K}_{\mathcal{K}}\left(\begin{bmatrix} u(k) \\ y(k) \end{bmatrix}\right) = \mathcal{N}_x \circ \mathcal{X}(u(k)) + \mathcal{N}_x \circ \mathcal{Y}(y(k))$$

is determined. To this end, it is, at first, equivalently written into

$$\mathcal{N}_x \circ \mathcal{X}(u(k)) + \mathcal{N}_x \circ \mathcal{Y}(y(k)) = \mathcal{N}_x^e \circ \mathcal{X}^e\left(u^e(k)\right) + \mathcal{N}_x^e \circ \mathcal{Y}^e(y(k)),$$
$$\mathcal{N}_x^e = \left(A_F(k), B^e(k), P(k), 0\right), \mathcal{X}^e = \left(A_L(k), -B_L^e(k), F^e(k), I\right),$$
$$\mathcal{Y}^e = \left(A_L(k), -L(k), F^e(k), 0\right), u^e(k) = \begin{bmatrix} u(k) \\ 0 \end{bmatrix}.$$

Here, the following extension is introduced without affecting the system dynamics,

$$B(k)F(k) = B^e(k)F^e(k),\, B^e(k) = \begin{bmatrix} B(k) & 0 \end{bmatrix},\, F^e(k) = \begin{bmatrix} F(k) \\ F_1(k) \end{bmatrix},$$
$$D^e(k) = \begin{bmatrix} D(k) & 0 \end{bmatrix}, rank\left(F^e(k)\right) = n,\, P(k) = P_0(k)F^e(k),$$

and correspondingly,

$$A_F(k) = A(k) + B(k)F(k) = A(k) + B^e(k)F^e(k),\, B_L^e(k) = B^e(k) - L(k)D^e(k),$$
$$B(k)u(k) = B^e(k)u^e(k),\, B_L(k)u(k) = B_L^e(k)u^e(k).$$

The motivation and the associated computations are well described in Sect. 3.3. With the same computation steps, we have

$$\mathcal{N}_x^e \circ \mathcal{X}^e\left(u^e(k)\right) + \mathcal{N}_x^e \circ \mathcal{Y}^e\left(y(k)\right) =$$
$$P_0(k)\left(\mathcal{M}^e \circ \mathcal{X}^e(u^e(k)) - \mathcal{X}^e(u^e(k)) + \mathcal{M}^e \circ \mathcal{Y}^e(y(k)) - \mathcal{Y}^e(y(k))\right),$$
$$\mathcal{M}^e = \left(A_F(k),\, B^e(k),\, F^e(k),\, I\right).$$

Introducing

$$\hat{\mathcal{Y}}^e = \left(A_F(k), -L(k)W^{-1}(k),\, F^e(k), 0\right),$$
$$\hat{\mathcal{N}}^e = \left(A_L(k),\, B_L^e(k),\, W(k)C(k),\, W(k)D^e(k)\right),$$
$$\hat{\mathcal{M}}^e = \hat{\mathcal{M}} = \left(A_L(k), -L(k),\, W(k)C(k),\, W(k)\right),$$

and by means of the Bezout identity that leads to

$$\mathcal{M}^e \circ \mathcal{X}^e - \mathcal{I} = -\hat{\mathcal{Y}}^e \circ \hat{\mathcal{N}}^e,\, \mathcal{M}^e \circ \mathcal{Y}^e = \hat{\mathcal{Y}}^e \circ \hat{\mathcal{M}}^e,$$

it is straightforward to obtain

$$\begin{aligned} P_0(k)\left(\mathcal{M}^e \circ \mathcal{X}^e(u^e(k)) - \mathcal{X}^e(u^e(k)) + \mathcal{M}^e \circ \mathcal{Y}^e(y(k)) - \mathcal{Y}^e(y(k))\right) \\ = P_0(k)\alpha(k) + P_0(k)\hat{\mathcal{Y}}^e\left(r_y(k)\right), \\ \alpha(k) = u^e(k) - \mathcal{X}^e(u^e(k)) - \mathcal{Y}^e(y(k)). \end{aligned} \tag{5.94}$$

Notice that the state space representation of the last equation is

$$\begin{cases} \hat{x}(k+1) = \left(A(k) - L(k)C(k)\right)\hat{x}(k) + \left(B(k) - L(k)D(k)\right)u(k) + L(k)y(k), \\ \alpha(k) = P_0(k)F^e(k)\hat{x}(k) = P(k)\hat{x}(k). \end{cases} \tag{5.95}$$

Summarising (5.93), (5.94) and (5.95), we finally have the following theorem.

Theorem 5.5 *Given the system model (5.1)–(5.2), all LDTV observers for $P(k)x(k)$ in sense of Definition 5.10 are parametrised by*

$$\begin{aligned}\hat{x}(k+1) &= (A(k) - L(k)C(k))\,\hat{x}(k) + (B(k) - L(k)D(k))\,u(k) + L(k)y(k),\\ x_p(k) &= P(k)\hat{x}(k) + r_Q(k),\\ r_Q(k) &= \mathcal{Q}\left(r_y(k)\right), r_y(k) = W(k)\left(y(k) - C(k)\hat{x}(k) - D(k)u(k)\right)\end{aligned}$$

with $\mathcal{Q}$ as the parameterisation system.

5.6 Notes and References

Our study in this chapter on system representations, analysis and synthesis of linear discrete time-varying systems is not a straightforward extension of the results presented in the previous chapters on LTI system. It is highly motivated by recent developments in research and application domains of control and detection. In the engineering world of real-time automatic control systems, most of control systems are in their nature time-varying. For instance, SD (sampling data) [226] and MSR (multi-sampling rate) systems [225, 239] are periodic and so time varying. Industrial automatic networked control systems (NCSs) [218] with TDMA (time division multiple access) protocol can be modelled as periodic systems [227]. The so-called event-triggered NCSs [234] or switching systems, which are recently receiving intensive research attention, are time-varying as well. Also, many linear parameter-varying (LPV) systems can be handled as LDTV systems. We notice that LDTV model-based fault detection technique has been successfully applied to deal with nonlinear fault detection issues, as reported in [113, 233, 237].

On the one hand, in the state space representation form, LDTV systems seem to be a natural extension of LTI models. On the other hand, some essential definitions and concepts like stability, stabilisability and detectability, which are of elemental importance for defining stable coprime factorisation system representations, need to be re-examined. To this aim, the definitions of uniformly exponential stability, uniform stabilisability and detectability have been introduced at the beginning of this chapter. The stability definition, Definition 5.1, is given in [240], while Definition 5.2 can be found in [144, 169]. For general knowledge of linear time-varying systems, the reader is referred to the monographs [17, 78].

Concerning coprime factorisations of LDTV systems, the results presented in the state space representation form are analogue to the corresponding ones for LTI systems, in particular, the identical control theoretical interpretations like

- RCF/SIR of the plant as a state feedback control system,
- LCF/SKR of the plant as an observer-based residual generator,
- LCF/RCF of the controller as an observer-based state feedback controller.

Recall that transfer functions are, on the one hand, a special system model form exclusively for LTI systems, they build, on the other hand, the cornerstone of the coprime factorisations of LTI systems. For instance, on their basis, normalised RCF and LCF are defined. In order to deal with LDTV systems alternatively and efficiently, operator theoretical model forms have been introduced. For necessary mathematical and knowledge, we refer the reader to [17, 55, 78]. In particular, for introducing normalised coprime factorisations, the concept of an adjoint system, also called dual system in the state space representation form, has been introduced. The results given in Theorem 5.1 on normalised RCF and LCF are well established, see, for instance, [144, 169].

The study in Sect. 5.3 on normalised right and left coprime factorisations is much more comprehensive than their computation formulas given in Theorem 5.1. For our work in the sequel, mathematically and control theoretically relevant properties and concepts have been thoroughly screened. These include

- isometric and co-isometric operators, which give a definition of a normalised factorisation in the context of operator theory,
- inner and co-inner systems, which are often investigated with regard to inner-outer/co-inner-outer factorisations with the use in designing optimal controller and estimator,
- LQ regulator and LS estimator, two standard optimal state feedback controller and observer, whose solutions can be achieved by means of inner-outer and co-inner-outer factorisations, as well as
- losslessness.

Analysis and synthesis of dissipative systems are an important thematic area of control theory. An inner system can be viewed as a special type of dissipative systems with energy lossless behaviour. In a recent study [39], it is demonstrated that an optimal observer-based fault detection system is a co-inner system. In this regard, the lossless property is interpreted as lossless information transfer.

The normalised coprime factorisations, their equivalences and related properties have been well summarised in Fig. 5.1. They are useful in our subsequent works.

Parameterisations of stabilising controllers and closed-loop dynamics for LDTV systems are the centerpiece of the control and detection unified framework. Although there exists a parameterisation form analogue to the standard Youla parameterisation given in (3.5) [54, 55], the observer-based parameterisation realisation of LDTV systems stands, bearing in mind the unified framework, at the centre of our endeavor. As the result, stabilising controllers are parametrised by (5.82)–(5.85), while the closed-loop dynamics are expressed by (5.86)–(5.87).

The last issue addressed in this chapter is parameterisation of all stable LDTV observers. To our best knowledge, there exist few works on this topic. Along the lines of the study in Sect. 3.3 on the parameterisation of LTI observers, a parameterisation form for LDTV systems has been proved. It is identical with the one for LTI observers, namely, an LDTV state observer together with a feedback of the residual vector, which parametrises the observers. These results are summarised in Theorem 5.5.

Chapter 6
Unified Framework for Nonlinear Systems

The main objective of this chapter is to present the unified control and detection framework for nonlinear dynamic systems. Like our study in the previous chapters, the system coprime factorisation technique serves as the major mathematical and control theoretical tool for this work. During the decades between the 80s and 90s, the factorisation technique for nonlinear systems and its applications to system stabilisation and control have been extensively investigated, which has ended in a well-established framework. In the sequel, we introduce those concepts, definitions and results in this framework, which are closely related to the unified framework. We align ourselves mainly to the monograph by van der Schaft. A literature review on the relevant issues is given at the end of this chapter.

6.1 Preliminaries

In this section, we review necessary preliminaries focusing on model and stability issues of nonlinear systems.

6.1.1 Notations

To begin with, we review and list notations below.

- $||\cdot||$ stands for the Euclidean norm of a vector in some Euclidean space. For $f \in \mathbb{R}^p$, F denotes a p-dimensional linear space with norm $||f||$.
- $\mathbb{R}_+ = [0, \infty)$.
- Let $f : \mathbb{R}_+ \rightarrow \mathbb{R}^p$. $\mathcal{L}_2(0, \infty)$ is the space of all functions f which are measurable and satisfy

$$\int_0^\infty ||f(t)||^2 dt < \infty.$$

S. X. Ding, *Unified Framework of Control and Detection*,
https://doi.org/10.1007/978-3-662-72417-0_6

- Endowed with the inner product,

$$\text{for } f(t), g(t) \in \mathcal{L}_2(0, \infty), \langle f, g \rangle = \int_0^\infty f^T(t) g(t) dt,$$

and associated with it, the $\mathcal{L}_2$-norm

$$||f||_2 = \langle f, f \rangle^{1/2},$$

$\mathcal{L}_2(0, \infty)$ is a Hilbert space.
- Let $f : \mathbb{R}_+ \to \mathbb{R}^p$ and F be the corresponding linear space. $\mathcal{F}^s = \mathcal{L}_2(F) \in \mathcal{L}_2(0, \infty)$ denotes the signal space with the inner product

$$\text{for } f_1, f_2 \in F, \langle f_1, f_2 \rangle = \int_0^\infty f_1^T(t) f_2(t) dt,$$

and the (bounded) norm

$$||f||_2 = \langle f, f \rangle^{1/2}.$$

Here, the superscript s stands for the stable signal subset.
- Let $f : \mathbb{R}_+ \to \mathbb{R}^p$. For $\tau \in \mathbb{R}_+$, $f_\tau : \mathbb{R}_+ \to \mathbb{R}^p$ is defined by

$$f_\tau(t) = \begin{cases} f(t), 0 \leq t < \tau \\ 0, t \geq \tau. \end{cases}$$

- $\mathcal{L}_{2,[0,\tau]}$-norm of $f(t)$ is defined by

$$\|f_\tau\|_2 = \left(\int_0^\tau ||f(t)||^2 dt \right)^{1/2}.$$

6.1.2 System Models and Stability

Given input and output signal spaces $\mathcal{U}$ and $\mathcal{Y}$, $u : \mathbb{R}_+ \to \mathbb{R}^p$, $y : \mathbb{R}_+ \to \mathbb{R}^m$, respectively, the mapping (operator) $\Sigma : \mathcal{U} \to \mathcal{Y}$ defines an input-output system model

$$y = \Sigma(u). \tag{6.1}$$

A causal system is defined by

$$y_\tau = (\Sigma(u))_\tau = (\Sigma(u_\tau))_\tau.$$

A state space model of Σ is described by

$$\Sigma : \begin{cases} \dot{x} = f(x, u) \\ y = h(x, u), \end{cases} \tag{6.2}$$

where $x \in \mathbb{R}^n$ denotes the state vector, $f(x, u)$ and $h(x, u)$ are continuously differentiable nonlinear functions with appropriate dimensions. The affine form of Σ,

$$\Sigma : \begin{cases} \dot{x} = a(x) + B(x)u \\ y = c(x) + D(x)u \end{cases} \tag{6.3}$$

with $a(x)$, $B(x)$, $c(x)$ and $D(x)$ being continuously differentiable and of appropriate dimensions, is a class of nonlinear systems which are widely adopted in nonlinear system research. This class of nonlinear systems can be considered as a natural extension of LTI systems studied in the previous chapters.

Stability is a fundamental issue of control theory, and plays naturally a substantial role in the unified framework of control and detection. For our purpose, known definitions related to stability issues, which will be adopted in our subsequent work on nonlinear systems, are introduced.

Definition 6.1 Given the input-output mapping (system) $\Sigma : \mathcal{U} \to \mathcal{Y}$, Σ is said to be $\mathcal{L}_2$-stable if

$$\forall u \in \mathcal{U}^s, y = \Sigma (u) \in \mathcal{Y}^s = \mathcal{L}_2(Y).$$

Remark 6.1 In our work, only $\mathcal{L}_2$-stability is under consideration. For brevity, in the sequel, an $\mathcal{L}_2$-stable system is called stable system. And stability is understood in the sense of Definition 6.1.

Remember that $\mathcal{U}^s, \mathcal{Y}^s$ are $\mathcal{L}_2$-norm bounded signal subspaces. Hence, the stability definition given above is in fact the widely adopted concept of bounded-input and bounded-output (BIBO) stability.

As known in our study on kernel and image subspaces for LTI systems, the system input-output relation is implicitly expressed by the relations

$$r_y = \begin{bmatrix} -\hat{N} & \hat{M} \end{bmatrix} \begin{bmatrix} u \\ y \end{bmatrix}, \begin{bmatrix} u \\ y \end{bmatrix} = \begin{bmatrix} M \\ N \end{bmatrix} v.$$

For the similar purpose, relations for nonlinear systems are defined. For input-output system $\Sigma : \mathcal{U} \to \mathcal{Y}$, R is a relation if $(u, y) \in R \subset \mathcal{U} \times \mathcal{Y}$.

Definition 6.2 Relation $R \subset \mathcal{U} \times \mathcal{Y}$ is said to be $\mathcal{L}_2$-stable if

$$(u, y) \in R, u \in \mathcal{U}^s \Longrightarrow y = \Sigma (u) \in \mathcal{Y}^s.$$

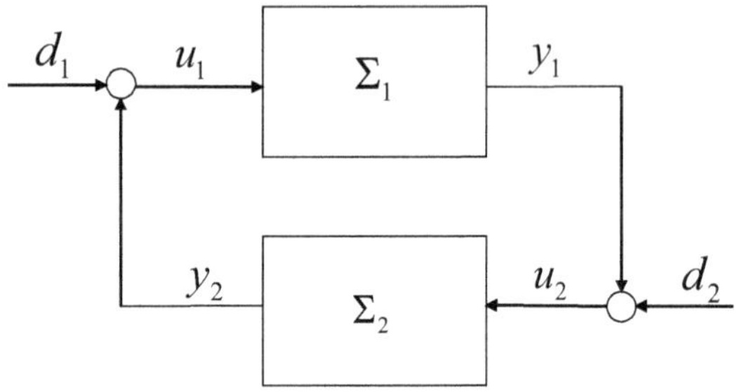

Fig. 6.1 A closed-loop system

Next, stability issues of closed-loop systems are addressed. Consider the closed-loop system depicted in Fig. 6.1, in which $u_1 \in \mathcal{U}_1, u_1 : \mathbb{R}_+ \rightarrow \mathbb{R}^p,\ y_1 \in \mathcal{Y}_1, y_1 : \mathbb{R}_+ \rightarrow \mathbb{R}^m$,

$$u_2 = y_1 + d_2 \in \mathcal{U}_2, y_2 = u_1 - d_1 \in \mathcal{Y}_2, \tag{6.4}$$

and signals $d_1 \in \mathcal{D}_1$ and $d_2 \in \mathcal{D}_2$ are external input vectors. Define

$$d = \begin{bmatrix} d_1 \\ d_2 \end{bmatrix} \in \mathcal{D},$$

and internal variable vectors

$$u = \begin{bmatrix} u_1 \\ u_2 \end{bmatrix} \in \mathcal{U}, y = \begin{bmatrix} y_1 \\ y_2 \end{bmatrix} \in \mathcal{Y}. \tag{6.5}$$

The stability definition of the above closed-loop system is analogous to the definition of internal stability of LTI systems.

Definition 6.3 Given the closed-loop system depicted in Fig. 6.1 and described by (6.4)–(6.5), the system is said to be stable if the input-output mappings (dynamics) from d to u and from d to y are stable, namely $\forall d \in \mathcal{D}^s$,

$$u \in \mathcal{U}^s, y \in \mathcal{Y}^s.$$

Note that

$$\begin{aligned} u &= \Sigma(u) + d, \Sigma(u) := \begin{bmatrix} \Sigma_2(u_2) \\ \Sigma_1(u_1) \end{bmatrix}, \\ y &= \Sigma(y + d), \Sigma(y + d) := \begin{bmatrix} \Sigma_1(d_1 + y_2) \\ \Sigma_2(d_2 + y_1) \end{bmatrix} \end{aligned}$$

define two relations. Thus, the above closed-loop system can be equivalently said to be stable, if the relations

$$R_{ud}(u,d) := \{(u,d) \in \mathcal{U} \times \mathcal{D} \,|\, u = \Sigma(u) + d\}, \tag{6.6}$$
$$R_{yd}(y,d) := \{(y,d) \in \mathcal{Y} \times \mathcal{D} \,|\, y = \Sigma(y+d)\} \tag{6.7}$$

are stable.

The following lemma is known in the literature.

Lemma 6.1 *Given the closed-loop system depicted in Fig. 6.1 with the relations defined by (6.6)–(6.7), then*

$$R_{ud}(u,d) \text{ is stable} \Longleftrightarrow R_{yd}(y,d) \text{ is stable.}$$

Concerning nonlinear systems presented in the state space model form (6.2), the following stability definition is adopted.

Definition 6.4 Given the state space model (6.2), the system is $\mathcal{L}_2$-stable if $\forall x(0) \in \mathcal{X}, u \in \mathcal{U}^s$,

$$y \in \mathcal{Y}^s.$$

6.2 System Coprime Factorisations

Like dealing with parameterisations of stabilising controllers and observers for LTI systems, coprime factorisation technique plays an essential role in addressing parameterisation issues for nonlinear systems. For the sake of simplicity, the affine form of Σ,

$$\Sigma^{x_0} : \begin{cases} \dot{x} = a(x) + B(x)u,\, x(0) = x_0 \\ y = c(x) + D(x)u, \end{cases} \tag{6.8}$$

is under consideration in our subsequent work.

Definition 6.5 Given system Σ^{x_0} defined by (6.8), a stable system $\mathcal{I}_\Sigma^{x_0} : \mathcal{V}^s \to \mathcal{U}^s \times \mathcal{Y}^s$ is an SIR of the system Σ^{x_0} if for all $u \in \mathcal{U}^s$ and $y = \Sigma(u) \in \mathcal{Y}^s$, there exists $v \in \mathcal{V}^s$ such that

$$\begin{bmatrix} u \\ y \end{bmatrix} = \mathcal{I}_\Sigma^{x_0}(v).$$

A state space representation of the SIR is given by

$$\mathcal{I}_\Sigma^{x_0} : \begin{cases} \dot{x} = a(x) + B(x)g(x) + B(x)V(x)v \\ \quad = a_I(x) + B_I(x)v \\ \begin{bmatrix} u \\ y \end{bmatrix} = c_I(x) + D_I(x)v, \end{cases} \tag{6.9}$$

$$a_I(x) = a(x) + B(x)g(x),\ B_I(x) = B(x)V(x),$$
$$c_I(x) = \begin{bmatrix} g(x) \\ c(x) + D(x)g(x) \end{bmatrix},\ D_I(x) = \begin{bmatrix} V(x) \\ D(x)V(x) \end{bmatrix},$$

where $g(x)$ is designed such that $a_I(x)$ is stable, and $V(x) \in \mathbb{R}^{p \times p}$ is invertible. In the context of feedback control systems,

$$u = g(x) + V(x)v$$

can be understood as a controller with a state feedback $g(x)$, feedforward controller $V(x)$ and v as the reference signal. Note that

$$D_I^T(x)D_I(x) = V^T(x)\left(I + D^T(x)D(x)\right)V(x)$$

is invertible.

Definition 6.6 Given system Σ^{x_0} defined by (6.8), a stable system $\mathcal{K}_\Sigma^{\hat{x}_0} : \mathcal{U}^s \times \mathcal{Y}^s \to \mathcal{R}^s$ is an SKR of the system Σ^{x_0} if for any $\hat{x}(0) = \hat{x}_0 = x_0$,

$$\mathcal{K}_\Sigma^{\hat{x}_0}\begin{pmatrix} u \\ y \end{pmatrix} = r_y = 0, r_y \in \mathbb{R}^m. \tag{6.10}$$

Remark 6.2 The state vector x is often not measurable. In that case, $\hat{x}(0)$ may be different from $x(0)$ and thus cause $r_y \neq 0$. $x(0) - \hat{x}(0) \neq 0$ will be handled as uncertainty.

The SKR (6.10) can be realised in a state space representation by

$$\mathcal{K}_\Sigma^{\hat{x}_0} : \begin{cases} \dot{\hat{x}} = a(\hat{x}) + B(\hat{x})u + L(\hat{x})\left(y - \hat{y}\right) \\ = a_K(\hat{x}) + B_K(\hat{x})\begin{bmatrix} u \\ y \end{bmatrix} \\ r_y = W(\hat{x})r_0 = c_K(\hat{x}) + D_K(\hat{x})\begin{bmatrix} u \\ y \end{bmatrix} \\ r_0 = y - \hat{y}, \hat{y} = c\left(\hat{x}\right) + D\left(\hat{x}\right)u, \end{cases} \tag{6.11}$$
$$a_K(\hat{x}) = a(\hat{x}) - L(\hat{x})c\left(\hat{x}\right), B_K(\hat{x}) = \left[\, B(\hat{x}) - L(\hat{x})D\left(\hat{x}\right)\ L(\hat{x}) \,\right],$$
$$c_K(\hat{x}) = -W(\hat{x})c\left(\hat{x}\right), D_K(\hat{x}) = \left[\, -W(\hat{x})D\left(\hat{x}\right)\ W(\hat{x}) \,\right],$$

where $L(\hat{x})$ is designed such that $a_K(\hat{x})$ is stable and $W(\hat{x}) \in \mathbb{R}^{m \times m}$ is invertible. In the detection context, r_0 builds a residual vector, W is a post-filter, and it holds

$$D_K^T(\hat{x})D_K(\hat{x}) = W(\hat{x})\left(I + D(\hat{x})D^T(\hat{x})\right)W^T(\hat{x})$$

being invertible. Like factorisations of LTI systems, the state equation in (6.11),

$$\dot{\hat{x}} = a(\hat{x}) + B(\hat{x})u + L(\hat{x})\left(y - \hat{y}\right), \tag{6.12}$$

plays a central role. It is a state observer.

It is obvious that for any $v \in \mathcal{V}^s$, $\hat{x}_0 = x_0$,

$$\mathcal{K}_{\Sigma}^{\hat{x}_0} \circ \mathcal{I}_{\Sigma}^{x_0}(v) = 0 \Longleftrightarrow \mathcal{K}_{\Sigma}^{\hat{x}_0} \circ \mathcal{I}_{\Sigma}^{x_0} = 0. \tag{6.13}$$

Now, coprime factorisations of Σ^{x_0} are introduced, which are, as we know from the framework for LTI systems, associated with feedback controllers and residual generators.

Definition 6.7 The SKR $\mathcal{K}_{\Sigma}^{\hat{x}_0}$ is said to be coprime if it has a stable right inverse $\mathcal{K}_{\Sigma}^{-} : \mathcal{R}^s \rightarrow \mathcal{U}^s \times \mathcal{Y}^s$ satisfying

$$\mathcal{K}_{\Sigma}^{\hat{x}_0} \circ \mathcal{K}_{\Sigma}^{-} = I. \tag{6.14}$$

Analogous to it, the SIR $\mathcal{I}_{\Sigma}^{x_0}$ is said to be coprime if it has a stable left inverse $\mathcal{I}_{\Sigma}^{-} : \mathcal{U}^s \times \mathcal{Y}^s \rightarrow \mathcal{V}^s$ satisfying

$$\mathcal{I}_{\Sigma}^{-} \circ \mathcal{I}_{\Sigma}^{x_0} = I. \tag{6.15}$$

Consistent with the LCF and RCF of linear systems, the state space realisation forms of $\mathcal{K}_{\Sigma}^{-}$ and $\mathcal{I}_{\Sigma}^{-}$ are now addressed. Substituting u in $\mathcal{K}_{\Sigma}^{\hat{x}_0}$ (6.11) by

$$u = g(\hat{x})$$

and y by

$$y = r_0 + \hat{y} = W^{-1}\left(\hat{x}\right) r_y + c\left(\hat{x}\right) + D(\hat{x})g(\hat{x})$$

leads to

$$\begin{cases} \dot{\hat{x}} = a_I(\hat{x}) + L(\hat{x})W^{-1}\left(\hat{x}\right) r_y, \\ \begin{bmatrix} u \\ y \end{bmatrix} = c_I(\hat{x}) + \begin{bmatrix} 0 \\ W^{-1}\left(\hat{x}\right) \end{bmatrix} r_y \\ r_y = W(\hat{x}) r_0 = c_K(\hat{x}) + D_K(\hat{x}) \begin{bmatrix} u \\ y \end{bmatrix}. \end{cases} \tag{6.16}$$

By defining

$$\begin{bmatrix} u \\ y \end{bmatrix} = \mathcal{K}_{\Sigma}^{-}\left(r_y\right) =: \begin{cases} \dot{\hat{x}} = a_I(\hat{x}) + L(\hat{x})W^{-1}\left(\hat{x}\right) r_y, \\ \begin{bmatrix} u \\ y \end{bmatrix} = c_I(\hat{x}) + \begin{bmatrix} 0 \\ W^{-1}\left(\hat{x}\right) \end{bmatrix} r_y, \end{cases} \tag{6.17}$$

system (6.16) is identical with

$$r_y = \mathcal{K}_{\Sigma}^{\hat{x}_0} \circ \mathcal{K}_{\Sigma}^{-}\left(r_y\right),$$

which holds for all $r_y \in \mathcal{R}^s$. Consequently, we have (6.14). Concerning Σ_I^-, let

$$u = g(\hat{x}) + V(\hat{x})v$$

and define, on account of the state observer (6.12),

$$\mathcal{I}_\Sigma^- := \begin{cases} \dot{\hat{x}} = a_K(\hat{x}) + B_K(\hat{x}) \begin{bmatrix} u \\ y \end{bmatrix} \\ v = V^{-1}(\hat{x}) \left(u - g(\hat{x})\right) = -V^{-1}(\hat{x})g(\hat{x}) + \left[\, V^{-1}(\hat{x}) \ 0 \,\right] \begin{bmatrix} u \\ y \end{bmatrix}. \end{cases} \tag{6.18}$$

Since for all $\hat{x}_0 = x_0$,

$$\mathcal{I}_\Sigma^- \circ \mathcal{I}_\Sigma^{x_0}(v) = \mathcal{I}_\Sigma^- \begin{pmatrix} u \\ y \end{pmatrix} = v,$$

which proves that (6.15) is true. Notice that the state equations of $\mathcal{K}_\Sigma^-$ and $\mathcal{I}_\Sigma^-$ given in (6.17) and (6.18) describe the same dynamic of a state observer. To a certain sense, $\mathcal{K}_\Sigma^-$ and $\mathcal{I}_\Sigma^-$ are related to an observer-based state feedback controller. Now, set $v = 0$, that is $u = g(\hat{x})$. It turns out that for all $r_y \in \mathcal{R}^s$,

$$\mathcal{I}_\Sigma^- \circ \mathcal{K}_\Sigma^-(r_y) = \mathcal{I}_\Sigma^- \begin{pmatrix} u \\ y \end{pmatrix} = 0.$$

As a result,

$$\mathcal{I}_\Sigma^- \circ \mathcal{K}_\Sigma^- = 0. \tag{6.19}$$

Identities (6.13)–(6.15) and (6.19), written in the following compact form

$$\begin{bmatrix} \mathcal{I}_\Sigma^- \circ \mathcal{I}_\Sigma^{x_0} & \mathcal{K}_\Sigma^{\hat{x}_0} \circ \mathcal{I}_\Sigma^{x_0} \\ \mathcal{I}_\Sigma^- \circ \mathcal{K}_\Sigma^- & \mathcal{K}_\Sigma^{\hat{x}_0} \circ \mathcal{K}_\Sigma^- \end{bmatrix} = \begin{bmatrix} I & 0 \\ 0 & I \end{bmatrix}, \tag{6.20}$$

seem to be similar to the Bezout identity. Unfortunately, it cannot be directly applied to the purpose of parameterising stabilising controllers. This is illustrated by the following observation. In the signal spaces, (6.20) leads to the following relation

$$\begin{bmatrix} v \\ r_y \end{bmatrix} = \begin{bmatrix} \mathcal{I}_\Sigma^- \begin{pmatrix} u \\ y \end{pmatrix} \\ \mathcal{K}_\Sigma^{\hat{x}_0} \begin{pmatrix} u \\ y \end{pmatrix} \end{bmatrix} =: \mathcal{K}_{\Sigma,C} \begin{pmatrix} u \\ y \end{pmatrix}. \tag{6.21}$$

On the other hand, the identity (6.20) does not imply the existence of an inverse mapping from (u, y) to (v, r_y), which is the core of proving the parameterisation of all stabilising controllers.

We end this section with the definition of image subspace.

Definition 6.8 Given system Σ, its SIR Σ_I, and $u \in \mathcal{U}^s$, $y \in \mathcal{Y}^s$,

$$\mathcal{I}_{\mathcal{I}_\Sigma} = \left\{ \begin{bmatrix} u \\ y \end{bmatrix} : \begin{bmatrix} u \\ y \end{bmatrix} = \Sigma_I(v), v \in \mathcal{V}^s \right\} \tag{6.22}$$

is called image subspace of Σ.

6.3 Hamiltonian and Inner Systems, and Normalised Coprime Factorisations

Due to its important role, we will address the issue of normalised coprime factorisations in this separate section. Recall that for LTI systems, the normalised coprime factorisations are defined in terms of SIR and SKR and their conjugates in the form of transfer functions, while for LTV systems the dual systems or adjoint operators of SIR and SKR are used for defining and computing the normalised coprime factorisations. Moreover, it has been showcased that a normalised SIR is an inner system. For nonlinear systems, the well-established Hamiltonian system technique can be applied to serve this purpose. In a broad sense, given a nonlinear system as an operator, the associated Hamiltonian system can be interpreted as a certain configuration of the operator and its adjoint. In this regard, we first introduce some essential concepts and definitions known in the literature.

6.3.1 *Hamiltonian Systems*

Consider the affine system (6.8) and assume that it is stable. The Hamiltonian extension of Σ is a dynamic system described by

$$\begin{cases} \dot{x} = a(x) + B(x)u \\ \dot{\lambda} = -\left(\frac{\partial a(x)}{\partial x} + \frac{\partial B(x)}{\partial x}u\right)^T \lambda - \left(\frac{\partial c(x)}{\partial x} + \frac{\partial D(x)}{\partial x}u\right)^T u_a \\ y = c(x) + D(x)u \\ y_a = B^T(x)\lambda + D^T(x)u_a, \; y_a \in \mathbb{R}^p, \; u_a \in \mathbb{R}^m, \end{cases} \tag{6.23}$$

with state variables (x, λ), input variables (u, u_a) and output variables (y, y_a). $\lambda \in \mathbb{R}^n$ is also called co-state variable. Connecting y and u_a,

$$u_a = y = c(x) + D(x)u,$$

and defining a Hamiltonian function

$$H\left(x,\lambda,u\right)=\frac{1}{2}\left(c(x)+D(x)u\right)^{T}\left(c(x)+D(x)u\right)+\lambda^{T}\left(a(x)+B(x)u\right)$$

lead to the following Hamiltonian system

$$y_{a}=\left(D\Sigma\right)^{T}\circ\Sigma\left(u\right),$$

whose state space representation can be written in a compact form

$$\left(D\Sigma\right)^{T}\circ\Sigma:\begin{cases}\dot{x}=\frac{\partial H}{\partial\lambda}\left(x,\lambda,u\right)\\ \dot{\lambda}=-\frac{\partial H}{\partial x}\left(x,\lambda,u\right)\\ y_{a}=\frac{\partial H}{\partial u}\left(x,\lambda,u\right)\end{cases}\tag{6.24}$$

with input vector u and output vector y_a.

Differently, if u and y_a of the Hamiltonian extension (6.23) are connected, namely

$$u=y_{a}=B^{T}(x)\lambda+D^{T}(x)u_{a},\tag{6.25}$$

(6.23) becomes

$$\Sigma\circ\left(D\Sigma\right)^{T}:\begin{cases}\dot{x}=a(x)+B(x)\left(B^{T}(x)\lambda+D^{T}(x)u_{a}\right)\\ \dot{\lambda}=-\left(\frac{\partial a(x)}{\partial x}+\frac{\partial\bar{B}(x,\lambda,u_{a})}{\partial x}\right)^{T}\lambda-\left(\frac{\partial c(x)}{\partial x}+\frac{\partial\bar{D}(x,\lambda,u_{a})}{\partial x}\right)^{T}u_{a}\\ y=c(x)+D(x)B^{T}(x)\lambda+D(x)D^{T}(x)u_{a},\end{cases}$$

$$\begin{aligned}\bar{B}(x,\lambda,u_{a})&=B(x)\left(B^{T}(x)\lambda+D^{T}(x)u_{a}\right),\\ \bar{D}(x,\lambda,u_{a})&=D(x)B^{T}(x)\lambda+D(x)D^{T}(x)u_{a}.\end{aligned}\tag{6.26}$$

Let

$$\begin{aligned}H\left(x,\lambda,u_{a}\right)=&\frac{1}{2}u_{a}^{T}D(x)D^{T}(x)u_{a}+c^{T}(x)u_{a}\\ &+\lambda^{T}\left(a(x)+\frac{1}{2}B(x)B^{T}(x)\lambda+B(x)D^{T}(x)u_{a}\right)\end{aligned}\tag{6.27}$$

be a Hamiltonian function. Accordingly, $\Sigma\circ\left(D\Sigma\right)^{T}$ can be compactly written as

$$\Sigma\circ\left(D\Sigma\right)^{T}:\begin{cases}\dot{x}=\frac{\partial H}{\partial\lambda}\left(x,\lambda,u_{a}\right)\\ \dot{\lambda}=-\frac{\partial H}{\partial x}\left(x,\lambda,u_{a}\right)\\ y=\frac{\partial H}{\partial u_{a}}\left(x,\lambda,u_{a}\right).\end{cases}\tag{6.28}$$

6.3.2 Inner and Co-inner

In Sect. 2.2, the concepts of inner and co-inner have been introduced for LTI systems. It has been showcased that normalised SIR and SKR are inner and co-inner, respectively. Before we define and address inner and co-inner issues for nonlinear systems, they are re-visited firstly for LTI systems, aiming at extending the definitions of inner and co-inner and highlighting some insightful interpretations.

To begin with, the LTI system modelled by

$$y_a(s) = G^{\sim}(s)G(s)u(s)$$

with the minimal state space representation $G = (A, B, C, D)$, is recast in the form of a Hamiltonian system,

$$G^{\sim}G : \begin{cases} \dot{x} = Ax + Bu \\ \dot{\lambda} = -A^T\lambda - C^TCx - C^TDu \\ y_a = D^TCx + B^T\lambda + D^TDu, \end{cases} \tag{6.29}$$

which can be, for Hamiltonian function

$$H(x, \lambda, u) = \frac{1}{2}(Cx + Du)^T (Cx + Du) + \lambda^T (Ax + Bu), \tag{6.30}$$

written into the compact form (6.24). For our purpose, the following lemma known in the literature is introduced.

Lemma 6.2 *Given $G \in \mathcal{RH}_\infty$ with the minimal state space representation (A, B, C, D), then G is an inner if and only if*

$$D^TC + B^TX = 0, D^TD = I.$$

Here, $X > 0$ is the solution of Lyapunov function

$$A^TX + XA + C^TC = 0,$$

Note that

$$\lambda = Xx$$

solves dynamic system equation (6.29). Now, define a positive definite function

$$V(x) = \frac{1}{2}x^TXx,$$

which can be interpreted as an (energy) storage function. Hence,

$$\lambda = \frac{\partial V(x)}{\partial x}, \lambda^T (Ax + Bu) = \left(\frac{\partial V(x)}{\partial x}\right)^T \dot{x} = \dot{V}(x).$$

Consequently, it turns out

$$H(x, \lambda, u) = \frac{1}{2} y^T y + \dot{V}(x) = x^T \left(\frac{1}{2} C^T C + XA\right) x + \frac{1}{2} u^T u.$$

Since

$$x^T \left(A^T X + XA + C^T C\right) x = 2x^T \left(\frac{1}{2} C^T C + XA\right) x = 0,$$

we finally have

$$H(x, \lambda, u) = \frac{1}{2} y^T y + \dot{V}(x) = \frac{1}{2} u^T u. \tag{6.31}$$

In the paradigm of dissipative systems, systems with property (6.31) or equivalently,

$$V(x(t_2)) - V(x(t_1)) = \int_{t_1}^{t_2} \left(\frac{1}{2} u^T u - \frac{1}{2} y^T y\right) d\tau, \tag{6.32}$$

are called lossless with respect to the supply rate

$$s(u, y) = \frac{1}{2} u^T u - \frac{1}{2} y^T y.$$

In this context, an inner Hamiltonian system, as given by (6.24), is defined as follows.

Definition 6.9 Given the stable nonlinear affine system (6.8) and the corresponding Hamiltonian system (6.24), Σ is inner if

$$y_a = u, \tag{6.33}$$

and there exists a storage function $V(x) \geq 0$, $V(0) = 0$ so that

$$V(x(t_2)) - V(x(t_1)) = \int_{t_1}^{t_2} \left(\frac{1}{2} u^T u - \frac{1}{2} y^T y\right) d\tau. \tag{6.34}$$

The following lemma is a known result and provides us with the existence conditions for a nonlinear inner system.

Lemma 6.3 *The nonlinear affine system (6.8) is inner, if there exists $V(x) \geq 0$ such that the following equations are feasible*

$$\begin{aligned} &V_x^T(x)\, a(x) + \frac{1}{2} c^T(x) c(x) = 0, \\ &V_x^T(x)\, B(x) + c^T(x) D(x) = 0, \\ &D^T(x) D(x) = I, \end{aligned}$$

where $V_x(x) = \frac{\partial V}{\partial x}(x)$. Moreover,

$$\lambda = V_x(x). \tag{6.35}$$

Next, issues related to co-inner are addressed. Analogous to Definition 6.9, we first introduce the definition of a co-inner.

Definition 6.10 Given the stable nonlinear affine system (6.8), Σ is co-inner if

$$y = u_a,$$

and there exists a storage function $V(x) \geq 0$, $V(0) = 0$ so that

$$V(x(t_2)) - V(x(t_1)) = \frac{1}{2} \int_{t_1}^{t_2} \left(u^T u - y^T y \right) d\tau.$$

Similar to our previous discussion, to pose insightful aspects of co-inner systems, we begin with LTI co-inner systems and consider $G(s)G^{\sim}(s)$. The corresponding Hamiltonian system of

$$y(s) = G(s)G^{\sim}(s) u_a(s)$$

is modelled by

$$GG^{\sim} : \begin{cases} \dot{x} = Ax + BB^T \lambda + BD^T u_a \\ \dot{\lambda} = -A^T \lambda - C^T C x - C^T u_a \\ y = Cx + DB^T \lambda + DD^T u_a, \end{cases} \tag{6.36}$$

which is, given the Hamiltonian function,

$$H(x, \lambda, u_a) = \frac{1}{2} u_a^T DD^T u_a + x^T C^T u_a + \lambda^T \left(Ax + \frac{1}{2} BB^T \lambda + BD^T u_a \right),$$

further written into (6.28). As the dual result of Lemma 6.2, it can be claimed that G is a co-inner if and only if

$$DB^T + CY = 0, \; DD^T = I, \tag{6.37}$$

where $Y > 0$ is the solution of Lyapunov function

$$AY + YA^T + BB^T = 0.$$

It is straightforward from (6.36) that

$$\lambda = Y^{-1}x,\ y = u_a.$$

Denoting by

$$V(x) = x^T Y^{-1} x > 0$$

a storage function, it follows from the Hamiltonian function

$$\begin{aligned} H(x, \lambda, u_a) &= \frac{1}{2}u_a^T DD^T u_a + x^T C^T u_a - \frac{1}{2}\lambda^T BB^T \lambda + V_x^T(x)\dot{x} \\ &= \frac{1}{2}u_a^T DD^T u_a - \lambda^T \left(BD^T u_a + \frac{1}{2}BB^T \lambda \right) + \dot{V}(x) \\ &= \frac{1}{2}u_a^T DD^T u_a = \frac{1}{2}y^T y, \end{aligned}$$

which, on account of the relation

$$u = y_a = B^T \lambda + D^T u_a,$$

can be written as

$$H(x, \lambda, u_a) = u_a^T DD^T u_a - \frac{1}{2}u^T u + \dot{V}(x) = \frac{1}{2}y^T y.$$

Consequently, the lossless property,

$$V(x(t_2)) - V(x(t_1)) = \frac{1}{2}\int_{t_1}^{t_2} \left(u^T u - \frac{1}{2}y^T y \right) d\tau,$$

is verified.

The existence conditions for a nonlinear co-inner system are summarised in the following lemma. Although the proof of these results can be found in the literature (refer to the references given at the end of this chapter), a slightly different proof is provided here with a focus on the lossless property.

Lemma 6.4 *The nonlinear affine system (6.8) is co-inner, if there exists $V(x) \geq 0$ such that the following equations are feasible*

$$V_x^T(x)\,a(x) + \frac{1}{2}V_x^T(x)\,B(x)B^T(x)V_x(x) = 0, \tag{6.38}$$

$$c^T(x) + V_x^T(x)\,B(x)D^T(x) = 0, \tag{6.39}$$

$$D(x)D^T(x) = I. \tag{6.40}$$

Furthermore,

$$\lambda = V_x(x)\,. \tag{6.41}$$

Proof For our purpose, a canonical transformation by means of the generating function approach is realised. To be specific, a so-called type 2 generating function,

$$G_2\left(x, \bar{\lambda}\right) = V(x) + \bar{\lambda}^T x,$$

is applied, which leads to

$$\lambda = V_x(x) + \bar{\lambda}, \bar{x} = x.$$

Here, $\bar{\lambda}$ and $\bar{x}$ are the transformed coordinates. Since $\frac{\partial G_2}{\partial t} = 0,$ the transformed Hamiltonian function

$$\bar{H}\left(x, \bar{\lambda}, u_a\right) = \frac{1}{2}u_a^T D(x)D^T(x)u_a + c^T(x)u_a$$
$$+ \left(V_x(x) + \bar{\lambda}\right)^T\left(a(x) + \frac{1}{2}B(x)B^T(x)\left(V_x(x) + \bar{\lambda}\right) + B(x)D^T(x)u_a\right)$$

is under consideration. Observe that, on account of (6.25) and (6.39), for $\bar{\lambda} = 0$

$$\bar{H}(x, 0, u_a) = \frac{1}{2}u_a^T u_a = u_a^T u_a - \frac{1}{2}u^T u + \dot{V}(x), \tag{6.42}$$

$$y = c(x) + D(x)B^T(x)V_x(x) + D(x)D^T(x)u_a \Longrightarrow y = u_a, \tag{6.43}$$

which implies

$$\bar{H}(x, 0, u_a) = \hat{H}(x, V_x(x), u_a) = \frac{1}{2}u_a^T u_a$$

is independent of the canonical coordinates. Accordingly, $\lambda = V_x(x)$ and x are the solutions of the differential equations in (6.23). Together with (6.42)–(6.43), the conclusion that system (6.8) is co-inner has been proved.

Remark 6.3 Canonical transformation and generating function approach are well-established methods in Hamiltonian mechanics. The reader is referred to the literature cited at the end of this chapter for a detailed introduction.

6.3.3 Normalised Right Coprime Factorisation

We are now in a position to introduce normalised coprime factorisations, and begin with normalised RCF/SIR. Motivated by the result that, for LTI systems, the normalised SIR is an inner, the definition of normalised SIR is firstly introduced.

Definition 6.11 The SIR (6.9) of the affine system (6.8) is said to be normalised, if it is inner.

Next, we study the conditions for an SIR to be inner. Consider SIR (6.9) of the affine system (6.8). The Hamiltonian extension of $\mathcal{I}_\Sigma$ is the dynamic system described by

$$\begin{cases} \dot{x} = a_I(x) + B_I(x)v \\ \dot{\lambda} = -\left(\frac{\partial a_I(x)}{\partial x} + \frac{\partial B_I(x)}{\partial x}v\right)^T \lambda - \left(\frac{\partial c_I(x)}{\partial x} + \frac{\partial D_I(x)}{\partial x}\right)^T v_a \\ z := \begin{bmatrix} u \\ y \end{bmatrix} = c_I(x) + D_I(x)v \\ z_a = B_I^T(x)\lambda + D_I^T(x)v_a, z_a \in \mathbb{R}^p, v_a \in \mathbb{R}^{m+p}, \end{cases}$$

where (v, v_a) and (z, z_a) are input and output vectors, respectively. Define the Hamiltonian function

$$H(x, \lambda, v) = \frac{1}{2}(c_I(x) + D_I(x)v)^T (c_I(x) + D_I(x)v) + \lambda^T (a_I(x) + B_I(x)v),$$

and connect z and v_a,

$$v_a = z = c_I(x) + D_I(x)v.$$

We have the following Hamiltonian system

$$z_a = (D\Sigma_\mathcal{I})^T \circ \Sigma_\mathcal{I}(v),$$

whose state space representation can be written in the compact form

$$(D\Sigma_\mathcal{I})^T \circ \Sigma_\mathcal{I} : \begin{cases} \dot{x} = \frac{\partial H}{\partial p}(x, p, v) \\ \dot{p} = -\frac{\partial H}{\partial x}(x, p, v) \\ z_a = \frac{\partial H}{\partial v}(x, p, v) \end{cases} \tag{6.44}$$

with input vector v and output vector z_a.

The existence conditions for the SIR (6.9) to be inner and thus normalised are summarised in the following theorem.

Theorem 6.1 *The nonlinear affine system (6.9) is inner, if*

$$g(x) = -\left(I + D^T(x)D(x)\right)^{-1}\left(B^T(x)P_x(x) + D^T(x)c(x)\right), \tag{6.45}$$

$$V(x) = V^T(x) = \left(I + D^T(x)D(x)\right)^{-1/2}, \tag{6.46}$$

and there exists $P(x) \geq 0$ such that the following equation is feasible

$$P_x^T(x)a(x) + \frac{1}{2}c^T(x)c(x) - \frac{1}{2}g^T(x)\left(I + D^T(x)D(x)\right)g(x) = 0, \tag{6.47}$$

where $P_x(x) = \frac{\partial P}{\partial x}(x)$.

Proof It follows from Lemma 6.3 that (6.9) is inner if the following three equations are solved by finding $g(x)$, $V(x)$ and $P(x)$:

$$P_x^T(x)a_I(x) + \frac{1}{2}c_I^T(x)c_I(x) = 0, \tag{6.48}$$

$$P_x^T(x)B_I(x) + c_I^T(x)D_I(x) = 0, \tag{6.49}$$

$$D_I^T(x)D_I(x) = I. \tag{6.50}$$

It is straightforward that $g(x)$ satisfying (6.45) and $V(x)$ satisfying (6.46) solve (6.49) and (6.50), respectively. Substituting them into (6.48) leads to

$$\begin{aligned}
&P_x^T(x)a(x) + \left(P_x^T(x)B(x) + c^T(x)D(x)\right)g(x) + \\
&\frac{1}{2}c^T(x)c(x) + \frac{1}{2}g^T(x)\left(I + D^T(x)D(x)\right)g(x) = \\
P_x^T(x)a(x) + \frac{1}{2}&c^T(x)c(x) - \frac{1}{2}g^T(x)\left(I + D^T(x)D(x)\right)g(x) = 0.
\end{aligned}$$

Hence, the theorem is proved.

Theorem 6.1 illustrates that the SIR (6.9) is an inner, when $g(x)$ and $V(x)$ are set according to (6.45) and (6.46), respectively. For this purpose, Hamilton-Jacobi equation (HJE) (6.47) is to be solved.

In the sequel, properties of the normalised (inner) SIR are investigated. To begin with, we would like to recall the definition of normalised SIR of LTI systems and some associated properties. An SIR,

$$I_G = \begin{bmatrix} M_0 \\ N_0 \end{bmatrix}, z := \begin{bmatrix} u \\ y \end{bmatrix} = I_G v,$$

is normalised, if

$$
\begin{aligned}
I_G^{\sim} I_G &= \begin{bmatrix} M_0^{\sim} & N_0^{\sim} \end{bmatrix} \begin{bmatrix} M_0 \\ N_0 \end{bmatrix} = I \Longrightarrow \\
z_a &:= I_G^{\sim} I_G v = v.
\end{aligned}
\tag{6.51}
$$

Moreover,

$$
\|z\|_2 = \|I_G v\|_2 = \|v\|_2 . \tag{6.52}
$$

It is apparent that if SIR (6.9) is inner, then it is of the properties given in (6.51) and (6.52) as well,

$$
\begin{aligned}
z_a &= (D\Sigma_{\mathcal{I}})^T \circ \Sigma_{\mathcal{I}}(v) = v, \\
\|z\|_2 &= \left\| \begin{bmatrix} u \\ y \end{bmatrix} \right\|_2 = \|\mathcal{I}_{\Sigma}(v)\|_2 = \|v\|_2 .
\end{aligned}
$$

Also, it is of interest to note the following comparison with the normalised SIR for LTI systems. Given a normalised SIR,

$$
\begin{gathered}
\begin{cases}
\dot{x} = A_{F_0} x + B V_0 v \\
\begin{bmatrix} u \\ y \end{bmatrix} = \begin{bmatrix} F_0 \\ C_{F_0} \end{bmatrix} x + \begin{bmatrix} V_0 \\ D V_0 \end{bmatrix} v,
\end{cases} \\
F_0 = -\left(I + D^T D\right)^{-1} \left(D^T C + B^T P\right), V_0 = \left(I + D^T D\right)^{-1/2}, \\
A_{F_0} = A + B F_0, C_{F_0} = C + D F_0,
\end{gathered}
$$

where $P > 0$ solves the following Riccati equation,

$$
A^T P + P A + C^T C - \left(D^T C + B^T P\right)^T R^{-1} \left(D^T C + B^T P\right) = 0,
$$

let

$$
\begin{aligned}
H(x, v) &= \dot{P}(x) + \frac{1}{2} \left(\begin{bmatrix} F_0 \\ C_{F_0} \end{bmatrix} x + \begin{bmatrix} V_0 \\ D V_0 \end{bmatrix} v \right)^T \left(\begin{bmatrix} F_0 \\ C_{F_0} \end{bmatrix} x + \begin{bmatrix} V_0 \\ D V_0 \end{bmatrix} v \right), \\
P(x) &:= \frac{x^T P x}{2}.
\end{aligned}
$$

It turns out

$$
H(x, v) = \dot{P}(x) + \frac{1}{2} v^T v.
$$

Comparing this result with the one for the inner $\mathcal{I}_{\Sigma}$ given in Theorem 6.1 apparently shows that

$$\lambda = P_x(x) = Px, \lambda^T (a_I(x) + B_I(x)v) = P_x(x)\dot{x} = \dot{P}(x) \Longrightarrow$$

$$H(x, \lambda, v) = \frac{1}{2}\left(\begin{array}{c} \lambda^T (a_I(x) + B_I(x)v) + \left(\begin{bmatrix} F_0 \\ C_{F_0} \end{bmatrix} x + \begin{bmatrix} V_0 \\ DV_0 \end{bmatrix} v \right)^T \\ \left(\begin{bmatrix} F_0 \\ C_{F_0} \end{bmatrix} x + \begin{bmatrix} V_0 \\ DV_0 \end{bmatrix} v \right) \end{array} \right)$$

$$= H(x, v),$$

when the affine system (6.8) is LTI. The analogue results for the nonlinear affine system (6.8) are, according to Lemma 6.3 and its proof,

$$\lambda = P_x(x), H(x, \lambda, v) = \dot{P}(x) + \frac{1}{2} v^T v.$$

Thus, in view of the system theoretical interpretation of $P(x)$ as an energy storage function, the co-state variable $\lambda = P_x(x)$ is an indicator for the variation of the storage energy with respect to the state variable x.

Remember that an inner-outer factorisation is related to the LQ optimal control scheme. For LTI systems, the solution of the LQ optimisation problem,

$$\min_u J(t) = \min_u \frac{1}{2} \int_t^\infty y^T(\tau) y(\tau) d\tau$$

$$\text{s.t. } \dot{x} = Ax + Bu, y = Cx + Du,$$

is given by

$$\min_u J(t) = \frac{1}{2} x^T(t) Px(t),$$

$$u(t) = Fx(t) = F_\lambda \lambda(t),$$

$$F = -R^{-1}\left(B^T P + D^T C\right), R = D^T D > 0, F_\lambda = FP^{-1},$$

$$A^T P + PA + C^T C - F^T RF = 0, P > 0.$$

Analogous to it, the existence conditions given in Theorem 6.1 can be interpreted as the solution of the following optimal control problem,

$$\min_u J(t) = \min_u \frac{1}{2} \int_t^\infty y^T(\tau) y(\tau) d\tau \tag{6.53}$$

$$\text{s.t. } \begin{cases} \dot{x} = a(x) + B(x)u \\ y = c(x) + D(x)u, \end{cases}$$

namely,

$$u = g(x),$$
$$\min_u \frac{1}{2} \int_t^{\infty} y^T(\tau)\, y(\tau) d\tau = P(x(t)).$$

Again, notice that

$$\lambda = P_x(x),$$

that is, the co-state variable λ is an indicator for the variation of the cost function with respect to the state variable x. In this context and observing

$$u = g(x) = -V^2(x) \left(B^T(x)\lambda + D^T(x)c(x) \right),$$

an interesting conclusion that the controller is driven by the co-state variable λ (and thus variation of the cost function) can be drawn.

6.3.4 Normalised Left Coprime Factorisation

We now consider issues of normalised SKR. Like handling issues of normalised SIR, we begin with the Hamiltonian extension of $\mathcal{K}_\Sigma$,

$$\begin{cases} \dot{\hat{x}} = a_K(\hat{x}) + B_K(\hat{x})z, \; z := \begin{bmatrix} u \\ y \end{bmatrix} \\ \dot{\lambda} = -\left(\frac{\partial a_K(\hat{x})}{\partial \hat{x}} + \frac{\partial B_K(\hat{x})}{\partial \hat{x}} z \right)^T \lambda - \left(\frac{\partial c_K(\hat{x})}{\partial \hat{x}} + \frac{\partial D_K(\hat{x})}{\partial \hat{x}} \right)^T z_a \\ r_y = c_K(\hat{x}) + D_K(\hat{x})z \\ r_{y,a} = B_K^T(\hat{x})\lambda + D_K^T(\hat{x})z_a, \; z_a \in \mathbb{R}^m, r_{y,a} \in \mathbb{R}^{m+p}. \end{cases} \tag{6.54}$$

Connecting z and $r_{y,a}$ builds

$$\Sigma_{\mathcal{K}} \circ (D\Sigma_{\mathcal{K}})^T : \begin{cases} \dot{\hat{x}} = a_K(\hat{x}) + \bar{B}(\hat{x}, \lambda, z_a) \\ \dot{\lambda} = -\left(\frac{\partial a_K(\hat{x})}{\partial \hat{x}} + \frac{\partial \bar{B}(\hat{x},\lambda,z_a)}{\partial \hat{x}} \right)^T \lambda - \left(\frac{\partial c_K(\hat{x})}{\partial \hat{x}} + \frac{\partial \bar{D}(\hat{x},\lambda,z_a)}{\partial \hat{x}} \right)^T z_a \\ r_y = c_K(\hat{x}) + \bar{D}(\hat{x}, \lambda, z_a), \end{cases} \tag{6.55}$$

$$\bar{B}(\hat{x}, \lambda, z_a) = B_K(\hat{x}) \left(B_K^T(\hat{x})\lambda + D_K^T(\hat{x})z_a \right),$$
$$\bar{D}(\hat{x}, \lambda, z_a) = D_K(\hat{x})B_K^T(\hat{x})\lambda + D_K(\hat{x})D_K^T(\hat{x})z_a.$$

Notice that the residual variable r_y contains exclusively uncertainties in the system under consideration. During nominal system operations, r_y is zero and independent of (u, y). So far, a direct adoption of the co-inner definition to characterise SKR is wrongly at the place. Another concern is missing control theoretic interpretations for SKR as a co-inner. These facts prompt us to recast the problem formulation for a normalised SKR.

Let us denote by (u_0, y_0) those process data that are generated during nominal system operations, that is,

$$\begin{bmatrix} u_0 \\ y_0 \end{bmatrix} = \mathcal{I}_\Sigma(v), \Sigma_\mathcal{K} \begin{bmatrix} u_0 \\ y_0 \end{bmatrix} = 0.$$

Assume that $\Sigma_\mathcal{K}$ is co-inner, and thus $r_y = \Sigma_\mathcal{K} \circ (D\Sigma_\mathcal{K})^T (z_a) = z_a$. A conclusion can be immediately drawn that z_a merely contains uncertainties in the system. In this context, we denote the (implicit) input of $\Sigma_\mathcal{K}$ by

$$\begin{bmatrix} \Delta u \\ \Delta y \end{bmatrix} = (D\Sigma_\mathcal{K})^T (z_a) = (D\Sigma_\mathcal{K})^T \left(r_y\right),$$

and $(\Delta u, \Delta y)$ are unknown input vectors. Note that

$$\Sigma_\mathcal{K} \begin{bmatrix} \Delta u \\ \Delta y \end{bmatrix} = r_y \neq 0 \Longrightarrow \begin{bmatrix} \Delta u \\ \Delta y \end{bmatrix} \neq \begin{bmatrix} u_0 \\ y_0 \end{bmatrix}.$$

To illustrate the meaning of such unknown input vectors, the following example is presented.

Example 6.1 Consider an LTI system

$$y = Gu + d.$$

Here, d represents an $\mathcal{L}_2$-bounded unknown input vector, and

$$G : \begin{cases} \dot{x} = Ax + Bu \\ y = Cx + Du \end{cases}, G = \hat{M}^{-1}\hat{N} = NM^{-1}$$

is the (minimal) state space realisation of the plant model, whose SKR is

$$\mathcal{K}_G : \begin{cases} \dot{\hat{x}} = (A - LC)\hat{x} + \begin{bmatrix} B - LD & L \end{bmatrix} \begin{bmatrix} u \\ y \end{bmatrix} \\ r_y = W\left(y - \hat{y}\right) = -WC\hat{x} + \begin{bmatrix} -WD & W \end{bmatrix} \begin{bmatrix} u \\ y \end{bmatrix} \end{cases}.$$

It is straightforward that, for $e = x - \hat{x}$, the dynamic of $\mathcal{K}_G$ is governed by

$$\begin{cases} \dot{e} = (A - LC)e - Ld \\ r_y = W(Ce + d). \end{cases} \tag{6.56}$$

Let

$$\begin{bmatrix} \Delta u \\ \Delta y \end{bmatrix} = \begin{bmatrix} 0 \\ d \end{bmatrix}.$$

The dynamic system (6.56) and thus $\mathcal{K}_G$ can be equivalently written into

$$\mathcal{K}_G : \begin{cases} \dot{e} = (A - LC)\, e + \begin{bmatrix} B - LD & L \end{bmatrix} \begin{bmatrix} \Delta u \\ \Delta y \end{bmatrix} \\ r_y = W\left(y - \hat{y}\right) = -WCe + \begin{bmatrix} -WD & W \end{bmatrix} \begin{bmatrix} \Delta u \\ \Delta y \end{bmatrix}. \end{cases}$$

In the case of existence of uncertainty (Δ_M, Δ_N) modelled by

$$G = (N_0 + \Delta_N)\,(M_0 + \Delta_M)^{-1}$$

the system dynamic is described by

$$\begin{bmatrix} u \\ y \end{bmatrix} = \begin{bmatrix} M_0 + \Delta_M \\ N_0 + \Delta_N \end{bmatrix} v$$

for some v as latent variable. Let

$$\hat{x} = \hat{x}_0 + \hat{x}_\Delta,$$

$$\begin{cases} \dot{\hat{x}}_0 = (A - LC)\, \hat{x}_0 + \begin{bmatrix} B - LD & L \end{bmatrix} \begin{bmatrix} M_0 \\ N_0 \end{bmatrix} v \\ \dot{\hat{x}}_\Delta = (A - LC)\, \hat{x}_\Delta + \begin{bmatrix} B - LD & L \end{bmatrix} \begin{bmatrix} \Delta_M \\ \Delta_N \end{bmatrix} v \\ r_y = W\left(y - \hat{y}\right) = -WC\left(\hat{x}_0 + \hat{x}_\Delta\right) + \begin{bmatrix} -WD & W \end{bmatrix} \begin{bmatrix} M_0 + \Delta_M \\ N_0 + \Delta_N \end{bmatrix} v. \end{cases}$$

Recall that

$$\begin{bmatrix} -\hat{N}_0 & \hat{M}_0 \end{bmatrix} \begin{bmatrix} M_0 \\ N_0 \end{bmatrix} = 0.$$

We have,

$$\begin{cases} \dot{\hat{x}}_0 = (A - LC)\, \hat{x}_0 + \begin{bmatrix} B - LD & L \end{bmatrix} \begin{bmatrix} M_0 \\ N_0 \end{bmatrix} v \\ r_{y,0} = W\left(y_0 - \hat{y}\right) = -WC\hat{x}_0 + \begin{bmatrix} -WD & W \end{bmatrix} \begin{bmatrix} M_0 \\ N_0 \end{bmatrix} v \end{cases}$$

$$\Longleftrightarrow r_{y,0} = \begin{bmatrix} -\hat{N}_0 & \hat{M}_0 \end{bmatrix} \begin{bmatrix} M_0 \\ N_0 \end{bmatrix} v = 0.$$

That implies,

$$r_y = W\left(y - \hat{y}\right) = -WC\hat{x}_\Delta + \begin{bmatrix} -WD & W \end{bmatrix} \begin{bmatrix} \Delta_M \\ \Delta_N \end{bmatrix} v.$$

As a result, the SKR dynamic is equivalent to

$$\mathcal{K}_G : \begin{cases} \dot{\hat{x}}_\Delta = (A - LC)\,\hat{x}_\Delta + \begin{bmatrix} B - LD & L \end{bmatrix} \begin{bmatrix} \Delta u \\ \Delta y \end{bmatrix} \\ r_y = -WC\hat{x}_\Delta + \begin{bmatrix} -WD & W \end{bmatrix} \begin{bmatrix} \Delta u \\ \Delta y \end{bmatrix}, \end{cases}$$

$$\begin{bmatrix} \Delta u \\ \Delta y \end{bmatrix} = \begin{bmatrix} \Delta_M \\ \Delta_N \end{bmatrix} v.$$

At the end of this example, it is worth to remark that closed-form expressions for the uncertainty $(\Delta u, \Delta y)$ like the ones presented above are generally only possible, when linear systems are concerned.

Following the aforementioned discussion and example, the SKR (6.11) can be equivalently expressed by

$$\mathcal{K}_\Sigma : \begin{cases} \dot{\hat{x}} = a_K(\hat{x}) + B_K(\hat{x}) \begin{bmatrix} \Delta u \\ \Delta y \end{bmatrix} \\ r_y = c_K(\hat{x}) + D_K(\hat{x}) \begin{bmatrix} \Delta u \\ \Delta y \end{bmatrix}, \end{cases} \tag{6.57}$$

where $\left(a_K(\hat{x}), B_K(\hat{x}), c_K(\hat{x}), D_K(\hat{x})\right)$ are functions of $L\left(\hat{x}\right)$ and $W\left(\hat{x}\right)$, as defined in (6.11). An emphasis is placed on the following two points:

- $\begin{bmatrix} \Delta u \\ \Delta y \end{bmatrix}$ is unknown $\mathcal{L}_2$-bounded, and there exist no further assumption on it;
- the residual vector r_y is generated by the SKR (6.11). The dynamic system (6.57) describes (implicitly) the dynamic of the SKR (6.11) with respect to the uncertainties in the system. As the input of the estimator, r_y is assumed to be an $\mathcal{L}_2$ -bounded signal.

We are now in a position to introduce the definition of the normalised SKR and its existence conditions.

Definition 6.12 The SKR (6.57) (equivalently (6.11)) of the affine system (6.8) is said to be normalised, if

$$r_y = z_a, \tag{6.58}$$

and there exists a storage function $V(x) \geq 0$, $V(0) = 0$ so that

$$V\left(x\left(t_2\right)\right) - V\left(x\left(t_1\right)\right) = \int_{t_1}^{t_2} \left(\frac{1}{2} \begin{bmatrix} \Delta u^T & \Delta y^T \end{bmatrix} \begin{bmatrix} \Delta u \\ \Delta y \end{bmatrix} - \frac{1}{2} r_y^T r_y \right) d\tau. \tag{6.59}$$

Remark 6.4 It should be emphasised that the system (6.57) and the SKR (6.11) are two closely related but different system dynamics. In fact, the system (6.57) describes the uncertain dynamic embedded in the SKR (6.11). Therefore, both systems have the identical output r_y that knowingly reflects the uncertainty in the SKR. The notation for the state variable in the system (6.57), $\hat{x}$, is, however, abused for representing the state variable of uncertain dynamic for the sake of simplicity. It is in general different from the one adopted in the SKR (6.11), as showcased by the above example.

The existence conditions for the SKR (6.11) satisfying (6.58)–(6.59) are given in the following theorem.

Theorem 6.2 *The SKR (6.11) is normalised, if there exists $V(\hat{x}) \geq 0$ solving*

$$V_x^T\left(\hat{x}\right) a(\hat{x}) + \frac{1}{2} V_x^T\left(\hat{x}\right) B(\hat{x}) B^T(\hat{x}) V_x\left(\hat{x}\right) - \frac{1}{2} L(\hat{x}) \left(I + D(\hat{x}) D^T(\hat{x})\right) L^T(\hat{x}) = 0, \tag{6.60}$$

and moreover,

$$V_x^T\left(\hat{x}\right) L(\hat{x}) = \left(c^T(\hat{x}) + V_x^T\left(\hat{x}\right) B(\hat{x}) D^T(\hat{x})\right) \left(I + D(\hat{x}) D^T(\hat{x})\right)^{-1}, \tag{6.61}$$

$$W(\hat{x}) = \left(I + D(\hat{x}) D^T(\hat{x})\right)^{-1/2}. \tag{6.62}$$

Proof It follows from Lemma 6.4 that the SKR (6.11) satisfies (6.58)–(6.59) if the following three equations are solved by finding $L(\hat{x})$, $W(\hat{x})$ and $V(\hat{x})$:

$$V_x^T\left(\hat{x}\right) a_K(\hat{x}) + \frac{1}{2} V_x^T\left(\hat{x}\right) B_K(\hat{x}) B_K^T(\hat{x}) V_x\left(\hat{x}\right) = 0, \tag{6.63}$$

$$c_K^T(\hat{x}) + V_x^T\left(\hat{x}\right) B_K(\hat{x}) D_K^T(\hat{x}) = 0, \tag{6.64}$$

$$D_K(\hat{x}) D_K^T(\hat{x}) = I, \tag{6.65}$$

where $a_K(\hat{x}), B_K(\hat{x}), c_K(\hat{x}), D_K(\hat{x})$ are given in (6.11). Furthermore, (6.64) and (6.65) lead to

$$V_x^T\left(\hat{x}\right) L(\hat{x}) = \left(c^T(\hat{x}) + V_x^T\left(\hat{x}\right) B(\hat{x}) D^T(\hat{x})\right) \left(I + D(\hat{x}) D^T(\hat{x})\right)^{-1},$$

$$W(\hat{x}) = \left(I + D(\hat{x}) D^T(\hat{x})\right)^{-1/2}.$$

According to them, (6.63) is equivalent with

$$V_x^T\left(\hat{x}\right) a(x) + \frac{1}{2} V_x^T\left(\hat{x}\right) B(\hat{x}) B^T(\hat{x}) V_x(\hat{x}) - \frac{1}{2} L(\hat{x}) \left(I + D(\hat{x}) D^T(\hat{x})\right) L^T(\hat{x}) = 0.$$

Hence, the theorem is proved.

A direct interpretation of the lossless property of the normalised SKR is the energy in the residual signal r_y is the sum of the storage energy in the system and the energy in the uncertainty signal $\begin{bmatrix} \Delta u \\ \Delta y \end{bmatrix}$.

6.4 Parameterisation of Observer-Based Stabilising Controllers

In this section, we introduce a parameterisation of observer-based stabilising controllers for the loop configuration depicted in Fig. 6.2. The plant dynamic is modelled by Σ with the state space representation (6.8). For brevity, we drop the superscripts x_0 and $\hat{x}_0$ in the sequel. C represents the controller that stabilises the loop dynamics. d and v are external input vectors.

To begin with, we study the stability conditions for the closed-loop. In compliance with the observer-based realisation of Youla parameterisation of LTI feedback loops, our study begins with (6.21), which is written, in the state space representation, as

$$\mathcal{K}_{\Sigma,C} := \begin{cases} \dot{\hat{x}} = a_K(\hat{x}) + B_K(\hat{x}) \begin{bmatrix} u \\ y \end{bmatrix} \\ \begin{bmatrix} v \\ r_y \end{bmatrix} = \begin{bmatrix} -V^{-1}(\hat{x}) g(\hat{x}) \\ c_K(\hat{x}) \end{bmatrix} + \begin{bmatrix} \bar{V}(\hat{x}) \\ D_K(\hat{x}) \end{bmatrix} \begin{bmatrix} u \\ y \end{bmatrix} \\ \bar{V}(\hat{x}) = \begin{bmatrix} V^{-1}(\hat{x}) \ 0 \end{bmatrix}. \end{cases} \tag{6.66}$$

Note that $\mathcal{K}_{\Sigma,C}$ can be viewed as the SKR of the closed-loop with (u, y) and (v, r_y) as its input and output vectors, respectively. On the other hand, the loop dynamic with the configuration depicted in Fig. 6.2 is governed by

$$\begin{bmatrix} v \\ d \end{bmatrix} = \begin{bmatrix} u - C(y) \\ y - \Sigma(u) \end{bmatrix}.$$

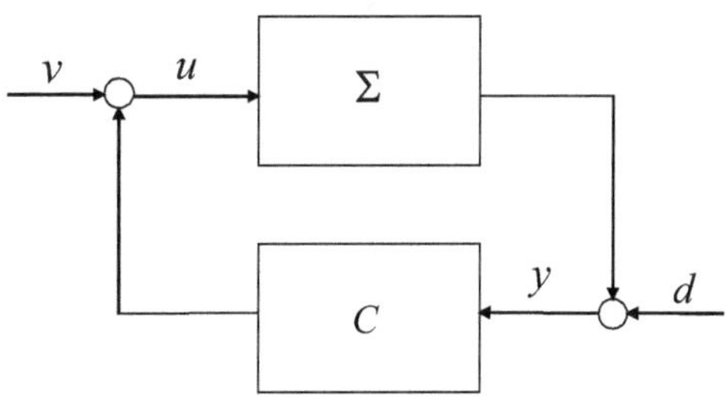

Fig. 6.2 Feedback control loop under consideration

Therefore, the loop is stable according to Lemma 6.1 only if the inverse mapping from (v, d) to (u, y) is stable. On account of (6.66) and recalling that r_y is driven by d, the closed-loop is stable if the SKR $\mathcal{K}_{\Sigma,C}$ has a stable right inverse.

Observe that the right inverse of the system (6.66) is equivalent to

$$\mathcal{K}_{\Sigma,C}^{-} = \begin{cases} \dot{\hat{x}} = a_I(\hat{x}) + \left[B_I(\hat{x}) \; L(\hat{x})W^{-1}\left(\hat{x}\right) \right] \begin{bmatrix} v \\ r_y \end{bmatrix} \\ \begin{bmatrix} u \\ y \end{bmatrix} = \begin{bmatrix} g(\hat{x}) \\ -\left(c(\hat{x}) + D(\hat{x})g(\hat{x})\right) \end{bmatrix} + \begin{bmatrix} V(\hat{x}) & 0 \\ -D(\hat{x})V(\hat{x}) & W^{-1}\left(\hat{x}\right) \end{bmatrix} \begin{bmatrix} v \\ r_y \end{bmatrix}. \end{cases} \tag{6.67}$$

Hence, the closed-loop is stable if and only if $\mathcal{K}_{\Sigma,C}^{-}$ is stable, namely $\mathcal{K}_{\Sigma,C}^{-}$ is input $\left(v, r_y\right)$ to output (u, y) stable.

It is clear that

$$\mathcal{K}_{\Sigma,C} \circ \mathcal{K}_{\Sigma,C}^{-} \begin{pmatrix} v \\ r_y \end{pmatrix} = \begin{bmatrix} v \\ r_y \end{bmatrix}, \mathcal{K}_{\Sigma,C}^{-} \circ \mathcal{K}_{\Sigma,C} \begin{pmatrix} u \\ y \end{pmatrix} = \begin{bmatrix} u \\ y \end{bmatrix},$$

and thus

$$\mathcal{K}_{\Sigma,C} \circ \mathcal{K}_{\Sigma,C}^{-} = \mathcal{K}_{\Sigma,C}^{-} \circ \mathcal{K}_{\Sigma,C} = I. \tag{6.68}$$

Identity (6.68) is an analogue form of the Bezout identity for nonlinear systems.

Next, we address the parameterisation issue. Let

$$\mathcal{Q} = \begin{cases} \dot{x}_Q = a_Q(x_Q) + B_Q(x_Q)r_y \\ v = c_Q(x_Q) + D_Q(x_Q)r_y \end{cases}$$

be a stable system that is driven by the residual vector r_y and output v. That is,

$$\begin{bmatrix} v \\ r_y \end{bmatrix} = \bar{\mathcal{Q}}\left(r_y\right) =: \begin{bmatrix} \mathcal{Q}\left(r_y\right) \\ r_y \end{bmatrix}$$

is a stable system. Substituting it into (6.67) yields

$$\begin{bmatrix} u \\ y \end{bmatrix} = \mathcal{K}_{\Sigma,C}^{-} \begin{pmatrix} v \\ r_y \end{pmatrix} = \mathcal{K}_{\Sigma,C}^{-} \circ \bar{\mathcal{Q}}\left(r_y\right) =: \mathcal{L}_{\Sigma,C}\left(r_y\right).$$

System $\mathcal{L}_{\Sigma,C}\left(r_y\right)$ is the loop dynamics, consisting of two sub-systems:

- feedback controller

$$\mathcal{C} := \begin{cases} \dot{\hat{x}} = a_I(\hat{x}) + B(\hat{x})v + L(\hat{x})W^{-1}\left(\hat{x}\right)r_y \\ v = \mathcal{Q}\left(r_y\right) \\ u = g(\hat{x}) + v, \end{cases} \tag{6.69}$$

- output response

$$\begin{cases} \dot{\hat{x}} = a_I(\hat{x}) + B(\hat{x})v + L(\hat{x})W^{-1}\left(\hat{x}\right)r_y \\ y = W^{-1}\left(\hat{x}\right)\left(r_y - c(\hat{x}) - D(\hat{x})g(\hat{x}) - D(\hat{x})v\right). \end{cases}$$

In the monograph by van der Schaft, it is proved that stabilising controllers can be parametrised by the system $\mathcal{C}$ with $\mathcal{Q}\left(r_y\right)$ as the (stable) parameterisation system. We summarise these results in the following theorem.

Theorem 6.3 *Given a feedback control system depicted in Fig. 6.2 with the plant dynamic Σ modelled by (6.8), and suppose that $\mathcal{K}_{\Sigma,C}^{-}$ given in (6.67) is stable, then stabilising controllers can be parametrised by (6.69) with $\mathcal{Q}\left(r_y\right)$ as the parameterisation system, and the loop dynamics are governed by*

$$\begin{cases} \dot{\hat{x}} = a_I(\hat{x}) + B(\hat{x})v + L(\hat{x})W^{-1}\left(\hat{x}\right)r_y \\ v = \mathcal{Q}\left(r_y\right) \\ \begin{bmatrix} u \\ y \end{bmatrix} = \begin{bmatrix} g(\hat{x}) \\ -\left(c(\hat{x}) + D(\hat{x})g(\hat{x})\right) \end{bmatrix} + \begin{bmatrix} I \\ -D(\hat{x}) \end{bmatrix} v + \begin{bmatrix} 0 \\ W^{-1}\left(\hat{x}\right) \end{bmatrix} r_y. \end{cases} \tag{6.70}$$

Theorem 6.3 is a nonlinear version of the Youla paremeterisation of observer-based stabilising controllers and thus is of considerable importance for controller design and optimisation. Beyond that, it verifies the known results for LTI systems that

- all feedback controllers are residual-driven, and
- they are observer-based state feedback controllers.

Moreover, (6.70) showcases the closed-loop dynamics and illustrates how uncertainties (in the form of r_y) affect the system dynamics.

6.5 Notes and References

Although the objective of this chapter is to present the unified control and detection framework for nonlinear dynamic systems, the main endeavours have been made to introduce and illustrate the major mathematical and control-theoretic tools needed for modelling, analysis and synthesis of nonlinear systems, including

- system factorisation technique,
- operator theory-based analysis and synthesis of nonlinear systems, and
- Hamiltonian systems.

Applications of these methods to system stability analysis, stabilising controller synthesis as well as stabilising controller parameterisation were extensively investigated during the decade between the 80s and 90s with rich results. The reader is referred to a series of publications on these issues [7, 62, 127, 153, 154, 162, 177].

In the section on preliminaries, we have mainly referred to the monograph by van der Schaft [194]. These include Definitions 6.1–6.4 on system stability and Lemma 6.1.

Section 6.2 addresses coprime factorisation methods for nonlinear systems. A detailed description about the related methods can be found primarily in [194]. Along with the lines of introducing the coprime factorisations for LTI systems in Chap. 2, the control theoretic interpretations, in particular, SKR of the plant as a residual generator and the observer-based state feedback controller for SKR and SIR of the corresponding controllers stand in the foreground.

The normalised SIR has been introduced in a separate section, Sect. 6.3. Our intention is not only to present the concept of normalised coprime factorisations, but also to introduce Hamiltonian system methods as an important mathematical tool useful for our subsequent works. The Hamiltonian extension and the related methods were introduced by Crouch and van der Schaft [27], and widely applied to the investigations on factorisations of nonlinear systems [7, 162, 177]. The definition of inner systems and the results in Lemma 6.3 and Theorem 6.2 can be found in [7, 177].

It is notable that, different from the study on Youla parameterisation for LTI systems, the parameterisation of all stabilising controllers for nonlinear systems has been developed directly on the basis of kernel representations [62, 63, 153, 154, 194] and thus presented in a similar form of the so-called observer-based realisation of Youla parameterisation. Theorem 6.3 has been proved in [194] (Theorems 6.2.5–6.2.6). A more general parameterisation form can be found in [62, 63].

It is remarkable that the parameterisation of all stabilising controllers for nonlinear systems has been successfully applied to the design of nonlinear fault-tolerant systems, as reported in [80, 118]. In [216, 217], application of nonlinear coprime factorisation technique to fault detection in nonlinear systems has been explored.

Chapter 7
Perspectives, Extensions, and Applications

In the previous five chapters, the basic ideas and the theoretic cornerstone of the unified framework of control and detection as well as the needed mathematical tools have been elaborately and comprehensively presented and illustrated. In the subsequent parts, Part II and Part III, applications in the unified framework aiming at fault diagnosis, fault-tolerant control and secure cyber-physical systems will be explored. Before that, we would like, in this chapter, to discuss and highlight potential extensions of the unified framework and the associated perspectives for its applications to control and detection.

7.1 Control Structure Embedded in Process Data

A cornerstone of the unified framework of control and detection is the parameterisation of the closed-loop response,

$$\begin{bmatrix} u \\ y \end{bmatrix} = \begin{bmatrix} M \\ N \end{bmatrix} v + \begin{bmatrix} -\hat{Y} + MQ \\ \hat{X} + NQ \end{bmatrix} r_y, \tag{7.1}$$

given in Theorem 3.1. This result is derived on the assumption of the plant model (2.1)–(2.3) and the control loop (3.1) whose configuration is sketched in Fig. 3.1. As the system response, it is natural to view the process data (u, y) as the causal result of the loop configuration and the applied controller. Below, we reveal that the relation in (7.1) can be interpreted as the data structure, which is embedded in the data (u, y) and determined exclusively by the nominal plant model (2.1)–(2.3). As such, it exists as well in case of an open-loop configuration.

Assumption In our subsequent analysis, it is assumed that

$$\begin{bmatrix} u \\ y \end{bmatrix} \in \mathcal{H}_2. \tag{7.2}$$

S. X. Ding, *Unified Framework of Control and Detection*,
https://doi.org/10.1007/978-3-662-72417-0_7

7.1.1 The Basic Form

To begin with, we consider a simplified form of (7.1) with $Q = 0$, namely

$$\begin{bmatrix} u \\ y \end{bmatrix} = \begin{bmatrix} M \\ N \end{bmatrix} v + \begin{bmatrix} -\hat{Y} \\ \hat{X} \end{bmatrix} r_y, \tag{7.3}$$

and recall the control-theoretic meanings of the two terms in (7.3). The first one describes the causal response of the system to the intended control action u (not necessarily within a closed-loop), while the second one reflects the influence of uncertainties on (u, y). Here, uncertainties include all those signals acting on the plant and possible variations in system parameters, that are addressed by the residual signal r_y. In this part of our analysis, it is supposed that only unknown inputs as uncertainties affect the system dynamics. Note that (7.3) follows immediately from the Bezout identity (2.31), when (v, r_y) are defined by

$$\begin{bmatrix} v \\ r_y \end{bmatrix} = \begin{bmatrix} X & Y \\ -\hat{N} & \hat{M} \end{bmatrix} \begin{bmatrix} u \\ y \end{bmatrix}.$$

More importantly, this relation is independent of any controller and its configuration.

Next, we explore the relation (7.3) from a geometric aspect of signal space. Remember that the image subspaces of the plant and controller are defined by

$$\mathcal{I}_G = \left\{ \begin{bmatrix} u \\ y \end{bmatrix} \in \mathcal{H}_2 : \begin{bmatrix} u \\ y \end{bmatrix} = \begin{bmatrix} M \\ N \end{bmatrix} v, v \in \mathcal{H}_2 \right\},$$

$$\mathcal{I}_K = \left\{ \begin{bmatrix} u \\ y \end{bmatrix} \in \mathcal{H}_2 : \begin{bmatrix} u \\ y \end{bmatrix} = \begin{bmatrix} -\hat{Y} \\ \hat{X} \end{bmatrix} r_y, r_y \in \mathcal{H}_2 \right\},$$

respectively. Hence, it can be concluded that the process data (u, y) are composed of two signal vectors respectively in the subspaces $\mathcal{I}_G$ and $\mathcal{I}_K$. While the geometric interpretation of $\mathcal{I}_G$ as the signal subspace representing the nominal dynamics is apparent, that of $\mathcal{I}_K$, in particular its relation to $\mathcal{I}_G$, remains unclear. This motivates us to take a close look at $\mathcal{I}_K$. The fact that

$$\begin{bmatrix} -\hat{N} & \hat{M} \end{bmatrix} \begin{bmatrix} -\hat{Y} \\ \hat{X} \end{bmatrix} = I, \begin{bmatrix} -\hat{N} & \hat{M} \end{bmatrix} \begin{bmatrix} M \\ N \end{bmatrix} = 0 \Longrightarrow$$

$$\forall \begin{bmatrix} u \\ y \end{bmatrix}, \begin{bmatrix} -\hat{N} & \hat{M} \end{bmatrix} \begin{bmatrix} u \\ y \end{bmatrix} = r_y \neq 0, \begin{bmatrix} -\hat{N} & \hat{M} \end{bmatrix} \begin{bmatrix} -\hat{Y} \\ \hat{X} \end{bmatrix} r_y = r_y \neq 0,$$

$$\forall \begin{bmatrix} u \\ y \end{bmatrix} \in \mathcal{I}_G, r_y = \begin{bmatrix} -\hat{N} & \hat{M} \end{bmatrix} \begin{bmatrix} M \\ N \end{bmatrix} v = 0 \Longrightarrow \begin{bmatrix} -\hat{Y} \\ \hat{X} \end{bmatrix} r_y = 0$$

suggests that

$$\mathcal{I}_K \cap \mathcal{R}_G \neq \mathcal{O},$$

where $\mathcal{R}_G$ denotes the orthogonal complement of $\mathcal{I}_G, \mathcal{I}_G^{\perp}$, because $\mathcal{R}_G$ is characterised by

$$\forall \begin{bmatrix} u \\ y \end{bmatrix} \in \mathcal{R}_G, r_y = \begin{bmatrix} -\hat{N} & \hat{M} \end{bmatrix} \begin{bmatrix} u \\ y \end{bmatrix} \neq 0.$$

The last claim is attributed to the definition of the kernel subspace $\mathcal{K}_G$,

$$\mathcal{K}_G = \left\{ \begin{bmatrix} u \\ y \end{bmatrix} \in \mathcal{H}_2 : \begin{bmatrix} -\hat{N} & \hat{M} \end{bmatrix} \begin{bmatrix} u \\ y \end{bmatrix} = 0 \right\},$$

given in Definition 2.5, and the relation $\mathcal{K}_G = \mathcal{I}_G$. We now assume that (M, N) and $\left(\hat{M}, \hat{N}\right)$ are the normalised RCP and LCP, as given in Theorem 2.1, and denote them and the corresponding RCP $\left(\hat{X}, \hat{Y}\right)$ by (M_0, N_0), $\left(\hat{M}_0, \hat{N}_0\right)$ and $\left(\hat{X}_0, \hat{Y}_0\right)$, respectively. For our purpose, the concept of orthogonal projection and the projection operator $\mathcal{P}_{\mathcal{I}_G}$ that projects (u, y) onto the image subspace $\mathcal{I}_G$ are borrowed from Chap. 9, where the application of projection theory to fault diagnosis is systematically and detailedly described. The operator $\mathcal{P}_{\mathcal{I}_G}$ is defined by

$$\mathcal{P}_{\mathcal{I}_G} = \mathcal{L}_{I_{G_0}} \mathcal{L}_{I_{G_0}}^*, \mathcal{L}_{I_{G_0}}^* : \mathcal{H}_2 \to \mathcal{H}_2, \mathcal{L}_{I_{G_0}}^* = \mathcal{P}_{\mathcal{H}_2} \mathcal{L}_{I_{G_0}^{\sim}}, \tag{7.4}$$

$$\mathcal{P}_{\mathcal{I}_G} \left(\begin{bmatrix} u \\ y \end{bmatrix} \right) = \mathcal{L}_{I_{G_0}} \mathcal{P}_{\mathcal{H}_2} \mathcal{L}_{I_{G_0}^{\sim}} \left(\begin{bmatrix} u \\ y \end{bmatrix} \right). \tag{7.5}$$

$\mathcal{L}_{I_{G_0}}$ and $\mathcal{L}_{I_{G_0}^{\sim}}$ are Laurent operators with the symbols $I_{G_0}, I_{G_0}^{\sim}$,

$$I_{G_0} = \begin{bmatrix} M_0 \\ N_0 \end{bmatrix}, I_{G_0}^{\sim} = \begin{bmatrix} M_0^T \left(z^{-1}\right) & N_0^T \left(z^{-1}\right) \end{bmatrix},$$

and $\mathcal{P}_{\mathcal{H}_2}$ is an orthogonal operator representing a projection onto $\mathcal{H}_2$. The subsequent properties of orthogonal projection $\mathcal{P}_{\mathcal{I}_G}$ are essential for our study,

$$\forall \begin{bmatrix} u \\ y \end{bmatrix} \in \mathcal{I}_G, \mathcal{P}_{\mathcal{I}_G} \left(\begin{bmatrix} u \\ y \end{bmatrix} \right) = \begin{bmatrix} u \\ y \end{bmatrix},$$

$$\forall \begin{bmatrix} u \\ y \end{bmatrix} \in \mathcal{I}_G^{\perp}, \mathcal{P}_{\mathcal{I}_G} \left(\begin{bmatrix} u \\ y \end{bmatrix} \right) = 0 \Longrightarrow \left(\mathcal{I} - \mathcal{P}_{\mathcal{I}_G}\right) \left(\begin{bmatrix} u \\ y \end{bmatrix} \right) = \begin{bmatrix} u \\ y \end{bmatrix},$$

$$\forall \begin{bmatrix} u \\ y \end{bmatrix} \in \mathcal{H}_2, \begin{bmatrix} u \\ y \end{bmatrix} = \mathcal{P}_{\mathcal{I}_G} \left(\begin{bmatrix} u \\ y \end{bmatrix} \right) + \left(\mathcal{I} - \mathcal{P}_{\mathcal{I}_G}\right) \left(\begin{bmatrix} u \\ y \end{bmatrix} \right),$$

$$\forall \begin{bmatrix} u \\ y \end{bmatrix} \in \mathcal{H}_2, \left\| \begin{bmatrix} u \\ y \end{bmatrix} \right\|_2^2 = \left\| \mathcal{P}_{\mathcal{I}_G} \left(\begin{bmatrix} u \\ y \end{bmatrix} \right) \right\|_2^2 + \left\| \left(\mathcal{I} - \mathcal{P}_{\mathcal{I}_G}\right) \left(\begin{bmatrix} u \\ y \end{bmatrix} \right) \right\|_2^2,$$

where $\mathcal{I}$ is a unit operator. Now, orthogonal projection $\mathcal{P}_{\mathcal{I}_G}$ is applied to

$$\mathcal{P}_{\mathcal{I}_G}\left(\begin{bmatrix} -\hat{Y}_0 \\ \hat{X}_0 \end{bmatrix} r_{y,0}\right) = \mathcal{L}_{I_{G_0}} \mathcal{P}_{\mathcal{H}_2}\left(\begin{bmatrix} M_0^{\sim} & N_0^{\sim} \end{bmatrix}\begin{bmatrix} -\hat{Y}_0 \\ \hat{X}_0 \end{bmatrix} r_{y,0}\right).$$

Herewith, $r_{y,0}$ represents the residual signal generated by the normalised SKR. Denote by $\mathcal{Q}_{\mathcal{H}_2}$ the projection

$$\mathcal{P}_{\mathcal{H}_2}\begin{bmatrix} M_0^{\sim} & N_0^{\sim} \end{bmatrix}\begin{bmatrix} -\hat{Y}_0 \\ \hat{X}_0 \end{bmatrix} =: \mathcal{Q}_{\mathcal{H}_2} \in \mathcal{RH}_\infty. \tag{7.6}$$

It yields

$$\begin{aligned} \mathcal{P}_{\mathcal{I}_G}\left(\begin{bmatrix} -\hat{Y}_0 \\ \hat{X}_0 \end{bmatrix} r_{y,0}\right) &= \begin{bmatrix} M_0 \\ N_0 \end{bmatrix} \mathcal{Q}_{\mathcal{H}_2} r_{y,0}, \\ (\mathcal{I} - \mathcal{P}_{\mathcal{I}_G})\left(\begin{bmatrix} -\hat{Y}_0 \\ \hat{X}_0 \end{bmatrix} r_{y,0}\right) &= \left(\begin{bmatrix} -\hat{Y}_0 \\ \hat{X}_0 \end{bmatrix} - \begin{bmatrix} M_0 \\ N_0 \end{bmatrix} \mathcal{Q}_{\mathcal{H}_2}\right) r_{y,0} \\ \Longrightarrow \begin{bmatrix} -\hat{Y}_0 \\ \hat{X}_0 \end{bmatrix} r_{y,0} &= \begin{bmatrix} M_0 \\ N_0 \end{bmatrix} \mathcal{Q}_{\mathcal{H}_2} r_{y,0} + (\mathcal{I} - \mathcal{P}_{\mathcal{I}_G})\left(\begin{bmatrix} -\hat{Y}_0 \\ \hat{X}_0 \end{bmatrix} r_{y,0}\right). \end{aligned}$$

Note that the above relation implies that $\left(\begin{bmatrix} -\hat{Y}_0 \\ \hat{X}_0 \end{bmatrix} - \begin{bmatrix} M_0 \\ N_0 \end{bmatrix} \mathcal{Q}_{\mathcal{H}_2}\right)$ spans the residual subspace $\mathcal{R}_G = \mathcal{I}_G^{\perp}$.

Theorem 7.1 *Given the image subspace $\mathcal{I}_G$, it holds*

$$\begin{aligned} &\forall \begin{bmatrix} u \\ y \end{bmatrix} \in \mathcal{I}_G^{\perp}, \exists r_y \in \mathcal{H}_2 \text{ s.t. } \begin{bmatrix} u \\ y \end{bmatrix} = \left(\begin{bmatrix} -\hat{Y}_0 \\ \hat{X}_0 \end{bmatrix} - \begin{bmatrix} M_0 \\ N_0 \end{bmatrix} \mathcal{Q}_{\mathcal{H}_2}\right) r_y, \\ &\forall r_y \in \mathcal{H}_2, \begin{bmatrix} u \\ y \end{bmatrix} = \left(\begin{bmatrix} -\hat{Y}_0 \\ \hat{X}_0 \end{bmatrix} - \begin{bmatrix} M_0 \\ N_0 \end{bmatrix} \mathcal{Q}_{\mathcal{H}_2}\right) r_y \in \mathcal{R}_G. \end{aligned}$$

Proof Recall that

$$\forall \begin{bmatrix} u \\ y \end{bmatrix} \in \mathcal{H}_2, \exists v, r_y \in \mathcal{H}_2, s.t. \begin{bmatrix} u \\ y \end{bmatrix} = \begin{bmatrix} M_0 \\ N_0 \end{bmatrix} v + \begin{bmatrix} -\hat{Y}_0 \\ \hat{X}_0 \end{bmatrix} r_y.$$

Hence,

$$\forall \begin{bmatrix} u \\ y \end{bmatrix} \in \mathcal{I}_G^{\perp}, \begin{bmatrix} u \\ y \end{bmatrix} = (\mathcal{I} - \mathcal{P}_{\mathcal{I}_G})\begin{bmatrix} -\hat{Y}_0 \\ \hat{X}_0 \end{bmatrix} r_y = \left(\begin{bmatrix} -\hat{Y}_0 \\ \hat{X}_0 \end{bmatrix} - \begin{bmatrix} M_0 \\ N_0 \end{bmatrix} \mathcal{Q}_{\mathcal{H}_2}\right) r_y.$$

For the same reason, it is apparent that $\forall r_y \in \mathcal{H}_2$,

$$\begin{bmatrix} u \\ y \end{bmatrix} = \left(\begin{bmatrix} -\hat{Y}_0 \\ \hat{X}_0 \end{bmatrix} - \begin{bmatrix} M_0 \\ N_0 \end{bmatrix} \mathcal{Q}_{\mathcal{H}_2} \right) r_y = (\mathcal{I} - \mathcal{P}_{\mathcal{I}_G}) \begin{bmatrix} -\hat{Y}_0 \\ \hat{X}_0 \end{bmatrix} r_y \in \mathcal{R}_G.$$

The theorem is thus proved.

An immediate result of this theorem is that the residual subspace $\mathcal{R}_G$ can be defined as

$$\mathcal{R}_G = \left\{ \begin{bmatrix} u \\ y \end{bmatrix}, \begin{bmatrix} u \\ y \end{bmatrix} = \left(\begin{bmatrix} -\hat{Y}_0 \\ \hat{X}_0 \end{bmatrix} - \begin{bmatrix} M_0 \\ N_0 \end{bmatrix} \mathcal{Q}_{\mathcal{H}_2} \right) r_y, r_y \in \mathcal{H}_2 \right\}.$$

Consequently, the data structure (7.3) is equivalently written as

$$\begin{bmatrix} u \\ y \end{bmatrix} = \begin{bmatrix} M_0 \\ N_0 \end{bmatrix} \left(v + \mathcal{Q}_{\mathcal{H}_2} r_{y,0} \right) + \left(\begin{bmatrix} -\hat{Y}_0 \\ \hat{X}_0 \end{bmatrix} - \begin{bmatrix} M_0 \\ N_0 \end{bmatrix} \mathcal{Q}_{\mathcal{H}_2} \right) r_{y,0}. \tag{7.7}$$

That means, the data (u, y) is decomposed into

$$\begin{bmatrix} M_0 \\ N_0 \end{bmatrix} \left(v + \mathcal{Q}_{\mathcal{H}_2} r_{y,0} \right) \in \mathcal{I}_G \text{ and } \left(\begin{bmatrix} -\hat{Y}_0 \\ \hat{X}_0 \end{bmatrix} - \begin{bmatrix} M_0 \\ N_0 \end{bmatrix} \mathcal{Q}_{\mathcal{H}_2} \right) r_{y,0} \in \mathcal{R}_G,$$

and it holds

$$\left\| \begin{bmatrix} u \\ y \end{bmatrix} \right\|_2^2 = \left\| \begin{bmatrix} M_0 \\ N_0 \end{bmatrix} \left(v + \mathcal{Q}_{\mathcal{H}_2} r_{y,0} \right) \right\|_2^2 + \left\| \left(\begin{bmatrix} -\hat{Y}_0 \\ \hat{X}_0 \end{bmatrix} - \begin{bmatrix} M_0 \\ N_0 \end{bmatrix} \mathcal{Q}_{\mathcal{H}_2} \right) r_{y,0} \right\|_2^2.$$

Now, consider the case (7.1) with $\mathcal{Q} \neq 0$, which leads to

$$\begin{bmatrix} u \\ y \end{bmatrix} = \begin{bmatrix} M_0 \\ N_0 \end{bmatrix} \left(v + \mathcal{Q} r_{y,0} + \mathcal{Q}_{\mathcal{H}_2} r_{y,0} \right) + \left(\begin{bmatrix} -\hat{Y}_0 \\ \hat{X}_0 \end{bmatrix} - \begin{bmatrix} M_0 \\ N_0 \end{bmatrix} \mathcal{Q}_{\mathcal{H}_2} \right) r_{y,0}, \tag{7.8}$$

$$\left\| \begin{bmatrix} u \\ y \end{bmatrix} \right\|_2^2 = \left\| \begin{bmatrix} M_0 \\ N_0 \end{bmatrix} \left(v + \mathcal{Q} r_{y,0} + \mathcal{Q}_{\mathcal{H}_2} r_{y,0} \right) \right\|_2^2 + \left\| \left(\begin{bmatrix} -\hat{Y}_0 \\ \hat{X}_0 \end{bmatrix} - \begin{bmatrix} M_0 \\ N_0 \end{bmatrix} \mathcal{Q}_{\mathcal{H}_2} \right) r_{y,0} \right\|_2^2.$$

It is apparent that setting

$$\mathcal{Q} = -\mathcal{Q}_{\mathcal{H}_2}$$

results in

$$\left\| \begin{bmatrix} u \\ y \end{bmatrix} \right\|_2^2 = \left\| \begin{bmatrix} M_0 \\ N_0 \end{bmatrix} v \right\|_2^2 + \left\| \left(\begin{bmatrix} -\hat{Y}_0 \\ \hat{X}_0 \end{bmatrix} - \begin{bmatrix} M_0 \\ N_0 \end{bmatrix} \mathcal{Q}_{\mathcal{H}_2} \right) r_{y,0} \right\|_2^2.$$

That means,

$$\min_{Q\in\mathcal{RH}_\infty}\left\|\begin{bmatrix}u\\y\end{bmatrix}\right\|_2^2=\left\|\begin{bmatrix}M_0\\N_0\end{bmatrix}v\right\|_2^2+\left\|\left(\begin{bmatrix}-\hat{Y}_0\\\hat{X}_0\end{bmatrix}-\begin{bmatrix}M_0\\N_0\end{bmatrix}Q_{\mathcal{H}_2}\right)r_{y,0}\right\|_2^2.$$

In other words, the norm (energy) in the data (u, y) varies depending on u and reaches its minimum when u is equivalent to a setting

$$u=F_0\hat{x}-Q_{\mathcal{H}_2}r_{y,0}+v.$$

We would like to emphasise that this result is independent of the system configuration, both for open- and closed-loop cases.

To conclude our analysis, we now remove the assumption on (M, N) and $\left(\hat{M}, \hat{N}\right)$ being the normalised RCP and LCP. To this end, recall Lemma 3.1 and Theorem 3.4, which allow us to express any given (M, N), $\left(\hat{Y}, \hat{X}\right)$ and $\left(\hat{M}, \hat{N}\right)$ as well as r_y in terms of (M_0, N_0), $\left(\hat{X}_0, \hat{Y}_0\right)$ and $\left(\hat{M}_0, \hat{N}_0\right)$ as well as $r_{y,0}$ as follows,

$$\begin{bmatrix}M\\N\end{bmatrix}=\begin{bmatrix}M_0\\N_0\end{bmatrix}Q_1,$$

$$Q_1=I+(F-F_0)(zI-A_F)^{-1}B\in\mathcal{RH}_\infty, A_F=A+BF, \tag{7.9}$$

$$\begin{bmatrix}-\hat{Y}\\\hat{X}\end{bmatrix}=\begin{bmatrix}-\hat{Y}_0\\\hat{X}_0\end{bmatrix}R+\begin{bmatrix}M_0\\N_0\end{bmatrix}Q_2, \tag{7.10}$$

$$R=I-C\left(zI-A_{L0}\right)^{-1}(L_0-L)\in\mathcal{RH}_\infty, A_{L_0}=A-L_0C,$$

$$Q_2=\begin{bmatrix}X_0 & Y_0\end{bmatrix}\begin{bmatrix}-\hat{Y}\\\hat{X}\end{bmatrix}\in\mathcal{RH}_\infty, \tag{7.11}$$

$$r_{y,0}=\begin{bmatrix}-\hat{N}_0 & \hat{M}_0\end{bmatrix}\begin{bmatrix}u\\y\end{bmatrix}=Rr_y=R\begin{bmatrix}-\hat{N} & \hat{M}\end{bmatrix}\begin{bmatrix}u\\y\end{bmatrix},$$

where (F, F_0) are the feedback gain matrices, while (L, L_0) denote the observer gain matrices applied in (M, N) and (M_0, N_0), $\left(\hat{M}, \hat{N}\right)$ and $\left(\hat{M}_0, \hat{N}_0\right)$, respectively. Sequentially, (7.1) is equivalent to

$$\begin{aligned}\begin{bmatrix}u\\y\end{bmatrix}&=\begin{bmatrix}M_0\\N_0\end{bmatrix}Q_1v+\begin{bmatrix}-\hat{Y}_0\\\hat{X}_0\end{bmatrix}r_{y,0}+\begin{bmatrix}M_0\\N_0\end{bmatrix}(Q_2+Q_1Q)r_y\\&=\begin{bmatrix}M_0\\N_0\end{bmatrix}Q_1v+\begin{bmatrix}-\hat{Y}_0\\\hat{X}_0\end{bmatrix}r_{y,0}+\begin{bmatrix}M_0\\N_0\end{bmatrix}\bar{Q}r_{y,0}\\&=\begin{bmatrix}M_0\\N_0\end{bmatrix}(Q_1v+Q_{\mathcal{H}_2}r_{y,0}+\bar{Q}r_{y,0})+\left(\begin{bmatrix}-\hat{Y}_0\\\hat{X}_0\end{bmatrix}-\begin{bmatrix}M_0\\N_0\end{bmatrix}Q_{\mathcal{H}_2}\right)r_{y,0},\end{aligned} \tag{7.12}$$

$$\bar{Q}=Q_1Q+Q_2\Longleftrightarrow Q=Q_1^{-1}\left(\bar{Q}-Q_2\right)\in\mathcal{RH}_\infty, \tag{7.13}$$

$$Q_1^{-1}=I+(F_0-F)\left(zI-A_{F0}\right)^{-1}B\in\mathcal{RH}_\infty.$$

Hence, defining

$$\bar{Q} = -Q_{\mathcal{H}_2} \Longrightarrow Q = -Q_1^{-1}\left(Q_2 + Q_{\mathcal{H}_2}\right), \hat{v} = Q_1^{-1} v \tag{7.14}$$

yields

$$\begin{aligned}
\begin{bmatrix} u \\ y \end{bmatrix} &= \begin{bmatrix} M_0 \\ N_0 \end{bmatrix} \hat{v} + \left(\begin{bmatrix} -\hat{Y}_0 \\ \hat{X}_0 \end{bmatrix} - \begin{bmatrix} M_0 \\ N_0 \end{bmatrix} Q_{\mathcal{H}_2} \right) r_{y,0} \\
&= \begin{bmatrix} M \\ N \end{bmatrix} v + \left(\begin{bmatrix} -\hat{Y}_0 \\ \hat{X}_0 \end{bmatrix} - \begin{bmatrix} M_0 \\ N_0 \end{bmatrix} Q_{\mathcal{H}_2} \right) r_{y,0}, \\
\left\| \begin{bmatrix} u \\ y \end{bmatrix} \right\|_2^2 &= \left\| \begin{bmatrix} M_0 \\ N_0 \end{bmatrix} \hat{v} \right\|_2^2 + \left\| \left(\begin{bmatrix} -\hat{Y}_0 \\ \hat{X}_0 \end{bmatrix} - \begin{bmatrix} M_0 \\ N_0 \end{bmatrix} Q_{\mathcal{H}_2} \right) r_{y,0} \right\|_2^2 .
\end{aligned}$$

In other words, it holds

$$\begin{aligned}
\min_{Q \in \mathcal{RH}_\infty} \left\| \begin{bmatrix} u \\ y \end{bmatrix} \right\|_2^2 &= \min_{Q \in \mathcal{RH}_\infty} \left\| \begin{bmatrix} M \\ N \end{bmatrix} v + \begin{bmatrix} -\hat{Y} + MQ \\ \hat{X} + NQ \end{bmatrix} r_y \right\|_2^2 \\
&= \left\| \begin{bmatrix} M_0 \\ N_0 \end{bmatrix} \hat{v} \right\|_2^2 + \left\| \left(\begin{bmatrix} -\hat{Y}_0 \\ \hat{X}_0 \end{bmatrix} - \begin{bmatrix} M_0 \\ N_0 \end{bmatrix} Q_{\mathcal{H}_2} \right) r_{y,0} \right\|_2^2 .
\end{aligned} \tag{7.15}$$

Summarising (7.12) and (7.15) proves the following theorem.

Theorem 7.2 *Given plant G with model (2.1)–(2.3), and suppose that (i) $(M, N), \left(\hat{M}, \hat{N}\right)$ are RCP, LCP of G, $\left(\hat{Y}, \hat{X}\right)$ is a RCP satisfying*

$$\begin{bmatrix} -\hat{N} & \hat{M} \end{bmatrix} \begin{bmatrix} -\hat{Y} \\ \hat{X} \end{bmatrix} = I,$$

(ii) $(M_0, N_0), \left(\hat{M}_0, \hat{N}_0\right)$ are the normalised RCP and LCP of G, and the RCP $\left(\hat{X}_0, \hat{Y}_0\right)$ is subject to

$$\begin{bmatrix} -\hat{N}_0 & \hat{M}_0 \end{bmatrix} \begin{bmatrix} -\hat{Y}_0 \\ \hat{X}_0 \end{bmatrix} = I,$$

then we have

$$\begin{aligned}
\begin{bmatrix} u \\ y \end{bmatrix} &= \begin{bmatrix} M \\ N \end{bmatrix} v + \begin{bmatrix} -\hat{Y} + MQ \\ \hat{X} + NQ \end{bmatrix} r_y \\
&= \begin{bmatrix} M_0 \\ N_0 \end{bmatrix} \left(\hat{v} + Q_{\mathcal{H}_2} r_{y,0} + \bar{Q} r_{y,0}\right) + \left(\begin{bmatrix} -\hat{Y}_0 \\ \hat{X}_0 \end{bmatrix} - \begin{bmatrix} M_0 \\ N_0 \end{bmatrix} Q_{\mathcal{H}_2} \right) r_{y,0},
\end{aligned} \tag{7.16}$$

$$\min_{Q\in\mathcal{RH}_\infty}\left\|\begin{bmatrix}u\\y\end{bmatrix}\right\|_2^2=\min_{Q\in\mathcal{RH}_\infty}\left\|\begin{bmatrix}M\\N\end{bmatrix}v+\begin{bmatrix}-\hat{Y}+MQ\\\hat{X}+NQ\end{bmatrix}r_y\right\|_2^2$$
$$=\left\|\begin{bmatrix}M_0\\N_0\end{bmatrix}\hat{v}\right\|_2^2+\left\|\left(\begin{bmatrix}-\hat{Y}_0\\\hat{X}_0\end{bmatrix}-\begin{bmatrix}M_0\\N_0\end{bmatrix}Q_{\mathcal{H}_2}\right)r_{y,0}\right\|_2^2 \tag{7.17}$$
$$=\left\|\begin{bmatrix}M\\N\end{bmatrix}v\right\|_2^2+\left\|\left(\begin{bmatrix}-\hat{Y}_0\\\hat{X}_0\end{bmatrix}-\begin{bmatrix}M_0\\N_0\end{bmatrix}Q_{\mathcal{H}_2}\right)r_{y,0}\right\|_2^2, \tag{7.18}$$
$$Q^*=\arg\min_{Q\in\mathcal{RH}_\infty}\left\|\begin{bmatrix}u\\y\end{bmatrix}\right\|_2^2=Q_1^{-1}\left(Q_2-Q_{\mathcal{H}_2}\right), \tag{7.19}$$

where $\hat{v}=Q_1^{-1}v$, *and* $Q_{\mathcal{H}_2}, \bar{Q}, Q_1$ *and* Q_2 *are defined in (7.6), (7.13), (7.9) and (7.11), and* $r_{y,0}$ *is the residual vector subject to*

$$r_{y,0}=\begin{bmatrix}-\hat{N}_0 & \hat{M}_0\end{bmatrix}\begin{bmatrix}u\\y\end{bmatrix}.$$

Equation (7.16) reveals that any process data (u, y) can be decomposed into signal vectors,

$$\begin{bmatrix}M_0\\N_0\end{bmatrix}\left(\hat{v}+Q_{\mathcal{H}_2}r_{y,0}+\bar{Q}r_{y,0}\right), \left(\begin{bmatrix}-\hat{Y}_0\\\hat{X}_0\end{bmatrix}-\begin{bmatrix}M_0\\N_0\end{bmatrix}Q_{\mathcal{H}_2}\right)r_{y,0},$$

respectively belonging to the image subspace $\mathcal{I}_G$ and the residual subspace $\mathcal{R}_G$. In our subsequent study, this decomposition is essential, and thus is called the basic form of the data structure. Bearing in mind that the residual signal reflects uncertainties in the system, the term $Q_{\mathcal{H}_2}r_{y,0}+\bar{Q}r_{y,0}$ causes direct deviation from the intended target defined by $\hat{v}$. Equation (7.17) highlights the reachable minimum influence of system uncertainties on the targeted dynamic

$$\begin{bmatrix}M_0\\N_0\end{bmatrix}\hat{v}=\begin{bmatrix}M\\N\end{bmatrix}v.$$

7.1.2 The Data Structure Under Uncertainties

We now address the data structure issue under existence of model uncertainties. Without loss of generality, it is supposed that the SIR of the plant G is described by

$$I_G=I_{G_0}+\Delta_{I_G}=\begin{bmatrix}M\\N\end{bmatrix}=\begin{bmatrix}M_0\\N_0\end{bmatrix}+\begin{bmatrix}\Delta_M\\\Delta_N\end{bmatrix}. \tag{7.20}$$

Here,

$$I_{G_0} = \begin{bmatrix} M_0 \\ N_0 \end{bmatrix}, \Delta_{I_G} = \begin{bmatrix} \Delta_M \\ \Delta_N \end{bmatrix} \in \mathcal{RH}_\infty,$$

represent the normalised nominal dynamics and the (unknown) uncertainties, respectively. As illustrated in Sect. 4.7, Δ_{I_G} is further written into

$$\Delta_{I_G} = \begin{bmatrix} M_0 \\ N_0 \end{bmatrix} \delta_{I_G} + \begin{bmatrix} -\hat{Y}_0 \\ \hat{X}_0 \end{bmatrix} \delta_{I_K}. \tag{7.21}$$

According to Theorem 4.6, (u, y) is subject to

$$\begin{aligned} \begin{bmatrix} u \\ y \end{bmatrix} &= \begin{bmatrix} M_0 \\ N_0 \end{bmatrix} v + \begin{bmatrix} M_0 \\ N_0 \end{bmatrix} Q\delta_{I_K} \left(I + \delta_{I_G} - Q\delta_{I_K}\right)^{-1} v \\ &+ \begin{bmatrix} -\hat{Y}_0 \\ \hat{X}_0 \end{bmatrix} \delta_{I_K} \left(I + \delta_{I_G} - Q\delta_{I_K}\right)^{-1} v. \end{aligned} \tag{7.22}$$

Considering that

$$\begin{aligned} r_{y,0} &= \begin{bmatrix} -\hat{N}_0 & \hat{M}_0 \end{bmatrix} \begin{bmatrix} u \\ y \end{bmatrix} = \begin{bmatrix} -\hat{N}_0 & \hat{M}_0 \end{bmatrix} \left(\begin{bmatrix} M_0 \\ N_0 \end{bmatrix} + \begin{bmatrix} \Delta_M \\ \Delta_N \end{bmatrix} \right) v_\delta \\ &= \delta_{I_K} \left(I + \delta_{I_G} - Q\delta_{I_K}\right)^{-1} v, \end{aligned} \tag{7.23}$$

$$v_\delta = \left(I + \begin{bmatrix} X_0 + Q\hat{N}_0 & Y_0 - Q\hat{M}_0 \end{bmatrix} \begin{bmatrix} \Delta_M \\ \Delta_N \end{bmatrix} \right)^{-1} v = \left(I + \delta_{I_G} - Q\delta_{I_K}\right)^{-1} v,$$

where the last equation is attributed to Lemma 4.2, (7.22) is further written as

$$\begin{bmatrix} u \\ y \end{bmatrix} = \begin{bmatrix} M_0 \\ N_0 \end{bmatrix} v + \left(\begin{bmatrix} -\hat{Y}_0 \\ \hat{X}_0 \end{bmatrix} + \begin{bmatrix} M_0 \\ N_0 \end{bmatrix} Q \right) r_{y,0}. \tag{7.24}$$

We would like to call the reader's attention to the fact that (7.22) or equivalently (7.24) is indeed an immediate result of the Bezout identity,

$$\begin{bmatrix} M_0 & -\hat{Y}_0 + M_0 Q \\ N_0 & \hat{X}_0 + N_0 Q \end{bmatrix} \begin{bmatrix} X_0 + Q\hat{N}_0 & Y_0 - Q\hat{M}_0 \\ -\hat{N}_0 & \hat{M}_0 \end{bmatrix} = \begin{bmatrix} I & 0 \\ 0 & I \end{bmatrix},$$

and independent of the loop configuration. To highlight this fact, let us define a transformation, $(u, y) \to \left(v, r_{y,0}\right)$,

$$\begin{aligned} u &= F_0 x + v + Q r_{y,0} \Longrightarrow v = X_0 u + Y_0 y - Q r_{y,0}, \\ r_{y,0} &= \hat{M}_0 y - \hat{N}_0 u, \end{aligned}$$

that is,

$$\begin{bmatrix} v \\ r_{y,0} \end{bmatrix} = \begin{bmatrix} X_0 + Q\hat{N}_0 & Y_0 - Q\hat{M}_0 \\ -\hat{N}_0 & \hat{M}_0 \end{bmatrix} \begin{bmatrix} u \\ y \end{bmatrix}.$$

Here, $(v, r_{y,0})$ serve as latent variables. Hence, by the above Bezout identity, we obtain

$$\begin{bmatrix} u \\ y \end{bmatrix} = \begin{bmatrix} M_0 \\ N_0 \end{bmatrix} v + \left(\begin{bmatrix} -\hat{Y}_0 \\ \hat{X}_0 \end{bmatrix} + \begin{bmatrix} M_0 \\ N_0 \end{bmatrix} Q \right) r_{y,0}.$$

As illustrated in the previous subsection, (7.24) can be expressed by (7.16), that is, (u, y) is decomposed into

$$\begin{bmatrix} u \\ y \end{bmatrix} = \begin{bmatrix} M_0 \\ N_0 \end{bmatrix} \left(v + Q_{\mathcal{H}_2} r_{y,0} + Q r_{y,0}\right) + \left(\begin{bmatrix} -\hat{Y}_0 \\ \hat{X}_0 \end{bmatrix} - \begin{bmatrix} M_0 \\ N_0 \end{bmatrix} Q_{\mathcal{H}_2} \right) r_{y,0}.$$

And it holds, for $Q = -Q_{\mathcal{H}_2}$,

$$\begin{bmatrix} u \\ y \end{bmatrix} = \begin{bmatrix} M_0 \\ N_0 \end{bmatrix} v + \left(\begin{bmatrix} -\hat{Y}_0 \\ \hat{X}_0 \end{bmatrix} - \begin{bmatrix} M_0 \\ N_0 \end{bmatrix} Q_{\mathcal{H}_2} \right) r_{y,0},$$

$$\left\| \begin{bmatrix} u \\ y \end{bmatrix} \right\|_2^2 = \left\| \begin{bmatrix} M_0 \\ N_0 \end{bmatrix} v \right\|_2^2 + \left\| \left(\begin{bmatrix} -\hat{Y}_0 \\ \hat{X}_0 \end{bmatrix} - \begin{bmatrix} M_0 \\ N_0 \end{bmatrix} Q_{\mathcal{H}_2} \right) r_{y,0} \right\|_2^2.$$

Note that $r_{y,0}$ is a function of Q, as shown in (7.23). Thus,

$$\left\| \begin{bmatrix} M_0 \\ N_0 \end{bmatrix} v \right\|_2^2 + \left\| \left(\begin{bmatrix} -\hat{Y}_0 \\ \hat{X}_0 \end{bmatrix} - \begin{bmatrix} M_0 \\ N_0 \end{bmatrix} Q_{\mathcal{H}_2} \right) r_{y,0} \right\|_2^2 \geq$$

$$\min_{Q \in \mathcal{RH}_\infty} \left(\left\| \begin{bmatrix} M_0 \\ N_0 \end{bmatrix} \left(v + Q_{\mathcal{H}_2} r_{y,0} + Q r_{y,0}\right) \right\|_2^2 + \left\| \left(\begin{bmatrix} -\hat{Y}_0 \\ \hat{X}_0 \end{bmatrix} - \begin{bmatrix} M_0 \\ N_0 \end{bmatrix} Q_{\mathcal{H}_2} \right) r_{y,0} \right\|_2^2 \right).$$

Nevertheless,

$$Q^* = -Q_{\mathcal{H}_2} = \arg \min_{Q \in \mathcal{RH}_\infty} \left\| \mathcal{P}_{\mathcal{I}_G} \left(\begin{bmatrix} u \\ y \end{bmatrix} \right) \right\|_2^2.$$

That means, setting $Q = -Q_{\mathcal{H}_2}$ assures that the image subspace $\mathcal{I}_G$ is free of uncertainties.

The following corollary is a summary of the main result achieved in this subsection.

Corollary 7.1 *Given plant G model (7.20) with the normalised RCP and LCP* (M_0, N_0) *and* $\left(\hat{M}_0, \hat{N}_0\right)$ *as well as the RCP* $\left(\hat{X}_0, \hat{Y}_0\right)$ *subject to*

$$\begin{bmatrix} -\hat{N}_0 & \hat{M}_0 \end{bmatrix} \begin{bmatrix} -\hat{Y}_0 \\ \hat{X}_0 \end{bmatrix} = I, \tag{7.25}$$

then we have

$$\begin{bmatrix} u \\ y \end{bmatrix} = \begin{bmatrix} M_0 \\ N_0 \end{bmatrix} \left(v + Q_{\mathcal{H}_2} r_{y,0} + Q r_{y,0}\right) + \left(\begin{bmatrix} -\hat{Y}_0 \\ \hat{X}_0 \end{bmatrix} - \begin{bmatrix} M_0 \\ N_0 \end{bmatrix} Q_{\mathcal{H}_2} \right) r_{y,0}, \quad (7.26)$$

$$\text{for } Q = -Q_{\mathcal{H}_2}, \left\| \begin{bmatrix} u \\ y \end{bmatrix} \right\|_2^2 = \left\| \begin{bmatrix} M_0 \\ N_0 \end{bmatrix} v \right\|_2^2 + \left\| \left(\begin{bmatrix} -\hat{Y}_0 \\ \hat{X}_0 \end{bmatrix} - \begin{bmatrix} M_0 \\ N_0 \end{bmatrix} Q_{\mathcal{H}_2} \right) r_{y,0} \right\|_2^2,$$

$$r_{y,0} = \begin{bmatrix} -\hat{N}_0 & \hat{M}_0 \end{bmatrix} \begin{bmatrix} u \\ y \end{bmatrix} = \delta_{I_K} \left(I + \delta_{I_G} - Q\delta_{I_K}\right)^{-1} v,$$

where v is a latent variable satisfying

$$v = X_0 u + Y_0 y - Q r_{y,0}, \quad (7.27)$$

and δ_{I_K}, δ_{I_G} are defined in (7.21).

On account of the equivalence of right and left coprime factor uncertainties proved in Sect. 4.7, we claim an interesting result: in an uncertain system, it is possible to assure that the (nominal) image subspace of the data (u, y) is free of the uncertainties addressed by the SKR and represented in terms of the residual $r_{y,0}$. Mathematically speaking, this implies

$$\mathcal{P}_{\mathcal{I}_G} \left(\begin{bmatrix} u \\ y \end{bmatrix} \right) = \begin{bmatrix} M_0 \\ N_0 \end{bmatrix} v.$$

Recall that the aforementioned results have been achieved on the assumption that only uncertainties addressed by the SKR and represented in terms of the residual are under consideration. Questions which naturally arise, are whether the impact of the uncertainties on the system dynamics is fully reflected in the data structure (7.26) and, if yes, how it is modelled. To answer these questions, it is sufficient to examine (7.22) and

$$u = F_0 x - Q r_{y,0} + v \Longleftrightarrow v = X_0 u + Y_0 y - Q r_{y,0},$$

which results in

$$v = \left(I + \delta_{I_G}\right) \left(I + \delta_{I_G} - Q\delta_{I_K}\right)^{-1} v - Q r_{y,0}. \quad (7.28)$$

Equation (7.28) poses that the impact of the uncertainties on the system image subspace and thus on the system dynamics is implicitly embedded in the latent variable v. In other words, the latent variable v might contain the targeted control action, a feedback of the residual $r_{y,0}$ to compensate the influence of $r_{y,0}$ on the image subspace, and uncertainties.

7.1.3 Discussions, New Insights, and Conclusions

The previous two subsections with Theorem 7.2 and Corollary 7.1 as major conclusions open a new perspective for essential control issues. The object in the focus of our investigation is process data, their structures and composition, rather than a controller and control system configuration. In this subsection, we attempt to explore hidden insights behind the results achieved in the previous subsections and hope to view control issues from the new perspective. For the simplicity of our discussion, the normalised RCP and LCP of G, (M_0, N_0), $\left(\hat{M}_0, \hat{N}_0\right)$ and, corresponding to them, $\left(\hat{X}_0, \hat{Y}_0\right)$, (X_0, Y_0) are considered.

7.1.3.1 Control, Uncertainties, and Latent Variables

To begin with, we examine the both signal subspaces, system image and residual subspaces, $\mathcal{I}_G$ and $\mathcal{R}_G$. They are exclusively defined by the nominal system dynamics and build the overall data space

$$\mathcal{H}_2 = \mathcal{I}_G \oplus \mathcal{R}_G.$$

In real-time implementation, all control actions are executed via the actuators represented by u. Due to the implicit relation of u and the latent variable v,

$$v = u - F_0 x - Q r_{y,0} \Longrightarrow v = X_0 u + Y_0 y - Q r_{y,0}, \tag{7.29}$$

the system responses (u, y) to all control actions build the image subspace $\mathcal{I}_G$,

$$\mathcal{I}_G = \left\{ \begin{bmatrix} u \\ y \end{bmatrix}, \begin{bmatrix} u \\ y \end{bmatrix} = I_G v = \begin{bmatrix} M \\ N \end{bmatrix} v, v \in \mathcal{H}_2 \right\}.$$

In this regard, $\mathcal{I}_G$ consists of all process data (u, y) as responses to control actions and represents the control capability of the actuators embedded in the system. However, it should be brought to mind that the latent variable v may be corrupted by uncertainties, which induces uncertainties in the image subspace $\mathcal{I}_G$.

Technically speaking, control actions consist of, in an extended sense, two tasks, (i) tracking a pre-defined trajectory, called target tracking, and (ii) suppressing the influence of all possible uncertainties that act on the plant and drive it to deviate from the target. The latter is called robust control. Let the target be described by (u_t, y_t). The task of target tracking can be schematically formulated as

$$\min_{v_t \in \mathcal{H}_2} \left\| \begin{bmatrix} u_t \\ y_t \end{bmatrix} - \begin{bmatrix} M \\ N \end{bmatrix} v_t \right\|.$$

Hereby, $\|\cdot\|$ represents a signal norm, v_t is a term in the latent variable v and dedicated to perform target tracking. This optimisation problem can be theoretically well solved, since all involved system models and variables, (M, N), (u_t, y_t) are available, and it is a standard signal matching problem. Differently, the robust control task is considerably challenging due to possible uncertainties. Thereby, numerous issues are to be addressed, among others, mapping uncertainties to the signal space, perceiving uncertainties, and compensating them in the signal space. In this regard, the residual subspace $\mathcal{R}_G$ plays an essential role. Signals in $\mathcal{R}_G$ represent uncertainties that cause the process data (u, y) to deviate from the targeted dynamics $\begin{bmatrix} M \\ N \end{bmatrix} v_t$. Variations in the process data induced by uncertainties in the residual subspace $\mathcal{R}_G$ can be perceived by the SKR of the system, $K_G = \begin{bmatrix} -\hat{N}_0 & \hat{M}_0 \end{bmatrix}$, because

$$r_{y,0} = \begin{bmatrix} -\hat{N}_0 & \hat{M}_0 \end{bmatrix} \begin{bmatrix} u \\ y \end{bmatrix} = \begin{cases} 0, \begin{bmatrix} u \\ y \end{bmatrix} \in \mathcal{I}_G \\ \neq 0, \begin{bmatrix} u \\ y \end{bmatrix} \neq \mathcal{I}_G. \end{cases} \tag{7.30}$$

From the system-theoretic viewpoint, $r_{y,0} = \begin{bmatrix} -\hat{N}_0 & \hat{M}_0 \end{bmatrix} \begin{bmatrix} u \\ y \end{bmatrix}$ is a residual generator driven by (u, y), whose role is the sensor-based perception of uncertainties. In the residual subspace $\mathcal{R}_G$, $r_{y,0}$ acts as a latent variable. Mathematically, mapping from the data (u, y) to the latent variables $(v, r_{y,0})$ is performed by

$$\begin{bmatrix} v \\ r_{y,0} \end{bmatrix} = \begin{bmatrix} X_0 + Q\hat{N}_0 & Y_0 - Q\hat{M}_0 \\ -\hat{N}_0 & \hat{M}_0 \end{bmatrix} \begin{bmatrix} u \\ y \end{bmatrix}.$$

Immediately, it follows from the Bezout identity that

$$\begin{aligned} \begin{bmatrix} u \\ y \end{bmatrix} &= \begin{bmatrix} X_0 + Q\hat{N}_0 & Y_0 - Q\hat{M}_0 \\ -\hat{N}_0 & \hat{M}_0 \end{bmatrix}^{-1} \begin{bmatrix} v \\ r_{y,0} \end{bmatrix} \\ &= \begin{bmatrix} M_0 \\ N_0 \end{bmatrix} \left(v + Qr_{y,0}\right) + \begin{bmatrix} -\hat{Y}_0 \\ \hat{X}_0 \end{bmatrix} r_{y,0}. \end{aligned}$$

Note that in the above equation, the term $Qr_{y,0}$ arises from (7.29) as a part of the relation of u and v, rather as a part of uncertainties. As such, $Qr_{y,0}$ parametrises the potential of the actuators to enhance the control robustness. In other words, the executing capability of the actuators is fully characterised by the sum $v_t + Qr_{y,0}$. Apparently, the impact of the system uncertainties in the residual subspace on the process data is modelled by $\begin{bmatrix} -\hat{Y}_0 \\ \hat{X}_0 \end{bmatrix} r_{y,0}$. It is of practical interest that, according to Theorem 7.2, $\begin{bmatrix} -\hat{Y}_0 \\ \hat{X}_0 \end{bmatrix} r_{y,0}$ might affect both $\mathcal{I}_G$ and $\mathcal{R}_G$, namely,

$$\begin{bmatrix} -\hat{Y}_0 \\ \hat{X}_0 \end{bmatrix} r_{y,0} = \begin{bmatrix} M_0 \\ N_0 \end{bmatrix} Q_{\mathcal{H}_2} r_{y,0} + \left(\begin{bmatrix} -\hat{Y}_0 \\ \hat{X}_0 \end{bmatrix} - \begin{bmatrix} M_0 \\ N_0 \end{bmatrix} Q_{\mathcal{H}_2} \right) r_{y,0},$$

$$\begin{bmatrix} M_0 \\ N_0 \end{bmatrix} Q_{\mathcal{H}_2} r_{y,0} \in \mathcal{I}_G, \left(\begin{bmatrix} -\hat{Y}_0 \\ \hat{X}_0 \end{bmatrix} - \begin{bmatrix} M_0 \\ N_0 \end{bmatrix} Q_{\mathcal{H}_2} \right) r_{y,0} \in \mathcal{R}_G,$$

Since $\mathcal{I}_G$ and $\mathcal{R}_G$ are perpendicular, $\left(\begin{bmatrix} -\hat{Y}_0 \\ \hat{X}_0 \end{bmatrix} - \begin{bmatrix} M_0 \\ N_0 \end{bmatrix} Q_{\mathcal{H}_2} \right) r_{y,0}$ is the uncertainty that exists in the process data and cannot be compensated, while the uncertainty $\begin{bmatrix} M_0 \\ N_0 \end{bmatrix} Q_{\mathcal{H}_2} r_{y,0}$ is compensable via the latent variable. Consequently, we have the result given in Theorem 7.2,

$$\begin{aligned} \left\| \begin{bmatrix} u \\ y \end{bmatrix} \right\|_2^2 &\geq \left\| \begin{bmatrix} M_0 \\ N_0 \end{bmatrix} v \right\|_2^2 + \left\| \left(\begin{bmatrix} -\hat{Y}_0 \\ \hat{X}_0 \end{bmatrix} - \begin{bmatrix} M_0 \\ N_0 \end{bmatrix} Q_{\mathcal{H}_2} \right) r_{y,0} \right\|_2^2 \\ &= \|v\|_2^2 + \left\| \left(\begin{bmatrix} -\hat{Y}_0 \\ \hat{X}_0 \end{bmatrix} - \begin{bmatrix} M_0 \\ N_0 \end{bmatrix} Q_{\mathcal{H}_2} \right) r_{y,0} \right\|_2^2 . \end{aligned} \tag{7.31}$$

Inequality (7.31) gives a lower bound of the energy (in the sense of $\mathcal{L}_2$-norm) in the process input and output signals (u, y). Thereby, $v = v_t + v_\Delta$ with v_Δ denoting possible uncertainties corrupted in v. $\|v\|_2^2$ represents the (energy) cost for the targeted control purpose, together with uncertainties in v, and $\left\| \left(\begin{bmatrix} -\hat{Y}_0 \\ \hat{X}_0 \end{bmatrix} - \begin{bmatrix} M_0 \\ N_0 \end{bmatrix} Q_{\mathcal{H}_2} \right) r_{y,0} \right\|_2^2$ is the energy in the non-compensable uncertainty. In this regard, the ratio,

$$R_{v/r} := \frac{\|v_t\|_2^2}{\left\| \left(\begin{bmatrix} -\hat{Y}_0 \\ \hat{X}_0 \end{bmatrix} - \begin{bmatrix} M_0 \\ N_0 \end{bmatrix} Q_{\mathcal{H}_2} \right) r_{y,0} \right\|_2^2 + \|v_\Delta\|_2^2}, \tag{7.32}$$

is a reasonable index for the data quality, and can be used, for instance, for learning-based data-driven control.

We adhere to $\begin{bmatrix} -\hat{Y}_0 \\ \hat{X}_0 \end{bmatrix} r_{y,0}$. Although uncertainties in the system are not explicitly modelled, their impact on the process data (u, y) is well perceived by $\begin{bmatrix} -\hat{Y}_0 \\ \hat{X}_0 \end{bmatrix} r_{y,0}$. That is, driven by the residual signal $r_{y,0}$, the system $\begin{bmatrix} -\hat{Y}_0 \\ \hat{X}_0 \end{bmatrix} r_{y,0}$ serves as a perception that delivers an estimate of variations in the process data caused by the uncertainties. As delineated in Sect. 2.1, $\begin{bmatrix} -\hat{Y}_0 \\ \hat{X}_0 \end{bmatrix}$ is a dynamic system with an observer at its core. As such, the computation of $\begin{bmatrix} -\hat{Y}_0 \\ \hat{X}_0 \end{bmatrix} r_{y,0}$ is trivial. This observation opens a new perspective to model dynamic systems with unmodelled uncertainties. The resulting model generates process input and output signals that perfectly fit the real process data (u, y) and, more importantly, is capable to run real-time. In fact, this model is a system realisation of the data structure,

$$\begin{bmatrix} u \\ y \end{bmatrix} = \begin{bmatrix} M_0 \\ N_0 \end{bmatrix} v + \begin{bmatrix} -\hat{Y}_0 \\ \hat{X}_0 \end{bmatrix} r_{y,0},$$

$$\begin{bmatrix} v \\ r_{y,0} \end{bmatrix} = \begin{bmatrix} X_0 & Y_0 \\ -\hat{N}_0 & \hat{M}_0 \end{bmatrix} \begin{bmatrix} u \\ y \end{bmatrix}.$$

A detailed introduction of this model and the related issues will be addressed in the next section.

7.1.3.2 On Separation Principle and a General Form

We now consecrate ourselves to the well-known separation principle, a fundamental result in control theory, and issues related to observer. Roughly speaking, the separation principle in its original form tells us, design of an observer-based controller towards closed-loop stabilisation can be performed by designing an observer and a state feedback controller separately. In an extended sense, it describes the fact that in a feedback control system, the dynamics of the controller and estimation error are decoupled. Note that, in general, the applicability of the separation principle is restricted to uncertainties in additive form.

The data structure (7.16) and the associated property (7.17) disclose a more general separation form in the process data (u, y). To be specific,

- the control actions executed by v and perception of uncertainties are naturally separated in data as shown by (7.1),
- control actions are fully decoupled from the uncertainties in the residual subspace $\mathcal{R}_G$, and
- the compensation control action $Qr_{y,0}$ is independent of the design of the residual generator,
- consequently, the design of the residual generator, independent of any executed control action, determines the impact of the uncertainties in the residual subspace $\mathcal{R}_G$ on the process data.

To put it in a nutshell, design of control actions and perception can be fully separately performed. This is a general form of separation principle that we can draw from the data structure introduced in this section.

The reader may notice that in the above discussion on separation principle a particular focus is on residual generators rather than observers. This is a natural consequence of the fact that any feedback control is a feedback of residual signals. This can be apparently observed from the data structure (7.1) or equivalently (7.16). In this context, the classical interpretation of separation principle as separated design of (state) feedback controller and (state) observer is misleading. In fact, the standard proof of separation principle is based on the decoupled relation of the dynamics of the (state) feedback control and the (state) estimation error. As known, the dynamic of the state estimation error is the dynamic of the residual generator, rather than that of the observer (state estimator).

As a closing remark, state observer and its role in observer-based feedback control schemes are concerned. A central conclusion of the unified framework of control and detection is that any (dynamic) output feedback controller is an observer-based state feedback controller. It is based on the result that feedback control and its response are schematically described by

$$\begin{bmatrix} u \\ y \end{bmatrix} = \begin{bmatrix} -\hat{Y} \\ \hat{X} \end{bmatrix} r_y \Longleftrightarrow$$

$$\hat{x}(k+1) = (A+BF)\hat{x}(k) + LW^{-1}r_y(k) \tag{7.33}$$

$$\begin{bmatrix} u(k) \\ y(k) \end{bmatrix} = \begin{bmatrix} F \\ C+DF \end{bmatrix} \hat{x}(k) + \begin{bmatrix} 0 \\ W^{-1} \end{bmatrix} r_y(k),$$

whose core is the state observer (7.33). It is worth noting that this state observer is driven by the residual r_y instead of the system input u. Accordingly, the state variable $\hat{x}$ reflects the influence of uncertainties expressed by r_y. In this regard, the state observer embedded in the feedback control serves perception of the uncertainties in the system.

7.1.3.3 On Optimal Robust Control and Data Structure

A classic optimal robust control problem is the MMP optimisation problem,

$$\min_{Q \in \mathcal{RH}_\infty} \left\| \begin{bmatrix} -\hat{Y} + MQ \\ \hat{X} + NQ \end{bmatrix} \right\|_\infty, \tag{7.34}$$

which apparently implies, in view of (7.1), a minimisation of the impact of uncertainties on (u, y). To highlight the relation between the MMP (7.34) and the compensation of uncertainties in the image subspace $\mathcal{I}_G$, formulated as

$$Q^* = -Q_{\mathcal{H}_2} = -\mathcal{P}_{\mathcal{H}_2}\left[M_0^{\sim} \; N_0^{\sim} \right] \begin{bmatrix} -\hat{Y}_0 \\ \hat{X}_0 \end{bmatrix}, \tag{7.35}$$

we firstly delineate the solution of MMP (7.34). Our discussion in the previous subsections has demonstrated that

$$\min_{Q \in \mathcal{RH}_\infty} \left\| \begin{bmatrix} -\hat{Y} + MQ \\ \hat{X} + NQ \end{bmatrix} \right\|_\infty = \min_{Q \in \mathcal{RH}_\infty} \left\| \begin{bmatrix} -\hat{Y}_0 + M_0 Q \\ \hat{X}_0 + N_0 Q \end{bmatrix} \right\|_\infty,$$

where (M_0, N_0), $\left(\hat{M}_0, \hat{N}_0\right)$ are normalised RCP and LCP and $\left(\hat{X}_0, \hat{Y}_0\right)$ satisfies (7.25). It can be further equivalently written as

$$\min_{Q \in \mathcal{RH}_\infty} \left\| \begin{bmatrix} -\hat{Y}_0 + M_0 Q \\ \hat{X}_0 + N_0 Q \end{bmatrix} \right\|_\infty = \min_{Q \in \mathcal{RH}_\infty} \left\| \begin{bmatrix} -\hat{Y}_{0,N} + M_0 Q \\ \hat{X}_{0,N} + N_0 Q \end{bmatrix} \right\|_\infty \tag{7.36}$$

with the normalised $\left(\hat{X}_{0,N}, \hat{Y}_{0,N}\right)$ satisfying

$$\begin{bmatrix} -\hat{Y}_{0,N} \\ \hat{X}_{0,N} \end{bmatrix} = \begin{bmatrix} -\hat{Y}_0 \\ \hat{X}_0 \end{bmatrix} V_0^{-1}, \ V_0^\sim V_0 = \begin{bmatrix} -\hat{Y}_0^\sim & \hat{X}_0^\sim \end{bmatrix} \begin{bmatrix} -\hat{Y}_0 \\ \hat{X}_0 \end{bmatrix},$$

$$V_0, V_0^{-1} \in \mathcal{RH}_\infty \Longrightarrow \begin{bmatrix} -\hat{Y}_{0,N}^\sim & \hat{X}_{0,N}^\sim \end{bmatrix} \begin{bmatrix} -\hat{Y}_{0,N} \\ \hat{X}_{0,N} \end{bmatrix} = I.$$

Remark 7.1 The computation of

$$V_0^\sim V_0 = \begin{bmatrix} -\hat{Y}_0^\sim & \hat{X}_0^\sim \end{bmatrix} \begin{bmatrix} -\hat{Y}_0 \\ \hat{X}_0 \end{bmatrix}$$

satisfying

$$V_0, V_0^{-1} \in \mathcal{RH}_\infty$$

can be achieved by a spectral factorisation.

By the subsequent computation

$$\begin{bmatrix} M_0^\sim & N_0^\sim \\ -\hat{N}_0 & \hat{M}_0 \end{bmatrix} \begin{bmatrix} -\hat{Y}_{0,N} + M_0 Q \\ \hat{X}_{0,N} + N_0 Q \end{bmatrix} = \begin{bmatrix} \begin{bmatrix} M_0^\sim & N_0^\sim \end{bmatrix} \begin{bmatrix} -\hat{Y}_{0,N} \\ \hat{X}_{0,N} \end{bmatrix} + Q \\ I \end{bmatrix},$$

and noticing that

$$\begin{bmatrix} M_0^\sim & N_0^\sim \\ -\hat{N}_0 & \hat{M}_0 \end{bmatrix} \begin{bmatrix} M_0^\sim & N_0^\sim \\ -\hat{N}_0 & \hat{M}_0 \end{bmatrix}^\sim = I \Longleftrightarrow \begin{bmatrix} M_0^\sim & N_0^\sim \\ -\hat{N}_0 & \hat{M}_0 \end{bmatrix}^\sim \begin{bmatrix} M_0^\sim & N_0^\sim \\ -\hat{N}_0 & \hat{M}_0 \end{bmatrix} = I,$$

the MMP (7.34) is equivalent to

$$\min_{Q \in \mathcal{RH}_\infty} \left\| \begin{bmatrix} -\hat{Y} + MQ \\ \hat{X} + NQ \end{bmatrix} \right\|_\infty \Longleftrightarrow \min_{Q \in \mathcal{RH}_\infty} \left(\begin{bmatrix} M_0^\sim & N_0^\sim \end{bmatrix} \begin{bmatrix} -\hat{Y}_{0,N} \\ \hat{X}_{0,N} \end{bmatrix} + Q \right).$$

Apparently, the solution of (7.35) can be equivalently achieved by setting

$$Q^* = -Q_{\mathcal{H}_2} = -Q_0 V_0, \tag{7.37}$$

$$Q_0 = \arg \min_{Q \in \mathcal{RH}_\infty} \left\| \left[M_0^\sim \; N_0^\sim \right] \begin{bmatrix} -\hat{Y}_{0,N} \\ \hat{X}_{0,N} \end{bmatrix} + Q \right\|_\infty^2 . \tag{7.38}$$

Remark 7.2 The claim (7.37) is well-founded as follows. It known from operator theory that the MMP (7.36) is equivalent to the computation of the so-called gap metric between the subspaces $\mathcal{I}_G$ and $\mathcal{I}_K$ spanned by

$$\begin{bmatrix} M_0 \\ N_0 \end{bmatrix} \text{ and } \begin{bmatrix} -\hat{Y}_{0,N} \\ \hat{X}_{0,N} \end{bmatrix},$$

respectively (refer to Sect. 3.1 for $\mathcal{I}_K$). As a result, the gap metric between the subspaces spanned by

$$\begin{bmatrix} M_0 \\ N_0 \end{bmatrix} \text{ and } \begin{bmatrix} -\hat{Y}_{0,N} \\ \hat{X}_{0,N} \end{bmatrix} - \begin{bmatrix} M_0 \\ N_0 \end{bmatrix} Q$$

is zero, which means these two subspaces are orthogonal. The reader is referred to Chap. 9 for the details about orthogonal projection and gap metric.

In combination with the discussion in the previous subsections, we notice that the equivalence (7.37) is not only to give an insightful interpretation of the robust control in the regard of signal space, but also to present an insightful aspect of robust control that results in signals in $\mathcal{I}_G$ and $\mathcal{R}_G$ being perpendicular. The significance of the latter property lies in its potential application to performance monitoring and fault detection.

7.1.3.4 Concluding Remarks

At the end of this section, we summarise the major results and exemplify their potential applications in fault detection, performance monitoring and fault-tolerant control. Some of the presented issues will be addressed in the subsequent sections.

- That any process data (u, y) is of the embedded control structure (7.1) or more precisely structure (7.16) forms the theoretical basis for data-driven learning of control systems.
  It is obvious that on this basis, both feedforward controller (in fact the latent variable v) and feedback controller, including the latent variable r_y, can be learnt, designed or optimised in various possible manners using existing powerful optimisation, statistic and machine learning algorithms and tools.
- The mapping between the process data (u, y) and the latent variables $\left(v, r_y\right)$ is one-to-one and determined by the Bezout identity.
  While the residual r_y contains information about uncertainties in the residual subspace exclusively, v might be composed of control action signals and uncertainties in the residual subspace.

- The system $\begin{bmatrix} -\hat{Y}_0 \\ \hat{X}_0 \end{bmatrix} r_{y,0}$ is a perceptive system of uncertainties in the process data (u, y).
  As such, its state space form,
$$\hat{x}(k+1) = (A + BF_0)\,\hat{x}(k) + L_0 W_0^{-1} r_{y,0}(k)$$
$$\begin{bmatrix} u(k) \\ y(k) \end{bmatrix} = \begin{bmatrix} F_0 \\ C + DF_0 \end{bmatrix} \hat{x}(k) + \begin{bmatrix} 0 \\ W_0^{-1} \end{bmatrix} r_{y,0}(k),$$
  can be applied for fault detection.
- The equivalence relation (7.37) offers an alternative way to optimise the feedback system Q in optimal robust control.
  To be specific, solving the following optimisation problem based on the data projection,
$$\sup_{r_y \in \mathcal{H}_2} \frac{\left\| \left(\mathcal{I} - \mathcal{P}_{\mathcal{I}_G}\right) \left(\begin{bmatrix} -\hat{Y}_0 \\ \hat{X}_0 \end{bmatrix} r_y \right) \right\|_2}{\left\| r_y \right\|_2}$$
  can improve the system robustness and fault-tolerance. Observing the result in Corollary 7.1, it can be claimed that the above conclusion holds for uncertain systems as well. Below is an interesting example to demonstrate such an application, for instance, in fault-tolerant control.

Example 7.1 Given a plant modelled by
$$G = \left(N_0 + \hat{X}_0 S\right) \left(M_0 - \hat{Y}_0 S\right)^{-1}, \tag{7.39}$$
where $S \in \mathcal{RH}_\infty$ represents an unknown (multiplicative) fault, and (M_0, N_0), $\left(\hat{X}_0, \hat{Y}_0\right)$ are the notations adopted in this section. According to Theorem 7.2, the data (u, y) can be structured into
$$\begin{aligned} \begin{bmatrix} u \\ y \end{bmatrix} &= \begin{bmatrix} M_0 \\ N_0 \end{bmatrix} v + \begin{bmatrix} -\hat{Y}_0 \\ \hat{X}_0 \end{bmatrix} r_{y,0} \\ &= \begin{bmatrix} M_0 \\ N_0 \end{bmatrix} \left(v + Q_{\mathcal{H}_2} r_{y,0}\right) + \left(\begin{bmatrix} -\hat{Y}_0 \\ \hat{X}_0 \end{bmatrix} - \begin{bmatrix} M_0 \\ N_0 \end{bmatrix} Q_{\mathcal{H}_2} \right) r_{y,0}, \\ r_{y,0} &= Sv. \end{aligned}$$
Let
$$v = v_t - Q_{\mathcal{H}_2} r_{y,0}. \tag{7.40}$$

It turns out

$$\begin{bmatrix} u \\ y \end{bmatrix} = \begin{bmatrix} M_0 \\ N_0 \end{bmatrix} v_t + \left(\begin{bmatrix} -\hat{Y}_0 \\ \hat{X}_0 \end{bmatrix} - \begin{bmatrix} M_0 \\ N_0 \end{bmatrix} Q_{\mathcal{H}_2} \right) r_{y,0}.$$

Thus, any control target will be approached by designing v_t. Note that (7.40) implicitly requires

$$v = v_t - Q_{\mathcal{H}_2} S v \Longrightarrow \left(I - Q_{\mathcal{H}_2} S\right)^{-1} \in \mathcal{RH}_\infty.$$

This is exactly the stability condition given in Theorem 4.4. On assumption of $\left(I - Q_{\mathcal{H}_2} S\right)^{-1} \in \mathcal{RH}_\infty$, controller,

$$v = v_t - Q_{\mathcal{H}_2} r_{y,0} \Longrightarrow u = F_0 \hat{x} + v_t - Q_{\mathcal{H}_2} r_{y,0},$$

is a fault-tolerant controller that can be implemented either in the passive (fault-tolerant control) or active style.

- The data structure,

$$\begin{bmatrix} u \\ y \end{bmatrix} = \begin{bmatrix} M_0 \\ N_0 \end{bmatrix} v + \left(\begin{bmatrix} -\hat{Y}_0 \\ \hat{X}_0 \end{bmatrix} - \begin{bmatrix} M_0 \\ N_0 \end{bmatrix} Q_{\mathcal{H}_2} \right) r_{y,0}, \tag{7.41}$$

  enables execution of any control action exclusively by tuning the latent variable v and is independent of uncertainties in the residual subspace.
  This is a natural conclusion drawn from the fact that

$$\begin{bmatrix} M_0 \\ N_0 \end{bmatrix} v \in \mathcal{I}_G, \left(\begin{bmatrix} -\hat{Y}_0 \\ \hat{X}_0 \end{bmatrix} - \begin{bmatrix} M_0 \\ N_0 \end{bmatrix} Q_{\mathcal{H}_2} \right) r_{y,0} \in \mathcal{R}_G,$$

  and they are orthogonal to each other. In this case,

$$u = F_0 \hat{x} + v - Q_{\mathcal{H}_2} r_{y,0}.$$

  This conclusion is useful to improve the existing model predictive control (MPC). A further potential application of the above data structure is robust and fault-tolerant control of multi-agent systems (MAS), where control actions of MAS, for instance, cooperative control, will be realised by designing the latent variable of each agent. The fault detection will be performed by means of information in

$$\left(\begin{bmatrix} -\hat{Y}_0 \\ \hat{X}_0 \end{bmatrix} - \begin{bmatrix} M_0 \\ N_0 \end{bmatrix} Q_{\mathcal{H}_2} \right) r_{y,0}$$

  of each agent.
- The relation (7.17) is useful for performance monitoring and control.
  For instance, for given control target, (u_t, y_t), real-time solving the optimisation problem

$$\min_{v \in \mathcal{H}_2} \left\| \begin{bmatrix} u_t \\ y_t \end{bmatrix} - \mathcal{P}_{\mathcal{I}_G} \left(\begin{bmatrix} u \\ y \end{bmatrix} \right) \right\|_2$$

leads to an optimal tracking of the target. Note that

$$\mathcal{P}_{\mathcal{I}_G} \left(\begin{bmatrix} u \\ y \end{bmatrix} \right) = \begin{bmatrix} M_0 \\ N_0 \end{bmatrix} v$$

is orthogonal to the uncertainty in the system,

$$\left(\begin{bmatrix} -\hat{Y}_0 \\ \hat{X}_0 \end{bmatrix} - \begin{bmatrix} M_0 \\ N_0 \end{bmatrix} Q_{\mathcal{H}_2} \right) r_{y,0}.$$

For performance monitoring, real-time computation of the ratio

$$R_{v/r} = \frac{\|v_t\|_2^2}{\left\| \left(\begin{bmatrix} -\hat{Y}_0 \\ \hat{X}_0 \end{bmatrix} - \begin{bmatrix} M_0 \\ N_0 \end{bmatrix} Q_{\mathcal{H}_2} \right) r_{y,0} \right\|_2^2 + \left\| \mathcal{P}_{\mathcal{I}_G} \left(\begin{bmatrix} u \\ y \end{bmatrix} \right) - \begin{bmatrix} M_0 \\ N_0 \end{bmatrix} v_t \right\|_2^2}$$

can be used. Or, a performance-based fault detection can be approached using

$$\left\| \begin{bmatrix} u \\ y \end{bmatrix} - \mathcal{P}_{\mathcal{I}_G} \left(\begin{bmatrix} u \\ y \end{bmatrix} \right) \right\|_2 = \left\| \left(\begin{bmatrix} -\hat{Y}_0 \\ \hat{X}_0 \end{bmatrix} - \begin{bmatrix} M_0 \\ N_0 \end{bmatrix} Q_{\mathcal{H}_2} \right) r_{y,0} \right\|_2$$

as an evaluation function, and the corresponding gap metric as the threshold.

7.2 A New Process Model Form: SIR-SKR-Based Model

The data structure (7.1) is a natural model form that encloses both the nominal system dynamics and the existing uncertainties in the system. Before we formally define it as a process model, the so-called observer-based unified system model introduced in Sect. 4.6 is firstly reviewed.

7.2.1 From SKR-Based to SIR-SKR-Based Model

The observer-based unified system model (4.37)–(4.38) can be equivalently expressed in terms of the system SKR,

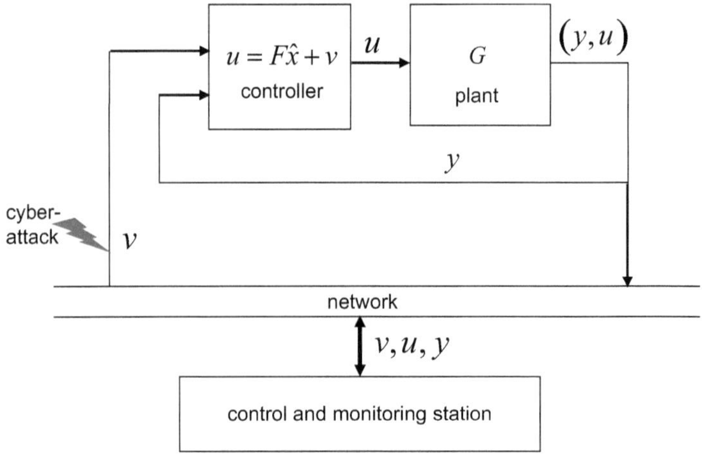

Fig. 7.1 A networked feedback control system under cyber-attack

$$\hat{x}(k+1) = A\hat{x}(k) + Bu(k) + LW^{-1}r_y(k), \tag{7.42}$$

$$y(k) = W^{-1}r_y(k) + C\hat{x}(k) + Du(k), \tag{7.43}$$

$$r_y(k) = W\left(y(k) - C\hat{x}(k) - Du(k)\right) \Longleftrightarrow r_y = \hat{M}y - \hat{N}u.$$

Hence, we also call the above form SKR-based input-output model. A major advantage of this model is that it is, independent of the existence of any model uncertainties, capable to run online. It is apparent that this model fully describes system input-output dynamics, also under existence of uncertainties. On the other hand, viewing from the data space (u, y), limitation of the SKR model can be identified. To exemplify this aspect, we look at the following example.

Example 7.2 We demonstrate some examples of uncertainties or anomalies resulting in variations in the kernel subspace (equivalently in the image subspace) of the plant. In the first example, a networked control system depicted in Fig. 7.1 is under consideration, in which the transmission of the reference signal v is under cyber-attacks. The attack model is

$$v^a(z) = \delta(z)v(z) + a_v(z),$$

where v^a is the attacked reference signal, and δ and a_v represent the multiplicative and additive changes in the dynamics of the transmission channel caused by the attack. On the assumption of the plant model $G = (A, B, C, D)$, the control input consists of two terms,

$$u(k) = F\hat{x}(k) + v^a(k),$$

and the observer-based state feedback term $F\hat{x}(k)$ is realised on the plant side. It is straightforward that

$$\hat{x}(k+1) = (A+BF)\hat{x}(k) + Bv^a(k) + Lr_y(k), \tag{7.44}$$

$$r_y(k) = r_0(k) = y(k) - \hat{y}(k), \hat{y}(k) = (C+DF)\hat{x}(k) + Dv^a(k), \tag{7.45}$$

$$\begin{bmatrix} u(k) \\ y(k) \end{bmatrix} = \begin{bmatrix} F \\ (C+DF) \end{bmatrix} \hat{x}(k) + \begin{bmatrix} I \\ D \end{bmatrix} v^a(k) + \begin{bmatrix} 0 \\ I \end{bmatrix} r_y(k), \tag{7.46}$$

which can be written as

$$\begin{bmatrix} u \\ y \end{bmatrix} = \begin{bmatrix} M \\ N \end{bmatrix} (\delta v + a_v) + \begin{bmatrix} -\hat{Y} \\ \hat{X} \end{bmatrix} r_y.$$

Since

$$\begin{bmatrix} -\hat{N} & \hat{M} \end{bmatrix} \begin{bmatrix} M \\ N \end{bmatrix} = 0,$$

it is apparent that r_y is decoupled from δ and a_v. The next example is related to the so-called covert attacks modelled by

$$x(k+1) = Ax(k) + B\left(u(k) + a_u(k)\right), \tag{7.47}$$

$$y^a(k) = Cx(k) + D\left(u(k) + a_u(k)\right) + a_y(k) \tag{7.48}$$

with a_u and a_y denoting attacks on the actuators and sensors, respectively, and satisfying

$$a_y(z) + G(z)a_u(z) = 0 \Longrightarrow \forall k \in [k_0, k_0 + N], y^a(k) = y(k).$$

Here, G is the transfer function matrix of the nominal plant dynamics and assumed to be stable. Observe that

$$\begin{bmatrix} -\hat{N} & \hat{M} \end{bmatrix} \begin{bmatrix} u \\ y^a \end{bmatrix} = \begin{bmatrix} -\hat{N} & \hat{M} \end{bmatrix} \begin{bmatrix} u + a_u - a_u \\ y^a - a_y + a_y \end{bmatrix},$$

$$\begin{bmatrix} -\hat{N} & \hat{M} \end{bmatrix} \begin{bmatrix} u + a_u \\ y^a - a_y \end{bmatrix} = 0 \Longrightarrow \begin{bmatrix} -\hat{N} & \hat{M} \end{bmatrix} \begin{bmatrix} -a_u \\ a_y \end{bmatrix} = 0.$$

Furthermore, remember that any signal $z \in \mathcal{H}_2^{m+p}$ satisfying

$$\begin{bmatrix} -\hat{N} & \hat{M} \end{bmatrix} z = 0$$

belongs to the system image subspace. Consequently, it can be claimed that the attack vector $\begin{bmatrix} -a_u \\ a_y \end{bmatrix}$ belongs to the image subspace of the plant.

In the second example, we consider a trivial case.

Example 7.3 Given a plant modelled by

$$\begin{aligned} x(k+1) &= Ax(k) + Bu(k) + E_d d(k), \\ y(k) &= Cx(k) + Du(k) + F_d d(k), \end{aligned}$$

where $d(k) \in \mathbb{R}^{k_d}$ is an $\mathcal{L}_{2,[0,N]}$ bounded unknown input vector that might be a fault vector. Equivalently, the input-output dynamic of this system can be presented by the SKR-based model, namely

$$\hat{x}(k+1) = A\hat{x}(k) + Bu(k) + L_0 W_0^{-1} r_{y,0}(k), \tag{7.49}$$

$$r_{y,0}(k) = W_0 \left(y(k) - C\hat{x}(k) - Du(k) \right), \tag{7.50}$$

$$y(k) = C\hat{x}(k) + Du(k) + W_0^{-1} r_{y,0}(k).$$

It is assumed that A is a Schur-matrix. The response of y to the unknown input d is subject to

$$\begin{aligned} y_d &= G_{yd} d = \left(I + C (zI - A)^{-1} L_0 \right) W_0^{-1} r_{y,0} = \hat{X} r_{y,0}, \\ r_{y,0} &= W_0 \left(C (zI - A + L_0 C)^{-1} (E_d - L_0 F_d) + F_d \right) d. \end{aligned}$$

Here, the normalised SKR is adopted. Now, let's have a close look at the response of (u, y) to d and its expression in the system image and residual subspaces. Write the response of (u, y) to d as

$$\begin{bmatrix} u_d \\ y_d \end{bmatrix} = \begin{bmatrix} 0 \\ \hat{X} \end{bmatrix} r_{y,0}.$$

Following our discussion in Sect. 7.1.1 and the results in Theorem 7.2, $\begin{bmatrix} 0 \\ \hat{X} \end{bmatrix} r_{y,0}$ can be decomposed into

$$\begin{aligned} \begin{bmatrix} 0 \\ \hat{X} \end{bmatrix} r_{y,0} &= \begin{bmatrix} M_0 \\ N_0 \end{bmatrix} \left(Q_{\mathcal{H}_2} + Q_2 \right) r_{y,0} + \left(\begin{bmatrix} -\hat{Y}_0 \\ \hat{X}_0 \end{bmatrix} - \begin{bmatrix} M_0 \\ N_0 \end{bmatrix} Q_{\mathcal{H}_2} \right) r_{y,0} \\ &= \begin{bmatrix} M_0 \\ N_0 \end{bmatrix} Q_2 r_{y,0} + \begin{bmatrix} -\hat{Y}_0 \\ \hat{X}_0 \end{bmatrix} r_{y,0}, \\ Q_2 &= Y_0 \hat{X} \in \mathcal{RH}_\infty. \end{aligned}$$

Recall that

$$\begin{bmatrix} M_0 \\ N_0 \end{bmatrix} \left(Q_{\mathcal{H}_2} + Q_2 \right) r_{y,0} = \mathcal{P}_{\mathcal{I}_G} \left(\begin{bmatrix} u_d \\ y_d \end{bmatrix} \right) \in \mathcal{I}_G,$$

$$\left(\begin{bmatrix} -\hat{Y}_0 \\ \hat{X}_0 \end{bmatrix} - \begin{bmatrix} M_0 \\ N_0 \end{bmatrix} \mathcal{Q}_{\mathcal{H}_2}\right) r_{y,0} = \mathcal{P}_{\mathcal{R}_G}\left(\begin{bmatrix} u_d \\ y_d \end{bmatrix}\right) \in \mathcal{R}_G,$$

$$\begin{bmatrix} -\hat{N}_0 & \hat{M}_0 \end{bmatrix} \begin{bmatrix} M_0 \\ N_0 \end{bmatrix} = 0.$$

As a result, it can be concluded that the unknown input d generally acts on the image and residual subspaces $\mathcal{I}_G$ and $\mathcal{R}_G$.

Examples 7.2 and 7.3 have not only helped us for identifying the deficit of the SKR-based model, but also implicitly suggested the possible solution, namely, SIR should be taken into account, in order to model the uncertainties in the image subspace.

Definition 7.1 Given the nominal plant model (2.1)–(2.3) and the associated RCPs and LCPs, (M, N), $\left(\hat{X}, \hat{Y}\right)$ and $\left(\hat{M}, \hat{N}\right)$, (X, Y) with the state space representations (2.6), (2.25) and (2.5), (2.26) and subject to the Bezout identity (2.31), the following form is called SIR-SKR-based model,

$$\hat{x}(k+1) = (A + BF)\,\hat{x}(k) + BVv(k) + LW^{-1}r_y(k), \tag{7.51}$$

$$\begin{bmatrix} u(k) \\ y(k) \end{bmatrix} = \begin{bmatrix} F \\ C + DF \end{bmatrix} \hat{x}(k) + \begin{bmatrix} V \\ DV \end{bmatrix} v(k) + \begin{bmatrix} 0 \\ W^{-1} \end{bmatrix} r_y(k), \tag{7.52}$$

$$r_y(k) = W\left(y(k) - C\hat{x}(k) - Du(k)\right) \tag{7.53}$$

$$v(k) = V^{-1}\left(u(k) - F\hat{x}(k)\right). \tag{7.54}$$

The above state space model form can be expressed compactly by

$$\begin{bmatrix} u \\ y \end{bmatrix} = \begin{bmatrix} M & -\hat{Y} \\ N & \hat{X} \end{bmatrix} \begin{bmatrix} v \\ r_y \end{bmatrix}, \tag{7.55}$$

$$\begin{bmatrix} v \\ r_y \end{bmatrix} = \begin{bmatrix} X & Y \\ -\hat{N} & \hat{M} \end{bmatrix} \begin{bmatrix} u \\ y \end{bmatrix}. \tag{7.56}$$

The reader may easily recognise that the SIR-SKR-based model (7.55)–(7.56) is coincident with the data structure (7.3). Apart from the advantage of the online capability under existence of any model uncertainties, similar to the SKR-based model, the SIR-SKR-based model

- apprehends all possible uncertainties in the whole signal space (u, y),
- guarantees a stable model implementation, when only unknown inputs as uncertainties act on the system,
- is applicable for both open-loop and closed-loop configurations, and
- can serve both as controller and detector, as described by (7.55) and (7.56), respectively.

In the next example, the application of the SIR-SKR-based model to the analysis of system dynamics under unknown input d is showcased. It is assumed that $x(0) = \hat{x}(0) = 0$.

Example 7.4 To begin with, the influence of d on (u, y) under the assumption $u = Fx + \upsilon_0$ is modelled. Hereby, υ_0 represents a targeted control action. It turns out

$$\begin{gathered}
x(k+1) = (A + BF)\,x(k) + BV\upsilon_0(k) + E_d d(k), \\
y(k) = (C + DF)\,x(k) + DV\upsilon_0(k) + F_d d(k) \Longrightarrow \\
\begin{bmatrix} u \\ y \end{bmatrix} = \begin{bmatrix} M \\ N \end{bmatrix} \upsilon_0 + \begin{bmatrix} G_{ud} \\ G_{yd} \end{bmatrix} d, \\
G_{ud} = F\,(zI - A - BF)^{-1}\,E_d,\, G_{yd} = (C + DF)\,(zI - A - BF)^{-1}\,E_d + F_d.
\end{gathered}$$

It yields

$$\begin{aligned}
\upsilon &= Xu + Yy = \upsilon_0 + \left(XG_{ud} + YG_{yd}\right) d, \\
r_y &= \hat{M}y - \hat{N}u = \left(\hat{M}G_{yd} - \hat{N}G_{ud}\right) d.
\end{aligned}$$

Observe that, according to the Bezout identity,

$$M\left(XG_{ud} + YG_{yd}\right) d = \left(\left(I - \hat{Y}\hat{N}\right) G_{ud} + \hat{Y}\hat{M}G_{yd}\right) d = G_{ud}d + \hat{Y}r_y, \quad (7.57)$$

$$N\left(XG_{ud} + YG_{yd}\right) d = \left(\hat{X}\hat{N}G_{ud} + \left(I - \hat{X}\hat{M}G_{yd}\right)\right) d = G_{yd}d - \hat{X}r_y. \quad (7.58)$$

Now, substituting $\left(\upsilon, r_y\right)$ into the SIR-SKR-based model (7.51)–(7.52) leads to

$$\begin{bmatrix} u \\ y \end{bmatrix} = \begin{bmatrix} M \\ N \end{bmatrix} \left(\upsilon_0 + \left(XG_{ud} + YG_{yd}\right) d\right) + \begin{bmatrix} -\hat{Y} \\ \hat{X} \end{bmatrix} r_y.$$

By means of (7.57)–(7.58), we finally obtain

$$\begin{aligned}
\begin{bmatrix} u \\ y \end{bmatrix} &= \begin{bmatrix} M \\ N \end{bmatrix} \left(\upsilon_0 + \left(XG_{ud} + YG_{yd}\right) d\right) + \begin{bmatrix} -\hat{Y} \\ \hat{X} \end{bmatrix} r_y \\
&= \begin{bmatrix} M \\ N \end{bmatrix} \upsilon_0 + \begin{bmatrix} G_{ud} \\ G_{yd} \end{bmatrix} d,
\end{aligned}$$

which proves that the SIR-SKR-based model exactly describes the system dynamics.

The reader may notice the assumption $x(0) = \hat{x}(0) = 0$, which is, in real applications, unrealistic. In the next example, we demonstrate how to model the uncertainty caused by the unknown initial state condition.

Example 7.5 Given the state space model

$$x(k+1) = Ax(k) + Bu(k),\, x(0) = x_0, \quad (7.59)$$

$$y(k) = Cx(k) + Du(k). \quad (7.60)$$

Let

$$E_d = Ax_0, F_d = Cx_0, d(k) = \delta(k), \delta(k) = \begin{cases} 1, k = 0 \\ 0, k \neq 0. \end{cases}$$

The model (7.59)–(7.60) is equivalent to

$$\begin{aligned} x(k+1) &= Ax(k) + Bu(k) + E_d d(k), x(0) = 0, \\ y(k) &= Cx(k) + Du(k) + F_d d(k). \end{aligned}$$

That means, the influence of the unknown initial condition can be modelled as an unknown input, and handled as shown in Example 7.4.

Remark 7.3 Based on the above results, it is assumed, in the remaining of this book, that $x(0) = \hat{x}(0) = 0$. The deviation in the system dynamics caused by $x(0) \neq 0$ will addressed by (v, r_y) as uncertainties.

It is noteworthy that the SIR-SKR-based model (7.51)–(7.52) describes a dynamic system with (v, r_y) as its input variables and (u, y) as the output variables. Differing from a system input-output model and an SKR-based model, substituting u by the latent variable v in the SIR-SKR-based model enables it to address the model uncertainties in the image subspace. More importantly, with u as an additional output vector, the system capability of perception and detection can be significantly enhanced. This specific feature is of particular importance for fault diagnosis. The mechanism for u serving as a sensor vector can be well illustrated by the control law adopted in the SIR,

$$u = F\hat{x} + Vv, \tag{7.61}$$

which simultaneously poses an output (sensor) model of the dynamic system (7.51) with F characterising the functionality of the sensor (vector). In other words, the relation (7.61) has a double function, acting as a controller and a sensor. It is of interest to observe that the output model (7.61) is a dual form of the output relation in the SKR-model,

$$y = C\hat{x} + W^{-1}r_y.$$

7.2.2 Canonical SIR-SKR-Based Model

In the SIR-SKR-based model (7.51)–(7.52), there are a number of selectable parameters. They are (L, W, F, V). To simplify and normalise the application of the model, the so-called canonical SIR-SKR- (C-SIR-SKR-) based model is introduced.

Definition 7.2 Given the nominal plant model (2.1)–(2.3) and the associated normalised RCP and LCP, (M_0, N_0) and $\left(\hat{M}_0, \hat{N}_0\right)$ as well as the corresponding RCP $\left(\hat{X}_0, \hat{Y}_0\right)$ and LCP (X_0, Y_0) subject to the Bezout identity (2.31), then

$$\begin{bmatrix} u \\ y \end{bmatrix} = \begin{bmatrix} M_0 & -\hat{Y}_0 \\ N_0 & \hat{X}_0 \end{bmatrix} \begin{bmatrix} v \\ r_{y,0} \end{bmatrix}, \tag{7.62}$$

$$\begin{bmatrix} v \\ r_{y,0} \end{bmatrix} = \begin{bmatrix} X_0 & Y_0 \\ -\hat{N}_0 & \hat{M}_0 \end{bmatrix} \begin{bmatrix} u \\ y \end{bmatrix} \tag{7.63}$$

is called C-SIR-SKR-based model.

Algorithm 7.1 Construction of C-SIR-SKR-based model

Step 0: *Input of the nominal model parameters* (A, B, C, D);
Step 1: *Solution of Riccati equations* (2.40) *and* (2.41);
Step 2: *Computation of* (L_0, W_0, L_0, V_0) *according to* (2.34), (2.35), (2.37) *and* (2.38), *respectively;*
Step 3: *Construction of* $\left(\hat{M}_0, \hat{N}_0\right)$, (M_0, N_0), $\left(\hat{X}_0, \hat{Y}_0\right)$ *and* (X_0, Y_0) *according to* (2.5), (2.6), (2.25) *and* (2.26), *respectively,*
Step 4: *Construction of C-SIR-SKR-based model according to* (7.62)–(7.63), *or alternatively, construction of its state space representation according to* (7.51)–(7.54).

The C-SIR-SKR-based model serves as a benchmark (reference) model. Furthermore, Lemma 3.1 and Theorem 3.4 provide the computation formulas for the necessary transformation of an existing observer or/and a controller in a real case to the normalised SKR and SIR as well as $\left(\hat{X}_0, \hat{Y}_0\right)$ and (X_0, Y_0).

From the system-theoretic point of view, the normalised SKR $\left(\hat{M}_0, \hat{N}_0\right)$ is a co-inner and delivers the LS residual signal $r_{y,0}$, while the normalised SIR is inner with the interpretation as an LQ controller. For details the reader is referred to Sect. 2.2.

7.2.3 Uncertainties and Latent Variables

In the previous chapters, uncertainty issues have been addressed under various aspects. Thereby, the major focus has been on the relation of the residual r_y and uncertainties. In this subsection, with respect to the SIR-SKR-based model (7.55), relations between (v, r_y) and uncertainties in the process data (u, y) are examined.

We begin with additive uncertainties in the form of an $\mathcal{L}_{2,[0,N]}$ bounded unknown input vector $d(k) \in \mathbb{R}^{k_d}$. The corresponding model is

$$\begin{aligned} x(k+1) &= Ax(k) + Bu(k) + E_d d(k), \\ y(k) &= Cx(k) + Du(k) + F_d d(k), \end{aligned}$$

which is further, setting $u = Fx + Vv_0$, written as

$$x(k+1) = (A + BF)x(k) + BVv_0(k) + E_d d(k),$$

$$y(k) = (C + DF)\,x(k) + DVv_0(k) + F_d d(k).$$

The uncertainty in (u, y) caused by d is modelled by

$$\begin{bmatrix} u_d \\ y_d \end{bmatrix} = \begin{bmatrix} G_{ud} \\ G_{yd} \end{bmatrix} d = \begin{bmatrix} F\,(zI - A - BF)^{-1}\,E_d \\ (C + DF)\,(zI - A - BF)^{-1}\,E_d + F_d \end{bmatrix} d.$$

It follows

$$\begin{bmatrix} v \\ r_y \end{bmatrix} = \begin{bmatrix} X & Y \\ -\hat{N} & \hat{M} \end{bmatrix} \begin{bmatrix} u \\ y \end{bmatrix} = \begin{bmatrix} v_0 + \left(XG_{ud} + YG_{yd}\right) d \\ \left(\hat{M}G_{yd} - \hat{N}G_{ud}\right) d \end{bmatrix}. \tag{7.64}$$

Next, consider an uncertain system modelled by

$$\begin{aligned} y &= Gu = NM^{-1}u, \\ \begin{bmatrix} M \\ N \end{bmatrix} &= \begin{bmatrix} M_0 \\ N_0 \end{bmatrix} + \begin{bmatrix} \Delta_M \\ \Delta_N \end{bmatrix} = I_{G_0} + \Delta_{I_G} \Longrightarrow \\ \begin{bmatrix} u \\ y \end{bmatrix} &= \left(\begin{bmatrix} M_0 \\ N_0 \end{bmatrix} + \begin{bmatrix} \Delta_M \\ \Delta_N \end{bmatrix} \right) v_\Delta. \end{aligned}$$

Hereby, Δ_{I_G} represents the coprime factor uncertainty with

$$\Delta_{I_G} = \begin{bmatrix} M_0 \\ N_0 \end{bmatrix} \delta_{I_G} + \begin{bmatrix} -\hat{Y}_0 \\ \hat{X}_0 \end{bmatrix} \delta_{I_K},$$

and v_Δ is the latent variable. When

$$u = F_0\hat{x} + v_0,$$

it holds

$$\begin{aligned} v &= \begin{bmatrix} X_0 & Y_0 \end{bmatrix} \begin{bmatrix} u \\ y \end{bmatrix} = v_0 = (I + \delta_{I_G})v_\Delta \Longrightarrow v_\Delta = (I + \delta_{I_G})^{-1}v_0, \\ r_{y,0} &= \begin{bmatrix} -\hat{N}_0 & \hat{M}_0 \end{bmatrix} \begin{bmatrix} u \\ y \end{bmatrix} = \delta_{I_K}v_\Delta = \delta_{I_K}(I + \delta_{I_G})^{-1}v_0. \end{aligned}$$

If

$$u = F_0\hat{x} + v_0 + Qr_{y,0},$$

it turns out

$$\begin{aligned} v &= \begin{bmatrix} X_0 & Y_0 \end{bmatrix} \begin{bmatrix} u \\ y \end{bmatrix} = v_0 + Qr_{y,0} = \left(I + \delta_{I_G}\right) v_\Delta \\ &\Longrightarrow v_\Delta = \left(I + \delta_{I_G}\right)^{-1} \left(v_0 + Qr_{y,0}\right), \end{aligned}$$

$$r_{y,0} = \begin{bmatrix} -\hat{N}_0 & \hat{M}_0 \end{bmatrix} \begin{bmatrix} u \\ y \end{bmatrix} = \delta_{I_K} \left(I + \delta_{I_G} \right)^{-1} \left(v_0 + Q r_{y,0} \right) \Longrightarrow$$

$$r_{y,0} = \delta_{I_K} \left(I + \delta_{I_G} - Q \delta_{I_K} \right)^{-1} v_0, \tag{7.65}$$

$$\begin{aligned} v_\Delta &= v_0 + Q \delta_{I_K} \left(I + \delta_{I_G} - Q \delta_{I_K} \right)^{-1} v_0 \\ &= \left(I + \delta_{I_G} \right) \left(I + \delta_{I_G} - Q \delta_{I_K} \right)^{-1} v_0. \end{aligned} \tag{7.66}$$

It is worth mentioning that analogue result can be achieved, when left coprime factor uncertainty Δ_{K_G},

$$\begin{aligned} G &= \left(\hat{M}_0 + \Delta_{\hat{M}} \right)^{-1} \left(\hat{N}_0 + \Delta_{\hat{N}} \right), \\ \Delta_{K_G} &= \begin{bmatrix} -\Delta_{\hat{N}} & \Delta_{\hat{M}} \end{bmatrix} \in \mathcal{RH}_\infty, \end{aligned}$$

is under consideration. To this end, the relation between Δ_{K_G} and Δ_{I_G} is useful. The reader is referred to Sect. 4.7 for details.

It seems that system

$$v = \begin{bmatrix} X & Y \end{bmatrix} \begin{bmatrix} u \\ y \end{bmatrix} \tag{7.67}$$

might serve as a detector for uncertainties in the image subspace $\mathcal{I}_G$ thanks to the relation

$$\begin{bmatrix} X & Y \end{bmatrix} \begin{bmatrix} M & -\hat{Y} \\ N & \hat{X} \end{bmatrix} = \begin{bmatrix} I & 0 \end{bmatrix}.$$

However, observe the known result,

$$\begin{aligned} \mathcal{P}_{\mathcal{R}_G} \left(\begin{bmatrix} -\hat{Y}_0 \\ \hat{X}_0 \end{bmatrix} r_{y,0} \right) &= \left(\begin{bmatrix} -\hat{Y}_0 \\ \hat{X}_0 \end{bmatrix} - \begin{bmatrix} M_0 \\ N_0 \end{bmatrix} Q_{\mathcal{H}_2} \right) r_{y,0} \\ \Longrightarrow \begin{bmatrix} X_0 & Y_0 \end{bmatrix} \mathcal{P}_{\mathcal{R}_G} \left(\begin{bmatrix} -\hat{Y}_0 \\ \hat{X}_0 \end{bmatrix} r_{y,0} \right) &= -Q_{\mathcal{H}_2} r_{y,0}. \end{aligned}$$

That implies, for $Q_{\mathcal{H}_2} \neq 0$,

$$\exists \begin{bmatrix} u \\ y \end{bmatrix} \in \mathcal{R}_G, \begin{bmatrix} X & Y \end{bmatrix} \begin{bmatrix} u \\ y \end{bmatrix} \neq 0.$$

In words, the latent variable v might be corrupted with uncertainties. At the same time, it should be noticed that all uncertainties acting on $\mathcal{I}_G$ through the system $\begin{bmatrix} -\hat{Y} \\ \hat{X} \end{bmatrix}$ cannot be captured by the detector (7.67). This observation motivates us to have a close look at the image subspace spanned by $\begin{bmatrix} -\hat{Y} \\ \hat{X} \end{bmatrix}$,

$$\mathcal{I}_K = \left\{ \begin{bmatrix} u \\ y \end{bmatrix} \in \mathcal{H}_2 : \begin{bmatrix} u \\ y \end{bmatrix} = \begin{bmatrix} -\hat{Y} \\ \hat{X} \end{bmatrix} r_y, r_y \in \mathcal{H}_2 \right\},$$

and to examine its relation to the image and residual subspaces $\mathcal{I}_G$ and $\mathcal{R}_G$.

Theorem 7.3 *Given the nominal plant model (2.1)–(2.3) and the associated normalised RCP and LCP,* (M_0, N_0) *and* $\left(\hat{M}_0, \hat{N}_0\right)$ *as well as the corresponding RCP* $\left(\hat{X}_0, \hat{Y}_0\right)$ *and LCP* (X_0, Y_0) *subject to the Bezout identity (2.31), then*

$$\mathcal{I}_K \cap \mathcal{R}_G = \left\{ \begin{bmatrix} u \\ y \end{bmatrix} : \begin{bmatrix} u \\ y \end{bmatrix} = \left(\begin{bmatrix} -\hat{Y}_0 \\ \hat{X}_0 \end{bmatrix} - \begin{bmatrix} M_0 \\ N_0 \end{bmatrix} \mathcal{Q}_{\mathcal{H}_2} \right) r_y, r_y \in \mathcal{H}_2 \right\} = \mathcal{R}_G, \tag{7.68}$$

$$\mathcal{I}_K \cap \mathcal{I}_G = \left\{ \begin{bmatrix} u \\ y \end{bmatrix} : \begin{bmatrix} u \\ y \end{bmatrix} = \begin{bmatrix} M_0 \\ N_0 \end{bmatrix} \mathcal{Q}_{\mathcal{H}_2} r_y, r_y \in \mathcal{H}_2 \right\}. \tag{7.69}$$

Proof On account of the Bezout identity,

$$\begin{bmatrix} M_0 & -\hat{Y}_0 \\ N_0 & \hat{X}_0 \end{bmatrix}^{-1} = \begin{bmatrix} X_0 & Y_0 \\ -\hat{N}_0 & \hat{M}_0 \end{bmatrix} \in \mathcal{RH}_\infty,$$

it is apparent that

$$\mathcal{I}_K \oplus \mathcal{I}_G = \mathcal{H}_2.$$

Since $\mathcal{R}_G = \mathcal{I}_G^\perp$, and $\mathcal{I}_G \otimes \mathcal{R}_G = \mathcal{H}_2$, it can be claimed that

$$\mathcal{I}_K \cap \mathcal{R}_G = \mathcal{R}_G.$$

In fact, the result in Sect. 7.1.1,

$$\begin{bmatrix} -\hat{Y}_0 \\ \hat{X}_0 \end{bmatrix} r_{y,0} = \begin{bmatrix} M_0 \\ N_0 \end{bmatrix} \mathcal{Q}_{\mathcal{H}_2} r_{y,0} + (\mathcal{I} - \mathcal{P}_{\mathcal{I}_G}) \left(\begin{bmatrix} -\hat{Y}_0 \\ \hat{X}_0 \end{bmatrix} r_{y,0} \right),$$

$$(\mathcal{I} - \mathcal{P}_{\mathcal{I}_G}) \left(\begin{bmatrix} -\hat{Y}_0 \\ \hat{X}_0 \end{bmatrix} r_{y,0} \right) = \left(\begin{bmatrix} -\hat{Y}_0 \\ \hat{X}_0 \end{bmatrix} - \begin{bmatrix} M_0 \\ N_0 \end{bmatrix} \mathcal{Q}_{\mathcal{H}_2} \right) r_{y,0},$$

implies (7.68) and (7.69). The theorem is thus proved.

7.3 From Behavioural Systems Theory to Finite Samples System Representations

7.3.1 About Behavioural Systems Theory

Behavioural systems theory established by Willems regards any dynamic system as a set of trajectories and calls such a set behaviour. Apart from the formal notations and conceptual definitions adopted in behavioural systems theory, we firstly introduce some basic features and concepts formlessly. They are at the core of behavioural systems theory and, more importantly, strongly related to our work.

Let ω be a discrete-time signal vector, representing a trajectory over a (finite) time interval. The behaviour of an LTI system $\mathcal{S}$ is the set of such trajectories denoted by $\Omega_{\mathcal{S}}$. Let (u, y) denote the input and output signals of $\mathcal{S}$. The following system representation forms are essential in behavioural systems theory.

- Kernel representation: the set of trajectories

$$\ker R_{\mathcal{S}} := \{\omega : R_{\mathcal{S}}(p)\,\omega = 0\},$$

  where $R_{\mathcal{S}}(p)$ represents a polynomial matrix with the unit shift operator, $(p\omega)(k) = \omega(k+1)$, i.e. $R_{\mathcal{S}}(p)\,\omega = 0$ is a difference equation;
- Image representation:

$$\forall\,(u, y)\,, \exists \omega \in \Omega_{\mathcal{S}},\ \text{s.t.} \begin{bmatrix} u \\ y \end{bmatrix} = I_{\mathcal{S}}(p)\omega. \tag{7.70}$$

  Here, $I_{\mathcal{S}}$ is a polynomial matrix satisfying

$$R_{\mathcal{S}}(p)\,I_{\mathcal{S}}(p) = 0. \tag{7.71}$$

  It follows from (7.70)–(7.71) that

$$\exists \hat{M}(p)\,, \hat{N}(p)\,,\ \text{s.t.}\ \hat{M}(p)\,y = \hat{N}(p)\,u,$$

  which gives another form of the kernel representation

$$\begin{bmatrix} -\hat{N}(p) & \hat{M}(p) \end{bmatrix} \begin{bmatrix} u \\ y \end{bmatrix} = \begin{bmatrix} -\hat{N}(p) & \hat{M}(p) \end{bmatrix} I_{\mathcal{S}}(p)\omega = 0. \tag{7.72}$$

- State space representation:

$$px = Ax + Bu,\ y = Cx + Du.$$

Although these representations are analogue to the SKR, SIR and the associated kernel and image subspaces, based on which the unified framework is established, they are different at one essential point: definition domain of the relevant signals. While signals are defined in Hilbert space $\mathcal{H}_2$ ($\mathcal{L}_2$) in the unified framework in the frequency domain or/and in the time domain $[0, \infty)$ ($(-\infty, \infty)$), trajectories are defined over a (finite) time interval in the behavioural system framework. In a certain sense, the aforementioned presentations of the behavioural approach can be interpreted as a data-driven version of the concepts SKR, SIR and kernel and image subspaces. From the mathematical point of view, the SKR and SIR are rational polynomial matrices in $\mathcal{RH}_\infty$, and $\left(\hat{M}(p), \hat{N}(p)\right)$ are polynomial matrices. On the other hand, it is known that the SKR and SIR in $\mathcal{RH}_\infty$ can be equivalently represented by the left and right coprime factorisations in the form of polynomial matrices. In the sequel, we provide a proof that the kernel and image representations in the behavioural system framework are a special realisation form the SKR and SIR in $\mathcal{RH}_\infty$.

7.3.2 Image Representation of Finite Samples Long Signals

Consider the system model

$$G : \begin{cases} x(k+1) = Ax(k) + Bu(k), x(0) = x_0 \\ y(k) = Cx(k) + Du(k) \end{cases}, x \in \mathbb{R}^n, y \in \mathbb{R}^m, u \in \mathbb{R}^p, \quad (7.73)$$

and introduce notations,

$$y_s(k) = \begin{bmatrix} y(k+1) \\ \vdots \\ y(k+s) \end{bmatrix} \in \mathbb{R}^{sm}, u_s(k) = \begin{bmatrix} u(k+1) \\ \vdots \\ u(k+s) \end{bmatrix} \in \mathbb{R}^{sp},$$
$$v_s(k) = \begin{bmatrix} v(k+1) \\ \vdots \\ v(k+s) \end{bmatrix} \in \mathbb{R}^{sp}.$$

Here, s is an integer that is not smaller than n. Define the set of all (u, y) over the time interval $[k+1, k+s]$,

$$\mathcal{S}_{G,s} = \left\{ \begin{bmatrix} u_s(k) \\ y_s(k) \end{bmatrix}, \begin{bmatrix} u(k+i) \\ y(k+i) \end{bmatrix} \in \mathcal{L}_{[0,\infty)}, i \in [1, s] \right\}. \quad (7.74)$$

Consider the SIR of the system (7.73). That is, set

$$\begin{aligned} u(k) &= Fx(k) + v(k) \Longrightarrow \\ x(k+1) &= A_F x(k) + Bv(k), A_F = A + BF, \\ y(k) &= C_F x(k) + Dv(k), C_F = C + DF. \end{aligned}$$

A routine computation leads to

$$\begin{bmatrix} u_s(k) \\ y_s(k) \end{bmatrix} = \begin{bmatrix} H_{u,x,F} \\ H_{y,x,F} \end{bmatrix} x(k+1) + \begin{bmatrix} H_{u,v,F} \\ H_{y,v,F} \end{bmatrix} v_s(k),$$

where

$$H_{u,x,F} = \begin{bmatrix} F \\ FA_F \\ \vdots \\ FA_F^{s-1} \end{bmatrix} \in \mathbb{R}^{sp\times n}, \ H_{u,v,F} = \begin{bmatrix} I & 0 & \cdots & 0 \\ FB & I & \ddots & \vdots \\ \vdots & \ddots & \ddots & 0 \\ FA_F^{s-2}B & \cdots & FB & I \end{bmatrix} \in \mathbb{R}^{sp\times sp},$$

$$H_{y,x,F} = \begin{bmatrix} C_F \\ C_FA_F \\ \vdots \\ C_FA_F^{s-1} \end{bmatrix} \in \mathbb{R}^{sm\times n}, \ H_{y,v,F} = \begin{bmatrix} D & 0 & \cdots & 0 \\ C_FB & D & \ddots & \vdots \\ \vdots & \ddots & \ddots & 0 \\ C_FA_F^{s-2}B & \cdots & C_FB & D \end{bmatrix} \in \mathbb{R}^{ms\times sp}.$$

Since

$$x(k+1) = A_F^n x(k-n+1) + \sum_{i=1}^{n} A_F^{n-i} Bv\,(k-n+i)\,,$$

setting $F = F_d$ so that all eigenvalues of $A + BF_d$ equal to zero, namely, $F_d x(k)$ is a deadbeat controller. It yields

$$A_F^n = 0, \sum_{i=1}^{n} A_F^{n-i} Bv\,(k-n+i) = H_{x,v,F} v_n(k-n),$$
$$H_{x,v,F} = \begin{bmatrix} A_F^{n-1}B & \cdots & B \end{bmatrix},$$

leading to

$$\begin{bmatrix} u_s(k) \\ y_s(k) \end{bmatrix} = \begin{bmatrix} H_{u,x,F} \\ H_{y,x,F} \end{bmatrix} H_{x,v,F} v_n(k-n) + \begin{bmatrix} H_{u,v,F} \\ H_{y,v,F} \end{bmatrix} v_s(k)$$
$$=: \begin{bmatrix} M_s \\ N_s \end{bmatrix} v_{s+n}(k-n), \ v_{s+n}(k-n) \in \mathbb{R}^{(s+n)p},$$
$$M_s = \begin{bmatrix} H_{u,x,F} H_{x,v,F} & H_{u,v,F} \end{bmatrix} \in \mathbb{R}^{sp\times(s+n)p}, \tag{7.75}$$
$$N_s = \begin{bmatrix} H_{y,x,F} H_{x,v,F} & H_{y,v,F} \end{bmatrix} \in \mathbb{R}^{sm\times(s+n)p}. \tag{7.76}$$

Recall Lemma 3.1 that provides us with the following relation between the SIR (M_d, N_d) with $F = F_d$ and another arbitrary SIR (M, N) with $F \neq F_d$,

$$\begin{bmatrix} M \\ N \end{bmatrix} = \begin{bmatrix} M_d \\ N_d \end{bmatrix} V, \tag{7.77}$$
$$V(z) = I + (F - F_d)(zI - A_F)^{-1} B \in \mathcal{RH}_\infty.$$

Note that V is invertible over $\mathcal{RH}_\infty$. As a result, the finite samples input-output model with $F \neq F_d$ can be written as

$$\forall F \neq F_d, \begin{bmatrix} u_s(k) \\ y_s(k) \end{bmatrix} = \begin{bmatrix} M_d \\ N_d \end{bmatrix} \bar{v}_{s+n}(k-n), \tag{7.78}$$
$$\bar{v}_{s+n}(k-n) = \begin{bmatrix} \bar{v}(k-n+1) \\ \vdots \\ \bar{v}(k+s) \end{bmatrix}, \bar{v}(z) = V(z)v(z).$$

It is of interest to notice that, given (M_s, N_s) satisfying (7.75)–(7.76) with $F \neq F_d$, it holds

$$\begin{bmatrix} u_s(k) \\ y_s(k) \end{bmatrix} = \begin{bmatrix} M_{s,d} \\ N_{s,d} \end{bmatrix} \bar{v}_{s+n}(k-n), \tag{7.79}$$
$$\bar{v}_{s+n}(k-n) = V_{s+n} v_{s+n}(k-n),$$
$$V_{s+n} = \begin{bmatrix} I & 0 & \cdots & \cdots & 0 \\ (F-F_d)B & I & 0 & \cdots & 0 \\ (F-F_d)A_F B & (F-F_d)B & I & \ddots & \vdots \\ \vdots & \vdots & \ddots & \ddots & 0 \\ (F-F_d)A_F^{s+n-2}B & (F-F_d)A_F^{s+n-3}B & \cdots & (F-F_d)B & I \end{bmatrix},$$

where $(M_{s,d}, N_{s,d})$ denote (M_s, N_s) with $F = F_d$. The relation (7.79) is a data-driven version of (7.78), whose proof is given in the following lemma and theorem.

Remark 7.4 In order to differentiate (M_s, N_s) corresponding to different feedback gains, the notations,

$$\begin{bmatrix} M_{s,d} \\ N_{s,d} \end{bmatrix}, \begin{bmatrix} M_s \\ N_s \end{bmatrix} = \begin{bmatrix} M_{s,d} \\ N_{s,d} \end{bmatrix} V_{s+n}$$

are adopted for the cases $F = F_d$ and $F \neq F_d$, respectively.

Lemma 7.1 *Given system matrices* $(\bar{A}, B, \bar{C}, \bar{D})$, $\bar{A}_F = \bar{A} + B\bar{F}$, $\bar{C}_F = \bar{C} + \bar{D}\bar{F}$ *with appropriate dimensions, then for* $j = 1, 2, \ldots$,

$$\begin{bmatrix} \bar{C}\bar{A}^j B & \cdots & \bar{C}B & \bar{D} \end{bmatrix} \begin{bmatrix} I \\ \bar{F}B \\ \vdots \\ \bar{F}\bar{A}_F^j B \end{bmatrix} = \bar{C}_F \bar{A}_F^j B. \tag{7.80}$$

Proof We firstly prove

$$\left[\,\bar{A}^j B \;\cdots\; B\,\right]\begin{bmatrix} I \\ \bar{F}B \\ \vdots \\ \bar{F}\bar{A}_F^{j-1}B \end{bmatrix} = \bar{A}_F^j B,\; j = 1, 2, \ldots, \tag{7.81}$$

using mathematical induction method. It is obvious that for $j = 1$, (7.81) is true. Suppose that (7.81) holds for $j = i$ and examine (7.81) for $j = i + 1$. Since for $j = i + 1$

$$\begin{aligned}\left[\,\bar{A}^{i+1} B \;\cdots\; B\,\right]\begin{bmatrix} I \\ \bar{F}B \\ \vdots \\ \bar{F}\bar{A}_F^{i}B \end{bmatrix} &= \bar{A}\left[\,\bar{A}^{i} B \;\cdots\; B\,\right]\begin{bmatrix} I \\ \bar{F}B \\ \vdots \\ \bar{F}\bar{A}_F^{i-1}B \end{bmatrix} + B\bar{F}\bar{A}_F^i B \\ &= \bar{A}\bar{A}_F^i B + B\bar{F}\bar{A}_F^i B = \bar{A}_F^{i+1} B,\end{aligned}$$

it is proved that (7.81) is true. Now, consider

$$\begin{aligned}&\left[\,\bar{C}\bar{A}^j B \;\cdots\; \bar{C}B \;\; \bar{D}\,\right]\begin{bmatrix} I \\ \bar{F}B \\ \vdots \\ \bar{F}\bar{A}_F^{j}B \end{bmatrix} \\ &= \bar{C}\left[\,\bar{A}^j B \;\cdots\; B\,\right]\begin{bmatrix} I \\ \bar{F}B \\ \vdots \\ \bar{F}\bar{A}_F^{j-1}B \end{bmatrix} + \bar{D}\bar{F}\bar{A}_F^j B.\end{aligned}$$

It follows from (7.81) that

$$\bar{C}\left[\,\bar{A}^j B \;\cdots\; B\,\right]\begin{bmatrix} I \\ \bar{F}B \\ \vdots \\ F\bar{A}_F^{j-1}B \end{bmatrix} + \bar{D}\bar{F}\bar{A}_F^j B = \bar{C}\bar{A}_F^j B + \bar{D}\bar{F}\bar{A}_F^j B$$

$$\Longrightarrow \left[\,\bar{C}\bar{A}^j B \;\cdots\; \bar{C}B \;\; \bar{D}\,\right]\begin{bmatrix} I \\ \bar{F}B \\ \vdots \\ \bar{F}\bar{A}_F^{j}B \end{bmatrix} = \bar{C}_F\bar{A}_F^j B.$$

The lemma is proved.

Theorem 7.4 *Given the system model (7.73), (7.79) holds.*

Proof Consider

$$\begin{bmatrix} M_{s,d} \\ N_{s,d} \end{bmatrix} = \begin{bmatrix} H_{u,x,F} H_{x,v,F} & H_{u,v,F} \\ H_{y,x,F} H_{x,v,F} & H_{y,v,F} \end{bmatrix},$$

and let

$$\begin{aligned} &\bar{F} = F - F_d, A_F = A + BF = A + BF_d + B\bar{F} := \bar{A}_F, \bar{D} = D, \\ &\bar{A} = A + BF_d, \bar{A}_F = \bar{A} + B\bar{F}, \bar{C} = C + DF_d, \bar{C}_F = \bar{C} + D\bar{F} = C_F. \end{aligned}$$

It yields

$$\begin{bmatrix} H_{y,x,F} H_{x,v,F} & H_{y,v,F} \end{bmatrix} = \begin{bmatrix} \bar{C}\bar{A}^{n-1}B & \cdots & D & 0 & \cdots & 0 \\ \bar{C}\bar{A}^{n}B & \cdots & \bar{C}B & D & \ddots & \vdots \\ \vdots & \vdots & \vdots & \ddots & \ddots & 0 \\ \bar{C}\bar{A}^{n+s-2}B & \cdots & \bar{C}\bar{A}^{s-2}B & \cdots & \bar{C}B & D \end{bmatrix},$$

$$V_{s+n} = \begin{bmatrix} I & 0 & \cdots & \cdots & 0 \\ \bar{F}B & I & 0 & \cdots & 0 \\ \bar{F}\bar{A}_F B & \bar{F}B & I & \ddots & \vdots \\ \vdots & \vdots & \ddots & \ddots & 0 \\ \bar{F}\bar{A}_F^{s+n-2}B & \bar{F}\bar{A}_F^{s+n-3}B & \cdots & \bar{F}B & I \end{bmatrix}.$$

Now, applying (7.80) to $\begin{bmatrix} H_{y,x,F} H_{x,v,F} & H_{y,v,F} \end{bmatrix} V_{s+n}$ gives

$$\begin{aligned} &\begin{bmatrix} H_{y,x,F} H_{x,v,F} & H_{y,v,F} \end{bmatrix} V_{s+n} = N_{s,d} V_{s+n} \\ &= \begin{bmatrix} \bar{C}_F\bar{A}_F^{n-1}B & \cdots & D & 0 & \cdots & 0 \\ \bar{C}_F\bar{A}_F^{n}B & \cdots & \bar{C}_F B & D & \ddots & \vdots \\ \vdots & \vdots & \vdots & \ddots & \ddots & 0 \\ \bar{C}_F\bar{A}_F^{n+s-2}B & \cdots & \bar{C}_F\bar{A}_F^{s-2}B & \cdots & \bar{C}_F B & D \end{bmatrix} \\ &= \begin{bmatrix} C_F A_F^{n-1}B & \cdots & D & 0 & \cdots & 0 \\ C_F A_F^{n}B & \cdots & C_F B & D & \ddots & \vdots \\ \vdots & \vdots & \vdots & \ddots & \ddots & 0 \\ C_F A_F^{n+s-2}B & \cdots & C_F A_F^{s-2}B & \cdots & C_F B & D \end{bmatrix}. \end{aligned}$$

Analogous to it, let

$$\bar{D} = I, \bar{C} = 0 + \bar{D}F_d, \bar{C}_F = \bar{C} + \bar{D}\bar{F} = F.$$

It turns out

$$\begin{bmatrix} H_{u,x,F} H_{x,v,F} & H_{u,v,F} \end{bmatrix} = \begin{bmatrix} \bar{C}\bar{A}^{n-1}B & \cdots & I & 0 & \cdots & 0 \\ \bar{C}\bar{A}^{n}B & \cdots & \bar{C}B & I & \ddots & \vdots \\ \vdots & \vdots & \vdots & \ddots & \ddots & 0 \\ \bar{C}\bar{A}^{n+s-2}B & \cdots & \bar{C}\bar{A}^{s-2}B & \cdots & \bar{C}B & I \end{bmatrix},$$

$$\begin{aligned} \begin{bmatrix} H_{u,x,F} H_{x,v,F} & H_{u,v,F} \end{bmatrix} V_{s+n} &= M_{s,d} V_{s+n} \\ &= \begin{bmatrix} \bar{C}_F\bar{A}_F^{n-1}B & \cdots & \bar{D} & 0 & \cdots & 0 \\ \bar{C}_F\bar{A}_F^{n}B & \cdots & \bar{C}_F B & \bar{D} & \ddots & \vdots \\ \vdots & \vdots & \vdots & \ddots & \ddots & 0 \\ \bar{C}_F\bar{A}_F^{n+s-2}B & \cdots & \bar{C}_F\bar{A}_F^{s-2}B & \cdots & \bar{C}_F B & \bar{D} \end{bmatrix} \\ &= \begin{bmatrix} FA_F^{n-1}B & \cdots & I & 0 & \cdots & 0 \\ FA_F^{n}B & \cdots & FB & I & \ddots & \vdots \\ \vdots & \vdots & \vdots & \ddots & \ddots & 0 \\ FA_F^{n+s-2}B & \cdots & FA_F^{s-2}B & \cdots & FB & I \end{bmatrix}. \end{aligned}$$

Hence, we have proved that

$$\begin{bmatrix} u_s(k) \\ y_s(k) \end{bmatrix} = \begin{bmatrix} M_{s,d} \\ N_{s,d} \end{bmatrix} \bar{v}_{s+n}(k-n),$$

$$M_s v_{s+n}(k-n) = \begin{bmatrix} C_F A_F^{n-1}B & \cdots & D & 0 & \cdots & 0 \\ C_F A_F^{n}B & \cdots & C_F B & D & \ddots & \vdots \\ \vdots & \vdots & \vdots & \ddots & \ddots & 0 \\ C_F A_F^{n+s-2}B & \cdots & C_F A_F^{s-2}B & \cdots & C_F B & D \end{bmatrix} v_{s+n}(k-n),$$

$$N_s v_{s+n}(k-n) = \begin{bmatrix} FA_F^{n-1}B & \cdots & I & 0 & \cdots & 0 \\ FA_F^{n}B & \cdots & FB & I & \ddots & \vdots \\ \vdots & \vdots & \vdots & \ddots & \ddots & 0 \\ FA_F^{n+s-2}B & \cdots & FA_F^{s-2}B & \cdots & FB & I \end{bmatrix} v_{s+n}(k-n).$$

An immediate application of the above results is the proof of the following two theorems.

Theorem 7.5 *Given matrices M_s, N_s defined by (7.75)–(7.76), it holds*

$$rank \begin{bmatrix} M_s \\ N_s \end{bmatrix} = sp + n. \tag{7.82}$$

Proof Observe that

$$\begin{bmatrix} M_s \\ N_s \end{bmatrix} = \left[\begin{bmatrix} H_{u,x,F} \\ H_{y,x,F} \end{bmatrix} H_{x,v,F} \begin{bmatrix} H_{u,v,F} \\ H_{y,v,F} \end{bmatrix} \right],$$

which implies that the number of the independent columns is not larger than $sp + n$. Hence,

$$rank \begin{bmatrix} M_s \\ N_s \end{bmatrix} \leq sp + n.$$

On the other hand, for $F = 0$, it holds, according to (7.79), that

$$\begin{bmatrix} M_s \\ N_s \end{bmatrix} = \begin{bmatrix} M_{s,d} \\ N_{s,d} \end{bmatrix} V_{s+n} = \begin{bmatrix} 0 & I \\ H_o H_c & H_y \end{bmatrix},$$

$$V_{s+n} = \begin{bmatrix} I & 0 & \cdots & \cdots & 0 \\ -F_d B & I & 0 & \cdots & 0 \\ -F_d AB & -F_d B & I & \ddots & \vdots \\ \vdots & \vdots & \ddots & \ddots & 0 \\ -F_d A^{s+n-2} B & -F_d A^{s+n-3} B & \cdots & -F_d B & I \end{bmatrix},$$

$$H_c = \begin{bmatrix} A^{n-1} B & \cdots & B \end{bmatrix},$$

$$H_0 = \begin{bmatrix} C \\ CA \\ \vdots \\ CA^{s-1} \end{bmatrix}, H_y = \begin{bmatrix} D & 0 & \cdots & 0 \\ CB & D & \ddots & \vdots \\ \vdots & \ddots & \ddots & 0 \\ CA^{s-2} B & \cdots & CB & D \end{bmatrix},$$

.

It is is apparent that

$$rank \begin{bmatrix} M_s \\ N_s \end{bmatrix} = rank \begin{bmatrix} 0 & I \\ H_o H_c & H_y \end{bmatrix} = sp + n.$$

This proves the theorem.

As the consequence of the above theorem,

$$\forall v_{s+n}(k-n), \exists \begin{bmatrix} u_s(k) \\ y_s(k) \end{bmatrix}, \text{ s.t. } \begin{bmatrix} u_s(k) \\ y_s(k) \end{bmatrix} = \begin{bmatrix} M_s \\ N_s \end{bmatrix} v_{s+n}(k-n), \tag{7.83}$$

$$\forall \begin{bmatrix} u_s(k) \\ y_s(k) \end{bmatrix} \in \mathcal{S}_{G,s}, \exists v_{s+n}(k-n), \text{ s.t. } \begin{bmatrix} u_s(k) \\ y_s(k) \end{bmatrix} = \begin{bmatrix} M_s \\ N_s \end{bmatrix} v_{s+n}(k-n). \tag{7.84}$$

Theorem 7.6 *Given the system model (7.73) and the corresponding input-output signal set (7.74), then we have*

$$\mathcal{S}_{G,s} = \mathcal{I}_{G,s}, \tag{7.85}$$

$$\mathcal{I}_{G,s} = \left\{ \begin{bmatrix} u_s(k) \\ y_s(k) \end{bmatrix}, \begin{bmatrix} u_s(k) \\ y_s(k) \end{bmatrix} = \begin{bmatrix} M_s \\ N_s \end{bmatrix} v_{s+n}(k-n), v_{s+n}(k-n) \in \mathcal{L}_{[0,\infty)} \right\}. \tag{7.86}$$

Proof The proof immediately follows (7.83), (7.84) and (7.79).

Analogous to the definitions of SIR and system image subspace, the concepts of image representation and image set or image subspace of finite samples long signals are introduced.

Definition 7.3 Given the system model (7.73), the set (7.86) is called image set or image subspace of s-samples long data. The equation

$$\begin{bmatrix} u_s(k) \\ y_s(k) \end{bmatrix} = I_{G,s} v_{s+n}(k-n), I_{G,s} = \begin{bmatrix} M_{s,d} \\ N_{s,d} \end{bmatrix} V_{s+n}, \tag{7.87}$$

is called image representation of s-samples long data. Vector $v_{s+n}(k-n) \in \mathbb{R}^{p(s+n)}$ is called latent variable.

It is apparent that the representation (7.87) is the image representation (7.70) and the latent variable $v_{s+n}(k-n)$ is the trajectory, defined in behaviour systems theory.

7.3.3 Kernel Representation of Finite Samples Long Signals

Next, the SKR of the system (7.73),

$$\hat{x}(k+1) = A_L \hat{x}(k) + B_L u(k) + L y(k), A_L = A - LC, B_L = B - LD, \tag{7.88}$$

$$r(k) = y(k) - C\hat{x}(k) - Du(k) \Longleftrightarrow y(k) = C\hat{x}(k) + Du(k) + r(k), \tag{7.89}$$

is considered. It leads to

$$r_s(k) = y_s(k) - H_{\hat{x},L}\hat{x}(k+1) - H_{r,u,L} u_s(k) - H_{r,y,L} y_s(k), \tag{7.90}$$

$$r_s(k) = \begin{bmatrix} r(k+1) \\ \vdots \\ r(k+s) \end{bmatrix} \in \mathbb{R}^{sm}, H_{\hat{x},L} = \begin{bmatrix} C \\ CA_L \\ \vdots \\ CA_L^{s-1} \end{bmatrix} \in \mathbb{R}^{sm \times n},$$

$$H_{r,u,L} = \begin{bmatrix} D & 0 & \cdots & 0 \\ CB_L & D & \ddots & \vdots \\ \vdots & \ddots & \ddots & 0 \\ CA_L^{s-2}B_L & \cdots & CB_L & D \end{bmatrix} \in \mathbb{R}^{sm \times sp},$$

$$H_{r,y,L} = \begin{bmatrix} 0 & 0 & \cdots & 0 \\ CL & 0 & \ddots & \vdots \\ \vdots & \ddots & \ddots & 0 \\ CA_L^{s-2}L & \cdots & CL & 0 \end{bmatrix} \in \mathbb{R}^{sm \times sm}. \tag{7.91}$$

Let $L = L_d$ be an observer gain matrix so that all eigenvalues of $A - L_d C$ equal to zero, namely, the above SKR is deadbeat. Consequently,

$$\begin{aligned} \hat{x}(k+1) &= A_L \hat{x}(k) + B_L u(k) + Ly(k) \\ = A_L^n \hat{x}(k-n+1) &+ \sum_{i=1}^{n} A_L^{n-i} B_L u(k-n+i) + \sum_{i=1}^{n} A_L^{n-i} Ly(k-n+i) \\ &= H_{\hat{x},u,L} u_n(k-n) + H_{\hat{x},y,L} y_n(k-n), \\ H_{\hat{x},u,L} &= \begin{bmatrix} A_L^{n-1} B_L & \cdots & B_L \end{bmatrix}, H_{\hat{x},y,L} = \begin{bmatrix} A_L^{n-1} L & \cdots & L \end{bmatrix}. \end{aligned}$$

It allows us to write (7.90) into

$$\begin{aligned} r_s(k) &= y_s(k) - H_{\hat{x},L} \begin{pmatrix} H_{\hat{x},u,L} u_n(k-n) \\ +H_{\hat{x},y,L} y_n(k-n) \end{pmatrix} - H_{r,u,L} u_s(k) - H_{r,y,L} y_s(k) \\ &= y_s(k) - H_{(u,y),p} \begin{bmatrix} u_n(k-n) \\ y_n(k-n) \end{bmatrix} - H_{r,u,L} u_s(k) - H_{r,y,L} y_s(k) \qquad (7.92) \\ &= \hat{M}_{s+n} y_{s+n}(k-n) - \hat{N}_{s+n} u_{s+n}(k-n), \\ H_{(u,y),p} &= \begin{bmatrix} H_{\hat{x},L} H_{\hat{x},u,L} & H_{\hat{x},L} H_{\hat{x},y,L} \end{bmatrix}, \qquad (7.93) \\ \hat{M}_{s+n} &= \begin{bmatrix} -H_{\hat{x},L} H_{\hat{x},y,L} & I - H_{r,y,L} \end{bmatrix}, \hat{N}_{s+n} = \begin{bmatrix} H_{\hat{x},L} H_{\hat{x},u,L} & H_{r,u,L} \end{bmatrix}. \qquad (7.94) \end{aligned}$$

It is evident that

$$rank \begin{bmatrix} -\hat{N}_{s+n} & \hat{M}_{s+n} \end{bmatrix} = sm. \tag{7.95}$$

In (7.92), the term

$$\hat{y}_s(k) = H_{(u,y),p} \begin{bmatrix} u_n(k-n) \\ y_n(k-n) \end{bmatrix} + H_{r,u,L} u_s(k) + H_{r,y,L} y_s(k)$$

is an estimate for $y_s(k)$, and thus

$$r_s(k) = y_s(k) - \hat{y}_s(k)$$

is a residual vector. For our purpose, the dynamic of $r_s(k)$ is examined. Let $e(k) = x(k) - \hat{x}(k)$. It is straightforward that

$$e(k+1) = A_L e(k) = A_L^n e(k-n+1) = 0 \Longrightarrow e(k+i) = 0, i = 1, \ldots,$$

which implies

$$\hat{x}(k+i) = x(k+i), i = 1, \ldots.$$

As a result, independent of the initial value of $x(k-n+1)$,

$$r(k+i) = Ce(k+i) = 0, i = 1, \ldots, s \Longrightarrow r_s(k) = 0. \tag{7.96}$$

Analogously to our discussion on SIR in the previous subsection, we consider an arbitrary SKR $\left(\hat{M}, \hat{N}\right)$ that is different from the SKR denoted by $\left(\hat{M}_d, \hat{N}_d\right)$ with the observer gain $L = L_d$. By means of the relation given in Lemma 3.1,

$$\begin{gathered}\left[\hat{M} \ \hat{N}\right] = R\left[\hat{M}_d \ \hat{N}_d\right], \\ R = I - C(zI - A_L)^{-1}(L - L_d) \in \mathcal{RH}_\infty, L \neq L_d, R^{-1} \in \mathcal{RH}_\infty,\end{gathered}$$

the residual generator (7.88)–(7.89) satisfies

$$r = \left[\hat{M} \ \hat{N}\right]\begin{bmatrix} u \\ y \end{bmatrix} = R\left[\hat{M}_d \ \hat{N}_d\right]\begin{bmatrix} u \\ y \end{bmatrix}.$$

Analogously to it, (7.90) for $L \neq L_d$ can be equivalently written as

$$r_s(k) = R_s\left(\hat{M}_{s+n} y_{s+n}(k-n) - \hat{N}_{s+n} u_{s+n}(k-n)\right) = 0, \tag{7.97}$$

$$R_s = \begin{bmatrix} I & 0 & \cdots & 0 \\ C(L_d - L) & I & \ddots & \vdots \\ \vdots & \ddots & \ddots & 0 \\ CA_L^{s-2}(L_d - L) & \cdots & C(L_d - L) & I \end{bmatrix} \in \mathbb{R}^{sm \times sm},$$

where R_s is invertible, and $\left(\hat{M}_{s+n}, \hat{N}_{s+n}\right)$ is defined in (7.94) with $L = L_d$. The above results are summarised in the subsequent theorem.

Theorem 7.7 *Given the system model (7.73) and the residual generator (7.88)–(7.89), then we have* $\forall \hat{x}(k+1)$,

$$\begin{gathered}r_s(k) = y_s(k) - \hat{y}_s(k) = \\ \left(I - H_{r,y,L}\right) y_s(k) - H_{\hat{x},L}\hat{x}(k+1) - H_{r,u,L} u_s(k) = 0.\end{gathered} \tag{7.98}$$

Proof The proof immediately follows (7.92), (7.96) and (7.97).

Now, we are in a position to introduce kernel representation, kernel space, residual generator and residual vector of $(s+n)$-samples long data.

Definition 7.4 Given the system model (7.73), the equation,

$$r_s(k) = K_{G,s+n} \begin{bmatrix} u_{s+n}(k-n) \\ y_{s+n}(k-n) \end{bmatrix}, K_{G,s+n} = R_s \begin{bmatrix} -\hat{N}_{s+n} & \hat{M}_{s+n} \end{bmatrix}, \tag{7.99}$$

is called kernel representation of $(s+n)$-samples long data, and the set

$$\mathcal{K}_{G,s+n} = \left\{ \begin{bmatrix} u_{s+n}(k-n) \\ y_{s+n}(k-n) \end{bmatrix} \in \mathcal{L}_{[0,\infty)}, K_{G,s+n} \begin{bmatrix} u_{s+n}(k-n) \\ y_{s+n}(k-n) \end{bmatrix} = 0 \right\} \tag{7.100}$$

is called kernel subspace of $(s+n)$-samples long data. The system

$$r_s(k) = \hat{M}_{s+n} y_{s+n}(k-n) - \hat{N}_{s+n} u_{s+n}(k-n)$$

is called residual generator with the residual vector $r_s(k)$.

Remark 7.5 When there is no risk of confusion, image/kernel representations and image/kernel subspaces will be used for simplicity.

Like the image representation, Theorem 7.7 demonstrates that the kernel representation (7.99) is a special realisation of the SKR in $\mathcal{H}_2$ space. The kernel representation (7.99) is a difference equation and a realisation of the kernel representation (7.72) of behaviour systems theory.

7.3.4 Discussion on Fundamental Lemma and Image, Kernel Representations

At the core of behavioural systems theory is the so-called fundamental lemma that, informally speaking, claims, any trajectories $(u_s(k), y_s(k))$ over the time interval $[k+1, k+s]$ can be represented by the image representation $I_{\mathcal{S}}\omega$. Hereby, $I_{\mathcal{S}}$ is a matrix consisting of process data (u, y), ω is the vector of latent variable. That means,

$$\forall \begin{bmatrix} u_s(k) \\ y_s(k) \end{bmatrix} \in \mathcal{S}_{G,s}, \exists \omega, \text{ s.t. } \begin{bmatrix} u_s(k) \\ y_s(k) \end{bmatrix} = I_{\mathcal{S}}\omega.$$

To be specific,

$$I_{\mathcal{S}} = \begin{bmatrix} u_s(j) & \cdots & u_s(j+M-1) \\ y_s(j) & \cdots & y_s(j+M-1) \end{bmatrix} \in \mathbb{R}^{(p+m)s \times M}, M \geq (s+n)\, p, \omega \in \mathbb{R}^M.$$

The matrix $I_{\mathcal{S}}$ is characterised by the distinct feature

$$rank\,(I_{\mathcal{S}}) = sp + n.$$

It is of interest to notice that the fundamental lemma can be alternatively proved by means of the image representation introduced in the previous subsection. The proof is indeed straightforward. Recall that

$$\forall \begin{bmatrix} u_s(k) \\ y_s(k) \end{bmatrix} \in \mathcal{S}_{G,s}, \exists v_{s+n}(k-n), \begin{bmatrix} u_s(k) \\ y_s(k) \end{bmatrix} = \begin{bmatrix} M_s \\ N_s \end{bmatrix} v_{s+n}(k-n).$$

Now, build

$$V_{s+n,M} = \begin{bmatrix} v_{s+n}(k-n) & \cdots & v_{s+n}(k-n+M-1) \end{bmatrix} \in \mathbb{R}^{(s+n)p \times M}$$

for $M \geq (s+n)\,p$ and satisfying

$$rank\left(V_{s+n,M}\right) = (s+n)\,p,$$

which is equivalent to the condition of sufficient excitation. As a result,

$$rank \begin{bmatrix} u_s(k) & \cdots & u_s(k+M-1) \\ y_s(k) & \cdots & y_s(k+M-1) \end{bmatrix} = rank \begin{bmatrix} M_s \\ N_s \end{bmatrix} = sp + n.$$

This implies, the matrix

$$I_{\mathcal{S}} := \begin{bmatrix} u_s(k) & \cdots & u_s(k+M-1) \\ y_s(k) & \cdots & y_s(k+M-1) \end{bmatrix} = \begin{bmatrix} M_s \\ N_s \end{bmatrix} V_{s+n,M}$$

defines the data matrix (set) described in the fundamental lemma.

Remark 7.6 It is worth pointing out that the image presentation defined by (7.87) is a causal (finite samples) system, while $I_{\mathcal{S}}$ given above might be a non-causal system.

Note that the fundamental lemma was proved with the aid of the fact that the system initial state $x(0)$ can be fully reconstructed using n input and output data. Apparently, the initial state $x(0)$ is not reconstructed in our proof. Instead, the system response to $x(0)$ is equal to zero after n sampling steps, thanks to the use of the deadbeat state feedback controller. It is worth mentioning that the kernel representation can be applied to prove the fundamental lemma as well. This is demonstrated by the following example.

Example 7.6 Given the kernel representation

$$r_s(k) = \hat{M}_{s+n} y_{s+n}(k-n) - \hat{N}_{s+n} u_{s+n}(k-n)$$

and keeping it in mind that

$$\hat{y}_s(k) = H_{(u,y),p} \begin{bmatrix} u_n(k-n) \\ y_n(k-n) \end{bmatrix} + H_{r,u,L} u_s(k) + H_{r,y,L} y_s(k),$$

we consider

$$
\begin{aligned}
y_s(k) &= H_{(u,y),p} \begin{bmatrix} u_n(k-n) \\ y_n(k-n) \end{bmatrix} + H_{r,u,L} u_s(k) + H_{r,y,L} y_s(k) \\
&= H_{u,s+n} u_{s+n}(k-n) + H_{y,s+n} y_{s+n}(k-n),
\end{aligned}
$$

where

$$
H_{u,s+n} = \begin{bmatrix} CA_L^{n-1}B_L & \cdots & D & 0 & \cdots & 0 \\ CA_L^{n}B_L & \cdots & CB_L & D & \ddots & \vdots \\ \vdots & \vdots & \vdots & \ddots & \ddots & 0 \\ CA_L^{n+s-2}B_L & \cdots & CA_L^{s-2}B_L & \cdots & CB_L & D \end{bmatrix},
$$

$$
H_{y,s+n} = \begin{bmatrix} CA_L^{n-1}L & \cdots & 0 & 0 & \cdots & 0 \\ CA_L^{n}L & \cdots & CL & 0 & \ddots & \vdots \\ \vdots & \vdots & \vdots & \ddots & \ddots & 0 \\ CA_L^{n+s-2}L & \cdots & CA_L^{s-2}L & \cdots & CL & 0 \end{bmatrix}.
$$

Note that $H_{u,s+n}$ and $H_{y,s+n}$ are of a dual structure to M_s and N_s of the image representation. Thus, multiplying by $H_{u,s+n}$ and $H_{y,s+n}$ by R_s

$$
R_s = \begin{bmatrix} I & 0 & \cdots & 0 \\ CL & I & \ddots & \vdots \\ \vdots & \ddots & \ddots & 0 \\ CA^{s-2}L & \cdots & CL & I \end{bmatrix},
$$

yields

$$
R_s H_{u,s+n} = \begin{bmatrix} CA^{n-1}B & \cdots & D & 0 & \cdots & 0 \\ CA^{n}B & \cdots & CB & D & \ddots & \vdots \\ \vdots & \vdots & \vdots & \ddots & \ddots & 0 \\ CA^{n+s-2}B & \cdots & CA^{s-2}B & \cdots & CB & D \end{bmatrix},
$$

$$
R_s H_{y,s+n} = 0.
$$

As a result,

$$
\begin{bmatrix} u_s(k) \\ y_s(k) \end{bmatrix} = \begin{bmatrix} 0 & I \\ H_{\hat{x},L} H_{\hat{x},u,L} & H_{r,u,L} \end{bmatrix} \begin{bmatrix} u_n(k-n) \\ u_s(k) \end{bmatrix},
$$

$$H_{\hat{x},L}H_{\hat{x},u,L}=\begin{bmatrix} C \\ CA \\ \vdots \\ CA^{s-1} \end{bmatrix}\begin{bmatrix} A^{n-1}B \cdots B \end{bmatrix},$$

$$H_{r,u,L}=\begin{bmatrix} D & 0 & \cdots & 0 \\ CB & D & \ddots & \vdots \\ \vdots & \ddots & \ddots & 0 \\ CA^{s-2}B & \cdots & CB & D \end{bmatrix}.$$

It is apparent that

$$rank\begin{bmatrix} 0 & I \\ H_{\hat{x},L}H_{\hat{x},u,L} & H_{r,u,L} \end{bmatrix}=sp+n.$$

7.3.5 Bezout Identity-Like Relations in the Form of Algebraic Matrices

We begin with a corollary that follows immediately from Theorems 7.6 and 7.7.

Corollary 7.2 *Given the system model (7.73) and*

$$I_{G,s+n}=\begin{bmatrix} M_{s+n} \\ N_{s+n} \end{bmatrix},\ K_{G,s+n}=\begin{bmatrix} -\hat{N}_{s+n} & \hat{M}_{s+n} \end{bmatrix},$$

defined in Definitions 7.3 and 7.4, respectively, as well as

$$\mathcal{I}_{G,s+n}=\left\{\begin{bmatrix} u_{s+n}(k-n) \\ y_{s+n}(k-n) \end{bmatrix}, \begin{array}{c}\begin{bmatrix} u_{s+n}(k-n) \\ y_{s+n}(k-n) \end{bmatrix}=I_{G,s+n}v_{s+2n}(k-2n), \\ v_{s+2n}(k-2n)\in\mathcal{L}_{[0,\infty)}\end{array}\right\},$$

$$\mathcal{K}_{G,s+n}=\left\{\begin{bmatrix} u_{s+n}(k-n) \\ y_{s+n}(k-n) \end{bmatrix}\in\mathcal{L}_{[0,\infty)}, K_{G,s+n}\begin{bmatrix} u_{s+n}(k-n) \\ y_{s+n}(k-n) \end{bmatrix}=0\right\},$$

then it holds

$$K_{G,s+n}I_{G,s+n}=0, \mathcal{I}_{G,s+n}\subseteq\mathcal{K}_{G,s+n}, \tag{7.101}$$

$$\dim\left(\mathcal{I}_{G,s+n}\right)=(s+n)\,p+n, \dim\left(\mathcal{K}_{G,s+n}\right)=(s+n)\,p+nm. \tag{7.102}$$

Proof That $K_{G,s+n}I_{G,s+n}=0$ follows from the fact that $\forall v_{s+2n}(k-2n)$,

$$K_{G,s+n}\begin{bmatrix} u_{s+n}(k-n) \\ y_{s+n}(k-n) \end{bmatrix}=K_{G,s+n}I_{G,s+n}v_{s+2n}(k-2n)=0.$$

This implies,

$$\forall \begin{bmatrix} u_{s+n}(k-n) \\ y_{s+n}(k-n) \end{bmatrix} \in \mathcal{I}_{G,s+n}, \begin{bmatrix} u_{s+n}(k-n) \\ y_{s+n}(k-n) \end{bmatrix} \in \mathcal{K}_{G,s+n},$$

i.e. $\mathcal{I}_{G,s+n} \subseteq \mathcal{K}_{G,s+n}$. Moreover, remember that

$$rank\left(I_{G,s+n}\right) = (s+n)\,p + n \Longrightarrow \dim\left(\mathcal{I}_{G,s+n}\right) = (s+n)\,p + n,$$
$$K_{G,s+n} \in \mathbb{R}^{ms \times (s+n)(p+m)}, rank\left(K_{G,s+n}\right) = sm.$$

Hence,

$$\dim\left(\mathcal{K}_{G,s+n}\right) = (s+n)(p+m) - sm = (s+n)\,p + nm.$$

Accordingly, we have, for single output systems,

$$\dim\left(\mathcal{I}_{G,s+n}\right) = \dim\left(\mathcal{K}_{G,s+n}\right) = (s+n)\,p + n.$$

Thus, in this case $\mathcal{I}_{G,s+n} = \mathcal{K}_{G,s+n}$. The corollary is proved.

This result is not surprising. Indeed, it is natural that $K_{G,s+n}$ and $I_{G,s+n}$, as special realisations of SKR K_G and SIR I_G in $\mathcal{H}_2$, inherit the property $K_G I_G = 0$.

Similarly, there exist matrices

$$I_{K,n+s} = \begin{bmatrix} -\hat{Y}_{s+n} \\ \hat{X}_{s+n} \end{bmatrix}, K_{K,n+s} = \begin{bmatrix} X_{s+n} & Y_{s+n} \end{bmatrix}$$

so that

$$K_{K,n+s} I_{K,n+s} = 0. \tag{7.103}$$

Below, (X_{s+n}, Y_{s+n}) and $\left(\hat{X}_{s+n}, \hat{Y}_{s+n}\right)$ are introduced and the relation given in (7.103) is showcased in a manner analogous to the study on $I_{G,s+n}$ and $K_{G,s+n}$.

Recall the control-theoretic interpretation of $I_K = \begin{bmatrix} -\hat{Y} \\ \hat{X} \end{bmatrix}$ and $K_K = \begin{bmatrix} X & Y \end{bmatrix}$,

$$I_K : \begin{cases} \hat{x}(k+1) = (A+BF)\,\hat{x}(k) + Lr(k), \hat{x}(0) = 0 \\ \begin{bmatrix} u(k) \\ y(k) \end{bmatrix} = \begin{bmatrix} F \\ C+DF \end{bmatrix} \hat{x}(k) + \begin{bmatrix} 0 \\ I \end{bmatrix} r(k), \end{cases} \tag{7.104}$$

$$K_K : \begin{cases} \hat{x}(k+1) = (A-LC)\,\hat{x}(k) + (B-LD)\,u(k) + Ly(k), \hat{x}(0) = 0 \\ v(k) = \begin{bmatrix} I & 0 \end{bmatrix} \begin{bmatrix} u(k) \\ y(k) \end{bmatrix} - F\hat{x}(k), \end{cases} \tag{7.105}$$

and the conditions $(u(i), y(i)) = (0,0)\,, i < 0$. Proceeding from that, we search for $I_{K,s+n}$ and $K_{K,s+n}$. For our purpose, set firstly

$$F = F_d, L = L_d.$$

It follows from (7.104) and (7.105) respectively that

$$\begin{bmatrix} u_s(k) \\ y_s(k) \end{bmatrix} = \begin{bmatrix} H_{u,\hat{x},F} \\ H_{y,\hat{x},F} \end{bmatrix} \hat{x}(k+1) + \begin{bmatrix} H_{u,r,F} \\ H_{y,r,F} \end{bmatrix} r_s(k), \tag{7.106}$$

$$\hat{x}(k+1) = H_{\hat{x},r,L} r_n(k-n) \Longrightarrow \begin{bmatrix} u_s(k) \\ y_s(k) \end{bmatrix} = \begin{bmatrix} -\hat{Y}_s \\ \hat{X}_s \end{bmatrix} r_{s+n}(k-n), \tag{7.107}$$

$$\hat{Y}_s = \begin{bmatrix} -H_{u,\hat{x},F} H_{\hat{x},r,L} & -H_{u,r,F} \end{bmatrix}, \hat{X}_s = \begin{bmatrix} H_{y,\hat{x},F} H_{\hat{x},r,L} & H_{y,r,F} \end{bmatrix}, \tag{7.108}$$

$$H_{\hat{x},r,L} = \begin{bmatrix} A_F^{n-1} L & \cdots & L \end{bmatrix}, H_{u,\hat{x},F} = H_{u,x,F}, H_{y,\hat{x},F} = H_{y,x,F},$$

$$H_{u,r,F} = \begin{bmatrix} 0 & 0 & \cdots & 0 \\ FL & 0 & \ddots & \vdots \\ \vdots & \ddots & \ddots & 0 \\ FA_F^{s-2} L & \cdots & FL & 0 \end{bmatrix}, H_{y,r,F} = \begin{bmatrix} I & 0 & \cdots & 0 \\ C_F L & I & \ddots & \vdots \\ \vdots & \ddots & \ddots & 0 \\ C_F A_F^{s-2} L & \cdots & C_F L & I \end{bmatrix},$$

as well as

$$v_s(k) = H_{u,F,L} u_s(k) + H_{y,F,L} y_s(k) - H_{\hat{x},F,L} \hat{x}(k+1), \tag{7.109}$$

$$\hat{x}(k+1) = H_{\hat{x},u,L} u_n(k-n) + H_{\hat{x},y,L} y_n(k-n) \Longrightarrow \tag{7.110}$$

$$v_s(k) = \begin{bmatrix} X_{s+n} & Y_{s+n} \end{bmatrix} \begin{bmatrix} u_{s+n}(k-n) \\ y_{s+n}(k-n) \end{bmatrix}, \tag{7.111}$$

$$X_{s+n} = \begin{bmatrix} -H_{\hat{x},F,L} H_{\hat{x},u,L} & H_{u,F,L} \end{bmatrix}, Y_{s+n} = \begin{bmatrix} -H_{\hat{x},F,L} H_{\hat{x},y,L} & H_{y,F,L} \end{bmatrix}, \tag{7.112}$$

$$H_{\hat{x},L,F} = \begin{bmatrix} F \\ FA_L \\ \vdots \\ FA_L^{s-1} \end{bmatrix}, H_{u,F,L} = \begin{bmatrix} I & 0 & \cdots & 0 \\ -FB_L & I & \ddots & \vdots \\ \vdots & \ddots & \ddots & 0 \\ -FA_L^{s-2} B_L & \cdots & -FB_L & I \end{bmatrix},$$

$$H_{y,F,L} = \begin{bmatrix} 0 & 0 & \cdots & 0 \\ -FL & 0 & \ddots & \vdots \\ \vdots & \ddots & \ddots & 0 \\ -FA_L^{s-2} L & \cdots & -FL & 0 \end{bmatrix}.$$

Let $v(k) = 0$ ($v_s(k) = 0$). It follows from (7.107) and (7.111) that $\forall r_{s+n}(k-2n)$,

$$0 = \begin{bmatrix} X_{s+n} & Y_{s+n} \end{bmatrix} \begin{bmatrix} u_{s+n}(k-n) \\ y_{s+n}(k-n) \end{bmatrix} = \begin{bmatrix} X_{s+n} & Y_{s+n} \end{bmatrix} \begin{bmatrix} -\hat{Y}_{s+n} \\ \hat{X}_{s+s} \end{bmatrix} r_{s+n}(k-2n)$$

$$\Longrightarrow \begin{bmatrix} X_{s+n} & Y_{s+n} \end{bmatrix} \begin{bmatrix} -\hat{Y}_{s+n} \\ \hat{X}_{s+s} \end{bmatrix} = K_{K,n+s} I_{K,n+s} = 0.$$

That is, (7.103) has been proved on the assumption that $F = F_d, L = L_d$. Next, this assumption is removed.

Recall Theorem 3.4,

$$\begin{bmatrix} -\hat{Y} \\ \hat{X} \end{bmatrix} = \begin{bmatrix} -\hat{Y}_1 \\ \hat{X}_1 \end{bmatrix} R + \begin{bmatrix} M_1 \\ N_1 \end{bmatrix} \bar{R},$$

$$\begin{bmatrix} X & Y \end{bmatrix} = V \begin{bmatrix} X_1 & Y_1 \end{bmatrix} + \bar{V} \begin{bmatrix} -\hat{N}_1 & \hat{M}_1 \end{bmatrix},$$

$$R = I - C\left(zI - A_{L_d}\right)^{-1}(L_d - L), \bar{R} = \begin{bmatrix} X_1 & Y_1 \end{bmatrix} \begin{bmatrix} -\hat{Y} \\ \hat{X} \end{bmatrix}, \tag{7.113}$$

$$V = I + (F_d - F)\left(zI - A_{F_d}\right)^{-1} B, \bar{V} = \begin{bmatrix} X & Y \end{bmatrix} \begin{bmatrix} -\hat{Y}_1 \\ \hat{X}_1 \end{bmatrix}, \tag{7.114}$$

where (M_1, N_1), $\left(\hat{M}_1, \hat{N}_1\right)$, $\left(\hat{X}_1, \hat{Y}_1\right)$ and (X_1, Y_1) represent the RCP and LCP of G and the RCP and LCP of K respectively with $F = F_d, L = L_d$, and (M, N), $\left(\hat{M}, \hat{N}\right)$, $\left(\hat{X}, \hat{Y}\right)$ and (X, Y) denote these coprime pairs with (F, L) different from (F_d, L_d). Furthermore, Example 3.4 delineates that

$$\bar{R} = (F_d - F)\left(zI - A_{F_d}\right)^{-1} L_d - F\left(zI - A_L\right)^{-1}(L - L_d),$$
$$\bar{V} = (F - F_d)\left(zI - A_F\right)^{-1} L - F_d\left(zI - A_{L_d}\right)^{-1}(L_d - L).$$

Note that $\left(-\hat{Y}, \hat{X}\right)$ builds an observer-based state feedback controller whose eigen dynamic is determined by $A + BF$. Hence, to model the SIR of the controller as a finite samples system, we only need consider possible variations of the state feedback gain matrix F. Analogous to it, variations of the observer gain matrix L are taken into account by modelling the SKR of the controller as a finite samples system. Accordingly, we have

$$\begin{bmatrix} u_s(k) \\ y_s(k) \end{bmatrix} = \begin{bmatrix} -\hat{Y}_s \\ \hat{X}_s \end{bmatrix} r_{s+n}(k-n) + \begin{bmatrix} M_s \\ N_s \end{bmatrix} \bar{R}_{s+n} r_{s+n}(k-n), \tag{7.115}$$

$$v_s(k) = \left(\begin{bmatrix} X_{s+n} & Y_{s+n} \end{bmatrix} + \bar{V}_s \begin{bmatrix} -\hat{N}_{s+n} & \hat{M}_{s+n} \end{bmatrix}\right) \begin{bmatrix} u_{s+n}(k-n) \\ y_{s+n}(k-n) \end{bmatrix}, \tag{7.116}$$

$$\bar{R}_{s+n} = \begin{bmatrix} 0 & 0 & \cdots & 0 \\ (F_d - F) L_d & 0 & \ddots & \vdots \\ \vdots & \ddots & \ddots & 0 \\ (F_d - F) A_{F_d}^{s+n-2} L_d & \cdots & (F_d - F) L_d & 0 \end{bmatrix},$$

$$\bar{V}_s = \begin{bmatrix} 0 & 0 & \cdots & 0 \\ F_d (L - L_d) & 0 & \ddots & \vdots \\ \vdots & \ddots & \ddots & 0 \\ F_d A_{L_d}^{s-2} (L - L_d) & \cdots & F_d (L - L_d) & 0 \end{bmatrix}.$$

Now, (7.103) is examined. Let $v_s(k) = 0$. It follows from (7.115) and (7.116) that $\forall r_{s+n}(k - 2n)$,

$$0 = \begin{bmatrix} V_s & \bar{V}_s \end{bmatrix} \begin{bmatrix} X_{s+n} & Y_{s+n} \\ -\hat{N}_{s+n} & \hat{M}_{s+n} \end{bmatrix} \begin{bmatrix} -\hat{Y}_{s+n} & M_{s+n} \\ \hat{X}_{s+n} & N_{s+n} \end{bmatrix} \begin{bmatrix} R_{s+2n} \\ \bar{R}_{s+2n} \end{bmatrix} r_{s+n}(k - 2n).$$

Hence,

$$\begin{bmatrix} V_s & \bar{V}_s \end{bmatrix} \begin{bmatrix} X_{s+n} & Y_{s+n} \\ -\hat{N}_{s+n} & \hat{M}_{s+n} \end{bmatrix} \begin{bmatrix} -\hat{Y}_{s+n} & M_{s+n} \\ \hat{X}_{s+n} & N_{s+n} \end{bmatrix} \begin{bmatrix} R_{s+2n} \\ \bar{R}_{s+2n} \end{bmatrix} = 0. \tag{7.117}$$

Theorem 7.8 *Given the system model (7.73), the corresponding image representation (7.87) and kernel representation (7.99), then there exist* $\left(\hat{X}_s, \hat{Y}_s\right)$ *and* (X_{s+n}, Y_{s+n}) *given by (7.108) and (7.112), respectively, so that*

$$\begin{bmatrix} u_s(k) \\ y_s(k) \end{bmatrix} = I_{K,s} r_{s+n}(k - n), \; I_{K,s+n} = \begin{bmatrix} -\hat{Y}_s & M_s \\ \hat{X}_s & N_s \end{bmatrix} \begin{bmatrix} I \\ \bar{R}_{s+n} \end{bmatrix},$$

$$v_s(k) = K_{K,s+n} \begin{bmatrix} u_{s+n}(k - n) \\ y_{s+n}(k - n) \end{bmatrix}, \; K_{K,s+n} = \begin{bmatrix} I & \bar{V}_s \end{bmatrix} \begin{bmatrix} X_{s+n} & Y_{s+n} \\ -\hat{N}_{s+n} & \hat{M}_{s+n} \end{bmatrix},$$

and $I_{K,s+n}$, $K_{K,s+n}$ *satisfy*

$$K_{K,s+n} I_{K,s+n} = 0.$$

Proof The proof immediately follows from the derivation of (7.108), (7.112), (7.115), (7.116) *and* (7.117).

In the sequel, we call

$$\begin{bmatrix} -\hat{Y}_s & M_s \\ \hat{X}_s & N_s \end{bmatrix} \begin{bmatrix} I \\ \bar{R}_{s+n} \end{bmatrix} \text{ and } \begin{bmatrix} I & \bar{V}_s \end{bmatrix} \begin{bmatrix} X_{s+n} & Y_{s+n} \\ -\hat{N}_{s+n} & \hat{M}_{s+n} \end{bmatrix}$$

image and kernel representation of controllers of finite samples long data, borrowing from the concepts of I_K and K_K.

At the core of the unified framework of control and detection are the Bezout identity and the associated system and residual dynamics,

$$\begin{bmatrix} K_K \\ K_G \end{bmatrix} \begin{bmatrix} I_G & I_K \end{bmatrix} = I \Longrightarrow \begin{bmatrix} u \\ y \end{bmatrix} = \begin{bmatrix} I_G & I_K \end{bmatrix} \begin{bmatrix} v \\ r \end{bmatrix}, \begin{bmatrix} v \\ r \end{bmatrix} = \begin{bmatrix} K_K \\ K_G \end{bmatrix} \begin{bmatrix} u \\ y \end{bmatrix}.$$

It is of considerable interest to examine the analogue relations expressed in terms of $\left(I_{G,s+n}, K_{G,s+n}\right)$ and $\left(I_{K,s+n}, K_{K,s+n}\right)$. To simplify the the comparison, (I_G, K_G) are expressed with respect to $\left(I_{d,G}, K_{d,G}\right)$, the SIR/SKR of the plant for $F = F_d$, $L = L_d$,

$$\begin{aligned}
&I_G = I_{d,G} V_d, K_G = R_d K_{d,G}, \\
&R_d = I - C\left(zI - A_L\right)^{-1}\left(L - L_d\right), V_d = I + \left(F - F_d\right)\left(zI - A_F\right)^{-1} B.
\end{aligned}$$

Similarly, (I_K, K_K) are represented by $\left(I_{d,G}, K_{d,G}\right)$ according to

$$\begin{aligned}
&I_K = I_{d,K} + I_{d,G}\bar{R}, K_K = K_{d,K} + \bar{V} K_{d,G}, \\
&\bar{R} = \left(F_d - F\right)\left(zI - A_{F_d}\right)^{-1} L_d, \bar{V} = F_d\left(zI - A_{L_d}\right)^{-1}\left(L - L_d\right).
\end{aligned}$$

Hereby, I_K is subject to (7.104) with $F = F_d$, while K_K is modelled by (7.105) with $L = L_d$. It yields

$$\begin{aligned}
\begin{bmatrix} u \\ y \end{bmatrix} &= \begin{bmatrix} I_{d,G} & I_{d,K} \end{bmatrix}\begin{bmatrix} V_d & \bar{R} \\ 0 & I \end{bmatrix}\begin{bmatrix} v \\ r \end{bmatrix} \qquad (7.118)\\
&= I_{d,G} v_d + \left(I_{d,K} + I_{d,G}\bar{R}\right) r, \; v_d = V_d v, \\
\begin{bmatrix} v \\ r \end{bmatrix} &= \begin{bmatrix} I & \bar{V} \\ 0 & R_d \end{bmatrix}\begin{bmatrix} K_{d,K} \\ K_{d,G} \end{bmatrix}\begin{bmatrix} u \\ y \end{bmatrix}. \qquad (7.119)
\end{aligned}$$

Now, we turn to the system dynamics in the form of finite samples long data and relations between $\left(I_{G,s+n}, K_{G,s+n}\right)$ and $\left(I_{K,s+n}, K_{K,s+n}\right)$. It follows from the SIR-SKR-based model (7.55)–(7.56) and the aforementioned discussions that

$$\begin{aligned}
\begin{bmatrix} u_s(k) \\ y_s(k) \end{bmatrix} &= I_{G,s}\bar{v}_{s+n}(k-n) + I_{K,s} r_{s+n}(k-n) \\
&= \begin{bmatrix} M_s \\ N_s \end{bmatrix}\bar{v}_{s+n}(k-n) + \left(\begin{bmatrix} -\hat{Y}_s \\ \hat{X}_s \end{bmatrix} + \begin{bmatrix} M_s \\ N_s \end{bmatrix}\bar{R}_{s+n}\right) r_{s+n}(k-n) \\
&= \begin{bmatrix} M_s & -\hat{Y}_s \\ N_s & \hat{X}_s \end{bmatrix}\begin{bmatrix} I & \bar{R}_{s+n} \\ 0 & I \end{bmatrix}\begin{bmatrix} \bar{v}_{s+n}(k-n) \\ r_{s+n}(k-n) \end{bmatrix}, \qquad (7.120)\\
\begin{bmatrix} v_s(k) \\ r_s(k) \end{bmatrix} &= \begin{bmatrix} K_{K,s+n} \\ K_{G,s+n} \end{bmatrix}\begin{bmatrix} u_{s+n}(k-n) \\ y_{s+n}(k-n) \end{bmatrix} \Longrightarrow \\
\begin{bmatrix} v_s(k) \\ \bar{r}_s(k) \end{bmatrix} &= \begin{bmatrix} I & \bar{V}_s \\ 0 & I \end{bmatrix}\begin{bmatrix} X_{s+n} & Y_{s+n} \\ -\hat{N}_{s+n} & \hat{M}_{s+n} \end{bmatrix}\begin{bmatrix} u_{s+n}(k-n) \\ y_{s+n}(k-n) \end{bmatrix}. \qquad (7.121)
\end{aligned}$$

It is obvious that

$$\begin{bmatrix} I & \bar{R}_{s+n} \\ 0 & I \end{bmatrix} \in \mathbb{R}^{(s+n)(m+p)\times(s+n)(p+m)},$$

$$rank \begin{bmatrix} I & \bar{R}_{s+n} \\ 0 & I \end{bmatrix} = (s+n)(p+m),$$

$$\begin{bmatrix} I & \bar{V}_s \\ 0 & I \end{bmatrix} \in \mathbb{R}^{s(m+p)\times s(p+m)}, rank \begin{bmatrix} I & \bar{V}_s \\ 0 & I \end{bmatrix} = s(p+m).$$

Moreover, for

$$\begin{bmatrix} M_s & -\hat{Y}_s \\ N_s & \hat{X}_s \end{bmatrix} \in \mathbb{R}^{s(m+p)\times(s+n)(p+m)}, \begin{bmatrix} X_{s+n} & Y_{s+n} \\ -\hat{N}_{s+n} & \hat{M}_{s+n} \end{bmatrix} \in \mathbb{R}^{s(m+p)\times(s+n)(p+m)},$$

$$\begin{bmatrix} M_s & -\hat{Y}_s \\ N_s & \hat{X}_s \end{bmatrix} = \begin{bmatrix} H_{u,x,F} H_{x,v,F} & H_{u,v,F} & H_{u,\hat{x},F} H_{\hat{x},r,L} & H_{u,r,F} \\ H_{y,x,F} H_{x,v,F} & H_{y,v,F} & H_{y,\hat{x},F} H_{\hat{x},r,L} & H_{y,r,F} \end{bmatrix},$$

$$\begin{bmatrix} X_{s+n} & Y_{s+n} \\ -\hat{N}_{s+n} & \hat{M}_{s+n} \end{bmatrix} = \begin{bmatrix} -H_{\hat{x},F,L} H_{\hat{x},u,L} & H_{u,F,L} & -H_{\hat{x},F,L} H_{\hat{x},y,L} & H_{y,F,L} \\ -H_{\hat{x},L} H_{\hat{x},u,L} & -H_{r,u,L} & -H_{\hat{x},L} H_{\hat{x},y,L} & I \end{bmatrix},$$

due to the special forms of $H_{u,v,F}$ and $H_{y,r,F}$ as well as $H_{r,u,L}$, it is apparent that

$$rank \begin{bmatrix} M_s & -\hat{Y}_s \\ N_s & \hat{X}_s \end{bmatrix} = rank\left(H_{u,v,F}\right) + rank\left(H_{y,r,F}\right) = s(m+p), \tag{7.122}$$

$$rank \begin{bmatrix} X_{s+n} & Y_{s+n} \\ -\hat{N}_{s+n} & \hat{M}_{s+n} \end{bmatrix} = rank\left(H_{u,F,L}\right) + rank\left(I_{sm\times sm}\right) = s(m+p). \tag{7.123}$$

Comparing (7.120)–(7.121) with (7.118)–(7.119) shows that both description forms have the identical structure, but are distinguished by the mappings from the process data to the residual/latent vectors and back. In $\mathcal{H}_2$ space, thanks to the Bezout identity, the mappings, $(u, y) \to (v, r)$ and $(v, r) \to (u, y)$, are one-to-one, while the system dynamics presented in the form of finite samples long signals are two mappings from $\mathbb{R}^{(s+n)(m+p)}$ to $\mathbb{R}^{s(m+p)}$. This difference is the consequence of modelling the system dynamics in terms of finite samples long signals. In order to "transfer" the state of the system dynamics before the beginning of a time interval, e.g. $[k+1, k+s]$, the initial state $x(k+1)$ is to be re-constructed using the data of the past n samples. Nevertheless, (7.122)–(7.123) tell us that two mappings from $\mathbb{R}^{(s+n)(m+p)}$ to $\mathbb{R}^{s(m+p)}$ are presented by two full row-rank matrices. The signal subspaces,

$$\mathcal{S}_{(u,y),,s} = \left\{ \begin{bmatrix} u_s(k) \\ y_s(k) \end{bmatrix}, \begin{bmatrix} u_s(k) \\ y_s(k) \end{bmatrix} = \begin{bmatrix} I_{G,s} & I_{K,s} \end{bmatrix} \begin{bmatrix} v_{s+n}(k-n) \\ r_{s+n}(k-n) \end{bmatrix}, \begin{matrix} \begin{bmatrix} v_{s+n}(k-n) \\ r_{s+n}(k-n) \end{bmatrix} \in \mathcal{L}_{[0,\infty)}^{(s+n)(m+p)} \end{matrix} \right\},$$

$$\mathcal{S}_{(v,r),s} = \left\{ \begin{bmatrix} v_s(k) \\ r_s(k) \end{bmatrix}, \begin{bmatrix} v_s(k) \\ r_s(k) \end{bmatrix} = \begin{bmatrix} K_{K,s+n} \\ K_{G,s+n} \end{bmatrix} \begin{bmatrix} u_{s+n}(k-n) \\ y_{s+n}(k-n) \end{bmatrix}, \begin{matrix} \begin{bmatrix} u_{s+n}(k-n) \\ y_{s+n}(k-n) \end{bmatrix} \in \mathcal{L}_{[0,\infty)}^{(s+n)(m+p)} \end{matrix} \right\},$$

are of dimension $s(m+p)$.

7.3.6 Minimal Order Realisations

In previous subsections, it has been delineated that n samples of the data in the past are used to re-construct the initial state $x(k+1)$ over the time interval $[k-n+1, k]$. It is known from the system realisation theory that for multiple input and multiple output (MIMO) systems less samples are necessary for this end. In the sequel, we investigate the minimum sample number needed for re-constructing the initial state, and call this issue minimal order realisation problem.

We begin with a short introduction of the concepts controllability and observability indices on the assumption that the state space realisation (A, B, C, D) of systems under consideration is minimum, where

$$B = \begin{bmatrix} b_1 \cdots b_p \end{bmatrix}, C = \begin{bmatrix} c_1 \\ \vdots \\ c_m \end{bmatrix}.$$

Thus, these systems are controllable and observable. The controllability indices are determined by the following algorithm.

Algorithm 7.2 Controllability indices

Step 0: *Set*

$$S_0 = \begin{bmatrix} s_{0,1} \cdots s_{0,p} \end{bmatrix} = \begin{bmatrix} b_1 \cdots b_p \end{bmatrix}, \mu_j = 1, j = 1, \ldots, p;$$

Step 1: *For* $i = 1 : n-1$

$S_{i,0} = \begin{bmatrix} s_{0,1} \cdots s_{0,p} \end{bmatrix} = S_{i-1};$
For $j = 1 : p$

- *if* $rank \begin{bmatrix} S_{i,j-1} \ A s_{j-1,j} \end{bmatrix} > rank \left(S_{i,j-1}\right)$, *then*

$$S_{i,j} = \begin{bmatrix} S_{i,j-1} \ A s_{j-1,j} \end{bmatrix}, \mu_j = \mu_j + 1$$

otherwise,

$$S_{i,j} = S_{i,i-1};$$

End
Set

$$S_i = S_{i,j};$$

If $S_i = S_{i-1}$, *go to Step 2;*
End;

Step 2: *Output* $S_i, \mu_j, j = 1, \ldots, p.$

This algorithm outputs the controllability matrix S_c,

$$S_c = \left[b_1 \cdots A^{\mu_1 - 1} b_1 \cdots b_p \cdots A^{\mu_p - 1} b_p \right] \in \mathbb{R}^{n \times n},$$

and μ_j, $j = 1, \ldots, p$, are controllability indices, satisfying

$$\sum_{j=1}^{p} \mu_j = n.$$

The controllability index is defined by

$$\mu_c = \max \left\{ \mu_j, j = 1, \ldots, p \right\}.$$

On account of the duality, given (C, A), the observability matrix S_o is

$$S_c = \begin{bmatrix} c_1 \\ \vdots \\ c_1 A^{\mu_1 - 1} \\ \vdots \\ c_m \\ \vdots \\ c_m A^{\mu_m - 1} \end{bmatrix} \in \mathbb{R}^{n \times n},$$

and μ_j, $j = 1, \ldots, m$, are observability indices, satisfying

$$\sum_{j=1}^{m} \mu_j = n.$$

The observability index is defined by

$$\mu_o = \max \left\{ \mu_j, j = 1, \ldots, m \right\}.$$

It is a well-known result in linear control theory that a deadbeat state feedback controller and an observer result in

$$A + BF_d \sim A_{c,0}, \; A - L_d C \sim A_{o,0},$$

$$A_{c,0} = diag\left(A_{c,1}, \ldots, A_{c,p} \right), \; A_{c,j} = \begin{bmatrix} 0 & 1 & 0 & \cdots & 0 \\ 0 & 0 & 1 & \cdots & 0 \\ \vdots & \ddots & \ddots & \ddots & \vdots \\ 0 & \cdots & 0 & 0 & 1 \\ 0 & \cdots & 0 & 0 & 0 \end{bmatrix} \in \mathbb{R}^{\mu_j \times \mu_j}, \; j = 1, \ldots, p,$$

$$A_{o,0} = diag\left(A_{o,1}, \ldots, A_{o,m}\right), A_{o,j} = \begin{bmatrix} 0 & 0 & 0 & \cdots & 0 \\ 1 & 0 & 0 & \cdots & 0 \\ \vdots & \ddots & \ddots & \ddots & \vdots \\ 0 & \cdots & 1 & 0 & 0 \\ 0 & 0 & \cdots & 1 & 0 \end{bmatrix} \in \mathbb{R}^{\mu_j \times \mu_j}, j = 1, \ldots, m,$$

where F_d and L_d are the feedback and observer gains of the deadbeat controller and observer, respectively. Therefore, for

$$\mu = \max\left\{\mu_o, \mu_c\right\},$$

it holds

$$(A + BF_d)^{\mu} = 0, (A - L_d C)^{\mu} = 0. \tag{7.124}$$

Equation (7.124) gives the minimal order of image and kernel representations of the plant and controller and allows us to rewrite them, as described in the following theorem.

Theorem 7.9 *Given the system model (7.73), the minimal order image and kernel representations of the plant and controller are respectively given by*

$$\begin{bmatrix} u_s(k) \\ y_s(k) \end{bmatrix} = I_{G,s} v_{s+\mu}(k-\mu), I_{G,s} = \begin{bmatrix} M_s \\ N_s \end{bmatrix} V_{s+\mu}, \tag{7.125}$$

$$r_s(k) = K_{G,s+\mu} \begin{bmatrix} u_{s+\mu}(k-\mu) \\ y_{s+\mu}(k-\mu) \end{bmatrix} = 0, K_{G,s+\mu} = R_s \begin{bmatrix} -\hat{N}_{s+\mu} & \hat{M}_{s+\mu} \end{bmatrix}, \tag{7.126}$$

$$\begin{bmatrix} u_s(k) \\ y_s(k) \end{bmatrix} = I_{K,s} r_{s+\mu}(k-\mu), I_{K,s} = \begin{bmatrix} -\hat{Y}_s & M_s \\ \hat{X}_s & N_s \end{bmatrix} \begin{bmatrix} R_{s+\mu} \\ \bar{R}_{s+\mu} \end{bmatrix}, \tag{7.127}$$

$$v_s(k) = K_{K,s+\mu} \begin{bmatrix} u_{s+\mu}(k-\mu) \\ y_{s+\mu}(k-\mu) \end{bmatrix}, K_{K,s+\mu} = \begin{bmatrix} V_s & \bar{V}_s \end{bmatrix} \begin{bmatrix} X_{s+\mu} & Y_{s+\mu} \\ -\hat{N}_{s+\mu} & \hat{M}_{s+\mu} \end{bmatrix}, \tag{7.128}$$

where (M_s, N_s), $\left(\hat{M}_{s+\mu}, \hat{N}_{s+\mu}\right)$, $\left(\hat{X}_s, \hat{Y}_s\right)$ *and* $\left(X_{s+\mu}, Y_{s+\mu}\right)$ *are respectively*

$$\begin{aligned}
M_s &= \begin{bmatrix} H_{u,x,F} H_{x,v,F} & H_{u,v,F} \end{bmatrix}, N_s = \begin{bmatrix} H_{y,x,F} H_{x,v,F} & H_{y,v,F} \end{bmatrix}, \\
H_{x,v,F} &= \begin{bmatrix} A_F^{\mu-1} B & \cdots & B \end{bmatrix}, \\
\hat{M}_{s+\mu} &= \begin{bmatrix} -H_{\hat{x},L} H_{\hat{x},y,L} & I \end{bmatrix}, \hat{N}_{s+\mu} = \begin{bmatrix} H_{\hat{x},L} H_{\hat{x},u,L} & H_{r,u,L} \end{bmatrix}, \\
H_{\hat{x},u,L} &= \begin{bmatrix} A_L^{\mu-1} B_L & \cdots & B_L \end{bmatrix}, H_{\hat{x},y,L} = \begin{bmatrix} A_L^{\mu-1} L & \cdots & L \end{bmatrix}, \\
\hat{Y}_s &= \begin{bmatrix} -H_{u,\hat{x},F} H_{\hat{x},r,L} & -H_{u,r,F} \end{bmatrix}, \hat{X}_s = \begin{bmatrix} H_{y,\hat{x},F} H_{\hat{x},r,L} & H_{y,r,F} \end{bmatrix}, \\
H_{\hat{x},r,L} &= \begin{bmatrix} A_F^{\mu-1} L & \cdots & L \end{bmatrix}, \\
X_{s+\mu} &= \begin{bmatrix} -H_{\hat{x},F,L} H_{\hat{x},u,L} & H_{u,F,L} \end{bmatrix}, Y_{s+\mu} = \begin{bmatrix} -H_{\hat{x},F,L} H_{\hat{x},y,L} & H_{y,F,L} \end{bmatrix}.
\end{aligned}$$

The proof of the theorem follows immediately from (7.124).

7.3.7 Concluding Remarks

The significance of the results in this section is, without a doubt, the equivalent expressions of the plant and controller dynamics in $\mathcal{H}_2$ Hilbert space in terms of algebraic matrices and vectors of finite samples long signals. The latter model and signal forms not only enable easy analysis and synthesis of the system and loop dynamics, but also build a control-theoretically well grounded bridge to system design and implementation in the data-driven fashion. It is worth emphasising that the system model forms of finite samples long signals introduced in this section are specific realisations of system and controller models, like SIR and SKR. Consequently, those essential properties like loop stability, the inherent data structure (7.1) or the structure of the loop dynamics are fully conserved. On the other hand, the properties of finite samples long signals and matrix-valued system models make it possible to solve a series of detection and control problems optimally without approximations. For instance,

- optimal fault detection
  A widely adopted type of residual evaluation functions is an $\mathcal{L}_2$-norm-based function over $[0, \infty\}$. Associated with it, $\mathcal{H}_2$- or $\mathcal{H}_\infty$-optimal design methods serve as design tools. In real-time applications, $[0, \infty\}$ is to be approximated by a finite time interval $[k_1, k_2]$, which may considerably degrade the detection performance. Alternatively, leveraging time-varying system theory to deal with such issues over finite evaluation time intervals can assure the targeted optimal performance, requires, on the other hand, considerably more online computation. The system representations of finite samples long signals and matrix-valued system models allow us, without any approximations, to address such issues in Euclidean space using algebraic methods.
- optimal control over finite time intervals
  Many control tasks are formulated as an optimisation problem over a finite time interval. MPC is a typical and widely applied example. The alternative system representations introduced in this section offer a new perspective for design, analysis and implementation of such control systems.
- data-driven control and detection
  At the core of most of so-called data-driven control and detection methods is an implicit system identification. The alternative system representation forms provide the designer with a comprehensive control-theoretic basis for a direct identification of primary controllers and residual generators.
- fault-tolerant control and performance recovery from degradation
  In engineering applications, it is expected that fault-tolerant control and performance recovery algorithms should be performed timely and reliably within a required (finite) time window. The proposed system representations fully meet such highly real-time and easy computation and implementation requirements.

7.4 Potential Applications in Detection and Control Domains

7.4.1 On Detection

We begin with the inherent data structure (7.1)

$$\begin{bmatrix} u \\ y \end{bmatrix} = \begin{bmatrix} M \\ N \end{bmatrix} v + \begin{bmatrix} -\hat{Y} + MQ \\ \hat{X} + NQ \end{bmatrix} r_y,$$

and assume that there is no uncertainty in the latent variable v. Since

$$\begin{bmatrix} r_u \\ r_y \end{bmatrix} = \begin{bmatrix} X & Y \\ -\hat{N} & \hat{M} \end{bmatrix} \begin{bmatrix} u \\ y \end{bmatrix} = \begin{bmatrix} v + Qr_y \\ r_y \end{bmatrix},$$

it is apparent that all uncertainties in the process signals (u, y) are driven by the residual r_y. On account of this fact, it can be claimed that it is sufficient to detect any uncertainty in the system or data by means of the residual generator/detector

$$r_y = \hat{M}y - \hat{N}u.$$

Example 7.7 In this example, we consider an uncertain system modelled by

$$\begin{gathered} y = G_\Delta u = N_\Delta M_\Delta^{-1} u, \\ \begin{bmatrix} M_\Delta \\ N_\Delta \end{bmatrix} = \begin{bmatrix} M \\ N \end{bmatrix} + \begin{bmatrix} \Delta_M \\ \Delta_N \end{bmatrix} = I_G + \Delta_{I_G} \Longrightarrow \\ \begin{bmatrix} u \\ y \end{bmatrix} = \left(\begin{bmatrix} M \\ N \end{bmatrix} + \begin{bmatrix} \Delta_M \\ \Delta_N \end{bmatrix} \right) v_\Delta. \end{gathered}$$

Hereby, Δ_{I_G} represents the coprime factor uncertainty that can be modelled by

$$\Delta_{I_G} = \begin{bmatrix} M \\ N \end{bmatrix} \delta_{I_G} + \begin{bmatrix} -\hat{Y} \\ \hat{X} \end{bmatrix} \delta_{I_K},$$

and v_Δ is the latent variable. It will be demonstrated that uncertainties in the data structure are driven by the residual $r_y = \hat{M}y - \hat{N}u$. Let us firstly define (any) $F\hat{x} + Qr_y$, $Q \in \mathcal{RH}_\infty$, $A + BF$ is a Schur matrix, and $\hat{x}$ is the state estimate. Now, introduce a latent variable v,

$$v = u - F\hat{x} - Qr_y = Xu + Yy - Qr_y.$$

Here, it is worth emphasising that also in open-loop configuration, $F\hat{x} + Qr_y$ and v can be (arbitrarily) defined. It turns out

$$r_u = \begin{bmatrix} X & Y \end{bmatrix} \begin{bmatrix} u \\ y \end{bmatrix} = v + Q r_y,$$

$$r_y = \begin{bmatrix} -\hat{N} & \hat{M} \end{bmatrix} \begin{bmatrix} u \\ y \end{bmatrix} = \delta_{I_K} \left(I + \delta_{I_G} \right)^{-1} \left(v + Q r_y \right) \Longrightarrow$$

$$r_y = \delta_{I_K} \left(I + \delta_{IG} - Q \delta_{I_K} \right)^{-1} v,$$

and, by means of the Bezout identity,

$$\begin{aligned} \begin{bmatrix} u \\ y \end{bmatrix} &= \begin{bmatrix} M \\ N \end{bmatrix} \left(v + Q r_y \right) + \begin{bmatrix} -\hat{Y} \\ \hat{X} \end{bmatrix} r_y \\ &= \begin{bmatrix} M \\ N \end{bmatrix} v + \begin{bmatrix} -\hat{Y} + MQ \\ \hat{X} + NQ \end{bmatrix} \delta_{I_K} \left(I + \delta_{IG} - Q \delta_{I_K} \right)^{-1} v. \end{aligned}$$

This demonstrates that all uncertain dynamics are driven by r_y.

Differently, when there exist uncertainties in the latent variable v, denoted by Δv, the data structure becomes

$$\begin{bmatrix} u \\ y \end{bmatrix} = \begin{bmatrix} M \\ N \end{bmatrix} (v + \Delta v) + \begin{bmatrix} -\hat{Y} + MQ \\ \hat{X} + NQ \end{bmatrix} r_y.$$

It is apparent that Δv cannot be detected generally by means of the residual generator $r_y = \hat{M} y - \hat{N} u$ due to the property

$$\begin{bmatrix} -\hat{N} & \hat{M} \end{bmatrix} \begin{bmatrix} M \\ N \end{bmatrix} = 0.$$

This motivates us to propose a fault detection scheme leveraging two residual generators,

$$\begin{bmatrix} r_u \\ r_y \end{bmatrix} = \begin{bmatrix} X & Y \\ -\hat{N} & \hat{M} \end{bmatrix} \begin{bmatrix} u \\ y \end{bmatrix}, \tag{7.129}$$

to detect uncertainties in the plant and model as well as in the latent variable v. Such a detection system will play an important role in our subsequent study on networked control systems (NCS) and cyber-physical systems (CPS) in Chaps. 14–16.

7.4.2 Tracking Control and MPC

Firstly, the tracking control problem under consideration is formulated. Given the nominal plant model

$$G : \begin{cases} x(k+1) = Ax(k) + Bu(k), x(0) = x_0 \\ y(k) = Cx(k) + Du(k), \end{cases}$$

$x \in \mathbb{R}^n$, $y \in \mathbb{R}^m$, $u \in \mathbb{R}^p$, and the reference model

$$y_r = G_r u_r, \; u_r \in \mathbb{C}^n, \; y_r \in \mathbb{C}^m, \tag{7.130}$$

whose minimal state space realisation is denoted by $G_r = (A_r, B_r, C_r, D_r)$, find u so that a pre-defined tracking performance, that will be defined in the sequel, is optimised. Suppose that there exist measurement and process noises or unknown inputs, but no model uncertainties. For our purpose, the canonical SIR-SKR-based model is applied,

$$\begin{bmatrix} u \\ y \end{bmatrix} = \begin{bmatrix} M_0 & -\hat{Y}_0 \\ N_0 & \hat{X}_0 \end{bmatrix} \begin{bmatrix} v + Q r_{y,0} \\ r_{y,0} \end{bmatrix}, \tag{7.131}$$

$$\begin{bmatrix} v \\ r_{y,0} \end{bmatrix} = \begin{bmatrix} X_0 & Y_0 \\ -\hat{N}_0 & \hat{M}_0 \end{bmatrix} \begin{bmatrix} u - Q r_{y,0} \\ y \end{bmatrix} \tag{7.132}$$

and the controller is set to be

$$u = F_0 \hat{x} + v + Q r_{y,0}, \; Q = -Q_{\mathcal{H}_2}.$$

According to Theorem 7.2, the closed-loop dynamic is governed by

$$\begin{bmatrix} u \\ y \end{bmatrix} = \begin{bmatrix} M_0 \\ N_0 \end{bmatrix} v + \left(\begin{bmatrix} -\hat{Y}_0 \\ \hat{X}_0 \end{bmatrix} - \begin{bmatrix} M_0 \\ N_0 \end{bmatrix} Q_{\mathcal{H}_2} \right) r_{y,0}. \tag{7.133}$$

Remember that

$$\begin{bmatrix} M_0 \\ N_0 \end{bmatrix} v \in \mathcal{I}_G, \left(\begin{bmatrix} -\hat{Y}_0 \\ \hat{X}_0 \end{bmatrix} - \begin{bmatrix} M_0 \\ N_0 \end{bmatrix} Q_{\mathcal{H}_2} \right) r_{y,0} \in \mathcal{R}_G.$$

Hence, the design of an optimal tracking controller is now formulated as minimising the difference between the reference model (7.130) and $\begin{bmatrix} M_0 \\ N_0 \end{bmatrix} v$. Below, we delineate a schematic solution.

Let $\left(M_{r,0}, N_{r,0}\right)$ be the normalised RCP and

$$I_{G_{r,0}} := \begin{bmatrix} M_{r,0} \\ N_{r,0} \end{bmatrix}$$

be the SIR of the reference model G_r. Projecting $I_{G_{r,0}} v_r$ to $\mathcal{I}_G$ by means of an orthogonal projection

$$\mathcal{P}_{\mathcal{I}_G} \left(\begin{bmatrix} M_{r,0} \\ N_{r,0} \end{bmatrix} v_r \right) = \mathcal{L}_{I_{G_0}} \mathcal{P}_{\mathcal{H}_2} \mathcal{L}_{I_{G_0}^{\sim}} \left(\begin{bmatrix} M_{r,0} \\ N_{r,0} \end{bmatrix} v_r \right) \tag{7.134}$$

yields, analogous to the discussion in Sect. 7.1,

$$\mathcal{P}_{\mathcal{I}_G}\left(\begin{bmatrix} M_{r,0} \\ N_{r,0} \end{bmatrix} v_r\right) = \mathcal{L}_{I_{G_0}} \mathcal{P}_{\mathcal{H}_2}\left(\begin{bmatrix} M_0^{\sim} & N_0^{\sim} \end{bmatrix} \begin{bmatrix} M_{r,0} \\ N_{r,0} \end{bmatrix} v_r\right) = \begin{bmatrix} M_0 \\ N_0 \end{bmatrix} Q_{r,\mathcal{H}_2} v_r,$$
$$Q_{r,\mathcal{H}_2} = \mathcal{P}_{\mathcal{H}_2}\left(\begin{bmatrix} M_0^{\sim} & N_0^{\sim} \end{bmatrix} \begin{bmatrix} M_{r,0} \\ N_{r,0} \end{bmatrix}\right) \in \mathcal{RH}_\infty.$$

Now, setting

$$v = Q_{r,\mathcal{H}_2} v_r + v_1$$

gives the tracking error

$$\begin{bmatrix} M_{r,0} \\ N_{r,0} \end{bmatrix} v_r - \begin{bmatrix} M_0 \\ N_0 \end{bmatrix} v = \left(\begin{bmatrix} M_{r,0} \\ N_{r,0} \end{bmatrix} - \begin{bmatrix} M_0 \\ N_0 \end{bmatrix} Q_{r,\mathcal{H}_2}\right) v_r - \begin{bmatrix} M_0 \\ N_0 \end{bmatrix} v_1. \quad (7.135)$$

On the basis (7.135), a design scheme is proposed: given v_r, solve the optimisation problem for v_1,

$$\min_{v_1} \left\| \left(\begin{bmatrix} M_{r,0} \\ N_{r,0} \end{bmatrix} - \begin{bmatrix} M_0 \\ N_0 \end{bmatrix} Q_{r,\mathcal{H}_2}\right) v_r - \begin{bmatrix} M_0 \\ N_0 \end{bmatrix} v_1 \right\|_{2,[k,k+s]} \quad (7.136)$$
$$= \min_{v} \left\| \begin{bmatrix} M_{r,0} \\ N_{r,0} \end{bmatrix} v_r - \begin{bmatrix} M_0 \\ N_0 \end{bmatrix} v \right\|_{2,[k,k+s]}. \quad (7.137)$$

Note that the tracking error is to be minimised over a time interval $[k, k+s]$. This problem can be well solved applying the system models of finite sample data introduced in the last section.

Remark 7.7 As mentioned in Sect. 7.1, handling with the projection (7.134) is equivalent to the solution of the MMP, i.e.

$$\gamma = \left\| \begin{bmatrix} M_{r,0} \\ N_{r,0} \end{bmatrix} - \begin{bmatrix} M_0 \\ N_0 \end{bmatrix} Q_{r,\mathcal{H}_2} \right\|_\infty = \min_{Q \in \mathcal{RH}_\infty} \left\| \begin{bmatrix} M_{r,0} \\ N_{r,0} \end{bmatrix} - \begin{bmatrix} M_0 \\ N_0 \end{bmatrix} Q \right\|_\infty.$$

In fact, the value γ is the so-called gap metric between the image subspaces spanned by $I_{G_{r,0}}$ and I_{G_0} respectively that measures the similarity between these two subspaces.

Now, we attend to MPC and investigate solutions in the unified framework of control and detection. Like tracking control, the canonical SIR-SKR model (7.131)–(7.132) of the plant and the reference model (7.130) are under consideration. In the context of MPC, the control performance is minimum prediction error. Suppose the prediction window $[k+1, k+s]$. To begin with, the loop dynamic (in Hilbert space) is considered,

$$\begin{bmatrix} u \\ y \end{bmatrix} = \begin{bmatrix} M_0 \\ N_0 \end{bmatrix} v + \left(\begin{bmatrix} -\hat{Y}_0 \\ \hat{X}_0 \end{bmatrix} - \begin{bmatrix} M_0 \\ N_0 \end{bmatrix} Q_{\mathcal{H}_2} \right) r_{y,0}.$$

Note that this model cannot be directly used for the prediction purpose, since $r_{y,0}$ over the future time interval $[k+1, k+s]$ is unknown due to the fact that no pre-knowledge of uncertainties in the system is available. Observe, on the other hand, that

$$\begin{cases} \hat{x}(k+1) = (A + BF_0)\,\hat{x}(k) + BV_0 v(k) + L_0 W_0^{-1} r_{y,0}(k) \\ y(k) = \hat{y}(k) + W_0^{-1} r_{y,0}(k),\, \hat{y}(k) = (C + DF_0)\,\hat{x}(k) + DV_0 v(k). \end{cases}$$

This implies, the impact of uncertainties in the past (until k) is reflected by $r_{y,0}(k)$. Taking this into account, the following prediction model is proposed under the condition that all unknown inputs in the system are set equal to zero over the time interval $[k+1, k+s]$,

$$\begin{aligned} \hat{x}_p(i+1) &= (A + BF_0)\,\hat{x}_p(i) + BV_0 v(i) + L_0 W_0^{-1} r_{y,p}(i),\, i = k, \ldots, k+s-1, \\ y_p(i) &= (C + DF_0)\,\hat{x}_p(i) + DV_0 v(i) + W_0^{-1} r_{y,p}(i),\, i = k+1, \ldots, k+s, \\ r_{y,p}(i) &= W_0 \left(y_p(i) - (C + DF_0)\,\hat{x}_p(i) - DV_0 v(i) \right),\, i = k+1, \ldots, k+s, \\ \hat{x}_p(k) &= \hat{x}(k),\, y_p(k) = y(k), \end{aligned}$$

where $\hat{x}_p(i)$, $y_p(i)$ denote the predicted estimate of the state vector and the output vector, and $r_{y,p}(i)$ models the influence of $r_{y,0}(k)$ on the dynamic of the predicted system and generated by

$$\begin{aligned} \hat{x}_r(i+1) &= (A + BF_0)\,\hat{x}_r(i) + L_0 W_0^{-1} r_{y,p}(i),\, i = k, \ldots, k+s-1,\, \hat{x}_r(k) = 0, \\ r_{y,p}(i) &= W_0\,(C + DF_0)\,\hat{x}_r(i),\, i = k+1, \ldots, k+s \end{aligned}$$

This allows us to model the prediction model of the closed-loop formally as

$$\begin{bmatrix} u_p \\ y_p \end{bmatrix} = \begin{bmatrix} M_0 \\ N_0 \end{bmatrix} v + \left(\begin{bmatrix} -\hat{Y}_0 \\ \hat{X}_0 \end{bmatrix} - \begin{bmatrix} M_0 \\ N_0 \end{bmatrix} Q_{\mathcal{H}_2} \right) r_{y,p}. \tag{7.138}$$

Analogously to the tracking control strategy described before, the MPC design is now formulated as

$$\begin{aligned} &\min_{v} \left\| \begin{bmatrix} M_{r,0} \\ N_{r,0} \end{bmatrix} v_r - \begin{bmatrix} u_p \\ y_p \end{bmatrix} \right\|_{2,[k+1,k+s]} \\ = &\min_{v} \left\| \begin{bmatrix} M_{r,0} \\ N_{r,0} \end{bmatrix} v_r - \begin{bmatrix} M_0 \\ N_0 \end{bmatrix} v - \left(\begin{bmatrix} -\hat{Y}_0 \\ \hat{X}_0 \end{bmatrix} - \begin{bmatrix} M_0 \\ N_0 \end{bmatrix} Q_{\mathcal{H}_2} \right) r_{y,p} \right\|_{2,[k+1,k+s]} \end{aligned} \tag{7.139}$$

$$= \min_{v_1} \left\| \begin{array}{c} \left(\begin{bmatrix} M_{r,0} \\ N_{r,0} \end{bmatrix} - \begin{bmatrix} M_0 \\ N_0 \end{bmatrix} Q_{r,\mathcal{H}_2} \right) v_r - \begin{bmatrix} M_0 \\ N_0 \end{bmatrix} v_1 \\ - \left(\begin{bmatrix} -\hat{Y}_0 \\ \hat{X}_0 \end{bmatrix} - \begin{bmatrix} M_0 \\ N_0 \end{bmatrix} Q_{\mathcal{H}_2} \right) r_{y,p} \end{array} \right\|_{2,[k+1,k+s]} \tag{7.140}$$

$$\text{s.t. } W \begin{bmatrix} u_{p,s}(k) \\ y_{p,s}(k) \end{bmatrix} + \zeta \leq 0. \tag{7.141}$$

Hereby, (7.141) denotes constraints on (u, y) formally for some defined matrix W and vector ζ. We would like to call the reader's attention to the difference between the two optimisation problems, (7.136) and (7.140). In the MPC scheme (7.140), the predicted uncertainty represented by $r_{y,p}$ is integrated into the optimisation. In other words, the determination of the latent variable v comprises both feedback $(r_{y,p})$ and feedforward $(r_{y,p})$ control signals. Concerning the solution and computation of optimisation problem (7.140), the system models of finite sample data introduced in the last section will be used.

7.4.3 Distributed Multi-agent Control Systems

In this subsection, it is attempted to address control and detection issues of multi-agent systems (MASs) in the unified framework of control and detection. It is to be mentioned that our purpose is not to solve control and detection issues in MASs. Instead, we would like to highlight some basic existing schemes from alternative aspects, viewed in the unified framework.

7.4.3.1 Preliminaries

Under consideration are a trivial type of LTI MASs modelled by

$$G_i : \begin{cases} x_i(k+1) = Ax_i(k) + Bu_i(k), x_i(0) = x_{i,0} \\ y_i(k) = Cx_i(k), \end{cases} \tag{7.142}$$

for $i = 1, \ldots, M$. G_i represents the dynamic of the ith agent with state, input and output vectors $x_i \in \mathbb{R}^n$, $u_i \in \mathbb{R}^p$ and $y_i \in \mathbb{R}^m$, and the MAS consists of M agents. They are connected by a network. Typical tasks of a MAS are cooperative work like consensus among the agents. For our exploration purpose, the following consensus requirement, $\forall i, j = 1, \ldots, M$,

$$\lim_{k \to \infty} \left\| x_i(k) - x_j(k) \right\| = 0, \tag{7.143}$$

is supposed to be the control objective.

Before presenting the known solutions to the above consensus control problem, some basic concepts, definitions and notations known in the literature about graph theory are shortly introduced. The reader is referred to the references given at the end of this chapter for detail. The topology of a network is defined by nodes in the network and their connections. Denote the set of the nodes by $\mathcal{N}$ and the set of connections, which are called edges in the graph theory, by E. Thus, the topology of a network with M nodes is expressed by

$$\mathcal{G} = (\mathcal{N}, E), \mathcal{N} = \{1, \ldots, M\},$$
$$E = \{(i, j) \,|i, j \in \mathcal{N}, i \neq j, \text{ they are networked}\}.$$

We call

$$\mathcal{G} = (\mathcal{N}, E)$$

graph. In a graph, edge (i, j) is called directed, when data transmissions between nodes i and j are performed in one direction. A graph is directed if its edges are directed, and a directed tree is a directed graph, in which every node has a single parent except for one node as the root, which has no parent and has directed paths to all other nodes. A directed tree is defined as spanning when it connects all the nodes in the graph. A graph is said to have a directed spanning tree if a subset of the edges forms a directed spanning tree. In our subsequent description, it is assumed that the network topology is modelled by a directed graph $\mathcal{G}$, which contains a directed spanning tree.

The following theorem is a summary of the existing results related to the aforementioned consensus problem, which are described in the monograph by Li and Duan, cited in the last section of this chapter. They build the basis for our endeavour.

Theorem 7.10 *Given the MAS (7.142) and the communication network with graph $\mathcal{G}$, the consensus control algorithms,*

$$\begin{aligned} K_{c,SIR} &: u_i(k) = F\hat{x}_{c_I,i}(k), i = 1, \ldots, M, \\ &: \hat{x}_{c_I,i}(k+1) = (A + BF)\,\hat{x}_{c_I,i}(k) \\ &+ L\sum_{j=1}^{M} w_{ij}\left(y_i(k) - y_j(k) - C\left(\hat{x}_{c_I,i}(k) - \hat{x}_{c_I,j}(k)\right)\right), \end{aligned} \tag{7.144}$$

$$\begin{aligned} K_{c,SKR} &: u_i(k) = F\hat{x}_{c_{II},i}(k), i = 1, \ldots, M, \\ &: \hat{x}_{c_{II},i}(k+1) = (A - LC)\,\hat{x}_{c_{II},i}(k) \\ &+ \sum_{j=1}^{M} w_{ij}\left(BF\left(\hat{x}_{c_{II},i}(k) - \hat{x}_{c_{II},j}(k)\right) - L\left(y_i(k) - y_j(k)\right)\right), \end{aligned} \tag{7.145}$$

where w_{ij} is the (i, j)th entry of the row-stochastic matrix $W \in \mathbb{R}^{M\times M}$ associated with the graph $\mathcal{G}$, assure the consensus (7.143), if, for controller $K_{c,SIR}$ (7.144), all the eigenvalues of matrices $A + BF$ and $A - \left(1 - \lambda_j\right)LC$, $j = 2, \ldots, M$, are

Schur, and for controller $K_{c,SKR}$ (7.145), all the eigenvalues of matrices $A - LC$ and $A + (1 - \lambda_j) BF$, $j = 2, \ldots, M$, are Schur, where λ_j, $j = 2, \ldots, M$, are the eigenvalues of W located in the open unit disk. Moreover, it holds

$$\lim_{k\to\infty} x_i(k+1) = \left(\alpha^T \otimes A^k\right) \begin{bmatrix} x_{1,0} \\ \vdots \\ x_{M,0} \end{bmatrix}, i = 1, \ldots, M. \tag{7.146}$$

Here, $\alpha \in \mathbb{R}^M$ satisfies

$$\alpha^T (I - W) = 0, \alpha^T \mathbf{1} = 1, \mathbf{1} = \begin{bmatrix} 1 \\ \vdots \\ 1 \end{bmatrix}.$$

The special notations $\left(K_{c,SIR}, K_{c,SKR}\right)$ will be explained in the sequel.

Remark 7.8 In the monograph by Li and Duan, the above two consensus controllers are called full-order observer-type protocol I and II, respectively.

7.4.3.2 Consensus Control Addressed in the Unified Framework of Control and Detection

To begin with, the MAS (7.142) is expressed in terms of the C-SIR-SKR-based model,

$$\begin{bmatrix} u_i \\ y_i \end{bmatrix} = \begin{bmatrix} M_0 & -\hat{Y}_0 \\ N_0 & \hat{X}_0 \end{bmatrix} \begin{bmatrix} v_i \\ r_{y_i,0} \end{bmatrix}, \begin{bmatrix} v_i \\ r_{y_i,0} \end{bmatrix} = \begin{bmatrix} X_0 & Y_0 \\ -\hat{N}_0 & \hat{M}_0 \end{bmatrix} \begin{bmatrix} u_i \\ y_i \end{bmatrix}, \tag{7.147}$$

$i = 1, \ldots M$, whose state space representation is,

$$G_i : \begin{cases} \hat{x}_i(k+1) = A_{F_0}\hat{x}_i(k) + BV_0^{-1}v_i(k) + L_0 W_0^{-1} r_{y_i,0}(k), \hat{x}_i(0) = 0 \\ y_i(k) = C\hat{x}_i(k) + W_0^{-1} r_{y_i,0}(k), A_{F_0} = A + BF_0. \end{cases} \tag{7.148}$$

It is supposed that only additive unknown inputs as uncertainties exist in the system. Note that the difference of the initial value $x_i(0) - \hat{x}_i(0)$ is modelled by the residual signal $r_{y_i,0}$ (refer to Example 7.5)). Hence, the consensus (7.143) is re-formulated as $\forall i, j = 1, \ldots, M$,

$$\lim_{k\to\infty} \left\| \hat{x}_i(k) - \hat{x}_j(k) \right\| = 0. \tag{7.149}$$

Observe that

$$\begin{bmatrix} u \\ y \end{bmatrix} = \left(I_M \otimes I_{G_0}\right) v + \left(I_M \otimes I_{K_0}\right) r_{y,0}, I_{G_0} = \begin{bmatrix} M_0 \\ N_0 \end{bmatrix}, I_{K_0} = \begin{bmatrix} -\hat{Y}_0 \\ \hat{X}_0 \end{bmatrix}. \tag{7.150}$$

$$u = \begin{bmatrix} u_1 \\ \vdots \\ u_M \end{bmatrix}, y = \begin{bmatrix} y_1 \\ \vdots \\ y_M \end{bmatrix}, v = \begin{bmatrix} v_1 \\ \vdots \\ v_M \end{bmatrix}, r_{y,0} = \begin{bmatrix} r_{y_1,0} \\ \vdots \\ r_{y_M,0} \end{bmatrix},$$

and there exist no physical coupling among the agents. Therefore, the consensus is to be reached exclusively by exchanging latent variables $v_i, i = 1, \ldots, M$. To this end, both consensus controllers in Theorem 7.10, $K_{c,SIR}$ in (7.144) and $K_{c,SKR}$ in (7.145), are useful.

Remark 7.9 In this subsection, notation I_M (I_m) is abused for representing an $M \times M$ ($m \times m$) dimensional unit matrix.

According to (7.144),

$$v_i(k) = F\hat{x}_{c_I,i}(k), i = 1, \ldots, M, \tag{7.151}$$

$$\hat{x}_{c_I,i}(k+1) = \left(A_{F_0} + BV_0^{-1}F\right)\hat{x}_{c_I,i}(k) + L\sum_{j=1}^{M} w_{ij}\left(y_i(k) - y_j(k) - C\left(\hat{x}_{c_I,i}(k) - \hat{x}_{c_I,j}(k)\right)\right). \tag{7.152}$$

Note that W is a row-stochastic matrix. It turns out

$$\sum_{j=1}^{M} w_{ij}\left(y_i(k) - y_j(k) - C\left(\hat{x}_{c_I,i}(k) - \hat{x}_{c_I,j}(k)\right)\right) = r_{y_i}(k) - \sum_{j=1, j\neq i}^{M} w_{ij}r_{y_j}(k),$$
$$r_{y_i}(k) = y_i(k) - C\hat{x}_{c_I,i}(k), i = 1, \ldots, M.$$

Let

$$\bar{r}_i(k) = \sum_{j=1, j\neq i}^{M} w_{ij}r_{y_j}(k). \tag{7.153}$$

The state equation (7.152) is equivalently written into

$$\hat{x}_{c_I,i}(k+1) = \left(A_{F_0} + BV_0^{-1}F\right)\hat{x}_{c_I,i}(k) + L\left(r_{y_i}(k) - \bar{r}_i(k)\right)$$
$$\Longrightarrow v_i = -\hat{Y}\left(r_{y_i} - \bar{r}_i\right), \hat{Y} = -F\left(zI - A_{F_0} - BV_0^{-1}F\right)^{-1}L,$$

$i = 1, \ldots, M$. Accordingly, the system (7.151)–(7.152) is an observer-based state feedback controller and driven by the residual. Now, we attend to (7.145), which is further written as

$$v_i(k) = F\hat{x}_{c_{II},i}(k), i = 1, \ldots, M, \tag{7.154}$$

$$\hat{x}_{c_{II},i}(k+1) = \left(A_{F_0} - LC\right)\hat{x}_{c_{II},i}(k) + BV_0^{-1}\left(v_i(k) - \bar{v}_i(k)\right) - L\left(y_i(k) - \bar{y}_i(k)\right) \Longrightarrow \tag{7.155}$$

$$X v_i + Y (y_i - \bar{y}_i) = (X - I) \bar{v}_i, \tag{7.156}$$

$$X = I - F\left(zI - A_{F_0} + LC\right)^{-1} B V_0^{-1}, Y = -F\left(zI - A_{F_0} + LC\right)^{-1} L,$$

$$\bar{v}_i(k) = \sum_{j=1, j\neq i}^{M} w_{ij} F \hat{x}_{c_{II},j}(k), \bar{y}_i(k) = \sum_{j=1, j\neq i}^{M} w_{ij} y_j(k). \tag{7.157}$$

The system (7.154)–(7.155) is an observer-based state feedback controller whose inputs are $y_i(k) - \bar{y}_i(k)$ and $v_i(k) - \bar{v}_i(k)$.

We now highlight alternative forms of the system descriptions (7.151)–(7.152) and (7.154)–(7.155), which might be useful to understand the consensus mechanisms from a feedback control-theoretic aspect. We firstly consider the system (7.151)–(7.152). Since

$$r_{y_i}(k) - \bar{r}_i(k) = y_i(k) - C\hat{x}_{c_I,i}(k) - \bar{r}_i(k)$$
$$\Longrightarrow y_i(k) - \bar{r}_i(k) = C\hat{x}_{c_I,i}(k) + \left(r_{y_i}(k) - \bar{r}_i(k)\right),$$

we obtain

$$y_i - \bar{r}_i = \hat{X}\left(r_{y_i} - \bar{r}_i\right), \hat{X} = I + C\left(zI - A_{F_0} - BV_0^{-1}F\right)^{-1} L.$$

As a result, $K_{c,SIR}$ in (7.144) can be written as a feedback controller of the form

$$v_i = -\hat{Y}\hat{X}^{-1}(y_i - \bar{r}_i) =: K_{SIR}(y_i - \bar{r}_i), \tag{7.158}$$
$$\hat{Y} = -F\left(zI - A_{F_0} - BV_0^{-1}F\right)^{-1} L.$$

Concerning the system (7.154)–(7.155), departing from the relation (7.156), the following relation holds true,

$$X(v_i - \bar{v}_i) + Y(y_i - \bar{y}_i) = -\bar{v}_i,$$
$$X = I - F\left(zI - A_{F_0} - LC\right)^{-1} BV_0^{-1}, Y = -F\left(zI - A_{F_0} - LC\right)^{-1} L,$$

which results in

$$v_i = -X^{-1}Y(y_i - \bar{y}_i) + \bar{v}_i - X^{-1}\bar{v}_i =: K_{SKR}(y_i - \bar{y}_i) + \bar{v}_i - X^{-1}\bar{v}_i. \tag{7.159}$$

Recall that $\left(\hat{X}, \hat{Y}\right)$ and (X, Y) are RCP and LCP respectively, and $\hat{Y}\hat{X}^{-1} = X^{-1}Y$. Hence, K_{SIR} and K_{SKR} are indeed the RCF and LCF of K, namely,

$$K := K_{SIR} = K_{SKR}.$$

On account of (7.158)–(7.159), Theorem 7.10 is now revised and re-formulated in the unified framework of control and detection.

Corollary 7.3 *Given the MAS (7.142) and the communication network with graph* $\mathcal{G}$ *and the associate row-stochastic matrix* $W \in \mathbb{R}^{M \times M}$*, then the controller*

$$v_i = K(y_i - \bar{r}_i) = -\hat{Y}\hat{X}^{-1}(y_i - \bar{r}_i) = -X^{-1}Y(y_i - \bar{r}_i) \tag{7.160}$$

assures the consensus (7.149), if all the eigenvalues of matrices

$$A + B\left(F_0 + V_0^{-1}F\right), A + BF_0 - \left(1 - \lambda_j\right)LC, j = 2, \ldots, M,$$

are Schur, where $\lambda_j, j = 2, \ldots, M$*, are the eigenvalues of* W *located in the open unit disk. The controller (7.160) is realised in two observer-based forms,*

$$\begin{aligned} K_{SIR} &: v_i(k) = F\hat{x}_i(k), v_i = \hat{Y}\left(\bar{r}_i - r_{y_i}\right), i = 1, \ldots, M, & (7.161)\\ &: \hat{x}_i(k+1) = \left(A_{F_0} + BV_0^{-1}F\right)\hat{x}_i(k) + L\left(r_{y_i}(k) - \bar{r}_i(k)\right), \\ K_{SKR} &: v_i(k) = F\hat{x}_i(k), Xv_i + Y\left(r_{y_i} - \bar{r}_i\right) = 0, i = 1, \ldots, M, & (7.162)\\ &: \hat{x}_i(k+1) = \left(A_{F_0} - LC\right)\hat{x}_i(k) + BV_0^{-1}v_i(k) - L\left(y_i(k) - \bar{r}_i(k)\right), \end{aligned}$$

where $\bar{r}_i$ *is defined by (7.153).*

Proof The proof of $v_i = K_{SIR}(y_i - \bar{r}_i) = -\hat{Y}\hat{X}^{-1}(y_i - \bar{r}_i)$ and state space realisation (7.161) follows the derivation of (7.158) given above. Because for $v_i(k) = F\hat{x}_i(k)$

$$\begin{aligned} \hat{x}_i(k+1) &= \left(A_{F_0} + BV_0^{-1}F\right)\hat{x}_i(k) + L\left(r_{y_i}(k) - \bar{r}_i(k)\right) \\ &= \left(A_{F_0} - LC\right)\hat{x}_i(k) + BV_0^{-1}v_i(k) - L\left(y_i(k) - \bar{r}_i(k)\right), \end{aligned}$$

(7.162) is equivalent to (7.161). Thanks to the known results about the LCF and RCF of a feedback controller, the relation $v_i = K_{SKR}(y_i - \bar{y}_i) = -X^{-1}Y(y_i - \bar{y}_i)$ becomes evident.

In the context of the above corollary, the system (7.160) and the state space realisations (7.161)–(7.162) are called SIR and SKR consensus controllers, respectively. They are observer-based state feedback controllers. It is noteworthy that for realising these two consensus controllers, only data

$$\begin{gathered} r_{y_j}(k), j \in \mathcal{N}_i, i = 1, \ldots, M, \\ \mathcal{N}_i = \{j \mid \text{node } j \text{ is networked with the } i\text{th node}, j \in \{1, \ldots, M\}\} \end{gathered}$$

are to be transmitted.

Finally, the loop dynamics are examined on the assumption that there exists no uncertainty in the system except for (unknown) initial values of the state variables. For the sake of simplicity, the SIR consensus controller is supposed being adopted. Departing from the C-SIR-SKR-based models (7.147), (7.150) and controller (7.160), the closed-loop dynamic of the ith agent is governed by

$$\begin{bmatrix} u_i \\ y_i \end{bmatrix} = \begin{bmatrix} M_0 \\ N_0 \end{bmatrix} \hat{Y} \left(\bar{r}_i - r_{y_i} \right) + \begin{bmatrix} -\hat{Y}_0 \\ \hat{X}_0 \end{bmatrix} r_{y_i,0}, \tag{7.163}$$

and the overall system dynamics are described by

$$\begin{bmatrix} u \\ y \end{bmatrix} = \left(I_M \otimes \left(I_{G_0} \hat{Y} \right) \right) (W \otimes I_m) r_y + \left(I_M \otimes I_{K_0} \right) r_{y,0}, r_y = \begin{bmatrix} r_{y_1} \\ \vdots \\ r_{y_M} \end{bmatrix}. \tag{7.164}$$

The final result is summarised in the following corollary.

Corollary 7.4 *Given the MAS (7.142) and the communication network with graph $\mathcal{G}$, and suppose that the consensus controller (7.160) is applied and the conditions in Corollary 7.1 are satisfied, then for $i = 1, \ldots, M$,*

$$\lim_{k \to \infty} \hat{x}_i (k+1) = x_i (k+1) = \left(\alpha^T \otimes A_{F_0}^k \right) \begin{bmatrix} x_{1,0} \\ \vdots \\ x_{M,0} \end{bmatrix}, \tag{7.165}$$

$$\lim_{k \to \infty} \begin{bmatrix} u_i (k+1) \\ y_i (k+1) \end{bmatrix} = \begin{bmatrix} F \\ C \end{bmatrix} \left(\alpha^T \otimes A_{F_0}^k \right) \begin{bmatrix} x_{1,0} \\ \vdots \\ x_{M,0} \end{bmatrix}. \tag{7.166}$$

Proof The proof is straightforward. Equation (7.165) is attributed to Theorem 7.10, Eq. (7.146), while (7.146) is the result of the state space representation (7.142) and the controller (7.161).

The aforementioned results reveal that

- a consensus controller is a feedback controller as well. As such, it follows the basic principle of feedback, namely, feedback of residual signals as the information carrier about uncertainties in the system. In case of consensus control, the uncertainty in the regard of consensus among the state variables is the difference between $x_i(k) - x_j(k), \forall i, j = 1, \ldots, M, j \neq i$, and the corresponding information carrier is the residual signal

$$r_{y_i} - \bar{r}_i = \sum_{j=1}^{M} w_{ij} \left(r_{y_i}(k) - r_{y_j}(k) \right),$$

  as indicated by (7.163);
- the closed-loop dynamic model (7.163) compasses three blocks with different functionalities, (i) the system eigen-dynamic modelled by

$$\begin{bmatrix} M_0 & -\hat{Y}_0 \\ N_0 & \hat{X}_0 \end{bmatrix},$$

and determined by the feedback and observer gain matrices (F_0, L_0) as well as pre- and post filters (V_0, W_0), (ii) the consensus controller (7.160) assuring the consensus of the state vectors of all agents, (iii) (local) residual generators,

$$r_{y_i,0} = \begin{bmatrix} -\hat{N}_0 & \hat{M}_0 \end{bmatrix} \begin{bmatrix} u_i \\ y_i \end{bmatrix}, i = 1, \ldots, M,$$

which is the information carrier of all uncertainties in the agents. It should be noticed that the two observers,

$$\begin{aligned} \hat{x}_{i,0}(k+1) &= (A - L_0 C)\, \hat{x}_{i,0}(k) + B u_i(k) + L_0 y_i(k) \\ &= (A + BF_0)\, \hat{x}_{i,0}(k) + B V_0^{-1} v_i(k) + L_0 W_0^{-1} r_{y_i,0}(k), \\ \hat{x}_i(k+1) &= \left(A_{F_0} + B V_0^{-1} F\right) \hat{x}_i(k) + L \left(r_{y_i}(k) - \bar{r}_i(k)\right) \\ &= \left(A_{F_0} - LC\right) \hat{x}_i(k) + B V_0^{-1} v_i(k) + L \left(y_i(k) - \bar{r}_i(k)\right), \end{aligned}$$

serve for distinguishing residual generation purposes. While the first one is applied to generating the residual

$$r_{y_i,0}(k) = y_i(k) - C\hat{x}_{i,0}(k)$$

as an indicator for the uncertainty in the ith agent, the latter generates the residual

$$r_{y_i}(k) - \bar{r}_i(k) = y_i(k) - C\hat{x}_i(k) - \bar{r}_i(k),$$

that carries the information about the consensus error;

- corresponding to these three functionalities, the control signal u_i acting on the ith agent G_i modelled by (7.142) is

$$\begin{aligned} u_i(k) &= F_0 \hat{x}_{i,0}(k) + F\hat{x}_i(k), \\ \hat{x}_{i,0}(k+1) &= (A - L_0 C)\, \hat{x}_{i,0}(k) + B u_i(k) + L_0 y_i(k), \\ \hat{x}_i(k+1) &= \left(A_{F_0} - LC\right) \hat{x}_i(k) + B V_0^{-1} v_i(k) + L \left(y_i(k) - \bar{r}_i(k)\right). \end{aligned}$$

At the end of this discussion, we would like to mention that a complete controller should include an additional term to compensate the uncertainties in the agents. It is of a general form described by

$$u_i = F_0 \hat{x}_{i,0} + F\hat{x}_i + Q_{i,0} r_{y_i,0},$$

where $Q_{i,0} \in \mathcal{RH}_\infty$ represents the solution of

$$\min_{Q_{i,0} \in \mathcal{RH}_\infty} \left\| \begin{bmatrix} -\hat{Y}_0 \\ \hat{X}_0 \end{bmatrix} - \begin{bmatrix} M_0 \\ N_0 \end{bmatrix} Q_{i,0} \right\|_\infty .$$

The control-theoretic meaning of the above MMP is well described in the previous section.

7.4.3.3 A Misunderstanding in Existing Research on Fault Diagnosis of MASs

Residual generation is the first step of an observer-based fault diagnosis system. For our purpose, the SKR-based model is adopted for addressing residual generation in the MAS (7.142),

$$G_i : \begin{cases} \hat{x}_i(k+1) = A\hat{x}_i(k) + Bu_i(k) + L_{0,i}r_{i,0}(k), \hat{x}_i(0) = 0 \\ r_{i,0}(k) = y_i(k) - C\hat{x}_i(k) \Longleftrightarrow y_i(k) = r_{i,0}(k) + C\hat{x}_i(k), \end{cases} \tag{7.167}$$

$i = 1, \ldots, M$. Herewith, $r_{i,0}$ represents any (additive) uncertainties in the ith agent, including possible faults. The global SKR model is

$$\begin{cases} \hat{x}(k+1) = (I_M \otimes A)\,\hat{x}(k) + (I_M \otimes B)\,u(k) + L_0 r_0(k) \\ r_0(k) = y(k) - (I_M \otimes C)\,\hat{x}(k), \end{cases} \tag{7.168}$$

$$u = \begin{bmatrix} u_1 \\ \vdots \\ u_M \end{bmatrix}, y = \begin{bmatrix} y_1 \\ \vdots \\ y_M \end{bmatrix}, \hat{x} = \begin{bmatrix} \hat{x}_1 \\ \vdots \\ \hat{x}_M \end{bmatrix}, r_0 = \begin{bmatrix} r_{1,0} \\ \vdots \\ r_{M,0} \end{bmatrix},$$

$$L_0 = diag\left(L_{0,1}, \ldots, L_{0,M}\right). \tag{7.169}$$

It is a popular method to consider the MAS (7.168) as an interconnected system and introduce the so-called relative output information defined by

$$\begin{aligned} r_{\mathcal{N}_i}(k) &= \sum_{j=1}^{M} w_{ij} \left(\left(y_i(k) - y_j(k)\right) - C\left(\hat{x}_i(k) - \hat{x}_j(k)\right)\right) \\ &= \sum_{j=1}^{M} w_{ij} \left(r_i(k) - r_j(k)\right), \end{aligned}$$

for constructing a residual generator as follows

$$\begin{cases} \hat{x}_i(k+1) = A\hat{x}_i(k) + Bu_i(k) + L_{\mathcal{N}_i} r_{\mathcal{N}_i}(k) \\ r_{\mathcal{N}_i}(k) = \displaystyle\sum_{j=1}^{M} w_{ij} \left(r_i(k) - r_j(k)\right), \end{cases}. \tag{7.170}$$

Accordingly, the corresponding residual generator of the overall system is given by

$$\begin{cases} \hat{x}(k+1) = (I_M \otimes A)\,\hat{x}(k) + (I_M \otimes B)\,u(k) + L_{\mathcal{N}} r_{\mathcal{N}}(k) \\ r_{\mathcal{N}}(k) = W_{\mathcal{N}} r(k), \end{cases} \tag{7.171}$$

$$r_{\mathcal{N}} = \begin{bmatrix} r_{\mathcal{N}1} \\ \vdots \\ r_{\mathcal{N}_M} \end{bmatrix}, L_{\mathcal{N}} = diag\left(L_{\mathcal{N}_1}, \ldots, L_{\mathcal{N}_M}\right), W_{\mathcal{N}} = W \otimes I_m.$$

Observe that the residual generator (7.171) is a trivial observer-based residual generator of the form

$$\begin{cases} \hat{x}(k+1) = A_{MAS}\hat{x}(k) + B_{MAS}u(k) + L_{\mathcal{N}} W_{\mathcal{N}} r(k) \\ r_{\mathcal{N}}(k) = W_{\mathcal{N}} r(k), \end{cases} \tag{7.172}$$

$$A_{MAS} = I_M \otimes A,\ B_{MAS} = I_M \otimes B,$$

whose dynamic is governed by

$$\begin{cases} e(k+1) = (A_{MAS} - L_{\mathcal{N}} W_{\mathcal{N}} C_{MAS})\,e(k) \\ r_{\mathcal{N}}(k) = W_{\mathcal{N}} C_{MAS} e(k) \end{cases}, \tag{7.173}$$

$$C_{MAS} = I_M \otimes C,\ e(k) = \begin{bmatrix} x_1 - \hat{x}_1 \\ \vdots \\ x_M - \hat{x}_M \end{bmatrix}.$$

An alternative, and popular as well, method is the use of the data from the neighbouring agents to construct the residual generator. To be specific, let

$$y_{\bar{\mathcal{N}}_i}(k) = \begin{bmatrix} y_{i_1}(k) \\ \vdots \\ y_{i_{\gamma_i}}(k) \end{bmatrix} =: C_{\bar{\mathcal{N}}_i} y(k) \in \mathbb{R}^{\gamma_i m}, i = 1, \ldots M,$$
$$\bar{\mathcal{N}_i} = \mathcal{N}_i \cup \{i\}, \gamma_i = \left|\bar{\mathcal{N}_i}\right|, i_j \in \bar{\mathcal{N}_i}, j = 1, \ldots, \gamma_i,$$

where $\left|\bar{\mathcal{N}_i}\right|$ denotes the cardinality of $\bar{\mathcal{N}_i}$. That is, $y_{\bar{\mathcal{N}}_i}$ is the model of all available data at the ith agent. As a result, a residual generator is constructed at the ith agent as

$$\begin{cases} \hat{x}_i(k+1) = A\hat{x}_i(k) + Bu_i(k) + L_{\bar{\mathcal{N}}_i} r_{\bar{\mathcal{N}}_i}(k) \\ r_{\bar{\mathcal{N}}_i}(k) = y_{\bar{\mathcal{N}}_i}(k) - C_{\bar{\mathcal{N}}_i}\hat{x}(k), \end{cases}$$

and thus

$$\begin{cases} \hat{x}(k+1) = A_{MAS}\hat{x}(k) + B_{MAS}u(k) + L_{\bar{\mathcal{N}}} r_{\bar{\mathcal{N}}}(k) \\ r_{\bar{\mathcal{N}}}(k) = y_{\bar{\mathcal{N}}}(k) - C_{\bar{\mathcal{N}}}\hat{x}(k), \end{cases}$$

$$y_{\bar{\mathcal{N}}} = \begin{bmatrix} y_{\bar{\mathcal{N}}_1} \\ \vdots \\ y_{\bar{\mathcal{N}}_M} \end{bmatrix}, C_{\bar{\mathcal{N}}} = diag\left(C_{\bar{\mathcal{N}}_1}, \ldots, C_{\bar{\mathcal{N}}_M}\right), r_{\bar{\mathcal{N}}} = \begin{bmatrix} r_{\bar{\mathcal{N}}_1} \\ \vdots \\ r_{\bar{\mathcal{N}}_M} \end{bmatrix},$$
$$L_{\bar{\mathcal{N}}} = diag\left(L_{\bar{\mathcal{N}}_1}, \ldots, L_{\bar{\mathcal{N}}_M}\right),$$

whose dynamic is subject to

$$e(k+1) = \left(A_{MAS} - L_{\bar{\mathcal{N}}} C_{\bar{\mathcal{N}}}\right) e(k), r_{\bar{\mathcal{N}}}(k) = C_{\bar{\mathcal{N}}} e(k).$$

Note that $r_{\bar{\mathcal{N}}_i}(k)$ can be written into

$$r_{\bar{\mathcal{N}}_i}(k) = W_{\bar{\mathcal{N}}_i}\left(y(k) - C_{MAS}\hat{x}(k)\right) = W_{\bar{\mathcal{N}}_i} r(k) = W_{\bar{\mathcal{N}}_i} C_{MAS} e(k)$$

for some $(\gamma_i m \times Mm)$-dimensional, full-row rank matrix $W_{\bar{\mathcal{N}}_i}$ whose entries are either one or zero, which yields

$$\begin{cases} \hat{x}(k+1) = A_{MAS}\hat{x}(k) + B_{MAS}u(k) + L_{\bar{\mathcal{N}}} W_{\bar{\mathcal{N}}} r(k) \\ r_{\bar{\mathcal{N}}}(k) = W_{\bar{\mathcal{N}}} r(k), \end{cases} \tag{7.174}$$
$$e(k+1) = \left(A_{MAS} - L_{\bar{\mathcal{N}}} W_{\bar{\mathcal{N}}} C_{MAS}\right) e(k), r_{\bar{\mathcal{N}}}(k) = W_{\bar{\mathcal{N}}} C_{MAS} e(k)$$

with $W_{\bar{\mathcal{N}}} = diag\left(W_{\bar{\mathcal{N}}_1}, \ldots, W_{\bar{\mathcal{N}}_M}\right)$. It is apparent that the residual generators (7.172) and (7.174) can be uniformly written as

$$\begin{cases} \hat{x}(k+1) = A_{MAS}\hat{x}(k) + B_{MAS}u(k) + L_{MAS} r(k) \\ e(k+1) = \left(A_{MAS} - L_{MAS} C_{MAS}\right) e(k) \\ r_{MAS,0}(k) = W_{MAS} r(k) = W_{MAS} C_{MAS} e(k), \end{cases} \tag{7.175}$$
$$\left(L_{MAS}, W_{MAS}\right) = \begin{cases} \left(L_{\mathcal{N}} W_{\mathcal{N}}, W_{\mathcal{N}}\right), \text{ residual generator (7.172)} \\ \left(L_{\bar{\mathcal{N}}} W_{\bar{\mathcal{N}}}, W_{\bar{\mathcal{N}}}\right), \text{ residual generator (7.174)}, \end{cases}$$

or in terms of the LCF of the MAS $G_{MAS} = (A_{MAS}, B_{MAS}, C_{MAS})$, as

$$r_{MAS,0} = \begin{bmatrix} -\hat{N}_{MAS} & \hat{M}_{MAS} \end{bmatrix} \begin{bmatrix} u \\ y \end{bmatrix}, \tag{7.176}$$
$$\hat{M}_{MAS} = \left(A_{MAS} - L_{MAS} C_{MAS}, -L_{MAS}, W_{MAS} C_{MAS}, W_{MAS}\right),$$
$$\hat{N}_{MAS} = \left(A_{MAS} - L_{MAS} C_{MAS}, B_{MAS}, W_{MAS} C_{MAS}, 0\right).$$

By means of the parameterisation of LTI observer-based residual generators, any observer-based residual generators for G_{MAS} can be written as

$$r_{MAS} = R_{MAS} r_{MAS,0} = R_{MAS} \begin{bmatrix} -\hat{N}_{MAS} & \hat{M}_{MAS} \end{bmatrix} \begin{bmatrix} u \\ y \end{bmatrix}, R_{MAS} \in \mathcal{RH}_\infty. \tag{7.177}$$

System R_{MAS} is a post-filter, whose design delivers any type of LTI residual generator, including a bank of unknown input observer-based residual generators, which are amazingly widely used for designing residual generators in MASs.

Now, we are in the position to illustrate that the use of the residual generator (7.176) is inefficient and unnecessary. To this end, we firstly consider an example system modelled by

$$G : \begin{cases} x(k+1) = Ax(k) + Bu(k), x(0) = x_0 \\ y(k) = Cx(k), \end{cases} \tag{7.178}$$

where the state, output and input variables are respectively composed of

$$x(k) = \begin{bmatrix} x_1(k) \\ \vdots \\ x_M(k) \end{bmatrix} \in \mathbb{R}^n, y(k) = \begin{bmatrix} y_1(k) \\ \vdots \\ y_M(k) \end{bmatrix} \in \mathbb{R}^m, u(k) = \begin{bmatrix} u_1(k) \\ \vdots \\ u_M(k) \end{bmatrix} \in \mathbb{R}^p,$$
$$x_i \in \mathbb{R}^{n_i}, y_i \in \mathbb{R}^{m_i}, u_i \in \mathbb{R}^{p_i}, i = 1, \ldots, M,$$
$$\sum_{i=1}^{M} n_i = n, \sum_{i=1}^{M} m_i = m, \sum_{i=1}^{M} p_i = p,$$

and the system matrices A, B, C are diagonal matrices of the form

$$A = diag\left(A_{1,} \cdots, A_M\right), B = diag\left(B_{1,} \cdots, B_M\right), C = diag\left(C_{1,} \cdots, C_M\right),$$
$$A_i \in \mathbb{R}^{n_i \times n_i}, B_i \in \mathbb{R}^{n_i \times p_i}, C_i \in \mathbb{R}^{m_i \times n_i}, i = 1, \ldots, M.$$

For our purpose, suppose that faults are modelled by

$$f(k) = \begin{bmatrix} f_1(k) \\ \vdots \\ f_M(k) \end{bmatrix}, f_i \in \mathbb{R}^{k_{f,i}}, i = 1, \ldots, M,$$

and affect the system dynamics in the form of

$$G : \begin{cases} x(k+1) = Ax(k) + Bu(k) + E_f f(k) \\ y(k) = Cx(k) + F_f f(k), \end{cases}$$
$$E_f = diag\left(E_{f,1,} \cdots, E_{f,M}\right), F_f = diag\left(F_{f,1,} \cdots, F_{f,M}\right),$$
$$E_{f,i} = \begin{bmatrix} E_i & 0 \end{bmatrix} \in \mathbb{R}^{n_i \times k_{f,i}}, F_{f,i} = \begin{bmatrix} 0 & F_i \end{bmatrix} \in \mathbb{R}^{m_i \times k_{f,i}}, i = 1, \ldots, M,$$

where E_f and F_f are known matrices. Now, design a special residual generator as constructed by

$$\begin{cases} \hat{x}(k+1) = A\hat{x}(k) + Bu(k) + L_O r_O(k) \\ r_O(k) = y(k) - \hat{y}(k), \hat{y}(k) = C\hat{x}(k), \end{cases}$$

$$L_0 = diag\left(L_{0,1,} \cdots, L_{0,M}\right), L_{0,i} \in \mathbb{R}^{n_i \times m_i}, i = 1, \ldots, M,$$

and $A_i - L_{0,i}C_i, i = 1, \ldots, M$, are Schur matrices. It is apparent that

$$r_0(k) = \begin{bmatrix} r_{0,1}(k) \\ \vdots \\ r_{0,M}(k) \end{bmatrix}, r_{0,i} = \hat{N}_{f_i} f_i, i = 1, \ldots, M,$$

$$\hat{N}_{f_i} = F_{f.i} + C_i \left(zI - A_i + L_{0,i}C_i\right)^{-1} \left(E_{f,i} - L_{0,i} F_{f.i}\right).$$

Next, construct an arbitrary residual generator as follows

$$\begin{cases} \hat{x}(k+1) = A\hat{x}(k) + Bu(k) + Lr(k) \\ r(k) = V\left(y(k) - \hat{y}(k)\right), \hat{y}(k) = C\hat{x}(k) \end{cases}$$

with the observer gain matrix assuring $A - LC$ being Schur. Accordingly,

$$r = V\hat{N}_f f, \hat{N}_f = F_f + C\left(zI - A + LC\right)^{-1}\left(E_f - LF_f\right). \tag{7.179}$$

Observe that (7.179) is equivalent to

$$r = \hat{N}_f f = Rr_0 = R \begin{bmatrix} \hat{N}_{f_i} f_1 \\ \vdots \\ \hat{N}_{f_M} f_M \end{bmatrix}, \tag{7.180}$$

$$R = V\left(I - C\left(zI - A + LC\right)^{-1}\left(L - L_0\right)\right) \in \mathcal{RH}_\infty.$$

Relation (7.180) implies that

- for detecting and estimating f_i, no residual generator is better than a (local) residual generator

$$\begin{cases} \hat{x}_i(k+1) = A_i\hat{x}_i(k) + B_iu_i(k) + L_{i,0}r_{i,0}(k) \\ r_{i,0} = R_{_i}\left(y_i - \hat{y}_i\right), \hat{y}_i = C_i\hat{x}_i, \end{cases}$$

$$\Longleftrightarrow r_{i,0} = R_{_i}\left[-\hat{N}_i \ \hat{M}_i\right]\begin{bmatrix} u_i \\ y_i \end{bmatrix} = R_{_i}\hat{N}_{f_i} f_i,$$

  where R_i might be a (stable) dynamic post-filter;
- for isolating f, r_0 offers the best solution.

Let us now return to our original issue of residual generation in MASs. Firstly, it is obvious that the MAS (7.168) is a special case of the system (7.178). As such, the corresponding residual generator (7.177) can be, according to (7.180), written into

$$r_{MAS} = R_{MAS} r_{MAS,0} = R_{MAS} R r_0,$$
$$R = \left(I - C_{MAS}\left(zI - A_{MAS} + L_{MAS} C_{MAS}\right)^{-1}\left(L_{MAS} - L_0\right)\right) \in \mathcal{RH}_\infty,$$

where r_0 is the residual vector generated by SKR (7.168), namely

$$r_0 = diag\left(K_{G_1}, \ldots, K_{G_M}\right)\begin{bmatrix} u \\ y \end{bmatrix}, K_{G_1} = \begin{bmatrix} -\hat{N}_i & \hat{M}_i \end{bmatrix}, \tag{7.181}$$
$$\hat{M}_i = \left(A - L_{0,i}C, -L_{0,i}, C, I\right), \hat{N}_i = \left(A - L_{0,i}C, B, C, 0\right). \tag{7.182}$$

Hence, for MASs with faults modelled by

$$G_i : \begin{cases} x_i(k+1) = Ax_i(k) + Bu_i(k) + E_{f,i} f_i \\ y_i(k) = Cx_i(k) + F_{f,i} f_i, \end{cases}$$
$$\Longrightarrow r_{0,i} = \hat{N}_{f_i} f_i, \hat{N}_{f_i} = F_{f.i} + C\left(zI - A + L_{0,i}C\right)^{-1}\left(E_{f,i} - L_{0,i} F_{f.i}\right),$$

it holds

$$r_{MAS} = R_{MAS} R r_0 = R_{MAS} R diag\left(\hat{N}_{f_1}, \ldots, \hat{N}_{f_M}\right) f, f = \begin{bmatrix} f_1 \\ \vdots \\ f_M \end{bmatrix}.$$

As a result, it can be claimed that the best and most practical residual generation scheme is the use of local (fully decentralised) residual generators.

7.4.3.4 On Uncertainties

In our previous study on MAS control and detection issues, no multiplicative uncertainties are under consideration. The major benefit of this handling is that the detection problems are fully decoupled from cooperative control tasks and can be purposefully addressed using individual residual generators running at individual agents. That multiplicative uncertainties, including faults, may yield coupling between the (output) residual and latent variable of the image subspace calls for a close look at the influence of existing (multiplicative) uncertainties in agents on detection and control performance. Without loss of generality, it is supposed that uncertainties in the ith agent are modelled by the (right) coprime factor uncertainty, namely

$$G_i = \left(N_0 + \Delta_{N,i}\right)\left(M_0 + \Delta_{M,i}\right)^{-1}, \begin{bmatrix} M_0 \\ N_0 \end{bmatrix} + \begin{bmatrix} \Delta_{M,i} \\ \Delta_{N,i} \end{bmatrix} =: I_{G_0} + \Delta_{I_{G_i}},$$
$$\Longrightarrow \begin{bmatrix} u_i \\ y_i \end{bmatrix} = \left(\begin{bmatrix} M_0 \\ N_0 \end{bmatrix} + \begin{bmatrix} \Delta_{M,i} \\ \Delta_{N,i} \end{bmatrix}\right) v_{i,\Delta}$$

with (normalised) RCP $\left(M_{0,}N_0\right)$, uncertainty $\Delta_{I_{G_i}}$ and the latent variable v_Δ. Let $\Delta_{I_{G_i}}$ be

$$\Delta_{I_G} = \begin{bmatrix} M_0 \\ N_0 \end{bmatrix} \delta_{I_{G_i}} + \begin{bmatrix} -\hat{Y}_0 \\ \hat{X}_0 \end{bmatrix} \delta_{I_{K_i}},$$

and suppose that consensus controller (7.161) is applied in the MAS, i.e.

$$u_i = F_0 \hat{x}_{i,0} + v_i + Q_{i,0} r_{y_i,0} = F_0 \hat{x}_{i,0} + F \hat{x}_i + Q_{i,0} r_{y_i,0}.$$

Hereby, the term $Q_{i,0} r_{y_i,0}$ is applied to enhance the robustness. According to our study in Sect. 7.2.3, it holds

$$\begin{aligned}
\left[\, X_0 \ Y_0 \,\right] \begin{bmatrix} u_i \\ y_i \end{bmatrix} &= v_i + Q_{i,0} r_{y_i,0} = \left(I + \delta_{I_{G_i}}\right) v_{i,\Delta} \\
\Longrightarrow v_{i,\Delta} &= \left(I + \delta_{I_{G_i}}\right)^{-1} \left(v_i + Q_{i,0} r_{y_i,0}\right) \\
&= \left(I - \left(I + \delta_{I_{G_i}}\right)^{-1} \delta_{I_{G_i}}\right) \left(v_i + Q_{i,0} r_{y_i,0}\right), \qquad (7.183) \\
r_{y_i,0} = \left[\, -\hat{N}_0 \ \hat{M}_0 \,\right] \begin{bmatrix} u_i \\ y_i \end{bmatrix} &= \delta_{I_{K_i}} \left(I + \delta_{I_{G_i}} - Q \delta_{I_{K_i}}\right)^{-1} \left(v_i + Q_{i,0} r_{y_i,0}\right) \\
\Longrightarrow r_{y_i,0} &= \delta_{I_{K_i}} \left(I + \delta_{I_{G_i}} - Q \delta_{I_{K_i}}\right)^{-1} v_i. \qquad (7.184)
\end{aligned}$$

And the closed-loop dynamic of the ith agent is subject to

$$\begin{aligned}
\begin{bmatrix} u_i \\ y_i \end{bmatrix} &= \begin{bmatrix} M_0 \\ N_0 \end{bmatrix} v_{i,\Delta} + \begin{bmatrix} -\hat{Y}_0 \\ \hat{X}_0 \end{bmatrix} r_{y_i,0} \\
&= \begin{bmatrix} M_0 \\ N_0 \end{bmatrix} \hat{Y} \left(\bar{r}_i - r_{y_i}\right) + \left(\begin{bmatrix} -\hat{Y}_0 \\ \hat{X}_0 \end{bmatrix} + \begin{bmatrix} M_0 \\ N_0 \end{bmatrix} Q_{i,0}\right) r_{y_i,0} \\
&\quad - \begin{bmatrix} M_0 \\ N_0 \end{bmatrix} \hat{Y} \left(I + \delta_{I_{G_i}}\right)^{-1} \delta_{I_{G_i}} \left(\left(\bar{r}_i - r_{y_i}\right) + Q_{i,0} r_{y_i,0}\right). \qquad (7.185)
\end{aligned}$$

Remember that the first two terms in (7.185) represent the closed-loop dynamics under the condition of being free of multiplicative uncertainties (refer to (7.163)). This implies that the third term describes the impact of the uncertainty Δ_{I_G} on the closed-loop performance. The corresponding overall system dynamics are then described by

$$\begin{aligned}
\begin{bmatrix} u \\ y \end{bmatrix} &= \left(I_M \otimes \left(I_{G_0} \hat{Y}\right)\right) \left(W \otimes I_m\right) r_y \\
&\quad + \left(I_M \otimes \left(I_{K_0} + I_{G_0} diag\left(Q_{1,0}, \ldots, Q_{M,0}\right)\right)\right) r_{y,0} \\
&\quad + \Pi \left(\left(W \otimes I_m\right) r_y + diag\left(Q_{1,0}, \ldots, Q_{M,0}\right) r_{y,0}\right), \qquad (7.186) \\
\Pi &= -\left(I_M \otimes \left(I_{G_0} \hat{Y}\right)\right) diag\left(\left(I + \delta_{I_{G_1}}\right)^{-1} \delta_{I_{G_1}}, \ldots, \left(I + \delta_{I_{G_M}}\right)^{-1} \delta_{I_{G_M}}\right).
\end{aligned}$$

Concerning fault diagnosis, the existence of couplings amongst the agents due to uncertainties degrades the capability of local residual-based (i.e. $r_{y_i,0}$ based) detection. Specifically, $r_{y_i,0}$ depends, according to (7.184), on r_{y_i}, a (local) residual vector at the ith agent and $\bar{r}_i$, i.e.

$$r_{y_i,0} = \delta_{I_{K_i}} \left(I + \delta_{I_{G_i}} - Q\delta_{I_{K_i}}\right)^{-1} \hat{Y} \left(\bar{r}_i - r_{y_i}\right).$$

Observe that

$$\begin{aligned} r_{y_i} &= \begin{bmatrix} -\hat{N} & \hat{M} \end{bmatrix} \begin{bmatrix} v_i \\ y_i - \bar{r}_i \end{bmatrix} = \begin{bmatrix} -\hat{N} & \hat{M} \end{bmatrix} \left(\begin{bmatrix} M_0 \\ N_0 \end{bmatrix} + \begin{bmatrix} \Delta_{M,i} \\ \Delta_{N,i} \end{bmatrix} \right) v_{i,\Delta} - \hat{M}\bar{r}_i \\ &= \Phi_{\Delta,i} \left(I + \delta_{I_{G_i}}\right)^{-1} \left(Q_{i,0} r_{y_i,0} - r_{y_i}\right) + \left(\Phi_{\Delta,i}\hat{Y} - \hat{M}\right)\bar{r}_i, \\ \Phi_{\Delta,i} &= \begin{bmatrix} -\hat{N} & \hat{M} \end{bmatrix} \left(\begin{bmatrix} M_0 \\ N_0 \end{bmatrix} + \begin{bmatrix} \Delta_{M,i} \\ \Delta_{N,i} \end{bmatrix} \right), \\ \hat{M} &= I - C\left(zI - A_{F_0} - LC\right)^{-1} L, \hat{N} = C\left(zI - A_{F_0} - LC\right)^{-1} BV_0^{-1}. \end{aligned}$$

By some routine but involved calculations, we obtain

$$\begin{bmatrix} r_{y_i,0} \\ r_{y_i} \end{bmatrix} = \Psi_{\Delta,i}^{-1} \begin{bmatrix} \delta_{I_{K_i}} \left(I + \delta_{I_{G_i}} - Q\delta_{I_{K_i}}\right)^{-1} \hat{Y} \\ \Phi_{\Delta,i}\hat{Y} - \hat{M} \end{bmatrix} \bar{r}_i, \tag{7.187}$$

$$\Psi_{\Delta,i} = \begin{bmatrix} I & \delta_{I_{K_i}} \left(I + \delta_{I_{G_i}} - Q\delta_{I_{K_i}}\right)^{-1} \hat{Y} \\ -\Phi_{\Delta,i} \left(I + \delta_{I_{G_i}}\right)^{-1} Q_{i,0} & I + \Phi_{\Delta,i} \left(I + \delta_{I_{G_i}}\right)^{-1} \end{bmatrix}.$$

Equation (7.187) depicts the dynamic relation of the ith (local) residuals and the residuals of the agents of its neighbouring set. Based on it, a fault detection scheme can be devised as follows,

- residual evaluation function:

$$\text{either } J_1 = \frac{\left\|r_{y_i,0}\right\|_2}{\left\|\bar{r}_i\right\|_2} \text{ or } J_2 = \frac{\left(\left\|r_{y_i,0}\right\|_2^2 + \left\|r_{y_i}\right\|_2^2\right)^{1/2}}{\left\|\bar{r}_i\right\|_2}$$

  can serve for this purpose,
- threshold setting:

$$\text{either } J_{1,th} := \left\| \begin{bmatrix} I \\ 0 \end{bmatrix} \Psi_{\Delta,i}^{-1} \begin{bmatrix} \delta_{I_{K_i}} \left(I + \delta_{I_{G_i}} - Q\delta_{I_{K_i}}\right)^{-1} \hat{Y} \\ \Phi_{\Delta,i}\hat{Y} - \hat{M} \end{bmatrix} \right\|_\infty,$$

$$\text{or } J_{2,th} := \left\| \Psi_{\Delta,i}^{-1} \begin{bmatrix} \delta_{I_{K_i}} \left(I + \delta_{I_{G_i}} - Q\delta_{I_{K_i}}\right)^{-1} \hat{Y} \\ \Phi_{\Delta,i}\hat{Y} - \hat{M} \end{bmatrix} \right\|_\infty.$$

7.4.3.5 Concluding Remarks

The aforementioned discussions and results reassure two essential principles in the unified framework of control and detection. One is that the nature of feedback control is the feedback of uncertainties in the form of residual. Hereby, the uncertainties are related the control task under consideration. In standard control framework, uncertainties are understood as deviations from the nominal dynamics. Dealing with MASs, whose tasks are in general cooperative control, consensus error dynamics expressed in terms of the difference of the residual signals of agents are the concerning uncertainties. This aspect can be clearly identified in the consensus controllers (7.161) and (7.162) in Corollary 7.3. Another essential principle deals with residual generation for fault detection. Roughly speaking, faults are referred to anomalies that destroy the parity relation of the nominal system dynamics, expressed, for instance, by the system kernel representation. Hereby, the nominal system dynamics comprise the sensors and actuators integrated in the system. Consequently, residual generation is to be achieved on the basis of the parity relation (model). In MASs, connections amongst agents are signal transmissions, which, generally speaking, do not trigger technical and physical faults in the agents. Hence, it is to be careful with handling of MASs as interconnected systems aiming at fault diagnosis. Unfortunately, this very basic principle is often disdained, as pointed out in the previous subsection.

A further interesting observation is that in an agent, the signals received from the agents of its neighbouring set for the consensus purpose acts like a reference signal. This can be recognised in the term $\hat{Y}\bar{r}_i$ in (7.163). In other words, the local dynamics of the ith agent are driven by $\bar{r}_i$. This fact is useful for fault diagnosis, in particular, when multiplicative uncertainties exist in the MAS, as depicted by (7.187).

Finally, we would like to address fault diagnosis issues for two classes of MASs. The first one is the so-called heterogeneous MASs, which differ from the homogeneous MASs modelled by (7.142) in that the agents may be of differing dynamics. In fact, the model (7.167) describes a typical heterogeneous MAS. Thus, also in this case, it is generally sufficient to generate residuals locally and directly in each agent. In special applications, sensors of agents often deliver relative information like relative position amongst neighbouring agents. To be specific, let $\mathcal{N}_i$ be the neighbouring set of the ith agent. The relative information can be generally modelled by

$$y_{i_j} = C\left(x_i - x_j\right), i_j \in \mathcal{N}_i \Longrightarrow y_{\mathcal{N}_i} = \begin{bmatrix} y_{i_1} \\ \vdots \\ y_{i_{\gamma_i}} \end{bmatrix}, \gamma_i = |\mathcal{N}_i| .$$

On the assumption that the ith agent is communicated with all agents in $\mathcal{N}_i$ and $u_{i_j}, i_j \in \mathcal{N}_i$, are transmitted to the ith agent, the ith agent cluster can be defined as

$$G_{\mathcal{N}_i} : \begin{cases} x_{\mathcal{N}_i}(k+1) = A_{\mathcal{N}_i} x_{\mathcal{N}_i}(k) + B_{\mathcal{N}_i} u_{\mathcal{N}_i}(k) \\ y_{\mathcal{N}_i}(k) = C_{\mathcal{N}_i} x_{\mathcal{N}_i}(k), \end{cases} \tag{7.188}$$

$$x_{\mathcal{N}_i} = \begin{bmatrix} x_i \\ x_{i_1} \\ \vdots \\ x_{i_{\gamma_i}} \end{bmatrix}, u_{\mathcal{N}_i} = \begin{bmatrix} u_i \\ u_{i_1} \\ \vdots \\ u_{i_{\gamma_i}} \end{bmatrix}, A_{\mathcal{N}_i} = I \otimes A \in \mathbb{R}^{n(\gamma_i+1)\times n(\gamma_i+1)},$$

$$B_{\mathcal{N}_i} = I \otimes B \in \mathbb{R}^{n(\gamma_i+1)\times p(\gamma_i+1)}, C_{\mathcal{N}_i} = diag\left(C_{i_1}, \ldots, C_{i_{\gamma_i}}\right) \in \mathbb{R}^{m\gamma_i \times n\gamma_i},$$
$$C_{i_j} = \left[I\ 0 \cdots 0\ -I\ 0 \cdots 0\right], i_j \in \mathcal{N}_i, \tag{7.189}$$

where the block $-I$ is in the position of the $(i_j + 1)$ block. Since any (additive) uncertainties in the ith agent and the agents belong to $\mathcal{N}_i$ exclusively depend on the uncertainties in the ith agent cluster, fault detection can be successfully achieved using an observer-based residual generator based on the model (7.188), as described in the last subsection. In such a manner, fault diagnosis problems can be well addressed using M such cluster residual generators. Note that $|\mathcal{N}_i|, i = 1, \ldots, M$, might be different. Therefore, such MASs are to be handled as heterogeneous MASs.

7.5 Notes and References

The four sections of this chapter are dedicated to four different issues. In a certain sense, they represent perspective applications of the unified framework to control and diagnosis issues.

The research object in Sect. 7.1 is the system data (u, y) belonging to $\mathcal{H}_2$ (signal) space. On the basis of the signal subspaces in $\mathcal{H}_2$, the image/kernel and residual subspaces, the data structure of (u, y) is defined. Clearly differing from the work in the previous chapters, the data structure is the inherent nature of the system data and independent of the loop configuration, namely, it exists both in closed- and open-loop configurations.

Given an LTI plant G, the data structure of (u, y) is characterised by the image/kernel and residual subspaces, $\mathcal{I}_G$ and $\mathcal{R}_G$, in $\mathcal{H}_2$ space, and

$$\mathcal{H}_2 = \mathcal{I}_G \otimes \mathcal{R}_G, \mathcal{R}_G = \mathcal{I}_G^{\perp}.$$

In order to identify the decomposition of (u, y) into $\mathcal{I}_G$ and $\mathcal{R}_G$, representing the nominal and uncertain system dynamics, two mathematical methods, being at the core of the intended data analysis, are used. The one is orthogonal projection operators and the associated computations. The projection of (u, y) onto $\mathcal{I}_G$ using the operator $\mathcal{P}_{\mathcal{I}_G}$ given in (7.4) is well-known in the literature [69, 199]. In Chaps. 9 and 10, we will introduce relevant orthogonal projection theory in more detail and leverage it for exploring alternative fault detection schemes. Another one is the Bezout identity (3.4),

$$\begin{bmatrix} M & -\hat{Y} + MQ \\ N & \hat{X} + NQ \end{bmatrix} \begin{bmatrix} X + Q\hat{N} & Y - Q\hat{M} \\ -\hat{N} & \hat{M} \end{bmatrix} = \begin{bmatrix} I & 0 \\ 0 & I \end{bmatrix}$$

which provides us with a decomposition of (u, y) in $\mathcal{I}_G$ and $\mathcal{I}_K, \mathcal{H}_2 = \mathcal{I}_G \otimes \mathcal{I}_K$,

$$\begin{bmatrix} u \\ y \end{bmatrix} = \begin{bmatrix} M \\ N \end{bmatrix} v + \begin{bmatrix} -\hat{Y} + MQ \\ \hat{X} + NQ \end{bmatrix} r_y = \begin{bmatrix} M \\ N \end{bmatrix} (v + Qr_y) + \begin{bmatrix} -\hat{Y} \\ \hat{X} \end{bmatrix} r_y.$$

$\mathcal{I}_K$ is the image subspace of a possible controller

$$K = -\left(\hat{Y} - MQ\right)\left(\hat{X} + NQ\right)^{-1} = -\left(X + Q\hat{N}\right)^{-1}\left(Y - Q\hat{M}\right).$$

It is remarkable that $\left(v, r_y\right)$ are latent variables and determine the coordinates of (u, y) in $\mathcal{I}_G$ and $\mathcal{I}_K$. In order to achieve the intended decomposition in $\mathcal{I}_G$ and $\mathcal{R}_G$, $\begin{bmatrix} -\hat{Y} \\ \hat{X} \end{bmatrix} r_y$ is, by means of the projection method, further decomposed into

$$\begin{bmatrix} -\hat{Y}_0 \\ \hat{X}_0 \end{bmatrix} r_{y,0} = \begin{bmatrix} M_0 \\ N_0 \end{bmatrix} Q_{\mathcal{H}_2} r_{y,0} + \left(\mathcal{I} - \mathcal{P}_{\mathcal{I}_G}\right)\left(\begin{bmatrix} -\hat{Y}_0 \\ \hat{X}_0 \end{bmatrix} r_{y,0}\right),$$

where (M_0, N_0) is the normalised (M, N) applied for the construction of $\mathcal{P}_{\mathcal{I}_G}$, and $\left(\hat{X}_0, \hat{Y}_0\right)$, $r_{y,0}$ are the associated SIR of K and the normalised residual r_y. Hereby, $Q_{\mathcal{H}_2} \in \mathcal{RH}_\infty$ is given by

$$Q_{\mathcal{H}_2} = \mathcal{P}_{\mathcal{H}_2}\left[M_0^\sim \; N_0^\sim \right] \begin{bmatrix} -\hat{Y}_0 \\ \hat{X}_0 \end{bmatrix}.$$

Now, setting $Q = Q_{\mathcal{H}_2}$ yields

$$\begin{aligned} \begin{bmatrix} u \\ y \end{bmatrix} &= \begin{bmatrix} M_0 \\ N_0 \end{bmatrix} \hat{v} + \left(\mathcal{I} - \mathcal{P}_{\mathcal{I}_G}\right)\left(\begin{bmatrix} -\hat{Y}_0 \\ \hat{X}_0 \end{bmatrix} r_{y,0}\right) \\ &= \begin{bmatrix} M_0 \\ N_0 \end{bmatrix} \hat{v} + \left(\begin{bmatrix} -\hat{Y}_0 \\ \hat{X}_0 \end{bmatrix} - \begin{bmatrix} M_0 \\ N_0 \end{bmatrix} Q_{\mathcal{H}_2}\right) r_{y,0}. \end{aligned}$$

This is a main result of this section, and highlighted from the following interesting aspects.

- It reveals the limitation of a feedback control, namely, the feedback control performance will never be better than

$$\left\| \left(\begin{bmatrix} -\hat{Y}_0 \\ \hat{X}_0 \end{bmatrix} - \begin{bmatrix} M_0 \\ N_0 \end{bmatrix} Q_{\mathcal{H}_2}\right) r_{y,0} \right\|.$$

- In this optimal case, the feedforward control represented by $\begin{bmatrix} M_0 \\ N_0 \end{bmatrix} \hat{v} \in \mathcal{I}_G$ and the feedback control $\left(\begin{bmatrix} -\hat{Y}_0 \\ \hat{X}_0 \end{bmatrix} - \begin{bmatrix} M_0 \\ N_0 \end{bmatrix} Q_{\mathcal{H}_2} \right) r_{y,0} \in \mathcal{R}_G$ are decoupled in the context that they act on two orthogonal subspaces.
- It gives a geometric interpretation of the optimal (robust) control problem,

$$\min_{Q \in \mathcal{RH}_\infty} \left\| \begin{bmatrix} -\hat{Y} + MQ \\ \hat{X} + NQ \end{bmatrix} \right\|_\infty .$$

  In Remark 7.2 and the related discussions, the equivalence of $Q_{\mathcal{H}_2}$ computation to the solution of the above MMP [57] has been explained in the context of the gap metric between the subspaces spanned by

$$\begin{bmatrix} -\hat{Y} \\ \hat{X} \end{bmatrix} \text{ and } \begin{bmatrix} M \\ N \end{bmatrix},$$

  which measures the similarity of two subspaces [39, 55, 102, 110]. The reader is referred to Chap. 9 for more details.
- As a by-product, the residual subspace $\mathcal{R}_G$ can be defined by

$$\mathcal{R}_G = \left\{ \begin{bmatrix} u \\ y \end{bmatrix} : \begin{bmatrix} u \\ y \end{bmatrix} = \left(\begin{bmatrix} -\hat{Y}_0 \\ \hat{X}_0 \end{bmatrix} - \begin{bmatrix} M_0 \\ N_0 \end{bmatrix} Q_{\mathcal{H}_2} \right) r_y, r_y \in \mathcal{H}_2 \right\},$$

  and the corresponding operator can be computed by means of a normalisation of $\begin{bmatrix} -\hat{Y}_0 \\ \hat{X}_0 \end{bmatrix} - \begin{bmatrix} M_0 \\ N_0 \end{bmatrix} Q_{\mathcal{H}_2}$. To this end, a spectral factorisation [243] is to be computed, as described in Remark 7.1.

The data structure explored in this section is control-theoretically informed, and provides us with new aspects and groundwork for data-driven control and detection research.

Section 7.2 deals with the extension of the kernel-based system model, the so-called observer-based uniform system model, introduced in Chap. 4. The motivation for this extension is twofold. From the control-theoretic viewpoint, the kernel-based model, driven by the residual r_y, is of the same eigen-dynamic like the original system. Accordingly, it may be unstable, which hinders a reliable online implementation of the model, for instance, towards process and performance monitoring. The major advantage of the kernel-based system model is the integration of the residual signal r_y in the system model, that reflects possible uncertainties in the system. However, uncertainties in the system image subspace or corrupted in the latent variable cannot be apprehended by the residual. While in classical control engineering systems such uncertainties are rare, they exist in networked control systems or cyber-physical systems, as demonstrated in Example 7.2. On the assumption that $(u, y) \in \mathcal{H}_2$, it is a natural extension to substitute u with

$$u = F\hat{x} + Vv,$$

which leads to the SIR-SKR-based model (7.55)–(7.56). We would like to call the reader's attention to the subsequent remarks about the SIR-SKR-based model,

- the overall SIR-SKR-based model comprises, besides the system (7.55) or its state space representation (7.51)–(7.52), the two residual generators described by (7.56). That means, the SIR-SKR-based model (7.55)–(7.56) is implicitly a high-dimensional system, including the nominal system dynamics, the dynamics corresponding to the system uncertainties including faults, and the dynamics induced by the uncertainties corrupted in the latent variable. On the other hand, for the online implementation, only an n-dimensional state observer is needed;
- the residual subspace $\mathcal{R}_K$, which consists of all signals (u, y) that result in

$$Xu + Yy \neq 0,$$

  is not orthogonal to the system image subspace $\mathcal{I}_G$. Instead, it is orthogonal to $\mathcal{I}_K$. As a result, the residual generated by $Xu + Yy$ only reflects those uncertainties corrupted in the latent variable v. Nevertheless, it is efficient and capable to detect cyber-attacks, as will be explored in Chaps. 14–16;
- in the SIR-SKR-based model, the unknown initial values of the state vector x are captured by the residual r_y. Hence, without loss of generality, the initial state vector in the SIR-SKR-based model, $\hat{x}(0)$, is assumed to be zero;
- the SIR-SKR-based model (7.51)–(7.52) is not unique, depends on the selection of the quadruple (L, W, F, V).

To simplify and normalise the application of the model, the C-SIR-SKR-based model has been introduced at the end of this section.

The behavioural systems theory was established by Willems in a series of publication in 1980s and 1990s. It is systematically and elaborately described in the monograph [163]. Emerging from behavioural systems theory as a system-theoretic basis, data-driven system analysis and control become a new research area in recent years [138]. Behavioural systems theory regards any dynamic system as a set of trajectories. At its core is the so-called fundamental lemma [207], which, informally speaking, tells us that any system input and output data (u, y) over a finite time interval can be expressed in terms of a linear combination of the data in an input-output data set (matrix), that can be formed, for instance, by recorded data in the past. Inspired by this result, we have studied, in Sect. 7.3, the image and kernel representations of plants and controllers over a finite time interval, and call them image and kernel representations of finite samples long signals/data. This exploration supplies us with a data-driven version of the unified framework of control and detection.

Theorem 7.6 presents the image representation of finite samples long data for a given plant. It is equivalent to the fundamental lemma with an alternative proof. The dual form of the kernel representation of finite samples long data is proved in Theorem 7.7. It is noteworthy that data-driven image and kernel representations have been firstly proposed by Ding et al. [46] in 2014 and, with the aid of the subspace identification technique, applied to data-driven fault detection and fault-tolerant control in automatic control systems [37].

Analogously to the unified framework with the plant and controller representations in Hilbert space, whose core is the Bezout identity, the kernel and image representations of finite samples long data for a "controller" are derived and summarised in Theorem 7.8. This enables us to describe the closed-loop dynamics over a finite time interval in terms of the image representations of the plant and controller $I_{G,s}$ and $I_{K,s}$ with the reference $v_{s+n}(k-n)$ and residual $r_{s+n}(k-n)$ of finite samples long as the latent variables, namely

$$\begin{bmatrix} u_s(k) \\ y_s(k) \end{bmatrix} = I_{G,s} v_{s+n}(k-n) + I_{K,s} r_{s+n}(k-n).$$

Different from the Bezout identity over $\mathcal{RH}_\infty$, the mappings between the system data $(u_s(k), y_s(k))$ and the latent variables $(v_s(k), r_s(k))$ are expressed with the aid of the following two signal/data sets,

$$\mathcal{S}_{(u,y),s} = \left\{ \begin{bmatrix} u_s(k) \\ y_s(k) \end{bmatrix}, \begin{array}{l} \begin{bmatrix} u_s(k) \\ y_s(k) \end{bmatrix} = \begin{bmatrix} I_{G,s} & I_{K,s} \end{bmatrix} \begin{bmatrix} v_{s+n}(k-n) \\ r_{s+n}(k-n) \end{bmatrix}, \\ \begin{bmatrix} v_{s+n}(k-n) \\ r_{s+n}(k-n) \end{bmatrix} \in \mathcal{L}_{[0,\infty)}^{(s+n)(m+p)} \end{array} \right\},$$

$$\mathcal{S}_{(v,r),s} = \left\{ \begin{bmatrix} v_s(k) \\ r_s(k) \end{bmatrix}, \begin{array}{l} \begin{bmatrix} v_s(k) \\ r_s(k) \end{bmatrix} = \begin{bmatrix} K_{K,s+n} \\ K_{G,s+n} \end{bmatrix} \begin{bmatrix} u_{s+n}(k-n) \\ y_{s+n}(k-n) \end{bmatrix}, \\ \begin{bmatrix} u_{s+n}(k-n) \\ y_{s+n}(k-n) \end{bmatrix} \in \mathcal{L}_{[0,\infty)}^{(s+n)(m+p)} \end{array} \right\}.$$

Applications of the unified framework with kernel and image representations of finite samples long data at its core to control and detection will be explored in Chap. 13.

At end of this section, the issue of the minimal order (samples long) image and kernel representations of the plant and controller has been addressed. Using the controllability and observerability canonical forms [128], it has been proved in Theorem 7.9 that the minimum order is given by

$$\mu = \max\{\mu_o, \mu_c\}$$

with μ_c and μ_o as the controllability and observability index respectively.

Three potential application domains have been shortly addressed in Section 7.4. The major attention has been paid to those possible new aspects of leveraging the control and detection methods in the unified framework to existing control and detection topics. Concerning detection issues, two substantial messages are conveyed, (i) any uncertainties, including faults, in the plant are captured by the residual signal r_y and they act on the system dynamics via r_y, (ii) the residual signal $Xy + Yy$ comprises uncertainties in the latent variable v and is useful for detecting cyber-attacks, which will be explored in Chaps. 14–16.

In the unified framework, tracking control and MPC can be alternatively and elegantly handled, where the latent variable v plays a key role. With the aid of the image representation and the latent variable, we are able to re-formulate tracking control problem as an optimisation problem with respect to the latent variable v as follows,

$$\min_{v \in \mathcal{H}_2} \left\| \begin{bmatrix} M_{r,0} \\ N_{r,0} \end{bmatrix} v_r - \begin{bmatrix} M_0 \\ N_0 \end{bmatrix} v \right\|_2 .$$

Moreover, it enables us to measure the "distance" between the image subspace of the reference model and the image subspace of the plant in terms of the so-called gap metric [69, 199], i.e.

$$\gamma = \left\| \begin{bmatrix} M_{r,0} \\ N_{r,0} \end{bmatrix} - \begin{bmatrix} M_0 \\ N_0 \end{bmatrix} Q_{r,\mathcal{H}_2} \right\|_\infty = \min_{Q \in \mathcal{RH}_\infty} \left\| \begin{bmatrix} M_{r,0} \\ N_{r,0} \end{bmatrix} - \begin{bmatrix} M_0 \\ N_0 \end{bmatrix} Q \right\|_\infty .$$

On this basis, the original optimisation problem can be rewritten as two optimisation sub-problems, the MMP given above that is generally solved off-line, and

$$\min_{v_1} \left\| \left(\begin{bmatrix} M_{r,0} \\ N_{r,0} \end{bmatrix} - \begin{bmatrix} M_0 \\ N_0 \end{bmatrix} Q_{r,\mathcal{H}_2} \right) v_r - \begin{bmatrix} M_0 \\ N_0 \end{bmatrix} v_1 \right\|_{2,[k,k+s]} ,$$

to be solved online. In other words, the latent variable is divided into

$$v = Q_{r,\mathcal{H}_2} v_r + v_1,$$

and v_1 is to be found by solving the optimisation given above. Considering that tracking control is often defined in terms of a reference trajectory over a finite time interval $[k, k+s]$, the concerned optimisation can be solved applying the system models of finite sample data introduced in Sect. 7.3.

Informally speaking, MPC can be viewed as an extended tracking control. Like tracking control, both reference and plant models are used for the prediction purpose. Differently, in the context of MPC, the control performance is minimum prediction error based on the present process data. In our study on MPC, a prediction model has been introduced to predict and compensate a part of the system uncertainties using control and detection methods developed in the unified framework.

The initial intention of our exploration to MASs is to explore the capability of the unified framework. The first part of our efforts has been devoted to this end. As an example, the trivial consensus problem among agents has been examined. The related preliminaries in this context, including those basic concepts, definitions and notations known in the graph theory and Theorem 7.10 are referred to the monograph by Li and Duan [119]. Hereby, Theorem 7.10 is a summary of Theorems 13–14, and Theorems 17 and 20 in [119]. The major results are given in Corollary 7.3, which show that

- the consensus controller at an agent can be equivalent written into the form of an output feedback controller,

$$v_i = K\left(y_i - \bar{r}_i\right) = K y_i - K\bar{r}_i = -\hat{Y}\hat{X}^{-1}\left(y_i - \bar{r}_i\right) = -X^{-1}Y\left(y_i - \bar{r}_i\right),$$

  where $\bar{r}_i$ comprises the residual signals received by the ith agent from its neighbours and $-K\bar{r}_i$ serves as a reference signal (latent variable) for the ith agent,
- equivalently, the consensus controller at the ith agent can be represented in the SIR and SKR forms

$$v_i = K\left(y_i - \bar{r}_i\right) = \hat{Y}\hat{X}^{-1}\left(y_i - \bar{r}_i\right) = -X^{-1}Y\left(y_i - \bar{r}_i\right),$$

  as well as their state space representations.

The significant message of this work is that the local feedback control in the agents are decoupled from the consensus controller. Consistently, the local controllers in the agents can be designed independent of the consensus algorithm to approach control and detection problems like robust control, fault detection and fault-tolerant control. This result motivates us to review the research literature of observer-based fault diagnosis in MASs, which has surprised us remarkably. It is state of the art that a great number of works in the past decade have dealt with observer-based residual generation based on a system model, in which the MAS under consideration is modelled as an interconnected system [95, 181]. As comprehensively illustrated in our study, such fault diagnosis schemes driven by the residual signals generated on the basis of an interconnected MAS model not only make the problem solution unnecessarily much complicated, but also might result in degraded fault detection performance. It has been demonstrated that based on a more general system model, the model (7.178), it is sufficient to generate residual signals using an observer-based residual generator located at an agent and without taking into account of the consensus algorithm. In this manner, also fault detection issues for heterogeneous MASs can be well addressed, and it is possible to extend this strategy to deal with fault detection in MASs with relative information [139].

Part II
Application of the Unified Framework to Fault Diagnosis and Control in Uncertain Dynamic Systems

Chapter 8
Re-visit of Fault Detection Methods, and Some New and Modified Schemes

Observer-based fault diagnosis methods, thanks to their intimate relations to automatic control, are widely accepted as a powerful technique in dealing with process monitoring and fault diagnosis in dynamic processes. In the unified framework of control and detection, it is natural to make use of the available detection capability in the observer-based feedback controller realisation to achieve fault detection and estimation. In this chapter, we firstly review some basic observer-based methods for residual generation, which can be applied in fact both for control and fault detection purposes. Our main focus is on the existing optimal fault detection schemes when additive disturbances (unknown input vectors) and multiplicative model uncertainties are present in the process under consideration. Thereby, some modifications and new observer-based schemes are introduced.

8.1 Basic Observer-Based Residual Generation Schemes

Residual generation is an essential step for detecting faults in dynamic processes driven by certain input variables. Roughly speaking, residual generation is

- to build software redundancy for process measurement variables and
- to create a residual vector by comparing the process measurement variables and their software redundancy.

In the ideal case, the residual vector is decoupled from the process input variables, and should contain information about faults to be detected. Unfortunately, in today's research works, the essential role of a residual generator to eliminate the influence of the process input on the residual vector is often ignored. Instead, much attention is paid to the robustness against unknown disturbances and model uncertainties. The most popular way to build software redundancy is the use of a nominal process model and, based on it, the construction of an observer, or in general, a residual generator.

S. X. Ding, *Unified Framework of Control and Detection*,
https://doi.org/10.1007/978-3-662-72417-0_8

In our subsequent introduction, the state space representation of the discrete-time LTI system model adopted in our previous work is as the nominal process model under consideration,

$$x(k+1) = Ax(k) + Bu(k), x(0) = x_0, \tag{8.1}$$

$$y(k) = Cx(k) + Du(k). \tag{8.2}$$

At first, we introduce two standard observer-based residual generation schemes.

8.1.1 Fault Detection Filter

Fault detection filter (FDF) is a type of observer-based residual generators proposed by Beard and Jones in the early 1970s. Their work marked the beginning of a stormy development of model-based FDI techniques.

The core of an FDF is a full-order state observer

$$\hat{x}(k+1) = A\hat{x}(k) + Bu(k) + L\left(y(k) - C\hat{x}(k) - Du(k)\right), \tag{8.3}$$

built upon which, the residual is simply defined by

$$r_0(k) = y(k) - \hat{y}(k) = y(k) - C\hat{x}(k) - Du(k).$$

The advantages of an FDF lie in its simple construction form and, for the reader who is familiar with the modern control theory, its close relationship with the state observer design and especially with the well-established robust control theory by designing robust residual generators.

The design of an FDF is the determination of the observer gain matrix L. To increase the degree of design freedom, an algebraic post-filter, W, is often added, that is

$$r_y(k) = Wr_0(k) = W\left(y(k) - \hat{y}(k)\right).$$

In order to preserve information about faults to be detected, W is typically selected as invertible.

8.1.2 Diagnostic Observer Scheme

A diagnostic observer (DO) is, thanks to its flexible structure and similarity to the Luenberger type observer, one of the mostly investigated observer-based residual generator forms.

The core of a DO is a Luenberger type (output) observer described by

$$\xi(k+1) = G\xi(k) + Hu(k) + Ly(k), \tag{8.4}$$

$$r_y(k) = Wy(k) - V\xi(k) - Qu(k), \tag{8.5}$$

where $\xi \in \mathbb{R}^s$, s denotes the observer order and can be equal to or lower or higher than the system order n. Although most contributions to the Luenberger type observer are focused on the case with lower order aiming at constructing a reduced order observer, higher order observers may play an important role in the optimisation of fault detection systems.

Given the process model (8.1)–(8.2), matrices G, H, L, Q, V and W together with a matrix $T \in \mathbb{R}^{s \times n}$ have to satisfy the so-called Luenberger conditions,

$$\begin{aligned} & I.G \text{ is a Schur matrix,} \\ & II.TA - GT = LC,\ H = TB - LD, \\ & III.WC - VT = 0,\ Q = WD, \end{aligned}$$

under which system (8.4)–(8.5) delivers a residual vector satisfying

$$\begin{aligned} e(k+1) = Ge(k), r_y(k) = We(k), e(k) = Tx(k) - \xi(k), \\ \Longrightarrow \forall u, x(0), \lim_{k \to \infty} r_y(k) = 0. \end{aligned}$$

8.1.3 Parity Space Approach

The parity space approach was initiated by Chow and Willsky in their pioneering work in the early 1980s. The parity space approach is commonly recognised as one of the important model-based residual generation approaches. Although a parity space based residual generator is in fact a finite impulse response filter (FIRF) and thus not an observer-based one, there exists a one-to-one mapping between, for instance, a DO and a parity vector, as will be briefly illustrated in this subsection. In this context, the parity space method is introduced in this section.

In the following, consider state space model (8.1)–(8.2) and, without loss of generality, assume

$$rank(C) = m.$$

Introducing the notations

$$y_s(k-s) = \begin{bmatrix} y(k-s+1) \\ \vdots \\ y(k) \end{bmatrix}, u_s(k-s) = \begin{bmatrix} u(k-s+1) \\ \vdots \\ u(k) \end{bmatrix}, \tag{8.6}$$

$$H_{o,s} = \begin{bmatrix} C \\ CA \\ \vdots \\ CA^{s-1} \end{bmatrix}, H_{u,s} = \begin{bmatrix} D & 0 & \cdots & 0 \\ CB & D & \ddots & \vdots \\ \vdots & \ddots & \ddots & 0 \\ CA^{s-2}B & \cdots & CB & D \end{bmatrix}, \tag{8.7}$$

a straightforward computation yields the following compact model form

$$y_s(k-s) = H_{o,s}x(k-s+1) + H_{u,s}u_s(k-s). \tag{8.8}$$

Note that (8.8) describes the input and output relationship in dependence on the past state vector $x(k-s+1)$, which is unknown (not measured). The underlying idea of the parity relation based residual generation lies in the utilisation of the fact, known from the linear control theory, that for $s > n$ the following rank condition holds:

$$rank\left(H_{o,s}\right) \leq n < \text{the row number of matrix } H_{o,s} = sm.$$

This ensures that for $s > n$ there exists at least a (row) vector

$$v_s(\neq 0) \in \mathbb{R}^{sm}$$

such that

$$v_s H_{o,s} = 0. \tag{8.9}$$

Hence, a parity relation based residual generator is constructed by

$$r_y(k) = v_s\left(y_s(k-s) - H_{u,s}u_s(k-s)\right), \tag{8.10}$$

whose dynamics is governed by

$$r_y(k) = v_s\left(y_s(k-s) - H_{u,s}u_s(k-s)\right) = v_s H_{o,s}x(k-s+1) = 0.$$

Vectors satisfying (8.9) are called parity vectors, the set of which,

$$P_s = \{v_s \mid v_s H_{o,s} = 0\},$$

is called the parity space of the s-th order.

One of the significant properties of parity relation based residual generators is their straightforward design that only deals with algebraic matrix computations. On the other hand, the implementation form (8.10) is disadvantageous, comparing with a recursive computation like an observer. The one-to-one relationship of parity vectors and observers given below is helpful to solve this problem.

Consider the process model (8.1)–(8.2) and a parity vector

$$v_s = \begin{bmatrix} v_{s,0} \; v_{s,1} \; \cdots \; v_{s,s-1} \end{bmatrix} \in \mathbb{R}^{sm}, \, v_s \begin{bmatrix} C \\ CA \\ \vdots \\ CA^{s-1} \end{bmatrix} = 0,$$

$v_{s,i} \in \mathbb{R}^m, i = 0, 1, \ldots, s-1$. It has been demonstrated that matrices

$$A_z = \begin{bmatrix} 0 & 0 & \cdots & 0 \\ 1 & 0 & \cdots & 0 \\ \vdots & \ddots & \ddots & \vdots \\ 0 & \cdots & 1 & 0 \end{bmatrix} \in \mathbb{R}^{s-1 \times s-1}, \, L_z = -\begin{bmatrix} v_{s,0} \\ v_{s,1} \\ \vdots \\ v_{s,s-2} \end{bmatrix}, \tag{8.11}$$

$$T = \begin{bmatrix} v_{s,1} & v_{s,2} & \cdots & v_{s,s-1} & v_{s,s-1} \\ v_{s,2} & \cdots & \cdots & v_{s,s-1} & 0 \\ \vdots & \cdots & \cdots & \vdots & \vdots \\ v_{s,s-1} & 0 & \cdots & \cdots & 0 \end{bmatrix} \begin{bmatrix} C \\ CA \\ \vdots \\ CA^{s-2} \end{bmatrix} \tag{8.12}$$

solve the Luenberger equations

$$TA - A_z T = L_z C, \, c_z T = gC, \, c_z = \begin{bmatrix} 0 \cdots 0 \; 1 \end{bmatrix}, g = v_{s,s-1}. \tag{8.13}$$

A direct application of this result is the construction of an observer-based residual generator for the given parity vector v_s as follows

$$z(k+1) = A_z z(k) + B_z u(k) + L_z y(k) \in \mathbb{R}^{s-1}, \, B_z = TB - L_z D, \tag{8.14}$$

$$r_y(k) = gy(k) - c_z z(k) - d_z u(k) \in \mathbb{R}, \, d_z = gD. \tag{8.15}$$

Note that

$$\begin{bmatrix} B_z \\ d_z \end{bmatrix} = \begin{bmatrix} v_{s,0} & v_{s,1} & \cdots & v_{s,s-1} & v_{s,s-1} \\ v_{s,1} & \cdots & \cdots & v_{s,s-1} & 0 \\ \vdots & \cdots & \cdots & \vdots & \vdots \\ v_{s,s-1} & 0 & \cdots & \cdots & 0 \end{bmatrix} \begin{bmatrix} D \\ CB \\ CAB \\ \vdots \\ CA^{s-2}B \end{bmatrix}$$

$$= \begin{bmatrix} v_s H_{u,s}(:, 1:p) \\ v_s H_{u,s}(:, p+1:2p) \\ \vdots \\ v_s H_{u,s}(:, (s-1)p+1:sp) \end{bmatrix}.$$

It is apparent that the system (8.14) is a dead-beat observer.

Based on the above result, a strategy named *parity space design, observer-based implementation* has been developed, which makes use of the computational advantage of parity space approaches for the system design (selection of a parity vector or matrix) and then realises the solution in the observer form to ensure a numerically stable and less consuming online computation. This strategy has been, for instance, successfully used in the sensor fault detection in vehicles and highly evaluated by engineers in industry. It is worth mentioning that the strategy of *parity space design, observer-based implementation* can also be applied to continuous-time systems.

8.1.4 Parameterisation of Observer-Based Residual Generators

It is a known result that all LTI observer-based residual generators can be represented by a full-order state observer and a stable dynamic post-filter, namely,

$$\hat{x}(k+1) = A\hat{x}(k) + Bu(k) + L\left(y(k) - \hat{y}(k)\right), \tag{8.16}$$

$$\hat{y}(k) = y(k) - C\hat{x}(k) - Du(k), \tag{8.17}$$

$$r_y(z) = R(z)\left(y(z) - \hat{y}(z)\right), \ R(z) \in \mathcal{RH}_\infty. \tag{8.18}$$

Such a general form of all types of LTI residual generators is called parameterisation of observer-based residual generators with R as the parameter system.

This parameterisation form is an immediate result of the parameterisation of LTI observers presented in Chap. 3. In fact, by definition any LTI observer-based residual generators can be written as

$$r_y(z) = Q_1(z)\left(y(z) - \hat{y}(z)\right)$$

for some $Q_1 \in \mathcal{RH}_\infty$. Recall that the parameterisation form of the estimate for y is given by

$$\hat{y}(z) = C\hat{x}(z) + Du(z) + Q_2(z)\left(y(z) - \hat{y}_0(z)\right)$$

with $Q_2 \in \mathcal{RH}_\infty$ being the parameterisation system. Here, $\hat{y}_0$ is the estimate of y delivered by the full-order state observer (8.16), namely

$$\hat{y}_0(k) = C\hat{x}(k) + Du(k).$$

As a result, it holds

$$\begin{aligned} r_y(z) &= Q_1(z)\left(y(z) - \hat{y}(z)\right) \\ &= Q_1(z)\left(y(z) - \hat{y}_0(z)\right) - Q_1(z)Q_2(z)\left(y(z) - \hat{y}_0(z)\right) \\ &= \left(Q_1(z) - Q_1(z)Q_2(z)\right) r_0(z) = R(z)\left(y(z) - \hat{y}_0(z)\right), \\ r_0(z) &= y(z) - \hat{y}_0(z), \ R(z) = Q_1(z) - Q_1(z)Q_2(z) \in \mathcal{RH}_\infty. \end{aligned}$$

Finally, the parameterisation form (8.16) can be expressed in terms of the SKR of the model (8.1)–(8.2) as

$$r_y(z) = R(z)\left(y(z) - \hat{y}(z)\right) = R(z)\left[\, \hat{M}(z) \; \hat{N}(z) \,\right] \begin{bmatrix} u(z) \\ y(z) \end{bmatrix}.$$

Note that, according to Lemma 3.1, the parameterisation of observer-based residual generators is invariant with regard to the observer gain matrix L. A major advantage of the parameterisation of observer-based residual generators is that analysis and synthesis with respect to an observer-based residual generator can be performed independent of the observer design.

8.2 Optimal Fault Detection Schemes for Processes with Additive Disturbances

Processes with additive unknown disturbances are a classic system model type that has been continuously receiving strong attention in the research domain since the beginning of research on observer-based fault detection technique. In this section, we introduce two standard fault detection schemes, the Kalman filter-based fault detection scheme and the so-called unified solution, developed for processes corrupted with stochastic and deterministic disturbances, respectively. Specifically, LDTV system models with process and measurement noises and unknown input vectors are under consideration. Although there exist various explanations and interpretations for optimal detection performance of these two schemes, we will discuss the optimality from a new aspect in the end of this section.

8.2.1 Fault Detection in Linear Stochastic Processes

Consider the process model

$$x(k+1) = A(k)x(k) + B(k)u(k) + E(k)w(k), \tag{8.19}$$

$$y(k) = C(k)x(k) + D(k)u(k) + v(k), \tag{8.20}$$

where $x \in \mathbb{R}^n$, $u \in \mathbb{R}^p$, $y \in \mathbb{R}^m$ are process state, input and output vectors, respectively, and all system matrices are of appropriate dimensions and known. $w \in \mathbb{R}^{k_w}$, v are process and measurement noise vectors. It is assumed that they are uncorrelated with the state and input vectors, and

$$w(k) \sim \mathcal{N}\left(0, \Sigma_w(k)\right), v(k) \sim \mathcal{N}\left(0, \Sigma_v(k)\right),$$

$$\mathbb{E}\left(\begin{bmatrix} w(i) \\ v(i) \\ x(0) \end{bmatrix}\begin{bmatrix} w(j) \\ v(j) \\ x(0) \end{bmatrix}^T\right) = \begin{bmatrix} \begin{bmatrix} \Sigma_w(i) & S_{wv}(i) \\ S_{wv}^T(i) & \Sigma_v(i) \end{bmatrix}\delta_{ij} & 0 \\ 0 & \Pi_0 \end{bmatrix}, \delta_{ij} = \begin{cases} 1, i = j \\ 0, i \neq j. \end{cases}$$

It is well-known that a Kalman filter with the recursive algorithm,

$$\begin{gathered} \hat{x}(k+1|k) = A(k)\hat{x}(k|k-1) + B(k)u(k) + K(k)r_0(k), \\ r_0(k) = y(k) - \hat{y}(k|k-1), \hat{x}(0) = 0, \\ \hat{y}(k|k-1) = C(k)\hat{x}(k|k-1) + D(k)u(k), \\ K(k) = \left(A(k)P(k|k-1)C^T(k) + E(k)S_{wv}(k)\right)\Sigma_r^{-1}(k), \\ P(k+1|k) = A(k)P(k|k-1)A^T(k) + E(k)\Sigma_w(k)E^T(k) \\ -K(k)\Sigma_r(k)K^T(k), \\ \Sigma_r(k) = C(k)P(k|k-1)C^T(k) + \Sigma_v(k) = \mathbb{E}\left(r_0(k)r_0^T(k)\right), \end{gathered}$$

delivers a residual vector $r_0(k) \in \mathbb{R}^m$, which is white and of minimum covariance matrix.

The whiteness of the residual vector allows us to approach the fault detection problem at each time instant equivalently to the following basic fault detection problem: given the model

$$r_0(k) = f(k) + \varepsilon(k), \varepsilon(k) \sim \mathcal{N}(0, \Sigma_r(k)), \tag{8.21}$$

where $f(k)$ represents any type of possible faults in the process or in the sensors and actuators, find a test statistic J and a threshold J_{th} so that the fault is optimally detected in the context

$$f(k) = 0 \Longrightarrow \text{fault-free}, \ f(k) \neq 0 \Longrightarrow \text{faulty}.$$

It is well-known that the optimal setting $\{J, J_{th}\}$ for the above detection problem is

$$J(k) = r_0^T(k)\Sigma_r^{-1}(k)r_0(k), \ J_{th} = \chi_\alpha^2(m). \tag{8.22}$$

Here, χ_α^2 is the value of $\chi^2(m)$-distribution corresponding to the false alarm rate (FAR) equal to α. This solution is optimal in the sense of *maximum fault detectability by* an acceptable FAR α. It is remarkable that the property with the minimum covariance matrix Σ_r results in the overall maximal fault detectability. That is, given system model (8.19)–(8.20), the above Kalman filter based fault detection system delivers, in comparison with all other potential (linear) fault detection systems, the best fault detectability at the FAR level α.

Remark 8.1 The property that the residual vector $r_0(k)$ is white Gaussian noise is a pre-condition for χ^2 statistic (8.22) to be the most powerful (optimal) test and hence widely applied in engineering application. Suppose that $\Sigma_r(k) = \Sigma_r, k = 0, 1, \ldots$.

Given $r_0(k_0+1), \ldots, r_0(k_0+N)$, a standard engineering solution is to build the average value

$$\bar{r}_0 = \frac{1}{N}\sum_{i=1}^{N} r_0(k_0+i) \sim \mathcal{N}\left(0, \frac{\Sigma_r}{N}\right) \tag{8.23}$$

and define the test statistic and the corresponding threshold

$$J(k) = N\bar{r}_0^T(k)\Sigma_r^{-1}\bar{r}_0(k),\ J_{th} = \chi_\alpha^2(m). \tag{8.24}$$

Thanks to the smaller covariance matrix Σ_r/N, the fault detectability increases considerably. Note that the distribution of $\bar{r}_0 \sim \mathcal{N}\left(0, \frac{\Sigma_r}{N}\right)$ only holds if $r_0(k)$ is white Gaussian noise, which can be interpreted as independent and identically distributed (i.i.d). It is noteworthy that in the literature the χ^2 statistic (8.24) is a widely used test statistic for fault detection in stochastic processes, even though the pre-condition that the residual signal should be i.i.d. random series is not satisfied.

8.2.2 Fault Detection in Linear Processes with Unknown Inputs

We now consider the process model

$$x(k+1) = A(k)x(k) + B(k)u(k) + E_d(k)d(k), \tag{8.25}$$
$$y(k) = C(k)x(k) + D(k)u(k) + F_d(k)d(k), \tag{8.26}$$

where $x(k), u(k), y(k)$ are as given in model (8.19)–(8.20), and all system matrices are of appropriate dimensions and known. $d(k) \in \mathbb{R}^{k_d}$ represents unknown input vector and is $l_{2,[0,N]}$ bounded with

$$\|d\|_{2,[0,N]} \le \delta_d. \tag{8.27}$$

It is assumed that $k_d \ge m$. The following FDF is an optimal observer-based residual generator, called unified solution,

$$\hat{x}(k+1) = A(k)\hat{x}(k) + B(k)u(k) + L(k)\, r_0(k), \tag{8.28}$$
$$r_0(k) = y(k) - \hat{y}(k),\ \hat{y}(k) = C(k)\hat{x}(k) + D(k)u(k), \tag{8.29}$$
$$L(k) = \left(E_d(k)F_d^T(k) + A(k)P(k)C^T(k)\right) R_r^{-1}(k), \tag{8.30}$$
$$R_r(k) = C(k)P(k)C^T(k) + F_d(k)F_d^T(k),$$

where $P(k)$ is the solution of Riccati recursion

$$P(k+1) = A(k)P(k)A^T(k) + E_d(k)E_d^T(k) - L(k)R_r(k)L^T(k),\ P(0) = 0.$$

Note that the dynamic of the above FDF is governed by

$$\mathcal{F}(d) : \begin{cases} e(k+1) = (A(k) - L(k)C(k))\, e(k) + (E_d(k) - L(k)F_d(k))\, d(k) \\ r_0(k) = C(k)e(k) + F_d(k)d(k), \end{cases}$$
$$e(k) = x(k) - \hat{x}(k). \tag{8.31}$$

In the literature, the FDF (8.28)–(8.29) is also called Least Squares (LS) observer or, analogous to its LTI version, H_2 observer. Let

$$r_{y,0}(k) = W(k)r_0(k),\ W(k) = R_r^{-1/2}(k). \tag{8.32}$$

It is easy to check that the system (8.31) and (8.32) is co-inner. Based on this property, it can be proved that for given additive faults modelled by

$$r_y(k) = W(k)\,(f(k) + \mathcal{F}(d)) \tag{8.33}$$

with $f(k)$ representing the influence of the faults on the system dynamic, the setting (J, J_{th}) defined by

$$J = \left\| r_{y,0} \right\|_{2,[0,N]},\ J_{th} = \delta_d$$

delivers the optimal fault detection performance.

Remark 8.2 It is noteworthy that the vector $f(k)$ is not limited to sensor fault modelling. It represents the influence of additive faults in the system (8.25)–(8.26) on the system output $y(k)$.

8.2.3 Discussion on the Optimality of the Fault Detection Performance

That Kalman filter and the unified solution deliver optimal performance of detecting additive faults has been intensively discussed and demonstrated, as reported in the references cited at the end of this chapter. In this subsection, we address the optimality issue from an alternative aspect. Since the detection mechanism of Kalman filter-based scheme is well explained from the stochastic viewpoint, only the system model (8.25)–(8.26) is under consideration. Recalling that the residual vector is decoupled from the input vector u, our discussion starts from the residual model

$$\begin{aligned} r_y(k) &= \hat{\mathcal{M}}\,(y(k)) - \hat{\mathcal{N}}\,(u(k)) = \hat{\mathcal{N}_d}\,(d(k))\,, \\ \hat{\mathcal{N}_d}\,(d(k)) &= (A_L(k), E_L(k), W(k)C(k), W(k)F_d(k))\,, \\ A_L(k) &= A_L(k) - L(k)C(k),\ E_L(k) = E_d(k) - L(k)F_d(k). \end{aligned}$$

As a result of condition (8.27), the set of the unknown input vector d is defined as

$$\mathcal{D} = \left\{ d(k), \ \|d\|_{2,[0,N]} \leq \delta_d \right\}.$$

Denote by $\hat{\mathcal{N}}_{d,0}\left(d(k)\right)$ the dynamic of the residual generator (8.31)–(8.32), and remember that $\hat{\mathcal{N}}_{d,0}$ is co-inner, that is,

$$\hat{\mathcal{N}}_{d,0} \circ \hat{\mathcal{N}}_{d,0}^{*} = \mathcal{I}, \tag{8.34}$$

where $\hat{\mathcal{N}}_{d,0}^{*}$ is the adjoint operator of $\hat{\mathcal{N}}_{d,0}$. Let

$$\mathcal{I}_{\hat{\mathcal{N}}_{d,0}} := \left\{ r_{y,0}(k), r_{y,0}(k) = \hat{\mathcal{N}}_{d,0}\left(d(k)\right), \ \|d\|_{2,[0,N]} \leq \delta_d \right\}$$

be the image subspace of $\hat{\mathcal{N}}_{d,0}$. It is apparent that, according to (8.34),

$$\begin{aligned} &\forall d \in \mathcal{D}, d \notin \ker\left(\mathcal{I}_{\hat{\mathcal{N}}_{d,0}}\right), \ \left\|r_{y,0}\right\|_{2,[0,N]} = \left\|\hat{\mathcal{N}}_{d,0}\left(d(k)\right)\right\|_{2,[0,N]} = \|d\|_{2,[0,N]} \\ &\Longrightarrow \sup_{r_{y,0} \in \mathcal{I}_{\hat{\mathcal{N}}_{d,0}}} \left\|r_{y,0}\right\|_{2,[0,N]} = \sup_{d \in \mathcal{D}} \|d\|_{2,[0,N]} = \delta_d. \end{aligned} \tag{8.35}$$

Consequently, the threshold setting, according to the definition

$$J_{th} = \sup_{d \in \mathcal{D}} \left\|r_{y,0}\right\|_{2,[0,N]} \quad \sup_{r_{y,0} \in \mathcal{I}_{\hat{\mathcal{N}}_{d,0}}} \left\|r_{y,0}\right\|_{2,[0,N]}$$

in the fault-free case, equals to

$$J_{th} = \sup_{d \in \mathcal{D}} \left\|r_{y,0}\right\|_{2,[0,N]} = \sup_{r_{y,0} \in \mathcal{I}_{\hat{\mathcal{N}}_{d,0}}} \left\|r_{y,0}\right\|_{2,[0,N]} = \delta_d.$$

Now, consider a residual vector $r_y(k)$, which is generated by the FDF (8.28)–(8.29) with the dynamic

$$\hat{\mathcal{N}}_d = \left(A_L(k), E_L(k), W(k)C(k), W(k)F_d(k)\right) \tag{8.36}$$

Suppose that the setting for $(L(k), W(k))$ is different from (8.30) and (8.32) and thus results in that $\hat{\mathcal{N}}_d$ is not co-inner. On the assumption that

$$\sup_{d \neq 0, d \in \mathcal{D}} \frac{\left\|\hat{\mathcal{N}}_d\left(d(k)\right)\right\|_{2,[0,N]}}{\|d(k)\|_{2,[0,N]}} = 1 = \sup_{d \neq 0, d \in \mathcal{D}} \frac{\left\|\hat{\mathcal{N}}_{d,0}\left(d(k)\right)\right\|_{2,[0,N]}}{\|d(k)\|_{2,[0,N]}}, \tag{8.37}$$

which leads to

$$\sup_{d\in\mathcal{D}}\left\|r_y\right\|_{2,[0,N]}=\sup_{r_y\in\mathcal{I}_{\hat{\mathcal{N}}_d}}\left\|r_y\right\|_{2,[0,N]}=\delta_d.$$

Remark 8.3 The assumption (8.37) loses no generality, since for any $\hat{\mathcal{N}}_d\left(d(k)\right)$ with

$$\sup_{d\neq 0,d\in\mathcal{D}}\frac{\left\|\hat{\mathcal{N}}_d\left(d(k)\right)\right\|_{2,[0,N]}}{\left\|d(k)\right\|_{2,[0,N]}}=\gamma,$$

the system $\hat{\mathcal{N}}_d\left(d(k)\right)/\gamma$ delivers the same fault detection performance like $\hat{\mathcal{N}}_d\left(d(k)\right)$. The introduction of this assumption enables us to compare $\hat{\mathcal{N}}_d$ and $\hat{\mathcal{N}}_{d,0}$.

For our study, the following lemma is useful, whose proof can be found in the reference cited at the end of this chapter.

Lemma 8.1 *Given residual vectors*

$$r_1(k)=\hat{\mathcal{N}}_{d,1}\left(d(k)\right),r_2(k)=\hat{\mathcal{N}}_{d,2}\left(d(k)\right),\tag{8.38}$$

where $r_1(k), r_2(k)$ are generated by two observer-based residual generators (8.28)–(8.29) with $(L_1(k), W_1(k))$ and $(L_2(k), W_2(k))$, respectively, then it holds

$$r_1(k)=\mathcal{Q}\left(r_2(k)\right),\tag{8.39}$$

$$\mathcal{Q}=\left(A(k)-L_1(k)C(k),\left(L_1(k)-L_2(k)\right)W_2^{-1}(k),W_1(k)C(k),W_1(k)W_2^{-1}(k)\right).$$

Moreover, the operator is invertible and the inverse is stable.

By Lemma 8.1, $\hat{\mathcal{N}}_d$ given in (8.36) can be written as

$$\hat{\mathcal{N}}_d=\mathcal{Q}_0\circ\hat{\mathcal{N}}_{d,0}$$

for some stable and invertible operator $\mathcal{Q}_0$. Define the image subspace of $\hat{\mathcal{N}}_d$,

$$\mathcal{I}_{\hat{\mathcal{N}}_d}:=\left\{r_y(k),r_y(k)=\hat{\mathcal{N}}_d\left(d(k)\right),\left\|d\right\|_{2,[0,N]}\leq\delta_d\right\}.$$

It follows from (8.35) and (8.39) that

$$\forall r_{y,0}(k)\in\mathcal{I}_{\hat{\mathcal{N}}_{d,0}},r_{y,0}(k)\in\mathcal{I}_{\hat{\mathcal{N}}_d}\Longrightarrow\mathcal{I}_{\hat{\mathcal{N}}_{d,0}}\subseteq\mathcal{I}_{\hat{\mathcal{N}}_d}.$$

It is known that $\mathcal{I}_{\hat{\mathcal{N}}_{d,0}}$ is a closed subspace in Hilbert space, and the operator,

$$\mathcal{P}_{\mathcal{I}_{\hat{\mathcal{N}}_{d,0}}}:=\hat{\mathcal{N}}_{d,0}^{*}\circ\hat{\mathcal{N}}_{d,0},$$

defines an orthogonal projection onto $\mathcal{I}_{\hat{\mathcal{N}}_{d,0}}$. An orthogonal projection is of the properties that

$$\begin{aligned}
\mathcal{P}_{\mathcal{I}_{\hat{\mathcal{N}}_{d,0}}} \circ \mathcal{P}_{\mathcal{I}_{\hat{\mathcal{N}}_{d,0}}} &= \mathcal{P}_{\mathcal{I}_{\hat{\mathcal{N}}_{d,0}}}, \\
\forall r_y(k), \left\| r_y(k) \right\|_{2,[0,N]}^2 &= \left\| \mathcal{P}_{\mathcal{I}_{\hat{\mathcal{N}}_{d,0}}} \left(r_y(k) \right) \right\|_{2,[0,N]}^2 \\
&+ \left\| r_y(k) - \mathcal{P}_{\mathcal{I}_{\hat{\mathcal{N}}_{d,0}}} \left(r_y(k) \right) \right\|_{2,[0,N]}^2, \\
\mathcal{P}_{\mathcal{I}_{\hat{\mathcal{N}}_{d,0}}} \left(r_y(k) \right) &= \hat{\mathcal{N}}_{d,0}^* \circ \hat{\mathcal{N}}_{d,0} \left(r_y(k) \right) \in \mathcal{I}_{\hat{\mathcal{N}}_{d,0}}.
\end{aligned} \tag{8.40}$$

Here, $\mathcal{P}_{\mathcal{I}_{\hat{\mathcal{N}}_{d,0}}} \left(r_y(k) \right)$ is the projection of $r_y(k)$ onto $\mathcal{I}_{\hat{\mathcal{N}}_{d,0}}$. Note that, due to (8.40), it holds

$$\left\| \mathcal{P}_{\mathcal{I}_{\hat{\mathcal{N}}_{d,0}}} \left(r_y(k) \right) \right\|_{2,[0,N]} = \left\| \hat{\mathcal{N}}_{d,0} \left(d(k) \right) \right\|_{2,[0,N]}.$$

Remark 8.4 In Chap. 9, orthogonal projections in Hilbert space and their applications to fault detection and isolation will be studied intensively, which also includes a detailed description of the relevant operator theoretical preliminaries.

Now, consider the residual dynamic in the faulty case and model it, without loss of generality, as

$$r_{y,o}(k) = \hat{\mathcal{N}}_{d,0} \left(d(k) \right) + f(k) \Longrightarrow r_y(k) = \mathcal{Q}_0 \left(\hat{\mathcal{N}}_{d,0} \left(d(k) \right) + f(k) \right),$$

where $f(k)$ represents the influence of an additive fault vector on the residual signals. By means of the orthogonal projection $\mathcal{P}_{\mathcal{I}_{\hat{\mathcal{N}}_{d,0}}}$, $r_{y,o}$ and r_y are further written into

$$\begin{aligned}
r_{y,o}(k) &= r_{y,o,\mathcal{I}_{\hat{\mathcal{N}}_{d,0}}}(k) + f_{\mathcal{I}^{\perp}_{\hat{\mathcal{N}}_{d,0}}}(k), \\
r_y(k) &= \mathcal{Q}_0 \left(r_{y,o,\mathcal{I}_{\hat{\mathcal{N}}_{d,0}}}(k) + f_{\mathcal{I}^{\perp}_{\hat{\mathcal{N}}_{d,0}}}(k) \right), \\
r_{y,o,\mathcal{I}_{\hat{\mathcal{N}}_{d,0}}}(k) &= \mathcal{P}_{\mathcal{I}_{\hat{\mathcal{N}}_{d,0}}} \left(\hat{\mathcal{N}}_{d,0} \left(d(k) \right) + f(k) \right) = \hat{\mathcal{N}}_{d,0} \left(d(k) \right) + \mathcal{P}_{\mathcal{I}_{\hat{\mathcal{N}}_{d,0}}} \left(f(k) \right), \\
f_{\mathcal{I}^{\perp}_{\hat{\mathcal{N}}_{d,0}}}(k) &= \left(\mathcal{I} - \mathcal{P}_{\mathcal{I}_{\hat{\mathcal{N}}_{d,0}}} \left(f(k) \right) \right).
\end{aligned}$$

Hence, the sets defined by

$$\begin{aligned}
\mathcal{NF}_0 &:= \left\{ r_{y,o}(k), \left\| r_{y,o} \right\|_{2,[0,N]} \leq J_{th} = \delta_d \right\}, \\
\mathcal{NF} &:= \left\{ r_y(k), \left\| r_y \right\|_{2,[0,N]} \leq J_{th} = \delta_d \right\}
\end{aligned}$$

are the undetectable set of the residual (generator) dynamics $r_{y,o}$ and r_y, respectively. It is evident that

$$\mathcal{NF}_0 \subseteq \mathcal{NF}. \tag{8.41}$$

Note that it follows from the property of orthogonal projection that

$$\left\| r_{y,o}(k) \right\|_{2,[0,N]}^2 = \left\| r_{y,o,\mathcal{I}_{\hat{\mathcal{N}}_{d,0}}}(k) \right\|_{2,[0,N]}^2 + \left\| f_{\mathcal{I}^{\perp}_{\hat{\mathcal{N}}_{d,0}}}(k) \right\|_{2,[0,N]}^2 .$$

Suppose that the unknown input vector d and fault vector f are such that

$$\left\| r_{y,o,\mathcal{I}_{\hat{\mathcal{N}}_{d,0}}}(k) \right\|_{2,[0,N]} = J_{th} = \delta_d, \left\| f_{\mathcal{I}^{\perp}_{\hat{\mathcal{N}}_{d,0}}}(k) \right\|_{2,[0,N]} = \varepsilon > 0, \tag{8.42}$$

which leads to

$$\forall \varepsilon > 0, \left\| r_{y,o}(k) \right\|_{2,[0,N]}^2 > J_{th} = \delta_d.$$

Recall that $\hat{\mathcal{N}}_d = \mathcal{Q}_0 \circ \hat{\mathcal{N}}_{d,0}$ is not a co-inner operator and satisfies condition (8.37), which implies that there exists a vector β so that

$$\left\| \mathcal{Q}_0 \left(\beta(k) \right) \right\|_{2,[0,N]} < \delta_d, \left\| \beta(k) \right\|_{2,[0,N]} = \delta_d. \tag{8.43}$$

Now, suppose that the vector $r_{y,o,\mathcal{I}_{\hat{\mathcal{N}}_{d,0}}}$ in (8.42) satisfies (8.43), and ε is sufficiently small so that

$$\begin{aligned} \left\| r_y(k) \right\|_{2,[0,N]} &= \left\| \mathcal{Q}_0 \left(r_{y,o,\mathcal{I}_{\hat{\mathcal{N}}_{d,0}}}(k) + f_{\mathcal{I}^{\perp}_{\hat{\mathcal{N}}_{d,0}}}(k) \right) \right\|_{2,[0,N]} \\ \leq \left\| \mathcal{Q}_0 \left(r_{y,o,\mathcal{I}_{\hat{\mathcal{N}}_{d,0}}}(k) \right) \right\|_{2,[0,N]} &+ \left\| \mathcal{Q}_0 \left(f_{\mathcal{I}^{\perp}_{\hat{\mathcal{N}}_{d,0}}}(k) \right) \right\|_{2,[0,N]} < \delta_d = J_{th}. \end{aligned} \tag{8.44}$$

Consequently, by means of the residual generator $\hat{\mathcal{N}}_d$ the fault under consideration cannot be detected. As a result, (8.41) can be further written as

$$\mathcal{NF}_0 \subset \mathcal{NF}. \tag{8.45}$$

In other words, the undetectable set $\mathcal{NF}_0$ is a sub-set of $\mathcal{NF}$, the undetectable set corresponding to the residual generator $\hat{\mathcal{N}}_d$. In this way, it is proved that the residual generator (8.28)–(8.29) with $(L(k), W(k))$-setting satisfying (8.30) and (8.32) is optimal in the sense of maximum fault detectability.

Remark 8.5 Given additive faults modelled by

$$x(k+1) = A(k)x(k) + B(k)u(k) + E_d(k)d(k) + E_f(k)f(k), \tag{8.46}$$

$$y(k) = Cx(k) + Du(k) + F_d(k)d(k) + F_f(k)f(k) \tag{8.47}$$

with $f(k) \in \mathbb{R}^{k_f}$, and assuming that

$$\dim\left\{r_{y,f}(k), r_{y,f}(k) = \hat{\mathcal{N}}_{f,0}\left(f(k)\right)\right\} = m,$$
$$\hat{\mathcal{N}}_{f,0} = \left(A_L(k), E_{L,f}(k), W(k)C(k), W(k)F_f(k)\right),$$
$$E_{L,f}(k) = E_f(k) - L(k)F_f(k),$$

the FDF with $(L(k), W(k))$ given by (8.30) and (8.32) delivers the optimal detection performance in the sense of maximum fault detectability as well. This conclusion has been proved in the reference cited at the end of this chapter.

8.3 Fault Detection Schemes for Processes with Multiplicative Uncertainties

Different from the well-established detection framework for processes with additive disturbances, it is state of the art in research on detecting faults in processes with multiplicative uncertainties that no standard fault detection schemes exist. The obvious difficulty herewith is handling of model uncertainties concerning observer-based residual generator design and, in particular, threshold setting. It is challenging to achieve an estimation of a tight upper-bound of the model uncertainty on the residual vector. Consequently, threshold might be set higher than needed, which may remarkably reduce the fault detectability. In this section, we briefly introduce some known methods reported in the literature.

8.3.1 *Problem Formulation*

Let

$$y(z) = G(z)u(z), G(z) = C\left(zI - A\right)^{-1}B + D, \tag{8.48}$$
$$G(z) = \left(\hat{M}(z) + \Delta_{\hat{M}}(z)\right)^{-1}\left(\hat{N}(z) + \Delta_{\hat{N}}(z)\right) \tag{8.49}$$
$$= \left(N(z) + \Delta_N(z)\right)\left(M(z) + \Delta_M(z)\right)^{-1} \tag{8.50}$$

be the LCF and RCF of the process model G under consideration. The LCP and RCP, $\left(\hat{M}, \hat{N}\right)$ and (M, N), are the nominal model, while $\left(\Delta_{\hat{M}}, \Delta_{\hat{N}}\right)$ and (Δ_M, Δ_N) denote the corrupted (multiplicative) uncertainty. It is supposed that

$$\left\|\begin{bmatrix} -\Delta_{\hat{N}} & \Delta_{\hat{M}} \end{bmatrix}\right\|_\infty \le \delta_{\Delta_K}, \left\|\begin{bmatrix} \Delta_M \\ \Delta_N \end{bmatrix}\right\|_\infty \le \delta_{\Delta_I}.$$

By an observer-based residual generator, the residual vector

$$r_y(z) = \hat{M}(z)y(z) - \hat{N}(z)u(z) \tag{8.51}$$

is generated, whose dynamic is governed by

$$r_y(z) = \begin{bmatrix} \Delta_{\hat{N}}(z) & -\Delta_{\hat{M}}(z) \end{bmatrix} \begin{bmatrix} u(z) \\ y(z) \end{bmatrix} \tag{8.52}$$

$$= \begin{bmatrix} -\hat{N}(z) & \hat{M}(z) \end{bmatrix} \begin{bmatrix} \Delta_M(z) \\ \Delta_N(z) \end{bmatrix} v(z). \tag{8.53}$$

In (8.53), vector v represents an unknown hidden variable that satisfies

$$v(z) = (M(z) + \Delta_M(z))^{-1} u(z).$$

Let

$$J = \|r_y\|_2$$

be the residual evaluation function. According to the commonly adopted definition of threshold setting,

$$J_{th} = \sup_{\left\|\begin{bmatrix} -\Delta_{\hat{N}} & \Delta_{\hat{M}} \end{bmatrix}\right\|_\infty \le \delta_{\Delta_K}} J, \; J_{th} = \sup_{\left\|\begin{bmatrix} \Delta_M \\ \Delta_N \end{bmatrix}\right\|_\infty \le \delta_{\Delta_I}} J$$

in the fault-free case, the threshold settings for the residual satisfying (8.52) or (8.53) are

$$J_{th} = \sup_{\left\|\begin{bmatrix} -\Delta_{\hat{N}} & \Delta_{\hat{M}} \end{bmatrix}\right\|_\infty \le \delta_{\Delta_K}} \|r_y\|_2, \tag{8.54}$$

$$J_{th} = \sup_{\left\|\begin{bmatrix} \Delta_M \\ \Delta_N \end{bmatrix}\right\|_\infty \le \delta_{\Delta_I}} \|r_y\|_2. \tag{8.55}$$

Note that different from the handling of additive uncertainties, the threshold here could be a function of the process input vector.

Now, we are in the position to formulate the design of fault detection systems consisting of (i) residual generator (8.51), and (ii) threshold setting given by (8.54) or/and (8.55). Given the process model with the model uncertainties (8.49) and (8.50), find (L, W) so that the threshold J_{th} is as tight as possible. In the context of one-class fault detection, a tighter threshold means a smaller set of undetectable faults. It is worth to mention that, under certain conditions, a solution of optimal one-class fault detection could also be the solution for a two-class optimal fault detection problem.

8.3.2 Threshold Setting Schemes

We begin with threshold setting schemes. By definition, a threshold should not be a function of the process output y, in order to ensure that it is an upper-bound of the impact of the model uncertainty on the residual vector in the fault-free case. Consider the residual dynamic (8.52). Since the process input u is independent of faults in the process, a logic step to estimate the upper-bound of the influence of the model uncertainty is to replace y by Gu, namely

$$r_y = \left[\Delta_{\hat{N}} \; -\Delta_{\hat{M}} \right] \begin{bmatrix} I \\ G \end{bmatrix} u,$$

which yields

$$J_{th} = \sup_{\left\| \left[-\Delta_{\hat{N}} \; \Delta_{\hat{M}} \right] \right\|_\infty \le \delta_{\Delta K}} \left\| \left[\Delta_{\hat{N}} \; -\Delta_{\hat{M}} \right] \begin{bmatrix} I \\ G \end{bmatrix} u \right\|_2 \le$$
$$\sup_{\left\| \left[-\Delta_{\hat{N}} \; \Delta_{\hat{M}} \right] \right\|_\infty \le \delta_{\Delta K}} \left\| \left[\Delta_{\hat{N}} \; -\Delta_{\hat{M}} \right] \begin{bmatrix} I \\ \left(\hat{M} + \Delta_{\hat{M}} \right)^{-1} \left(\hat{N} + \Delta_{\hat{N}} \right) \end{bmatrix} \right\|_\infty \|u\|_2 .$$

It is straightforward that

$$\left\| \left[\Delta_{\hat{N}} \; -\Delta_{\hat{M}} \right] \begin{bmatrix} I \\ \left(\hat{M} + \Delta_{\hat{M}} \right)^{-1} \left(\hat{N} + \Delta_{\hat{N}} \right) \end{bmatrix} \right\|_\infty$$
$$\le \delta_{\Delta K} \sqrt{1 + \left\| \left(I + \hat{M}^{-1} \Delta_{\hat{M}} \right)^{-1} \left(\hat{M}^{-1} \hat{N} + \hat{M}^{-1} \Delta_{\hat{N}} \right) \right\|_\infty^2} .$$

Suppose that

$$G_0 = \hat{M}^{-1} \hat{N} \in \mathcal{RH}_\infty, \gamma_{G_0} = \|G_0\|_\infty ,$$
$$\hat{M}^{-1} \left[\Delta_{\hat{N}} \; \Delta_{\hat{M}} \right] \in \mathcal{H}_\infty, \left\| \hat{M}^{-1} \left[\Delta_{\hat{N}} \; \Delta_{\hat{M}} \right] \right\|_\infty \le \bar{\delta}_{\Delta K} < 1.$$

Then, it turns out

$$\left\| \left[\Delta_{\hat{N}} \; -\Delta_{\hat{M}} \right] \begin{bmatrix} I \\ \left(\hat{M} + \Delta_{\hat{M}} \right)^{-1} \left(\hat{N} + \Delta_{\hat{N}} \right) \end{bmatrix} \right\|_\infty \le \delta_{\Delta K} \sqrt{1 + \left(\frac{\gamma_{G_0} + \bar{\delta}_{\Delta K}}{1 - \bar{\delta}_{\Delta K}} \right)^2} .$$

There is no doubt that

$$J_{th} = \delta_{\Delta_K} \sqrt{1 + \left(\frac{\gamma_{G_0} + \bar{\delta}_{\Delta_K}}{1 - \bar{\delta}_{\Delta_K}} \right)^2} \|u\|_2 \tag{8.56}$$

is a considerably conservative threshold setting. In the literature, efforts have been reported to achieve a tighter threshold setting. For our purpose of finding a tight threshold setting, the following known inequality is introduced.

Lemma 8.2 *Let $\Delta_1, \Delta_2 \in \mathcal{H}_\infty$ be such that*

$$\left\| \begin{bmatrix} \Delta_1 \\ \Delta_2 \end{bmatrix} \right\|_\infty \leq b < 1.$$

Then,

$$\left\| \Delta_1 (I + \Delta_2)^{-1} \right\|_\infty \leq \frac{b}{\sqrt{1 - b^2}}. \tag{8.57}$$

It is evident that the dual form of this lemma,

$$\left\| \begin{bmatrix} \Delta_1 & \Delta_2 \end{bmatrix} \right\|_\infty \leq b < 1 \Longrightarrow \left\| (I + \Delta_2)^{-1} \Delta_1 \right\|_\infty \leq \frac{b}{\sqrt{1 - b^2}}, \tag{8.58}$$

also holds.

Rewriting (8.52) into

$$r_y = \begin{bmatrix} \Delta_{\hat{N}} & -\Delta_{\hat{M}} \end{bmatrix} \begin{bmatrix} I \\ \left(I + \hat{M}^{-1} \Delta_{\hat{M}} \right)^{-1} \left(G_0 + \hat{M}^{-1} \Delta_{\hat{N}} \right) \end{bmatrix} u,$$

we obtain, by means of Lemma 8.2,

$$\begin{aligned} \left\| r_y \right\|_2 &\leq \delta_{\Delta_K} \sqrt{1 + \left(\frac{\bar{\delta}_{\Delta_K}}{\sqrt{1 - \bar{\delta}_{\Delta_K}^2}} + \frac{\gamma_{G_0}}{1 - \bar{\delta}_{\Delta_K}} \right)^2} \|u\|_2 \\ &= \frac{\delta_{\Delta_K}}{1 - \bar{\delta}_{\Delta_K}} \sqrt{\left(1 - \bar{\delta}_{\Delta_K}\right)^2 + \left(\sqrt{\frac{1 - \bar{\delta}_{\Delta_K}}{1 + \bar{\delta}_{\Delta_K}}} + \gamma_{G_0} \right)^2} \|u\|_2 \\ \Longrightarrow J_{th} &= \frac{\delta_{\Delta_K}}{1 - \bar{\delta}_{\Delta_K}} \sqrt{\left(1 - \bar{\delta}_{\Delta_K}\right)^2 + \left(\sqrt{\frac{1 - \bar{\delta}_{\Delta_K}}{1 + \bar{\delta}_{\Delta_K}}} + \gamma_{G_0} \right)^2} \|u\|_2 . \end{aligned} \tag{8.59}$$

Alternatively, we now consider the residual dynamic expression (8.53) and substitute (Δ_M, Δ_N) by Δ_{I,K_G} given in (2.68) in Example 2.1 yields

$$\begin{aligned} r_y &= -\begin{bmatrix} -\hat{N} & \hat{M} \end{bmatrix}\begin{bmatrix} -\hat{Y} \\ \hat{X} \end{bmatrix}\left(I + \Delta_{K_G}\begin{bmatrix} -\hat{Y} \\ \hat{X} \end{bmatrix}\right)^{-1}\Delta_{K_G}\begin{bmatrix} M \\ N \end{bmatrix} v, \\ &= -\left(I + \Delta_{K_G}\begin{bmatrix} -\hat{Y} \\ \hat{X} \end{bmatrix}\right)^{-1}\Delta_{K_G}\begin{bmatrix} M \\ N \end{bmatrix} v, \\ \Delta_{K_G} &= -\begin{bmatrix} \Delta_{\hat{N}} & -\Delta_{\hat{M}} \end{bmatrix}. \end{aligned} \tag{8.60}$$

On the assumption that

$$\left\| \begin{bmatrix} \Delta_{\hat{N}} & -\Delta_{\hat{M}} \end{bmatrix}\begin{bmatrix} M & -\hat{Y} \\ N & \hat{X} \end{bmatrix} \right\|_\infty = \tilde{\delta}_{\Delta_K} < 1,$$

we have, according to Lemma 8.2,

$$\begin{aligned} \left\| r_y \right\|_2 &= \left\| \left(I + \Delta_{K_G}\begin{bmatrix} -\hat{Y} \\ \hat{X} \end{bmatrix}\right)^{-1}\Delta_{K_G}\begin{bmatrix} M \\ N \end{bmatrix} v \right\|_2 \\ &\le \frac{\tilde{\delta}_{\Delta_K}}{\sqrt{1-\tilde{\delta}^2_{\Delta_K}}} \left\| v \right\|_2 . \end{aligned} \tag{8.61}$$

The value $\frac{\tilde{\delta}_{\Delta_K}}{\sqrt{1-\tilde{\delta}^2_{\Delta_K}}} \left\| v \right\|_2$ is a tight upper-bound of the residual vector. Unfortunately, it can only serve as a threshold if $\left\| v \right\|_2$ is known online or offline. This could be the case, for instance, when a feedback control system is under consideration with v as the known reference vector.

We now consider the uncertain system (8.50). The corresponding residual dynamic is given by (8.53), which yields

$$\left\| r_y \right\|_2 \le \left\| \begin{bmatrix} -\hat{N} & \hat{M} \end{bmatrix}\begin{bmatrix} \Delta_M \\ \Delta_N \end{bmatrix} v \right\|_2 .$$

In general, v is unknown and thus, aiming at threshold setting, is substituted by a function of u. To this end, the relation

$$\begin{aligned} u &= (M + \Delta_M)\, v = \left(I + \Delta_M M^{-1}\right) M v \Longrightarrow \\ M v &= \left(I + \Delta_M M^{-1}\right)^{-1} u \end{aligned}$$

is used, which, on the assumption

$$\begin{bmatrix} \Delta_M \\ \Delta_N \end{bmatrix} M^{-1} \in \mathcal{H}_\infty, \left\| \begin{bmatrix} \Delta_M \\ \Delta_N \end{bmatrix} M^{-1} \right\|_\infty \leq \bar{\delta}_{\Delta_I} < 1,$$

results in

$$\|Mv\|_2 \leq \frac{1}{1-\bar{\delta}_{\Delta_I}} \|u\|_2 .$$

Suppose that $\begin{bmatrix} -\hat{N} & \hat{M} \end{bmatrix}$ is a normalised SKR. It turns out

$$\left\|r_y\right\|_2 \leq \left\| \begin{bmatrix} \Delta_M \\ \Delta_N \end{bmatrix} M^{-1} \right\|_\infty \|Mv\|_2 \leq \frac{\bar{\delta}_{\Delta_I}}{1-\bar{\delta}_{\Delta_I}} \|u\|_2 =: J_{th}. \tag{8.62}$$

As a summary of the aforementioned discussion, it is remarked that the threshold setting laws (8.56), (8.59) and (8.62) are based on an upper-bound estimation of the process input u on the residual r_u. Due to the existence of conservativeness, the resulted thresholds are not tight.

8.3.3 Design of Observer-Based Residual Generators

In the literature, there are a great number of methods dealing with observer design for the residual generation purpose, mainly in the context of robustness against model uncertainties and, when possible, simultaneously increasing sensitivity for the faults to be detected. The achieved upper-bound of the influence of the uncertainty on the residual vector is then adopted for the threshold setting. Consequently, most of these methods result in conservative threshold setting. Below, an alternative method is introduced.

We firstly consider the uncertain system (8.50) and write the residual dynamic as

$$r_y = R \begin{bmatrix} -\hat{N}_o & \hat{M}_o \end{bmatrix} \begin{bmatrix} u \\ y \end{bmatrix} = R \begin{bmatrix} -\hat{N}_o & \hat{M}_o \end{bmatrix} \begin{bmatrix} \Delta_M \\ \Delta_N \end{bmatrix} v$$

for some $\mathcal{L}_2$ bounded signal v, where $\left(\hat{M}_o, \hat{N}_o\right)$ is the normalised LCP and $R \in \mathcal{RH}_\infty$ is an invertible post-filter with $R^{-1} \in \mathcal{RH}_\infty$. Due to the uncertainty $\begin{bmatrix} \Delta_M \\ \Delta_N \end{bmatrix}$,

$$d := \begin{bmatrix} \Delta_M \\ \Delta_N \end{bmatrix} v$$

is also an $\mathcal{L}_2$ bounded unknown vector as far as $v \in \mathcal{H}_2$. Hence,

$$r_y = R \begin{bmatrix} -\hat{N}_o & \hat{M}_o \end{bmatrix} d. \tag{8.63}$$

Recall our discussion in Sect. 8.2 on the optimal fault detection for systems with additive disturbances and notice the facts (i) d is an additive disturbance, (ii) the normalised $\left[-\hat{N}_o \; \hat{M}_o \right]$ is inner. As a result, the normalised $\left[-\hat{N}_o \; \hat{M}_o \right]$ delivers the optimal solution. That implies, on the other hand,

$$R = I$$

is the optimal setting for the post-filter.

Next, consider the uncertain system (8.49) and the corresponding residual dynamic given by

$$r_y = R \left[-\hat{N}_o \; \hat{M}_o \right] \begin{bmatrix} u \\ y \end{bmatrix} = R \left[\Delta_{\hat{N}} \; -\Delta_{\hat{M}} \right] \begin{bmatrix} u \\ y \end{bmatrix}.$$

Analogous to the handling of $\left[\Delta_{\hat{N}} \; -\Delta_{\hat{M}} \right]$ by its equivalent uncertainty in the SIR Δ_{I,K_G}, namely by means of (2.68) in Example 2.1, we obtain

$$\begin{aligned} r_y &= -R \left[-\hat{N}_o \; \hat{M}_o \right] \begin{bmatrix} -\hat{Y}_0 \\ \hat{X}_0 \end{bmatrix} \left(I + \Delta_{K_G} \begin{bmatrix} -\hat{Y}_0 \\ \hat{X}_0 \end{bmatrix} \right)^{-1} \Delta_{K_G} \begin{bmatrix} M_0 \\ N_0 \end{bmatrix} v \\ &= -R \left[-\hat{N}_o \; \hat{M}_o \right] \left(I + \begin{bmatrix} -\hat{Y}_0 \\ \hat{X}_0 \end{bmatrix} \Delta_{K_G} \right)^{-1} \begin{bmatrix} -\hat{Y}_0 \\ \hat{X}_0 \end{bmatrix} \Delta_{K_G} \begin{bmatrix} M_0 \\ N_0 \end{bmatrix} v, \end{aligned}$$

where (M_0, N_0) is the normalised RCP and $\left(\hat{X}_0, \hat{Y}_0 \right)$ satisfies the corresponding Bezout identity. Assume that

$$\begin{bmatrix} -\hat{Y}_0 \\ \hat{X}_0 \end{bmatrix} \Delta_{K_G} \in \mathcal{H}_\infty, \left\| \begin{bmatrix} -\hat{Y}_0 \\ \hat{X}_0 \end{bmatrix} \Delta_{K_G} \right\|_\infty < 1.$$

Thus

$$d := \left(I + \begin{bmatrix} -\hat{Y}_0 \\ \hat{X}_0 \end{bmatrix} \Delta_{K_G} \right)^{-1} \begin{bmatrix} -\hat{Y}_0 \\ \hat{X}_0 \end{bmatrix} \Delta_{K_G} \begin{bmatrix} M_0 \\ N_0 \end{bmatrix} v$$

is an $\mathcal{L}_2$ bounded unknown vector for $v \in \mathcal{H}_2$. It holds

$$r_y = R \left[-\hat{N}_o \; \hat{M}_o \right] d.$$

With the same above-mentioned argument, it can be concluded that $R = I$ is the optimal setting for the post-filter.

8.4 Notes and References

Observer-based fault diagnosis is the state of the art technique in dealing with fault detection, isolation and identification in dynamic systems [13, 20, 34, 71, 157]. Beginning 50 years ago with the pioneering work [9, 99], the observer-based fault diagnosis technique has undergone a rapid development over a couple of decades and become today well established as an active research area in control theory and engineering [58, 60, 64, 93, 136, 197, 238]. Even in the era when data-driven and machine learning methods become the most dominant research domain in technical fault diagnosis, observer-based fault diagnosis technique is still the major and efficient tool applied to addressing fault diagnosis issues for dynamic and particularly automatic control systems [39]. The recent development of data-driven design of observer-based fault detection systems [37, 38] has contributed to the extension of this technique so that it could meet practical demands for industrial applications.

An observer-based fault diagnosis system consists of two major functional blocks: residual (feature) generation and decision making. While decisions for fault detection or isolation are made on the basis of residual processing algorithms, the major task of residual generation is the construction of an observer serving as residual generator. Thanks to this intimate relation to control theory, most of investigations on observer-based fault diagnosis systems are dedicated to observer design issues. In this chapter, we have reviewed the basics of the model-based fault detection methods for LTI systems. Most of the results can be found in the monographs [13, 20, 36, 71, 77, 158, 184] and in the early survey papers [60, 93, 136, 224]. The first work on FDF and DO has been reported by Beard and Jones [9, 99], and Chow and Willsky have proposed the first optimal solution using the parity space scheme [24]. The one-to-one mapping of parity vectors and observers presented in Sect. 8.1.3 has been reported in [34, 52] and applied for the purpose of the data-driven design of observer-based residual generators [37, 38, 47].

Keeping in mind that any fault detection (or isolation) problem is in its core to distinguish two different system operations, i.e. the fault (to be detected) versus uncertainties as fault detection, or two different classes of faults as fault isolation, research on observer-based residual generation was, right from the beginning, shaped into the trade-off framework of sensitivity (e.g. to fault) vs. robustness (against uncertainty) [50, 58, 59, 89, 156]. This development is a natural result, as robust control dominated control theory and engineering between 80's and 90's, and has stamped the progress of observer-based fault diagnosis technique since then. Although not addressed in our review, the mostly representative research in this regard is the (optimal) index-based observer-based residual generator design. The $\mathcal{H}_\infty/\mathcal{H}_\infty$ design problem was first proposed and solved in [49], lately in [59, 166, 175]. In 1993, $\mathcal{H}_-/\mathcal{H}_\infty$ design problem was proposed and handled in [50]. It has been extensively studied after the publication of the LMI (linear matrix inequality) solution to this problem [89]. The most significant contributions to this issue are [83, 89, 120, 123, 167, 168, 202]. It is worth mentioning the work by Zhong et al. [235], Wang and Yang [200] and Chadli et al. [18], in which the

$\mathcal{H}_-/\mathcal{H}_\infty$ design scheme has been applied to (i) uncertain LTI systems, (ii) the solution in a finite frequency range, and (iii) nonlinear systems modelled by means of T-S fuzzy technique, respectively. In [34, 41], it has been proved that the unified solution offers a simultaneous solution to the multi-objective $\mathcal{H}_i/\mathcal{H}_\infty$ optimisation problem, and the $\mathcal{H}_-/\mathcal{H}_\infty$, $\mathcal{H}_\infty/\mathcal{H}_\infty$ as well as $\mathcal{H}_-/\mathcal{H}_\infty$ in a finite frequency range are only special cases of the $\mathcal{H}_i/\mathcal{H}_\infty$ optimisation. The unified solution was derived using the co-inner-outer factorisation technique, which is, in comparison with the LMI solutions, computational less involved. The solution to the optimal fault detection introduced in Sect. 8.2.2 is an extension of the unified solution to the LDTV systems and has been reported in [39], including Lemma 8.1.

Efficient and optimal fault detection in dynamic systems with model uncertainties is an open issue. Among numerous reported solutions, the LMI-aided scheme [235] is the mostly popular one. In Sect. 8.3, we have introduced some recent results achieved in the factorisation framework and reported in [39, 110, 111]. Lemma 8.2 is a known result given in [70]. In fact, the lack of systematic and efficient design schemes for optimal fault detection in uncertain dynamic systems is one of major motivations for developing alternative fault detection techniques. The next chapters will be devoted to this issue.

Chapter 9
A New Paradigm of Fault Diagnosis for Dynamic Systems

9.1 Motivation

The short re-visit of the observer-based fault detection technique and the state of the art review in the previous chapter disclose that there exist a number of open and challenging issues even after extensive studies over past decades. These include:

- efficient and optimal detection of parametric (multiplicative) faults, in particular in the presence of model uncertainties as well as in feedback control systems,
- definition and introduction of convincing and mathematically established metrics to measure the distance between nominal and faulty operations for fault detection purpose or the distance between two different classes of faults towards fault isolation. Although the widely accepted optimisation criteria like $\mathcal{H}_\infty/\mathcal{H}_\infty$, $\mathcal{H}_-/\mathcal{H}_\infty$ and $\mathcal{H}_2$ are applied for optimal design of observer-based residual generators, they only reflect the impact of faults and uncertainties on system dynamics under consideration, and do not serve as a metric for measuring the distance of different system operations,
- residual generation and threshold setting in an integrated manner. This is indeed a consequence of the above-mentioned problem with missing metric for distance measurement, and
- a uniform design of fault diagnosis systems, both in the model-based and data-driven fashions.

These facts have considerably motivated us to explore potential alternative strategies in recent years.

In the previous chapters, we have learnt that system dynamics can be modelled as a subspace in Hilbert space $\mathcal{H}$, like image and kernel subspaces. Hilbert space is a vector space endowed with an inner product $\langle \cdot, \cdot \rangle$. In this regard, Hilbert space is a complete metric space. Accordingly, a classification problem can be well defined in Hilbert space as: given a subspace $\mathcal{V} \in \mathcal{H}$, check if $\alpha \in \mathcal{H}$ belongs to $\mathcal{V}$. The solution is straightforward and consists of two steps: (i) projecting α onto $\mathcal{V}$, and

S. X. Ding, *Unified Framework of Control and Detection*,
https://doi.org/10.1007/978-3-662-72417-0_9

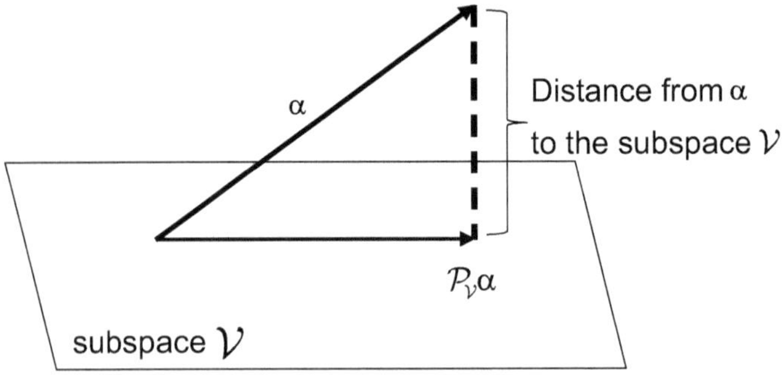

Fig. 9.1 Schematic description of projection-based classification ($\mathcal{P}_{\mathcal{V}}\alpha$ denotes the projection of α onto $\mathcal{V}$)

then (ii) calculating the distance between α and its projection using the induced norm. Figure 9.1 schematically illustrates this projection-based classification problem and its solution.

This conceptional formulation of classification problem in terms of projection and distance in a Hilbert subspace, and above all, the facts that

- all signals in dynamic systems can be represented by vectors in a Hilbert subspace, and
- system dynamics can be modelled as subspaces in Hilbert space as well,

are convincing arguments for us to study fault diagnosis issues using projection-based methods in the above described context. To this end, it is needed firstly to formulate fault detection and isolation issues, in particular those open ones like detecting parametric faults in uncertainty systems or fault detection in feedback control systems, as classification problems in Hilbert subspaces, and then to solve them using projection-based methods. This is the essential step towards establishing an alternative paradigm for fault diagnosis in dynamic systems. It should be furthermore demonstrated that, in this framework, fault detection and isolation issues can be systematically handled in a uniform manner. To be specific, residual generation and threshold setting will be addressed by determining (i) the distance of a (signal) vector to a (system) subspace and (ii) the distance between two (system) subspaces.

As a part of our study, relations between the proposed projection-based methods and the well-established observer-based framework will be investigated and comparison will be made when possible. It will be demonstrated that the proposed projection-based methods result in higher fault detectability. Moreover, a projection-based fault detection system will be proposed, which is comparable with the existing observer-based systems with regard to the needed online computation, but delivers better fault detection performance.

9.2 Mathematical Preliminaries

In this section, mathematical preliminaries related to orthogonal projection in Hilbert space and gap metric as a distance measurement between two closed Hilbert subspaces are introduced.

An orthogonal projection onto a (closed) subspace $\mathcal{V}$, denoted by $\mathcal{P}_{\mathcal{V}}$, in Hilbert space $\mathcal{H}$ endowed with an inner product, denoted by $\langle x, y\rangle, x, y \in \mathcal{H}$, is a linear operator satisfying

$$x, y \in \mathcal{H}, \mathcal{P}_{\mathcal{V}}^2 = \mathcal{P}_{\mathcal{V}}, \langle \mathcal{P}_{\mathcal{V}} x, y\rangle = \langle x, \mathcal{P}_{\mathcal{V}} y\rangle . \tag{9.1}$$

The first relation is called idempotent, while the second one is self-adjoint. In our subsequent work, the following definitions of inner products corresponding to different signal spaces are considered,

- Hilbert space $\mathcal{L}_2, \mathcal{L}_2 = \mathcal{L}_2(-\infty, 0] \oplus \mathcal{L}_2[0, \infty)$:

$$\langle x, y\rangle = \sum_{k=-\infty}^{\infty} x^T(k)y(k), x, y \in \mathcal{L}_2, \tag{9.2}$$

- Hilbert space $\mathcal{L}_2[0, \infty)$:

$$\langle x, y\rangle = \sum_{k=0}^{\infty} x^T(k)y(k), x, y \in \mathcal{L}_2, \tag{9.3}$$

- Hilbert space $\mathcal{L}_{2,[k_1,k_2]}, k_2 > k_1 \geq 0$:

$$\langle x, y\rangle = \sum_{k=k_1}^{k_2} x^T(k)y(k), x, y \in \mathcal{L}_{2,[k_1,k_2]}. \tag{9.4}$$

The following well-known properties of an orthogonal projection are of importance for our subsequent study:

- given $x \in \mathcal{H}$, the (orthogonal) projection of x onto $\mathcal{V}$, $\mathcal{P}_{\mathcal{V}} x$, satisfies

$$\langle \mathcal{P}_{\mathcal{V}} x, x - \mathcal{P}_{\mathcal{V}} x\rangle = 0. \tag{9.5}$$

  Often, $\mathcal{P}_{\mathcal{V}} x$ serves as an estimate for x, and in this context, $x - \mathcal{P}_{\mathcal{V}} x$ is understood as the estimation error. That means, the projection-induced estimate is orthogonal to the estimation error;
- relation (9.5) can also be equivalently expressed by

$$x = \mathcal{P}_{\mathcal{V}} x + \mathcal{P}_{\mathcal{V}^{\perp}} x, \tag{9.6}$$

  where $\mathcal{V}^{\perp}$ is the orthogonal complement of $\mathcal{V}$;

- Consequently, it holds

$$\|x\|_2^2 = \|\mathcal{P}_{\mathcal{V}} x\|^2 + \|\mathcal{P}_{\mathcal{V}^{\perp}} x\|^2, \tag{9.7}$$

the well-known Pythagorean equation,
- and given $y \in \mathcal{H}, \forall x \in \mathcal{V} \in \mathcal{H}$,

$$\langle y - x, y - x\rangle^{1/2} = \|y - x\|_2 \geq \|y - \mathcal{P}_{\mathcal{V}} y\|_2 . \tag{9.8}$$

Given a closed subspace $\mathcal{V} \in \mathcal{H}$ and a vector $y \in \mathcal{H}$, the distance between y and $\mathcal{V}$, $dist\,(y, \mathcal{V})$, is defined as

$$dist\,(y, \mathcal{V}) = \inf_{x \in \mathcal{V}} \|y - x\|_2 ,$$

which, following (9.8), can be computed as

$$dist\,(y, \mathcal{V}) = \|(\mathcal{I} - \mathcal{P}_{\mathcal{V}})\, y\|_2 = \|\mathcal{P}_{\mathcal{V}^{\perp}} y\|_2 .$$

Here, $\mathcal{I}$ is the unit operator.

In order to measure the distance between two (closed) subspaces in Hilbert space $\mathcal{H}$, the concept of gap metric is established. Given two closed subspaces $\mathcal{V}, \mathcal{U} \in \mathcal{H}$, the gap metric between them is defined by

$$\delta\,(\mathcal{V}, \mathcal{U}) = \max\left\{\boldsymbol{\delta}\,(\mathcal{V}, \mathcal{U}), \boldsymbol{\delta}\,(\mathcal{U}, \mathcal{V})\right\}, \tag{9.9}$$

$$\boldsymbol{\delta}\,(\mathcal{V}, \mathcal{U}) = \sup_{\substack{x \in \mathcal{V} \\ \|x\|_2 = 1}} dist\,(x, \mathcal{U}) = \|(\mathcal{I} - \mathcal{P}_{\mathcal{U}})\, \mathcal{P}_{\mathcal{V}}\| = \sup_{x \in \mathcal{V}} \inf_{y \in \mathcal{U}} \frac{\|x - y\|_2}{\|x\|_2}, \tag{9.10}$$

$$\boldsymbol{\delta}\,(\mathcal{U}, \mathcal{V}) = \sup_{\substack{y \in \mathcal{U} \\ \|y\|_2 = 1}} dist\,(y, \mathcal{V}) = \|(\mathcal{I} - \mathcal{P}_{\mathcal{V}})\, \mathcal{P}_{\mathcal{U}}\| = \sup_{y \in \mathcal{U}} \inf_{x \in \mathcal{V}} \frac{\|y - x\|_2}{\|y\|_2}. \tag{9.11}$$

Here, $\boldsymbol{\delta}\,(\mathcal{V}, \mathcal{U})$ and $\boldsymbol{\delta}\,(\mathcal{U}, \mathcal{V})$ are called directed gap, and $\|(\mathcal{I} - \mathcal{P}_{\mathcal{U}})\, \mathcal{P}_{\mathcal{V}}\|$ and $\|(\mathcal{I} - \mathcal{P}_{\mathcal{V}})\, \mathcal{P}_{\mathcal{U}}\|$ are norms of operators $(\mathcal{I} - \mathcal{P}_{\mathcal{U}})\, \mathcal{P}_{\mathcal{V}}$ as well as $(\mathcal{I} - \mathcal{P}_{\mathcal{V}})\, \mathcal{P}_{\mathcal{U}}$. The norm of an operator $\mathcal{P}: \mathcal{U} \to \mathcal{V}$, is defined by

$$\|\mathcal{P}\| = \sup\left\{\mathcal{P}x, x \in \mathcal{U}, \|x\|_2 \leq 1\right\}.$$

The following properties are well-known and useful for our subsequent investigation:

$$\begin{gathered} 0 \leq \delta\,(\mathcal{V}, \mathcal{U}) \leq 1, \\ \text{for } \delta\,(\mathcal{V}, \mathcal{U}) < 1, \boldsymbol{\delta}\,(\mathcal{V}, \mathcal{U}) = \boldsymbol{\delta}\,(\mathcal{U}, \mathcal{V}) = \delta\,(\mathcal{V}, \mathcal{U}), \\ \text{for } \delta\,(\mathcal{V}, \mathcal{U}) = 0, \mathcal{V} = \mathcal{U}, \text{ and for } \delta\,(\mathcal{V}, \mathcal{U}) = 1, \mathcal{V} \perp \mathcal{U}. \end{gathered}$$

Let $K \in \mathcal{RL}_\infty$ define a function from $\mathcal{L}_2$ to $\mathcal{L}_2$. The operator,

$$\mathcal{L}_K x := K x, x \in \mathcal{L}_2$$

is called with symbol K. It is obvious that

$$\|\mathcal{L}_K\| = \|K\|_\infty .$$

9.3 Problem Formulation

With the aid of the results on

- system SIR/SKR and image/kernel subspaces introduced in Chap. 2,
- orthogonal projection operator as well as the associated properties, in particular,
- the distance and gap metric definitions and their computations,

we are now in a position to concretise fault diagnosis problems in terms of projection-based classification, as sketched in Fig. 9.1. They are:

- given a system measurement vector $\begin{bmatrix} u \\ y \end{bmatrix}$, find its orthogonal projection onto image subspaces $\mathcal{I}_G^{\mathcal{L}_2}$ and $\mathcal{I}_G^{\mathcal{H}_2}$, respectively,
- determine the distance $dist\left(\begin{bmatrix} u \\ y \end{bmatrix}, \mathcal{I}_G^{\mathcal{L}_2}\right)$ and $dist\left(\begin{bmatrix} u \\ y \end{bmatrix}, \mathcal{I}_G^{\mathcal{H}_2}\right)$, which build the so-called projection-based residual vector,
- find threshold setting schemes based on a gap metric. In our work, the threshold will be set schematically to guarantee that model uncertainties do not cause false alarms.
- In addition, online realisation issues towards constructing a fault detection system based on $dist\left(\begin{bmatrix} u \\ y \end{bmatrix}, \mathcal{I}_G\right)$ will be discussed.
- The aforementioned fault detection implies a one-class classification. That is, the fault detection is addressed on the assumption that only knowledge of fault-free operations of the process under consideration is available for the fault detection purpose. Fault detection and isolation issues will be, as well, formulated as binary and multi-class classification problems and handled in the projection-based framework.

Remark 9.1 As described, the definition of the adopted inner product is the basis to define an orthogonal projection. In this chapter, the inner products defined by (9.2) and (9.3) are considered.

In our work, attention will be paid to comparison study with the well-established observer-based fault diagnosis methods when possible. In this context, we will illustrate that the projection-based methods are comparable with an observer-based fault detection schemes with regard to online computation but deliver better detection performance.

9.4 A Fault Detection Method Based on Orthogonal Projection onto $\mathcal{I}_G^{\mathcal{L}_2}$

We begin with applying the orthogonal projection onto $\mathcal{I}_G^{\mathcal{L}_2}$ technique to achieving fault detection. Our major focus is on optimally detecting faults in dynamic systems with model uncertainty.

9.4.1 *Orthogonal Projection-Based Residual Generation*

It is a known result that the orthogonal projection onto the image subspace $\mathcal{I}_G^{\mathcal{L}_2}$ is given by

$$\mathcal{P}_{\mathcal{I}_G^{\mathcal{L}_2}} = \mathcal{L}_{I_{G_0}} \mathcal{L}_{I_{G_0}}^*, \tag{9.12}$$

where $\mathcal{L}_{I_{G_0}} : \mathcal{L}_2 \to \mathcal{L}_2$ is Laurent operator with symbol I_{G_0},

$$\mathcal{L}_{I_{G_0}} v = I_{G_0} v, \, v \in \mathcal{L}_2,$$

with the normalised SIR I_{G_0}, $\mathcal{L}_{I_{G_0}}^* : \mathcal{L}_2 \to \mathcal{L}_2$ as the adjoint of $\mathcal{L}_{I_{G_0}}$ is given by

$$\mathcal{L}_{I_{G_0}}^* \begin{bmatrix} u \\ y \end{bmatrix} = I_{G_0}^{\sim} \begin{bmatrix} u \\ y \end{bmatrix} = \mathcal{L}_{I_{G_0}^{\sim}} \begin{bmatrix} u \\ y \end{bmatrix}, \begin{bmatrix} u \\ y \end{bmatrix} \in \mathcal{L}_2.$$

Recall that

$$I_{G_0}^{\sim} I_{G_0} = \begin{bmatrix} M_0 \\ N_0 \end{bmatrix}^{\sim} \begin{bmatrix} M_0 \\ N_0 \end{bmatrix} = I,$$

which yields

$$\forall \begin{bmatrix} u \\ y \end{bmatrix} \in \mathcal{L}_2, \mathcal{P}_{\mathcal{I}_G^{\mathcal{L}_2}} \mathcal{P}_{\mathcal{I}_G^{\mathcal{L}_2}} \begin{bmatrix} u \\ y \end{bmatrix} = I_{G_0} I_{G_0}^{\sim} \begin{bmatrix} u \\ y \end{bmatrix}$$
$$\Longrightarrow \mathcal{P}_{\mathcal{I}_G^{\mathcal{L}_2}} \mathcal{P}_{\mathcal{I}_G^{\mathcal{L}_2}} = \mathcal{P}_{\mathcal{I}_G^{\mathcal{L}_2}}.$$

Moreover,

$$\forall \begin{bmatrix} u_1 \\ y_1 \end{bmatrix}, \begin{bmatrix} u_2 \\ y_2 \end{bmatrix} \in \mathcal{L}_2, \left\langle \mathcal{P}_{\mathcal{I}_G^{\mathcal{L}_2}} \begin{bmatrix} u_1 \\ y_1 \end{bmatrix}, \begin{bmatrix} u_2 \\ y_2 \end{bmatrix} \right\rangle = \left\langle \begin{bmatrix} u_1 \\ y_1 \end{bmatrix}, \mathcal{P}_{\mathcal{I}_G^{\mathcal{L}_2}} \begin{bmatrix} u_2 \\ y_2 \end{bmatrix} \right\rangle,$$

where the adopted inner product is given by (9.2). Hence, $\mathcal{P}_{\mathcal{I}_G^{\mathcal{L}_2}}$ is an operator of an orthogonal projection. The projection

$$\mathcal{P}_{\mathcal{I}_G^{\mathcal{L}_2}} \begin{bmatrix} u \\ y \end{bmatrix} =: \begin{bmatrix} \hat{u} \\ \hat{y} \end{bmatrix}$$

can be viewed as an estimate of process data $\begin{bmatrix} u \\ y \end{bmatrix}$ by means of the nominal model (i.e. fault- and uncertainty-free). Consequently, the difference between $\begin{bmatrix} u \\ y \end{bmatrix}$ and $\begin{bmatrix} \hat{u} \\ \hat{y} \end{bmatrix}$ builds the residual vector,

$$r_{\mathcal{I}_G^{\mathcal{L}_2}} = \begin{bmatrix} u \\ y \end{bmatrix} - \begin{bmatrix} \hat{u} \\ \hat{y} \end{bmatrix} = \left(\mathcal{I} - \mathcal{P}_{\mathcal{I}_G^{\mathcal{L}_2}}\right) \begin{bmatrix} u \\ y \end{bmatrix} = \left(\mathcal{I} - \mathcal{L}_{I_{G_0}} \mathcal{L}_{I_{\tilde{G}_0}}^{\sim}\right) \begin{bmatrix} u \\ y \end{bmatrix}. \tag{9.13}$$

Residual vector $r_{\mathcal{I}_G^{\mathcal{L}_2}}$ indicates how far the measurement (data) vector $\begin{bmatrix} u \\ y \end{bmatrix}$ deviates from the nominal system dynamics expressed by the system SIR. Thus, its $\mathcal{L}_2$-norm is the distance between the data vector and the image subspace. In this regard, we introduce the following definition.

Definition 9.1 Given the orthogonal projection onto the image subspace $\mathcal{I}_G^{\mathcal{L}_2}$ defined by (9.12), the dynamic system (9.13) is called projection-based residual generator. The $\mathcal{L}_2$-norm of its output $r_{\mathcal{I}_G^{\mathcal{L}_2}}$,

$$\left\| r_{\mathcal{I}_G^{\mathcal{L}_2}} \right\|_2 = dist\left(\begin{bmatrix} u \\ y \end{bmatrix}, \mathcal{I}_G^{\mathcal{L}_2} \right) =: J \tag{9.14}$$

is the distance from $\begin{bmatrix} u \\ y \end{bmatrix}$ to $\mathcal{I}_G^{\mathcal{L}_2}$, and defined as the residual evaluation function J.

It is of interest to note that $\mathcal{I} - \mathcal{P}_{\mathcal{I}_G^{\mathcal{L}_2}}$ defines the orthogonal projection onto the orthogonal complement subspace of $\mathcal{I}_G^{\mathcal{L}_2}$. That is,

$$\begin{aligned} \mathcal{I}_G^{\mathcal{L}_2,\perp} &= \left\{ r_{\mathcal{I}_G^{\mathcal{L}_2}} : r_{\mathcal{I}_G^{\mathcal{L}_2}} = \left(\mathcal{I} - \mathcal{L}_{I_{G_0}} \mathcal{L}_{I_{\tilde{G}_0}}^{\sim}\right) \begin{bmatrix} u \\ y \end{bmatrix}, \begin{bmatrix} u \\ y \end{bmatrix} \in \mathcal{L}_2 \right\}, \\ \mathcal{L}_2 &= \mathcal{I}_G^{\mathcal{L}_2} \oplus \mathcal{I}_G^{\mathcal{L}_2,\perp}. \end{aligned}$$

In other words, the residual subspace is in fact $\mathcal{I}_G^{\mathcal{L}_2,\perp}$. Under consideration of our study purpose, we prefer the term "residual" over projection onto the orthogonal complement of $\mathcal{I}_G^{\mathcal{L}_2}$. In this context, we introduce the concept of projection-based residual subspace.

Definition 9.2 Given the orthogonal projection onto the image subspace $\mathcal{I}_G^{\mathcal{L}_2}$ defined by (9.12), the projection-based residual generator (9.13), we call the subspace

$$\mathcal{R}_{\mathcal{I}_G^{\mathcal{L}_2}} = \left\{ r_{\mathcal{I}_G^{\mathcal{L}_2}} : r_{\mathcal{I}_G^{\mathcal{L}_2}} = \left(\mathcal{I} - \mathcal{L}_{I_{G_0}} \mathcal{L}_{I_{G_0}^{\sim}} \right) \begin{bmatrix} u \\ y \end{bmatrix}, \begin{bmatrix} u \\ y \end{bmatrix} \in \mathcal{L}_2 \right\} \tag{9.15}$$

projection-based residual subspace.

Remembering the natural relation between SKR and the residual generation as well as the known fact that $\mathcal{K}_G = \mathcal{I}_G$, it is of considerable interest to investigate relations among the operator $\mathcal{L}_{I_G}$, the orthogonal projection $\mathcal{P}_{\mathcal{I}_G}$ and the system SKR.

Lemma 9.1 *Let $\mathcal{L}_{K_{G_0}}$ and $\mathcal{L}_{K_{G_0}^{\sim}}$ be Laurent operators with symbol K_{G_0} and $K_{G_0^{\sim}}$, respectively, where K_{G_0} is the normalised SKR. It holds*

$$\mathcal{L}_{K_{G_0}^{\sim}} \mathcal{L}_{K_{G_0}} + \mathcal{L}_{I_{G_0}} \mathcal{L}_{I_{G_0}^{\sim}} = \mathcal{I}. \tag{9.16}$$

Proof. Identity (9.16) follows from

$$\begin{bmatrix} M_0^{\sim} & N_0^{\sim} \\ -\hat{N}_0 & \hat{M}_0 \end{bmatrix} \begin{bmatrix} M_0 & -\hat{N}_0^{\sim} \\ N_0 & \hat{M}_0^{\sim} \end{bmatrix} = \begin{bmatrix} M_0 & -\hat{N}_0^{\sim} \\ N_0 & \hat{M}_0^{\sim} \end{bmatrix} \begin{bmatrix} M_0^{\sim} & N_0^{\sim} \\ -\hat{N}_0 & \hat{M}_0 \end{bmatrix} = \begin{bmatrix} I & 0 \\ 0 & I \end{bmatrix},$$

since

$$I_{G_0} = \begin{bmatrix} M_0 \\ N_0 \end{bmatrix}, K_{G_0} = \begin{bmatrix} -\hat{N}_0 & \hat{M}_0 \end{bmatrix}$$

are normalised SIR and SKR, respectively.

An immediate result of this lemma is that $\mathcal{P}_{\mathcal{K}_G^{\mathcal{L}_2}} : \mathcal{L}_2 \to \mathcal{L}_2$,

$$\mathcal{P}_{\mathcal{R}_G^{\mathcal{L}_2}} := \mathcal{L}_{K_{G_0}^{\sim}} \mathcal{L}_{K_{G_0}} \tag{9.17}$$

is an operator of the orthogonal projection onto residual subspace $\mathcal{R}_{\mathcal{I}_G^{\mathcal{L}_2}}$.

9.4.2 Real-Time Implementation Algorithm of the Projection-Based Residual Generator

In this subsection, online computation issues of $J = \left\| r_{\mathcal{I}_G^{\mathcal{L}_2}} \right\|_2$ are addressed. It is clear that computation of $r_{\mathcal{I}_G^{\mathcal{L}_2}}$ in its original form,

$$r_{\mathcal{I}_G^{\mathcal{L}_2}} = \left(\mathcal{I} - \mathcal{L}_{I_{G_0}} \mathcal{L}_{I_{G_0}}^* \right) \begin{bmatrix} u \\ y \end{bmatrix},$$

demands considerable computational power, for instance, in comparison with an observer-based residual generator, due to

- its higher dimension,

$$r_{\mathcal{I}_G^{\mathcal{L}_2}} \in \mathbb{R}^{m+p},$$

- computation of $K_{G_0}^{\sim}$.

The following discussion demonstrates that this problem can be well solved thanks to the relation given in Lemma 9.1. It follows from (9.16) that

$$\left\| r_{\mathcal{I}_G^{\mathcal{L}_2}} \right\|_2 = \left\| \mathcal{L}_{K_{G_0}^{\sim}} \mathcal{L}_{K_{G_0}} \begin{bmatrix} u \\ y \end{bmatrix} \right\|_2 = \left\| K_{G_0} \begin{bmatrix} u \\ y \end{bmatrix} \right\|_2, \tag{9.18}$$

since $\mathcal{L}_{K_{G_0}^{\sim}} \mathcal{L}_{K_{G_0}}$ is an orthogonal projection operator and K_{G_0} is a normalised SKR. Recall that

$$r_y = K_{G_0} \begin{bmatrix} u \\ y \end{bmatrix} = \begin{bmatrix} -\hat{N}_0 & \hat{M}_0 \end{bmatrix} \begin{bmatrix} u \\ y \end{bmatrix}$$

is an observer-based residual generator. Here, the setting for the observer gain and post-filter, (L, W), is as given in Theorem 2.1 for the normalised SKR. As a result, computation of K_{G_0} instead of $\mathcal{P}_{\mathcal{K}_G^{\mathcal{L}_2}}$ is sufficient for our purpose of fault detection. In other words, the projection-based residual generation can be equivalently realised by means of an observer-based residual generator.

9.4.3 *Threshold Setting Schemes*

We now consider threshold setting issues for systems with model uncertainty described by

$$G = NM^{-1} = (N_0 + \Delta_N)(M_0 + \Delta_M)^{-1} = \left(\hat{M}_0 + \Delta_{\hat{M}}\right)^{-1} \left(\hat{N}_0 + \Delta_{\hat{N}}\right), \tag{9.19}$$

$$I_G = \begin{bmatrix} M \\ N \end{bmatrix} = I_{G_0} + \Delta_{I_G}, I_{G_0} = \begin{bmatrix} M_0 \\ N_0 \end{bmatrix}, \Delta_{I_G} = \begin{bmatrix} \Delta_M \\ \Delta_N \end{bmatrix} \in \mathcal{RH}_\infty, \tag{9.20}$$

$$K_G = \begin{bmatrix} -\hat{N} & \hat{M} \end{bmatrix} = \begin{bmatrix} -\left(\hat{N}_0 + \Delta_{\hat{N}}\right) & \hat{M}_0 + \Delta_{\hat{M}} \end{bmatrix} = K_{G_0} + \Delta_{K_G}, \tag{9.21}$$

$$K_{G_0} = \begin{bmatrix} -\hat{N}_0 & \hat{M}_0 \end{bmatrix}, \Delta_{K_G} = \begin{bmatrix} -\Delta_{\hat{N}} & \Delta_{\hat{M}} \end{bmatrix} \in \mathcal{RH}_\infty$$

with normalised SIR I_{G_0} and SKR K_{G_0}.

On the assumption

$$\sup \left\| \Delta_{I_G} \right\|_\infty = \delta_{\Delta_{I_G}} < 1, \sup \left\| \Delta_{K_G} \right\|_\infty = \delta_{\Delta_{K_G}} < 1, \tag{9.22}$$

the threshold is set to prevent false alarms, which implies that, during fault-free operations,

$$J = \left\| r_{\mathcal{I}_G^{\mathcal{L}_2}} \right\|_2 \leq J_{th} \tag{9.23}$$

should hold.

Recall that for some Δ_{I_G},

$$\mathcal{I}_G^{\mathcal{L}_2} = \left\{ \begin{bmatrix} u \\ y \end{bmatrix} : \begin{bmatrix} u \\ y \end{bmatrix} = \begin{bmatrix} M \\ N \end{bmatrix} v, v \in \mathcal{L}_2 \right\}$$

defines an image subspace that is obviously different from $\mathcal{I}_{G_0}$,

$$\mathcal{I}_{G_0}^{\mathcal{L}_2} = \left\{ \begin{bmatrix} u \\ y \end{bmatrix} : \begin{bmatrix} u \\ y \end{bmatrix} = \begin{bmatrix} M_0 \\ N_0 \end{bmatrix} v, v \in \mathcal{L}_2 \right\}.$$

To measure the distance between the subspaces $\mathcal{I}_G^{\mathcal{L}_2}$ and $\mathcal{I}_{G_0}^{\mathcal{L}_2}$, the concept of gap metric defined in (9.9)–(9.10) is adopted. Against this background, the following definition is introduced.

Definition 9.3 Given system model (9.19) and the projection-based residual generator (9.13), the threshold J_{th} is defined by $\forall \begin{bmatrix} u \\ y \end{bmatrix} \in \mathcal{I}_G^{\mathcal{L}_2}$,

$$J_{th} = \sup_{\|\Delta_{I_G}\|_\infty \leq \delta_{\Delta_{I_G}}} \left\| r_{\mathcal{I}_G^{\mathcal{L}_2}} \right\|_2 = \sup_{\|\Delta_{I_G}\|_\infty \leq \delta_{\Delta_{I_G}}} \delta_{\mathcal{I}} \left(\mathcal{I}_G^{\mathcal{L}_2}, \mathcal{I}_{G_0}^{\mathcal{L}_2} \right) \left\| \begin{bmatrix} u \\ y \end{bmatrix} \right\|_2. \tag{9.24}$$

Next, we will study the threshold setting issues based on (9.24). To this end, the following definition and computation algorithms related to gap metric are introduced.

Definition 9.4 Let

$$\mathcal{I}_{G_i}^{\mathcal{L}_2} = \left\{ \begin{bmatrix} u_i \\ y_i \end{bmatrix} : \begin{bmatrix} u_i \\ y_i \end{bmatrix} = \begin{bmatrix} M_i \\ N_i \end{bmatrix} v, v \in \mathcal{L}_2 \right\}, i = 1, 2, \tag{9.25}$$

be two image subspaces. The directed gap $\vec{\delta}_{\mathcal{I}} \left(\mathcal{I}_{G_1}^{\mathcal{L}_2}, \mathcal{I}_{G_2}^{\mathcal{L}_2} \right)$ and gap metric $\delta_{\mathcal{I}} \left(\mathcal{I}_{G_1}^{\mathcal{L}_2}, \mathcal{I}_{G_2}^{\mathcal{L}_2} \right)$ are respectively defined as

$$\vec{\delta}_{\mathcal{I}} \left(\mathcal{I}_{G_1}^{\mathcal{L}_2}, \mathcal{I}_{G_2}^{\mathcal{L}_2} \right) = \sup_{\begin{bmatrix} u_1 \\ y_1 \end{bmatrix} \in \mathcal{I}_{G_1}^{\mathcal{L}_2}} \inf_{\begin{bmatrix} u_2 \\ y_2 \end{bmatrix} \in \mathcal{I}_{G_2}^{\mathcal{L}_2}} \frac{\left\| \begin{bmatrix} u_1 \\ y_1 \end{bmatrix} - \begin{bmatrix} u_2 \\ y_2 \end{bmatrix} \right\|_2}{\left\| \begin{bmatrix} u_1 \\ y_1 \end{bmatrix} \right\|_2} \tag{9.26}$$

$$= \sup_{\begin{bmatrix} u_1 \\ y_1 \end{bmatrix} \in \mathcal{I}_{G_1}^{\mathcal{L}_2}, \left\| \begin{bmatrix} u_1 \\ y_1 \end{bmatrix} \right\|_2 = 1} dist \left(\begin{bmatrix} u_1 \\ y_1 \end{bmatrix}, \mathcal{I}_{G_2}^{\mathcal{L}_2} \right), \tag{9.27}$$

$$\delta_{\mathcal{I}}\left(\mathcal{I}_{G_1}^{\mathcal{L}_2},\mathcal{I}_{G_2}^{\mathcal{L}_2}\right) = \max\left\{\vec{\delta}_{\mathcal{I}}\left(\mathcal{I}_{G_1}^{\mathcal{L}_2},\mathcal{I}_{G_2}^{\mathcal{L}_2}\right), \vec{\delta}_{\mathcal{I}}\left(\mathcal{I}_{G_2}^{\mathcal{L}_2},\mathcal{I}_{G_1}^{\mathcal{L}_2}\right)\right\} \tag{9.28}$$

$$= \left\|\mathcal{P}_{\mathcal{I}_{G_1}^{\mathcal{L}_2}} - \mathcal{P}_{\mathcal{I}_{G_2}^{\mathcal{L}_2}}\right\|, \tag{9.29}$$

where $\mathcal{P}_{\mathcal{I}_{G_i}^{\mathcal{L}_2}}$, $i = 1, 2$, is the operator of the orthogonal projection onto $\mathcal{I}_{G_i}^{\mathcal{L}_2}$, and $\|\cdot\|$ is the operator norm.

Remark 9.2 In the above definition, different definition forms of gap metric are presented. Although they are equivalent, they can be used for different application purposes. In particular, from (9.29) the computation algorithm of the gap metric can be derived in a straightforward manner.

The computation of $\delta_{\mathcal{I}}\left(\mathcal{I}_{G_i}^{\mathcal{L}_2},\mathcal{I}_{G_j}^{\mathcal{L}_2}\right)$ has been intensively investigated, and one of the key results is summarised in the following theorem.

Theorem 9.1 *Given $\mathcal{I}_{G_i}^{\mathcal{L}_2}$, $i = 1, 2$, as defined in (9.25), and the normalised SIR and SKR of G_i,*

$$I_{G_{i,0}} = \begin{bmatrix} M_{i,0} \\ N_{i,0} \end{bmatrix}, K_{G_{i,0}} = \begin{bmatrix} -\hat{N}_{i,0} & \hat{M}_{i,0} \end{bmatrix},$$

then

$$\delta_{\mathcal{I}}\left(\mathcal{I}_{G_i}^{\mathcal{L}_2},\mathcal{I}_{G_j}^{\mathcal{L}_2}\right) = \left\|K_{G_{i,0}} I_{G_{j,0}}\right\|_\infty = \left\|K_{G_{j,0}} I_{G_{i,0}}\right\|_\infty, \tag{9.30}$$

$$\left\|K_{G_{i,0}} I_{G_{j,0}}\right\|_\infty = \left\|\hat{M}_{i,0} N_{j,0} - \hat{N}_{i,0} M_{j,0}\right\|_\infty, i, j = 1, 2, i \neq j.$$

The result (9.30) and its proof are well-known in the literature, see for instance Vinnicombe's book cited at the end of this chapter. Below, we outline the key steps of the proof, which are helpful to gain a deeper insight into gap metric computation.

It follows from (9.29) in Definition 9.4 that

$$\delta_{\mathcal{I}}\left(\mathcal{I}_{G_i}^{\mathcal{L}_2},\mathcal{I}_{G_j}^{\mathcal{L}_2}\right) = \left\|\mathcal{P}_{\mathcal{I}_{G_i}^{\mathcal{L}_2}} - \mathcal{P}_{\mathcal{I}_{G_j}^{\mathcal{L}_2}}\right\| = \left\|I_{G_{i,0}} I_{G_{i,0}}^{\sim} - I_{G_{j,0}} I_{G_{j,0}}^{\sim}\right\|_\infty.$$

Note that

$$\begin{bmatrix} K_{G_{i,0}} \\ I_{G_{i,0}}^{\sim} \end{bmatrix} \begin{bmatrix} K_{G_{i,0}} \\ I_{G_{i,0}}^{\sim} \end{bmatrix}^{\sim} = \begin{bmatrix} K_{G_{i,0}} \\ I_{G_{i,0}}^{\sim} \end{bmatrix}^{\sim} \begin{bmatrix} K_{G_{i,0}} \\ I_{G_{i,0}}^{\sim} \end{bmatrix} = \begin{bmatrix} I & 0 \\ 0 & I \end{bmatrix},$$

$$\begin{bmatrix} K_{G_{i,0}} \\ I_{G_{i,0}}^{\sim} \end{bmatrix} I_{G_{i,0}} I_{G_{i,0}}^{\sim} = \begin{bmatrix} 0 \\ I_{G_{i,0}}^{\sim} \end{bmatrix}, I_{G_{i,0}} I_{G_{i,0}}^{\sim} \begin{bmatrix} I_{G_{i,0}}^{\sim} \\ K_{G_{i,0}} \end{bmatrix}^{\sim} = \begin{bmatrix} I_{G_{i,0}} & 0 \end{bmatrix}.$$

It turns out

$$\begin{aligned}\left\| I_{G_{i,0}} I_{G_{i,0}}^{\sim} - I_{G_{j,0}} I_{G_{j,0}}^{\sim} \right\|_{\infty} &= \left\| \begin{bmatrix} K_{G_{i,0}} \\ I_{G_{i,0}}^{\sim} \end{bmatrix} \left(I_{G_{i,0}} I_{G_{i,0}}^{\sim} - I_{G_{j,0}} I_{G_{j,0}}^{\sim} \right) \begin{bmatrix} I_{G_{j,0}}^{\sim} \\ K_{G_{j,0}} \end{bmatrix}^{\sim} \right\|_{\infty} \\ &= \left\| \begin{bmatrix} -K_{G_{i,0}} I_{G_{j,0}} & 0 \\ I_{G_{i,0}}^{\sim} I_{G_{j,0}} - I_{G_{i,0}}^{\sim} I_{G_{j,0}} & I_{G_{i,0}}^{\sim} K_{G_{j,0}}^{\sim} \end{bmatrix} \right\|_{\infty} \\ &= \max \left\{ \left\| K_{G_{i,0}} I_{G_{j,0}} \right\|_{\infty}, \left\| K_{G_{j,0}} I_{G_{i,0}} \right\|_{\infty} \right\}.\end{aligned}$$

In the final step, it is to show

$$\left\| K_{G_{i,0}} I_{G_{j,0}} \right\|_{\infty} = \left\| K_{G_{j,0}} I_{G_{i,0}} \right\|_{\infty}.$$

For this purpose, the following lemma is applied.

Lemma 9.2 *Given*

$$Z = \begin{bmatrix} X \\ Y \end{bmatrix} \in \mathbb{C}^{q \times p}, X \in \mathbb{C}^{p \times p}, \bar{Z} = \begin{bmatrix} \bar{X} & \bar{Y} \end{bmatrix} \in \mathbb{C}^{p \times q}, \bar{X} \in \mathbb{C}^{p \times p}$$

satisfying

$$Z^{\sim} Z = I, \bar{Z} \bar{Z}^{\sim} = I,$$

then it holds

$$\bar{\sigma}^2 (Y) = 1 - \underline{\sigma}^2 (X), \bar{\sigma}^2 \left(\bar{Y} \right) = 1 - \underline{\sigma}^2 \left(\bar{X} \right).$$

Now, let

$$Z_1 = \begin{bmatrix} I_{G_{i,0}}^{\sim} I_{G_{j,0}} \\ K_{G_{i,0}} I_{G_{j,0}} \end{bmatrix}, Z_2 = \begin{bmatrix} I_{G_{i,0}}^{\sim} I_{G_{j,0}} & I_{G_{i,0}}^{\sim} K_{G_{j,0}}^{\sim} \end{bmatrix}.$$

Because

$$\begin{aligned} Z_1^{\sim} Z_1 &= I_{G_{j,0}}^{\sim} I_{G_{i,0}} I_{G_{i,0}}^{\sim} I_{G_{j,0}} + I_{G_{j,0}}^{\sim} K_{G_{i,0}}^{\sim} K_{G_{i,0}} I_{G_{j,0}} = I_{G_{j,0}}^{\sim} I_{G_{j,0}} = I, \\ Z_2 Z_2^{\sim} &= I_{G_{i,0}}^{\sim} I_{G_{j,0}} I_{G_{j,0}}^{\sim} I_{G_{i,0}} + I_{G_{i,0}}^{\sim} K_{G_{j,0}}^{\sim} K_{G_{j,0}} I_{G_{i,0}} = I_{G_{i,0}}^{\sim} I_{G_{i,0}} = I, \end{aligned}$$

it holds, by Lemma 9.2, that

$$\begin{aligned} &\bar{\sigma}^2 \left(K_{G_{i,0}} I_{G_{j,0}} \right) = 1 - \underline{\sigma}^2 \left(I_{G_{i,0}}^{\sim} I_{G_{j,0}} \right) = \bar{\sigma}^2 \left(I_{G_{i,0}}^{\sim} K_{G_{j,0}}^{\sim} \right) \Longrightarrow \\ &\left\| K_{G_{i,0}} I_{G_{j,0}} \right\|_{\infty} = \left\| I_{G_{i,0}}^{\sim} K_{G_{j,0}}^{\sim} \right\|_{\infty} = \left\| K_{G_{j,0}} I_{G_{i,0}} \right\|_{\infty}. \end{aligned}$$

It is of interest to mention that $\left\|K_{G_{i,0}} I_{G_{j,0}}\right\|_\infty$ and $\left\|K_{G_{j,0}} I_{G_{i,0}}\right\|_\infty$ are also the solution of the following MMP

$$\begin{aligned}
\left\|K_{G_{j,0}} I_{G_{i,0}}\right\|_\infty &= \min_{Q \in \mathcal{L}_\infty} \left\| \begin{bmatrix} M_{i,0} \\ N_{i,0} \end{bmatrix} - \begin{bmatrix} M_{j,0} \\ N_{j,0} \end{bmatrix} Q \right\|_\infty = \\
\left\|K_{G_{i,0}} I_{G_{j,0}}\right\|_\infty &= \min_{Q \in \mathcal{L}_\infty} \left\| \begin{bmatrix} M_{j,0} \\ N_{j,0} \end{bmatrix} - \begin{bmatrix} M_{i,0} \\ N_{i,0} \end{bmatrix} Q \right\|_\infty .
\end{aligned} \tag{9.31}$$

This conclusion becomes obvious, observing the facts below

$$\begin{gathered}
\begin{bmatrix} K_{G_{i,0}} \\ I_{G_{i,0}}^{\sim} \end{bmatrix} \left(\begin{bmatrix} M_{j,0} \\ N_{j,0} \end{bmatrix} - \begin{bmatrix} M_{i,0} \\ N_{i,0} \end{bmatrix} Q \right) = \begin{bmatrix} K_{G_{i,0}} I_{G_{j,0}} \\ I_{G_{i,0}}^{\sim} I_{G_{j,0}} \end{bmatrix} - \begin{bmatrix} 0 \\ I \end{bmatrix} Q, \\
\left\| \begin{bmatrix} K_{G_{i,0}} \\ I_{G_{i,0}}^{\sim} \end{bmatrix} \left(\begin{bmatrix} M_{j,0} \\ N_{j,0} \end{bmatrix} - \begin{bmatrix} M_{i,0} \\ N_{i,0} \end{bmatrix} Q \right) \right\|_\infty = \left\| \begin{bmatrix} M_{j,0} \\ N_{j,0} \end{bmatrix} - \begin{bmatrix} M_{i,0} \\ N_{i,0} \end{bmatrix} Q \right\|_\infty, \\
\therefore \left\| \begin{bmatrix} M_{j,0} \\ N_{j,0} \end{bmatrix} - \begin{bmatrix} M_{i,0} \\ N_{i,0} \end{bmatrix} Q \right\|_\infty = \left\| \begin{bmatrix} K_{G_{i,0}} I_{G_{j,0}} \\ I_{G_{i,0}}^{\sim} I_{G_{j,0}} \end{bmatrix} - \begin{bmatrix} 0 \\ I \end{bmatrix} Q \right\|_\infty \Longrightarrow \\
Q^* = \arg \min_{Q \in \mathcal{L}_\infty} \left\| \begin{bmatrix} M_{j,0} \\ N_{j,0} \end{bmatrix} - \begin{bmatrix} M_{i,0} \\ N_{i,0} \end{bmatrix} Q \right\|_\infty = I_{G_{i,0}}^{\sim} I_{G_{j,0}} \Longrightarrow \\
\left\|K_{G_{i,0}} I_{G_{j,0}}\right\|_\infty = \min_{Q \in \mathcal{L}_\infty} \left\| \begin{bmatrix} M_{j,0} \\ N_{j,0} \end{bmatrix} - \begin{bmatrix} M_{i,0} \\ N_{i,0} \end{bmatrix} Q \right\|_\infty, i, j = 1, 2, i \neq j.
\end{gathered} \tag{9.32}$$

Now, we are in a position to solve the threshold setting problem defined by (9.24). Applying Theorem 9.1 to (9.24) leads to

$$\begin{aligned}
\forall \begin{bmatrix} u \\ y \end{bmatrix} \in \mathcal{I}_G^{\mathcal{L}_2}, &\sup_{\|\Delta_{I_G}\|_\infty \le \delta_{\Delta_{I_G}}} \delta_{\mathcal{I}} \left(\mathcal{I}_G^{\mathcal{L}_2}, \mathcal{I}_{G_0}^{\mathcal{L}_2} \right) \left\| \begin{bmatrix} u \\ y \end{bmatrix} \right\|_2 \\
&= \sup_{\|\Delta_{I_G}\|_\infty \le \delta_{\Delta_{I_G}}} \left\|K_{G_0} I_G\right\|_\infty \left\| \begin{bmatrix} u \\ y \end{bmatrix} \right\|_2 .
\end{aligned}$$

Since

$$I_G = I_{G_0} + \Delta_{I_G}, \; K_{G_0} I_{G_0} = 0,$$

and K_{G_0} is a normalised SKR, it turns out

$$\begin{aligned}
&\left\|K_{G_0} I_G\right\|_\infty = \left\|K_{G_0} \Delta_{I_G}\right\|_\infty \Longrightarrow \\
\sup_{\|\Delta_{I_G}\|_\infty \le \delta_{\Delta_{I_G}}} \left\|K_{G_0} I_G\right\|_\infty \left\| \begin{bmatrix} u \\ y \end{bmatrix} \right\|_2 &\le \sup_{\|\Delta_{I_G}\|_\infty \le \delta_{\Delta_{I_G}}} \left\|\Delta_{I_G}\right\|_\infty \left\| \begin{bmatrix} u \\ y \end{bmatrix} \right\|_2 \\
&\le \delta_{\Delta_{I_G}} \left\| \begin{bmatrix} u \\ y \end{bmatrix} \right\|_2 .
\end{aligned} \tag{9.33}$$

On the other hand, as a result of Theorem 9.1, it holds as well

$$\begin{aligned}
\forall \begin{bmatrix} u \\ y \end{bmatrix} \in \mathcal{I}_G^{\mathcal{L}_2}, & \sup_{\|\Delta_{K_G}\|_\infty \leq \delta_{\Delta_{K_G}}} \delta_{\mathcal{I}}\left(\mathcal{I}_G^{\mathcal{L}_2}, \mathcal{I}_{G_0}^{\mathcal{L}_2}\right) \left\| \begin{bmatrix} u \\ y \end{bmatrix} \right\|_2 \\
= & \sup_{\|\Delta_{K_G}\|_\infty \leq \delta_{\Delta_{K_G}}} \left\| K_G I_{G_0} \right\|_\infty \left\| \begin{bmatrix} u \\ y \end{bmatrix} \right\|_2 \\
\leq & \sup_{\|\Delta_{K_G}\|_\infty \leq \delta_{\Delta_{K_G}}} \left\| \Delta_{K_G} \right\|_\infty \left\| \begin{bmatrix} u \\ y \end{bmatrix} \right\|_2 \leq \delta_{\Delta_{K_G}} \left\| \begin{bmatrix} u \\ y \end{bmatrix} \right\|_2 .
\end{aligned} \tag{9.34}$$

Summarising (9.33)–(9.34) and for the sake of simplicity, the threshold is set to be

$$J_{th} = \min\left\{\delta_{\Delta_{K_G}}, \delta_{\Delta_{I_G}}\right\} \left\| \begin{bmatrix} u \\ y \end{bmatrix} \right\|_2 . \tag{9.35}$$

We would like to remark that the threshold setting law (9.35) is relatively conservative. Moreover, the dependence of J_{th} on y can lead to a larger threshold value when, for instance, measurement noises are strong, which, in turn, reduces the fault detectability. In order to enhance the fault detectability and simultaneously system robustness, further improvement of threshold setting is made.

Consider the relation during fault-free operations

$$\left\| r_{\mathcal{I}_G^{\mathcal{L}_2}} \right\|_2^2 \leq \delta_{\Delta_{I_G}}^2 \left\| \begin{bmatrix} u \\ y \end{bmatrix} \right\|_2^2$$

and substitute (u, y) by its orthogonal decomposition

$$\begin{aligned}
& \left\| r_{\mathcal{I}_G^{\mathcal{L}_2}} \right\|_2^2 \leq \delta_{\Delta_{I_G}}^2 \left(\left\| r_{\mathcal{I}_G^{\mathcal{L}_2}} \right\|_2^2 + \left\| \mathcal{P}_{\mathcal{I}_G^{\mathcal{L}_2}} \begin{bmatrix} u \\ y \end{bmatrix} \right\|_2^2 \right) \Longrightarrow \\
& \left\| r_{\mathcal{I}_G^{\mathcal{L}_2}} \right\|_2^2 \leq \frac{\delta_{\Delta_{I_G}}^2}{1 - \delta_{\Delta_{I_G}}^2} \left\| \mathcal{P}_{\mathcal{I}_G^{\mathcal{L}_2}} \begin{bmatrix} u \\ y \end{bmatrix} \right\|_2^2 .
\end{aligned} \tag{9.36}$$

This result enables us to prove the following theorem.

Theorem 9.2 *Given system model (9.19) and the projection-based residual generator (9.13), then it holds*

$$\left\| r_{\mathcal{I}_G^{\mathcal{L}_2}} \right\|_2 \leq \frac{\delta_{\Delta_{I_G}}}{\sqrt{1 - \delta_{\Delta_{I_G}}^2}} \left(\left\| \begin{bmatrix} u \\ y \end{bmatrix} \right\|_2^2 - \left\| r_{\mathcal{I}_G^{\mathcal{L}_2}} \right\|_2^2 \right)^{1/2} . \tag{9.37}$$

Proof. *Since* (9.37) *holds for all* $\begin{bmatrix} u \\ y \end{bmatrix} \in I_G$, *it is straightforward that*

$$J_{th} = \sup_{\|\Delta_I\|_\infty \le \delta_{\Delta_I}} \left\| r_{\mathcal{I}_G^{\mathcal{L}_2}} \right\|_2 = \frac{\delta_{\Delta_{IG}}}{\sqrt{1-\delta_{\Delta_{IG}}^2}} \left\| \mathcal{P}_{\mathcal{I}_G^{\mathcal{L}_2}} \begin{bmatrix} u \\ y \end{bmatrix} \right\|_2 .$$

It follows from the relation

$$\left\| \begin{bmatrix} u \\ y \end{bmatrix} \right\|_2^2 = \left\| r_{\mathcal{I}_G^{\mathcal{L}_2}} \right\|_2^2 + \left\| \mathcal{P}_{\mathcal{I}_G^{\mathcal{L}_2}} \begin{bmatrix} u \\ y \end{bmatrix} \right\|_2^2$$

that the inequality (9.36) *can be written as*

$$\begin{aligned} \left\| r_{\mathcal{I}_G^{\mathcal{L}_2}} \right\|_2^2 &\le \frac{\delta_{\Delta_{IG}}^2}{1-\delta_{\Delta_{IG}}^2} \left\| \mathcal{P}_{\mathcal{I}_G^{\mathcal{L}_2}} \begin{bmatrix} u \\ y \end{bmatrix} \right\|_2^2 \\ &= \frac{\delta_{\Delta_{IG}}^2}{1-\delta_{\Delta_{IG}}^2} \left(\left\| \begin{bmatrix} u \\ y \end{bmatrix} \right\|_2^2 - \left\| r_{\mathcal{I}_G^{\mathcal{L}_2}} \right\|_2^2 \right) \\ \Longleftrightarrow \left\| r_{\mathcal{I}_G^{\mathcal{L}_2}} \right\|_2 &\le \frac{\delta_{\Delta_{IG}}}{\sqrt{1-\delta_{\Delta_{IG}}^2}} \left(\left\| \begin{bmatrix} u \\ y \end{bmatrix} \right\|_2^2 - \left\| r_{\mathcal{I}_G^{\mathcal{L}_2}} \right\|_2^2 \right)^{1/2} . \end{aligned}$$

Let the relative residual evaluation function be defined by

$$J_{rel} = \frac{\left\| r_{\mathcal{I}_G^{\mathcal{L}_2}} \right\|_2}{\left\| \begin{bmatrix} u \\ y \end{bmatrix} \right\|_2} .$$

The upper-bound of J_{rel} on the right-hand side of (9.37) varies over the value interval $\left(0, \frac{\delta_{\Delta_{IG}}}{\sqrt{1-\delta_{\Delta_{IG}}^2}}\right)$. To be specific:

- during fault-free operations, namely

$$\frac{\left\| r_{\mathcal{I}_G^{\mathcal{L}_2}} \right\|_2}{\left\| \begin{bmatrix} u \\ y \end{bmatrix} \right\|_2} \le \delta_{\Delta_{IG}},$$

J_{rel} decreases from $\frac{\delta_{\Delta_{IG}}}{\sqrt{1-\delta_{\Delta_{IG}}^2}}$ (without uncertainty) to $\delta_{\Delta_{IG}}$ (maximal uncertainty),

- once the fault causes J_{rel} larger than $\delta_{\Delta_{I_G}}$, it holds

$$J_{rel} = \frac{\left\| r_{\mathcal{I}_G^{\mathcal{L}_2}} \right\|_2}{\left\| \begin{bmatrix} u \\ y \end{bmatrix} \right\|_2} > \delta_{\Delta_{I_G}} > \sqrt{\frac{1 - J_{rel}^2}{1 - \delta_{\Delta_{I_G}}^2}} \delta_{\Delta_{I_G}},$$

- in case that $(u, y) \in \mathcal{I}_G^{\mathcal{L}_2, \perp}$,

$$J_{rel} = 1 \Longrightarrow \sqrt{\frac{1 - J_{rel}^2}{1 - \delta_{\Delta_{I_G}}^2}} \delta_{\Delta_{I_G}} = 0.$$

It is of interest to observe that the upper-bound of the relative residual evaluation function J_{rel} monotonically decreases with the increasing uncertainties expressed in terms of $\frac{\left\| r_{\mathcal{I}_G^{\mathcal{L}_2}} \right\|_2}{\left\| \begin{bmatrix} u \\ y \end{bmatrix} \right\|_2}$. Furthermore, the following typical cases are often observed during fault-free process operations,

- due to process and measurement noises, $\left\| r_{\mathcal{I}_G^{\mathcal{L}_2}} \right\|_2$ will constantly deviate from zero, but keep at a lower level, namely $0 < \left\| r_{\mathcal{I}_G^{\mathcal{L}_2}} \right\|_2 \ll \delta_{\Delta_{I_G}}$,
- the upper-bound $\delta_{\Delta_{I_G}}$ will be reached at a (very) low probability, and
- parametric faults arise in a continuous process from a (considerably large) parameter deviation to an incipient fault.

On account of these facts, a turning parameter can be introduced, which provides us with certain degree of design freedom to set an adaptive threshold. Let

$$\gamma \in (0, 1)$$

be the turning parameter and set the threshold equal to

$$J_{rel,th} = \begin{cases} \gamma \delta_{\Delta_{I_G}}, \ J_{rel} < \gamma \delta_{\Delta_{I_G}}, \\ \gamma \delta_{\Delta_{I_G}} \sqrt{\frac{1 - J_{rel}^2}{1 - \gamma^2 \delta_{\Delta_{I_G}}^2}}, \ \text{otherwise}. \end{cases} \tag{9.38}$$

It is evident that for $J_{rel} \leq \gamma \delta_{\Delta_{I_G}}$, the threshold $J_{rel,th}$ is a constant equal to $\gamma \delta_{\Delta_{I_G}}$. Once $J_{rel} > \gamma \delta_{\Delta_{I_G}}$, $J_{rel,th}$ becomes smaller than $\gamma \delta_{\Delta_{I_G}}$, and monotonically decreases with a further increase of J_{rel}. Consequently, in case of the existence of (strong) uncertainties, incipient faults can be detected. Such an adaptive threshold reflects the typical emergence process of parametric faults. Note that

$$J_{rel} \leq \gamma \delta_{\Delta_{I_G}} \Longrightarrow J_{rel} \leq J_{rel,th} \Longrightarrow \text{fault-free}, \tag{9.39}$$

$$J_{rel} > \gamma\delta_{\Delta_{I_G}} \Longrightarrow J_{rel} > J_{rel,th} \Longrightarrow \text{alarm}. \tag{9.40}$$

Since there would exist uncertainties that lead to $\gamma\delta_{\Delta_{I_G}} < J_{rel} \left(\leq \delta_{\Delta_{I_G}}\right)$, condition (9.40) means that false alarms are possible. On the other hand, this condition enables to detect those faults that cause $J_{rel} > \gamma\delta_{\Delta_{I_G}}$ but are bounded by $J_{rel} \leq \delta_{\Delta_{I_G}}$ and thus could not be detected when the threshold setting $J_{rel,th} = \delta_{\Delta_{I_G}}$ is adopted. In other words, the turning parameter γ determines the trade-off between the fault detectability rate and false alarm rate.

9.4.4 Application to a Class of Uncertain Systems

In the first part of the book, we have introduced a model form for systems with uncertainties, which can be stabilised by a (dynamic) output feedback controller. According to Theorem 4.3, such uncertain systems can be modelled by

$$G = NM^{-1} = \left(N_0 + \hat{X}_0 S\right)\left(M_0 - \hat{Y}_0 S\right)^{-1}, \tag{9.41}$$

$$I_G = I_{G_0} + \Delta_{I_G} = \begin{bmatrix} M_0 - \hat{Y}_0 S \\ N_0 + \hat{X}_0 S \end{bmatrix}. \tag{9.42}$$

Here, $\left(\hat{X}_0, \hat{Y}_0\right)$ is a RCP satisfying (2.25) and the Bezout identity (2.31), and $S \in \mathcal{H}_\infty$ is unknown.

It follows from the Bezout identity that

$$r_{\mathcal{I}_G^{\mathcal{L}_2}} = \left(\mathcal{I} - \mathcal{L}_{I_{G_0}}\mathcal{L}_{I_{G_0}}\right)\begin{bmatrix} u \\ y \end{bmatrix} = \mathcal{L}_{K_{G_0}^{\sim}}\mathcal{L}_{K_{G_0}}\begin{bmatrix} u \\ y \end{bmatrix} = K_{G_0}^{\sim} K_{G_0}\begin{bmatrix} M_0 - \hat{Y}_0 S \\ N_0 + \hat{X}_0 S \end{bmatrix} v$$
$$\Longrightarrow \forall \begin{bmatrix} u \\ y \end{bmatrix} \in \mathcal{I}_G^{\mathcal{L}_2}, \left\| r_{\mathcal{I}_G^{\mathcal{L}_2}} \right\|_2 = \left\| \mathcal{L}_{K_{G_0}^{\sim}}\mathcal{L}_{K_{G_0}}\begin{bmatrix} u \\ y \end{bmatrix} \right\|_2 = \|Sv\|_2, v \in \mathcal{L}_2.$$

Consequently,

$$\sup_{S\in\mathcal{H}_\infty} \left\| r_{\mathcal{I}_G^{\mathcal{L}_2}} \right\|_2 = \sup_{S\in\mathcal{H}_\infty} dist\left(\begin{bmatrix} u \\ y \end{bmatrix}, \mathcal{I}_{G_0}^{\mathcal{L}_2}\right) = \sup \|S\|_\infty \|v\|_2.$$

Recall

$$u = F_0 x + Vv \Longleftrightarrow v = V^{-1}(u - F_0 x), \tag{9.43}$$

which allows to compute v in terms of u and x (or its estimate). An adaptive threshold is set to be

$$J_{th} = \sup \|S\|_\infty \|v\|_2. \tag{9.44}$$

9.5 Fault Detection Based on Orthogonal Projection onto $\mathcal{I}_G^{\mathcal{H}_2}$

Recall that the process data (u, y) used for detection and control purposes in fact belong to the $\mathcal{H}_2$ subspace. It would be more reasonable and convincing to deal with fault detection issues by means of an orthogonal projection onto $\mathcal{I}_G^{\mathcal{H}_2}$ instead of onto $\mathcal{I}_G^{\mathcal{L}_2}$ as described in the previous section. The objective of this section is to introduce a fault detection approach based on orthogonal projection onto $\mathcal{I}_G^{\mathcal{H}_2}$. The obvious and legitimate question why the method based on the orthogonal projection onto $\mathcal{I}_G^{\mathcal{L}_2}$ has been firstly studied will be answered at the end of this chapter by a comparison of both schemes.

Remark 9.3 As mentioned before, for the sake of simplicity, the image and kernel subspaces $\mathcal{I}_G^{\mathcal{H}_2}$ and $\mathcal{K}_G^{\mathcal{H}_2}$ are denoted by $\mathcal{I}_G$ and $\mathcal{K}_G$. Moreover, it is assumed in this section

$$\begin{bmatrix} u \\ y \end{bmatrix} \in \mathcal{H}_2.$$

Therefore, Laurent operators used in this section are restricted to $\mathcal{H}_2$, for instance, Laurent operator with symbol I_{G_0} with restriction to $\mathcal{H}_2$, $\mathcal{L}_{I_{G_0}} |\mathcal{H}_2 : \mathcal{H}_2 \to \mathcal{H}_2$ or Laurent operator with symbol $I_{G_0^{\sim}}$ with restriction to $\mathcal{H}_2$, $\mathcal{L}_{I_{G_0^{\sim}}} |\mathcal{H}_2 : \mathcal{H}_2 \to \mathcal{L}_2$. For the sake of simplicity, the notation $\mathcal{L}_{I_{G_0}}$ is abused in the sequel Laurent operators restricted to $\mathcal{H}_2$, when there is no risk of confusion.

9.5.1 On Orthogonal Projection onto $\mathcal{H}_2$ Subspaces

That the orthogonal projection onto the image subspace $\mathcal{I}_G$ is given by

$$\mathcal{P}_{\mathcal{I}_G} = \mathcal{L}_{I_{G_0}} \mathcal{L}_{I_{G_0}}^*, \mathcal{L}_{I_{G_0}}^* : \mathcal{H}_2 \to \mathcal{H}_2, \mathcal{L}_{I_{G_0}}^* = \mathcal{P}_{\mathcal{H}_2} \mathcal{L}_{I_{\tilde{G}_0}}, \tag{9.45}$$

is a well-known result. Hereby, $\mathcal{P}_{\mathcal{H}_2} : \mathcal{L}_2 \to \mathcal{H}_2$ is an operator representing an orthogonal projection onto $\mathcal{H}_2$. Correspondingly, the projection of a data vector $\begin{bmatrix} u \\ y \end{bmatrix} \in \mathcal{H}_2$ onto the image subspace is

$$p_{\mathcal{I}_G} = \mathcal{P}_{\mathcal{I}_G} \begin{bmatrix} u \\ y \end{bmatrix}.$$

The difference between $\begin{bmatrix} u \\ y \end{bmatrix}$ and its estimate $p_{\mathcal{I}_G}$,

$$r_{\mathcal{I}_G} = \begin{bmatrix} u \\ y \end{bmatrix} - p_{\mathcal{I}_G} = \left(\mathcal{I} - \mathcal{L}_{I_{G_0}} \mathcal{P}_{\mathcal{H}_2} \mathcal{L}_{I_{\tilde{G}_0}} \right) \begin{bmatrix} u \\ y \end{bmatrix}, \tag{9.46}$$

gives the projection-based residual, denoted by $r_{\mathcal{I}_G}$, and indicates how far the measurement (data) vector $\begin{bmatrix} u \\ y \end{bmatrix}$ deviates from the nominal system dynamics expressed by the system SIR. Thus, its $\mathcal{L}_2$-norm is the distance between the data vector and the image subspace.

Note that the residual generator (9.46) is different from the one described by (9.13). Correspondingly, the nice property of (9.18), namely the projection-based residual can be equivalently realised using a trivial observer-based residual generator with a lower online computation load, does not hold.

Next, computation issues of the projection-based residual generator (9.46) are addressed. Remembering the natural relation between SKR and the residual generation as well as the known fact that $\mathcal{K}_G = \mathcal{I}_G$, it is of considerable interest to investigate relations among the operator $\mathcal{L}_{I_{G_0}}$, the orthogonal projection $\mathcal{P}_{\mathcal{I}_G}$ and the system SKR.

Lemma 9.3 *Let $\mathcal{L}_{K_{G_0}}$ and $\mathcal{L}_{K_{\tilde{G}_0}}$ be Laurent operators (restricted to $\mathcal{H}_2$) with symbol K_{G_0} and $K_{\tilde{G}_0}$, respectively. It holds*

$$\mathcal{P}_{\mathcal{H}_2}\left(\mathcal{I} - \mathcal{L}_{K_{\tilde{G}_0}}\mathcal{L}_{K_{G_0}}\right) = \mathcal{P}_{\mathcal{I}_G} + \mathcal{P}_{\mathcal{H}_2}\mathcal{L}_{I_{G_0}}\mathcal{P}_{\mathcal{H}_2^\perp}\mathcal{L}_{I_{\tilde{G}_0}}. \tag{9.47}$$

Proof. It follows from Lemma 9.1 *that*

$$\mathcal{P}_{\mathcal{H}_2}\mathcal{L}_{K_{\tilde{G}_0}}\mathcal{L}_{K_{G_0}} = \mathcal{P}_{\mathcal{H}_2}\left(\mathcal{I} - \mathcal{L}_{I_{G_0}}\mathcal{L}_{I_{\tilde{G}_0}}\right),$$

which yields

$$\begin{aligned}\mathcal{P}_{\mathcal{H}_2}\left(\mathcal{I} - \mathcal{L}_{K_{\tilde{G}_0}}\mathcal{L}_{K_{G_0}}\right) &= \mathcal{P}_{\mathcal{H}_2}\mathcal{L}_{I_{G_0}}\mathcal{L}_{I_{\tilde{G}_0}} \\ = \mathcal{P}_{\mathcal{H}_2}\mathcal{L}_{I_{G_0}}\mathcal{P}_{\mathcal{H}_2}\mathcal{L}_{I_{\tilde{G}_0}} &+ \mathcal{P}_{\mathcal{H}_2}\mathcal{L}_{I_{G_0}}\mathcal{P}_{\mathcal{H}_2^\perp}\mathcal{L}_{I_{\tilde{G}_0}} \\ &= \mathcal{P}_{\mathcal{I}_G} + \mathcal{P}_{\mathcal{H}_2}\mathcal{L}_{I_{G_0}}\mathcal{P}_{\mathcal{H}_2^\perp}\mathcal{L}_{I_{\tilde{G}_0}}.\end{aligned}$$

It follows from (9.7) and (9.46) and that

$$\begin{aligned}\left\|\begin{bmatrix} u \\ y \end{bmatrix}\right\|_2^2 &= \left\|r_{\mathcal{I}_G}\right\|_2^2 + \left\|\mathcal{L}_{I_G}\mathcal{P}_{\mathcal{H}_2}\mathcal{L}_{I_{\tilde{G}}}\begin{bmatrix} u \\ y \end{bmatrix}\right\|_2^2 \\ \Longrightarrow \left\|r_{\mathcal{I}_G}\right\|_2^2 &= \left\|\begin{bmatrix} u \\ y \end{bmatrix}\right\|_2^2 - \left\|\mathcal{L}_{I_G}\mathcal{P}_{\mathcal{H}_2}\mathcal{L}_{I_{\tilde{G}}}\begin{bmatrix} u \\ y \end{bmatrix}\right\|_2^2.\end{aligned}$$

Notice that the relation

$$\mathcal{L}_{I_{\tilde{G}}} = \mathcal{P}_{\mathcal{H}_2}\mathcal{L}_{I_{\tilde{G}}} + \mathcal{P}_{\mathcal{H}_2^\perp}\mathcal{L}_{I_{\tilde{G}}}$$

leads to

$$\left\| \mathcal{P}_{\mathcal{H}_2} \mathcal{L}_{I_{\tilde{G}}} \begin{bmatrix} u \\ y \end{bmatrix} \right\|_2^2 = \left\| \mathcal{L}_{I_G} \mathcal{L}_{I_{\tilde{G}}} \begin{bmatrix} u \\ y \end{bmatrix} \right\|_2^2 - \left\| \mathcal{L}_{I_G} \mathcal{P}_{\mathcal{H}_2^{\perp}} \mathcal{L}_{I_{\tilde{G}}} \begin{bmatrix} u \\ y \end{bmatrix} \right\|_2^2$$

$$\Longrightarrow \left\| r_{\mathcal{I}_G} \right\|_2^2 = \left\| \begin{bmatrix} u \\ y \end{bmatrix} \right\|_2^2 - \left\| \mathcal{L}_{I_G} \mathcal{L}_{I_{\tilde{G}}} \begin{bmatrix} u \\ y \end{bmatrix} \right\|_2^2 + \left\| \mathcal{L}_{I_G} \mathcal{P}_{\mathcal{H}_2^{\perp}} \mathcal{L}_{I_{\tilde{G}}} \begin{bmatrix} u \\ y \end{bmatrix} \right\|_2^2.$$

On the other hand, according to Lemma 9.1,

$$\left\| \begin{bmatrix} u \\ y \end{bmatrix} \right\|_2^2 - \left\| \mathcal{L}_{I_G} \mathcal{L}_{I_{\tilde{G}}} \begin{bmatrix} u \\ y \end{bmatrix} \right\|_2^2 = \left\| \mathcal{L}_{K_{\tilde{G}_0}} \mathcal{L}_{K_{G_0}} \begin{bmatrix} u \\ y \end{bmatrix} \right\|_2^2,$$

which results in

$$\left\| r_{\mathcal{I}_G} \right\|_2^2 = \left\| \mathcal{L}_{K_{G_0}} \begin{bmatrix} u \\ y \end{bmatrix} \right\|_2^2 + \left\| \mathcal{L}_{I_{G_0}} \mathcal{P}_{\mathcal{H}_2^{\perp}} \mathcal{L}_{I_{\tilde{G}_0}} \begin{bmatrix} u \\ y \end{bmatrix} \right\|_2^2. \tag{9.48}$$

Recalling that $\mathcal{L}_{K_{G_0}} \begin{bmatrix} u \\ y \end{bmatrix}$ is equivalent to an observer-based residual generator given by

$$r_y = \mathcal{L}_{K_{G_0}} \begin{bmatrix} u \\ y \end{bmatrix} = K_{G_0} \begin{bmatrix} u \\ y \end{bmatrix} = \begin{bmatrix} -\hat{N}_0 & \hat{M}_0 \end{bmatrix} \begin{bmatrix} u \\ y \end{bmatrix},$$

it holds

$$\left\| r_{\mathcal{I}_G} \right\|_2^2 = \left\| r_y \right\|_2^2 + \left\| \mathcal{L}_{I_{G_0}} \mathcal{P}_{\mathcal{H}_2^{\perp}} \mathcal{L}_{I_{\tilde{G}_0}} \begin{bmatrix} u \\ y \end{bmatrix} \right\|_2^2. \tag{9.49}$$

Let us have a close look at the term

$$\left\| \mathcal{L}_{I_{G_0}} \mathcal{P}_{\mathcal{H}_2^{\perp}} \mathcal{L}_{I_{\tilde{G}_0}} \begin{bmatrix} u \\ y \end{bmatrix} \right\|_2 = \left\| \mathcal{P}_{\mathcal{H}_2^{\perp}} \mathcal{L}_{I_{\tilde{G}_0}} \begin{bmatrix} u \\ y \end{bmatrix} \right\|_2$$

that marks the difference between $\left\| r_{\mathcal{I}_G} \right\|_2$ and $\left\| r_{\mathcal{I}_G^{\mathcal{L}_2}} \right\|_2 = \left\| r_y \right\|_2$. Note that, on the one hand,

$$\mathcal{L}_{I_{G_0}} \mathcal{P}_{\mathcal{H}_2^{\perp}} \mathcal{L}_{I_{\tilde{G}_0}} \begin{bmatrix} u \\ y \end{bmatrix} \in \mathcal{L}_2 \text{ and } \mathcal{L}_{K_{G_0}} \mathcal{L}_{I_{G_0}} = 0, \tag{9.50}$$

and furthermore,

$$\forall \begin{bmatrix} u \\ y \end{bmatrix} \in \mathcal{I}_G, \exists v \in \mathcal{H}_2, \text{ s.t. } \begin{bmatrix} u \\ y \end{bmatrix} = \mathcal{L}_{I_{G_0}} v,$$

which leads to

$$\mathcal{L}_{I_{G_0}} \mathcal{P}_{\mathcal{H}_2^{\perp}} \mathcal{L}_{I_{\tilde{G}_0}} \begin{bmatrix} u \\ y \end{bmatrix} = \mathcal{L}_{I_{G_0}} \mathcal{P}_{\mathcal{H}_2^{\perp}} \mathcal{L}_{I_{\tilde{G}_0}} \mathcal{L}_{I_{G_0}} v = 0.$$

On the other hand, for $\begin{bmatrix} u \\ y \end{bmatrix} \in \mathcal{I}_{\tilde{G}}^{\perp} \subset \mathcal{H}_2$, it is possible that

$$\mathcal{L}_{I_G} \mathcal{P}_{\mathcal{H}_2^{\perp}} \mathcal{L}_{I_{\tilde{G}}} \begin{bmatrix} u \\ y \end{bmatrix} \neq 0. \tag{9.51}$$

Relations (9.50)–(9.51) showcase that

- changes caused by $\begin{bmatrix} u \\ y \end{bmatrix} \in \mathcal{I}_{\tilde{G}}^{\perp} \subset \mathcal{H}_2$ and satisfying (9.51) cannot be detected using an observer-based residual generator, since they do not lead to any variation in $r_{\mathcal{I}_G^{\mathcal{L}_2}} \left(r_y \right)$;
- against, it is possible to detect these changes using $r_{\mathcal{I}_G}$, as described by (9.49).

In the context of fault detection, it can thus be claimed that the residual signal $r_{\mathcal{I}_G}$ is more sensitive to faults than $r_{\mathcal{I}_G^{\mathcal{L}_2}}$ and r_y.

Next, we address the operator $\mathcal{P}_{\mathcal{H}_2^{\perp}} \mathcal{L}_{I_{\tilde{G}_0}}$ using the results presented in Francis's monograph. The operator

$$\mathcal{H}_{I_{\tilde{G}_0}} := \mathcal{P}_{\mathcal{H}_2^{\perp}} \mathcal{L}_{I_{\tilde{G}_0}} : \mathcal{H}_2 \to \mathcal{H}_2^{\perp}$$

is the so-called Hankel operator with symbol $I_{\tilde{G}_0}$. Given the state space representation of $I_{\tilde{G}_0}$,

$$\xi(k-1) = \bar{A}\xi(k) + \bar{B} \begin{bmatrix} u(k) \\ y(k) \end{bmatrix}, \xi \in \mathbb{R}^n, \tag{9.52}$$

$$\varsigma(k) = \bar{C}\xi(k) + \bar{D} \begin{bmatrix} u(k) \\ y(k) \end{bmatrix} \in \mathbb{R}^p, \tag{9.53}$$

$$\bar{A} = (A + BF_0)^T, \bar{B} = \begin{bmatrix} F_0^T & (C + DF_0)^T \end{bmatrix}, \bar{C} = B^T, \bar{D} = \begin{bmatrix} V_0^T & (DV_0)^T \end{bmatrix}$$

with $\varsigma(k)$ as the output of system $I_{\tilde{G}_0} \begin{bmatrix} u \\ y \end{bmatrix}$, the output

$$\varsigma_{\mathcal{H}} := \mathcal{H}_{I_{\tilde{G}_0}} \begin{bmatrix} u \\ y \end{bmatrix}$$

can be computed by means of discrete convolution as follows

$$\varsigma_{\mathcal{H}}(k) = -\sum_{i=1}^{\infty} \bar{C}\bar{A}^{-k+i-1}\bar{B} \begin{bmatrix} u(i) \\ y(i) \end{bmatrix}, k \in (-\infty, 0].$$

As demonstrated, $\varsigma_{\mathcal{H}}$ can also be written as

$$\varsigma_{\mathcal{H}}(k) = \left(\Psi_o \Psi_c \begin{bmatrix} u \\ y \end{bmatrix}\right)(k), k \in (-\infty, 0], \tag{9.54}$$

where Ψ_o and Ψ_c,

$$\Psi_c \begin{bmatrix} u \\ y \end{bmatrix} = -\sum_{i=1}^{\infty} \bar{A}^{i-1} \bar{B} \begin{bmatrix} u(i) \\ y(i) \end{bmatrix}, (\Psi_o \xi)(k) = \bar{C} \bar{A}^{-k} \xi(k), k \in (-\infty, 0],$$
$$\Longrightarrow \mathcal{H}_{I_{\tilde{G}_0}} = \Psi_o \Psi_c,$$

are controllability and observability operators, respectively.

We now discuss the interpretation of $\left\| \mathcal{P}_{\mathcal{H}_2^\perp} \mathcal{L}_{I_{\tilde{G}_0}} \begin{bmatrix} u \\ y \end{bmatrix} \right\|_2$ from the fault detection aspect. Consider the relation

$$\left\| \mathcal{H}_{I_{\tilde{G}_0}} \begin{bmatrix} u \\ y \end{bmatrix} \right\|_2 = \left\| \mathcal{H}_{I_{\tilde{G}_0}}^* \varsigma_{\mathcal{H}} \right\|_2, \varsigma_{\mathcal{H}} \in \mathcal{H}_2^\perp,$$

where $\mathcal{H}_{I_{\tilde{G}_0}}^*$ is the adjoint of $\mathcal{H}_{I_{\tilde{G}_0}}$, and can be, based on (9.54), written as

$$\mathcal{H}_{I_{\tilde{G}_0}}^* : \mathcal{H}_2^\perp \to \mathcal{H}_2, \mathcal{H}_{I_{\tilde{G}_0}}^* = (\Psi_o \Psi_c)^* = \Psi_c^* \Psi_o^*,$$

and correspondingly $\mathcal{H}_{I_{\tilde{G}_0}}^* \varsigma_{\mathcal{H}}$ yields

$$x_0 = \Psi_o^* \varsigma_{\mathcal{H}} = \sum_{k=-\infty}^{0} \left(\bar{A}^T\right)^{-k} \bar{C}^T \varsigma_{\mathcal{H}}(k), \tag{9.55}$$

$$\begin{bmatrix} u(k) \\ y(k) \end{bmatrix} = \left(\Psi_c^* x_0\right)(k) = -\bar{B}^T \left(\bar{A}^T\right)^k x_0, k \in [0, \infty). \tag{9.56}$$

In this regard, $\left\| \mathcal{P}_{\mathcal{H}_2^\perp} \mathcal{L}_{I_{\tilde{G}_0}} \begin{bmatrix} u \\ y \end{bmatrix} \right\|_2$ can be interpreted as the influence of deviations from the (nominal) image subspace in the past (i.e. over the time interval $(-\infty, 0]$), which affects the dynamics of the residual generator $\left(\mathcal{I} - \mathcal{P}_{\mathcal{I}_G}\right) \begin{bmatrix} u \\ y \end{bmatrix}$ in form of the response to the corresponding changes in initial condition (i.e. via x_0).

Summarising the discussion on relation (9.49) and the interpretation of $\left\| \mathcal{P}_{\mathcal{H}_2^\perp} \mathcal{L}_{I_{\tilde{G}_0}} \begin{bmatrix} u \\ y \end{bmatrix} \right\|_2$, it can be concluded that the projection-based residual generator $\left(\mathcal{I} - \mathcal{P}_{\mathcal{I}_G}\right) \begin{bmatrix} u \\ y \end{bmatrix}$ is not only efficient in detecting existing faults (i.e. over the time interval $[0, \infty)$) like the observer-based residual generator $\mathcal{L}_{K_{G_0}} \begin{bmatrix} u \\ y \end{bmatrix}$, but also capable of detecting faults in the past (i.e. over the time interval $(-\infty, 0]$).

9.5.2 Residual Generators and Implementation Issues

In this subsection, online computation issues of $\|r_{\mathcal{I}_G}\|_2$ are addressed. It follows from (9.49) that the residual evaluation function

$$J = \|r_{\mathcal{I}_G}\|_2 = \sqrt{\|r_y\|_2^2 + \left\| \mathcal{H}_{I_{\tilde{G}_0}} \begin{bmatrix} u \\ y \end{bmatrix} \right\|_2^2} \tag{9.57}$$

can be realised using two dynamic systems, namely

- an observer-based residual generator,

$$\begin{aligned} \hat{x}(k+1) &= A\hat{x}(k) + Bu(k) + L_0 \left(y(k) - \hat{y}(k) \right), \\ r_y(k) &= W_0 \left(y(k) - \hat{y}(k) \right) = W_0 \left(y(k) - C\hat{x}(k) - Du(k) \right), \end{aligned}$$

  where (L_0, W_0) is the observer gain and post-filter setting that leads to the normalised SKR, and
- the filter (9.54) for the computation of Hankel operator $\mathcal{H}_{I_{\tilde{G}_0}}$.

These two dynamic systems can independently run in parallel. Recall that

$$\mathcal{P}_{\mathcal{H}_2} \mathcal{L}_{K_{\tilde{G}_0}} \mathcal{L}_{K_{G_0}} \begin{bmatrix} u \\ y \end{bmatrix} \text{ and } \mathcal{P}_{\mathcal{H}_2} \mathcal{L}_{I_{G_0}} \mathcal{P}_{\mathcal{H}_2^\perp} \mathcal{L}_{I_{\tilde{G}_0}} \begin{bmatrix} u \\ y \end{bmatrix}$$

are orthogonal, which implies that r_y with

$$\|r_y\|_2 = \left\| \mathcal{P}_{\mathcal{H}_2} \mathcal{L}_{K_{\tilde{G}_0}} \mathcal{L}_{K_{G_0}} \begin{bmatrix} u \\ y \end{bmatrix} \right\|_2$$

and $\mathcal{H}_{I_{\tilde{G}_0}} \begin{bmatrix} u \\ y \end{bmatrix}$,

$$\left\| \mathcal{H}_{I_{\tilde{G}_0}} \begin{bmatrix} u \\ y \end{bmatrix} \right\|_2 = \left\| \mathcal{P}_{\mathcal{H}_2} \mathcal{L}_{I_{G_0}} \mathcal{P}_{\mathcal{H}_2^\perp} \mathcal{L}_{I_{\tilde{G}_0}} \begin{bmatrix} u \\ y \end{bmatrix} \right\|_2,$$

contain uncorrelated information.

At the end of this subsection, an example is given that provides an algorithm for the online realisation of $\left\| \mathcal{H}_{I_{\tilde{G}_0}} \begin{bmatrix} u \\ y \end{bmatrix} \right\|_2$.

Example 9.1 According to

$$\left\| \mathcal{H}_{I_{\tilde{G}_0}} \begin{bmatrix} u \\ y \end{bmatrix} \right\|_2 = \left\| \mathcal{H}^*_{I_{\tilde{G}_0}} \varsigma_{\mathcal{H}} \right\|_2 ,$$

$\left\| \mathcal{H}_{I_{\tilde{G}_0}} \begin{bmatrix} u \\ y \end{bmatrix} \right\|_2$ can be computed online based on (9.55)–(9.56). To this end, write the space space equation of (9.55) as

$$\xi(k+1) = \bar{A}^T \xi(k) + \bar{C}^T u(k) = (A + BF_0)\,\xi(k) + B\varsigma_{\mathcal{H}}(k), \tag{9.58}$$

$$\varsigma_{\mathcal{H}}(k) = V_0 v(k) = u\,(k) - F_0 \xi(k),\, k \in (-\infty, -1]\,, \xi(-\infty) = 0 \tag{9.59}$$

$$\Longrightarrow x_0 = \varsigma(0) = A\xi(-1) + Bu(-1), \tag{9.60}$$

$$\xi(k+1) = (A + BF_0)\,\xi(k),\, k \in [0, \infty)\,, \tag{9.61}$$

and the output equation as

$$\begin{bmatrix} u(k) \\ y(k) \end{bmatrix} = -\bar{B}^T \left(\bar{A}^T\right)^k x_0 = \begin{bmatrix} F_0 \\ C + DF_0 \end{bmatrix} \varsigma(k),\, k \in [0, \infty)\,, \tag{9.62}$$

respectively. For the online implementation, $\left\| \mathcal{H}_{I_{\tilde{G}_0}} \begin{bmatrix} u \\ y \end{bmatrix} \right\|_2$ is then calculated in three steps

- run system (9.58);
- for a detection over the time interval $[k_0, k_1]$, run system (9.61) and (9.62), and
- calculate

$$\left\| \begin{bmatrix} u \\ y \end{bmatrix} \right\|_2 = \left(\sum_{k=k_0}^{k_1} \left(u^T(k)u(k) + y^T(k)y(k) \right) \right)^{1/2}$$

as an approximation of $\left\| \mathcal{H}_{I_{\tilde{G}_0}} \begin{bmatrix} u \\ y \end{bmatrix} \right\|_2$.

It is worth remarking that this algorithm can also run with a moving window.

9.5.3 *Threshold Setting*

As described in the previous section, threshold setting for projection-based fault detection systems can be formulated as determining the gap metric between two subspaces. To be specific, given the plant model (9.19) with

$$\mathcal{I}_{G_0} = \left\{ \begin{bmatrix} u \\ y \end{bmatrix} : \begin{bmatrix} u \\ y \end{bmatrix} = \begin{bmatrix} M_0 \\ N_0 \end{bmatrix} v,\, v \in \mathcal{H}_2 \right\},$$

$$\mathcal{I}_G = \left\{ \begin{array}{l} \begin{bmatrix} u \\ y \end{bmatrix} : \begin{bmatrix} u \\ y \end{bmatrix} = \left(\begin{bmatrix} M_0 \\ N_0 \end{bmatrix} + \begin{bmatrix} \Delta_M \\ \Delta_N \end{bmatrix} \right) v, \\ \sup \left\| \begin{bmatrix} \Delta_M \\ \Delta_N \end{bmatrix} \right\|_\infty = \delta_{\Delta_{I_G}} < 1, v \in \mathcal{H}_2 \end{array} \right\},$$

the projection-based residual generator (9.49) and the residual evaluation function $J = \left\| r_{\mathcal{I}_G} \right\|_2$, the threshold J_{th} is defined by, $\forall \begin{bmatrix} u \\ y \end{bmatrix} \in \mathcal{I}_G$,

$$J_{th} = \sup_{\left\| \Delta_{I_G} \right\|_\infty \leq \delta_{\Delta_{I_G}}} \left\| r_{\mathcal{I}_G} \right\|_2 = \sup_{\left\| \Delta_{I_G} \right\|_\infty \leq \delta_{\Delta_{I_G}}} \delta_{\mathcal{I}} \left(\mathcal{I}_G, \mathcal{I}_{G_0} \right) \left\| \begin{bmatrix} u \\ y \end{bmatrix} \right\|_2. \tag{9.63}$$

For the computation of $\delta_{\mathcal{I}} \left(\mathcal{I}_G, \mathcal{I}_{G_0} \right)$, some relevant known results are introduced below. The references are given in the end of the chapter.

Theorem 9.3 *Given G_i and, associated with it, $\mathcal{I}_{G_i}, i = 1, 2$,*

$$\mathcal{I}_{G_i} = \left\{ \begin{bmatrix} u_i \\ y_i \end{bmatrix} : \begin{bmatrix} u_i \\ y_i \end{bmatrix} = \begin{bmatrix} M_i \\ N_i \end{bmatrix} v, v \in \mathcal{H}_2 \right\}, i = 1, 2, \tag{9.64}$$

with the normalised SIR and SKR of G_i,

$$I_{G_{i,0}} = \begin{bmatrix} M_{i,0} \\ N_{i,0} \end{bmatrix}, K_{G_{i,0}} = \begin{bmatrix} -\hat{N}_{i,0} & \hat{M}_{i,0} \end{bmatrix},$$

then

$$\delta_{\mathcal{I}} \left(\mathcal{I}_{G_i}, \mathcal{I}_{G_j} \right) = \min_{Q \in \mathcal{H}_\infty} \left\| \begin{bmatrix} M_{i,0} \\ N_{i,0} \end{bmatrix} - \begin{bmatrix} M_{j,0} \\ N_{j,0} \end{bmatrix} Q \right\|_\infty, i, j = 1, 2, i \neq j, \tag{9.65}$$

$$\delta_{\mathcal{I}} \left(\mathcal{I}_{G_i}, \mathcal{I}_{G_j} \right) = \max \left\{ \delta_{\mathcal{I}} \left(\mathcal{I}_{G_1}, \mathcal{I}_{G_2} \right), \delta_{\mathcal{I}} \left(\mathcal{I}_{G_2}, \mathcal{I}_{G_1} \right) \right\}. \tag{9.66}$$

It is apparent and of interest to notice that

$$\begin{aligned} \delta_{\mathcal{I}} \left(\mathcal{I}_{G_i}, \mathcal{I}_{G_j} \right) \geq \delta_{\mathcal{I}} \left(\mathcal{I}_{G_i}^{\mathcal{L}_2}, \mathcal{I}_{G_j}^{\mathcal{L}_2} \right) &= \min_{Q \in \mathcal{L}_\infty} \left\| \begin{bmatrix} M_{i,0} \\ N_{i,0} \end{bmatrix} - \begin{bmatrix} M_{j,0} \\ N_{j,0} \end{bmatrix} Q \right\|_\infty \\ &= \left\| K_{G_{i,0}} I_{G_{j,0}} \right\|_\infty = \left\| K_{G_{j,0}} I_{G_{i,0}} \right\|_\infty, i, j = 1, 2, i \neq j. \end{aligned} \tag{9.67}$$

In fact, relation (9.67) is the consequence of the relation

$$\begin{aligned} \left\| r_{\mathcal{I}_G} \right\|_2^2 &= \left\| \mathcal{L}_{K_{G_0}} \begin{bmatrix} u \\ y \end{bmatrix} \right\|_2^2 + \left\| \mathcal{P}_{\mathcal{H}_2^\perp} \mathcal{L}_{I_{G_0}^\sim} \begin{bmatrix} u \\ y \end{bmatrix} \right\|_2^2 \\ &= \left\| r_{\mathcal{I}_G}^{\mathcal{L}_2} \right\|_2^2 + \left\| \mathcal{P}_{\mathcal{H}_2^\perp} \mathcal{L}_{I_{G_0}^\sim} \begin{bmatrix} u \\ y \end{bmatrix} \right\|_2^2, \end{aligned}$$

and $\mathcal{L}_{K_{G_0}}\begin{bmatrix} u \\ y \end{bmatrix}, \mathcal{P}_{\mathcal{H}_2^{\perp}}\mathcal{L}_{I_{\widetilde{G}_0}}\begin{bmatrix} u \\ y \end{bmatrix}$ are orthogonal. Below, we give an alternative proof of Theorem 9.3 by means of the above relation.

Proof. By definition of the gap metric

$$\delta_{\mathcal{I}}\left(\mathcal{I}_{G_i}, \mathcal{I}_{G_j}\right) = \sup_{\substack{z\in\mathcal{I}_{G_i} \\ \|z\|_2=1}} dist\left(z, \mathcal{I}_{G_j}\right), z = \begin{bmatrix} u \\ y \end{bmatrix},$$

we have

$$\delta_{\mathcal{I}}\left(\mathcal{I}_{G_i}, \mathcal{I}_{G_j}\right) = \sup_{\substack{z\in\mathcal{I}_{G_i} \\ \|z\|_2=1}} \left\|r_{\mathcal{I}_{G_j}}\right\|_2, K_{G_{j,0}}z = \begin{bmatrix} -\hat{N}_{j,0} & \hat{M}_{j,0} \end{bmatrix}\begin{bmatrix} u \\ y \end{bmatrix}.$$

Since

$$\forall \begin{bmatrix} u \\ y \end{bmatrix} \in \mathcal{I}_{G_i}, \left\|\begin{bmatrix} u \\ y \end{bmatrix}\right\|_2 = 1 \Longleftrightarrow \begin{bmatrix} u \\ y \end{bmatrix} = I_{G_i}v, \|v\|_2 = 1,$$

it turns out, $\forall z \in I_{G_i}, \|z\|_2 = 1,$

$$\begin{aligned}
\left\|r_{\mathcal{I}_{G_j}}\right\|_2^2 &= \left\|K_{G_{j,0}}\begin{bmatrix} u \\ y \end{bmatrix}\right\|_2^2 + \left\|\mathcal{P}_{\mathcal{H}_2^{\perp}}\mathcal{L}_{I_{\widetilde{G}_{j,0}}}\begin{bmatrix} u \\ y \end{bmatrix}\right\|_2^2 \\
&= \left\|K_{G_{j,0}}I_{G_{i,0}}v\right\|_2^2 + \left\|\mathcal{P}_{\mathcal{H}_2^{\perp}}\mathcal{L}_{I_{\widetilde{G}_{j,0}}}I_{G_{i,0}}v\right\|_2^2 \Longrightarrow \\
\sup_{\substack{z\in\mathcal{I}_{G_i} \\ \|z\|_2=1}} \left\|r_{\mathcal{I}_{G_j}}\right\|_2^2 &= \sup_{\|v\|_2=1}\left(\left\|K_{G_{j,0}}I_{G_{i,0}}v\right\|_2^2 + \left\|\mathcal{P}_{\mathcal{H}_2^{\perp}}\mathcal{L}_{I_{\widetilde{G}_{j,0}}}\mathcal{L}_{I_{G_{i,0}}}v\right\|_2^2\right) \\
&= \left\|K_{G_{j,0}}I_{G_{i,0}}\right\|_\infty^2 + \left\|\mathcal{H}_{I_{\widetilde{G}_{j,0}}, I_{G_{i,0}}}\right\|^2,
\end{aligned}$$

where

$$\mathcal{H}_{I_{\widetilde{G}_{j,0}}, I_{G_{i,0}}} := \mathcal{P}_{\mathcal{H}_2^{\perp}}\mathcal{L}_{I_{\widetilde{G}_{j,0}}}\mathcal{L}_{I_{G_{i,0}}} : \mathcal{H}_2 \to \mathcal{H}_2^{\perp}$$

is the Hankel operator with symbol $I_{\widetilde{G}_{j,0}}I_{G_{i,0}}$. *By Nehari's Theorem (see Francis's monograph, Theorem 1 in Sect. 5.2), it holds*

$$\left\|\mathcal{H}_{I_{\widetilde{G}_{j,0}}, I_{G_{i,0}}}\right\| = \min_{Q\in\mathcal{H}_\infty}\left\|I_{\widetilde{G}_{j,0}}I_{G_{i,0}} - Q\right\|_\infty.$$

Remember that

$$\begin{aligned}
&\left\|I_{G_{i,0}} - I_{G_{j,0}}Q\right\|_\infty = \left\|\begin{bmatrix} K_{G_{j,0}} \\ I_{\widetilde{G}_{j,0}} \end{bmatrix}\left(I_{G_{i,0}} - I_{G_{j,0}}Q\right)\right\|_\infty = \left\|\begin{bmatrix} K_{G_{j,0}}I_{G_{i,0}} \\ I_{\widetilde{G}_{j,0}}I_{G_{i,0}} - Q \end{bmatrix}\right\|_\infty \\
&\Longrightarrow \min_{Q\in\mathcal{H}_\infty}\left\|I_{G_{i,0}} - I_{G_{j,0}}Q\right\|_\infty^2 = \left\|K_{G_{j,0}}I_{G_{i,0}}\right\|_\infty^2 + \min_{Q\in\mathcal{H}_\infty}\left\|I_{\widetilde{G}_{j,0}}I_{G_{i,0}} - Q\right\|_\infty^2
\end{aligned}$$

$$= \min_{Q \in \mathcal{L}_\infty} \left\| I_{G_{i,0}} - I_{G_{j,0}} Q \right\|_\infty^2 + \min_{Q \in \mathcal{H}_\infty} \left\| I_{\tilde{G}_{j,0}} I_{G_{i,0}} - Q \right\|_\infty^2 .$$

The theorem is thus proved.

The following theorem is well-known and widely applied in dealing with robustness issues of control systems with uncertainties.

Theorem 9.4 *Given plant model (9.19) with* $\sup \left\| \Delta_{I_G} \right\|_\infty = \delta_{\Delta_{I_G}} < 1$, *and let*

$$\mathcal{I}_{G,\delta} = \left\{ \mathcal{I}_G : \left\| \Delta_{I_G} \right\|_\infty = \left\| I_G - I_{G_0} \right\|_\infty \leq \delta_{\Delta_{I_G}} \right\}, \tag{9.68}$$

then it holds

$$\mathcal{I}_{G,\delta} = \left\{ \mathcal{I}_G : \delta_{\mathcal{I}} \left(\mathcal{I}_G, \mathcal{I}_{G_0} \right) \leq \delta_{\Delta_{I_G}} \right\}.$$

An immediate result from Theorem 9.4 is that

$$J_{th} = \sup_{\left\| \Delta_{I_G} \right\|_\infty \leq \delta_{\Delta_{I_G}}} \delta_{\mathcal{I}} \left(\mathcal{I}_G, \mathcal{I}_{G_0} \right) \left\| \begin{bmatrix} u \\ y \end{bmatrix} \right\|_2 = \delta_{\Delta_{I_G}} \left\| \begin{bmatrix} u \\ y \end{bmatrix} \right\|_2, \tag{9.69}$$

$$J_{th,rel} = \sup_{\left\| \Delta_{I_G} \right\|_\infty \leq \delta_{\Delta_{I_G}}} J_{rel} = \sup_{\left\| \Delta_{I_G} \right\|_\infty \leq \delta_{\Delta_{I_G}}} \delta_{\mathcal{I}} \left(\mathcal{I}_G, \mathcal{I}_{G_0} \right) = \delta_{\Delta_{I_G}}, \tag{9.70}$$

$$J_{rel} := \frac{\left\| r_{\mathcal{I}_G} \right\|_2}{\left\| \begin{bmatrix} u \\ y \end{bmatrix} \right\|_2}, \begin{bmatrix} u \\ y \end{bmatrix} \neq 0.$$

Analogous to the adaptively residual-triggered threshold introduced in the previous section, the following threshold algorithm can be alternatively adopted,

$$J_{rel,th} = \begin{cases} \gamma \delta_{\Delta_{I_G}}, \; J_{rel} < \gamma \delta_{\Delta_{I_G}}, \\ \gamma \delta_{\Delta_{I_G}} \sqrt{\frac{1 - J_{rel}^2}{1 - \gamma^2 \delta_{\Delta_{I_G}}^2}}, \; \text{otherwise}, \end{cases} \tag{9.71}$$

with the turning parameter $\gamma \in (0, 1)$.

9.6 A Generalisation of Fault Detection as One-Class Classification

The fault detection issues addressed in this chapter can be equivalently formulated as a one-class classification problem. That is, given information about process nominal operations, including the process model as offline information and real-time measurements as online information, which defines the class fault-free, a decision is to be made on if (online) received measurements belong to fault-free or faulty operation. Like the well-established model- or observer-based fault detection technique, the proposed projection-based method consists of two separate steps, (i) residual generation, and (ii) residual evaluation, on which a "zero or one", "fault-free or faulty" decision is made. Observe that, in the machine learning research domain, one-class classification problems are typically solved by means of identifying the set membership of the received measurement data. For fault detection, two sets, fault-free and faulty, are defined. This section is dedicated to an attempt to give a "one-step" solution of the fault detection problem in the context of set membership identification with the aid of projection technique.

9.6.1 Recasting the Projection-Based Fault Detection Solution

Recall the detection logic,

$$\begin{cases} J_{rel} \leq J_{rel,th} \Longrightarrow \text{fault-free} \\ J_{rel} > J_{rel,th} \Longrightarrow \text{alarm}, \end{cases} \tag{9.72}$$

$$J_{rel} = \frac{\left\| r_{\mathcal{I}_G} \right\|_2}{\left\| \begin{bmatrix} u \\ y \end{bmatrix} \right\|_2}, \ J_{th,rel} = \delta < 1. \tag{9.73}$$

For our purpose, the residual can be generated by projection either onto $\mathcal{I}_G^{\mathcal{H}_2}$ or onto $\mathcal{I}_G^{\mathcal{L}_2}$. To simplify the notation and without loss of generality, $r_{\mathcal{I}_G}$ is abused to represent a projection-based residual vector, and

$$\mathcal{S} \in \{\mathcal{H}_2, \mathcal{L}_2\}$$

for the subspace under consideration. The above detection logic divides the (relative) residual subspace,

$$\mathcal{R}_{\mathcal{I}_G,rel} = \left\{ r_{\mathcal{I}_G,rel} : r_{\mathcal{I}_G,rel} = \frac{\left(\mathcal{I} - \mathcal{P}_{I_G}\right) \begin{bmatrix} u \\ y \end{bmatrix}}{\left\| \begin{bmatrix} u \\ y \end{bmatrix} \right\|_2}, \begin{bmatrix} u \\ y \end{bmatrix} \in \mathcal{S} \right\}, \tag{9.74}$$

into two subspaces $\mathcal{R}_{\mathcal{I}_G,ff}$ and $\mathcal{R}_{\mathcal{I}_G,f}$,

$$\mathcal{R}_{\mathcal{I}_G,rel} = \mathcal{R}_{\mathcal{I}_G,ff} \oplus \mathcal{R}_{\mathcal{I}_G,f}, \tag{9.75}$$

$$\mathcal{R}_{\mathcal{I}_G,ff} = \left\{ r_{\mathcal{I}_G,rel} : \left\| r_{\mathcal{I}_G,rel} \right\|_2 \leq \delta \right\}, \tag{9.76}$$

$$\mathcal{R}_{\mathcal{I}_G,f} = \left\{ r_{\mathcal{I}_G,rel} : \left\| r_{\mathcal{I}_G,rel} \right\|_2 > \delta \right\}. \tag{9.77}$$

It is obvious that $\mathcal{R}_{\mathcal{I}_G,ff}$ (fault-free) is a (closed) $\mathcal{L}_2$ ball, which represents the residual signal generated during fault-free operations. Accordingly, an operator for a projection onto $\mathcal{R}_{\mathcal{I}_G,ff}$ can be defined as follows,

$$\begin{aligned}
\mathcal{P}_{\mathcal{R}_{\mathcal{I}_G,ff}}\left(r_{\mathcal{I}_G,rel}\right) &:= \frac{r_{\mathcal{I}_G,rel}}{\max\left\{1, \left\|r_{\mathcal{I}_G,rel}\right\|_2/\delta\right\}} \\
&= \begin{cases} r_{\mathcal{I}_G,rel}, r_{\mathcal{I}_G,rel} \in \mathcal{R}_{\mathcal{I}_G,ff}, \\ \frac{r_{\mathcal{I}_G,rel}}{\left\|r_{\mathcal{I}_G,rel}\right\|_2/\delta}, r_{\mathcal{I}_G,rel} \notin \mathcal{R}_{\mathcal{I}_G,ff} \end{cases} \Longrightarrow \\
\mathcal{P}_{\mathcal{R}_{\mathcal{I}_G,ff}} &= \frac{I}{\max\left\{1, \left\|r_{\mathcal{I}_G,rel}\right\|_2/\delta\right\}}.
\end{aligned}$$

Now, define the detection signal (label) as

$$S_{\det} := \begin{cases} 0, \text{ fault-free}, \\ 1, \text{ faulty}. \end{cases}$$

Noticing that

$$\begin{aligned}
&\left\|r_{\mathcal{I}_G,rel}\right\|_2 \leq 1, \\
&\left\|r_{\mathcal{I}_G,rel} - \mathcal{P}_{\mathcal{R}_{\mathcal{I}_G,ff}}\left(r_{\mathcal{I}_G,rel}\right)\right\|_2 = \left(1 - \frac{1}{\max\left\{1, \left\|r_{\mathcal{I}_G,rel}\right\|_2/\delta\right\}}\right)\left\|r_{\mathcal{I}_G,rel}\right\|_2 \\
&\Longrightarrow \left\|r_{\mathcal{I}_G,rel} - \mathcal{P}_{\mathcal{R}_{\mathcal{I}_G,ff}}\left(r_{\mathcal{I}_G,rel}\right)\right\|_2 = \begin{cases} 0, r_{\mathcal{I}_G,rel} \in \mathcal{R}_{\mathcal{I}_G,ff}, \\ \eta, r_{\mathcal{I}_G,rel} \notin \mathcal{R}_{\mathcal{I}_G,ff}, \end{cases} \\
&\eta = \left\|r_{\mathcal{I}_G,rel}\right\|_2 - \delta \in (0, 1),
\end{aligned}$$

the detection logic can be compactly written as

$$S_{\det} = \left\lceil \left\|r_{\mathcal{I}_G,rel} - \mathcal{P}_{\mathcal{R}_{\mathcal{I}_G,ff}}\left(r_{\mathcal{I}_G,rel}\right)\right\|_2 \right\rceil. \tag{9.78}$$

Here, the symbol $\lceil\cdot\rceil$ denotes the ceiling function. Expression (9.78) gives an abstract solution of the (optimal) fault detection as a one-class classification problem. It is noteworthy that the solution (9.78), although being equivalent to the projection-based one, is helpful for us to gain a deep insight into optimal fault detection from the aspect of classification and, based on it, to design advanced fault detection systems with enhancing fault detection performance.

9.6.2 Interpretations

In this subsection, we examine the signal $r_{\mathcal{I}_G,rel} - \mathcal{P}_{\mathcal{R}_{\mathcal{I}_G,ff}}\left(r_{\mathcal{I}_G,rel}\right)$ and reveal its properties. On this basis, the projection-based fault detection will be extended.

The operator $\mathcal{P}_{\mathcal{R}_{\mathcal{I}_G,ff}}$ gives a projection of the residual vector $r_{\mathcal{I}_G,rel}$ onto $\mathcal{R}_{\mathcal{I}_G,ff}$ and can be further written as

$$\mathcal{P}_{\mathcal{R}_{\mathcal{I}_G,ff}}\left(r_{\mathcal{I}_G,rel}\right) = \mathcal{P}_{\mathcal{R}_{\mathcal{I}_G,ff}}\left(\mathcal{I} - \mathcal{P}_{I_G}\right)\frac{\begin{bmatrix} u \\ y \end{bmatrix}}{\left\|\begin{bmatrix} u \\ y \end{bmatrix}\right\|_2}.$$

Observe that

$$\begin{aligned}
\mathcal{P}_{\mathcal{R}_{\mathcal{I}_G,ff}} &= \frac{I}{\max\left\{1, \left\|r_{\mathcal{I}_G,rel}\right\|_2/\delta\right\}} \Longrightarrow \\
\mathcal{P}_{\mathcal{R}_{\mathcal{I}_G,ff}}\left(\mathcal{I} - \mathcal{P}_{I_G}\right)\begin{bmatrix} u \\ y \end{bmatrix} &= \left(\mathcal{I} - \mathcal{P}_{I_G}\right)\mathcal{P}_{\mathcal{R}_{\mathcal{I}_G,ff}}\begin{bmatrix} u \\ y \end{bmatrix}, \\
\left(\mathcal{P}_{\mathcal{R}_{\mathcal{I}_G,ff}}\left(\mathcal{I} - \mathcal{P}_{I_G}\right)\right)^2 &= \mathcal{P}_{\mathcal{R}_{\mathcal{I}_G,ff}}\left(\mathcal{I} - \mathcal{P}_{I_G}\right)^2\mathcal{P}_{\mathcal{R}_{\mathcal{I}_G,ff}} \\
&= \mathcal{P}^2_{\mathcal{R}_{\mathcal{I}_G,ff}}\left(\mathcal{I} - \mathcal{P}_{I_G}\right) = \mathcal{P}_{\mathcal{R}_{\mathcal{I}_G,ff}}\left(\mathcal{I} - \mathcal{P}_{I_G}\right), \\
\left\langle \mathcal{P}_{\mathcal{R}_{\mathcal{I}_G,ff}}\left(\mathcal{I} - \mathcal{P}_{I_G}\right)\begin{bmatrix} u_1 \\ y_1 \end{bmatrix}, \begin{bmatrix} u_2 \\ y_2 \end{bmatrix}\right\rangle &= \left\langle \begin{bmatrix} u_1 \\ y_1 \end{bmatrix}, \mathcal{P}_{\mathcal{R}_{\mathcal{I}_G,ff}}\left(\mathcal{I} - \mathcal{P}_{I_G}\right)\begin{bmatrix} u_2 \\ y_2 \end{bmatrix}\right\rangle.
\end{aligned}$$

Hence, operator

$$\begin{aligned}
\mathcal{P}_{\mathcal{R}_\delta} &:= \mathcal{P}_{\mathcal{R}_{\mathcal{I}_G,ff}}\left(\mathcal{I} - \mathcal{P}_{I_G}\right) : \mathcal{S} \to \mathcal{S}, \\
\mathcal{P}_{\mathcal{R}_\delta}\begin{bmatrix} u \\ y \end{bmatrix} &= \xi\left(\mathcal{I} - \mathcal{P}_{I_G}\right)\begin{bmatrix} u \\ y \end{bmatrix}, \xi = \frac{1}{\max\left\{1, \left\|r_{\mathcal{I}_G,rel}\right\|_2/\delta\right\}}
\end{aligned}$$

is idempotent, self-adjoint and defines a projection from $(u, y) \in \mathcal{B}_{(u,y)} \subset \mathcal{S}$ to the residual subspace $\mathcal{R}_\delta \subset \mathcal{S}$,

$$\begin{aligned}
\mathcal{B}_{(u,y)} &= \left\{\begin{bmatrix} u \\ y \end{bmatrix}, \left\|\begin{bmatrix} u \\ y \end{bmatrix}\right\|_2 = 1, \begin{bmatrix} u \\ y \end{bmatrix} \in \mathcal{S}\right\}, \\
\mathcal{R}_\delta &= \left\{r_{\mathcal{I}_G,rel} : r_{\mathcal{I}_G,rel} = \mathcal{P}_{\mathcal{R}_\delta}\begin{bmatrix} u \\ y \end{bmatrix}, \begin{bmatrix} u \\ y \end{bmatrix} \in \mathcal{B}_{(u,y)}\right\}.
\end{aligned}$$

This reveals two meaningful points:

- the fault detection problem under consideration can be interpreted as finding the distance from a vector $\begin{bmatrix} u \\ y \end{bmatrix}$ in the unit ball $\mathcal{B}_{(u,y)}$ to the subspace $\mathcal{R}_\delta$, and consequently,
- an optimal solution of the addressed problem can be achieved by means of the projection $\mathcal{P}_{\mathcal{R}_\delta}\begin{bmatrix} u \\ y \end{bmatrix}$.

Recall that for $\mathcal{S} = \mathcal{L}_2$,

$$\left\| r_{\mathcal{I}_G,rel} \right\|_2 = \left\| \mathcal{L}_{K_{\tilde{G}_0}} \mathcal{L}_{K_{G_0}} \begin{bmatrix} u \\ y \end{bmatrix} \right\|_2 = \left\| K_{G_0} \begin{bmatrix} u \\ y \end{bmatrix} \right\|_2 .$$

This allows us to write the operator $\mathcal{P}_{\mathcal{R}_\delta}$ into

$$\mathcal{P}_{\mathcal{R}_\delta} = \mathcal{L}_{K_{\tilde{G}_0}} \mathcal{P}_{\mathcal{R}_{y,ff}} \mathcal{L}_{K_{G_0}}, \tag{9.79}$$

$$\mathcal{P}_{\mathcal{R}_{y,ff}} = \frac{I_{m \times m}}{\max\left\{1, \left\| K_{G_0} \begin{bmatrix} u \\ y \end{bmatrix} \right\|_2 / \delta \right\}}, \begin{bmatrix} u \\ y \end{bmatrix} \in \mathcal{B}_{(u,y)}, \tag{9.80}$$

and moreover,

$$\left\| r_{\mathcal{I}_G,rel} - \mathcal{P}_{\mathcal{R}_{\mathcal{I}_G,ff}} \left(r_{\mathcal{I}_G,rel} \right) \right\|_2 = \xi \left\| K_{G_0} \begin{bmatrix} u \\ y \end{bmatrix} \right\|_2 = \begin{cases} 0, & \text{fault-free}, \\ \eta, & \text{faulty} \end{cases},$$

$$\xi = 1 - \frac{1}{\max\left\{1, \left\| K_{G_0} \begin{bmatrix} u \\ y \end{bmatrix} \right\|_2 / \delta \right\}}, \begin{bmatrix} u \\ y \end{bmatrix} \in \mathcal{B}_{(u,y)},$$

$$\eta = \left\| K_{G_0} \begin{bmatrix} u \\ y \end{bmatrix} \right\|_2 - \delta \in (0, 1) .$$

Accordingly, our original optimal fault detection problem can be recast into

$$S_{\text{det}} = \left[\xi \left\| K_{G_0} \begin{bmatrix} u \\ y \end{bmatrix} \right\|_2 \right], \begin{bmatrix} u \\ y \end{bmatrix} \in \mathcal{B}_{(u,y)}, \tag{9.81}$$

and thus solved (and realised) by an observer-based algorithm.

9.6.3 An Extension

Considering the wide application of observer-based residual generation schemes, the observer-based fault detector (9.81) is of significant interest. It motivates us to deepen our work along this line.

Let $\mathcal{P}_{\mathcal{R}_y} : \mathbb{R}^m \to \mathbb{R}^m$ be an operator that defines a projection onto a convex set in the (observer-based) residual subspace, that satisfies

$$\mathcal{P}_{\mathcal{R}_y} r_y(k) = \mathcal{P}^2_{\mathcal{R}_y} r_y(k) = \mathcal{P}^*_{\mathcal{R}_y} r_y(k), \left\langle r_y(k_1), r_y(k_2) \right\rangle = r_y^T(k_1) r_y(k_2),$$

namely idempotent and self-adjoint. As will be demonstrated in the example given below, *a priori* available information about residual dynamics can be well modelled by such a convex set. Inspired by (9.79), we firstly prove the following theorem.

Theorem 9.5 *Given $\mathcal{L}_{K_{G_0}}, \mathcal{L}_{K_{\tilde{G}_0}}$, and operator $\mathcal{P}_{\mathcal{R}_y}$, then the operator*

$$\mathcal{P}_{\mathcal{R}} : \mathcal{L}_2 \to \mathcal{L}_2, \mathcal{P}_{\mathcal{R}} = \mathcal{L}_{K_{\tilde{G}_0}} \mathcal{P}_{\mathcal{R}_y} \mathcal{L}_{K_{G_0}} \tag{9.82}$$

is idempotent and self-adjoint.

Proof. *It is straightforward that*

$$\begin{aligned}
\mathcal{P}_{\mathcal{R}}^2 &= \mathcal{L}_{K_{\tilde{G}_0}} \mathcal{P}_{\mathcal{R}_y} \mathcal{L}_{K_{G_0}} \mathcal{L}_{K_{\tilde{G}_0}} \mathcal{P}_{\mathcal{R}_y} \mathcal{L}_{K_{G_0}} = \mathcal{L}_{K_{\tilde{G}_0}} \mathcal{P}_{\mathcal{R}_y} \mathcal{L}_{K_{G_0}}, \\
\mathcal{P}_{\mathcal{R}}^* &= \left(\mathcal{L}_{K_{\tilde{G}_0}} \mathcal{P}_{\mathcal{R}_y} \mathcal{L}_{K_{G_0}} \right)^* = \mathcal{L}_{K_{\tilde{G}_0}} \mathcal{P}_{\mathcal{R}_y} \mathcal{L}_{K_{G_0}}.
\end{aligned}$$

On the basis of Theorem 9.5, we are able to integrate *a priori* information of the residual signal generated during fault-free operations in terms of the operator $\mathcal{P}_{\mathcal{R}_y}$ in the construction of an optimal fault detector. Note that

$$\begin{aligned}
\mathcal{L}_{K_{\tilde{G}_0}} \mathcal{L}_{K_{G_0}} &= \mathcal{L}_{K_{\tilde{G}_0}} \mathcal{P}_{\mathcal{R}_y} \mathcal{L}_{K_{G_0}} + \mathcal{L}_{K_{\tilde{G}_0}} \left(\mathcal{I} - \mathcal{P}_{\mathcal{R}_y} \right) \mathcal{L}_{K_{G_0}} \Longrightarrow \\
r_{\mathcal{I}_G,rel} &= \mathcal{L}_{K_{\tilde{G}_0}} \mathcal{P}_{\mathcal{R}_y} \mathcal{L}_{K_{G_0}} \begin{bmatrix} u \\ y \end{bmatrix} + \mathcal{L}_{K_{\tilde{G}_0}} \left(\mathcal{I} - \mathcal{P}_{\mathcal{R}_y} \right) \mathcal{L}_{K_{G_0}} \begin{bmatrix} u \\ y \end{bmatrix} \\
&= r_{rel} + \left(r_{\mathcal{I}_G,rel} - r_{rel} \right), \begin{bmatrix} u \\ y \end{bmatrix} \in \mathcal{B}_{(u,y)}, \\
r_{rel} &= \mathcal{P}_{\mathcal{R}} \begin{bmatrix} u \\ y \end{bmatrix}, \left\langle r_{rel}, r_{\mathcal{I}_G,rel} - r_{rel} \right\rangle = 0 \Longrightarrow \\
\mathcal{R}_{\mathcal{I}_G,rel} &= \mathcal{R}_{rel} \oplus \mathcal{R}_{rel}^{\perp}, \\
\mathcal{R}_{rel} &= \left\{ r_{rel} : r_{rel} = \mathcal{P}_{\mathcal{R}} \begin{bmatrix} u \\ y \end{bmatrix}, \begin{bmatrix} u \\ y \end{bmatrix} \in \mathcal{B}_{(u,y)} \right\}, \\
\mathcal{R}_{rel}^{\perp} &= \left\{ r_{\mathcal{I}_G,rel} - r_{rel} \right\}.
\end{aligned}$$

Corresponding to fault-free and faulty operations, $\mathcal{R}_{\mathcal{I}_G,rel}$ is divided into $\mathcal{R}_{\mathcal{I}_G,ff}$ and $\mathcal{R}_{\mathcal{I}_G,f}$ as follows

$$\mathcal{R}_{\mathcal{I}_G,rel} = \mathcal{R}_{\mathcal{I}_G,ff} \otimes \mathcal{R}_{\mathcal{I}_G,f}, \tag{9.83}$$

$$\mathcal{R}_{\mathcal{I}_G,ff} = \left\{ r_{\mathcal{I}_G,rel}, r_{\mathcal{I}_G,rel} = r_{rel} : \| r_{rel} \|_2 \le \delta \right\}, \tag{9.84}$$

$$\mathcal{R}_{\mathcal{I}_G,f} = \left\{ r_{\mathcal{I}_G,rel}, r_{\mathcal{I}_G,rel} = r_{rel} : \| r_{rel} \|_2 > \delta \right\} \cup \mathcal{R}_{rel}^{\perp}. \tag{9.85}$$

Let

$$\mathcal{P}_{\mathcal{R}_{rel}}\left(r_{\mathcal{I}_G,rel}\right) := \begin{cases} r_{\mathcal{I}_G,rel},\ r_{\mathcal{I}_G,rel} = r_{rel} \in \mathcal{R}_{\mathcal{I}_G,ff} \\ \frac{r_{\mathcal{I}_G,rel}}{\left\|r_{\mathcal{I}_G,rel}\right\|_2/\delta},\ r_{\mathcal{I}_G,rel} = r_{rel},\ \|r_{rel}\|_2 > \delta \\ 0,\ r_{\mathcal{I}_G,rel} \in \mathcal{R}_{rel}^{\perp}, \end{cases} \tag{9.86}$$

define a projection onto $\mathcal{R}_{\mathcal{I}_G,ff}$. It yields

$$\left\|r_{\mathcal{I}_G,rel} - \mathcal{P}_{\mathcal{R}_{rel}}\left(r_{\mathcal{I}_G,rel}\right)\right\|_2 = \begin{cases} 0,\ r_{\mathcal{I}_G,rel} \in \mathcal{R}_{\mathcal{I}_G,ff} \\ \left\|r_{\mathcal{I}_G,rel}\right\|_2 - \delta,\ r_{\mathcal{I}_G,rel} = r_{rel},\ \|r_{rel}\|_2 > \delta \\ \left\|r_{\mathcal{I}_G,rel}\right\|_2,\ r_{\mathcal{I}_G,rel} \in \mathcal{R}_{rel}^{\perp}. \end{cases}$$

Accordingly,

$$S_{\det} = \left\lceil \left\|r_{\mathcal{I}_G,rel} - \mathcal{P}_{\mathcal{R}_{rel}}\left(r_{\mathcal{I}_G,rel}\right)\right\|_2 \right\rceil \tag{9.87}$$

is an optimal fault detector with the detection signal as its output,

$$S_{\det} = \left\lceil \left\|r_{\mathcal{I}_G,rel} - \mathcal{P}_{\mathcal{R}_{rel}}\left(r_{\mathcal{I}_G,rel}\right)\right\|_2 \right\rceil = \begin{cases} 0, & \text{fault-free}, \\ 1, & \text{faulty}. \end{cases}$$

Since

$$r_{\mathcal{I}_G,rel} - \mathcal{P}_{\mathcal{R}_{rel}}\left(r_{\mathcal{I}_G,rel}\right) = \mathcal{L}_{K_{\tilde{G}_0}}\left(\mathcal{I} - \mathcal{P}_{\mathcal{R}_y}\right)\mathcal{L}_{K_{G_0}}\begin{bmatrix} u \\ y \end{bmatrix},$$

$$\left\|r_{\mathcal{I}_G,rel} - \mathcal{P}_{\mathcal{R}_{rel}}\left(r_{\mathcal{I}_G,rel}\right)\right\|_2 = \left\|\left(\mathcal{I} - \mathcal{P}_{\mathcal{R}_y}\right)K_{G_0}\begin{bmatrix} u \\ y \end{bmatrix}\right\|_2,\ \begin{bmatrix} u \\ y \end{bmatrix} \in \mathcal{B}_{(u,y)},$$

we finally have

$$S_{\det} = \left\lceil \left\|\left(\mathcal{I} - \mathcal{P}_{\mathcal{R}_y}\right)K_{G_0}\begin{bmatrix} u \\ y \end{bmatrix}\right\|_2 \right\rceil,\ \begin{bmatrix} u \\ y \end{bmatrix} \in \mathcal{B}_{(u,y)}. \tag{9.88}$$

That is, the fault detector (9.88) is an observer-based system with the (normalised) process data (u, y) as its input and the detection signal as its output.

We summarise the above results as a theorem.

Theorem 9.6 *Given $\mathcal{L}_{K_{G_0}}$, $\mathcal{L}_{K_{\tilde{G}_0}}$, and operator $\mathcal{P}_{\mathcal{R}_y}$, the fault detector (9.88) is an optimal observer-based fault detector.*

Comparing with fault detector (9.81), the fault detector (9.88) is of enhanced fault detectability. The key to the performance improvement lies in the additional information about process nominal operations modelled by $\mathcal{P}_{\mathcal{R}_y}$. Integration of such information into the data projection results in a tighter closure with the residual subspace as a subset, which corresponds to the fault-free operations (the white box shown in Fig. 9.2). Thus, it will reduce missing detection, as delineated by (9.83)–(9.86). Figure 9.2 gives a schematic comparison without and with embedding $\mathcal{P}_{\mathcal{R}_y}$,

Fig. 9.2 Schematic description of the extension: **a** without the integration of projection $\mathcal{P}_{\mathcal{R}_y}$ **b** with the integration of projection $\mathcal{P}_{\mathcal{R}_y}$. Here, the white box represents the subspace corresponding to the fault-free operation

from which it can be seen that the subspace $\mathcal{R}_{\mathcal{I}_G,f}$ is extended to include $\mathcal{R}_{rel}^{\perp}$ when $\mathcal{P}_{\mathcal{R}_y}$ is embedded. In a nutshell, the convex set represented by $\mathcal{P}_{\mathcal{R}_y}$ bounds the residual subspace, and the tighter this closure becomes, the better the fault detection performance becomes.

Below is an example of operator $\mathcal{P}_{\mathcal{R}_y}$ for a given convex set.

Example 9.2 Two convex sets are considered here. The first one is described by

$$\begin{aligned} \mathcal{R}_y &= \mathcal{R}_{y,1} \cap \mathcal{R}_{y,2}, \\ \mathcal{R}_{y,1} &:= \left\{ r_y(k) : \sqrt{r_y^T(k) r_y(k)} = \left\| r_y(k) \right\| \leq \Delta \right\}, \\ \mathcal{R}_{y,2} &:= \left\{ r_y(k) : \alpha^T r_y(k) \leq \upsilon \right\}, \end{aligned}$$

where $\alpha \in \mathbb{R}^m, \alpha^T \alpha = 1, \upsilon \in \mathbb{R}$ are given. The corresponding operator $\mathcal{P}_{\mathcal{R}_y}$ is defined by

$$\mathcal{P}_{\mathcal{R}_y}\left(r_y(k)\right) = \begin{cases} r_y(k), r_y(k) \in \mathcal{R}_y, \\ \frac{r_y(k)\Delta}{\left\| r_y(k) \right\|}, r_y(k) \in \mathcal{R}_{y,2}, r_y(k) \notin \mathcal{R}_{y,1} \\ \frac{\left(I - \alpha\alpha^T\right) r_y(k) + \alpha\upsilon}{\max\left\{1, \left\| r_y(k) \right\| / \Delta\right\}}, r_y(k) \notin \mathcal{R}_{y,2}. \end{cases} \tag{9.89}$$

To examine (9.89), consider $\mathcal{P}_{\mathcal{R}_y}^2\left(r_y(k)\right)$. Note that

$$\begin{aligned} \left\| \frac{r_y(k)}{\left\| r_y(k) \right\|} \Delta \right\| = \Delta &\Longrightarrow \frac{r_y(k)}{\left\| r_y(k) \right\|} \Delta \in \mathcal{R}_{y,1}, \\ \frac{\alpha^T \left(\left(I - \alpha\alpha^T\right) r_y(k) + \alpha\upsilon \right)}{\max\left\{1, \left\| r_y(k) \right\| / \Delta\right\}} &= \frac{\upsilon}{\max\left\{1, \left\| r_y(k) \right\| / \Delta\right\}} \in \mathcal{R}_y. \end{aligned}$$

Hence, it holds

$$\mathcal{P}_{\mathcal{R}_y}^2 r_y(k) = \mathcal{P}_{\mathcal{R}_y}\left(r_y(k)\right).$$

We now address $\mathcal{I} - \mathcal{P}_{\mathcal{R}_y}$. It is apparent that

$$\left(\mathcal{I} - \mathcal{P}_{\mathcal{R}_y}\right)\left(r_y(k)\right) = \begin{cases} 0, r_y(k) \in \mathcal{R}_y, \\ \left(1 - \frac{\Delta}{\|r_y(k)\|}\right) r_y(k), r_y(k) \in \mathcal{R}_{y,2}, r_y(k) \notin \mathcal{R}_{y,1} \\ \left(1 - \frac{\left(I-\alpha\alpha^T\right)}{\kappa}\right) r_y(k) + \frac{\alpha v}{\kappa}, r_y(k) \notin \mathcal{R}_{y,2}, \end{cases} \tag{9.90}$$
$$\kappa = \max\left\{1, \left\|r_y(k)\right\| / \Delta\right\}.$$

Thus, the online implementation consists of two steps, (i) generation of $r_y(k)$ using an observer, (ii) computation of $\left(\mathcal{I} - \mathcal{P}_{\mathcal{R}_y}\right)\left(r_y(k)\right)$ according to (9.90).

The second case is described by

$$\mathcal{R}_y = \mathcal{R}_{y,1} \cap \mathcal{R}_{y,2},$$
$$\mathcal{R}_{y,1} := \left\{r_y(k) : \sqrt{r_y^T(k) r_y(k)} = \left\|r_y(k)\right\| \leq \Delta\right\},$$
$$\mathcal{R}_{y,2} := \left\{r_y(k) : r_{y,LB} \preceq r_y(k) \preceq r_{y,UB}\right\},$$

where $\preceq$ denotes componentwise inequality between vectors, and

$$r_{y,LB} = \begin{bmatrix} r_{y,1,LB} \\ \vdots \\ r_{y,m,LB} \end{bmatrix}, r_{y,UB} = \begin{bmatrix} r_{y,1,UB} \\ \vdots \\ r_{y,m,UB} \end{bmatrix}$$

are lower and upper bounds, respectively. The corresponding projection operator $\mathcal{P}_{\mathcal{R}_y}$ is given by

$$\mathcal{P}_{\mathcal{R}_y}\left(r_y(k)\right) = \begin{cases} r_y(k), r_y(k) \in \mathcal{R}_y, \\ \frac{r_y(k)\Delta}{\|r_y(k)\|}, r_y(k) \in \mathcal{R}_{y,2}, r_y(k) \notin \mathcal{R}_{y,1} \\ \frac{\mathcal{P}_{\mathcal{R}_{y,2}}\left(r_y(k)\right)}{\max\left\{1, \|r_y(k)\|/\Delta\right\}}, r_y(k) \notin \mathcal{R}_{y,2}, \end{cases} \tag{9.91}$$

$$\mathcal{P}_{\mathcal{R}_{y,2}}\left(r_y(k)\right) := \begin{bmatrix} \hat{r}_{y,1}(k) \\ \vdots \\ \hat{r}_{y,m}(k) \end{bmatrix}, \hat{r}_{y,i}(k) = \begin{cases} r_{y,i}(k), r_{y,i,LB} \leq r_{y,i}(k) \leq r_{y,i,UB}, \\ r_{y,i,LB}, r_{y,i}(k) \leq r_{y,i,LB}, \\ r_{y,i,UB}, r_{y,i}(k) \geq r_{y,i,UB}, \end{cases} \tag{9.92}$$

and finally it turns out

$$\left(\mathcal{I} - \mathcal{P}_{\mathcal{R}_y}\right)\left(r_y(k)\right) = \begin{cases} 0, r_y(k) \in \mathcal{R}_y, \\ \left(1 - \frac{\Delta}{\|r_y(k)\|}\right) r_y(k), r_y(k) \in \mathcal{R}_{y,2}, r_y(k) \notin \mathcal{R}_{y,1} \\ \left(1 - \frac{\mathcal{P}_{\mathcal{R}_{y,2}}\left(r_y(k)\right)}{\kappa}\right) r_y(k), r_y(k) \notin \mathcal{R}_{y,2}, \end{cases} \tag{9.93}$$
$$\kappa = \max\left\{1, \left\|r_y(k)\right\| / \Delta\right\}.$$

Remark 9.4 It is noteworthy that the operators (9.90)–(9.92) are idempotent and self-adjoint, but non-linear. Nevertheless, it holds true that

$$\left\| \left(\mathcal{I} - \mathcal{P}_{\mathcal{R}_y}\right)\left(r_y(k)\right) \right\| = dist\left(r_y(k), \mathcal{R}_y\right) = \inf_{r \in \mathcal{R}_y} \left\| r_y(k) - r \right\|.$$

that is, $\left\| \left(\mathcal{I} - \mathcal{P}_{\mathcal{R}_y}\right)\left(r_y(k)\right) \right\|$ is the distance between $r_y(k)$ and $\mathcal{R}_y$.

9.7 Notes and References

In this chapter, we have presented a fault detection scheme using the method of orthogonal projection onto a system subspace, including image and kernel subspaces, in Hilbert space. The design of the corresponding fault detection system follows a uniform procedure with three major steps: (i) definition and determination of an orthogonal projection operator, (ii) construction of the projection-based residual generator and realisation of the online implementation algorithm, and (iii) gap metric-based threshold setting. This scheme provides, on the one hand, an alternative framework for fault detection in dynamic systems to the well-established observer-based methods, and is, on the other hand, of evident similarities to the observer-based schemes, amongst others,

- residual generation is realised on the basis of an observer,
- the residual evaluation function is norm-based, and
- the threshold is set equal to the maximal value of the residual norm in the fault-free case.

From the system theoretic viewpoint, a significant feature of this framework is that the parameter settings are, up to a certain degree, well determined by the orthogonal projection and the associated gap (metric) computation, and can be calculated using the existing algorithms. This distinguishes the projection-based framework from the observer-based one essentially. It is the state of the art in the observer-based fault detection framework, even after five decades of development, that observer design remains the major object of this research domain, and there exists no commonly recognised and accepted design (optimisation) criterion. The underlying cause of this difference consists in the basic idea of dealing with the fault detection task. While the projection-based method formulates fault detection as a one-class classification problem, the observer-based one addresses it from the control-theoretic viewpoint and formulate it as a trade-off between robustness (against uncertainties) and sensitivity (to the faults). Consequently, in the projection-based framework, (i) identifying how far the process (measurement) data (u, y) is from the nominal process operation, and (ii) determining whether the identified deviation lies in the tolerant interval, build the two major tasks. Differently, the main task of designing an observer-based fault detection system mainly consists of (i) decision for an optimisation criterion that reflects the trade-off requirement for the system under consideration, and (ii) applying an existing or developing an algorithm to solve the trade-off optimisation problem. In general, established control-theoretic methods serve as tools for this purpose.

Comparing the both frameworks from the "geometric" viewpoint may illustrate their major difference. Given process data (u, y), by an orthogonal projection onto the image subspace $\mathcal{I}_G$ of the system under consideration, (u, y) can be uniquely decomposed into

$$\begin{bmatrix} u \\ y \end{bmatrix} = \begin{bmatrix} \hat{u} \\ \hat{y} \end{bmatrix} + r_{\mathcal{I}_G}, \begin{bmatrix} \hat{u} \\ \hat{y} \end{bmatrix} = \mathcal{P}_{\mathcal{I}_G} \begin{bmatrix} u \\ y \end{bmatrix}$$
$$\Longrightarrow \left\| \begin{bmatrix} u \\ y \end{bmatrix} \right\|_2^2 = \left\| \begin{bmatrix} \hat{u} \\ \hat{y} \end{bmatrix} \right\|_2^2 + \left\| r_{\mathcal{I}_G} \right\|_2^2,$$

where $(\hat{u}, \hat{y})$ is the projection value (estimate) of (u, y) in the image subspace, and the residual $r_{\mathcal{I}_G}$ is the distance between (u, y) and the image subspace $\mathcal{I}_G$, namely

$$\left\| r_{\mathcal{I}_G} \right\|_2 = dist\left(\begin{bmatrix} u \\ y \end{bmatrix}, \mathcal{I}_G \right) = \inf_{z \in \mathcal{I}_G} \left\| \begin{bmatrix} u \\ y \end{bmatrix} - z \right\|_2.$$

On this basis, the "distance" between a (closed) subspace, say $\mathcal{I}_{G_\Delta}$ representing the image subspace of the corresponding system with uncertainties, and the image subspace $\mathcal{I}_G$ can be well measured by the gap metric,

$$\left\| r_{\mathcal{I}_G} \right\|_2 = dist\left(\begin{bmatrix} u \\ y \end{bmatrix}, \mathcal{I}_G \right) = \inf_{z \in \mathcal{I}_G} \left\| \begin{bmatrix} u \\ y \end{bmatrix} - z \right\|_2,$$
$$\delta\left(\mathcal{I}_{G_\Delta}, \mathcal{I}_G\right) = \sup_{\substack{z \in \mathcal{I}_{G_\Delta} \\ \|z\|_2 = 1}} dist\left(\begin{bmatrix} u \\ y \end{bmatrix}, \mathcal{I}_G \right) = \sup_{\substack{z \in \mathcal{I}_{G_\Delta} \\ \|z\|_2 = 1}} \left\| r_{\mathcal{I}_G} \right\|_2,$$

which serves as the (relative) threshold.

We now turn to the observer-based residual generation. As known,

$$r_y = K_G \begin{bmatrix} u \\ y \end{bmatrix} = W\left(y - C\hat{x} - Du\right),$$

is a residual vector generated by an observer with K_G denoting the SKR of the process G. Let

$$\begin{bmatrix} \hat{u}_{obs} \\ \hat{y}_{obs} \end{bmatrix} = \begin{bmatrix} u \\ C\hat{x} + Du \end{bmatrix} = \begin{bmatrix} u \\ K_G \begin{bmatrix} u \\ y \end{bmatrix} - y \end{bmatrix} =: \mathcal{P}_{obs} \begin{bmatrix} u \\ y \end{bmatrix},$$
$$r_{\mathcal{P}_{obs}} := \begin{bmatrix} u \\ y \end{bmatrix} - \begin{bmatrix} \hat{u}_{obs} \\ \hat{y}_{obs} \end{bmatrix} = \begin{bmatrix} 0 \\ r_y \end{bmatrix} \Longleftrightarrow \begin{bmatrix} u \\ y \end{bmatrix} = \begin{bmatrix} \hat{u}_{obs} \\ \hat{y}_{obs} \end{bmatrix} + r_{\mathcal{P}_{obs}},$$

$\mathcal{P}_{obs}$ is an operator representing a mapping of $\begin{bmatrix} u \\ y \end{bmatrix}$. Here, without loss of generality, it is assumed that post-filter $W = I$. Observe that

$$K_G \begin{bmatrix} \hat{u}_{obs} \\ \hat{y}_{obs} \end{bmatrix} = 0,$$
$$\begin{bmatrix} \hat{u}_{obs} \\ \hat{y}_{obs} \end{bmatrix} = \mathcal{P}_{obs} \begin{bmatrix} u \\ y \end{bmatrix} = \mathcal{P}_{obs} \begin{bmatrix} \hat{u}_{obs} \\ \hat{y}_{obs} \end{bmatrix} = \mathcal{P}_{obs}^2 \begin{bmatrix} u \\ y \end{bmatrix}.$$

Hence, $\mathcal{P}_{obs}$ is an operator of a projection onto $\mathcal{I}_G$. On the other hand, in general,

$$\left\| \begin{bmatrix} u \\ y \end{bmatrix} \right\|_2^2 \leq \left\| \begin{bmatrix} \hat{u}_{obs} \\ \hat{y}_{obs} \end{bmatrix} \right\|_2^2 + \left\| r_{\mathcal{P}_{obs}} \right\|_2^2. \tag{9.94}$$

Therefore, $\mathcal{P}_{obs}$ represents an oblique projection. Inequality (9.94) reveals that

- the residual $r_{\mathcal{P}_{obs}}$ contains redundant information, and
- $\left\| r_{\mathcal{P}_{obs}} \right\|_2$ is not a measurement of the distance between $\begin{bmatrix} u \\ y \end{bmatrix}$ and the image subspace $\mathcal{I}_G$.

This result explains, as well, that an observer-based residual generator could be optimal only under certain trade-off cost function, rather than in the "geometric" sense.

Remark 9.5 It is noteworthy that, when the setting of an observer-based residual generator follows the normalised SKR setting, namely

$$(L, W) = (L_0, W_0) \Longrightarrow K_G = K_{G_0},$$

then

$$\left\| r_{\mathcal{P}_{obs}} \right\|_2 = \left\| r_y \right\|_2 = \left\| r_{\mathcal{I}_G}^{\mathcal{L}_2} \right\|_2.$$

Nevertheless,

$$\begin{bmatrix} \hat{u} \\ \hat{y} \end{bmatrix} = \mathcal{P}_{\mathcal{I}_G^{\mathcal{L}_2}} \begin{bmatrix} u \\ y \end{bmatrix} = I_{G_0} I_{G_0}^{\sim} \begin{bmatrix} u \\ y \end{bmatrix} \neq \begin{bmatrix} \hat{u}_{obs} \\ \hat{y}_{obs} \end{bmatrix} = \mathcal{P}_{obs} \begin{bmatrix} u \\ y \end{bmatrix} = \begin{bmatrix} u \\ C\hat{x} + Du \end{bmatrix}$$
$$\Longrightarrow r_{\mathcal{P}_{obs}} \neq r_{\mathcal{I}_G}^{\mathcal{L}_2}.$$

We would like to call the reader's attention to two essential mathematical concepts, the inner product and the Hilbert subspace on which the system image subspace is defined. Their definitions induce the form and computation algorithms of the corresponding operator, projection, distance and gap. In the fault detection framework, both of them play an important role and could significantly influence the fault detection performance and computation demands. To be specific, the inner product, $\langle \cdot, \cdot \rangle$, induces the definition of the residual norm as the residual evaluation function. In our study, inner products of (9.2) and (9.3) have been adopted. Their implementation in practical applications implies residual evaluation within a (very) large time window. This would lead to a large detection delay, but considerably save online computations thanks to the LTI system dynamics. On the other hand, the adoption of inner product (9.4) enables the selection of the evaluation time window to meet the practical requirements on fault detection time. Unfortunately, in this case, the system dynamics are LDTV and thus system realisation requires remarkably more online computations. In this chapter, we didn't address projection-based fault detection for LDTV-systems. Some relevant results and algorithms are reported in [109].

In Sects. 9.4 and 9.5, two fault detection schemes have been presented, based on orthogonal projections onto $\mathcal{L}_2$ and $\mathcal{H}_2$ image subspaces, respectively. In spite of the fact that all process data (u, y) in our control engineering consideration belong to $\mathcal{H}_2$ subspace, that is

$$\begin{bmatrix} u(k) \\ y(k) \end{bmatrix} \in \mathcal{L}_2 [0, \infty),$$

our study on projection onto $\mathcal{L}_2$ subspace is useful and helpful from the following aspects:

- the needed mathematical handling is less demanding and thus helpful to understand the underlying ideas towards optimal fault detection,
- the equivalence to the observer- (SKR-) based residual generation is a useful conclusion for practical applications, which allows, for instance,
- to realise an orthogonal projection-based optimal fault detection using an observer-based fault detection system.

On the other hand, this simplified online realisation is achieved at the cost of fault detectability, as demonstrated in our study on the projection onto $\mathcal{H}_2$ subspace. It has been delineated that the residual subspace of the $\mathcal{H}_2$ image subspace, namely,

$$\mathcal{I}_G^{\mathcal{H}_2,\perp} = \left\{ r_{\mathcal{I}_G}, r_{\mathcal{I}_G} = \left(\mathcal{I} - \mathcal{P}_{\mathcal{I}_G^{\mathcal{H}_2}} \right) \begin{bmatrix} u \\ y \end{bmatrix}, \begin{bmatrix} u \\ y \end{bmatrix} \in \mathcal{H}_2 \right\},$$

consists of two orthogonal subspaces,

$$\begin{aligned}
\mathcal{I}_G^{\mathcal{H}_2,\perp} &= \mathcal{R}_K^{\mathcal{H}_2} \oplus \mathcal{R}_I^{\mathcal{H}_2}, \\
\mathcal{R}_K^{\mathcal{H}_2} &= \left\{ r_K, r_K := \mathcal{P}_{\mathcal{H}_2} \mathcal{L}_{K_{\tilde{G}_0}^{\sim}} \mathcal{L}_{K_{G_0}} \begin{bmatrix} u \\ y \end{bmatrix}, \begin{bmatrix} u \\ y \end{bmatrix} \in \mathcal{H}_2 \right\}, \\
\mathcal{R}_I^{\mathcal{H}_2} &= \left\{ r_I, r_I := \mathcal{P}_{\mathcal{H}_2} \mathcal{L}_{I_{G_0}} \mathcal{P}_{\mathcal{H}_2^{\perp}} \mathcal{L}_{I_{\tilde{G}_0}^{\sim}} \begin{bmatrix} u \\ y \end{bmatrix}, \begin{bmatrix} u \\ y \end{bmatrix} \in \mathcal{H}_2 \right\}.
\end{aligned}$$

Notice that $\mathcal{I}_G^{\mathcal{H}_2}$ is the subspace composed of the nominal process data, and $\mathcal{I}_G^{\mathcal{H}_2,\perp}$ is the subspace consisting of the process data during operations with uncertainties and possible faults. Due to the relations,

$$\forall \begin{bmatrix} u \\ y \end{bmatrix} \in \mathcal{H}_2, \left\| r_{\mathcal{I}_G}^{\mathcal{L}_2} \right\|_2 = \left\| r_y \right\|_2 = \| r_K \|_2 ,$$

$$\| r_{\mathcal{I}_G} \|_2^2 = \left\| \mathcal{L}_{K_{G_0}} \begin{bmatrix} u \\ y \end{bmatrix} \right\|_2^2 + \left\| \mathcal{P}_{\mathcal{H}_2^{\perp}} \mathcal{L}_{I_{\tilde{G}_0}} \begin{bmatrix} u \\ y \end{bmatrix} \right\|_2^2 = \| r_K \|_2^2 + \| r_I \|_2^2 ,$$

it is apparent that the residual $r_{\mathcal{I}_G}$ generated by means of the projection onto $\mathcal{H}_2$ subspace delivers higher fault detectability. To this end, the computation of

$$\left\| \mathcal{P}_{\mathcal{H}_2^{\perp}} \mathcal{L}_{I_{\tilde{G}_0}} \begin{bmatrix} u \\ y \end{bmatrix} \right\|_2 = \left\| \mathcal{H}_{I_{\tilde{G}_0}} \begin{bmatrix} u \\ y \end{bmatrix} \right\|_2$$

is necessary, which, as discussed in Sect. 9.5, is computationally intensive during online operations.

Projection methods are standard techniques in machine and statistic learning and extensively applied in data-driven fault diagnosis. The most popular ones are principal component analysis (PCA) method [23, 37, 98] and partial least squares (PLS) algorithm [23, 37, 108]. Although their applications are limited to detecting faults in statistic processes (including steady dynamic systems), PCA and PLS are, thanks to their uniform formulation and simple computation algorithms, and above all, their clear geometric interpretation of projections onto subspaces, not only widely accepted in industry, but also recognised in research as a conceptional basis for the development of advanced fault diagnosis methods. This example also inspires our efforts towards a projection-based fault diagnosis framework. Noticing that projections adopted in PCA and PLS deal with linear algebraic computations, and thus are limited to solving fault diagnosis problems in statistic processes, different mathematical and control theoretical tools are applied in our study on fault diagnosis in dynamic systems. We consider system signals (data) defined in Hilbert spaces like Lebesgue space or Hardy space, and model the dynamic systems under consideration as subspaces in Hilbert space, for instance, image and kernel subspaces [57, 199]. In order to deal with orthogonal projections and distance between (signal) vector and (system) subspace or between two subspaces, basics of operator theory, gap metric as well as their computations and realisations in the form of dynamic systems serve as the major tools for our study.

The mathematical preliminaries for the following topics are summarized below:

- the concepts and definitions related to orthogonal projections in Hilbert space and gap metric introduced in Sect. 9.2 are mainly referred to [55, 102],
- function (signal) spaces, operator theory and its application to orthogonal projections are from [55, 57, 199],
- the orthogonal projections onto the image subspaces $\mathcal{I}_G^{\mathcal{H}_2}$ and $\mathcal{I}_G^{\mathcal{L}_2}$, $\mathcal{P}_{\mathcal{I}_G^{\mathcal{H}_2}}$, $\mathcal{P}_{\mathcal{I}_G^{\mathcal{L}_2}}$ given in (9.45) and (9.12) respectively, are well known and can be found, for instance, in [69, 199],
- the definitions and concepts concerning Hankel operator, the computations algorithms for the Hankel operator $\mathcal{H}_{I_{\tilde{G}_0}} = \mathcal{P}_{\mathcal{H}_2^{\perp}} \mathcal{L}_{I_{\tilde{G}_0}}$ as well as the relation to Nehari's Theorem are referred to [57].

The results presented in Theorem 9.1 and Lemma 9.2, as well as the discussion related to MMP (9.31) are referred to [199]. The detailed proofs of Theorems 9.3 and 9.4 are given in [70]. It is worth mentioning that the discussions provided in [199] about the relevant results and derivations are helpful for understanding the issues. Also, the alternative proof of Theorem 9.3 demonstrates the important role of Lemma 9.3 and the relation,

$$\left\| r_{\mathcal{I}_G} \right\|_2^2 = \left\| \mathcal{L}_{K_{G_0}} \begin{bmatrix} u \\ y \end{bmatrix} \right\|_2^2 + \left\| \mathcal{P}_{\mathcal{H}_2^\perp} \mathcal{L}_{I_{G_0}^\sim} \begin{bmatrix} u \\ y \end{bmatrix} \right\|_2^2,$$

in the projection-based fault detection framework.

The objective of Sect. 9.6 is twofold. It serves as an extension of the projection-based fault detection methods introduced in the previous sections, and, maybe more importantly, highlights a uniform formulation of projection-based fault detection of dynamic systems as a one-class classification problem. Based on it, a perspective solution in this framework is proposed. The first result is expressed by (9.78), which provides us with a uniform description of an optimal fault detection problem as finding the distance between process data (u, y) in its unit ball, $\mathcal{B}_{(u,y)}$, and the residual subspace $\mathcal{R}_\delta$. Accordingly, the orthogonal projection, $\mathcal{P}_{\mathcal{R}_\delta} \begin{bmatrix} u \\ y \end{bmatrix}$, gives a reasonable theoretical solution. From the implementation viewpoint, the observer-based realisation algorithm (9.81) is of practical significance. It demonstrates once more that observer algorithms are a powerful tool in dealing with fault detection issues for dynamic systems. On the other hand, we would like to emphasise the significant difference to the traditional observer-based fault detection technique, whose core consists in optimal design algorithms of observer-based residual generators.

The first application of the aforementioned uniform problem formulation yields a promising result. By means of (orthogonal) projection onto convex sets in the residual subspace, which represent typical engineering *a priori* information and requirements, we are able to integrate such useful information into the projection-based fault detection algorithms, and consequently to enhance the fault detectability. The theoretical basis of this application is Theorem 9.5, a simple mathematical claim, but fundamental and useful from the following two aspects:

- the operator $\mathcal{P}_{\mathcal{R}}$ defined by (9.82) describes an orthogonal projection, in which the operator $\mathcal{P}_{\mathcal{R}_y}$ describes an orthogonal projection onto the intersection of those convex sets that reflect *a priori* information;
- the special configuration of $\mathcal{P}_{\mathcal{R}}$ preserves the property that the detector can be well realised in form of an observer-based algorithm, as given by (9.88).

In our work and Example 9.2, some projection operators for $\mathcal{P}_{\mathcal{R}_y}$ are included. They are simple and known examples. The reader is referred to [15] for details. It is noteworthy that in general there exists no closed analytical form of operator for an orthogonal projection onto an intersection of convex sets. Established solutions are numerical algorithms, for instance, using alternating projection method or convex optimisation [15].

At the end of this chapter, we would like to emphasise that the projection-based fault diagnosis framework is established on the basis of assessment of the system dynamics in terms of distances between the process variables (data) and a signal subspace as well as between two signal subspaces. These signal subspaces are the process data sets corresponding to different process operation modes, amongst others, nominal operations, operations under disturbances and varying operating conditions, and faulty operations. In other words, the core of the solution strategy is to classify the system dynamics according to their belonging to the defined data sets. In this context, this framework can be applied uniformly both in the model-based and data-driven fashions. Concerning the data-driven version, the existing methods and algorithms for data-driven realisations of SKR and SIR [37] and gap computation [104] are useful. Recently, explainable design of autoencoders for fault detection in nonlinear dynamic systems using the projection-based method has been reported in [112].

Chapter 10
Projection-Based Fault Classification

In the observer-based fault diagnosis framework, fault detection and isolation are two major tasks, mainly handled with the aid of system analysis and observer design. In the previous chapter, a projection-based approach was proposed, whose core is to assess to which (signal) subspace the acquired process data belong. This is a standard fault classification strategy. Roughly speaking, the task of fault classification is to determine to which class a fault belongs. In this chapter, projection-based methods and algorithms are applied and extended to deal with fault classification issues.

10.1 Fault Detection: A Binary Classification Scheme

In the previous chapter, we have studied a basic projection-based fault detection method whose basis is the nominal system model and (limited) information about model uncertainties. This is a typical one-class classification problem. When a system model (or data) describing faulty operations exists, in addition to the nominal model, fault detection can be achieved using both models. This is a binary fault classification task. It is evident that binary classification solutions, thanks to additional information, could considerably enhance the fault detection performance. Additionally, a multi-class classification problem towards fault isolation can be reformulated as a bank of binary fault classification sub-problems. In this section, binary fault classification issues are addressed in the projection-based framework.

S. X. Ding, *Unified Framework of Control and Detection*,
https://doi.org/10.1007/978-3-662-72417-0_10

10.1.1 Problem Formulation

Given $G_i \in \mathcal{RL}_\infty^{m\times p}, i = 0, 1$, let

$$I_{G_{i,0}} = \begin{bmatrix} M_{i,0} \\ N_{i,0} \end{bmatrix} = \left(A_i + B_i F_i, B_i V_i, \begin{bmatrix} F_i \\ C_i + D_i F_i \end{bmatrix}, \begin{bmatrix} V_i \\ D_i V_i \end{bmatrix} \right) \tag{10.1}$$

denote the corresponding normalised SIR of G_i and its state space realisation, and

$$\mathcal{I}_{G_i} = \left\{ \begin{bmatrix} u \\ y \end{bmatrix} : \begin{bmatrix} u \\ y \end{bmatrix} = I_{G_{i,0}} v, v \in \mathcal{L}_2 \right\} \tag{10.2}$$

be the image subspace. It is supposed that the system operates either in the nominal or faulty state, and the nominal and faulty models are represented by sub-index 0 and sub-index 1, respectively. In other words, $\mathcal{I}_{G_0}$ and $\mathcal{I}_{G_1}$ are complementary subspaces and

$$\mathcal{I}_G := \mathcal{I}_{G_0} \oplus \mathcal{I}_{G_1} = \mathcal{L}_2.$$

Remark 10.1 For simplifying notation, we drop the superscript $\mathcal{L}_2$ in $\mathcal{I}_{G_i}^{\mathcal{L}_2}$.

In the sequel, aiming at addressing practical cases, it is assumed that

$$0 < \delta\left(\mathcal{I}_{G_0}, \mathcal{I}_{G_1}\right) = \delta\left(\mathcal{I}_{G_1}, \mathcal{I}_{G_0}\right) = \delta\left(\mathcal{I}_{G_0}, \mathcal{I}_{G_1}\right) < 1. \tag{10.3}$$

The task of fault detection is to make a decision to which subspace, $\mathcal{I}_{G_0}$ or $\mathcal{I}_{G_1}$, the collected process data (u, y) belong. Below, a fault detector (classifier) is proposed, whose design and realisation consist of two steps:

- design and implementation of two projection-based dynamic estimators,
- design and construction of a classification function.

While the first step is trivial, the second one is demanding and will be intensively addressed.

10.1.2 Projection-Based Estimators

For the fault detection purpose, we now construct two projection-based estimators,

$$\mathcal{P}_{\mathcal{I}_{G_i}} = \mathcal{L}_{I_{G_{i,0}}} \mathcal{L}_{I_{G_{i,0}}^{\sim}}, \ p_{\mathcal{I}_{G_i}} = \mathcal{P}_{\mathcal{I}_{G_i}} \begin{bmatrix} u \\ y \end{bmatrix}, i = 0, 1. \tag{10.4}$$

Consequently,

$$\left\| p_{\mathcal{I}_{G_i}} \right\|_2 = \left\| \mathcal{P}_{\mathcal{I}_{G_i}} \begin{bmatrix} u \\ y \end{bmatrix} \right\|_2 = \left\| I_{G_{i,0}^{\sim}} \begin{bmatrix} u \\ y \end{bmatrix} \right\|_2. \tag{10.5}$$

For the online implementation, the state space realisation of $I_{G_{i,0}^{\sim}}$,

$$x_{i,0}^{\sim}(k) = (A_i + B_i F_i)^T x_{i,0}^{\sim}(k+1) + \begin{bmatrix} F_i^T & (C_i + D_i F_i)^T \end{bmatrix} \begin{bmatrix} u(k) \\ y(k) \end{bmatrix}, \tag{10.6}$$

$$v_{i,0}^{\sim}(k) = V_i^T B_i^T x_{i,0}^{\sim}(k+1) + \begin{bmatrix} V_i^T & V_i^T D_i^T \end{bmatrix} \begin{bmatrix} u(k) \\ y(k) \end{bmatrix}, \tag{10.7}$$

will be computed. The computation amount is similar to the one of an observer.

For our application purpose, the normalised projection,

$$p_{\mathcal{I}_{G_i},n} := \frac{p_{\mathcal{I}_{G_i}}}{\left\| \begin{bmatrix} u \\ y \end{bmatrix} \right\|_2},$$

will be adopted, which can also be written as

$$p_{\mathcal{I}_{G_i},n} = \mathcal{P}_{\mathcal{I}_{G_i}} \begin{bmatrix} u \\ y \end{bmatrix}, \begin{bmatrix} u \\ y \end{bmatrix} \in \mathcal{B}_{(u,y)},$$

$$\mathcal{B}_{(u,y)} = \left\{ \begin{bmatrix} u \\ y \end{bmatrix}, \left\| \begin{bmatrix} u \\ y \end{bmatrix} \right\|_2 = 1, \begin{bmatrix} u \\ y \end{bmatrix} \in \mathcal{L}_2 \right\}.$$

It seems that if

$$\left\| p_{\mathcal{I}_{G_0},n} \right\|_2 = \begin{cases} 1, & \begin{bmatrix} u \\ y \end{bmatrix} \in \mathcal{I}_{G_0} \\ 0, & \begin{bmatrix} u \\ y \end{bmatrix} \in \mathcal{I}_{G_1} \end{cases}, \left\| p_{\mathcal{I}_{G_1},n} \right\|_2 = \begin{cases} 1, & \begin{bmatrix} u \\ y \end{bmatrix} \in \mathcal{I}_{G_1} \\ 0, & \begin{bmatrix} u \\ y \end{bmatrix} \in \mathcal{I}_{G_0}, \end{cases}$$

a perfect fault detection would be achieved. Unfortunately, under the practical assumption (10.3),

$$\forall \begin{bmatrix} u \\ y \end{bmatrix} \in \mathcal{I}_{G_1}, \left\| p_{\mathcal{I}_{G_0},n} \right\|_2 > 0 \text{ and } \forall \begin{bmatrix} u \\ y \end{bmatrix} \in \mathcal{I}_{G_0} \left\| p_{\mathcal{I}_{G_1},n} \right\|_2 > 0.$$

Next, this issue is thoroughly addressed.

10.1.3 Similarity of $\mathcal{I}_{G_0}$ and $\mathcal{I}_{G_1}$

To begin with, we analyse relations between $\mathcal{P}_{\mathcal{I}_{G_i}}, i = 0, 1$, and, associated with them, the dynamics of $p_{\mathcal{I}_{G_i},n}$. The following theorem given in Feintuch's monograph is essential for our study.

Theorem 10.1 *Suppose that* $\mathcal{V}_1$ *and* $\mathcal{V}_2$ *are two closed subspaces of* $\mathcal{H}$. *Then* $\delta(\mathcal{V}_1, \mathcal{V}_2) < 1$ *if and only if operator* $\mathcal{P}_{\mathcal{V}_1} : \mathcal{V}_2 \to \mathcal{V}_1$ *is one-to-one and onto.*

Remark 10.2 The proof given in Feintuch's monograph shows that

$$\begin{aligned} \left\| \mathcal{P}_{\mathcal{V}_1} - \mathcal{P}_{\mathcal{V}_2} \right\| = \delta(\mathcal{V}_1, \mathcal{V}_2) < 1 &\Longrightarrow \text{operator } \left(\mathcal{I} + \mathcal{P}_{\mathcal{V}_2} - \mathcal{P}_{\mathcal{V}_1}\right) \text{ is invertible} \\ &\Longrightarrow \mathcal{P}_{\mathcal{V}_1}\mathcal{H} = \mathcal{P}_{\mathcal{V}_1}\left(\mathcal{I} + \mathcal{P}_{\mathcal{V}_2} - \mathcal{P}_{\mathcal{V}_1}\right)\mathcal{H} = \mathcal{P}_{\mathcal{V}_1}\mathcal{P}_{\mathcal{V}_2}\mathcal{H}. \end{aligned}$$

As a result, $\mathcal{P}_{\mathcal{V}_1}$ maps $\mathcal{V}_2$ to $\mathcal{V}_1$ and it is onto. Furthermore,

$$\forall z \neq 0, z \in \mathcal{V}_2, \mathcal{P}_{\mathcal{V}_1} z \neq 0,$$

otherwise,

$$\begin{aligned} \|z\|_2 &= \left\| \left(\mathcal{I} - \mathcal{P}_{\mathcal{V}_1}\right) z \right\|_2 = \left\| \left(\mathcal{I} - \mathcal{P}_{\mathcal{V}_1}\right) \mathcal{P}_{\mathcal{V}_2} z \right\|_2 \\ &\leq \left\| \left(\mathcal{I} - \mathcal{P}_{\mathcal{V}_1}\right) \mathcal{P}_{\mathcal{V}_2} \right\| \|z\|_2 < \|z\|_2 \end{aligned}$$

leads to a contradiction to $\delta(\mathcal{V}_1, \mathcal{V}_2) < 1$. Let $\mathcal{M} : \mathcal{V}_2 \to \mathcal{V}_1$ denote the mapping from $\mathcal{P}_{\mathcal{V}_2}\mathcal{H}$ onto $\mathcal{P}_{\mathcal{V}_1}\mathcal{H}$ and assume it is one-to-one. And, it is showcased that

$$\begin{aligned} \delta(\mathcal{V}_1, \mathcal{V}_2) &= \sqrt{1 - \mu^2} < 1, \\ \mu &= \inf\left\{ \left\| \mathcal{P}_{\mathcal{V}_2} z \right\|_2, \|z\|_2 = 1, z \in \mathcal{P}_{\mathcal{V}_1}\mathcal{H} \right\} \\ &= \inf\left\{ \left\| \mathcal{P}_{\mathcal{V}_1} z \right\|_2, \|z\|_2 = 1, z \in \mathcal{P}_{\mathcal{V}_2}\mathcal{H} \right\} \neq 0. \end{aligned} \tag{10.8}$$

Thus, $\mathcal{P}_{\mathcal{V}_1}$ maps $\mathcal{V}_2$ to $\mathcal{V}_1$ one-to-one.

Applying this theorem to our case results in the following corollary.

Corollary 10.1 *Given* $\mathcal{I}_{G_i}, i = 0, 1$, *and the corresponding projection operators* $\mathcal{P}_{\mathcal{I}_{G_i}}$, *then*

$$\forall \begin{bmatrix} u \\ y \end{bmatrix} \in \mathcal{I}_{G_j}, j \neq i, 0 < \left\| r_{\mathcal{I}_{G_i}} \right\|_2 = dist\left(\begin{bmatrix} u \\ y \end{bmatrix}, \mathcal{I}_{G_i} \right) \tag{10.9}$$

$$= \sqrt{ \left\| \begin{bmatrix} u \\ y \end{bmatrix} \right\|_2^2 - \left\| \mathcal{P}_{\mathcal{I}_{G_i}} \mathcal{P}_{\mathcal{I}_{G_j}} \begin{bmatrix} u \\ y \end{bmatrix} \right\|_2^2 } \leq \sqrt{1 - \mu^2} < \left\| \begin{bmatrix} u \\ y \end{bmatrix} \right\|_2 \tag{10.10}$$

with μ *defined by (10.8).*

Proof The equation in (10.10) is straightforward, because

$$\begin{aligned} \left\| \begin{bmatrix} u \\ y \end{bmatrix} \right\|_2^2 &= \left\| \mathcal{P}_{\mathcal{I}_{G_i}} \begin{bmatrix} u \\ y \end{bmatrix} \right\|_2^2 + \left\| r_{\mathcal{I}_{G_i}} \right\|_2^2, \\ \forall \begin{bmatrix} u \\ y \end{bmatrix} &\in \mathcal{I}_{G_j}, \begin{bmatrix} u \\ y \end{bmatrix} = \mathcal{P}_{\mathcal{I}_{G_j}} \begin{bmatrix} u \\ y \end{bmatrix}. \end{aligned}$$

Note that

$$\mathcal{I}_{G_j} = \left(\mathcal{I} - \mathcal{P}_{\mathcal{I}_{G_i}}\right)\mathcal{I}_{G_j} + \mathcal{P}_{\mathcal{I}_{G_i}}\mathcal{I}_{G_j}.$$

The inequality in (10.10) is due to the fact that

$$\forall \begin{bmatrix} u \\ y \end{bmatrix}, \mathcal{P}_{\mathcal{I}_{G_i}}\mathcal{P}_{\mathcal{I}_{G_j}}\begin{bmatrix} u \\ y \end{bmatrix} \neq 0 \text{ and } \left\| \mathcal{P}_{\mathcal{I}_{G_i}}\mathcal{P}_{\mathcal{I}_{G_j}}\begin{bmatrix} u \\ y \end{bmatrix} \right\|_2^2 \geq \mu^2,$$

while the inequality in (10.9) is attributed to the assumption that $\mathcal{I}_{G_i}$ and $\mathcal{I}_{G_j}, i \neq j$, are complementary, which implies

$$\forall \begin{bmatrix} u \\ y \end{bmatrix} \in \mathcal{I}_{G_j}, \mathcal{I}_{G_j}\begin{bmatrix} u \\ y \end{bmatrix} \neq \mathcal{P}_{\mathcal{I}_{G_i}}\mathcal{I}_{G_j}\begin{bmatrix} u \\ y \end{bmatrix}$$

$$\Longrightarrow \left(\mathcal{I} - \mathcal{P}_{\mathcal{I}_{G_i}}\right)\mathcal{I}_{G_j}\begin{bmatrix} u \\ y \end{bmatrix} \neq 0 \Longrightarrow \left\| r_{\mathcal{I}_{G_i}} \right\|_2 = dist\left(\begin{bmatrix} u \\ y \end{bmatrix}, \mathcal{I}_{G_i}\right) > 0.$$

The corollary is proved.

It follows from Corollary 10.1 that

$$\forall \begin{bmatrix} u \\ y \end{bmatrix} \in \mathcal{I}_{G_0}, \left\| p_{\mathcal{I}_{G_0},n} \right\|_2 = 1, \mu \leq \left\| p_{\mathcal{I}_{G_1},n} \right\|_2 < 1,$$
$$\forall \begin{bmatrix} u \\ y \end{bmatrix} \in \mathcal{I}_{G_1}, \left\| p_{\mathcal{I}_{G_1},n} \right\|_2 = 1, \mu \leq \left\| p_{\mathcal{I}_{G_0},n} \right\|_2 < 1.$$

Accordingly, it is suggested to split the data space $\mathcal{I}_G = \mathcal{I}_{G_0} \oplus \mathcal{I}_{G_1}$ into three sets

$$\mathcal{I}_G = \mathcal{S}_0 \oplus \mathcal{S}_1 \oplus \mathcal{S}_{01}, \tag{10.11}$$
$$\mathcal{S}_0 = \left\{ \begin{bmatrix} u \\ y \end{bmatrix}, \left\| p_{\mathcal{I}_{G_0},n} \right\|_2 = 1, J_{th,0} \leq \left\| r_{\mathcal{I}_{G_1},n} \right\|_2 \right\},$$
$$\mathcal{S}_1 = \left\{ \begin{bmatrix} u \\ y \end{bmatrix}, \left\| p_{\mathcal{I}_{G_1},n} \right\|_2 = 1, J_{th,1} \leq \left\| r_{\mathcal{I}_{G_0},n} \right\|_2 \right\},$$
$$\mathcal{S}_{01} = \bar{\mathcal{S}}_0 \cup \bar{\mathcal{S}}_1, \tag{10.12}$$
$$\bar{\mathcal{S}}_0 := \left\{ \begin{bmatrix} u \\ y \end{bmatrix}, \left\| p_{\mathcal{I}_{G_0},n} \right\|_2 = 1, \left\| r_{\mathcal{I}_{G_1},n} \right\|_2 < J_{th,0} \right\}, \tag{10.13}$$
$$\bar{\mathcal{S}}_1 := \left\{ \begin{bmatrix} u \\ y \end{bmatrix}, \left\| p_{\mathcal{I}_{G_1},n} \right\|_2 = 1, \left\| r_{\mathcal{I}_{G_0},n} \right\|_2 < J_{th,1} \right\}, \tag{10.14}$$

where $J_{th,0}, J_{th,1}$ are constants satisfying

$$0 << J_{th,0} \leq \sqrt{1-\mu^2}, 0 << J_{th,1} \leq \sqrt{1-\mu^2},$$

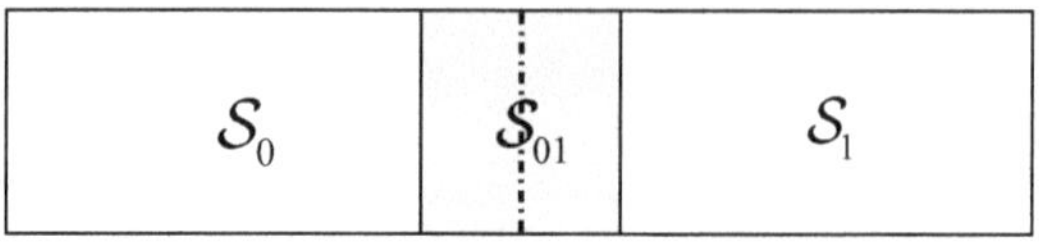

Fig. 10.1 Schematic description of the composition of the data subspace

Figure 10.1 depicts the above composition of the data subspace schematically. Now, we explain the meaning and selection of $J_{th,0}$, $J_{th,1}$ as well as the interpretation of the sets $\mathcal{S}_0, \mathcal{S}_{01}, \mathcal{S}_1$, which are denominated as Operation Mode (OM) I, II and III, respectively. It is apparent that

- in OM I, $\forall \begin{bmatrix} u \\ y \end{bmatrix} \in \mathcal{S}_0$,

$$\begin{bmatrix} u \\ y \end{bmatrix} \in \mathcal{I}_{G_0}, dist\left(\begin{bmatrix} u \\ y \end{bmatrix}, \mathcal{I}_{G_1}\right) = \left\| r_{\mathcal{I}_{G_1}} \right\|_2 \geq J_{th,0}.$$

  This implies, the system is under the nominal operation condition and distant from the faulty dynamic. Hence, OM I indicates fault-free operation.

- As against, in OM III, $\forall \begin{bmatrix} u \\ y \end{bmatrix} \in \mathcal{S}_1$,

$$\begin{bmatrix} u \\ y \end{bmatrix} \in \mathcal{I}_{G_1}, dist\left(\begin{bmatrix} u \\ y \end{bmatrix}, \mathcal{I}_{G_0}\right) = \left\| r_{\mathcal{I}_{G_0}} \right\|_2 \geq J_{th,1},$$

  that is, the system operates under the faulty condition and is far away from the nominal dynamic. Accordingly, OM III means faulty operation.

OM II raises our interest for a reasonable and convincing interpretation. It is well-known that formation of faults, for instance, caused by ageing, is a longtime process. It begins with incipient degradation that does not affect the system dynamics significantly, and thus is interpreted as the transitional phase from the fault-free to faulty operation. Accordingly, the operations in this range can be called incipient fault. This is one interpretation of OM II. A further possible explanation for this case is the so-called intermittent faults which repeatedly occur in the process over a time interval and then disappear. Their emergence and disappearance are triggered under certain system operation conditions. That means, in OM II, for some input signals (and so the corresponding outputs), the system may operate normally, and then, as a response to the change of input signals, works in the faulty operation. In summary, it can be concluded that, if the system operates in OM II, warning should be triggered to call operator's attention for possible faults. Figure 10.2 sketches the system operation ranges and states. In order to gain a deeper insight into the aforementioned discussions, we would like to introduce some useful system theoretical aspects in the sequel. The following theorem is a summary of some relevant results in Feintuch's monograph.

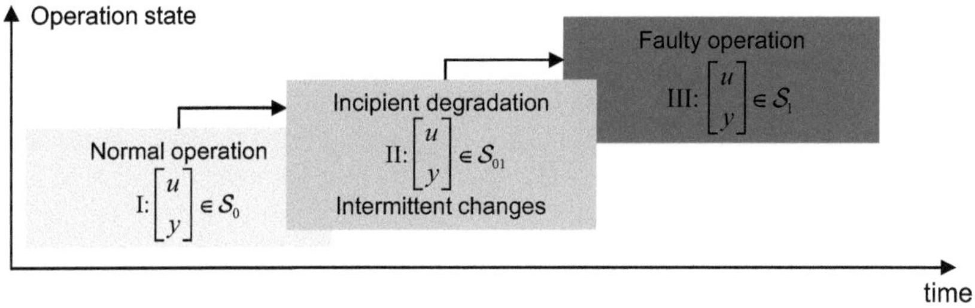

Fig. 10.2 Schematic description of operation states

Theorem 10.2 *Suppose that $\mathcal{V}_1$ and $\mathcal{V}_2$ are two closed subspaces of $\mathcal{H}$, $\delta(\mathcal{V}_1, \mathcal{V}_2) < 1$, and the operators,*

$$\mathcal{M}_{12} : \mathcal{V}_2 \to \mathcal{V}_1, \forall x \in \mathcal{V}_2, \mathcal{M}_{12} x = \mathcal{P}_{\mathcal{V}_1} x,$$
$$\mathcal{M}_{21} : \mathcal{V}_1 \to \mathcal{V}_2, \forall x \in \mathcal{V}_1, \mathcal{M}_{21} x = \mathcal{P}_{\mathcal{V}_2} x,$$

are thus invertible. Then, it holds

$$\mathcal{V}_2 = \mathcal{M}_{21}\mathcal{V}_1 = (\mathcal{I}(\mathcal{V}_1) + \mathcal{X})\mathcal{V}_1, \mathcal{X} = \left(\mathcal{I} - \mathcal{P}_{\mathcal{V}_1}\right)\mathcal{M}_{12}^{-1},$$

where $\mathcal{I}(\mathcal{V}_1)$ is the identity operator restricted to $\mathcal{V}_1$.

According to this theorem, the two image subspaces in our binary fault classification setting have the following relation,

$$\mathcal{I}_{G_1} = \mathcal{M}_{10}\mathcal{I}_{G_0} = \left(\mathcal{I}\left(\mathcal{I}_{G_0}\right) + \mathcal{X}\right)\mathcal{I}_{G_0}, \mathcal{X} = \left(\mathcal{I} - \mathcal{P}_{\mathcal{I}_{G_0}}\right)\mathcal{M}_{01}^{-1}, \tag{10.15}$$

where $\mathcal{M}_{10} : \mathcal{I}_{G_0} \to \mathcal{I}_{G_1}$, $\mathcal{M}_{01} : \mathcal{I}_{G_1} \to \mathcal{I}_{G_0}$ are two invertible operators. In fact, as given in the proof of Theorem 10.2,

$$\mathcal{M}_{10}\mathcal{I}_{G_0} = \left(\mathcal{P}_{\mathcal{I}_{G_0}}\mathcal{M}_{10} + \mathcal{M}_{10} - \mathcal{P}_{\mathcal{I}_{G_0}}\mathcal{M}_{10}\right)\mathcal{I}_{G_0}$$
$$= \mathcal{P}_{\mathcal{I}_{G_0}}\mathcal{M}_{10}\mathcal{I}_{G_0} + \left(\mathcal{I} - \mathcal{P}_{\mathcal{I}_{G_0}}\right)\mathcal{M}_{10}\mathcal{I}_{G_0} = \mathcal{P}_{\mathcal{I}_{G_0}}\mathcal{I}_{G_1} + \left(\mathcal{I} - \mathcal{P}_{\mathcal{I}_{G_0}}\right)\mathcal{I}_{G_1}.$$

On the other hand,

$$\begin{aligned} &\mathcal{P}_{\mathcal{I}_{G_0}}\mathcal{M}_{10}\mathcal{I}_{G_0} + \left(\mathcal{I} - \mathcal{P}_{\mathcal{I}_{G_0}}\right)\mathcal{M}_{10}\mathcal{I}_{G_0} \\ &= \left(\mathcal{P}_{\mathcal{I}_{G_0}}\mathcal{M}_{01}^{-1} + \left(\mathcal{I} - \mathcal{P}_{\mathcal{I}_{G_0}}\right)\mathcal{M}_{01}^{-1}\right)\mathcal{M}_{01}\mathcal{M}_{10}\mathcal{I}_{G_0} \\ &= \left(\mathcal{P}_{\mathcal{I}_{G_0}}\mathcal{M}_{01}^{-1} + \left(\mathcal{I} - \mathcal{P}_{\mathcal{I}_{G_0}}\right)\mathcal{M}_{01}^{-1}\right)\mathcal{I}_{G_0}, \end{aligned}$$

which finally leads to (10.15). Therefore, (10.15) is an alternative expression of the decomposition of $\mathcal{I}_{G_1}$. It illustrates that the gap between faulty and fault-free subspaces can be expressed, for instance, by the fault-free subspace and an operator defined on this subspace, as far as these two subspaces are similar up to a certain degree (i.e. the gap metric is not equal to one). This makes the binary fault classification problem, in spite of the additional information, hard to solve. The primary cause for this difficultly consists in the transitional phase from the fault-free to faulty operation represented by the set $\mathcal{S}_{01}$. Although it is difficult to specify $\mathcal{S}_{01}$ analytically, it is possible to characterise this subspace by means of simulation or data. This would be helpful to understand the mechanism of incipient and intermittent faults, and thus to detect them reliably.

10.1.4 Classification Function

Next, we propose, based on the aforementioned results, a classification function $f(u, y)$ that discriminates fault-free and faulty data and is defined by

$$f(u, y) = \frac{\left\| p_{\mathcal{I}_{G_1},n} \right\|_2^2 - \left\| p_{\mathcal{I}_{G_0},n} \right\|_2^2}{1 - J_{th}^2}, \begin{bmatrix} u \\ y \end{bmatrix} \in \mathcal{B}_{(u,y)}. \tag{10.16}$$

Hereby, it is assumed that $J_{th,0} = J_{th,1} = J_{th}$. Observe that $\forall \begin{bmatrix} u \\ y \end{bmatrix} \in \mathcal{I}_{G_j}$,

$$\left\| p_{\mathcal{I}_{G_i},n} \right\|_2^2 = 1 - \left\| r_{\mathcal{I}_{G_i}} \right\|_2^2 \Longrightarrow \begin{cases} 1 - J_{th}^2 \geq \left\| p_{\mathcal{I}_{G_i},n} \right\|_2^2, \left\| r_{\mathcal{I}_{G_i}} \right\|_2 \geq J_{th} \\ 1 - J_{th}^2 < \left\| p_{\mathcal{I}_{G_i},n} \right\|_2^2, \left\| r_{\mathcal{I}_{G_i}} \right\|_2 < J_{th}, \end{cases}$$

$i = 0, 1$. It turns out

$$\begin{cases} f(u, y) \leq -\varepsilon, \begin{bmatrix} u \\ y \end{bmatrix} \in \mathcal{S}_0 \\ -\varepsilon < f(u, y) < \varepsilon, \begin{bmatrix} u \\ y \end{bmatrix} \in \mathcal{S}_{01} \\ f(u, y) \geq \varepsilon, \begin{bmatrix} u \\ y \end{bmatrix} \in \mathcal{S}_1 \end{cases}, \varepsilon = \frac{J_{th}^2}{1 - J_{th}^2}, \tag{10.17}$$

which allows us to define the detection logic as

$$\begin{cases} f(u, y) \leq -\varepsilon, \text{ no alarm } \Longrightarrow \text{fault-free} \\ -\varepsilon < f(u, y) < \varepsilon, \text{ warning for incipient/intermittent degradations} \\ f(u, y) \geq \varepsilon, \text{ alarm } \Longrightarrow \text{faulty.} \end{cases} \tag{10.18}$$

10.1.5 A Summary

Beginning with a formulation of fault detection problem as a binary (fault) classification, our work ends with a solution consisting of

- two projection-based dynamic systems (10.5) that can be realised using the dynamic system given in (10.6)–(10.7),
- the classification function (10.16), and
- the decision rule (10.18).

It is somewhat surprising that this solution for a binary classification problem is finally given in the form of three classes, as described by the decision rule (10.18). The reason is obvious, namely, the two classes defined by the corresponding system models $G_i, i = 0, 1$, are inseparable in the sense that both system models could generate similar process data (u, y). From the control-theoretic viewpoint, this conclusion is explained by the condition $0 < \delta\left(\mathcal{I}_{G_0}, \mathcal{I}_{G_1}\right) < 1$. On the other hand, we have given a plausible engineering explanation for the similarity of $\mathcal{I}_{G_1}$ and $\mathcal{I}_{G_0}$, that is, it models incipient and/or intermittent faults. In the traditional research of observer-based detection of incipient and/or intermittent faults, major efforts focus on how to detect these types of faults. No study has been reported on modelling of their evolution process from control theoretic point of view.

Another point is the dependence of the fault detectability on the value of the gap metric $\delta\left(\mathcal{I}_{G_0}, \mathcal{I}_{G_1}\right)$. The volumes of $\mathcal{S}_0$ and $\mathcal{S}_1$ become larger and the one of $\mathcal{S}_{01}$ becomes smaller, as the gap metric $\delta\left(\mathcal{I}_{G_0}, \mathcal{I}_{G_1}\right)$ increases. The reason is obvious, namely, the gap metric is a similarity measurement.

Figure 10.3 visualises the relations between the system subspaces $\mathcal{I}_{G_0}, \mathcal{I}_{G_1}$, the corresponding classification function and the detection signal.

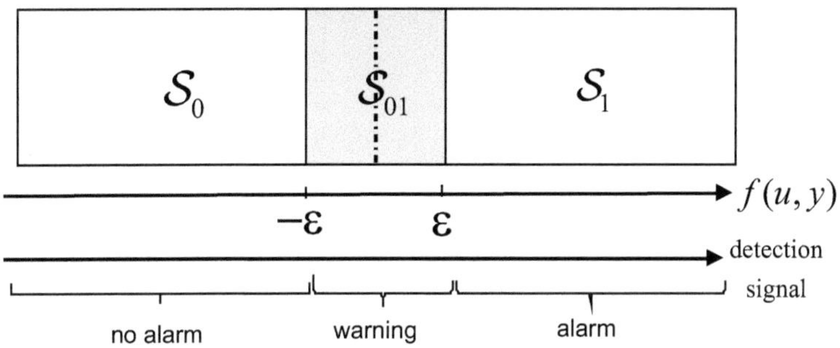

Fig. 10.3 System subspaces and the corresponding classification function as well as detection signal

10.2 Modified Binary Fault Detection Schemes

In Sect. 9.6, we have introduced an approach to incorporate a priori process knowledge into the system projection aiming at enhancing the fault detectability. Analogously to this handling and the result in Theorem 9.5, we propose algorithms to improve the binary classification scheme studied in the previous section.

10.2.1 On Nominal System Dynamics

In practice, besides process models like SIRs, there often exists additional a priori information about the system nominal dynamics, for instance, a nominal value region of the hidden variable v. Suppose that such information is represented by a convex set $\mathcal{V}_{conv} \subset \mathcal{V} = \{v : v \in \mathcal{L}_2\}$. Let $\mathcal{P}_{\mathcal{V}_{conv}} : \mathcal{L}_2 \to \mathcal{L}_2$ be an operator orthogonally projecting $v \in \mathcal{L}_2$ onto $\mathcal{V}_{conv}$. Analogous to Theorem 9.5, it is straightforward to prove that

$$\mathcal{P}_{\mathcal{I}_{0},\mathcal{V}_{conv}} : \mathcal{L}_2 \to \mathcal{L}_2, \mathcal{P}_{\mathcal{I}_{0},\mathcal{V}_{conv}} = \mathcal{L}_{I_{G_0}} \mathcal{P}_{\mathcal{V}_{conv}} \mathcal{L}_{I_{\tilde{G}_0}} \tag{10.19}$$

is an orthogonal operator.

Example 10.1 Suppose that

$$\mathcal{V}_{conv} = \left\{ v(k) \in \mathbb{R}^p : v_{LB} \preceq v(k) \preceq v_{UB} \right\},$$

where $\preceq$ denotes componentwise inequality between vectors, and

$$v_{LB} = \begin{bmatrix} v_{1,LB} \\ \vdots \\ v_{p,LB} \end{bmatrix}, v_{UB} = \begin{bmatrix} v_{1,UB} \\ \vdots \\ v_{p,UB} \end{bmatrix}$$

are lower and upper bounds, respectively. As shown in Example 9.2, the corresponding projection operator $\mathcal{P}_{\mathcal{V}_{conv}}$ is given by

$$\mathcal{P}_{\mathcal{V}_{conv}}(v(k)) = \begin{bmatrix} \hat{v}_1(k) \\ \vdots \\ \hat{v}_p(k) \end{bmatrix},$$

$$\hat{v}_i(k) = \begin{cases} v_i(k), v_{i,LB} \leq v_i(k) \leq v_{i,UB}, \\ v_{i,LB}, v_i(k) < v_{i,LB}, \\ v_{i,UB}, v_i(k) > v_{i,UB}. \end{cases}, i = 1, \cdots, p.$$

10.2.2 An Improved Detection Scheme

It is plain that embedding the projection onto $\mathcal{V}_{conv}$ as additional information in the detection system will enhance the fault detectability and reduce false alarms. To this end, we examine $\mathcal{S}_0$ and $\mathcal{S}_{01}$ under embedding the projection onto $\mathcal{V}_{conv}$. It is obvious that they can be split into

$$
\begin{aligned}
\mathcal{S}_0 &= \mathcal{S}_{0,ff} \oplus \mathcal{S}_{0,f},\\
\mathcal{S}_{0,ff} &= \left\{ \begin{bmatrix} u \\ y \end{bmatrix}, \left\| \mathcal{P}_{\mathcal{I}_0,\mathcal{V}_{conv}} \begin{bmatrix} u \\ y \end{bmatrix} \right\|_2 = 1, \begin{bmatrix} u \\ y \end{bmatrix} \in \mathcal{S}_0 \right\},\\
\mathcal{S}_{0,f} &= \left\{ \begin{bmatrix} u \\ y \end{bmatrix}, \left\| \mathcal{P}_{\mathcal{I}_0,\mathcal{V}_{conv}} \begin{bmatrix} u \\ y \end{bmatrix} \right\|_2 = 0, \begin{bmatrix} u \\ y \end{bmatrix} \in \mathcal{S}_0 \right\},
\end{aligned}
$$

as well as

$$
\begin{aligned}
\mathcal{S}_{01} &= \bar{\mathcal{S}}_{01} \oplus \mathcal{S}_{01,f}, \bar{\mathcal{S}}_{01} = \bar{\mathcal{S}}_{0,0} \cup \bar{\mathcal{S}}_{1,0}, \mathcal{S}_{01,f} = \bar{\mathcal{S}}_{0,f} \cup \bar{\mathcal{S}}_{1,f},\\
\bar{\mathcal{S}}_{0,0} &= \left\{ \begin{bmatrix} u \\ y \end{bmatrix}, \left\| \mathcal{P}_{\mathcal{I}_0,\mathcal{V}_{conv}} \begin{bmatrix} u \\ y \end{bmatrix} \right\|_2 = 1, \begin{bmatrix} u \\ y \end{bmatrix} \in \bar{\mathcal{S}}_0 \right\},\\
\bar{\mathcal{S}}_{1,0} &= \left\{ \begin{bmatrix} u \\ y \end{bmatrix}, \left\| \mathcal{P}_{\mathcal{I}_1,\mathcal{V}_{conv}} \begin{bmatrix} u \\ y \end{bmatrix} \right\|_2 = 1, \begin{bmatrix} u \\ y \end{bmatrix} \in \bar{\mathcal{S}}_1 \right\},\\
\bar{\mathcal{S}}_{0,f} &= \left\{ \begin{bmatrix} u \\ y \end{bmatrix}, \left\| \mathcal{P}_{\mathcal{I}_0,\mathcal{V}_{conv}} \begin{bmatrix} u \\ y \end{bmatrix} \right\|_2 = 0, \begin{bmatrix} u \\ y \end{bmatrix} \in \bar{\mathcal{S}}_0 \right\},\\
\bar{\mathcal{S}}_{1,f} &= \left\{ \begin{bmatrix} u \\ y \end{bmatrix}, \left\| \mathcal{P}_{\mathcal{I}_1,\mathcal{V}_{conv}} \begin{bmatrix} u \\ y \end{bmatrix} \right\|_2 = 0, \begin{bmatrix} u \\ y \end{bmatrix} \in \bar{\mathcal{S}}_1 \right\},
\end{aligned}
$$

where $\mathcal{S}_{0,f}$ and $\mathcal{S}_{01,f}$ denote the data sets consisting of the data generated during faulty operations, $\mathcal{S}_{0,ff}$ indicates the data set with fault-free data. Consequently, the system signal space $\mathcal{I}_G$ can now be rewritten as

$$
\begin{aligned}
\mathcal{I}_G &= \mathcal{S}_{ff} \oplus \bar{\mathcal{S}}_{01} \oplus \mathcal{S}_f, \\
\mathcal{S}_{ff} &= \mathcal{S}_{0,ff}, \mathcal{S}_f = \mathcal{S}_1 \cup \mathcal{S}_{0,f} \cup \mathcal{S}_{01,f}.
\end{aligned} \tag{10.20}
$$

Comparing (10.20) with (10.11) makes it clear that embedding the projection $\mathcal{P}_{\mathcal{V}_{conv}}$ enhances the fault detectability and reduce false alarms. To be specific, it follows from the definition of $\mathcal{I}_{G_1}$ and $\mathcal{I}_{G_0}$ by (10.2) that the alternative projection $\mathcal{P}_{\mathcal{I}_0,\mathcal{V}_{conv}} = \mathcal{L}_{I_{G_0}} \mathcal{P}_{\mathcal{V}_{conv}} \mathcal{L}_{I_{G_0}^{\sim}}$ results in

- reducing the set $\mathcal{S}_{01}$, from $\mathcal{S}_{01}$ to $\bar{\mathcal{S}}_{01}$, representing OM II,
- reducing the set $\mathcal{S}_0$, from $\mathcal{S}_0$ to $\mathcal{S}_{0,ff}$, representing OM I, and
- increasing the set $\mathcal{S}_1$,

$$\text{from } \mathcal{S}_1 \text{ to } \mathcal{S}_f = \mathcal{S}_1 \cup \mathcal{S}_{0,f} \cup \mathcal{S}_{01,f},$$

which indicates the faulty operations.

10.2.3 A Binary Fault Detection Scheme for a Class of Faulty Dynamics

The fault detection schemes introduced in the previous subsections are based on the assumption of the faulty model given in (10.1). In this subsection, we consider a class of faulty dynamic systems modelled by

$$G_1 = \left(N_{0,0} + N_f \Theta\right)\left(M_{0,0} + M_f \Theta\right)^{-1}, \tag{10.21}$$

$$I_{G_1} = I_{G_0} + \Delta_f = \begin{bmatrix} M_1 \\ N_1 \end{bmatrix} = \begin{bmatrix} M_{0,0} + M_f \Theta \\ N_{0,0} + N_f \Theta \end{bmatrix}, I_f = \begin{bmatrix} M_f \\ N_f \end{bmatrix}. \tag{10.22}$$

It is assumed that

- $M_f, N_f \in \mathcal{RH}_\infty$ are known and

$$\mathcal{I}_{G_0} \cap \mathcal{I}_f \neq \mathcal{O}, \mathcal{I}_{G_0}^{\perp} \cap \mathcal{I}_f \neq \mathcal{O}, \tag{10.23}$$

$$\mathcal{I}_f := \left\{ \begin{bmatrix} u \\ y \end{bmatrix} : \begin{bmatrix} u \\ y \end{bmatrix} = \begin{bmatrix} M_f \\ N_f \end{bmatrix} \alpha, \alpha \in \mathcal{L}_2 \right\}, \tag{10.24}$$

- $\Theta \in \mathcal{H}_\infty$ is unknown, but bounded, and
- the data (u, y) generated during nominal operations build a set $\mathcal{I}_{G_0,n} \subset \mathcal{I}_{G_0}$,

$$\mathcal{I}_{G_0,n} = \left\{ \begin{bmatrix} u \\ y \end{bmatrix} : \begin{bmatrix} u \\ y \end{bmatrix} = I_{G_0} v, v \in \mathcal{L}_2, \|v\|_2 \leq \delta_v \right\}, \tag{10.25}$$

and the upper-bound δ_v is known.

It is of interest to notice that, according to Theorem 4.3, any faulty system can be modeled by

$$G = NM^{-1} = \left(N_{0,0} + \hat{X}_0 S\right)\left(M_{0,0} - \hat{Y}_0 S\right)^{-1} \Longrightarrow I_{G_1} = \begin{bmatrix} M_{0,0} - \hat{Y}_0 S \\ N_{0,0} + \hat{X}_0 S \end{bmatrix},$$

if it is stabilisable by an output feedback controller. Here, $\left(\hat{X}_0, \hat{Y}_0\right)$ is a RCP satisfying (2.25) and the Bezout identity (2.31), and $S \in \mathcal{H}_\infty$ is unknown. This is a special case of faulty model (10.22). Observe that in this case, by an orthogonal projection onto the residual subspace,

$$r_{\mathcal{I}_G^{\mathcal{L}_2}} = \mathcal{L}_{K_{\tilde{G}_0}} \mathcal{L}_{K_{G_0}} \begin{bmatrix} u \\ y \end{bmatrix} = K_{\tilde{G}_0} K_{G_0} \begin{bmatrix} M_{0,0} - \hat{Y}_0 S \\ N_{0,0} + \hat{X}_0 S \end{bmatrix} v$$
$$\Longrightarrow \forall Sv \neq 0, \left\| r_{\mathcal{I}_G^{\mathcal{L}_2}} \right\|_2 = \left\| K_{\tilde{G}_0} K_{G_0} \begin{bmatrix} u \\ y \end{bmatrix} \right\|_2 = \|Sv\|_2 \neq 0,$$
$$K_{G_0} = \begin{bmatrix} -\hat{N}_{0,0} & \hat{M}_{0,0} \end{bmatrix},$$

the fault can be perfectly (without considering possible uncertainties) detected. This result is indeed not surprising, considering the Bezout identity,

$$\begin{bmatrix} X_0 & Y_0 \\ -\hat{N}_{0,0} & \hat{M}_{0,0} \end{bmatrix} \begin{bmatrix} M_{0,0} & -\hat{Y}_0 \\ N_{0,0} & \hat{X}_0 \end{bmatrix} = \begin{bmatrix} I & 0 \\ 0 & I \end{bmatrix},$$

which indicates dissimilarity between $\begin{bmatrix} M_{0,0} \\ N_{0,0} \end{bmatrix}$ and $\begin{bmatrix} -\hat{Y}_0 \\ \hat{X}_0 \end{bmatrix}$. The relation

$$K_{G_0} \begin{bmatrix} u \\ y \end{bmatrix} = Sv$$

means a full recovery of information in the subspace spanned by $\begin{bmatrix} -\hat{Y}_0 \\ \hat{X}_0 \end{bmatrix}$.

As a more general form, the assumption (10.23) reflects practical process operations. In order to achieve a reliable fault detection, the intersection set of $\mathcal{I}_{G_0}$ and $\mathcal{I}_f$ should be taken into account as well. In the sequel, we focus on this task. Thanks to the Bezout identity given above, $\begin{bmatrix} M_f \\ N_f \end{bmatrix}$ can be written into

$$\begin{bmatrix} M_f \\ N_f \end{bmatrix} = \begin{bmatrix} M_{0,0} & -\hat{Y}_0 \\ N_{0,0} & \hat{X}_0 \end{bmatrix} \begin{bmatrix} \bar{I}_{f,1} \\ I_{f,2} \end{bmatrix},$$

for some $\bar{I}_{f,1}, I_{f,2} \in \mathcal{RH}_\infty$. Recall the discussion in Sect. 7.1 that

$$\begin{bmatrix} -\hat{Y}_0 \\ \hat{X}_0 \end{bmatrix} = \begin{bmatrix} M_{0,0} \\ N_{0,0} \end{bmatrix} Q_{\mathcal{H}_2} + \left(\begin{bmatrix} -\hat{Y}_0 \\ \hat{X}_0 \end{bmatrix} - \begin{bmatrix} M_{0,0} \\ N_{0,0} \end{bmatrix} Q_{\mathcal{H}_2} \right),$$

where $Q_{\mathcal{H}_2} \in \mathcal{RH}_\infty$ is the solution of the following MMP,

$$Q_{\mathcal{H}_2} = \arg\inf_{Q} \left\| \begin{bmatrix} -\hat{Y}_0 \\ \hat{X}_0 \end{bmatrix} - Q \begin{bmatrix} M_{0,0} \\ N_{0,0} \end{bmatrix} \right\|_\infty .$$

Thus,

$$\Delta_f = \begin{bmatrix} M_f \\ N_f \end{bmatrix} \Theta = \begin{bmatrix} M_{0,0} \\ N_{0,0} \end{bmatrix} I_{f,1}\Theta + \left(\begin{bmatrix} -\hat{Y}_0 \\ \hat{X}_0 \end{bmatrix} - \begin{bmatrix} M_{0,0} \\ N_{0,0} \end{bmatrix} Q_{\mathcal{H}_2} \right) I_{f,2}\Theta, \tag{10.26}$$
$$I_{f,1} = \bar{I}_{f,1} + Q_{\mathcal{H}_2} I_{f,2}.$$

Accordingly, the intersection set of $\mathcal{I}_{G_0}$ and $\mathcal{I}_f$, denoted by $\mathcal{S}_{\mathcal{I}_{G_0}\cap\mathcal{I}_f}$, can be modelled by

$$\mathcal{S}_{\mathcal{I}_{G_0}\cap\mathcal{I}_f} = \left\{ \begin{bmatrix} u \\ y \end{bmatrix} : \begin{bmatrix} u \\ y \end{bmatrix} = \begin{bmatrix} M_{0,0} \\ N_{0,0} \end{bmatrix} I_{f,1}\beta, \beta \in \mathcal{L}_2 \right\}. \tag{10.27}$$

Apart from the assumption (10.25), the following algorithm yields a classification of $\begin{bmatrix} u \\ y \end{bmatrix} \in \mathcal{S}_{\mathcal{I}_{G_0}\cap\mathcal{I}_f}$:

- computation of the orthogonal projection of $\begin{bmatrix} u \\ y \end{bmatrix}$ onto $\mathcal{I}_{G_0,n}$ and its relative norm,

$$p_{\mathcal{P}_{\mathcal{I}_{G_0,n}}} = \mathcal{P}_{\mathcal{I}_{G_0,n}} \begin{bmatrix} u \\ y \end{bmatrix} = \frac{\mathcal{P}_{\mathcal{I}_{G_0}} \begin{bmatrix} u \\ y \end{bmatrix}}{\max\left\{1, \left\| \begin{bmatrix} u \\ y \end{bmatrix} \right\|_2 / \delta_v \right\}}, \tag{10.28}$$

$$\frac{\left\| p_{\mathcal{P}_{\mathcal{I}_{G_0,n}}} \right\|_2}{\left\| \begin{bmatrix} u \\ y \end{bmatrix} \right\|_2} = \frac{\left\| \mathcal{P}_{\mathcal{I}_{G_0}} \begin{bmatrix} u \\ y \end{bmatrix} \right\|_2}{\left\| \begin{bmatrix} u \\ y \end{bmatrix} \right\|_2 \max\left\{1, \left\| \begin{bmatrix} u \\ y \end{bmatrix} \right\|_2 / \delta_v \right\}} \tag{10.29}$$

$$= \begin{cases} 1, \begin{bmatrix} u \\ y \end{bmatrix} \in \mathcal{I}_{G_0,n}, \\ < 1, \begin{bmatrix} u \\ y \end{bmatrix} \notin \mathcal{I}_{G_0,n}, \end{cases} \tag{10.30}$$

- implementation of detection logic

$$f_1(u, y) := 1 - \frac{\left\| p_{\mathcal{P}_{\mathcal{I}_{G_0,n}}} \right\|_2}{\left\| \begin{bmatrix} u \\ y \end{bmatrix} \right\|_2} - \varepsilon_1,$$

$$\text{detection logic} \begin{cases} f_1(u, y) < 0, \begin{bmatrix} u \\ y \end{bmatrix} \in \mathcal{I}_{G_0,n}, \\ f_1(u, y) > 0, \begin{bmatrix} u \\ y \end{bmatrix} \notin \mathcal{I}_{G_0,n}, \end{cases}$$

where $0 < \varepsilon_1 << 1$ represents an upper-bound for system uncertainties like noises or model parameter variations.

It is easy to examine that

- $\mathcal{P}_{\mathcal{I}_{G_0,n}}$ given in (10.28) is an orthogonal projection, and
- (10.28) is true, because

$$\forall \begin{bmatrix} u \\ y \end{bmatrix} \in \mathcal{I}_{G_0}, \left\| \begin{bmatrix} u \\ y \end{bmatrix} \right\|_2 = \|v\|_2 .$$

Now, combining the both algorithms of classifying $\begin{bmatrix} u \\ y \end{bmatrix} \in \mathcal{I}_{G_0}^{\perp}$ and $\begin{bmatrix} u \\ y \end{bmatrix} \notin \mathcal{I}_{G_0,n}$ provides us an algorithm to detect $\begin{bmatrix} u \\ y \end{bmatrix} \in \mathcal{I}_{G_1}$, which is summarised as follows: (i) let

$$f_1(u, y) = 1 - \frac{\left\| p_{\mathcal{P}_{\mathcal{I}_{G_0,n}}} \right\|_2}{\left\| \begin{bmatrix} u \\ y \end{bmatrix} \right\|_2} - \varepsilon_1, \tag{10.31}$$

$$f_2(u, y) = J_{rel} - \varepsilon_2, \tag{10.32}$$

$$J_{rel} = \frac{\left\| r_{\mathcal{I}_G^{\mathcal{L}_2}} \right\|_2}{\left\| \begin{bmatrix} u \\ y \end{bmatrix} \right\|_2} = \frac{\left\| K_{G_0} \begin{bmatrix} u \\ y \end{bmatrix} \right\|_2}{\left\| \begin{bmatrix} u \\ y \end{bmatrix} \right\|_2} \tag{10.33}$$

be classification functions, where $0 < \varepsilon_1 << 1, 0 < \varepsilon_2 << 1$, represent the upper bounds of influences of noises on $\left\| p_{\mathcal{P}_{\mathcal{I}_{G_0,n}}} \right\|_2$ and J_{rel}, respectively, (ii) define

$$\text{Detection logic}: \begin{cases} f_1(u, y) < 0 \text{ and } f_2(u, y) < 0 \Longrightarrow \text{fault-free} \\ f_1(u, y) > 0 \text{ or/and } f_2(u, y) > 0 \Longrightarrow \text{faulty}, \end{cases} \tag{10.34}$$

which allows to discriminate fault-free and faulty data. To verify this claim, we examine the detection performance of algorithm (10.34). For the sake of simplicity, the influence of the noises are neglected. It can be obviously seen from (10.32) that

$$K_{G_0} \begin{bmatrix} u \\ y \end{bmatrix} = K_{G_0} I_f \alpha \Longrightarrow K_{G_0} I_f \alpha \neq 0 \Longrightarrow f_2(u, y) > 0,$$

that is, the fault can be observed in the residual subspace and thus directly detected. For $f_2(u, y) < 0$ and hence

$$\begin{bmatrix} u \\ y \end{bmatrix} \in \mathcal{I}_{G_0} \Longrightarrow K_{G_0} I_f = 0,$$

there exist two scenarios: (i)

$$\begin{bmatrix} u \\ y \end{bmatrix} \in \mathcal{S}_{\mathcal{I}_{G_0} \cap \mathcal{I}_f},$$

which yields, according to (10.26) and (10.27),

$$\begin{bmatrix} u \\ y \end{bmatrix} = \begin{bmatrix} M_{0,0} \\ N_{0,0} \end{bmatrix} \left(I + I_{f,1} \Theta \right) v.$$

Consequently, once

$$\left\| \begin{bmatrix} u \\ y \end{bmatrix} \right\|_2 = \left\| \left(I + I_{f,1} \Theta \right) v \right\|_2 > \delta_v,$$

that is, the fault $I_{f,1} \Theta v$ is sufficient large,

$$\frac{\left\| p_{\mathcal{P}_{\mathcal{I}_{G_0,n}}} \right\|_2}{\left\| \begin{bmatrix} u \\ y \end{bmatrix} \right\|_2} < 1 \Longrightarrow f_1(u, y) > 0,$$

and (ii)

$$I_{f,1} \Theta = 0 \Longrightarrow \frac{\left\| p_{\mathcal{P}_{\mathcal{I}_{G_0,n}}} \right\|_2}{\left\| \begin{bmatrix} u \\ y \end{bmatrix} \right\|_2} = 1 \Longrightarrow f_1(u, y) < 0.$$

Therefore, the claim holds true.

10.3 Fault Isolation: Multi-class Classification

A natural extension of the binary classification methods introduced in the previous sections to multi-class classification will provide us with a projection-based scheme for fault isolation. To this end, we first define the problem setting.

10.3.1 Problem Setting

The starting point is a detected fault. It is to decide to which class the arising fault belongs. Let the faulty dynamics with M fault classes be modelled by

$$\begin{bmatrix} u \\ y \end{bmatrix} = \begin{bmatrix} M_i \\ N_i \end{bmatrix} \Theta_i f_i, i = 1, \ldots, M. \tag{10.35}$$

Here, $\begin{bmatrix} M_i \\ N_i \end{bmatrix} \in \mathcal{RH}_{\infty}^{(m+p)\times p}$ is a RCF, $\Theta_i \in \mathcal{RH}_{\infty}^{p\times k_{f_i}}$, and $f_i \in \mathcal{L}_2^{k_{f_i}}$ denotes the i-th fault vector. M_i, N_i, Θ_i are given. We would like to call the reader's attention that the model (10.35) is not an SIR of a transfer function matrix, as defined and adopted before, since k_{f_i} can be different from p and Θ_i is not invertible, although a transfer function could be formally defined by

$$v_i := \Theta_i f_i, y = N_i v_i, u = M_i v_i \Longrightarrow G_i = N_i M_i^{-1}.$$

Example 10.2 Consider a feedback control system whose dynamic, as introduced in the first part of this book and given in Theorem 3.1, is governed by

$$\begin{bmatrix} u \\ y \end{bmatrix} = \begin{bmatrix} M_0 \\ N_0 \end{bmatrix} \bar{v} + \begin{bmatrix} -\hat{Y}_0 + M_0 Q \\ \hat{X}_0 + N_0 Q \end{bmatrix} r_y, \tag{10.36}$$

$$\bar{v} = \left(X_0 + Q\hat{N}_0 \right) v, \tag{10.37}$$

where $v \in \mathbb{R}^p, r_y \in \mathbb{R}^m$ are the reference vector and (output) residual vector, respectively, (M_0, N_0), $\left(\hat{X}_0, \hat{Y}_0\right)$, $\hat{N}_0$ and X_0 are given in (2.5)–(2.6) and (2.25)–(2.26), and $Q \in \mathcal{RH}_\infty$ is the parameter transfer function matrix. Now, two cases are included here to illustrate the use of the model (10.35) to describe faulty dynamics. Firstly, consider an additive fault vector described by

$$\begin{aligned} x(k+1) &= Ax(k) + Bu(k) + Ef(k), \\ y(k) &= Cx(k) + Du(k) + Ff(k), \end{aligned}$$

where $f \in \mathbb{R}^{k_f}$ represents the fault vector. It is well-known that

$$\begin{aligned} &r_y = \hat{M}_0 y - \hat{N}_0 u = \hat{N}_f f, \\ &\hat{M}_0 = W - WC(zI - A_L)^{-1} L, \hat{N}_f = WF + WC(zI - A_L)^{-1} E_L, \\ &A_L = A - LC, E_L = E - LF. \end{aligned}$$

Hence, (10.36) can be rewritten as

$$\begin{bmatrix} \Delta u \\ \Delta y \end{bmatrix} = \begin{bmatrix} u \\ y \end{bmatrix} - \begin{bmatrix} M_0 \\ N_0 \end{bmatrix} \bar{v} = \begin{bmatrix} -\hat{Y}_0 + M_0 Q \\ \hat{X}_0 + N_0 Q \end{bmatrix} \hat{N}_f f.$$

The latter is of the same form given in (10.35). We now address a multiplicative fault modelled by the plant dynamic

$$\begin{bmatrix} u \\ y \end{bmatrix} = \begin{bmatrix} M_0 + \Delta_{M_f} \\ N_0 + \Delta_{N_f} \end{bmatrix} v,$$

where $\left(\Delta_{M_f}, \Delta_{N_f}\right) \in \mathcal{H}_\infty$ denote the multiplicative fault. The corresponding closed-loop dynamic is

$$\begin{bmatrix} u \\ y \end{bmatrix} = \begin{bmatrix} M_0 \\ N_0 \end{bmatrix} \bar{v} + \begin{bmatrix} -\hat{Y}_0 + M_0 Q \\ \hat{X}_0 + N_0 Q \end{bmatrix} \begin{bmatrix} -\hat{N}_0 & \hat{M}_0 \end{bmatrix} \begin{bmatrix} \Delta_{M_f} \\ \Delta_{N_f} \end{bmatrix} v.$$

Let

$$f = \begin{bmatrix} \Delta_{M_f} \\ \Delta_{N_f} \end{bmatrix} v, \begin{bmatrix} \Delta u \\ \Delta y \end{bmatrix} = \begin{bmatrix} u \\ y \end{bmatrix} - \begin{bmatrix} M_0 \\ N_0 \end{bmatrix} \bar{v}.$$

It holds

$$\begin{bmatrix} \Delta u \\ \Delta y \end{bmatrix} = \begin{bmatrix} -\hat{Y}_0 + M_0 Q \\ \hat{X}_0 + N_0 Q \end{bmatrix} \begin{bmatrix} -\hat{N}_0 & \hat{M}_0 \end{bmatrix} f,$$

which is of the same form like (10.35).

Definition 10.1 The system under consideration is said to be in faulty operation with fault class i, if the system dynamic is governed by (10.35) and $f_i \neq 0$.

It is assumed that during faulty operations, the system dynamic is governed by a single fault class. In fact, this assumption loses no generality, as any so-called simultaneous faults can be defined as a fault class.

Let $\mathcal{I}_i$ denote the image subspace defined by

$$\mathcal{I}_i = \left\{ \begin{bmatrix} u \\ y \end{bmatrix} : \begin{bmatrix} u \\ y \end{bmatrix} = I_i v_i, v_i \in \mathcal{L}_2 \right\},$$
$$I_i = \begin{bmatrix} M_i \\ N_i \end{bmatrix}, i = 1, \ldots, M.$$

It is assumed that

$$0 < \delta_{\mathcal{I}} \left(\mathcal{I}_{G_i}, \mathcal{I}_{G_j} \right) < 1, i, j \in \{1, \ldots, M\}, i \neq j. \tag{10.38}$$

10.3.2 A Primitive Classification Algorithm

In Sects. 9.6 and 10.1.4, we have introduced one-class and binary classification approaches, respectively. The basic ideas can be adopted to deal with the multi-class case. To begin with, orthogonal projections are to be defined.

Note that I_i is generally not normalised, but satisfies

$$\begin{bmatrix} M_i\left(e^{j\theta}\right) \\ N_i\left(e^{j\theta}\right) \end{bmatrix}^{\sim} \begin{bmatrix} M_i\left(e^{j\theta}\right) \\ N_i\left(e^{j\theta}\right) \end{bmatrix} > 0(p.d.) \text{ for all } \theta. \tag{10.39}$$

In order to define the operator for an orthogonal projection onto $\mathcal{I}_i$, a spectral factorisation is performed,

$$I_i^{\sim} I_i = \bar{I}_i^{\sim} \bar{I}_i,$$

where $\bar{I}_i$ is (right) invertible with a stable right inverse $\bar{I}_i^{-}$, namely,

$$\left(\bar{I}_i^{\sim} \bar{I}_i\right)^{-1} = \bar{I}_i^{-} \left(\bar{I}_i^{\sim}\right)^{-}.$$

Spectral factorisation is a well-established technique. The reader is referred to the references given at the end of this chapter for the relevant computation algorithms.

Now, we are in a position to define operator

$$\mathcal{P}_{\mathcal{I}_i} : \mathcal{L}_2 \rightarrow \mathcal{L}_2, \mathcal{P}_{\mathcal{I}_i} = \mathcal{L}_{I_i} \mathcal{L}_{\left(\bar{I}_i^{\sim} \bar{I}_i\right)^{-1}} \mathcal{L}_{I_i^{\sim}},$$

which projects the process data onto $\mathcal{I}_i$. To verify the orthogonality, observe

$$\mathcal{P}_{\mathcal{I}_i}^* = \left(\mathcal{L}_{I_i} \mathcal{L}_{\left(\bar{I}_i^{\sim} \bar{I}_i\right)^{-1}} \mathcal{L}_{I_i^{\sim}}\right)^* = \mathcal{L}_{I_i} \mathcal{L}_{\left(\bar{I}_i^{\sim} \bar{I}_i\right)^{-1}} \mathcal{L}_{I_i^{\sim}} = \mathcal{P}_{\mathcal{I}_i},$$
$$\mathcal{P}_{\mathcal{I}_i}^2 = \mathcal{L}_{I_i} \mathcal{L}_{\left(\bar{I}_i^{\sim} \bar{I}_i\right)^{-1}} \mathcal{L}_{I_i^{\sim}} \mathcal{L}_{I_i} \mathcal{L}_{\left(\bar{I}_i^{\sim} \bar{I}_i\right)^{-1}} \mathcal{L}_{I_i^{\sim}} = \mathcal{L}_{I_i} \mathcal{L}_{\left(\bar{I}_i^{\sim} \bar{I}_i\right)^{-1}} \mathcal{L}_{I_i^{\sim}}.$$

Next, construct M classifiers,

$$p_{\mathcal{I}_i,n} = \mathcal{P}_{\mathcal{I}_i} \begin{bmatrix} u \\ y \end{bmatrix}, \begin{bmatrix} u \\ y \end{bmatrix} \in \mathcal{B}_{(u,y)}, i = 1, \ldots M,$$
$$\left\| p_{\mathcal{I}_i,n} \right\|_2 = \left\| \mathcal{P}_{\mathcal{I}_i} \begin{bmatrix} u \\ y \end{bmatrix} \right\|_2 = \left\| \left(\bar{I}_i^{\sim}\right)^{-} I_i^{\sim} \begin{bmatrix} u \\ y \end{bmatrix} \right\|_2, i = 1, \ldots M.$$

And set (lower) thresholds $0 < J_{th,i} < 1, i = 1, \ldots M$, to deal with possible noises and uncertainties. As a result, we have the following classification logic,

$$\begin{cases} \left\| p_{\mathcal{I}_i,n} \right\|_2 \geq J_{th,i} \Longrightarrow \text{the } i\text{-th fault class}, \\ \left\| p_{\mathcal{I}_i,n} \right\|_2 < J_{th,i}, \text{ free of the } i\text{-th fault class}, \end{cases} i = 1, \ldots M.$$

This simply appearing algorithm has a flaw that a number of classifiers (corresponding to numerous fault classes) will very probably indicate faulty, although only a single fault (in the sense of Definition 10.1) would arise. The reason is obvious. As claimed in Theorems 10.1 and 10.2, the assumption (10.38) implies the existence of an intersection of $\mathcal{I}_i$ and $\mathcal{I}_j, i \neq j$. In order to make good this deficiency, additional information would be helpful, as delineated in the advanced classification algorithm proposed in the sequel.

10.3.3 An Advanced Classification Algorithm

We begin with a short analysis of the intersection set of two image subspaces $\mathcal{I}_i$ and $\mathcal{I}_j, i \neq j$. Given $I_i = \begin{bmatrix} M_i \\ N_i \end{bmatrix} \in \mathcal{RH}_\infty^{p\times(m+p)}$, there exist $K_i \in \mathcal{RH}_\infty^{(m+p)\times m}$ so that

$$\forall \begin{bmatrix} u \\ y \end{bmatrix} \in \mathcal{I}_i,\ K_i \begin{bmatrix} u \\ y \end{bmatrix} = 0,$$

that is, K_i is a kernel of the system I_i. The existing relation

$$I - I_i \left(\bar{I}_i^\sim \bar{I}_i\right)^{-1} I_i^\sim = K_{i,0}^\sim K_{i,0}$$

gives the normalised kernel $K_{i,0}$. Let

$$\mathcal{I}_i^\perp = \left\{\begin{bmatrix} u \\ y \end{bmatrix} : \mathcal{P}_{\mathcal{I}_i} \begin{bmatrix} u \\ y \end{bmatrix} = 0, \begin{bmatrix} u \\ y \end{bmatrix} \in \mathcal{L}_2\right\}$$

be the orthogonal complement of $\mathcal{I}_i$. It is straightforward that the operator

$$\mathcal{P}_{\mathcal{I}_i^\perp} : \mathcal{L}_2 \to \mathcal{L}_2, \mathcal{P}_{\mathcal{I}_i^\perp} = \mathcal{L}_{K_{i,0}^\sim} \mathcal{L}_{K_{i,0}}$$

defines an orthogonal projection onto $\mathcal{I}_i^\perp$. Now, consider $\begin{bmatrix} u \\ y \end{bmatrix} \in \mathcal{I}_j, j \neq i$, which can be split into

$$\begin{bmatrix} u \\ y \end{bmatrix} = \mathcal{P}_{\mathcal{I}_i} \begin{bmatrix} u \\ y \end{bmatrix} + \mathcal{P}_{\mathcal{I}_i^\perp} \begin{bmatrix} u \\ y \end{bmatrix}.$$

Accordingly,

$$\forall \begin{bmatrix} u \\ y \end{bmatrix} \in \mathcal{I}_j, \exists v \in \mathcal{L}_2, \text{ s.t. } \mathcal{P}_{\mathcal{I}_i} \begin{bmatrix} u \\ y \end{bmatrix} = I_i v. \tag{10.40}$$

Equation (10.40) reveals that tightly bordering v is helpful to distinguish the i-th fault class from the j-th one. Making use of this fact, the following classification algorithm is proposed.

Consider the faulty model (10.35). It is assumed that

- $\Theta_i \in \mathcal{RH}_\infty^{p \times k_{f_i}}$ can be inner-outer factorised into

$$\Theta_i = \Theta_i^i \Theta_i^o, \tag{10.41}$$

with the outer $\Theta_i^o \in \mathcal{RH}_\infty^{k_{f_i} \times k_{f_i}}$ and inner $\Theta_i^i \in \mathcal{RH}_\infty^{p \times k_{f_i}}$,
- the fault vector f_i belongs to the set

$$\left\{ f_i(k) \in \mathbb{R}^{k_{f_i}} : f_{i,LB} \preceq f_i(k) \preceq f_{i,UB} \right\}, \tag{10.42}$$

where

$$f_{i,LB} = \begin{bmatrix} f_{i1,LB} \\ \vdots \\ f_{ik_{f_i},LB} \end{bmatrix}, f_{i,UB} = \begin{bmatrix} f_{i1,UB} \\ \vdots \\ f_{ik_{f_i},UB} \end{bmatrix}$$

are lower and upper bounds, respectively.

Remark 10.3 The assumption (10.41) requires $p \geq k_{f_i}$. This constraint can be relaxed by modifying the faulty model (10.35) to

$$\begin{bmatrix} u \\ y \end{bmatrix} = \begin{bmatrix} M_i \\ N_i \end{bmatrix} \Theta_i f_i, \begin{bmatrix} M_i \\ N_i \end{bmatrix} \in \mathcal{RH}_\infty^{(m+p) \times k_i},$$

where $k_i > 0$ is an integer smaller than $m + p$. Correspondingly, the assumption (10.38) is reformulated as

$$\mathcal{I}_{G_i} \cap \mathcal{I}_{G_j} \neq \mathcal{Q}, \mathcal{I}_{G_i} \cap \mathcal{I}_{G_j}^{\perp} \neq \mathcal{Q}, i, j \in \{1, \ldots, M\}, i \neq j.$$

Remark 10.4 The assumption of the set (10.42) is widely adopted in the set-membership-based fault detection research. It is well-known that such a set can be modelled as a zonotope.

Summarising the assumptions (10.41)–(10.42) gives rise to the definition of the following faulty set,

$$\mathcal{F}_i = \left\{ \begin{bmatrix} u \\ y \end{bmatrix} : \begin{bmatrix} u \\ y \end{bmatrix} = \begin{bmatrix} M_i \\ N_i \end{bmatrix} \Theta_i f_i, f_i \in \mathcal{L}_2 \cap \left\{ f_{i,LB} \preceq f_i(k) \preceq f_{i,UB} \right\} \right\}. \tag{10.43}$$

Define

$$\mathcal{P}_{\mathcal{F}_i} := \mathcal{L}_{I_i} \mathcal{L}_{\bar{I}_i^-} \mathcal{L}_{\Theta_i} \mathcal{L}_{(\Theta_i^o)^{-1}} \mathcal{P}_{f_i} \mathcal{L}_{(\Theta_i^{o\sim})^{-1}} \mathcal{L}_{\Theta_i^\sim} \mathcal{L}_{(\bar{I}_i^\sim)^-} \mathcal{L}_{I_i^\sim}, \tag{10.44}$$

$$\mathcal{P}_{f_i}(f_i(k)) = \begin{bmatrix} \hat{f}_{i1}(k) \\ \vdots \\ \hat{f}_{ik_{f_i}}(k) \end{bmatrix},$$

$$\hat{f}_{ij}(k) = \begin{cases} f_{ij}(k), f_{ij,LB} \leq f_{ij}(k) \leq f_{ij,UB}, \\ f_{ij,LB}, f_{ij}(k) < f_{ij,LB}, & , j = 1, \ldots, k_{f_i}. \\ f_{ij,UB}, f_{ij}(k) > f_{ij,UB}. \end{cases}$$

Note that

$$\begin{aligned} &\Theta_i^{\sim}\Theta_i = \Theta_i^{o\sim}\Theta_i^{i\sim}\Theta_i^{i}\Theta_i^{o} = \Theta_i^{o\sim}\Theta_i^{o} \Longrightarrow \left(\Theta_i^{o\sim}\right)^{-1}\Theta_i^{\sim}\Theta_i\left(\Theta_i^{o}\right)^{-1} = I \\ &\Longrightarrow \mathcal{L}_{\left(\Theta_i^{o\sim}\right)^{-1}}\mathcal{L}_{\Theta_i^{\sim}}\mathcal{L}_{\left(\bar{I}_i^{\sim}\right)^{-}}\mathcal{L}_{I_i^{\sim}}\mathcal{L}_{I_i}\mathcal{L}_{\bar{I}_i^{-}}\mathcal{L}_{\Theta_i}\mathcal{L}_{\left(\Theta_i^{o}\right)^{-1}} = \mathcal{I} \\ &\Longrightarrow \mathcal{P}_{\mathcal{F}_i}^2 = \mathcal{L}_{I_i}\mathcal{L}_{\bar{I}_i^{-}}\mathcal{L}_{\Theta_i}\mathcal{L}_{\left(\Theta_i^{o}\right)^{-1}}\mathcal{P}_{f_i}\mathcal{L}_{\left(\Theta_i^{o\sim}\right)^{-1}}\mathcal{L}_{\Theta_i^{\sim}}\mathcal{L}_{\left(\bar{I}_i^{\sim}\right)^{-}}\mathcal{L}_{I_i^{\sim}} = \mathcal{P}_{\mathcal{F}_i}, \end{aligned}$$

and it is apparent that

$$\mathcal{P}_{\mathcal{F}_i}^{*} = \mathcal{P}_{\mathcal{F}_i}.$$

Hence, $\mathcal{P}_{\mathcal{F}_i}$ defines an orthogonal projection of the process data (u, y) onto the faulty set $\mathcal{F}_i$. Now, by means of the $\mathcal{L}_2$-norm of the projection, $\left\| p_{\mathcal{F}_i} \right\|_2$,

$$p_{\mathcal{F}_i} = \mathcal{P}_{\mathcal{F}_i}\begin{bmatrix} u \\ y \end{bmatrix}, i = 1, \ldots M,$$

we are able to classify the fault based on the following classification logic

$$\text{fault class} = \arg \max_{i \in \{1,\ldots,M\}} \left\{ \left\| p_{\mathcal{F}_i} \right\|_2, i = 1, \ldots, M \right\}.$$

In order to take possible uncertainties into account, the above classification logic is finally extended as follows: the i-th fault class is identified if

$$\forall j \in \{1, \ldots, M\}, j \neq i, \left\| p_{\mathcal{F}_i} \right\|_2 - \left\| p_{\mathcal{F}_j} \right\|_2 - \varepsilon \geq 0, \tag{10.45}$$

where $\varepsilon > 0$ represents an upper-bound of the possible uncertainties.

Concerning the online implementation of $\left\| p_{\mathcal{F}_i} \right\|_2$, observe that

$$\left\| p_{\mathcal{F}_i} \right\|_2 = \left\| \mathcal{P}_{f_i}\left(\Theta_i^{o\sim}\right)^{-1}\Theta_i^{\sim}\left(\bar{I}_i^{\sim}\right)^{-} I_i^{\sim}\begin{bmatrix} u \\ y \end{bmatrix} \right\|_2.$$

Accordingly, before the online implementation, a spectral factorisation of I_i and an inner-outer factorisation of Θ_i are to be performed (offline). To this end, the reader can refer to the algorithms given in the references cited at the chapter end. It is followed by the (offline) computation of $\left(\Theta_i^o\right)^{-}$ and $\bar{I}_i^{-}$. Without loss of generality, denote the state space representation of $I_i \bar{I}_i^{-}\Theta_i\left(\Theta_i^o\right)^{-1}$ by

$$\begin{aligned} x^{f_i}(k+1) &= A^{f_i}x^{f_i}(k) + B^{f_i}f_i(k), \\ \begin{bmatrix} u(k) \\ y(k) \end{bmatrix} &= C^{f_i}x^{f_i}(k) + D^{f_i}f_i(k), \end{aligned}$$

where $x^{f_i}(k) \in \mathbb{R}^{n_i}$ denotes the state vector of the system $I_i \bar{I}_i^- \Theta_i \left(\Theta_i^o\right)^{-1}$. This enables us to (online) compute the conjugate of $I_i \bar{I}_i^- \Theta_i \left(\Theta_i^o\right)^{-1}$ using the following system algorithm

$$\xi^{f_i}(k-1) = \left(A^{f_i}\right)^T \xi^{f_i}(k) + \left(C^{f_i}\right)^T \begin{bmatrix} u(k) \\ y(k) \end{bmatrix}, \xi^{f_i}(k) \in \mathbb{R}^{n_i},$$

$$f_i^{\sim}(k) = \left(B^{f_i}\right) \xi^{f_i}(k) + \left(D^{f_i}\right)^T \begin{bmatrix} u(k) \\ y(k) \end{bmatrix}.$$

Finally, at each time instance k,

$$\mathcal{P}_{f_i} \left(\Theta_i^{o\sim}\right)^{-1} \Theta_i^{\sim} \left(\bar{I}_i^{\sim}\right)^{-} I_i^{\sim} \begin{bmatrix} u(k) \\ y(k) \end{bmatrix} = \hat{f}_i^{\sim}(k) = \begin{bmatrix} \hat{f}_{i1}^{\sim}(k) \\ \vdots \\ \hat{f}_{ik_{f_i}}^{\sim}(k) \end{bmatrix},$$

$$\hat{f}_{ij}^{\sim}(k) = \begin{cases} \hat{f}_{ij}^{\sim}(k), \; f_{ij,LB} \le \hat{f}_{ij}^{\sim}(k) \le f_{ij,UB}, \\ f_{ij,LB}, \; \hat{f}_{ij}^{\sim}(k) < f_{ij,LB}, \\ f_{ij,UB}, \; \hat{f}_{ij}^{\sim}(k) > f_{ij,UB}. \end{cases}, j = 1, \ldots, k_{f_i}.$$

And, with sufficient number of $\hat{f}_i^{\sim}(k)$,

$$\left\| p_{\mathcal{F}_i} \right\|_2 = \left\| \mathcal{P}_{f_i} \left(\Theta_i^{o\sim}\right)^{-1} \Theta_i^{\sim} \left(\bar{I}_i^{\sim}\right)^{-} I_i^{\sim} \begin{bmatrix} u \\ y \end{bmatrix} \right\|_2 = \left\| \hat{f}_i^{\sim}(k) \right\|_2,$$

will be (online) approximately computed.

10.4 An Integrated Method

The last part of our work on projection-based fault classification is devoted to a combination of the projection-based fault classification scheme with data-driven methods like support vector data description (SVDD) and support vector machine (SVM), which are well-established data-driven algorithms for classification in machine learning. The endeavour is motivated by the cognition that the model-based projection methods might be less efficient in practical applications, for instance, when the gap metric-based threshold setting is too conservative. There are two major reasons. Firstly, the upper-bound of uncertainties

$$\delta_{\Delta_{I_G}} = \sup_{I_\Delta \in \mathcal{H}_\infty} \left\| I_\Delta \right\|_\infty$$

can only be conservatively estimated. Secondly, the threshold setting rule,

$$J_{th} = \sup_{I_\Delta \in \mathcal{H}_\infty} J,$$

implies demand for zero FAR, which is in engineering practice too strict and often unrealistic. The basic idea behind the endeavour in this section is to exemplified apply SVDD and SVM algorithms to set a tighter threshold. For this end, we briefly introduce the standard SVDD and SVM algorithms at first.

10.4.1 SVDD and SVM for One-Class and Binary Classifications

SVDD and SVM are pure data-driven algorithms, originally developed for one-class and binary classification, respectively. Below is a short summary. For details, the reader is referred to the references given at the chapter end.

We begin with the SVDD algorithm. Given the data set, $x(i) \in \mathbb{R}^n, i = 1, \ldots, N$, find $c \in \mathbb{R}^n$, $R \in \mathbb{R}$ and $\xi_i \in \mathbb{R}, i = 1, \ldots, N$, that solve the following minimisation problem:

$$\begin{aligned} &\min_{R,c,\xi_i} J\left(R, c, \xi_i\right) = \min_{R,c,\xi_i} \left(R^2 + C \sum_{i=1}^{N} \xi_i\right) \\ &\text{s.t. } \|x(i) - c\|^2 \leq R^2 + \xi_i, \xi_i \geq 0, \end{aligned} \tag{10.46}$$

where $C \geq 0$ is a design parameter and $\xi_i, i = 1, \ldots, N$, are slack variables representing tolerable outliers. For the real-time fault detection with the online measurement data $x(k)$, the detection logic is

$$\begin{cases} \|x(k) - c\|^2 \leq R^2 \Longrightarrow \text{fault-free} \\ \|x(k) - c\|^2 > R^2 \Longrightarrow \text{faulty.} \end{cases} \tag{10.47}$$

It is apparent from the detection logic (10.47) that R^2 serves as the detection threshold. Geometrically, c and R are the center and radius of the sphere $S_R(c) = \{x : \|x - c\| = R\}$. Now, we learn the solution of optimisation problem (10.46).

Introducing Lagrange multipliers $\alpha_i, \gamma_i, i = 1, \ldots, N$, the constrained optimisation problem (10.46) is rewritten as

$$\min_{R,c,\xi_i,\alpha_i,\gamma_i} L\left(R, c, \xi_i, \alpha_i, \gamma_i\right), \tag{10.48}$$

$$L\left(R, c, \xi_i, \alpha_i, \gamma_i\right) = R^2 + C \sum_{i=1}^{N} \xi_i + \sum_{i=1}^{N} \left(\alpha_i \left(\|x(i) - c\|^2 - R^2 - \xi_i\right) - \gamma_i \xi_i\right).$$

The following relations give the necessary conditions for the solution,

$$\begin{aligned}
\frac{\partial L}{\partial R} &= 0 \Longrightarrow \sum_{i=1}^{N} \alpha_i = 1, \\
\frac{\partial L}{\partial c} &= 0 \Longrightarrow c = \frac{\sum_{i=1}^{N} \alpha_i x(i)}{\sum_{i=1}^{N} \alpha_i} = \sum_{i=1}^{N} \alpha_i x(i), \qquad (10.49)\\
\frac{\partial L}{\partial \xi_i} &= 0 \Longrightarrow C - \alpha_i - \gamma_i = 0, i = 1, \ldots, N, \\
\alpha_i &= C - \gamma_i, \gamma_i \geq 0 \Longrightarrow 0 \leq \alpha_i \leq C. \qquad (10.50)
\end{aligned}$$

Conditions (10.49) and (10.50) are of considerable significance for our comparison. It is apparent that $\forall x(i), i \in \{1, \ldots, N\}$, satisfying

$$\|x(i) - c\|^2 < R^2 + \xi_i,$$

the corresponding Lagrange multiplier α_i is reasonably set equal to zero. Solving the following optimisation problem, also called dual form of the primal problem (10.46), finally results in α_i,

$$\max_{\alpha_i} \left(\sum_{i=1}^{N} \alpha_i x^T(i) x(i) - \sum_{i=1}^{N} \sum_{j=1}^{N} \alpha_i \alpha_j x^T(i) x(j) \right) \qquad (10.51)$$

$$\text{s.t. } 0 \leq \alpha_i \leq C, \sum_{i=1}^{N} \alpha_i = 1.$$

It is of interest to notice the following relations concerning α_i,

$$\alpha_i = \begin{cases} 0, \|x(i) - c\| < R \\ \alpha_i \in (0, C), \|x(i) - c\| = R \\ C, \|x(i) - c\| > R. \end{cases}$$

The last equation implies that $x(i)$ is an outlier, which would trigger a (false) alarm. For our detection purpose, the set

$$\mathbb{I}_{>R} := \{j, \|x(j) - c\| > R, j \in \{1, \ldots, N\}\} \qquad (10.52)$$

is defined.

Next, we summarise the SVM algorithm for binary classification. Similar to the SVDD algorithm, SVM is formulated and solved in the primal and dual forms. Given the data set $x(i) \in \mathbb{R}^n, i = 1, \ldots, N$, that comprises data collected during fault-free and faulty operations, which are labelled by $y(i), i = 1, \ldots, N$,

$$y(i)=\begin{cases}-1, \text{ fault-free}\\ 1, \text{ faulty.}\end{cases}$$

Accordingly, the data set is divided into two groups,

$$\Theta_{-}=\{x(i), y(i)=-1, i\in\{1,\ldots,N\}\},$$
$$\Theta_{+}=\{x(i), y(i)=1, i\in\{1,\ldots,N\}\}.$$

Roughly speaking, the basic idea of SVM algorithms is to find a hyperplane,

$$w^Tx-b=0, w\in\mathbb{R}^n, b\in\mathbb{R},$$

that separates the group of Θ_- from the group Θ_+. Concerning the real-time fault detection with online data $x(k)$, the detection logic is

$$\begin{cases}w^Tx(k)-b\leq -1\Longrightarrow \text{fault-free}\\ w^Tx(k)-b\geq 1\Longrightarrow \text{faulty.}\end{cases} \tag{10.53}$$

The goal of SVM algorithms is to find (w, b) so that the data sets Θ_- and Θ_+ are separated by the hyperplane with the maximal distance. The corresponding optimisation problem is primally formulated as

$$\min_{w,b,\xi_i} J(w,b,\xi_i)=\min_{w,b,\xi_i}\left(\lambda\left\|w\right\|^2+\frac{1}{N}\sum_{i=1}^{N}\xi_i\right) \tag{10.54}$$

$$\text{s.t. } y(i)\left(w^Tx(i)-b\right)\geq 1-\xi_i, \xi_i\geq 0, i=1,\ldots,N,$$

with the slack variables $\xi_i, i=1,\ldots,N$, and weighting constant $\lambda>0$. For our study, the dual form of the above primal problem is of interest. Similar to the SVDD algorithm, solving the Lagrangian dual of the above problem results in

$$\max_{\alpha_i}\left(\sum_{i=1}^{N}\alpha_i-\frac{1}{2}\sum_{i=1}^{N}\sum_{j=1}^{N}y(i)\alpha_i x^T(i)x(j)y(j)\alpha_j\right) \tag{10.55}$$

$$\text{s.t. } \sum_{i=1}^{N}\alpha_i y(i)=0, 0\leq\alpha_i\leq\frac{1}{2N\lambda}, i=1,\ldots,N.$$

Once $\alpha_i, i=1,\ldots,N$, are determined, (w,b) are calculated as

$$w=\sum_{i=1}^{N}\alpha_i y(i)x(i), b=w^Tx(i)-y(i).$$

Moreover, it holds the relations,

$$\alpha_i = \begin{cases} 0, \ y(i)\left(w^T x(i) - b\right) > 1 \\ \alpha_i \in \left(0, \frac{1}{2N\lambda}\right), \ y(i)\left(w^T x(i) - b\right) = 1 \\ \frac{1}{2N\lambda}, \ y(i)\left(w^T x(i) - b\right) < 1. \end{cases}$$

Analogous to the SVDD algorithm, the set

$$\mathbb{I}_{FC} := \left\{i, \ y(i)\left(w^T x(i) - b\right) < 1, \ y(i) = -1, i \in \{1, \ldots, N\}\right\}, \tag{10.56}$$

is defined, which comprises the miss-classified fault-free data.

10.4.2 Combination of SVDD, SVM and Projection-Based Classification

We begin with SVDD-aided one-class classification on the assumption that sufficient number of projection-based residual data $r_{\mathcal{I}_G}(i), i = 1, \ldots, N$, are available. Suppose that the following data ball of $r_{\mathcal{I}_G}(k)$ is constructed using the SVDD algorithm,

$$B_{R, r_{\mathcal{I}_G}} := \left\{ \begin{array}{c} r_{\mathcal{I}_G}(k) : \left\| r_{\mathcal{I}_G}(k) - c \right\| \le R, \\ r_{\mathcal{I}_G}(k) = \left(\mathcal{I} - \mathcal{L}_{I_{G_0}} \mathcal{L}_{I_{G_0}^{\sim}}\right) \begin{bmatrix} u(k) \\ y(k) \end{bmatrix} \end{array} \right\} \in \mathbb{R}^{p+m},$$

with the identified (c, R) and the set $\mathbb{I}_{>R}$ defined by (10.52). With an expected FAR,

$$FAR = \frac{|\mathbb{I}_{>R}|}{N},$$

the following detection logic is suggested for fault detection,

$$J = \left\| r_{\mathcal{I}_G}(k) - c \right\|, \ J_{th} = R,$$

$$\begin{cases} J \le J_{th}, \ \text{fault-free} \\ J > J_{th}, \ \text{faulty.} \end{cases}$$

Hereby, $|\mathbb{I}_{>R}|$ is the cardinality of the set $\mathbb{I}_{>R}$.

Concerning SVM-based binary classification, sufficient the residual $r_{\mathcal{I}_G}(i), i = 1, \ldots, N$, are generated and divided into two groups,

$$\Theta_- = \left\{r_{\mathcal{I}_G}(i), l(i) = -1, i \in \{1, \ldots, N\}\right\},$$
$$\Theta_+ = \left\{r_{\mathcal{I}_G}(i), l(i) = 1, i \in \{1, \ldots, N\}\right\},$$

with the labels $l(i), i = 1, \ldots, N$. Applying the SVM algorithm, the parameters (w, b) of the hyperplane and the set $\mathbb{I}_{FC}$ are identified, which enable us to define the detection logic as follows,

$$\begin{cases} w^T r_{\mathcal{I}_G}(k) - b \leq -1 \Longrightarrow \text{fault-free} \\ w^T r_{\mathcal{I}_G}(i) - b \geq 1 \Longrightarrow \text{faulty}. \end{cases} \tag{10.57}$$

Notice that

$$FAR = \frac{|\mathbb{I}_{FC}|}{N}$$

is an estimate of the expected FAR.

Remark 10.5 In engineering practice, faulty data are often lacking. In this case, synthetic fault data, for instance, generated by means of the so-called randomised algorithm technique, can be used for the classification purpose.

Finally, we present the generalised projection-based fault detection scheme that describes the above two schemes in a unified form. Along the lines of the projection-based detection scheme presented in Example 9.2, the convex set

$$\mathcal{R}_{ff} = \begin{cases} \left\{ r_{\mathcal{I}_G}(k) : \left\| r_{\mathcal{I}_G}(k) - c \right\| \leq R \right\}, \text{ SVDD-based detection} \\ \left\{ r_{\mathcal{I}_G}(k) : \bar{w}^T r_{\mathcal{I}_G}(k) \leq \bar{b} \right\}, \text{ SVM-based detection,} \end{cases}$$
$$\bar{w} = \frac{w}{\|w\|}, \bar{b} = \frac{b-1}{\|w\|},$$

and the corresponding operator $\mathcal{P}_{\mathcal{R}_{ff}}$,

$$\mathcal{P}_{\mathcal{R}_{ff}}\left(r_{\mathcal{I}_G}(k)\right) = \begin{cases} r_{\mathcal{I}_G}(k), r_{\mathcal{I}_G}(k) \in \mathcal{R}_{ff} \\ c + \frac{R\left(r_{\mathcal{I}_G}(k) - c\right)}{\left\|r_{\mathcal{I}_G}(k) - c\right\|}, r_{\mathcal{I}_G}(k) \notin \mathcal{R}_{ff}, \text{ SVDD-based} \\ \left(I - \bar{w}\bar{w}^T\right) r_{\mathcal{I}_G}(k) + \bar{w}\bar{b}, r_{\mathcal{I}_G}(k) \notin \mathcal{R}_{ff}, \text{ SVM-based} \end{cases} \tag{10.58}$$

are defined. That means, $\mathcal{R}_{ff}$ is the residual data set of fault-free operations. It is apparent that

$$J = \begin{cases} \left\| \left(\mathcal{I} - \mathcal{P}_{\mathcal{R}_{ff}}\right) \left(r_{\mathcal{I}_G}(k)\right) \right\|, \text{ SVDD-based} \\ \bar{w}^T \left(\mathcal{I} - \mathcal{P}_{\mathcal{R}_{ff}}\right) \left(r_{\mathcal{I}_G}(k)\right). \text{ SVM-based} \end{cases} = \begin{cases} 0, r_{\mathcal{I}_G}(k) \in \mathcal{R}_{ff}, \\ > 0, r_{\mathcal{I}_G}(k) \notin \mathcal{R}_{ff}. \end{cases}$$

Recall that in the case of the SVD-based binary classification, the evaluation function $J > 0$ indicates that the operation is not fault-free. To assure a reliable fault detection, further examining

$$w^T r_{\mathcal{I}_G}(k) - b \geq 1 \Longrightarrow \text{faulty}$$

can remarkably reduce FAR. For $r_{\mathcal{I}_G}(k)$ satisfying

$$w^T r_{\mathcal{I}_G}(k) - b \geq 1 \Longrightarrow \bar{w}^T r_{\mathcal{I}_G}(k) \geq \bar{b} + \frac{2}{\|w\|},$$

it holds,

$$\bar{w}^T \left(\mathcal{I} - \mathcal{P}_{\mathcal{R}_{ff}}\right) \left(r_{\mathcal{I}_G}(k)\right) \geq \frac{2}{\|w\|}.$$

In summary, we have

$$J = \begin{cases} \left\| \left(\mathcal{I} - \mathcal{P}_{\mathcal{R}_{ff}}\right) \left(r_{\mathcal{I}_G}(k)\right) \right\|, \text{ SVDD-based} \\ \bar{w}^T \left(\mathcal{I} - \mathcal{P}_{\mathcal{R}_{ff}}\right) \left(r_{\mathcal{I}_G}(k)\right), \text{ SVM-based} \end{cases}$$
$$= \begin{cases} 0, \text{ fault-free} \\ > 0, \text{ faulty for SVDD-based one-class FD} \\ \geq \frac{2}{\|w\|}, \text{ faulty for SVM-based binary FD} \end{cases} \tag{10.59}$$

with the expected FAR,

$$FAR = \begin{cases} \frac{|\mathbb{I}_{>R}|}{N}, \text{ for SVDD-based one-class FD} \\ \frac{|\mathbb{I}_{FC}|}{N}, \text{ for SVM-based binary FD.} \end{cases} \tag{10.60}$$

10.5 Notes and References

This chapter continues the work on projection-based fault diagnosis and is mainly focused on

- binary classification, which serves as the basis for the extension to
- multi-class classification towards isolation, and
- extension and generalisation of projection-based fault detection methods.

Dealing with projection-based binary and multi-class classification, the residual generation is a trivial work. For instance, for binary classification, two process models $G_{i,0}, i = 0, 1$, are given with the corresponding SIRs,

$$I_{G_{i,0}} = \begin{bmatrix} M_{i,0} \\ N_{i,0} \end{bmatrix}, i = 0, 1, \tag{10.61}$$

representing nominal and faulty system dynamic respectively. Understandably, two projection systems are to be constructed and, based on them, classification decision is then made. The crucial scenario is the similarity between those projections respectively generated by both projection-based residual generators. This may remarkably hinder a correct classification. In operator theory, such similarity is measured by the gap metric between two subspaces. In our case, the gap metric,

$$0 < \boldsymbol{\delta}\left(\mathcal{I}_{G_0}, \mathcal{I}_{G_1}\right) = \boldsymbol{\delta}\left(\mathcal{I}_{G_1}, \mathcal{I}_{G_0}\right) = \delta\left(\mathcal{I}_{G_0}, \mathcal{I}_{G_1}\right) < 1,$$

is adopted, where

$$\mathcal{I}_{G_i} = \left\{ \begin{bmatrix} u \\ y \end{bmatrix} : \begin{bmatrix} u \\ y \end{bmatrix} = I_{G_{i,0}} v, \, v \in \mathcal{L}_2 \right\}, i = 0, 1,$$

are image subspaces of the process data corresponding to the fault-free and faulty operations.

In order to gain insightful view of this crucial issue, similarity of $\mathcal{I}_{G_0}$ and $\mathcal{I}_{G_1}$ has been investigated in Sect. 10.1.3. To begin with, Theorem 10.1 given in Feintuch's monograph [55] (Theorem 9.1.4) has been presented, which tells us, the data in $\mathcal{I}_{G_i}$ can be generated by the data in $\mathcal{I}_{G_j}$, $j \neq i$. This implies that it is hard to classify the process operation condition when the data lie in certain region. Theorem 10.2, which can be found in Feintuch's monograph [55] (Sect. 9.1), further illustrates that the gap between faulty and fault-free subspaces can be expressed by the fault-free subspace and an operator defined on this subspace. In the classification context, this result reveals that, without additional information, the binary fault classification problem is hard to solve. In fact, the primary cause for this difficulty consists in the transitional phase from the fault-free to faulty operations. Although it is difficult to give an analytical form of the similarity, it is possible to characterise it by means of simulation or data, for instance, using the so-called randomised algorithm technique [43]. This would be helpful to understand the mechanism of incipient and intermittent faults.

Based on the aforementioned theoretic results, a classification function $f(u, y)$ and the associated decision logic have been proposed in Sect. 10.1.4, which enable to unambiguously classify the data in three sets for the final classification aim.

It is noteworthy that the computation aspect of a projection system like

$$\mathcal{P}_{\mathcal{I}_{G_i}} \begin{bmatrix} u \\ y \end{bmatrix} = I_{G_{i,0}} I_{G_{i,0}}^{\sim} \begin{bmatrix} u \\ y \end{bmatrix},$$

has not been addressed in our study. It is obvious that, in engineering applications, the computation of

$$\left\| \mathcal{P}_{\mathcal{I}_{G_i}} \begin{bmatrix} u \\ y \end{bmatrix} \right\|_2 = \left\| I_{G_{i,0}^{\sim}} \begin{bmatrix} u \\ y \end{bmatrix} \right\|_2$$

can only be approximated over a finite time interval using the data generated by the state space realisation of $I_{G_{i,0}^{\sim}}$ given by (10.6)–(10.7). This requires collecting and processing data batchwise. Alternatively, time-varying projection systems proposed in [109] or the finite samples representations introduced in Chap. 7 offers capable solutions.

Aiming at enhancing fault detectability and reducing false alarms, modification schemes have been investigated in Sect. 10.2. To this end, two approaches have been proposed. The first one is to leverage available knowledge of the latent variable, for instance, presented in the form of a convex set, to embed a further projection in the projection-based detection system, as described by

$$p_{\mathcal{I}_{G_0},\mathcal{V}_{conv},n} = \mathcal{L}_{I_{G_0}} \mathcal{P}_{\mathcal{V}_{conv}} \mathcal{L}_{I_{G_0}^{\sim}} \begin{bmatrix} u \\ y \end{bmatrix}.$$

This modification is similar to the work presented in Sect. 9.6. Another approach deals with the issue by modelling the faulty dynamic in terms of

$$G_1 = \left(N_{0,0} + N_f \Theta\right) \left(M_{0,0} + M_f \Theta\right)^{-1},$$
$$I_{G_1} = I_{G_0} + \Delta_f = \begin{bmatrix} M_1 \\ N_1 \end{bmatrix} = \begin{bmatrix} M_{0,0} + M_f \Theta \\ N_{0,0} + N_f \Theta \end{bmatrix},$$
$$I_{G_0} = \begin{bmatrix} M_{0,0} \\ N_{0,0} \end{bmatrix}, \Delta_f = I_f \Theta, I_f = \begin{bmatrix} M_f \\ N_f \end{bmatrix},$$

under the assumption that I_f is known. In fact, I_f can be interpreted as the structural information about the faulty dynamic Δ_f. On account of such additional information, a classification approach has been elaborated.

Extensions of binary classification-based fault detection approaches presented in Sects. 13.1 and 10.2 to fault isolation have been explored in Sect. 10.3.

Notice that the projection-based classification methods might become less efficient, when the gap metric-based threshold setting is too conservative, like most of model-based fault detection algorithms. Two reasons are mainly responsible for this case: (i) the conservative estimate of the upper-bound of uncertainties, and (ii) the demand for zero FAR. To solve this problem, an integrated method has been proposed in Sect. 10.4, which comprises a combination of the projection-based fault classification schemes with data-driven methods SVDD [178, 188] and SVM [30, 91]. SVDD and SVM are well-established data-driven algorithms in machine learning. They map the data onto context sets and formulate the classifications under consideration as standard optimisation problems. For our application, the dual forms of the primal optimisation problems are of a special interest. They enable us to estimate the FAR using the collected data. The technical descriptions of the SVDD and SVM algorithms have been referred to [178, 188] and [30, 91]. Concerning the theoretical details of the optimisation problems addressed by the algorithms and the related projection operators used for residual evaluation and threshold setting, the reader is referred to [15] as well. As a result of our endeavour, we are able to set considerably tighter threshold and validate the possible FAR of the designed fault detection system.

Chapter 11
Projection-Based Fault Diagnosis Schemes for Nonlinear Dynamic Systems

11.1 Introduction

Having undergone a rapid development over a couple of decades, the model-based fault diagnosis technique, in particular the observer-based one, has become well established as an active research area in control theory and engineering. As is well known, an observer-based fault detection system consists of two major units, a residual generator and evaluator. For their design and analysis system input-output and the associated state space models build a common basis, and advanced control-theoretic methods serve as the major theoretical tool. Specifically, the observer-based residual generator design is mainly achieved by means of advanced observer theory, while the construction of the evaluator with a focus on the threshold setting is performed on the basis of norm-based system analysis. To put it in a nutshell, state of the art of observer-based fault diagnosis technique is marked by the well-established framework for fault diagnosis in LTI systems, while systematically dealing with fault diagnosis issues is missing for nonlinear systems.

In their survey, Alcorta-Garcia and Frank reviewed state of the art of nonlinear observer-based fault detection in the 90's. The major methods include the application of feedback-based linearisation, differential algebra techniques to observer-based residual generator design, and the geometric approach for nonlinear fault detection. Later, considerable attention has been paid to the application of those special techniques that enable dealing with analysis and synthesis of nonlinear dynamic systems more efficiently. These are, amongst others, linear matrix inequalities (LMI) technique to address a special class of nonlinear systems with, for instance, Lipschitz nonlinearity or sector bounded nonlinearity, fuzzy technique based fault detection, adaptive fault diagnosis for nonlinear systems, linear parameter-varying fault detection systems, and sliding mode observer-based fault detection. Although all these efforts are successful in dealing with fault detection of the corresponding system classes, a common theoretical framework for design and analysis of nonlinear fault detection systems seems to be missing. In our early endeavours towards such a

S. X. Ding, *Unified Framework of Control and Detection*,
https://doi.org/10.1007/978-3-662-72417-0_11

framework, we noticed that little attention was paid to essential issues of nonlinear observer-based fault detection systems like existence conditions, detectability and parameterisation. With the aid of nonlinear system theory as a major tool, these issues have been examined and the first results were reported.

Fault detection is a classification problem, namely classifying those potential anomalies from the data collected during system operations. The use of a system model and, based on it, an observer allows to distinguish anomalies from the normal input-output dynamics and thus leads to an efficient classification. On the other hand, in industrial practice, the use of a mathematical model is generally associated with model uncertainties. Coupled with external disturbances (unknown inputs), they considerably complicate a reliable classification, in particular when nonlinear dynamic systems are under consideration. In fact, distinguishing anomalies/faults from the uncertainties existing during fault-free operations is a most crucial issue in dealing with classification/fault detection. A further challenging issue is to detect process faults, also called component faults in the literature. As reported in innumerable publications, observer-based fault detection methods are capable to detect sensor and actuator faults in dynamic systems. Two plausible reasons for such successful applications are,

- an input-output model gives a natural modelling of sensor and actuator faults,
- the norm-based evaluation of residual signals builds an explicit indicator for these two types of faults.

Differently, process faults are corrupted with model uncertainties, and as a consequence of nonlinear system dynamics, process data build a manifold that is a non-Euclidean space. It is known that usual Euclidean distances or inner products may not be appropriate to measure the dissimilarity between the data points on a manifold, which is the core purpose of classification. That process faults as well as uncertainties can result in remarkable system performance degradation but are hardly detected using norm-based evaluation function is the further aspects that motivate our work.

In the era of artificial intelligence and big data, machine learning (ML) based classification methods have become now the mainstream in the research area of engineering fault diagnosis. Classification-based fault diagnosis is a technique that uses ML-classifiers to detect faults in a system based on the features extracted from the process data. As a crucial step towards reliable and explainable fault diagnosis, dimensionality reduction technique (DRT), also known as manifold learning, serves as a capable tool. The basic idea behind DRT is to reduce irrelevant and redundant information in the data, leading to not only lower computation complexity but also, and more importantly, better classification and fault diagnosis performance. In order to capture nonlinear and nonstationary characteristics of the data, nonlinear dimensionality reduction methods have been developed, among them, the so-called Isomap (Isometric mapping) and LLE (Locally-linear embedding) are two representative ones. Informally speaking, an essential assumption in manifold learning is that the data have been generated from a lower-dimensional manifold that is embedded inside of a higher-dimensional space composed of redundant and irrelevant information (uncertainties). The basic idea of Isomap is the use of a geodesic distance, which enables incorporation of the manifold structure in the resulting embedding. The objective of the LLE algorithm is to create a data mapping from the higher-dimensional manifold to the lower-dimensional one and preserving local neighbourhoods at every point of the underlying manifold. In recent years, the so-called autoencoder (AE) technique, a capable ML-based algorithm for dimensionality reduction and feature learning, is widely and successfully applied to fault diagnosis and anomaly detection. An AE consists of an encoder and a decoder, where the encoder compresses the data in the higher-dimensional space into a lower-dimensional feature manifold represented by the so-called latent variable, and the decoder is driven by the latent variable and delivers the process variables reflecting, in case of fault diagnosis, nominal process operations. The encoder and decoder are constructed using neural networks (NNs) that are learnt in such a way that the latent variable (the lower-dimensional feature) contains exclusive information of the nominal process dynamics, from which the process variables in the nominal state are then fully reconstructed. To put it in a nutshell, dimensionality reduction is a capable strategy to remove redundant and irrelevant informations in the data and thus recover exclusive information about the nominal process operations by means of projecting the data in the higher-dimensional space to the lower-dimensional feature manifold.

A comparison of the observer-based and dimensionality reduction-based ML diagnosis methods reveals the following insightful differences of these fault detection strategies, in spite of the evidently different assumptions on the types of available process knowledge and information,

- the input-output/state space models describing the nominal system dynamics build the system core of observer-based fault detection, while the ML-based methods handle the nominal system dynamics as a sub-manifold (feature manifold) embedded in the (higher-dimensional) data space composed of all the process data with uncertainties,
- accordingly, the observer-based methods handle system uncertainties by means of residual generation, and the ML-based methods make use of projection techniques,
- and consequently, in the framework of the observer-based fault detection, the detection decision is made on the basis of a norm-based residual evaluation, while for the ML-based methods, the decision is based on the distance metric between the current data vector and the lower-dimensional feature manifold. It is noteworthy that in non-Euclidean space the concept of a geodesic distance is often applied in order to capture the nonlinear structure of the data manifold.

In a short form, these major differences are summarised as

$$\begin{gathered}\langle \text{I/O-model, residual, norm-based evaluation} \rangle \ vs. \\ \langle \text{feature manifold, projection, geodesic distance} \rangle\,,\end{gathered}$$

whose core is handling of nominal dynamics and uncertainties.

Handling of model uncertainties using the projection technique in a Hilbert subspace is nothing new in control theory. In the 90s, the concept of gap metric was widely applied in robust control theory to measure influences of model uncertainties on a feedback control system. Roughly speaking, a gap is a similarity metric between subspaces in Hilbert space based on an orthogonal projection. Applied to robust control theory, a gap metric gives, for instance, the "distance" between the system dynamics with and without uncertainties. The use of gap metrics to fault detection has been comprehensively investigated by various research groups. Motivated by the promising results and inspired by the aforementioned ML-based fault detection methods, an initial project has been launched to investigate application of orthogonal projection technique to fault detection in LTI systems. The primary objective of that project is to establish a projection-based fault detection technique as a potential alternative framework to the observer-based one and to compare the capacity of the two detection schemes. Below are two distinct conclusions,

- in the projection-based fault detection framework, residual generation and evaluation can be well realised in an optimal and integrated manner, and
- a projection-based fault detection system can be, under certain conditions, equivalently realised by a class of optimal observer-based fault detection systems.

As described in the previous chapters, the theoretical basis for projection-based fault detection is the orthogonal projection in Hilbert space which allows an orthogonal decomposition of a process data vector into an optimal estimate (projection) and the residual. As a result, the norms of the process data, the residual vector and the projection satisfy, according to the Hilbert Projection Theorem, the Pythagorean equation, which leads to an optimal fault detection.

Motivated and driven by the aforementioned preliminary results, the main project has started with the objective of establishing a projection-based fault diagnosis framework, in which nonlinear dynamic systems with uncertainties are considered, and an optimal system performance-oriented fault detection as well as fault estimation are defined and approached. This is also the major objective of this chapter reporting the main results of our endeavours. We would like to mention that the overall goal of our efforts is to build a uniform framework, in which both control theory based and ML-based methods can be developed for fault diagnosis in dynamic control systems. To be specific, the design of ML-based fault diagnosis systems could be well guided by known control-theoretic knowledge and, on the other hand, the design and realisation of model-based fault diagnosis systems are supported by data-driven and ML-algorithms. Our recent work on control theory-informed application of autoencoder methods to fault detection is an example of the former case, while the latter is inspired by current research on physics-informed machine learning (PINN).

To approach the above described objective, we focus on the following four major tasks.

- Definition of nominal system dynamics as a manifold in the process data space and realisation of normalised image and kernel representations.

Let the process input and output data build a data space. As known, for linear systems SIR and SKR are system model forms that are equivalent to the input-output (transfer function) and the associated state space models. In particular, the normalised SIR and SKR, as inner and co-inner systems, play an important role in the (orthogonal) projection related research. By means of SIR and SKR, two equivalent subspaces, the image and kernel subspaces, are defined in Hilbert space. The analogue concepts are also well defined for nonlinear systems. Nevertheless, the lossless properties of inner and co-inner are of special importance in our subsequent study and motivate us to follow the works by van der Schaft and his co-worker on nonlinear systems. The core of these works is various configurations of the so-called Hamiltonian extension as Hamiltonian systems, which serve as the major system-theoretic tool in our work as well.

- Design and analysis of projection systems.
  For LTI systems, the orthogonal projection onto the image space is a series system composed of the normalised SIR and its conjugate. Analogously to it, we propose to configure the projection as a Hamiltonian system using the normalised SIR. The centrepiece in this work is undoubtedly the Hamiltonian function of the system, which describes the system dynamics in the sense of energy transfer in the system, and is thus interpreted as a system performance function. Associated with it, a key concept, the so-called Legendre transform, is introduced, which gives a dual relation of the input and output vectors of the Hamiltonian system as well as a dual form of the Hamiltonian system. For our purpose of fault detection, the dual pair of the Hamiltonian functions is analysed corresponding to the nominal system operation and operations when uncertainties or even faults exist. Although this task is primarily dedicated to constructing projection systems, it serves as the system-theoretic basis for the next task as well.
- Bregman divergence-based performance-oriented fault detection.
  While the first two tasks are, more or less, extended applications of the existing control-theoretic methods, this task is crucially different from the projection-based scheme for LTI systems, where the residual is built by means of the difference of the data vector and its projection and further norm-based evaluated. This task is devoted to three issues: (i) handling of process data in non-Euclidean space aiming at classification, (ii) performance-oriented fault detection, and (iii) interpretation of the projection-based method. To this end, Bregman divergences are introduced. A Bregman divergence is defined in terms of a convex function and serves as a measure of difference between two points in a space. We propose to define Bregman divergences in terms of Hamiltonian functions. The idea behind that is to measure the difference between the data vector and its projection with respect to the change in the Hamiltonian function as a performance function under consideration. Observe that Bregman divergences form an important class of divergences and are widely used in information geometry. In particular, a Bregman divergence together with a Legendre transform may induce a dually flat Riemannian manifold in the data space that may be regarded as a dualistic extension of the Euclidean space. In this context, the interpretation of the Hamiltonian system based projection as a geodesic projection will be studied.

- Modelling of data uncertainty, manifold of uncertain data, and fault detection and estimation.
  In the projection-based fault diagnosis framework for LTI systems, thanks to the Pythagorean equation, projections onto the image subspace and its orthogonal complement are dual problems and can be addressed, in a certain sense, in a uniform manner. For process data in a non-Euclidean space, they cannot be decomposed into two components representing nominal and uncertain/faulty data, respectively. In other words, nominal data are corrupted with uncertain data and they are inseparable. For our purposes, the concepts of (data) uncertainty model and manifold of uncertain data are introduced on the basis of SKR. While the data in the image manifold are created under nominal system dynamics, the data in the manifold of uncertain data are assumed to contain uncertainties exclusively. In this sense, the SKR-based uncertainty model is a dual form of the SIR-based model for nominal dynamics. Here, we place a greater emphasis on the possible interpretation of uncertainties as faults. In such a system setting, Hamiltonian systems, the associated Hamiltonian functions and the corresponding Bregman divergences, dual to the ones defined in previous tasks, are defined, designed, and then applied for the fault detection purpose. In addition, on the assumption that the uncertainty is caused by faulty operations, the SKR-based projection onto the manifold of uncertain data delivers an estimate for the faulty data. The performance of such an estimation will finally be analysed.

The chapter is organised as follows. At first, necessary preliminaries are introduced. The control-theoretic preliminaries include image and kernel representations of nonlinear dynamic systems, Hamiltonian systems, inner and co-inner systems. It is followed by an introduction to definitions and computation of Bregman divergences and Legendre transform. They are used not only as divergences for performance-oriented fault detection, but also for inducing a dual Riemannian structure in the data space, which admits a conjugate pair of affine connections. In this regard, the concepts of divergences from a (data) vector to a manifold and geodesic projections are described. On the basis of the introduced preliminaries, problems to be addressed are specified. Two schemes will be investigated in this chapter. The first one is devoted to the projection and Bregman divergence based fault detection. This work consists of three parts. The first part deals with construction of a projection system by means of the normalised SIR and its associated Hamiltonian systems, and the analysis of the corresponding Hamiltonian functions under different operation conditions. It is followed by the second part focusing on (i) computation of the Bregman divergence from the data vector to its projection, which is derived from the (dual) Hamiltonian function, (ii) its use for fault detection, and (iii) its interpretation as a geodesic projection. The last part discusses implementations issues, including definition of a reliable evaluation function over an evaluation time interval and determination of a threshold. In the second scheme, two different but strongly related issues are addressed. The first one is an SKR-based projection, as a dual fault detection scheme to the SIR-based projection. Fault estimation is the second issue. The focus of this study is on the application of an SKR-based projection as a fault estimator and analysis of its estimation performance. For both parts, the so-called uncertainty model and the manifold of uncertain data serve as a common basis and are introduced at the beginning of this section.

11.2 Preliminaries

11.2.1 Kernel and Image Representations of Nonlinear Systems, and Manifolds

Coprime factorisations of nonlinear affine systems and some related concepts described in this subsection have been introduced in Chap. 6 in considerable details. In order to help the reader to have an easy start with the challenging study in this chapter, we firstly give a short summary of relevant results needed for our subsequent work.

We consider nonlinear affine systems of the form

$$\Sigma : \begin{cases} \dot{x} = a(x) + B(x)u, x(0) = x_0 \\ y = c(x) + D(x)u, \end{cases} \tag{11.1}$$

where $x \in \mathbb{R}^n$ denotes the state vector, $u \in \mathbb{R}^p$ and $y \in \mathbb{R}^m$ represent system input and output vectors, respectively, x_0 represents the initial value of $x(t)$, and $a(x)$, $B(x)$, $c(x)$ and $D(x)$ are continuously differentiable and of appropriate dimensions.

Given system Σ, the state space representation of the SIR $\Sigma_{\mathcal{I}}$ is an essential model form used in our subsequent work and is given by

$$\Sigma_{\mathcal{I}} : \begin{cases} \dot{x} = a_I(x) + B_I(x)v \\ \begin{bmatrix} u \\ y \end{bmatrix} = c_I(x) + D_I(x)v, \end{cases} \tag{11.2}$$

$$a_I(x) = a(x) + B(x)g(x),\ B_I(x) = B(x)V(x),$$

$$c_I(x) = \begin{bmatrix} g(x) \\ c(x) + D(x)g(x) \end{bmatrix},\ D_I(x) = \begin{bmatrix} V(x) \\ D(x)V(x) \end{bmatrix},$$

where $g(x)$ is designed such that $a_I(x)$ is stable, and $V(x) \in \mathbb{R}^{p \times p}$ is invertible. In the context of feedback control systems,

$$u = g(x) + V(x)v$$

is understood as a controller with a state feedback $g(x)$, feed-forward controller $V(x)$ and reference signal v. Note that

$$D_I^T(x)D_I(x) = V^T(x)\left(I + D^T(x)D(x)\right)V(x)$$

is invertible.

The SKR of Σ is another system model form and useful in our subsequent study as well. The SKR of Σ, denoted by $\Sigma_{\mathcal{K}}$, is a stable system that satisfies, technically speaking,

$$r_y = \Sigma_{\mathcal{K}}\begin{pmatrix} u \\ y \end{pmatrix} = 0, r_y \in \mathbb{R}^m, \tag{11.3}$$

when there exists no uncertainty in the system. The state space representation of SKR $\Sigma_{\mathcal{K}}$ is described by

$$\Sigma_{\mathcal{K}}: \begin{cases} \dot{\hat{x}} = a(\hat{x}) + B(\hat{x})u + L(\hat{x})\left(y - \hat{y}\right), \hat{x}(0) = \hat{x}_0 \\ \quad = a_K(\hat{x}) + B_K(\hat{x})\begin{bmatrix} u \\ y \end{bmatrix} \\ r_y = W(\hat{x})r_0 = c_K(\hat{x}) + D_K(\hat{x})\begin{bmatrix} u \\ y \end{bmatrix} \\ r_0 = y - \hat{y}, \hat{y} = c\left(\hat{x}\right) + D\left(\hat{x}\right)u, \end{cases} \tag{11.4}$$

$$a_K(\hat{x}) = a(\hat{x}) - L(\hat{x})c\left(\hat{x}\right), B_K(\hat{x}) = \left[\, B(\hat{x}) - L(\hat{x})D\left(\hat{x}\right) \;\; L(\hat{x}) \,\right],$$
$$c_K(\hat{x}) = -W(\hat{x})c\left(\hat{x}\right), D_K(\hat{x}) = \left[\, -W(\hat{x})D\left(\hat{x}\right) \;\; W(\hat{x}) \,\right],$$

where $L(\hat{x})$ is designed such that $a_K(\hat{x})$ is stable and $W(\hat{x}) \in \mathbb{R}^{m\times m}$ is invertible. In the detection context, r_0 builds a residual vector, W is an invertible post-filter, and it holds

$$D_K^T(\hat{x})D_K(\hat{x}) = W(\hat{x})\left(I + D(\hat{x})D^T(\hat{x})\right)W^T(\hat{x}).$$

Note that for $\hat{x}(0) = x(0) = x_0$, $r_y = r_0 = 0$. For this case, we denote $\Sigma_{\mathcal{K}}$ by $\Sigma_{\mathcal{K}}^{x_0}$. It is apparent that $\forall v$,

$$\Sigma_{\mathcal{K}}^{x_0} \circ \Sigma_{\mathcal{I}}(v) = \Sigma_{\mathcal{K}}^{x_0}\begin{pmatrix} u \\ y \end{pmatrix} = 0 \Longrightarrow \Sigma_{\mathcal{K}}^{x_0} \circ \Sigma_{\mathcal{I}} = 0. \tag{11.5}$$

In terms of the concepts of SIR and SKR, we are in a position to introduce the definitions of image and kernel manifolds. They are lower-dimensional manifolds embedded in the data space of (u, y). This manner of modelling dynamic systems enables us to address fault detection/classification issues directly in the data space, rather than on the basis of the system input-output relations.

Definition 11.1 Given system (11.1), its SIR (11.2) and SKR (11.4),

$$\mathcal{I}_\Sigma = \left\{ \begin{bmatrix} u \\ y \end{bmatrix} : \begin{bmatrix} u \\ y \end{bmatrix} = \Sigma_\mathcal{I}(v), v \in \mathcal{L}_2 \right\}, \tag{11.6}$$

$$\mathcal{K}_\Sigma = \left\{ \begin{bmatrix} u \\ y \end{bmatrix} \in \mathcal{L}_2 : \Sigma_\mathcal{K}^{x_0}\left(\begin{bmatrix} u \\ y \end{bmatrix} \right) = 0 \right\} \tag{11.7}$$

are called image and kernel sub-manifold of Σ, respectively.

Note that, due to relation (11.5), it holds $\mathcal{I}_\Sigma = \mathcal{K}_\Sigma$.

11.2.2 Hamiltonian, Inner Systems, and Normalised SIR and SKR

Recall that for LTI systems, the normalised SIR and SKR are defined in terms of the transfer functions of the RCF and LCF as well as their conjugates. For nonlinear systems, the well-established Hamiltonian system technique can be applied for this purpose. In a broad sense, given a nonlinear system as an operator, the associated Hamiltonian system can be interpreted as a certain configuration of the operator and its adjoint. In this regard, we first introduce some essential concepts and definitions known in the literature and also introduced in Chap. 6.

Consider the affine system (11.1) and assume that it is stable. The Hamiltonian extension of Σ is a dynamic system modelled by

$$\begin{cases} \dot{x} = a(x) + B(x)u \\ \dot{\lambda} = -\left(\frac{\partial a(x)}{\partial x} + \frac{\partial B(x)}{\partial x} u \right)^T \lambda - \left(\frac{\partial c(x)}{\partial x} + \frac{\partial D(x)}{\partial x} u \right)^T u_a \\ y = c(x) + D(x)u \\ y_a = B^T(x)\lambda + D^T(x)u_a,\ y_a \in \mathbb{R}^p,\ u_a \in \mathbb{R}^m, \end{cases} \tag{11.8}$$

with state variables (x, λ), input variables (u, u_a) and output variables (y, y_a). $\lambda \in \mathbb{R}^n$ is called co-state variable. Connecting y and u_a, i.e.

$$u_a = y = c(x) + D(x)u,$$

and defining a Hamiltonian function

$$H(x, \lambda, u) = \frac{1}{2} (c(x) + D(x)u)^T (c(x) + D(x)u) + \lambda^T (a(x) + B(x)u)$$

lead to the Hamiltonian system $y_a = (D\Sigma)^T \circ \Sigma(u)$, whose state space representation can be written in a compact form

$$(D\Sigma)^T \circ \Sigma : \begin{cases} \dot{x} = \frac{\partial H}{\partial \lambda}(x, \lambda, u) \\ \dot{\lambda} = -\frac{\partial H}{\partial x}(x, \lambda, u) \\ y_a = \frac{\partial H}{\partial u}(x, \lambda, u) \end{cases} \tag{11.9}$$

with input vector u and output vector y_a. Differently, connecting u and y_a of the Hamiltonian extension (11.8) results in

$$\Sigma \circ (D\Sigma)^T : \begin{cases} \dot{x} = a(x) + B(x)\left(B^T(x)\lambda + D^T(x)u_a\right) \\ \dot{\lambda} = -\left(\frac{\partial a(x)}{\partial x} + \frac{\partial \bar{B}(x,\lambda,u_a)}{\partial x}\right)^T \lambda - \left(\frac{\partial c(x)}{\partial x} + \frac{\partial \bar{D}(x,\lambda,u_a)}{\partial x}\right)^T u_a \\ y = c(x) + D(x)B^T(x)\lambda + D(x)D^T(x)u_a, \end{cases}$$
$$\bar{B}(x, \lambda, u_a) = B(x)\left(B^T(x)\lambda + D^T(x)u_a\right),$$
$$\bar{D}(x, \lambda, u_a) = D(x)B^T(x)\lambda + D(x)D^T(x)u_a. \tag{11.10}$$

Let

$$H(x, \lambda, u_a) =$$
$$\frac{1}{2}u_a^T D(x)D^T(x)u_a + c^T(x)u_a + \lambda^T\left(a(x) + \frac{1}{2}B(x)B^T(x)\lambda + B(x)D^T(x)u_a\right) \tag{11.11}$$

be a Hamiltonian function. Accordingly, $\Sigma \circ (D\Sigma)^T$ can be compactly written as

$$\Sigma \circ (D\Sigma)^T : \begin{cases} \dot{x} = \frac{\partial H}{\partial \lambda}(x, \lambda, u_a) \\ \dot{\lambda} = -\frac{\partial H}{\partial x}(x, \lambda, u_a) \\ y = \frac{\partial H}{\partial u_a}(x, \lambda, u_a). \end{cases} \tag{11.12}$$

Considering that the normalised SIR and SKR are inner and co-inner, respectively, we firstly review the concepts and existence conditions of nonlinear inner and co-inner systems.

Definition 11.2 Consider the stable nonlinear affine system (11.1). Σ is inner if the Hamiltonian system (11.9) satisfies

$$y_a = u, \tag{11.13}$$

and lossless , and it is co-inner, when it holds, for the Hamiltonian system (11.12),

$$y = u_a \tag{11.14}$$

and lossless. Here, Σ is called lossless, if there exists a storage function $V(x) \geq 0$, $V(0) = 0$ so that

$$V(x(t_2)) - V(x(t_1)) = \frac{1}{2}\int_{t_1}^{t_2} \left(u^T u - y^T y\right) d\tau. \tag{11.15}$$

Lemma 11.1 *The nonlinear affine system (11.1) is inner, if there exists $V(x) \geq 0$ such that the following equations are feasible*

$$V_x^T(x)\, a(x) + \frac{1}{2} c^T(x) c(x) = 0, \tag{11.16}$$
$$V_x^T(x)\, B(x) + c^T(x) D(x) = 0, \tag{11.17}$$
$$D^T(x) D(x) = I, \tag{11.18}$$

where $V_x(x) = \frac{\partial V(x)}{\partial x}$. And, it is co-inner, if there exists $V(x) \geq 0$ such that the following equations are feasible

$$V_x^T(x)\, a(x) + \frac{1}{2} V_x^T(x)\, B(x) B^T(x) V_x(x) = 0, \tag{11.19}$$
$$c^T(x) + V_x^T(x)\, B(x) D^T(x) = 0, \tag{11.20}$$
$$D(x) D^T(x) = I. \tag{11.21}$$

Definition 11.3 The SIR $\Sigma_{\mathcal{I}}$ and SKR $\Sigma_{\mathcal{K}}$ given in (11.2) and (11.4) are, respectively, normalised, when $(g(x), V(x))$ and $(L(x), W(x))$ are set such that $\Sigma_{\mathcal{I}}$ is inner and $\Sigma_{\mathcal{K}}$ is co-inner.

Theorem 11.1 *The SIR (11.2) is normalised, if there exists $P(x) \geq 0$ such that the following equation is feasible*

$$P_x^T(x) a(x) + \frac{1}{2} c^T(x)\, c(x) - \frac{1}{2} g^T(x) \left(I + D^T(x) D(x)\right) g(x) = 0, \tag{11.22}$$

and $(g(x), V(x))$ are set to be

$$g(x) = -\left(I + D^T(x) D(x)\right)^{-1} \left(B^T(x) P_x(x) + D^T(x) c(x)\right), \tag{11.23}$$
$$V(x) = V^T(x) = \left(I + D^T(x) D(x)\right)^{-1/2}. \tag{11.24}$$

And the SKR (11.4) is normalised, if there exists $V(\hat{x}) \geq 0$ solving

$$\begin{gathered} V_{\hat{x}}^T(\hat{x})\, a(\hat{x}) + \tfrac{1}{2} V_{\hat{x}}^T(\hat{x})\, B(\hat{x}) B^T(\hat{x}) V_{\hat{x}}(\hat{x}) - \\ \tfrac{1}{2}\left(c^T(\hat{x}) + V_{\hat{x}}^T(\hat{x})\, B(\hat{x}) D^T(\hat{x})\right) W^2(\hat{x}) \left(c^T(\hat{x}) + V_{\hat{x}}^T(\hat{x})\, B(\hat{x}) D^T(\hat{x})\right)^T = 0 \end{gathered} \tag{11.25}$$

and $(L(x), W(x))$ are set to be

$$V_{\hat{x}}^T(\hat{x})\, L(\hat{x}) = \left(c^T(\hat{x}) + V_{\hat{x}}^T(\hat{x})\, B(\hat{x}) D^T(\hat{x})\right) \left(I + D(\hat{x}) D^T(\hat{x})\right)^{-1}, \tag{11.26}$$
$$W(\hat{x}) = \left(I + D(\hat{x}) D^T(\hat{x})\right)^{-1/2}. \tag{11.27}$$

The proof of this theorem follows immediately from Lemma 11.1 and thus is omitted.

Remark 11.1 The normalised SIR can be interpreted as an optimal controller under a quadratic cost function associated to the Hamiltonian functions. A normalised SKR based estimator results in an LS estimation of uncertainties corrupted in data (u, y).

11.2.3 Bregman Divergence, Legendre Transform, and Geodesic Projection

For our purpose, we restrict our attention to the following definition of a Bregman divergence. Let $\varphi : \mathbb{R}^n \to \mathbb{R}$ be a continuously-differentiable, strictly convex function. Bregman divergence from $\alpha \in \mathbb{R}^n$ to $\beta \in \mathbb{R}^n$, denoted by $D_\varphi [\alpha : \beta]$, is defined by

$$D_\varphi [\alpha : \beta] = \varphi (\alpha) - \varphi (\beta) - \left(\frac{\partial \varphi (\beta)}{\partial \beta} \right)^T (\alpha - \beta). \tag{11.28}$$

The function φ is also called generating function. Bregman divergences were introduced by Bregman as a measure of difference between two points in a space. A Bregman divergence is of all properties of a divergence. In our subsequent work, the dual divergence to $D_\varphi [\alpha : \beta]$ plays an important role. To define it, we firstly consider a Legendre transform. Let

$$\alpha^\times = \frac{\partial \varphi (\alpha)}{\partial \alpha} \in \mathbb{R}^n.$$

The function

$$\varphi^\times \left(\alpha^\times\right) := \alpha^T \alpha^\times - \varphi (\alpha)$$

is called the Legendre dual of φ. It is known that

$$\alpha^\times = \frac{\partial \varphi (\alpha)}{\partial \alpha}, \alpha = \frac{\partial \varphi^\times \left(\alpha^\times\right)}{\partial \alpha^\times},$$

form a dualistic structure, and $\varphi^\times \left(\alpha^\times\right)$ is convex with respect to $\alpha^\times$. By means of the Legendre dual of φ, the dual divergence to $D_\varphi [\alpha : \beta]$ is defined by

$$D_{\varphi^\times} \left[\alpha^\times : \beta^\times\right] = \varphi \left(\alpha^\times\right) - \varphi \left(\beta^\times\right) - \left(\frac{\partial \varphi^\times \left(\beta^\times\right)}{\partial \beta^\times} \right)^T \left(\alpha^\times - \beta^\times\right).$$

By some routine computations, we obtain

$$D_{\varphi}[\alpha:\beta]=\varphi(\alpha)+\varphi^{\times}\left(\beta^{\times}\right)-\alpha^{T}\beta^{\times}, \tag{11.29}$$

$$D_{\varphi^{\times}}\left[\alpha^{\times}:\beta^{\times}\right]=D_{\varphi}[\beta:\alpha]. \tag{11.30}$$

The introduction of the dual divergence is not only useful from the computational point of view. It is the theoretical basis for the definition and computation of the so-called geodesic projection onto a manifold. Informally speaking, a geodesic projection can be understood as an extension of the concept of an orthogonal projection onto a vector subspace. Below, we shortly highlight the basic ideas and some results on the Bregman divergence-based geodesic projection described in the literature, and restrict the introduction only to the so-called projection theorem that is needed for our fault detection study. The reader is referred to references given in the last section of this chapter for more details.

In his monograph on Information geometry, Amari begins the introduction to a dually flat Riemannian manifold with the following statement, "*we can establish a new edifice of differential geometry, by treating a pair of affine connections which are dually coupled with respect to the Riemannian metric. Such a structure ... is the heart of information geometry. ... As an important special case, we study a dually flat Riemannian manifold. It may be regarded as a dualistic extension of the Euclidean space.*" Given two convex functions connected by the Legendre transform $\alpha^{\times}=\nabla F(\alpha), \alpha=\nabla F^{\times}\left(\alpha^{\times}\right)$, Bregman and dual Bregman divergences on a manifold $\mathcal{M}$ are defined,

$$D_{\varphi}[\alpha:\beta]=:F(\alpha), D_{\varphi^{\times}}\left[\alpha^{\times}:\beta^{\times}\right]=:F^{\times}\left(\alpha^{\times}\right).$$

On this basis, metric tensors

$$G:=\nabla^{2}F(\alpha), G^{\times}:=\nabla^{2}F^{\times}\left(\alpha^{\times}\right)$$

provide $\mathcal{M}$ with a dual Riemannian structure that admits a conjugate pair of affine connections. It is proved that these two affine connections are curvature- and torsion-free, and thus the corresponding Riemannian structure is called dually flat. Consequently, any pair of points in $\mathcal{M}$ can be linked by a straight line either in the coordinate system α or in its dual coordinate system $\alpha^{\times}$. A main result related to a dually flat structured manifold is the proof of the Generalised Pythagorean Theorem by Nagaoka and Amari. Based on it, the Projection Theorem given below was proved.

Let $\mathcal{M}\subset\mathbb{R}^{n}$ denote a dually flat manifold.

Definition 11.4 Given $\alpha\in\mathcal{M}$ and a submanifold $\mathcal{S}\subset\mathcal{M}$, the divergence from α to $\mathcal{S}$ is defined by

$$D_{\varphi}[\alpha:\mathcal{S}]=\min_{\beta\in\mathcal{S}}D_{\varphi}[\alpha:\beta].$$

A curve is said to be orthogonal to $\mathcal{S}$ when its tangent vector is orthogonal to any tangent vectors of $\mathcal{S}$ at the intersection.

Definition 11.5 Vector $\hat{\alpha}_{\mathcal{S}} \in \mathcal{S}$ is called geodesic projection of α to $\mathcal{S}$ when the geodesic connecting α and $\hat{\alpha}_{\mathcal{S}}$ is orthogonal to $\mathcal{S}$. Dually, $\hat{\alpha}_{\mathcal{S}}^{\times} \in \mathcal{S}$ is the dual geodesic projection of α to $\mathcal{S}$, when the dual geodesic connecting α and $\hat{\alpha}_{\mathcal{S}}^{\times}$ is orthogonal to $\mathcal{S}$.

Theorem 11.2 (Projection theorem) *Given $\alpha \in \mathcal{M}$ and a smooth submanifold $\mathcal{S}$ in a dually flat manifold $\mathcal{M}$, the vector that minimises the divergence $D_{\varphi}[\alpha : \beta], \beta \in \mathcal{S}$, is the dual geodesic projection of α to $\mathcal{S}$. The vector that minimises the dual divergence $D_{\varphi^{\times}}[\alpha : \beta], \beta \in \mathcal{S}$, is the geodesic projection of α to $\mathcal{S}$.*

11.2.4 Problem Formulation

The main intention of introducing the system coprime factorisations and the SIR/SKR concepts is to model the (nominal) system dynamic as a subspace/sub-manifold in the data space of (u, y). Although these model forms are equivalent to the input-output models widely applied in control-theoretic research, they provide us with the alternative methodology to deal with fault diagnosis problems, namely directly in data space and by means of an orthogonal projection. As delineated by the projection-based fault detection scheme for LTI systems presented in the previous chapters, an orthogonal projection of (u, y) onto the system image (kernel) subspace helps us to extract information from the data optimally using the distance of the data vector to the system image subspace as the (nominal) system dynamics. In this regard, the normalised SIR and SKR serve for two purposes. Firstly, it is a mathematical tool for the computation of the projection and the distance. Secondly, it connects the projection-based fault detection with the observer-based one. The latter implies that the projection-based fault detection can be realised using an observer-based detection system and thus, on the other hand, enables an optimal design of an observer-based detection system.

A direct extension of the projection-based fault detection to nonlinear dynamic systems poses obviously a number of problems. While the SIR and SKR can be, analogous to LTI systems, straightforwardly introduced, concepts like projection and distance definitions have to be addressed with the aid of different mathematical tools and knowledge. Concerning projection issues, the well-established theoretic framework of Hamiltonian systems may serve as the system foundation, and as an analysis and computational tool, where the Hamiltonian systems corresponding to the normalised SIR and SKR and their solutions as inner and co-inner systems play an important role in our subsequent work. The centrepiece in our framework is the associated Hamiltonian functions, by which the Bregman divergences are introduced for our fault detection study.

To be specific, for LTI systems, the system image (kernel) is a subspace in Hilbert space, in which an inner product and the corresponding norm of a vector are well defined. In such a Euclidean space, the shortest distance between two points (vectors) is a straight line, and thus the norm of a residual vector is efficient for measuring the distance from a data vector to the system image subspace that represents the nominal dynamics. This is the theoretic basis for an optimal fault detection. In a non-Euclidean space, as the manifold formed by process data from a nonlinear system, the metric tensor at each point of the manifold is different from the Euclidean metric tensor like an inner product defined on Hilbert space. Therefore, the norm-based distance defined in Hilbert space may not be compatible with the metric tensor of the non-Euclidean space. In other words, the norm-based metric is not suitable to measure the distance from a data vector to the system image manifold, based on which a fault detection should be made. To solve this problem, we propose to use Bregman divergences to measure deviations in the system dynamics and data from the nominal ones. Initially, this idea has been inspired by wide applications of Bregman divergences in machine learning and, in particular, the Kullback-Leibler (KL) divergence as a difference measure of two probability distributions. In the course of our extensive investigation, three distinguishing aspects of applying Bregman divergences to fault detection have drawn our attention: (i) realisation of performance-oriented fault detection, (ii) simplified computation and realisation of Bregman divergences making use of the dual Hamiltonian system pair as a Legendre transform, and (iii) interpretation of the SIR-based projection as a geodesic projection in terms of Bregman divergence from a data vector to the image manifold. To be specific,

- performance-oriented fault detection has been proposed in our earlier works, dedicated to reliably detecting performance degradation in automatic control systems. It has been observed that considerable engineering efforts are often needed to set the parameters of the performance evaluation function. Bregman divergences are induced by convex functions and measure the difference of two points (vectors) defined in terms of this convex function. In optimal control, Hamiltonian functions act as (control) performance cost functions. In a broad sense, Hamiltonian functions can be interpreted as the energy balance (parity relation) in dynamic systems. This fact motivates us to define the Hamiltonian function of the system under consideration as a system performance and, based on it, to induce Bregman divergences towards a performance-oriented fault detection.
- It is known that for a given system, there exists a dual Hamiltonian system pair that forms a Legendre transform. As delineated in Sect. 11.2.3, the Legendre dual of the Hamiltonian functions is useful for the Bregman divergence computations. More importantly, the Bregman divergence endows the data manifold with a dually flat structured Riemannian manifold, which can be regarded as a dualistic extension of the Euclidean space.
- As a result, the projection theorem can be used to characterise the Hamiltonian function-based projection and to highlight its insight from the viewpoint of information geometry.

As described in Chap. 9, a distinguishing feature for projections in Hilbert space is that the data vector can be decomposed as the orthogonal projection of the data vector onto the image subspace and its complementary. It leads to the useful relations

$$\begin{bmatrix} \hat{u} \\ \hat{y} \end{bmatrix} = \mathcal{P}_{\mathcal{I}_G^{\mathcal{L}_2}} \left(\begin{bmatrix} u \\ y \end{bmatrix} \right), r_{\mathcal{I}_G^{\mathcal{L}_2}} = \begin{bmatrix} u \\ y \end{bmatrix} - \begin{bmatrix} \hat{u} \\ \hat{y} \end{bmatrix} = \left(\mathcal{I} - \mathcal{P}_{\mathcal{I}_G^{\mathcal{L}_2}} \right) \left(\begin{bmatrix} u \\ y \end{bmatrix} \right),$$
$$\left\| \begin{bmatrix} u \\ y \end{bmatrix} \right\|_2^2 = \left\| \begin{bmatrix} \hat{u} \\ \hat{y} \end{bmatrix} \right\|_2^2 + \left\| r_{\mathcal{I}_G^{\mathcal{L}_2}} \right\|_2^2, \tag{11.31}$$

that enable the realisation of a projection-based fault detection system in the well-established observer-based detection framework. Here, the second equation is known as Pythagorean equation as well. Moreover, they pose a novel scheme to design observer-based fault detection systems that are capable of dealing with systems with model uncertainties and multiplicative faults. Unfortunately, for nonlinear systems such relations are generally not valid. Bearing this fact in mind, we will study the (nonlinear) SIR- and SKR-based projections separately. The SIR-based projection, similar to the application to LTI systems, concerns the projection of process data (u, y) to the system image manifold representing the nominal dynamics. By means of a Hamiltonian function induced Bregman divergence, the projection is evaluated, which delivers a distance measure between the data and the image manifold and is used for fault detection. As known from the previous chapters, an SKR is an observer-based residual generator that reflects influences of uncertainties in the system, including faults, on the system dynamics. Correspondingly, the SKR-based projection is an estimate for uncertainties. For the fault detection purpose, the evaluation of the SKR-based projection in terms of a Hamiltonian function induced Bregman divergence will be dedicated to a measure of the distance between the data and the uncertainty-free operation.

In course of our study on the SKR-based projection, the imperative of insightfully understanding the SKR-based projection as an estimate for uncertainties has been recognised. Remember that the essential reason why the Pythagorean equation doesn't hold for nonlinear systems is that the data (u, y) are inextricably corrupted with uncertainties. Consequently, the influences of the system input (representing the nominal dynamic) and uncertainties in the system are not separable in the data. As will be shown in our subsequent study, an SKR-based projection is a Hamiltonian system with the system data (u, y) (equivalent with an observer-based residual generator) as its input and the projection as its output. Our endeavour, to be specific, is to figure out in which context the projection delivers an estimate of the uncertainty corrupted in the data. To this end, an uncertainty model is introduced, which uniformly models deviations in the data from their nominal values and caused by all types of possible uncertainties, including faults. In this regard, it will be delineated that the SKR-based projection is an LS estimate of the uncertainty corrupted in the data. This result not only highlights the insightful interpretation of the SKR-based projection, but also underlines its potential use, for instance, for fault estimation.

To put it in a nutshell, the main problems to be addressed in this work are formulated as follows:

- configuration of a normalised SIR-based projection system that projects (u, y) onto $\mathcal{I}_\Sigma$. The work will be performed in the framework of Hamiltonian systems and their dual form as a Legendre transform.

- introduction of a Bregman divergence induced by a Hamiltonian function and computation solution of the Bregman divergence from (u, y) to its projection. This solution will serve as an evaluation function for the detection (decision) logic.
- study on interpretations of the SIR-based projection and Bregman divergence as a fault detection system. Hereby, two aspects are of special interest: (i) the control-theoretic insight, i.e. understanding of the Hamiltonian function induced Bregman divergence in the context of system performance-oriented fault detection, (ii) the interpretation from the viewpoint of information geometry and in the context of the projection theorem.
- discussion on the realisation and implementation issues. In order to achieve a reliable fault detection, it is of considerable practical importance to reduce false alarms. To this end, the defined Bregman divergence will be evaluated over an evaluation time interval. Moreover, a threshold should be set to ensure an optimal fault detection decision. At the end of this work, a one-class fault detection scheme for nonlinear dynamic systems is completed.
- investigation on a scheme of a normalised SKR-based projection system. Although this work is analogue to the one of the SIR-based projection and fault detection system, it is significantly different in (i) its construction as an observer-based residual generator, (ii) the definition of the Bregman divergence, and in particular (iii) the interpretation of the Bregman divergence as a distance measure of (u, y) and the nominal system dynamics.
- study on the SKR-based projection system as an estimator for the uncertainty corrupted in (u, y) with a focus on (i) definition of an uncertainty model that should be independent of the possible types of uncertainties in the system, including faults, and thus lays the basis for interpreting and evaluating the estimation performance, (ii) evaluation of the projection-based estimation.

11.3 A Projection and Bregman Divergence Based Fault Detection Scheme

In this section, the first part of our work is presented. It is a projection-based scheme for detecting faults in the affine system (11.1). To be specific, a projection of process data (u, y) onto the image subspace $\mathcal{I}_\Sigma$ is realised using the normalised SIR $\Sigma_\mathcal{I}$. Then, the introduction of a Hamiltonian function induced Bregman divergence and the associated Riemannian structure of the data manifold enables the definition and computation of a geodesic projection onto $\mathcal{I}_\Sigma$. Based on these theoretical results, fault detection algorithms are presented. To this end, we will firstly study the normalised SIR and its properties in the framework of Hamiltonian systems. It is followed by defining and analysing a Bregman divergence with the Hamiltonian function as a generating function. The proof that the normalised SIR-based projection is the geodesic projection gives an interesting interpretation of the projection as an important result of this study. At the end of this section, implementation issues towards a practical fault detection and the associated algorithm are discussed.

11.3.1 Normalised SIR and the Associated Hamiltonian Projection Systems

Consider the affine system (11.1) and its SIR $\Sigma_{\mathcal{I}}$ given by (11.2). The corresponding Hamiltonian extension of $\Sigma_{\mathcal{I}}$ is the dynamic system described by

$$\begin{cases} \dot{x} = a_I(x) + B_I(x)v \\ \dot{\lambda} = -\left(\frac{\partial a_I(x)}{\partial x} + \frac{\partial B_I(x)}{\partial x}v\right)^T \lambda - \left(\frac{\partial c_I(x)}{\partial x} + \frac{\partial D_I(x)}{\partial x}\right)^T v_a \\ \hat{z} := c_I(x) + D_I(x)v \in \mathbb{R}^{m+p} \\ z_a := B_I^T(x)\lambda + D_I^T(x)v_a \in \mathbb{R}^p. \end{cases}$$

For our purpose of projecting (u, y) onto the image subspace $\mathcal{I}_\Sigma$, connect v and z_a, i.e.

$$v = z_a = B_I^T(x)\lambda + D_I^T(x)v_a, \tag{11.32}$$

and let the data (u, y) be the input and $\hat{z}$ the output of $\Sigma_{\mathcal{I}} \circ (D\Sigma_{\mathcal{I}})^T$,

$$v_a = \begin{bmatrix} u \\ y \end{bmatrix} =: z, \hat{z} = \Sigma_{\mathcal{I}} \circ (D\Sigma_{\mathcal{I}})^T(z).$$

Then, we have

$$\Sigma_{\mathcal{I}} \circ (D\Sigma_{\mathcal{I}})^T : \begin{cases} \dot{x} = a_I(x) + B_I(x)B_I^T(x)\lambda + B_I(x)D_I^T(x)z \\ \dot{\lambda} = -\left(\frac{\partial a_I(x)}{\partial x} + \frac{\partial \bar{B}_I(x,\lambda,z)}{\partial x}\right)^T \lambda - \left(\frac{\partial\left(c_I(x)+\bar{D}_I(x,\lambda,z)\right)}{\partial x}\right)^T z \\ \hat{z} = c_I(x) + D_I(x)B_I^T(x)\lambda + D_I(x)D_I^T(x)z, \end{cases}$$

$$\bar{B}_I(x,\lambda,z) = B_I(x)B_I^T(x)\lambda + B_I(x)D_I^T(x)z, \tag{11.33}$$

$$\bar{D}_I(x,\lambda,z) = D_I(x)B_I^T(x)\lambda + D_I(x)D_I^T(x)z.$$

The configuration of $\Sigma_{\mathcal{I}} \circ (D\Sigma_{\mathcal{I}})^T$ is schematically depicted in Fig. 11.1.

Define the Hamiltonian function

$$\begin{aligned} H(x,\lambda,z) = &\frac{1}{2}z^T D_I(x)D_I^T(x)z + c_I^T(x)z \\ &+ \lambda^T\left(a_I(x) + \frac{1}{2}B_I(x)B_I^T(x)\lambda + B_I(x)D_I^T(x)z\right). \end{aligned} \tag{11.34}$$

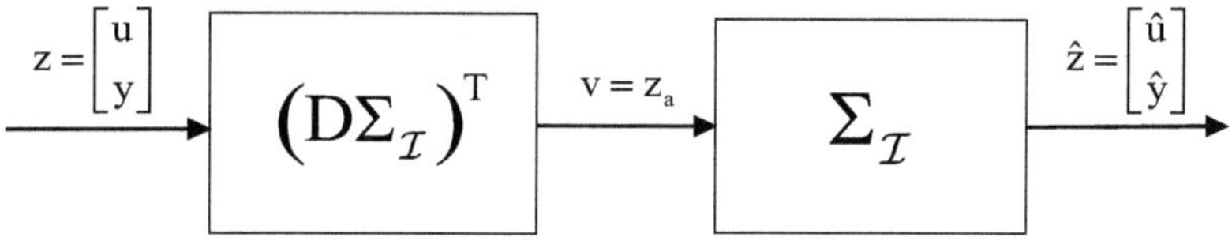

Fig. 11.1 SIR-based projection system $\Sigma_{\mathcal{I}} \circ (D\Sigma_{\mathcal{I}})^T$

It is straightforward that the Hamiltonian system (11.33) can then be written in the compact form

$$\Sigma_{\mathcal{I}} \circ (D\Sigma_{\mathcal{I}})^T : \begin{cases} \dot{x} = \frac{\partial H}{\partial \lambda}(x, \lambda, z) \\ \dot{\lambda} = -\frac{\partial H}{\partial x}(x, \lambda, z) \\ \hat{z} = \frac{\partial H}{\partial z}(x, \lambda, z) . \end{cases} \tag{11.35}$$

Notice that the Hamiltonian function (11.34) can be, after some routine calculations, written as

$$\begin{aligned} H(x, \lambda, z) &= -\frac{1}{2} z^T D_I(x) D_I^T(x) z + \hat{z}^T z + \lambda^T \left(a_I(x) + \frac{1}{2} B_I(x) B_I^T(x) \lambda \right) \\ &= -\frac{1}{2} v^T v + \hat{z}^T z + \lambda^T \dot{x}, \\ \dot{x} &= a_I(x) + B_I(x) B_I^T(x) \lambda + B_I(x) D_I^T(x) z. \end{aligned} \tag{11.36}$$

Let

$$H^{\times}\left(x, \lambda, \hat{z}\right) = \hat{z}^T z - H(x, \lambda, z) = \frac{1}{2} v^T v - \lambda^T \dot{x}. \tag{11.37}$$

Note that

$$z = \frac{\partial H^{\times}\left(x, \lambda, \hat{z}\right)}{\partial \hat{z}},$$

and the Hamiltonian system (11.35) can be expressed in terms of $H^{\times}\left(x, \lambda, \hat{z}\right)$ as

$$\begin{cases} \dot{x} = -\frac{\partial H^{\times}}{\partial \lambda}\left(x, \lambda, \hat{z}\right) \\ \dot{\lambda} = \frac{\partial H^{\times}}{\partial x}\left(x, \lambda, \hat{z}\right) \\ z = \frac{\partial H^{\times}}{\partial \hat{z}}\left(x, \lambda, \hat{z}\right) \end{cases} \tag{11.38}$$

with $\hat{z}$ as the input and z as the output. The above Hamiltonian system (11.38) has been investigated in the context of inner-outer factorisation, and $H^{\times}\left(x, \lambda, \hat{z}\right)$ is the Legendre dual of $H(x, \lambda, z)$. In other words, the Hamiltonian functions $H(x, \lambda, z)$ and $H^{\times}\left(x, \lambda, \hat{z}\right)$ with

$$\hat{z} = \frac{\partial H}{\partial z}(x, \lambda, z) =: z^{\times}, z = \frac{\partial H^{\times}}{\partial \hat{z}}\left(x, \lambda, \hat{z}\right),$$

form a Legendre transform and give the system data a dualistic structure. Here, $\hat{z}$ is dual to z and thus is denoted by $z^{\times}$ as well. It is obvious from (11.36) that

$$\begin{aligned} H^{\times}\left(x, \lambda, z^{\times}\right) &= \frac{1}{2} v^T v - \lambda^T \dot{x} \\ &= \frac{1}{2} z^T D_I(x) D_I^T(x) z - \lambda^T \left(a_I(x) + \frac{1}{2} B_I(x) B_I^T(x) \lambda \right). \end{aligned} \tag{11.39}$$

Now, consider the normalised SIR and the corresponding Hamiltonian systems (11.35) and (11.38). The following theorem showcases that $\Sigma_{\mathcal{I}} \circ (D\Sigma_{\mathcal{I}})^T$ is an idempotent operator and the normalised SIR projects the process data (u, y) to $\mathcal{I}_{\Sigma}$.

Theorem 11.3 *Given the normalised SIR and the corresponding Hamiltonian system (11.33), then $\Sigma_{\mathcal{I}} \circ (D\Sigma_{\mathcal{I}})^T$ is idempotent, and*

$$\forall z = \begin{bmatrix} u \\ y \end{bmatrix} \in \mathcal{L}_2, \hat{z} \in \mathcal{I}_{\Sigma}, \tag{11.40}$$

$$\forall z \neq 0, \hat{z} = \begin{bmatrix} u \\ y \end{bmatrix} \text{ iff } \begin{bmatrix} u \\ y \end{bmatrix} \in \mathcal{I}_{\Sigma}. \tag{11.41}$$

Proof That $\Sigma_{\mathcal{I}} \circ (D\Sigma_{\mathcal{I}})^T$ is idempotent immediately follows from the normalised SIR, i.e.

$$\Sigma_{\mathcal{I}} \circ (D\Sigma_{\mathcal{I}})^T \circ \Sigma_{\mathcal{I}} \circ (D\Sigma_{\mathcal{I}})^T = \Sigma_{\mathcal{I}} \circ (D\Sigma_{\mathcal{I}})^T .$$

The proof of (11.40) is apparent, since

$$(D\Sigma_{\mathcal{I}})^T \left(\begin{bmatrix} u \\ y \end{bmatrix} \right) =: v \in \mathcal{L}_2 \Longrightarrow \Sigma_{\mathcal{I}} \circ (D\Sigma_{\mathcal{I}})^T \left(\begin{bmatrix} u \\ y \end{bmatrix} \right) = \Sigma_{\mathcal{I}}(v) \in \mathcal{I}_{\Sigma}.$$

To prove (11.41), observe that

$$\begin{gathered} \forall \begin{bmatrix} u \\ y \end{bmatrix} \in \mathcal{I}_{\Sigma}, \exists v \in \mathcal{L}_2, \text{ s.t. } \begin{bmatrix} u \\ y \end{bmatrix} = \Sigma_{\mathcal{I}}(v) \Longrightarrow \\ \Sigma_{\mathcal{I}} \circ (D\Sigma_{\mathcal{I}})^T \left(\begin{bmatrix} u \\ y \end{bmatrix} \right) = \Sigma_{\mathcal{I}} \circ (D\Sigma_{\mathcal{I}})^T \circ \Sigma_{\mathcal{I}}(v) = \Sigma_{\mathcal{I}}(v) = \begin{bmatrix} u \\ y \end{bmatrix}. \end{gathered}$$

On the other hand, it follows from (11.40) that

$$\hat{z} = \begin{bmatrix} u \\ y \end{bmatrix} \Longrightarrow \begin{bmatrix} u \\ y \end{bmatrix} \in \mathcal{I}_{\Sigma}.$$

Thus, (11.41) is proved.

Remark 11.2 That $\Sigma_{\mathcal{I}} \circ (D\Sigma_{\mathcal{I}})^T$ is idempotent is an important property that ensures the projection preserving local neighbourhoods at every point of the underlying manifold.

As a projection of (u, y) to $\mathcal{I}_\Sigma$, $\hat{z}$ can be interpreted as an estimate of (u, y) as well. In this context, the notation

$$\hat{z} = \begin{bmatrix} \hat{u} \\ \hat{y} \end{bmatrix}$$

will also be adopted in the sequel.

The following theorem is useful for us to examine the interpretation of both Hamiltonian functions as generalised energy functions and, based on it, to establish our fault detection strategy.

Theorem 11.4 *Given the normalised SIR and the corresponding Hamiltonian system (11.33),*

$$\forall \begin{bmatrix} u \\ y \end{bmatrix} \in \mathcal{I}_\Sigma, H^\times\left(x, \lambda, \hat{z}\right) = H\left(x, \lambda, z\right) = \frac{1}{2}\hat{z}^T\hat{z} = \frac{1}{2}z^T z = \frac{1}{2}v^T v - \dot{P}(x),$$

where $P(x) \geq 0$ is the solution of the HJE (11.22).

Proof It follows from Theorem 11.3 that

$$\forall z = \begin{bmatrix} u \\ y \end{bmatrix} \in \mathcal{I}_\Sigma, \hat{z} = z,$$

leading to

$$z_a = \left(D\Sigma_\mathcal{I}\right)^T\left(\begin{bmatrix} u \\ y \end{bmatrix}\right) = \left(D\Sigma_\mathcal{I}\right)^T \circ \Sigma_\mathcal{I}\left(v\right) \Longrightarrow z_a = v \tag{11.42}$$

for some v. Subsequently,

$$\Sigma_\mathcal{I}\left(z_a\right) = \Sigma_\mathcal{I}\left(v\right). \tag{11.43}$$

Equations (11.42)–(11.43) imply that the subsystems in (11.33), $\Sigma_\mathcal{I}(z_a)$ and $\left(D\Sigma_\mathcal{I}\right)^T(z)$, have the identical dynamics like the subsystems $\Sigma_\mathcal{I}(v)$ and $\left(D\Sigma_\mathcal{I}\right)^T(v_a)$ in

$$z_a = \left(D\Sigma_\mathcal{I}\right)^T \circ \Sigma_\mathcal{I}\left(v\right) = \left(D\Sigma_\mathcal{I}\right)^T\left(z\right).$$

Hence, following Lemma 11.1, it holds

$$P_x^T\left(x\right) a_I(x) + \frac{1}{2}c_I^T(x)c_I(x) = 0,\ P_x^T\left(x\right) B_I(x) + c_I^T(x)D_I(x) = 0, \tag{11.44}$$

$$D_I^T(x)D_I(x) = I, \lambda = P_x\left(x\right). \tag{11.45}$$

Equation (11.36) and the relation

$$v = z_a = B_I^T(x)\lambda + D_I^T(x)z,$$

yield

$$H(x,\lambda,z) = -\frac{1}{2}\left(B_I^T(x)\lambda + D_I^T(x)z\right)^T\left(B_I^T(x)\lambda + D_I^T(x)z\right) + \hat{z}^T z$$
$$+ \lambda^T\left(a_I(x) + B_I(x)B_I^T(x)\lambda + B_I(x)D_I^T(x)z\right). \tag{11.46}$$

After some routine computations with the aid of equations (11.44)–(11.45), it turns out

$$H(x,\lambda,z) = -\frac{1}{2}\hat{z}^T\hat{z} + \hat{z}^T z = \frac{1}{2}z^T z = \frac{1}{2}\hat{z}^T\hat{z}, \tag{11.47}$$

where the last two identities are attributed to Theorem 11.3. Because

$$H^\times\left(x,\lambda,\hat{z}\right) = \hat{z}^T z - H(x,\lambda,z),$$

it is apparent that

$$H^\times\left(x,\lambda,\hat{z}\right) = H(x,\lambda,z) = \frac{1}{2}z^T z = \frac{1}{2}\hat{z}^T\hat{z}.$$

Finally, attributed to

$$H^\times\left(x,\lambda,\hat{z}\right) = \frac{1}{2}\upsilon^T\upsilon - P_x^T(x)\dot{x} = \frac{1}{2}\upsilon^T\upsilon - \dot{P}(x),$$

the theorem is proved.

Theorem 11.4 tells us, corresponding to the nominal system dynamics, both Hamiltonian functions (11.34) and (11.37) as energy functions are equivalent. Moreover, from the viewpoint of the energy balance, the lossless property, as defined in Definition 11.2 and in the sense of (11.15), holds. When uncertainties or faults are present in the system, i.e. $z \notin \mathcal{I}_\Sigma$, such an energy balance will be disturbed. In this context, the fault detection problem can be formulated as detecting deviations of the Hamiltonian functions from the nominal value $\frac{1}{2}\hat{z}^T\hat{z} = \frac{1}{2}z^T z$. To this end, it is plain that a well-founded evaluation function should be defined in such a way that it is possible to examine the energy balance relations real-time and thus to detect possible changes in the parity relations reliably and timely. This is the problem to be addressed in the next subsection.

11.3.2 A Bregman Divergence-Based Evaluation Function, and Geodesic Projection

In the well-established observer-based fault detection framework, building (i) a residual vector as the difference between the measurement and its estimate, (ii) evaluating the residual vector by means of a signal norm, typically $\mathcal{L}_2$ norm, are the two essential steps to a successful fault detection. In our divergence-based detection framework, we propose, alternatively, to evaluate possible changes in the system dynamic in terms of a divergence between the measurement (data) and its estimate (projection). The basic idea behind this approach is twofold, as motivated and described in the section of problem formulation,

- to achieve a system performance-oriented fault detection by means of a Hamiltonian function induced Bregman divergence, and
- to detect deviations in a non-Euclidean space using a Bregman divergence as a distance measure.

Consider the normalised SIR (11.2) and the corresponding Hamiltonian system (11.33) with the input and output vectors, z and $\hat{z}$, where z is the collected data and $\hat{z}$ is its estimate (projection) delivered by (11.33). In the sequel, for the sake of notation simplicity, the Hamiltonian functions (11.34) and (11.39) are expressed respectively by

$$H(z) := H(x, \lambda, z), \; H^{\times}(z^{\times}) := H^{\times}\left(x, \lambda, z^{\times}\right).$$

Since our objective is to detect faults by checking deviations from the nominal dynamics, we examine, according to Theorem 11.4, the condition

$$H^{\times}(z) = H(z) = \frac{1}{2}\hat{z}^T\hat{z} = \frac{1}{2}z^T z. \tag{11.48}$$

Let the Bregman divergence,

$$D_H\left[z : \hat{z}\right] := H(z) - H(\hat{z}) - \left(\frac{\partial H(\hat{z})}{\partial \hat{z}}\right)^T (z - \hat{z}), \tag{11.49}$$

be the test (evaluation) function, where $H(z)$ is the Hamiltonian function (11.34). The computation of $D_H\left[z : \hat{z}\right]$ is given in following theorem, which also provides us with an interesting interpretation of $D_H\left[z : \hat{z}\right]$.

Theorem 11.5 *Given the normalised SIR (11.2), the corresponding Hamiltonian system (11.33) and Hamiltonian functions (11.34) and (11.39), then the Bregman divergence (11.49) is given by*

$$D_H\left[z : \hat{z}\right] = H\left(\hat{z}\right) - H^{\times}\left(\hat{z}\right). \tag{11.50}$$

Proof According to Theorems 11.3 and 11.4,

$$\hat{z} \in \mathcal{I}_{\Sigma}, \; H\left(\hat{z}\right) = \frac{1}{2}\hat{z}^T\hat{z},$$

which yields, according to (11.28),

$$D_H\left[z:\hat{z}\right] = H(z) - \frac{1}{2}\hat{z}^T\hat{z} - \hat{z}^T\left(z-\hat{z}\right) = \frac{1}{2}\hat{z}^T\hat{z} + H(z) - \hat{z}^T z.$$

Recall that z and $\hat{z}$ are a Legendre dual pair,

$$\hat{z} = \frac{\partial H\left(x,\lambda,z\right)}{\partial z}, z = \frac{\partial H^{\times}\left(x,\lambda,\hat{z}\right)}{\partial \hat{z}}.$$

Hence,

$$H^{\times}\left(z^{\times}\right) = \hat{z}^T z - H(z) \Longleftrightarrow H(z) - \hat{z}^T z = -H^{\times}\left(\hat{z}\right)$$
$$\Longrightarrow D_H\left[z:\hat{z}\right] = H(\hat{z}) - H^{\times}\left(\hat{z}\right).$$

The theorem is proved.

Theorem 11.5 demonstrates that the Bregman divergence $D_H\left[z:\hat{z}\right]$ as a distance measure between process data z and its projection on to $\mathcal{I}_{\Sigma}$, $\hat{z}$, is equal to the difference between the Hamiltonian functions (11.34) and its Legendre dual (11.39) with $\hat{z}$ as their input.

It is worth emphasising that in a non-Euclidean space, the relation (11.31) for an orthogonal projection does not hold. Our intention of introducing the Bregman divergence as a distance measure is to find an alternative solution. In this regard, it is of considerably theoretic interest to study the projection $\hat{z} = \Sigma_{\mathcal{I}} \circ \left(D\Sigma_{\mathcal{I}}\right)^T(z)$ from the viewpoint of information geometry. To this end, the projection theorem, Theorem 11.2, is applied.

Consider the Bregman divergence from the process data z to the system image manifold $\mathcal{I}_{\Sigma}$ derived by $D_{H^{\times}}$,

$$D_{H^{\times}}\left[z:\mathcal{I}_{\Sigma}\right] = \min_{z_0\in\mathcal{I}_{\Sigma}} D_{H^{\times}}\left[z:z_0\right]. \tag{11.51}$$

The following theorem provides us with the interpretation that the projection $\hat{z}$ delivered by the Hamiltonian system (11.33) is a geodesic projection.

Theorem 11.6 *Consider the normalised SIR (11.2) and the Bregman divergence $D_{H^{\times}}\left[z:\mathcal{I}_{\Sigma}\right]$. The projection $\hat{z}$ delivered by the corresponding Hamiltonian system (11.33) is a geodesic projection of z onto $\mathcal{I}_{\Sigma}$.*

Proof It follows from Theorem 11.2 that the geodesic projection of z onto $\mathcal{I}_{\Sigma}$ is the vector $z^* \in \mathcal{I}_{\Sigma}$ that minimises $D_{H^{\times}}\left[z:z_0\right], \forall z_0 \in \mathcal{I}_{\Sigma}$, i.e.

$$D_{H^{\times}}\left[z:\mathcal{I}_{\Sigma}\right] = \min_{z_0\in\mathcal{I}_{\Sigma}} D_{H^{\times}}\left[z:z_0\right] = D_{H^{\times}}\left[z:z^*\right].$$

Accordingly, it is to prove

$$\hat{z} = z^* = \arg\min_{z_0\in\mathcal{I}_{\Sigma}} D_{H^{\times}}\left[z:z_0\right]. \tag{11.52}$$

Following (11.30), it holds

$$D_{H^{\times}}\left[z:z_{0}\right]=D_{H}\left[z_{0}:z^{\times}\right].$$

Moreover, by Theorems 11.3 and 11.4, we obtain

$$\begin{aligned}D_{H}\left[z_{0}:z^{\times}\right]&=H\left(z_{0}\right)-H\left(z^{\times}\right)-\left(z^{\times}\right)^{T}\left(z_{0}-z^{\times}\right)\\&=\frac{1}{2}z_{0}^{T}z_{0}+\frac{1}{2}\hat{z}^{T}\hat{z}-\hat{z}^{T}z_{0}.\end{aligned}$$

It is evident that

$$z^{*}=\arg\min_{z_{0}\in\mathcal{I}_{\Sigma}}D_{H^{\times}}\left[z:z_{0}\right]=\hat{z}.$$

The theorem is proved.

This result gives a nice interpretation of the projection

$$\hat{z}=\Sigma_{\mathcal{I}}\circ\left(D\Sigma_{\mathcal{I}}\right)^{T}(z).$$

Thanks to the Bregman divergence $D_{H^{\times}}$ and the Legendre transform, the data manifold is equipped with a dual Riemannian structure and dually flat. In this context, the projection $\hat{z}$ is a geodesic projection of z onto $\mathcal{I}_{\Sigma}$, and the distance defined by the Bregman divergence $D_{H^{\times}}$ is at minimum. This result can be interpreted as a generalisation of the orthogonal projection $\mathcal{P}_{\mathcal{I}_{G}^{\mathcal{L}_{2}}}$ defined in Hilbert space.

As a summary of this subsection, it is concluded that the two major objectives of our work, (i) system performance-oriented fault detection by means of a Hamiltonian function induced Bregman divergence, (ii) detection of deviations in a non-Euclidean space using a Bregman divergence as a distance measure, have been satisfactorily achieved.

11.3.3 Implementation Issues

Before the theoretical results presented in the previous subsections can be applied to a reliable fault detection, two fault detection specified issues should be clarified. They are,

- definition of a reliable (robust) evaluation function J and, based on it,
- determination of a threshold J_{th}.

The first issue is of considerable importance to reduce false alarms. For instance, the Bregman divergence $D_H\left[z:\hat{z}\right]$ given in (11.50) is a time function. When it is directly used as an evaluation function, detection decision will be made solely depending on the data received at a single time instant. Due to possible transient disturbances during system operations, such handling would probably result in a flood of false alarms. Concerning the second issue, it is natural that, under the detection logic,

$$\text{detection logic: } \begin{cases} J \leq J_{th} \Longrightarrow \text{fault-free} \\ J > J_{th} \Longrightarrow \text{faulty,} \end{cases}$$

the selection of J_{th} would have a crucial influence on the detection performance. An integrated design of (J, J_{th}) is state of the art in model-based fault detection towards an optimal trade-off between the false alarm rate and fault detection rate. Below, these two issues are addressed jointly.

Let $[t_0, t_1]$ be the evaluation time window and define

$$J = \int_{t_0}^{t_1} D_H\left[z:\hat{z}\right] d\tau = \int_{t_0}^{t_1} \left(H\left(\hat{z}\right) - H^{\times}\left(\hat{z}\right)\right) d\tau \tag{11.53}$$

as the evaluation function. In the literature, the divergence given by (11.53) is called pointwise Bregman divergence. Consider that, in real applications, digital data are collected and recorded. Suppose that over the time interval $[t_0, t_1]$, $z\left(k_i\right), i = 1, \ldots, M$, are collected, where k_i denotes the sampling instant. For our purpose of fault detection, an approximation of function (11.53) is implemented using the available data $z\left(k_i\right)$ as follows,

$$J = \frac{1}{M}\sum_{i=1}^{M} D_H\left[z\left(k_i\right):\hat{z}\left(k_i\right)\right] = \frac{1}{M}\sum_{i=1}^{M}\left(H\left(\hat{z}\left(k_i\right)\right) - H^{\times}\left(\hat{z}\left(k_i\right)\right)\right). \tag{11.54}$$

It is of interest to note that the above evaluation function J is equivalent to a Bregman divergence. To delineate this claim, let us define

$$z_M := \begin{bmatrix} z\left(k_1\right) \\ \vdots \\ z\left(k_M\right) \end{bmatrix} \in \mathbb{R}^{M(m+p)}, \hat{z}_M := \begin{bmatrix} \hat{z}\left(k_1\right) \\ \vdots \\ \hat{z}\left(k_M\right) \end{bmatrix} \in \mathbb{R}^{M(m+p)},$$

$$H_{\Sigma}\left(z_M\right) := \frac{1}{M}\sum_{i=1}^{M} H\left(z\left(k_i\right)\right), H_{\Sigma}\left(\hat{z}_M\right) := \frac{1}{M}\sum_{i=1}^{M} H\left(\hat{z}\left(k_i\right)\right). \tag{11.55}$$

Observe that

$$
\begin{aligned}
&H_\Sigma(z_M) - H_\Sigma(\hat{z}_M) - \left(\frac{\partial H_\Sigma(\hat{z}_M)}{\partial \hat{z}_M}\right)^T (z_M - \hat{z}_M) \\
&= \frac{1}{M}\sum_{i=1}^{M} H(z(k_i)) - \frac{1}{M}\sum_{i=1}^{M} H(\hat{z}(k_i)) - \frac{1}{M}\sum_{i=1}^{M}\left(\frac{\partial H(\hat{z}(k_i))}{\partial \hat{z}(k_i)}\right)^T (z(k_i) - \hat{z}(k_i)) \\
&= \frac{1}{M}\sum_{i=1}^{M} D_H\left[z(k_i) : \hat{z}(k_i)\right] = \frac{1}{M}\sum_{i=1}^{M}\left(H(\hat{z}(k_i)) - H^{\times}(\hat{z}(k_i))\right).
\end{aligned} \tag{11.56}
$$

Thus, the value,

$$
J = D_{H_\Sigma}\left[z_M : \hat{z}_M\right] =: H_\Sigma(z_M) - H_\Sigma(\hat{z}_M) - \left(\frac{\partial H_\Sigma(\hat{z}_M)}{\partial \hat{z}_M}\right)^T (z_M - \hat{z}_M), \tag{11.57}
$$

is a Bregman divergence.

Theorem 11.7 *Consider the normalised SIR (11.2), the corresponding Hamiltonian system (11.33), Hamiltonian function (11.39), and data* $z(k_i), i = 1, \ldots, M$, *then the evaluation function (11.53) is equal to the Bregman divergence (11.57) with the generating function (11.55) and satisfies*

$$
D_{H_\Sigma}\left[z_M : \hat{z}_M\right] = J = \frac{1}{M}\sum_{i=1}^{M}\left(H(\hat{z}(k_i)) - H^{\times}(\hat{z}(k_i))\right).
$$

Proof The proof follows from (11.56).

Theorem 11.7 not only reveals that the evaluation function J is a Bregman divergence, but also poses the basis of a potential framework of data-driven fault detection. The centrepiece of this framework is the Bregman divergence

$$
D_{H_\Sigma}\left[z_M : \hat{z}_M\right]
$$

and the Legendre transform

$$
z_M = \frac{\partial H_M^{\times}(\hat{z}_M)}{\partial \hat{z}_M} = \begin{bmatrix} \frac{\partial H^{\times}(\hat{z}(k_1))}{\partial \hat{z}(k_1)} \\ \vdots \\ \frac{\partial H^{\times}(\hat{z}(k_M))}{\partial \hat{z}(k_M)} \end{bmatrix}, \hat{z}_M = \frac{\partial H_M(z_M)}{\partial z_M} = \begin{bmatrix} \frac{\partial H(z(k_1))}{\partial z(k_1)} \\ \vdots \\ \frac{\partial H(z(k_M))}{\partial z(k_M)} \end{bmatrix},
$$

$$
H_M^{\times}(\hat{z}_M) = \begin{bmatrix} H^{\times}(\hat{z}(k_1)) \\ \vdots \\ H^{\times}(\hat{z}(k_M)) \end{bmatrix}, H_M(z_M) = \begin{bmatrix} H(z(k_1)) \\ \vdots \\ H(z(k_M)) \end{bmatrix},
$$

which would enable us to endow the data set

$$\mathcal{M} = \left\{ z_M = \begin{bmatrix} z(k_1) \\ \vdots \\ z(k_M) \end{bmatrix} \in \mathcal{L}_2 \right\}$$

with a dual Riemannian structure and thus a projection-based fault detection. This issue is a part of our future work and will not be discussed here.

Before introducing the threshold setting scheme, we would like to place a greater emphasis on the basic task of one-class fault detection. Simply speaking, it is to detect anomalies arising during normal system operations. The challenge is a reliable detection of those incipient anomalies as early as possible in spite of possibly existing uncertainties in the system. Now, we are in the position to study the threshold setting issue on the assumption of the evaluation function J given by (11.54).

Recall that the Bregman divergence $D_{H_\Sigma}\left[z_M : \hat{z}_M\right]$ gives the difference between the nominal function as a reference and the real function $H_M(z_M)$ over the time interval $[t_0, t_1]$. During the ideal operation without faults and uncertainty,

$$H\left(\hat{z}(k_i)\right) = H^\times\left(\hat{z}(k_i)\right) \Longrightarrow$$

$$J = D_{H_\Sigma}\left[z_M : \hat{z}_M\right] = \frac{1}{M}\sum_{i=1}^{M}\left(H\left(\hat{z}(k_i)\right) - H^\times\left(\hat{z}(k_i)\right)\right) = 0.$$

To meet the needs of engineering applications, the evaluation function is modified as

$$J_{rel} := \frac{J}{\frac{1}{M}\sum_{i=1}^{M} H\left(\hat{z}(k_i)\right)} = \frac{\sum_{i=1}^{M}\left(H\left(\hat{z}(k_i)\right) - H^\times\left(\hat{z}(k_i)\right)\right)}{\sum_{i=1}^{M} H\left(\hat{z}(k_i)\right)}. \tag{11.58}$$

With respect to J_{rel}, the threshold is set to be

$$J_{rel,th} = \gamma << 1. \tag{11.59}$$

Here, γ can be determined from the real fault-free operations or simulation data.

At the end of this section, the online detection algorithm is summarised as follows:

- Collection of data $z(k_i)$ and computation of $\hat{z}(k_i), i = 1, \ldots, M$, according to (11.48);
- Computation of the evaluation function J and J_{rel} according to (11.56) and (11.58), respectively;
- Detection decision logic:

$$\begin{cases} J_{rel} \leq J_{rel,th} \Longrightarrow \text{fault-free} \\ J_{rel} > J_{rel,th} \Longrightarrow \text{faulty.} \end{cases}$$

11.4 An SKR-Based Fault Detection Scheme, and Uncertainty/Fault Estimation

Thanks to the dual relation of SKR and SIR, most of the schemes and ideas of our study in this section on the application of the normalised SKR-based projection to fault detection are analogue to the ones presented in the previous section. Hence, our descriptions and discussions on the related issues will be shortened. On the other hand, intensive endeavours will be made to highlight the projection-based estimation of system uncertainties and faults aiming at insightfully understanding and interpreting the normalised SKR-based projection.

11.4.1 *On the Uncertainty Corrupted in Process Data and Its Estimation*

In this subsection, we take a closer look at system $(D\Sigma_{\mathcal{K}})^T \circ \Sigma_{\mathcal{K}}$ and delineate its use for estimating uncertainties/faults corrupted in the process data (u, y). We begin with the Hamiltonian extension of $\Sigma_{\mathcal{K}}$,

$$\begin{cases} \dot{\hat{x}} = a_K(\hat{x}) + B_K(\hat{x})z, \; z := \begin{bmatrix} u \\ y \end{bmatrix} \\ \dot{\lambda} = -\left(\frac{\partial a_K(\hat{x})}{\partial \hat{x}} + \frac{\partial B_K(\hat{x})}{\partial \hat{x}} z\right)^T \lambda - \left(\frac{\partial c_K(\hat{x})}{\partial \hat{x}} + \frac{\partial D_K(\hat{x})}{\partial \hat{x}}\right)^T z_a \\ r_y = c_K(\hat{x}) + D_K(\hat{x})z \\ r_{y,a} = B_K^T(\hat{x})\lambda + D_K^T(\hat{x})z_a, \; z_a \in \mathbb{R}^m, \; r_{y,a} \in \mathbb{R}^{m+p}. \end{cases} \tag{11.60}$$

Connecting r_y and z_a, and defining the Hamiltonian function,

$$\begin{aligned} H\left(\hat{x}, \lambda, z\right) = {} & \frac{1}{2}\left(c_K(\hat{x}) + D_K(\hat{x})z\right)^T \left(c_K(\hat{x}) + D_K(\hat{x})z\right) \\ & + \lambda^T \left(a_K(\hat{x}) + B_K(\hat{x})z\right), \end{aligned} \tag{11.61}$$

give the Hamiltonian system $(D\Sigma_{\mathcal{K}})^T \circ \Sigma_{\mathcal{K}}$,

$$(D\Sigma_{\mathcal{K}})^T \circ \Sigma_{\mathcal{K}} : \begin{cases} \dot{\hat{x}} = \frac{\partial H(\hat{x},\lambda,z)}{\partial \lambda} \\ \dot{\lambda} = -\frac{\partial H(\hat{x},\lambda,z)}{\partial \hat{x}} \\ r_{y,a} = \frac{\partial H(\hat{x},\lambda,z)}{\partial z}. \end{cases} \tag{11.62}$$

System $(D\Sigma_{\mathcal{K}})^T \circ \Sigma_{\mathcal{K}}$ is an $\mathcal{L}_2 \to \mathcal{L}_2$ mapping which can be interpreted as an estimation of the uncertainty corrupted in the data z, as will be discussed in the sequel. For this reason, we adopt the notation

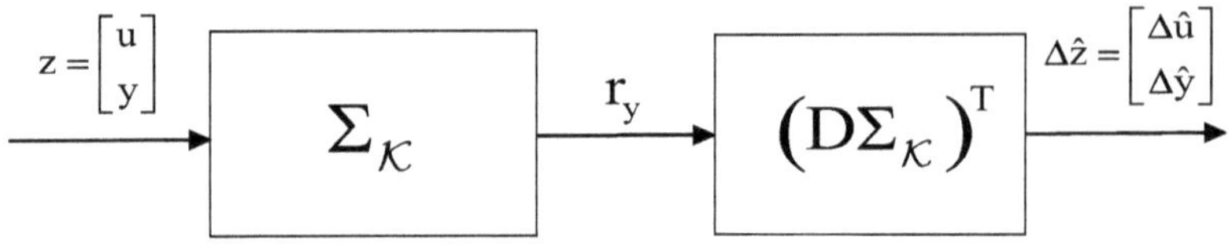

Fig. 11.2 SKR-based projection system $(D\Sigma_{\mathcal{K}})^T \circ \Sigma_{\mathcal{K}}$

$$\hat{z}_{\Delta} := r_{y,a} = (D\Sigma_{\mathcal{K}})^T \circ \Sigma_{\mathcal{K}}(z).$$

Figure 11.2 sketches the configuration of $(D\Sigma_{\mathcal{K}})^T \circ \Sigma_{\mathcal{K}}$ schematically.

11.4.1.1 Co-inner and Uncertainties

Recall that the normalised SKR is a co-inner. That means, by definition, the system

$$r_y = \Sigma_{\mathcal{K}} \circ (D\Sigma_{\mathcal{K}})^T (z_a),$$

configured by connecting z and $r_{y,a}$ in the Hamiltonian extension of $\Sigma_{\mathcal{K}}$ (11.60) is of the lossless property, namely,

$$\begin{aligned} r_y &= z_a, \\ \exists V(x) &\geq 0,\ V(0) = 0, \\ \text{s.t. } V(x(t_2)) - V(x(t_1)) &= \frac{1}{2}\int_{t_1}^{t_2} \left(r_{y,a}^T r_{y,a} - r_y^T r_y\right) d\tau. \end{aligned}$$

It leads to

$$r_{y,a} = (D\Sigma_{\mathcal{K}})^T (z_a) = (D\Sigma_{\mathcal{K}})^T (r_y). \tag{11.63}$$

Since the residual vector r_y contains information about uncertainties in the system, relation (11.63) indicates that the output of the system $(D\Sigma_{\mathcal{K}})^T (r_y)$ is of the character of uncertainties. This observation draws our attention to the (normalised) SKR and system uncertainties.

11.4.1.2 System Kernel and Uncertainty Models

In advanced control and model-based fault diagnosis theories, handling of system uncertainties is an essential issue. There are numerous ways to model various types of model uncertainties and, based on them, the influences (often in form of an upper-bound) of system uncertainties on the system output or the residual signal can be estimated and used, e.g. for robust controller design or threshold setting in a fault detection scheme. Below, we introduce a model that describes uncertainties corrupted in the system data (u, y). This model underlines the insightful interpretation of the projection $\hat{z}_\Delta = (D\Sigma_\mathcal{K})^T \circ \Sigma_\mathcal{K}(z)$ as an estimator for uncertainties/faults.

To begin with, a modified form of the SKR $\Sigma_\mathcal{K}$ (11.4),

$$\begin{cases} \dot{\hat{x}} = a_K(\hat{x}) + B_K(\hat{x}) \begin{bmatrix} u \\ y \end{bmatrix} \\ r_y = c_K(\hat{x}) + D_K(\hat{x}) \begin{bmatrix} u \\ y \end{bmatrix} \\ y = \hat{y} + W^{-1}(\hat{x}) r_y, \ \hat{y} = c(\hat{x}) + D(\hat{x}) u, \end{cases} \tag{11.64}$$

is introduced. Model (11.64) describes the I/O-dynamics of the system (11.1), and thus is called kernel-based I/O model. Apparently, it models the I/O dynamics both in the nominal and uncertain/faulty operations, and is independent of possible types of system uncertainties. A plausible explanation of this distinguishing capability is the fact that system uncertainties are implicitly corrupted in the process data (u, y) and represented by the residual vector r_y. This raises the question of whether it is possible to (i) model the uncertainties corrupted in (u, y), and (ii) estimate them. An apparent answer in case of a linear system, as shown below, motivates our work.

Example 11.1 Consider an LTI system G. Let

$$\begin{bmatrix} u \\ y \end{bmatrix} = \begin{bmatrix} u_0 \\ y_0 \end{bmatrix} + \begin{bmatrix} \Delta u \\ \Delta y \end{bmatrix}, \tag{11.65}$$

where $\begin{bmatrix} u_0 \\ y_0 \end{bmatrix} \in \mathcal{I}_G$ denotes the nominal operation, and $\begin{bmatrix} \Delta u \\ \Delta y \end{bmatrix}$ represents the uncertainties, for instance, caused by unknown input vectors in the system. As is known, we have

$$\begin{cases} \dot{\hat{x}} = A_L \hat{x} + \begin{bmatrix} B_L & -L \end{bmatrix} \begin{bmatrix} u \\ y \end{bmatrix} \\ r_y = -W \left(C\hat{x} + \begin{bmatrix} -D & I \end{bmatrix} \begin{bmatrix} u \\ y \end{bmatrix} \right) \end{cases} = \begin{cases} \dot{\hat{x}}_\Delta = A_L \hat{x}_\Delta + \begin{bmatrix} B_L & -L \end{bmatrix} \begin{bmatrix} \Delta u \\ \Delta y \end{bmatrix} \\ r_y = -W \left(C\hat{x}_\Delta + \begin{bmatrix} -D & I \end{bmatrix} \begin{bmatrix} \Delta u \\ \Delta y \end{bmatrix} \right), \end{cases}$$

$$A_L = A - LC, \ B_L = B - LD,$$

since it holds $K_G I_G = 0$, where K_G and I_G are the SKR and SIR of G. Note that the state variables $\hat{x}$ and $\hat{x}_\Delta$ are generally different, and the above two systems are equal in the context that they deliver the identical output.

Although for nonlinear systems the data corrupted with uncertainties cannot be written in the additive form like (11.65), this example inspires the introduction of the following uncertainty model.

Definition 11.6 Given the normalised SKR (11.4), the state space representation of $\Delta\Sigma_\mathcal{K}$,

$$\Delta\Sigma_K : \begin{cases} \dot{\hat{x}}_\Delta = a_K(\hat{x}_\Delta) + B_K(\hat{x}_\Delta)\begin{bmatrix} \Delta u \\ \Delta y \end{bmatrix} \\ r_y = c_K(\hat{x}_\Delta) + D_K(\hat{x}_\Delta)\begin{bmatrix} \Delta u \\ \Delta y \end{bmatrix}, \end{cases} \hat{x}_\Delta(0) = 0, \begin{bmatrix} \Delta u \\ \Delta y \end{bmatrix} \in \mathcal{L}_2, \tag{11.66}$$

subject to

$$\begin{cases} \dot{\hat{x}}_\Delta = a_K(\hat{x}_\Delta) + B_K(\hat{x}_\Delta)\begin{bmatrix} \Delta u \\ \Delta y \end{bmatrix} \\ r_y = c_K(\hat{x}_\Delta) + D_K(\hat{x}_\Delta)\begin{bmatrix} \Delta u \\ \Delta y \end{bmatrix} \end{cases} = \begin{cases} \dot{\hat{x}} = a_K(\hat{x}) + B_K(\hat{x})\begin{bmatrix} u \\ y \end{bmatrix} \\ r_y = c_K(\hat{x}) + D_K(\hat{x})\begin{bmatrix} u \\ y \end{bmatrix}, \end{cases} \tag{11.67}$$

is called uncertainty model of the SKR with $(\Delta u, \Delta y)$ denoting the uncertainties corrupted in the data (u, y).

The uncertainty model (11.66) and the condition (11.67) imply that,

- on the one hand, the residual vector r_y exclusively contains the information about uncertainties in the system,
- on the other hand, r_y can be equivalently generated by the SKR (11.4) driven by (u, y).

Although $(\Delta u, \Delta y)$ is in general unknown, its value can be specified in two (extreme) cases, which play an essential role in our subsequent study:

- corresponding to the nominal dynamics,

$$\forall \begin{bmatrix} u \\ y \end{bmatrix} \in \mathcal{I}_\Sigma, \begin{bmatrix} \Delta u \\ \Delta y \end{bmatrix} = 0; \tag{11.68}$$

- when it holds

$$\begin{aligned} \hat{z}_\Delta &= (D\Sigma_K)^T \circ \Sigma_K\left(\begin{bmatrix} u \\ y \end{bmatrix}\right) = (D\Sigma_K)^T \circ \Sigma_K\left(\begin{bmatrix} \Delta u \\ \Delta y \end{bmatrix}\right) \\ &= \begin{bmatrix} u \\ y \end{bmatrix} = \begin{bmatrix} \Delta u \\ \Delta y \end{bmatrix}. \end{aligned} \tag{11.69}$$

The first case is true due to the relation (6.13) and reflects the system response during nominal operations. The second one means that the data (u, y) solely contains uncertainties, which, despite being rarely the case in real operations, characterises a crucial data set and inspires us to introduce the concept of the manifold of uncertain data.

Definition 11.7 Given the normalised SKR (11.4), the manifold $\mathcal{U}_\Sigma \subset \mathcal{L}_2$,

$$\mathcal{U}_\Sigma = \left\{ \begin{bmatrix} u \\ y \end{bmatrix}, \begin{bmatrix} u \\ y \end{bmatrix} = (D\Sigma_\mathcal{K})^T (r), r \in \mathcal{L}_2^m \right\}, \tag{11.70}$$

is called manifold of uncertain data.

It is apparent that $\forall \begin{bmatrix} u \\ y \end{bmatrix} \in \mathcal{U}_\Sigma$,

$$\hat{z}_\Delta = (D\Sigma_\mathcal{K})^T \circ \Sigma_\mathcal{K} \left(\begin{bmatrix} u \\ y \end{bmatrix} \right) = \begin{bmatrix} u \\ y \end{bmatrix},$$

i.e. the relation (11.69) holds. Thus, the data in $\mathcal{U}_\Sigma$ solely contain uncertainties. Note that

$$\forall \begin{bmatrix} u \\ y \end{bmatrix} \in \mathcal{L}_2, \begin{bmatrix} \Delta\hat{u} \\ \Delta\hat{y} \end{bmatrix} = (D\Sigma_\mathcal{K})^T \circ \Sigma_\mathcal{K} \left(\begin{bmatrix} u \\ y \end{bmatrix} \right) \in \mathcal{U}_\Sigma \tag{11.71}$$

indicates that the operator $(D\Sigma_\mathcal{K})^T \circ \Sigma_\mathcal{K}$ projects the data (u, y) to the manifold $\mathcal{U}_\Sigma$. Observe that, as a result of the uncertainty model (11.66),

$$\begin{bmatrix} \hat{u}_\Delta \\ \hat{y}_\Delta \end{bmatrix} := (D\Sigma_\mathcal{K})^T \circ \Sigma_\mathcal{K} \left(\begin{bmatrix} u \\ y \end{bmatrix} \right) = (D\Sigma_\mathcal{K})^T \circ \Sigma_\mathcal{K} \left(\begin{bmatrix} \Delta u \\ \Delta y \end{bmatrix} \right). \tag{11.72}$$

That is, $\begin{bmatrix} \hat{u}_\Delta \\ \hat{y}_\Delta \end{bmatrix}$ is a projection of the uncertainty $\begin{bmatrix} \Delta u \\ \Delta y \end{bmatrix}$ corrupted in the system data and thus serves as an estimate of $\begin{bmatrix} \Delta u \\ \Delta y \end{bmatrix}$ in the context of the uncertainty model (11.66). It can be, for instance, applied for re-constructing the detected faults or examining the data quality. To this end, we will study the normalised SKR and the associated Hamiltonian projection system,

$$\hat{z}_\Delta = \begin{bmatrix} \hat{u}_\Delta \\ \hat{y}_\Delta \end{bmatrix} = (D\Sigma_\mathcal{K})^T \circ \Sigma_\mathcal{K} \left(\begin{bmatrix} u \\ y \end{bmatrix} \right), \tag{11.73}$$

in the next subsection.

Remark 11.3 We would like to emphasise that $\hat{z}_\Delta$ is an estimation of the uncertainty $\begin{bmatrix} \Delta u \\ \Delta y \end{bmatrix}$ corrupted in the system data. And the Hamiltonian projection system (11.73) is a projection of the data (u, y) onto the manifold of uncertain data $\mathcal{U}_\Sigma$.

11.4.2 Normalised SKR and the Associated Hamiltonian Projection System

We begin with the state space description of system (11.73). Let

$$H\left(\hat{x}, \lambda, z\right) = \frac{1}{2}\left(c_K(\hat{x}) + D_K(\hat{x})z\right)^T\left(c_K(\hat{x}) + D_K(\hat{x})z\right) + \lambda^T\left(a_K(\hat{x}) + B_K(\hat{x})z\right) \tag{11.74}$$

be the Hamiltonian function, where, for the sake of simplicity, notation $\hat{x}$ instead of $\hat{x}_\Delta$ is abused. Accordingly, we have

$$(D\Sigma_{\mathcal{K}})^T \circ \Sigma_{\mathcal{K}} : \begin{cases} \dot{\hat{x}} = \frac{\partial H(\hat{x},\lambda,z)}{\partial \lambda} \\ \dot{\lambda} = -\frac{\partial H(\hat{x},\lambda,z)}{\partial \hat{x}} \\ \hat{z}_\Delta = \frac{\partial H(\hat{x},\lambda,z)}{\partial z}. \end{cases} \tag{11.75}$$

By some routine calculations, the Hamiltonian function (11.74) is further written into

$$\begin{aligned} H\left(\hat{x}, \lambda, z\right) &= \frac{1}{2}r_y^T r_y + \lambda^T \dot{\hat{x}} \\ &= \hat{z}_\Delta^T \hat{z}_\Delta + \frac{1}{2}c_K^T(\hat{x})c_K(\hat{x}) - \frac{1}{2}z^T D_K^T(\hat{x})D_K(\hat{x})z + \lambda^T a_K(\hat{x}). \end{aligned}$$

For our study, we are interested in the values of Hamiltonian function corresponding to the two extreme cases: (i) $z \in \mathcal{I}_\Sigma$, (ii) $z \in \mathcal{U}_\Sigma$. The following theorem gives the solutions.

Theorem 11.8 *Given the normalised SKR (11.4) and the corresponding Hamiltonian system (11.62), then*

$$\forall z = \begin{bmatrix} u \\ y \end{bmatrix} \in \mathcal{I}_\Sigma, \; H^\times\left(x, \lambda, \hat{z}_\Delta\right) = H\left(x, \lambda, z\right) = 0, \tag{11.76}$$

$$\forall z = \begin{bmatrix} u \\ y \end{bmatrix} \in \mathcal{U}_\Sigma, \; H\left(\hat{x}, \lambda, z\right) = H^\times\left(\hat{x}, \lambda, \hat{z}_\Delta\right) = \frac{1}{2}\hat{z}_\Delta^T \hat{z}_\Delta = \frac{1}{2}z^T z. \tag{11.77}$$

Proof Equation (11.76) is attributed to (6.13). The proof of (11.76) is analogous to the proof of Theorem 11.4 and thus omitted here.

Theorem 11.8 describes two special operation situations, in which

$$H^\times\left(x, \lambda, \hat{z}_\Delta\right) = H\left(x, \lambda, z\right).$$

The first one given by (11.76) corresponds to the nominal dynamics, while the second one with (11.77) reflects the situation when the data exclusively consist of uncertainties.

11.4.3 Applications to Fault Detection and Uncertainty Estimation

Based on the results presented in the previous subsections, Bregman divergence-based fault detection schemes as well as a further study on uncertainty estimation are addressed in this subsection.

11.4.3.1 Bregman Divergence-Based Fault Detection

To begin with, we discuss about the application of the Bregman divergence from Δz to the manifold $\mathcal{I}_{\Sigma}$ for the purpose of fault detection. Consider the uncertainty model (11.66) and recall that

$$\forall z = \begin{bmatrix} u \\ y \end{bmatrix} \in \mathcal{I}_{\Sigma}, \begin{bmatrix} \Delta u \\ \Delta y \end{bmatrix} = 0.$$

To detect possible uncertainties in the data (u, y), the Hamiltonian functions

$$H^{\times}\left(\hat{z}_{\Delta}\right) := H^{\times}\left(\hat{x}, \lambda, \hat{z}_{\Delta}\right), H(z) = H\left(\hat{x}, \lambda, z\right)$$

are applied as the generating function of the Bregman divergence from $\hat{z}_{\Delta}$ to $z \in \mathcal{I}_{\Sigma}$ given by

$$D_{H^{\times}}\left[\hat{z}_{\Delta} : z\right] = D_H\left[z^{\times} : \hat{z}_{\Delta}^{\times}\right], , H(z) = H\left(\hat{x}, \lambda, z\right), \tag{11.78}$$

where the last equation is attributed to (11.30). Recall that

$$\hat{z}_{\Delta} \in \mathcal{U}_{\Sigma} \Longrightarrow \hat{z}_{\Delta} = \hat{z}_{\Delta}^{\times}, z \in \mathcal{I}_{\Sigma} \Longrightarrow z^{\times} = 0.$$

It turns out

$$D_{H^{\times}}\left[\hat{z}_{\Delta} : z\right] = D_H\left[z^{\times} : \hat{z}_{\Delta}^{\times}\right] = H(0) - \frac{1}{2}\hat{z}_{\Delta}^T\hat{z}_{\Delta} - \hat{z}_{\Delta}^T\left(0 - \hat{z}_{\Delta}\right) = \frac{1}{2}\hat{z}_{\Delta}^T\hat{z}_{\Delta}. \tag{11.79}$$

It is of interest to notice that the Bregman divergence (11.78) is true for all (u, y) belonging to $\mathcal{I}_{\Sigma}$. Moreover, $\hat{z}_{\Delta} = \left(D\Sigma_{\mathcal{K}}\right)^T \circ \Sigma_{\mathcal{K}}(z)$. In this sense, $D_{H^{\times}}\left[\hat{z}_{\Delta} : z\right]$ can be interpreted as a divergence from the uncertainty in the data to $\mathcal{I}_{\Sigma}$.

Analogous to the discussion on the implementation issues in Sect. 11.3.3, we shortly summarise the major steps to apply the Bregman divergence (11.78) for a practical fault detection. Let $z(k_i), i = 1, \ldots, M$, be the data collected over the time interval $[t_0, t_1]$ and

$$J = D_{H_M^{\times}}\left[\hat{z}_{\Delta M} : z_M\right] = \frac{1}{2M}\sum_{i=1}^{M}\hat{z}_{\Delta}^T(k_i)\,\hat{z}_{\Delta}(k_i) = \frac{1}{2M}\hat{z}_{\Delta M}^T\hat{z}_{\Delta M}, \tag{11.80}$$

$$\hat{z}_{\Delta M}=\begin{bmatrix}\hat{z}_{\Delta}(k_1)\\ \vdots \\ \hat{z}_{\Delta}(k_M)\end{bmatrix}\in\mathbb{R}^{M(m+p)},\ z_M=\begin{bmatrix}z(k_1)\\ \vdots \\ z(k_M)\end{bmatrix}\in\mathbb{R}^{M(m+p)},$$

$z(k_i)\in\mathcal{I}_{\Sigma}, i=1,\ldots,M$, be the corresponding evaluation function. It is straightforward that the evaluation function (11.80) is a Bregman divergence with the generating function

$$H_M^{\times}\left(\hat{z}_{\Delta M}\right)=\frac{1}{M}\sum_{i=1}^{M}H^{\times}\left(\hat{z}_{\Delta}(k_i)\right).$$

To set the threshold, consider the ratio

$$J_{rel}=\frac{\hat{z}_{\Delta M}^{T}\hat{z}_{\Delta M}}{z_M^T z_M} \tag{11.81}$$

that gives an estimate of the relative change in the process data z_M. Let $\alpha\in[0,1]$ be the tolerant limit so that it is accepted as fault-free operation when

$$J_{rel}=\frac{\hat{z}_{\Delta M}^{T}\hat{z}_{\Delta M}}{z_M^T z_M}\leq\alpha.$$

Accordingly, the threshold is set to be

$$J_{rel,th}=\alpha. \tag{11.82}$$

Below is the online detection algorithm:

- Collection of data $z(k_i)$ and computation of $\hat{z}(k_i), i=1,\ldots,M$, according to (11.62);
- Computation of the evaluation function J_{rel} according to (11.81);
- Detection decision logic:

$$\begin{cases}J_{rel}\leq J_{rel,th}\Longrightarrow \text{fault-free}\\ J_{rel}>J_{rel,th}\Longrightarrow \text{faulty.}\end{cases}$$

11.4.3.2 On the Projection-Based Estimation

Having studied the use of the projection,

$$\hat{z}_{\Delta}=\left(D\Sigma_{\mathcal{K}}\right)^{T}\left(r_y\right)=\left(D\Sigma_{\mathcal{K}}\right)^{T}\circ\Sigma_{\mathcal{K}}(z),$$

for the fault detection purpose, we now comprehend $\hat{z}_{\Delta}$ as an estimate of uncertainties corrupted in the data. To this end, the issue of examining $\hat{z}_{\Delta}$ as a projection of the system data on the uncertain data manifold $\mathcal{U}_{\Sigma}$ is addressed from the viewpoint of information geometry.

The following theorem is a dual result of Theorem 11.6 and gives $\hat{z}_{\Delta}$ an insightful interpretation from the information geometric viewpoint.

Theorem 11.9 *Consider the normalised SKR (11.4) and the Bregman divergence* $D_{H^{\times}}[z : \mathcal{U}_{\Sigma}]$,

$$D_{H^{\times}}[z : \mathcal{U}_{\Sigma}] = \min_{\Delta z_0 \in \mathcal{U}_{\Sigma}} D_{H^{\times}}[z : \Delta z_0],$$

$$D_{H^{\times}}[z : \Delta z_0] := H^{\times}(z) - H^{\times}(\Delta z_0) - \left(\frac{\partial H^{\times}(\Delta z_0))}{\partial \Delta z_0}\right)^T (z - \Delta z_0),$$

$$H^{\times}(z) = \hat{z}_{\Delta}^T z - H\left(\hat{x}, \lambda, z\right) = \frac{1}{2}\hat{z}_{\Delta}^T \hat{z}_{\Delta}.$$

The projection $\hat{z}_{\Delta}$ delivered by the corresponding Hamiltonian system (11.75) is a geodesic projection of z onto $\mathcal{U}_{\Sigma}$.

Proof The proof is similar to the one of Theorem 11.6. Thus, only key steps are described. According to Theorem 11.2, the geodesic projection of z onto U_{Σ} is the vector Δz^* leading to

$$\min_{\Delta z_0 \in \mathcal{U}_{\Sigma}} D_{H^{\times}}[z : \Delta z_0] = D_{H^{\times}}\left[z : \Delta z^*\right].$$

Following (11.30), Theorem 11.8 and noticing $\forall \Delta z_0 \in U_{\Sigma}$, $\Delta z_0^{\times} = \Delta z_0$, $z^{\times} = \Delta \hat{z}$, it turns out

$$\begin{aligned} D_{H^{\times}}[z : \Delta z_0] &= D_H\left[\Delta z_0^{\times} : z^{\times}\right] = H(\Delta z_0) - H\left(\Delta \hat{z}\right) - \Delta \hat{z}^T \left(\Delta z_0 - \Delta \hat{z}\right) \\ &= \frac{1}{2}\Delta z_0^T \Delta z_0 - \Delta \hat{z}^T \Delta z_0 + \frac{1}{2}\Delta \hat{z}^T \Delta \hat{z}. \end{aligned}$$

As a result,

$$\begin{aligned} \min_{\Delta z_0 \in \mathcal{U}_{\Sigma}} D_{H^{\times}}[z : \Delta z_0] &= \min_{\Delta z_0 \in \mathcal{U}_{\Sigma}} \left(\frac{1}{2}\Delta z_0^T \Delta z_0 - \Delta \hat{z}^T \Delta z_0 + \frac{1}{2}\Delta \hat{z}^T \Delta \hat{z}\right) \\ &\Longrightarrow \arg \min_{\Delta z_0 \in \mathcal{U}_{\Sigma}} \left(\frac{1}{2}\Delta z_0^T \Delta z_0 - \Delta \hat{z}^T \Delta z_0 + \frac{1}{2}\Delta \hat{z}^T \Delta \hat{z}\right) \\ &= \Delta \hat{z} = (D\Sigma_{\mathcal{K}})^T \left(r_y\right) = (D\Sigma_{\mathcal{K}})^T \circ \Sigma_{\mathcal{K}}(z). \end{aligned}$$

The theorem is thus proved.

Theorem 11.9 underlines $\hat{z}_\Delta$ as an optimal estimate for Δz in the context of a geodesic projection of z onto $\mathcal{U}_\Sigma$. Here, the manifold $\mathcal{U}_\Sigma$ is equipped with a dual Riemannian structure by means of the Bregman divergence $D_{H^\times}$ and it is dually flat.

11.5 Notes and References

In this chapter, performance-oriented one-class fault detection and estimation in nonlinear dynamic systems with uncertainties have been addressed. Bearing in mind that the objective of this work is to establish a framework, in which not only model-based but also data-driven and ML-based fault diagnosis strategies can be uniformly handled, our work has followed a paradigm that is different from the conventional control theoretically oriented model-based framework. To be specific, our work is based on the SIR and SKR model forms of dynamic systems, rather than on the well-established input-output and the associated state space models. This change enables us to model the nominal system dynamics as a lower-dimensional manifold embedded in the process data space (u, y). The main idea behind this handling is, along the line of data-driven and ML-based fault detection schemes, to deal with fault detection as a classification problem in the process data space. To this end, projection technique provides us with a capable mathematical tool. In this regard, our work has focused on projection-based fault detection and estimation in nonlinear dynamic systems.

Before starting with our work, a short literature review about state of the art in model- and observer-based fault diagnosis techniques has been presented. The survey paper by Alcorta-Garcia and Frank [2] reviewed the major nonlinear observer-based fault detection schemes developed in the 90s, including the feedback-based linearisation and differential algebra techniques [79, 100, 180], and the geometric approach to nonlinear fault detection [159]. The advanced observer-based fault detection schemes developed in the past two decades are, amongst others, LMI technique to address fault detection nonlinear systems with, for instance, Lipschitz nonlinearity [161, 231] or sector bounded nonlinearity [82], fuzzy technique based fault detection [18, 113–117, 146, 183], adaptive fault diagnosis for nonlinear systems [183, 211, 230], linear parameter-varying fault detection systems [5, 14], and sliding mode observer-based fault detection [56, 215]. With the aid of nonlinear system theory as a major tool, issues like existence conditions, detectability and parameterisation of nonlinear observer-based fault detection systems have been examined in [39, 216, 217].

Encouraged by our previous work on application of the orthogonal projection technique to fault detection in LTI systems [44], we have started our effort with constructing projection systems. As an extension of the well-established projections onto the image and kernel subspaces of LTI systems [44, 69, 199], normalised SIR and SKR of nonlinear systems have been adopted for our aim. In order to address nonlinear system specified issues, we have focused on the lossless properties of the normalised SIR and SKR as inner and co-inner, and introduced various configurations of Hamiltonian extensions [27] as Hamiltonian systems [7, 162, 177]. They serve as the system theoretical tool throughout our work. For instance, Definition 11.2, Lemma 11.1 and Theorem 11.1 can be found in [7, 162, 177].

In order to deal with projection-based one-class fault detection as a classification problem, we have studied projections onto system image $\mathcal{I}_{\Sigma}$ manifold. Inspired by the manifold learning methods Isomap [191] and LLE [173], the construction of the projection systems is based on the normalised SIR. Note that an essential assumption in manifold learning is that the data have been generated from a lower-dimensional manifold that is embedded inside of a higher-dimensional space composed of redundant and irrelevant information (uncertainties). Moreover, the basic idea behind the LLE algorithm is to create a data mapping from the higher-dimensional manifold to the lower-dimensional one and preserving local neighbourhoods at every point of the underlying manifold. The projection system constructed by the normalised SIR is idempotent and thus preserves the local neighbourhoods of the underlying manifold, as described by Theorem 11.3. It is of considerable interest to notice that the projection system proposed in our work, (11.33), whose configuration is sketched in Fig. 11.1, can be interpreted as autoencoders. AE technique is a capable ML-based algorithm for dimensionality reduction and feature learning [73, 84] and widely applied to fault diagnosis and anomaly detection [1, 92, 96, 124, 172].

The theoretical basis for a successful application of the orthogonal projection technique to fault detection in LTI systems is the Hilbert Projection Theorem. Concretely, the normalised SIR-based orthogonal projection of the process data (u, y) onto the image subspace in Hilbert space, describing the nominal system dynamics, allows an orthogonal decomposition of (u, y) into an optimal estimate (projection) and the residual. By means of the Pythagorean equation, a norm-based evaluation of (u, y) leads to a unique separation of the nominal dynamic and the dynamics corresponding to uncertainties/faults. As a result, a classification and thus an optimal fault detection is achieved. For nonlinear dynamic systems, (process) data processing should be performed in non-Euclidean space. Consequently, the norm-based distance defined in Hilbert space may not be compatible with the metric tensor of the non-Euclidean space and hence is not suitable to measure the distance from a data vector to the system image manifold. As one of the major results of our work, we have proposed to use a Bregman divergence [16], a measure of difference between two points in a space, to solve the problem of distance measure in non-Euclidean space. A Bregman divergence is defined in terms of a convex function. When the system performance function under consideration is convex, it is plain that we are able to use the Bregman divergence defined by such a performance to achieve performance-oriented fault detection. A further distinguishing contribution of our work is the proposed scheme of combining the normalised SIR as Hamiltonian systems and the Bregman divergence induced by the corresponding Hamiltonian functions. This scheme not only enables us to realise the objective of performance-oriented fault detection, but also uncovers the information geometric aspect of our work. It is known that the input and output of a Hamiltonian system build a Legendre transform and there exists a Hamiltonian system as the Legendre dual system. As a result, the computation of the corresponding Bregman divergence can be expressed in terms of the dual Hamiltonian functions. In particular, a Bregman divergence together with a Legendre transform induces a dually flat Riemannian manifold in the data space that is regarded as a dualistic extension of the Euclidean space [3, 147]. In this context, the Hamiltonian system based projection can be interpreted as a geodesic projection. The preliminary knowledge of information geometry like the concepts, Bregman divergence, Legendre transform, and geodesic projection etc. introduced in Sect. 11.2.3 are mainly referred to [3, 147]. Definitions 11.4 and 11.5 are from [3]. The proof of the Projection Theorem, Theorem 11.2, is given in [3], which is based on the Generalised Pythagorean Theorem by Nagaoka and Amari [145]. So far, this part of work is an integration of control theory and information geometry, whose main results are summarised as follows:

- Construction of the normalised SIR-based projection system that projects (u, y) onto $\mathcal{I}_\Sigma$, the image manifold describing the nominal system dynamics,
- Determination of the Hamiltonian functions with respect to the nominal dynamics,
- Determination and computation of the Bregman divergence for performance-oriented fault detection, and based on it,
- Definition of the evaluation function over a time interval, which has been proved to be a Bregman divergence as well, and the corresponding threshold setting,
- Demonstration of the interpretation that the projection onto the image manifold is a geodesic projection.

As reported in our previous work [44], for LTI systems the orthogonal projection onto the image subspace is equivalent to a projection onto the kernel subspace. Thanks to the Pythagorean equation, the projection-based fault detection can then be equivalently realised using an observer-based residual generator. A further aspect of studying the SKR-based projection is the uncertainty issue. In robust control theory, there are numerous ways to model various types of model uncertainties [57, 194, 199, 243]. And in the observer-based fault detection, uncertainty analysis is essential for threshold setting [34, 39]. These conclusions and results have motivated and inspired our study on the normalised SKR induced Hamiltonian system $(D\Sigma_\mathcal{K})^T \circ \Sigma_\mathcal{K}$ as a projection. Since the SKR $\Sigma_\mathcal{K}$ is a residual generator and the residual contains uncertainty information, we have firstly studied $(D\Sigma_\mathcal{K})^T \circ \Sigma_\mathcal{K}$ and its co-inner property from the information (uncertainty) viewpoint. To be specific, the uncertainty model (11.66) and Definition 11.7, the manifold of uncertain data $\mathcal{U}_\Sigma$, have been introduced. It has been proved that $(D\Sigma_\mathcal{K})^T \circ \Sigma_\mathcal{K}$ represents a projection of the process data onto $\mathcal{U}_\Sigma$. Analogous to our work in the first part, a fault detection scheme has been proposed, consisting of

- design of $(D\Sigma_\mathcal{K})^T \circ \Sigma_\mathcal{K}$ and the corresponding dual Hamiltonian system,
- determination of the Hamiltonian functions with respect to the two extreme cases, $z \in \mathcal{I}_\Sigma$ and $z \in \mathcal{U}_\Sigma$,
- definition and computation of the Hamiltonian function induced Bregman divergence and its use for fault detection,
- analysis and interpretation of $(D\Sigma_\mathcal{K})^T \circ \Sigma_\mathcal{K}$ as a projection onto $\mathcal{U}_\Sigma$, and
- study on the realisation issues.

The further result of our study on $(D\Sigma_{\mathcal{K}})^T \circ \Sigma_{\mathcal{K}}$ is the proof that the projection $\hat{z}_{\Delta}$ is an optimal estimate for Δz in the context of a geodesic projection of z onto $\mathcal{U}_{\Sigma}$. An immediate application of this result is fault estimation. In other words, variations in the process data caused by faults, in particular those process faults, are modelled like uncertainties by means of the uncertainty model and the manifold $\mathcal{U}_{\Sigma}$ represents the corresponding set in the data space. And $(D\Sigma_{\mathcal{K}})^T \circ \Sigma_{\mathcal{K}}$ serves as an optimal estimator for such variations in the process data.

We would like to call the reader's attention to the fact that both projection systems studied here, $\Sigma_{\mathcal{I}} \circ (D\Sigma_{\mathcal{I}})^T$ and $(D\Sigma_{\mathcal{K}})^T \circ \Sigma_{\mathcal{K}}$, are of the typical configuration of an AE, where the latent variables are v and r_y respectively. In our previous work [112], we have proposed to train $\Sigma_{\mathcal{I}} \circ (D\Sigma_{\mathcal{I}})^T$ as neural networks in such a way that $\Sigma_{\mathcal{I}} \circ (D\Sigma_{\mathcal{I}})^T$ is idempotent and of the lossless property. This work can be interpreted as control theoretically learning of AE for optimal and data-driven fault diagnosis. A further interesting perspective is the application of PINN methods [101]. In our theoretical work, we notice that solving HJEs like (11.22) and (11.25) analytically is a hard work. The PINN methods offer a capable tool to deal with such problems by means of data-driven and learning algorithms.

Chapter 12
Estimation, Estimation-Based Control and Performance Recovery

The objective of this chapter is to re-visit and study estimation, including unknown inputs, faults, uncertainties and state estimation, and the estimation-based control and detection issues. Observer-based state estimation is a fully developed technique in control theory and engineering. Nonetheless, it becomes considerably difficult, when the system under consideration is corrupted with disturbances, uncertainties and faults. The challenge consists in, undoubtedly, mutual interference of unknown and uncertain variables like state, disturbances, faults and parameter variations. For our purpose, the attention is paid to the following issues

- unknown input, fault (as unknown inputs) and state estimation,
- uncertainty estimation,
- disturbance observer-based control (DOBC) schemes, as well as
- learning-based fault-tolerant observer and control schemes.

12.1 Fault, Unknown Input and State Estimation

Fault estimation provides us with information about faults which is useful for fault-tolerant control. There are various techniques well-established for fault estimation. In particular, for those additive faults, which can be addressed as unknown inputs, there exist well-developed estimation methods. Considerably differing from it, estimation of multiplicative faults is mainly explored in the framework of parameter identification. In this section, we firstly review existing unknown input estimation methods. In this context, fault estimation is investigated, where faults are bounded $\mathcal{L}_2$ signals and addressed as unknown inputs. We will introduce a method for optimally estimating unknown inputs, the least squares (LS) method. Finally, we will leverage the application of the projection method to optimal estimation of faulty data.

S. X. Ding, *Unified Framework of Control and Detection*,
https://doi.org/10.1007/978-3-662-72417-0_12

12.1.1 Unknown Input Observers and Estimation

The unknown input observer (UIO) is a well-established technique widely applied for fault detection, robust and fault-tolerant control. A state unknown input observer outputs a state estimate whose estimation error is decoupled from the unknown input vector, while a UIO-based residual generator delivers a residual signal decoupled from the unknown input. In fact, a UIO is based on the principle of a full compensation of the unknown input vector, which implicitly means that the unknown input vector is re-constructable using the system measurement y. To delineate this aspect, let us examine the subsequent example.

Example 12.1 In this example, two trivial UIOs are reviewed. Consider

$$x(k+1) = Ax(k) + Bu(k) + E_d d(k), \tag{12.1}$$

$$y(k) = Cx(k) + Du(k) + F_d d(k), \tag{12.2}$$

where $d(k) \in \mathbb{R}^{k_d}$ is an $\mathcal{L}_2$ bounded unknown input vector that might be a fault vector, E_d, F_d are known and F_d satisfies

$$rank\,(F_d) = k_d.$$

Suppose that

$$\dim\,(y) = m > k_d.$$

Therefore,

$$\begin{aligned} &\exists \begin{bmatrix} F_d^- \\ F_d^\perp \end{bmatrix}, F_d^- F_d = I, F_d^\perp F_d = 0, \\ &\text{s.t.} \begin{bmatrix} F_d^- \\ F_d^\perp \end{bmatrix} y\,(k) = \begin{bmatrix} F_d^- \\ F_d^\perp \end{bmatrix} (Cx(k) + Du(k)) + \begin{bmatrix} d(k) \\ 0 \end{bmatrix}. \end{aligned}$$

Now, construct an observer as

$$\begin{aligned} \hat{x}(k+1) &= A\hat{x}(k) + Bu(k) + E_d F_d^- \left(y\,(k) - C\hat{x}(k) - Du(k)\right) \\ &\quad + LF_d^\perp \left(y\,(k) - C\hat{x}(k) - Du(k)\right), \end{aligned} \tag{12.3}$$

which yields an state estimate $\hat{x}(k)$ satisfying, for $e(k) = x(k) - \hat{x}(k)$,

$$e(k+1) = \left(A - E_d F_d^- C - LF_d^\perp C\right) e(k).$$

On the assumption that $\exists\, L$ s.t. matrix $\left(A - E_d F_d^- C - LF_d^\perp C\right)$ is Schur, it holds

$$\lim_{k\to\infty} e(k) = 0.$$

Hence, the above observer is a UIO. Let

$$\hat{d}(k) = F_d^- r_y(k), r_y = y(k) - C\hat{x}(k) - Du(k).$$

It is apparent that

$$d(k) - \hat{d}(k) = -F_d^- Ce(k) \to 0.$$

Thus, $\hat{d}(k)$ is an unbiased estimate for $d(k)$.

Next, consider a modified model

$$\begin{aligned} x(k+1) &= Ax(k) + Bu(k) + E_d d(k), \\ y(k) &= Cx(k) + Du(k), \end{aligned}$$

and suppose

$$rank\,(CE_d) = k_d. \tag{12.4}$$

Now, design two observers. The first one is described by

$$\begin{aligned} \hat{x}(k+1) &= A\hat{x}(k) + Bu(k) + L\left(y(k) - C\hat{x}(k) - Du(k)\right), \\ r(k) &= y(k) - C\hat{x}(k) - Du(k), \end{aligned}$$

resulting in

$$\begin{aligned} e(k+1) &= (A - LC)\,e(k) + E_d d(k), r(k) = Ce(k), \\ r(k+1) &= Ce(k+1) = C\,(A - LC)\,e(k) + CE_d d(k) \Longrightarrow \\ d(k) &= P\,(r(k+1) - C\,(A - LC)\,e(k))\,, PCE_d = I. \end{aligned}$$

On this basis, the second observer is constructed as follows,

$$\begin{aligned} &\hat{e}(k+1) = (A - LC)\,\hat{e}(k) + E_d P\left(r(k+1) - C\,(A - LC)\,\hat{e}(k)\right) \\ &\Longrightarrow \hat{e}(k+1) - E_d Pr(k+1) = (I - E_d PC)\,(A - LC)\,\hat{e}(k). \end{aligned}$$

Defining $\hat{z}(k+1) = \hat{e}(k+1) - E_d Pr(k+1)$ leads to the state space representation

$$\hat{z}(k+1) = (I - E_d PC)\,(A - LC)\,\hat{z}(k) + (I - E_d PC)\,(A - LC)\,E_d Pr(k),$$

whose dynamics are governed by

$$e_z(k+1) = (I - E_d PC)\,(A - LC)\,e_z(k), e_z(k) = e(k) - E_d Pr(k) - \hat{z}(k).$$

Assume that there exists L so that the matrix $(I - E_d PC)\,(A - LC)$ is Schur, then

$$\lim_{k\to\infty} e_z(k) = 0 \Longleftrightarrow \lim_{k\to\infty}\left(e(k) - \hat{e}(k)\right) = 0,$$

and hence for

$$\hat{d}(k) = P\left(r(k+1) - C(A - LC)\hat{e}(k)\right), \tag{12.5}$$

it holds

$$d(k) - \hat{d}(k) = -C(A - LC)\left(e(k) - \hat{e}(k)\right) \to 0.$$

Since

$$\begin{aligned} e(k+1) &= (A - LC)e(k) + E_d P(r(k+1) - C(A - LC)e(k)) \\ (I - E_d PC)e(k+1) &= (I - E_d PC)(A - LC)e(k) \Longrightarrow \lim_{k\to\infty} e(k) = 0, \end{aligned}$$

the aforementioned two observers form a UIO and delivers an estimate for $d(k)$.

Remark 12.1 It is to be denoted that the matrix $(I - E_d PC)$ is singular, since

$$(I - E_d PC)E_d = 0.$$

Consequently, the system

$$(I - E_d PC)e(k+1) = (I - E_d PC)(A - LC)e(k)$$

is indeed a singular system. A reasonable and also practical solution is to construct a reduced observer, for instance, solving the Luenberger equations,

$$\begin{aligned} T(A - LC) - FT &= L_z C, TE_d = 0, T \in \mathcal{R}^{s\times n}, s = n - k_d, \\ WT + VC &= I, \end{aligned}$$

for T, a Schur matrix F, and matrices L, V, W. The UIO is constructed as

$$\begin{aligned} \hat{z}(k+1) &= F\hat{z}(k) + Hu(k) + L_z y(k), \hat{x}(k) = W\hat{z}(k) + Vy(k), \\ \hat{z}(k) &= T\hat{x}(k), H = T(B - L_z D). \end{aligned}$$

Example 12.1 demonstrates that the well-known UIOs are an unknown input estimator as well. Note that the estimate (12.5) is achieved by a (one step) backward computation.

In the DOBC research, a further class of UIOs are widely investigated on the assumption that prior knowledge of disturbances in the form of a model is available. Suppose that

$$\begin{cases} \xi(k+1) = A_d\xi(k) + B_d\varsigma(k), \xi \in \mathcal{R}^{n_d}, \varsigma \in \mathcal{R}^{p_d} \\ d(k) = C_d\xi(k), \end{cases} \tag{12.6}$$

where the model matrices (A_d, B_d, C_d) are known, A_d is Schur, ς represents the unmodellable uncertainty. The integers (n_d, p_d) reflect, in a certain sense, the degree of availability of prior knowledge of d. In particular, p_d is smaller than k_d. In engineering applications, it is often (and reasonably) assumed that $\xi(0) = \xi_0$ is unknown and $\varsigma(k) = 0$. This case can be equivalently modelled by

$$\begin{aligned} \xi(k+1) &= A_d \xi(k) + B_d \varsigma(k), \xi(0) = 0, \\ B_d \varsigma(k) &= A_d \xi_0 \delta(k), \delta(k) = \begin{cases} 1, k = 0 \\ 0, k > 0. \end{cases} \end{aligned}$$

Given the plant model (12.1)–(12.2) and disturbance model (12.6), the overall system dynamics are modelled by

$$\begin{bmatrix} x(k+1) \\ \xi(k+1) \end{bmatrix} = \begin{bmatrix} A & E_d C_d \\ 0 & A_d \end{bmatrix} \begin{bmatrix} x(k) \\ \xi(k) \end{bmatrix} + \begin{bmatrix} B \\ 0 \end{bmatrix} u(k) + \begin{bmatrix} 0 \\ B_d \end{bmatrix} \varsigma(k),$$
$$y(k) = \begin{bmatrix} C & F_d C_d \end{bmatrix} \begin{bmatrix} x(k) \\ \xi(k) \end{bmatrix}.$$

Note that when

$$rank\,(F_d C_d B_d) = p_d,$$

a UIO can be constructed for a perfect decoupling and estimation of d, as illustrated in Example 12.1. In general, the observer is constructed as follows,

$$\begin{bmatrix} \hat{x}(k+1) \\ \hat{\xi}(k+1) \end{bmatrix} = \begin{bmatrix} A & E_d C_d \\ 0 & A_d \end{bmatrix} \begin{bmatrix} \hat{x}(k) \\ \hat{\xi}(k) \end{bmatrix} + \begin{bmatrix} B \\ 0 \end{bmatrix} u(k) + \begin{bmatrix} L \\ L_d \end{bmatrix} \left(y(k) - \hat{y}(k)\right),$$
$$\hat{y}(k) = C\hat{x}(k) + F_d C_d \hat{\xi}(k) + Du(k),$$
$$\hat{d}(k) = C_d \hat{\xi}(k),$$

whose dynamic is subject to

$$\begin{bmatrix} e_x(k+1) \\ e_\xi(k+1) \end{bmatrix} = \begin{bmatrix} A - LC & (E_d - LF_d) C_d \\ -L_d C & A_d - L_d F_d C_d \end{bmatrix} \begin{bmatrix} e_x(k) \\ e_\xi(k) \end{bmatrix} + \begin{bmatrix} 0 \\ B_d \end{bmatrix} \varsigma(k),$$
$$e_x(k) = x(k) - \hat{x}(k), e_\xi(k) = \xi(k) - \hat{\xi}(k).$$

Hence, (L, L_d) are to be designed, when possible, to ensure the matrix

$$\begin{bmatrix} A - LC & (E_d - LF_d) C_d \\ -L_d C & A_d - L_d F_d \end{bmatrix}$$

being Schur. As a result, the disturbance estimator can be equivalently described by

$$\hat{\xi}(k+1) = A_d \hat{\xi}(k) + L_d \left(y(k) - \hat{y}(k)\right), \hat{d}(k) = C_d \hat{\xi}(k) \Longrightarrow$$
$$\hat{d} = R_d r_y, r_y = \hat{N}_d C_d e_\xi = \hat{N}_d \left(d - \hat{d}\right),$$

$$R_d = C_d (zI - A_d)^{-1} L_d, \hat{N}_d = (A - LC, E_d - LF_d, C, F_d),$$

from which it follows immediately

$$\hat{d} = \left(I + R_d\hat{N}_d\right)^{-1} R_d\hat{N}_d d \Longleftrightarrow d - \hat{d} = \left(I + R_d\hat{N}_d\right)^{-1} G_\varsigma \varsigma \qquad (12.7)$$
$$\Longrightarrow \hat{d} = R_d\hat{N}_d \left(I + R_d\hat{N}_d\right)^{-1} G_\varsigma \varsigma, G_\varsigma = C_d (zI - A_d)^{-1} B_d.$$

According to (12.7), the design of (L, L_d), besides the stability requirement, is formulated as

$$\min_{\substack{L, L_d \\ \left(I + R_d\hat{N}_d\right)^{-1} \in \mathcal{RH}_\infty}} \left\| \left(I + R_d\hat{N}_d\right)^{-1} G_\varsigma \right\|_\infty .$$

12.1.2 LS Unknown Input Estimation

The assumption (12.4) considerably limits the application of UIOs for fault detection and estimation. In fact, fault and unknown input estimation is an inverse problem, and (12.4) is one of the basic conditions for the existence of the inverse (estimate). For a more general case, for instance, when (12.4) does not hold, a solution will be achieved by solving an optimisation problem. The so-called Least Squares (LS) estimation is such a solution. In the sequel, plants modelled by

$$\begin{cases} x(k+1) = Ax(k) + Bu(k) + E_d d(k), x(0) = x_0 \\ y(k) = Cx(k) + Du(k) + F_d d(k), \end{cases} \qquad (12.8)$$

are considered, where $d(k) \in \mathbb{R}^{k_d}$ is an $\mathcal{L}_{2,[0,N]}$ bounded unknown input vector that might be a fault vector, E_d, F_d are known with F_d satisfying

$$rank\,(F_d) = m < k_d, \qquad (12.9)$$

and x_0 is the unknown initial state vector. Before specifying the problem to be addressed, two remarks on the model (12.8) and the associated assumptions are made.

- The unknown input vector d may comprehend disturbances/faults in the process (state) model and the output model independently. To be specific, d and the associated matrices E_d and F_d are of the general form

$$d = \begin{bmatrix} d_p \\ d_o \end{bmatrix}, \begin{bmatrix} E_d \\ F_d \end{bmatrix} = \begin{bmatrix} E_{d_p} & 0 \\ 0 & F_{d_o} \end{bmatrix};$$

- If
$$rank\left(F_d\right) < m,$$
i.e. the condition (12.8) is not satisfied, it is then assumed that
$$rank\begin{bmatrix} F_d \\ CE_d \end{bmatrix} = m.$$
Analogous to Example 12.1, in this case $y(k+1)$ or a linear mapping of $y(k+1)$ are used as (a part of) the output for the estimation purpose.

Like Example 12.1, it is, without loss of generality, supposed that an observer-based residual generator,
$$r_y = \hat{M}y - \hat{N}u, \tag{12.10}$$
delivers an m-dimensional residual vector that is independent of u and driven by d,
$$\begin{aligned} r_y &= \hat{N}_d d, \hat{N}_d = W\left(F_d + C\left(zI - A_L\right)^{-1} E_{d,L}\right), \\ A_L &= A + LC, E_{d,L} = E_d - LF_d. \end{aligned} \tag{12.11}$$
Notice that the system (12.11) is equivalent to
$$\begin{cases} x_d(k+1) = Ax_d(k) + E_d d(k), x_d(0) = x_0 \\ y_d(k) = Cx(k) + F_d d(k), \end{cases} \tag{12.12}$$
$$r_y = \hat{M}y_d,$$
where y_d is the response of y to d. In this context, generation of the residual signal r_y can be equivalently realised by
$$\hat{x}_d(k+1) = A\hat{x}_d(k) + Lr_y(k), \hat{x}_d\left(0\right) := 0, \tag{12.13}$$
whose dynamic is governed by
$$\begin{cases} e_d(k+1) = A_L e_d(k) + E_{d,L} d(k), e_d\left(0\right) = x_0, e_d(k) = x_d(k) - \hat{x}_d(k) \\ r_y(k) = W\left(Ce_d(k) + F_d d(k)\right). \end{cases} \tag{12.14}$$

Definition 12.1 Given the system (12.8) and data $r_y(k)$ generated by the residual generator (12.10), $k \in [0, N]$, find estimates for $d(k), k \in [0, N-1]$, and x_0 which minimise the cost function,
$$\begin{aligned} J &= \left\|x\left(0\right)\right\|^2_{P^{-1}(0)} + \left\|d\right\|^2_{[0,N-1]}, \\ &\text{s.t. Eq. (12.8)}, \end{aligned} \tag{12.15}$$

where $P(0) > 0$ is a pre-defined weighting matrix. The resulted estimates are called LS estimates.

Before starting with the solution, following notations are introduced. Let

$$\hat{x}(0|N), \hat{d}(k|N), k \in \{0, 1, \ldots, N-1\},$$

denote estimates of $x(0), d(k)$ given data r_y over the time interval $[0, N]$. The optimal estimation problem (12.15) is called smoothing as well. Also, we would like to remind the reader about the results in Sect. 5.3, where it has been illustrated that the normalised LCF is a co-inner system that delivers an LS estimate of $x(0), d(k)$.

Below, we summarise the estimation algorithm and prove that it solves the LS estimation problem (12.15).

Algorithm 12.1 *LS unknown input estimation*

Step 0: *Set* $\hat{x}(0) = 0, P(0)$;
Step 1: *Run the observer-based residual generator*

$$\hat{x}(k+1) = A\hat{x}(k) + Bu(k) + L_0(k)r_y(k), \tag{12.16}$$

$$r_y(k) = W_0(k)\left(y(k) - \hat{y}(k)\right), \hat{y}(k) = C\hat{x}(k) + Du(k), \tag{12.17}$$

and output $x(k), r_y(k), k \in [0, N]$, *where*

$$L_0(k) = \left(E_d F_d^T + AP(k)C^T\right) W_0^2(k), W_0(k) = \left(F_d F_d^T + CP(k)C^T\right)^{-1/2},$$
$$P(k+1) = A_{L_0}(k)P(k)A_{L_0}^T(k) + E_{d,L_0}(k)E_{d,L_0}^T(k),$$
$$A_{L_0}(k) = A - L_0(k)C, E_{d,L_0}(k) = E_d - L_0(k)F_d;$$

Step 2: *Run the unknown input estimator backwards*

$$\begin{cases} \bar{e}_d(k) = A_{L_0}^T(k)\bar{e}_d(k+1) + (W_0(k)C)^T r_y(k), \bar{e}_d(N) = 0 \\ \hat{d}(k) = E_{L_0}^T(k)\bar{e}_d(k+1) + (W_0(k)F_d)^T r_y(k), \end{cases} \tag{12.18}$$

and output $\bar{e}_d(k), \hat{d}(k|N) := \hat{d}(k), k \in [0, N-1]$;
Step 3: *Calculate*

$$\hat{x}(k|N) = P(k)\bar{e}_d(k) + \hat{x}(k),$$

and output $\hat{x}(k|N), k \in [0, N]$.

In the following theorem, we prove that $\hat{d}(k|N), k = 1, \ldots, N-1, \hat{x}(0|N)$ solve the optimisation problem (12.15).

Theorem 12.1 *Given the system (12.14) and data* $r_y(k), k \in [0, N]$, *then* $\hat{d}(k|N), k = 0, \ldots, N-1, \hat{x}(0|N)$ *delivered by Algorithm 12.1 solve the optimisation problem (12.15). Moreover, the optimal value of the cost function* J^* *is given by*

$$J^* = \left\| \hat{x}(0\,|N\,) \right\|_{P^{-1}(0)}^2 + \left\| \hat{d}(k\,|N\,) \right\|_{[0,N-1]}^2 = \left\| r_y\,(k) \right\|_{[0,N-1]}^2 . \tag{12.19}$$

Proof Our proof is based on the property that

$$\begin{aligned} r_y &= \hat{\mathcal{N}_0}\,(d)\,, \\ \hat{\mathcal{N}_0} &= \left(A_{L_0}(k),\, E_{d,L_0}(k),\, W_0(k)C,\, W_0(k)F_d\right), \end{aligned}$$

where $\hat{\mathcal{N}_0}$ is a co-inner time-varying system (refer to Sect. 5.3) and defines a linear operator that maps $(x(0), d)$ to r_y, $\hat{\mathcal{N}_0} : R^n \times \mathcal{L}_{2,[0,N]} \to \mathcal{L}_{2,[0,N]}$. To prove this property, let

$$\begin{aligned} e(k) = x(k) - \hat{x}(k),\, e_d(k) = x_d(k) - \hat{x}_d(k), \\ \hat{x}_d(k+1) = A\hat{x}_d(k) + L_0(k)r_y(k),\, \hat{x}_d\,(0) = 0. \end{aligned}$$

It is straightforward that

$$\begin{cases} e(k+1) = A_{L_0}(k)e(k) + E_{d,L_0}(k)d(k),\, e\,(0) = x\,(0) = x_0 \\ r_y(k) = W_0(k)\,(Ce(k) + F_d d(k))\,, \end{cases}$$

which is apparently equivalent to (12.14), i.e. $e(k) = e_d(k)$. As a result, we have

$$r_y = \hat{\mathcal{M}}_0\,(y_d) = \hat{\mathcal{N}_0}\,(d)\,,$$

where $\hat{\mathcal{M}}_0$ is a time-varying system given by

$$\hat{\mathcal{M}}_0 = \left(A_{L_0}(k),\, -L_0(k),\, W_0(k)C,\, W_0(k)\right).$$

As described in Theorems 5.2 and 5.4, the adjoint of the system (12.14),

$$\begin{aligned} \bar{e}_d(k) &= A_{L_0}^T(k)\bar{e}_d(k+1) + (W_0(k)C)^T\, r_y(k), \\ \hat{d}(k) &= E_{d,L_0}^T(k)\bar{e}_d(k+1) + (W_0(k)F_d)^T\, r_y(k) \end{aligned}$$

delivers an (LS) estimate $\hat{d}(k)$ for $d(k)$. Recall that $\bar{e}_d(k) = P^{-1}(k)e_d(k) = P^{-1}(k)e(k)$. It turns out

$$\hat{d}(k) = E_{d,L_0}^T(k)P^{-1}(k+1)e_d(k+1) + (W_0(k)F_d)^T\, r_y(k).$$

Notice that $\hat{d}(k) = \hat{d}(k\,|N\,)$ can be written as

$$\begin{aligned} \hat{d}(k) = E_{L_0}^T(k)P^{-1}\,(k+1)\left(\hat{x}(k+1\,|N\,) - \hat{x}(k+1)\right) \\ + (W_0(k)F_d)^T\, r_y(k),\, k = N-1, N-2, \cdots, 0, \end{aligned} \tag{12.20}$$

and $\left(\hat{d}(k\,|N\,), \hat{x}_d(k\,|N\,)\right)$ are subject to

$$\hat{x}(k+1\,|N\,) = A\hat{x}(k\,|N\,) + Bu(k) + E_d\hat{d}(k\,|N\,), \hat{x}(N\,|N\,) = \hat{x}(N), \quad k = N-1, N-2, \ldots, 0. \tag{12.21}$$

Thus, they are a solution of (12.8) for

$$x(N) = \hat{x}(N\,|N\,) = \hat{x}(N), x(0) = \hat{x}(0\,|N\,), d(k) = \hat{d}(k) = \hat{d}(k\,|N\,).$$

Now, with

$$e_d(k) = e(k) = \hat{x}(k\,|N\,) - \hat{x}(k), d(k) = \hat{d}(k\,|N\,)$$

we obtain

$$\begin{cases} e_d(k+1) = A_{L_0}(k)e_d(k) + E_{d,L_0}(k)\hat{d}(k), e_d\,(0) = \hat{x}_d(0\,|N\,), e_d(N) = 0. \\ r_y(k) = W_0(k)\left(Ce_d(k) + F_d\hat{d}(k)\right). \end{cases}$$

As a result of Theorem 5.4, we have the unknown input estimator (12.21)–(12.20). Moreover, it holds

$$J^* = \left\|\hat{x}(0\,|N\,)\right\|^2_{P^{-1}(0)} + \left\|\hat{d}(k\,|N\,)\right\|^2_{[0,N-1]} = \left\|r_y\,(k)\right\|^2_{[0,N-1]}.$$

Since for $\forall x(0), d(k)$,

$$J = \|x(0)\|^2_{P^{-1}(0)} + \|d(k)\|^2_{[0,N-1]} \geq \left\|r_y\,(k)\right\|^2_{[0,N-1]},$$

it can be concluded that $x(0) = \hat{x}(0\,|N\,), d(k) = \hat{d}(k\,|N\,)$ subject to (12.21)–(12.20) solve the optimisation problem (12.15). The theorem has been proved.

12.1.3 Projection-Based Uncertainty Estimation

In this subsection, projection-based estimation of uncertainties in process data (u, y) is researched. We consider the following SKR-based process model,

$$G: \begin{cases} \hat{x}(k+1) = A\hat{x}(k) + Bu(k) + Lr_y(k), \hat{x}(0) = 0 \\ y(k) = C\hat{x}(k) + Du(k) + r_y(k), \end{cases} \tag{12.22}$$

and suppose that no a priori knowledge of uncertainties is available. For our purpose of estimating uncertainties corrupted in the process data (u, y), an SKR-based projection system is constructed as

$$K_{G_0}^{\sim} K_{G_0} : \begin{cases} \hat{x}(k+1) = A_K \hat{x}(k) + B_K \begin{bmatrix} u(k) \\ y(k) \end{bmatrix} \\ r_y(k) = C_K \hat{x}(k) + D_K \begin{bmatrix} u(k) \\ y(k) \end{bmatrix} \\ \lambda(k) = A_K^T \lambda(k+1) + C_K^T r_y(k+1) \\ \begin{bmatrix} \Delta u(k) \\ \Delta y(k) \end{bmatrix} = B_K^T \lambda(k) + D_K^T r_y(k), \end{cases} \tag{12.23}$$

$$A_K = A - L_0 C,\; B_K = \begin{bmatrix} B - L_0 D & L_0 \end{bmatrix},\; C_K = -W_0 C,\; D_K = \begin{bmatrix} -W_0 D & W_0 \end{bmatrix},$$

where K_{G_0} is the normalised SKR of G with its conjugate $K_{G_0}^{\sim}$, $\lambda \in \mathbb{R}^n$ is the co-state vector, and (L_0, W_0) are the observer gain matrix and post-filter of the normalised SKR (refer to Theorem 2.1). The output of the projection system $K_{G_0}^{\sim} K_{G_0}$ is $(\Delta u(k), \Delta y(k))$ that represents an estimate of uncertainties corrupted in the data. In fact, the dynamics modelled by the normalised SKR describe the influence of $(\Delta u(k), \Delta y(k))$ on the system. In other words, it models the uncertain dynamics. Indeed, recall that

$$\begin{bmatrix} \Delta u(k) \\ \Delta y(k) \end{bmatrix} \in \mathcal{R}_G.$$

Hence, it holds

$$\begin{bmatrix} \Delta u \\ \Delta y \end{bmatrix} = K_{G_0}^{\sim} K_{G_0} \begin{bmatrix} \Delta u \\ \Delta y \end{bmatrix},$$

namely, $K_{G_0}^{\sim} K_{G_0}$ can be equivalently written into

$$K_{G_0}^{\sim} K_{G_0} : \begin{cases} \tilde{x}(k+1) = A_K \tilde{x}(k) + B_K \begin{bmatrix} \Delta u(k) \\ \Delta y(k) \end{bmatrix} \\ r_y(k) = C_K \tilde{x}(k) + D_K \begin{bmatrix} \Delta u(k) \\ \Delta y(k) \end{bmatrix} \\ \lambda(k) = A_K^T \lambda(k+1) + C_K^T r_y(k) \\ \begin{bmatrix} \Delta u(k) \\ \Delta y(k) \end{bmatrix} = B_K^T \lambda(k+1) + D_K^T r_y(k) \end{cases}$$

with $\tilde{x}$ representing the state vector of the uncertain dynamics. Note that for the nominal system, i.e. $\Delta u = 0$, $\Delta y = 0$, it holds

$$r_y(k) = C_K \tilde{x}(k) = -W_0 C \tilde{x}(k).$$

This implies,

$$\tilde{x}(k) = \hat{x}(k) - x(k),$$

as illustrated in the following example.

Example 12.2 Consider the nominal system model

$$\begin{cases} x(k+1) = Ax(k) + Bu(k), x(0) = x_0 \\ y(k) = Cx(k) + Du(k). \end{cases}$$

The corresponding normalised SKR is given by

$$K_{G_0}: \begin{cases} \hat{x}(k+1) = A_K \hat{x}(k) + B_K \begin{bmatrix} u(k) \\ y(k) \end{bmatrix} \\ r_y(k) = C_K \hat{x}(k) + D_K \begin{bmatrix} u(k) \\ y(k) \end{bmatrix}. \end{cases}$$

Let $e(k) = x(k) - \hat{x}(k)$, and consider $\Delta u = 0$, $\Delta y = 0$. It turns out, after routine computation,

$$e(k+1) = A_K e(k) \Longrightarrow$$

$$K_{G_0}: \begin{cases} e(k+1) = A_K e(k), e(0) = x_0 \\ r_y(k) = W_0 e(k) = -W_0 C \tilde{x}(k) = C_K \tilde{x}(k). \end{cases}$$

The aforementioned results reveal two elemental properties of the SKR-based projection system (12.23):

- there exists a one-to-one mapping between the co-state vector $\lambda(k)$ and $\tilde{x}(k)$, namely
$$P\lambda(k) = \tilde{x}(k), \tag{12.24}$$
where $P > 0$ is the solution of the Riccati equation given in Theorem 2.3;
- in this context, $P\lambda(k-i)$ gives an estimate of $\tilde{x}(k-i), i = 0, 1, \ldots, k$. This result is useful for data-driven learning of an observer and residual generator.

The claim (12.24) can be informally proved as follows. Since K_{G_0} is co-inner, it holds

$$K_{G_0} K_{G_0}^{\sim} = I, P\lambda(k) = \tilde{x}(k),$$
$$V(\lambda(k-1)) - V(\lambda(k)) = \frac{1}{2}\left(r_y^T(k) r_y(k) - \Delta z^T(k) \Delta z(k)\right), \tag{12.25}$$
$$V(\lambda(k)) = \frac{1}{2}\lambda^T(k) P \lambda(k), z(k) = \begin{bmatrix} \Delta u \\ \Delta y \end{bmatrix}.$$

Equation (12.25) describes the so-called lossless property with $V(\lambda(k))$ as the "storage information" function, the analogue form of the storage energy function. The reader is referred to Theorem 5.4 for details. Subsequently,

$$K_{G_0} \begin{bmatrix} \Delta u \\ \Delta y \end{bmatrix} = K_{G_0} \begin{bmatrix} u \\ y \end{bmatrix} = r_y.$$

On account of the fact that both systems,

$$\begin{bmatrix} \Delta u \\ \Delta y \end{bmatrix} = K_{\tilde{G}_0} r_y, \begin{bmatrix} \Delta u \\ \Delta y \end{bmatrix} = K_{\tilde{G}_0} K_{G_0} \begin{bmatrix} u \\ y \end{bmatrix} = K_{\tilde{G}_0} r_y,$$

have the identical storage information, it can be concluded that (12.24) holds for the projection system as well. Notice that in this aspect,

$$\tilde{x}(k) = P\lambda(k) = \frac{\partial V(\lambda(k))}{\partial \lambda(k)}.$$

This implies, $\tilde{x}(k)$ is the change of the storage information with respect to $\lambda(k)$.

Concerning the computation of $(\Delta u, \Delta y)$, according to the space space representation (12.23), given $r_y(k)$ and setting $\lambda(k+1) = 0$, values of $\Delta u(k-i), \Delta y(k-i), i = 0, 1, \ldots, k$, are computed backwards as follows,

$$\begin{cases} \lambda(k) = A_K^T \lambda(k+1) + C_K^T r_y(k), \lambda(k+1) = 0 \\ \begin{bmatrix} \Delta u(k-i) \\ \Delta y(k-i) \end{bmatrix} = B_K^T \lambda(k-i+1) + D_K^T r_y(k-i). \end{cases} \tag{12.26}$$

It is remarkable that A_K^T is a Schur matrix. Therefore, the boundary setting $\lambda(k+1) = 0$ has negligible influence on the estimation performance of $\lambda(k-i)$ and (equivalently) $\tilde{x}(k-i) = P\lambda(k-i)$. It is noteworthy that

- the algorithm,
  - Step 1: residual generation,

$$\begin{cases} \hat{x}(k+1) = A_K \hat{x}(k) + B_K \begin{bmatrix} u(k) \\ y(k) \end{bmatrix} \\ r_y(k) = C_K \hat{x}(k) + D_K \begin{bmatrix} u(k) \\ y(k) \end{bmatrix}, \end{cases} \tag{12.27}$$

  - Step 2: run (12.26),

  is in fact a general form of Algorithm 12.1 for the LS unknown input estimation. In this sense, it is called LS estimation of uncertainty as well;
- this algorithm can be applied to improve the estimation performance of the state vector $\hat{x}(k-i)$ as well. To be specific,
  - run (12.27) as an observer for $\hat{x}(k)$ and residual generator for $r_y(k)$, and save $(\hat{x}(k), r_y(k)), k \in [0, N]$;
  - run (12.26) for $\lambda(k), k = N, N-1, \ldots, 0$ under the boundary condition $\lambda(N+1) = 0$;
  - set

$$\bar{x}(k) = \hat{x}(k) - P\lambda(k), k = N, N-1, \ldots, 0.$$

Hereby, $\bar{x}(k)$ is the improved estimate of the state vector $x(k)$.

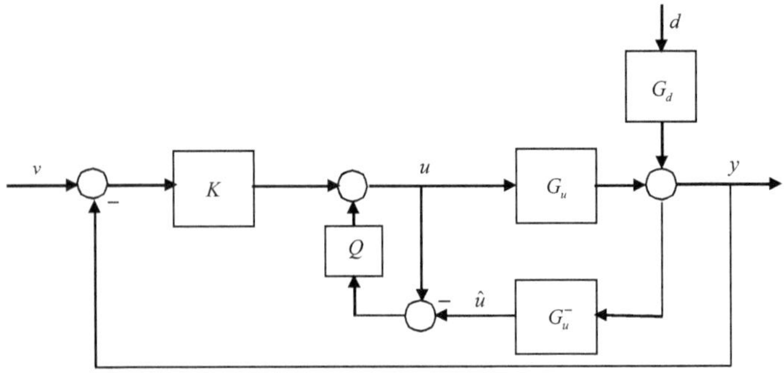

Fig. 12.1 A schematic description of DOBC system

12.2 Disturbance Observer-Based Control

Disturbance observer-based control (DOBC) is a popular technique widely applied in engineering systems for enhancing system robustness and fault tolerance. A DOBC basically consists of two loops. At its core is a disturbance compensator located in the inner loop. The outer loop is a trivial (output) feedback control. Research efforts reported in the literature are mainly focused on the design of UIO aiming at unknown input estimation. In this section, DOBC schemes will be explored in the unified framework of control and detection, which will help us to get some insightful aspects of DOBC.

12.2.1 *Re-visit of DOBC*

Figure 12.1 sketches the configuration of a DOBC system schematically, where d represents an unknown input vector, being possibly faults, G_u^- is a notation for an approximation of the inverse of G_u, and $K(v-y)$ builds the outer loop. The dynamic of the inner loop is described by

$$\begin{aligned} y = G_u u + G_d d, \hat{u} = G_u^- y = G_u^- G_u u + G_u^- G_d d \Longrightarrow \\ u - \hat{u} = \left(I - G_u^- G_u\right) u - G_u^- G_d d \Longrightarrow \\ y = G_u \left(K (v - y) + Q \left(I - G_u^- G_u\right) u - Q G_u^- G_d d\right) + G_d d. \end{aligned} \tag{12.28}$$

Hence, on the assumption that $\exists G_u^-, Q \in \mathcal{RH}_\infty$ s.t.

$$G_u^- G_u \approx I, \left(I - G_u Q G_u^-\right) G_d \approx 0, \tag{12.29}$$

it holds

$$y = G_u K (v - y) \Longrightarrow y = (I + G_u K)^{-1} v,$$

namely, the system response is decoupled from the unknown input d. The brilliant idea behind this perfect (decoupling) result is building of the residual signal $u - \hat{u}$ given in (12.28), which, at the first glance, is different from the residual generation using the system (measurement) output y, as adopted through our work. On the other hand, (12.29) imposes very hard conditions for the realisation of the idea, in particular the existence of G_u^- is the centerpiece of the DOBC scheme. It is somewhat amazing that considerable reported research efforts have been dedicated to issues on unknown input estimation and design of UIOs, and very few attention has been paid to the realisation of the initial idea of generating the residual $u - \hat{u}$.

12.2.2 Analysis of DOBC in the Unified Framework

We now explore the DOBC scheme in the unified framework of control and detection. For the sake of simplicity but without loss of generality, the system model under consideration is described by

$$G : \begin{cases} x(k+1) = Ax(k) + Bu(k) + E_d d(k) \\ y(k) = Cx(k) + Du(k) + F_d d(k) \end{cases} \tag{12.30}$$

with d as the unknown input vector. The corresponding DOBC is subject to

$$\begin{aligned} u &= K(v - y) + Q\left(u - G_u^- y\right) \\ &\Longrightarrow u = -(I - Q)^{-1}\left(\left(QG_u^- + K\right)y + Kv\right), \\ K &= \hat{U}\hat{V}^{-1} = V^{-1}U, \hat{U} = \hat{Y} - MQ_0, \hat{V} = \hat{X} + NQ_0, \\ V &= X + Q_0\hat{N}, U = Y - Q_0\hat{M}, \end{aligned} \tag{12.31}$$

where (M, N), $\left(\hat{M}, \hat{N}\right)$, $\left(\hat{X}, \hat{Y}\right)$ and (X, Y) are the RCPs and LCPs of the plant $G_u = (A, B, C, D)$ and the controller K, Q_0 is the Youla parameter matrix, and $\left(Q, G_u^-\right)$ are to be determined. It turns out

$$\begin{aligned} \begin{bmatrix} u \\ y \end{bmatrix} &= \begin{bmatrix} I - Q & QG_u^- + K \\ -G_u & I \end{bmatrix}^{-1} \begin{bmatrix} Kv \\ G_d d \end{bmatrix} \\ &= \begin{bmatrix} V - VQ & VQG_u^- + U \\ -\hat{N} & \hat{M} \end{bmatrix}^{-1} \begin{bmatrix} Uv \\ \hat{N}_d d \end{bmatrix}, \\ G_u &= \hat{M}^{-1}\hat{N}, G_d = (A, E_d, C, F_d) = \hat{M}^{-1}\hat{N}_d. \end{aligned}$$

On account of the conditions (12.29), G_u^-, Q are parametrised by

$$\begin{aligned} G_u^- G_u &= I \Longrightarrow G_u^- := \hat{N}^- \hat{M}, G_u Q G_u^- = I \Longrightarrow Q = M Q_1 \Longrightarrow \\ G_u^- G_u &= \hat{N}^- \hat{N}, G_u Q G_u^- = N Q_1 G_u^-, \hat{N}^-, Q_1 \in \mathcal{RH}_\infty. \end{aligned}$$

Thus, the conditions (12.29) are rewritten as $\exists \hat{N}^-$, $Q_1 \in \mathcal{RH}_\infty$, s.t.

$$\hat{N}^- \hat{N} = I, \hat{M} N Q_1 \hat{N}^- \hat{N}_d = \hat{N}_d. \tag{12.32}$$

By the Bezout identity and routine calculations,

$$\begin{gathered} \begin{bmatrix} V & U \\ -\hat{N} & \hat{M} \end{bmatrix}^{-1} = \begin{bmatrix} M & -\hat{Y} + M Q_0 \\ N & \hat{X} + N Q_0 \end{bmatrix} = \begin{bmatrix} M & -\hat{U} \\ N & \hat{V} \end{bmatrix}, \\ \begin{bmatrix} V - V Q & V Q G_u^- + U \\ -\hat{N} & \hat{M} \end{bmatrix}^{-1} \\ = \begin{bmatrix} V & U \\ -\hat{N} & \hat{M} \end{bmatrix}^{-1} \left(I + \begin{bmatrix} -V Q & V Q G_u^- \\ 0 & 0 \end{bmatrix} \begin{bmatrix} V & U \\ -\hat{N} & \hat{M} \end{bmatrix}^{-1} \right)^{-1} = \\ \begin{bmatrix} M & -\hat{Y} + M Q_0 \\ N & \hat{X} + N Q_0 \end{bmatrix} \begin{bmatrix} \Pi & -V Q \left(-\hat{N}^- + \left(I - \hat{N}^- \hat{N} \right) \left(-\hat{Y} + M Q_0 \right) \right) \\ 0 & I \end{bmatrix}^{-1} = \\ \begin{bmatrix} M & -\hat{Y} + M Q_0 \\ N & \hat{X} + N Q_0 \end{bmatrix} \begin{bmatrix} \Pi^{-1} & \Pi^{-1} V Q \left(-\hat{N}^- + \left(I - \hat{N}^- \hat{N} \right) \left(-\hat{Y} + M Q_0 \right) \right) \\ 0 & I \end{bmatrix}, \\ \Pi = I - V Q \left(I - \hat{N}^- \hat{N} \right) M, \end{gathered}$$

we obtain

$$\begin{aligned} \begin{bmatrix} u \\ y \end{bmatrix} &= \begin{bmatrix} M \\ N \end{bmatrix} \Pi^{-1} U v + \begin{bmatrix} -\hat{Y} + M Q_0 \\ \hat{X} + N Q_0 \end{bmatrix} \hat{N}_d d \\ &+ \begin{bmatrix} M \\ N \end{bmatrix} \Pi^{-1} V Q \left(-\hat{N}^- + \left(I - \hat{N}^- \hat{N} \right) \left(-\hat{Y} + M Q_0 \right) \right) \hat{N}_d d. \end{aligned} \tag{12.33}$$

Hence, when $\hat{N}^- \hat{N} = I$, the transfer function from d to y, G_{yd}, is subject to

$$G_{yd} = \left(\hat{X} + N Q_0 - N V Q \hat{N}^- \right) \hat{N}_d.$$

Note that for $Q_0 = 0$,

$$G_{yd} = \hat{X} \left(I - \hat{N} Q \right) \hat{N}_d.$$

We summarise the above results in a theorem.

Theorem 12.2 *Given system (12.30), controller (12.31) and an observer-based residual generator whose dynamic is described by (12.11), then y is decoupled from from d if and only if*

$$\left(\hat{X} + N\left(Q_0 + Q^*\right)\right)\hat{N}_d = 0, \tag{12.34}$$

$$Q^* = \Pi^{-1} V Q \left(-\hat{N}^- + \left(I - \hat{N}^-\hat{N}\right)\left(-\hat{Y} + M Q_0\right)\right). \tag{12.35}$$

Recall that the transfer function $\left(\hat{X} + N\left(Q_0 + Q^*\right)\right)$ is invertible. Hence, as a consequence of (12.34), the following condition becomes apparent.

Corollary 12.1 *Given system (12.30), controller (12.31) and an observer-based residual generator whose dynamic is described by (12.11), then y is decoupled from from d if and only if*

$$\exists L, W, \ s.t. \ \hat{N}_d = 0. \tag{12.36}$$

It is well-known that the condition (12.36) implies a (perfect) UIO, as described in Example 12.1.

The summary of the initial idea of the DOBC and the results from the analysis of DOBC prompt us to make the following remarks on DOBC research.

- The extensive efforts dedicated to UIO schemes are, on account of the condition (12.36), are well-grounded.
- It can be observed that the reported works on UIOs are mainly focused on unknown input state observers, namely, the dynamic of the state estimation error is to be decoupled from the unknown input vector d. It is well-known that the existence of such an observer is equivalent to an inverse problem, whose existence condition are expressed by, for $F_d \neq 0$,

$$rank \begin{bmatrix} A - zI & E_d \\ C & F_d \end{bmatrix} = n + k_d, \forall z_0, rank \begin{bmatrix} A - z_0 I & E_d \\ C & F_d \end{bmatrix} < n + k_d, |z_0| < 1,$$

  or for $F_d = 0$,

$$rank\left(C E_d\right) = k_d, \forall z_0, rank \begin{bmatrix} A - z_0 I & E_d \\ C & 0 \end{bmatrix} < n + k_d, |z_0| < 1,$$

  where z_0 is a transmission zero. In other words, the transmission zeros should be located inside the unit disk. On the other hand, the loop dynamic (12.33) and the condition (12.36) reveal that it is sufficient to achieve a full decoupling of y from d if the residual r_y,

$$r_y = \hat{N}_d d,$$

  is decoupled from d.

- Note that $\hat{N}_d = 0$ is a necessary condition as well for the realisation of those DOBC systems (12.31) designed based on the initial idea of generating the residual $u - \hat{u}$.
- The last fact motivates us to have a close look at the residual $u - \hat{u}$ and compare its role with the one of the residual r_y used in the unified framework. Recall that

$$u - \hat{u} = u - G_u^- y = u - \hat{N}^- \hat{M} y,$$

where G_u^- is factorised, on the account of the condition $G_u^- G_u = I$, as $G_u^- = \hat{N}^- \hat{M}$. The residual vector r_y is generated by

$$r_y = \hat{M} y - \hat{N} u.$$

Since the condition $G_u^- G_u = I$ requires

$$G_u^- G_u = \hat{N}^- \hat{N} = I,$$

it turns out

$$\hat{N}^- r_y = \hat{N}^- \hat{M} y - u = - \left(u - \hat{u} \right).$$

Viewing $\hat{N}^-$ as a post-filter, the residual vector $u - \hat{u}$ is in fact a special case of a function of r_y. In this context, it becomes clear that a direct use of r_y for unknown input decoupling control would significantly simplify the problem solution.

12.2.3 DOBC Design in the Unified Framework

Motivated by the aforementioned observations and analysis, we now explore DOBC design in the unified framework. A major advantage of DOBC technique is its capability to deal with uncertain systems in a straightforward manner. For instance, for an uncertain system,

$$\begin{cases} x(k+1) = (A + \Delta A)\, x(k) + (B + \Delta B)\, u(k) + d_p(k) \\ y(k) = (C + \Delta C)\, x(k) + (D + \Delta D)\, u(k) + d_s(k), \end{cases} \tag{12.37}$$

where $(\Delta A, \Delta B, \Delta C, \Delta D)$ represent uncertainties of the model parameters and d_p, d_s are process and sensor disturbances, the model (12.37) is rewritten as

$$\begin{cases} x(k+1) = Ax(k) + Bu(k) + E_{\mathrm{d}} d(k) \\ y(k) = Cx(k) + Du(k) + F_{\mathrm{d}} d(k), \end{cases} \tag{12.38}$$

$$E_{\mathrm{d}} d(k) = \Delta Ax(k) + \Delta Bu(k) + d_p(k),\ F_{\mathrm{d}} d(k) = \Delta Cx(k) + \Delta Du(k) + d_s(k). \tag{12.39}$$

Note that the unknown input vector d is now a function of (x, u) and thus is multiplicative uncertainties. For our purpose, we firstly formulate the design problem of DOBC in the unified framework.

Suppose that a UIO is applied for estimating d, and the state vector x, denoted by $\hat{d}$ and $\hat{x}$ respectively. The corresponding residual generator and the estimate $\hat{d}$ are, without loss of generality, expressed by

$$r_y = \hat{M}y - \hat{N}u, \hat{d} = Q_d r_y,$$

respectively. The controller consists of an observer-based state feedback and a disturbance compensator, written in a general form

$$u = F\hat{x} + Q_c\hat{d} = F\hat{x} + Qr_y, Q = Q_c Q_d \in \mathcal{RH}_\infty. \tag{12.40}$$

Here, for the sake of simplicity, a reference signal is not taken into account. Controller (12.37) is a standard form of an observer-based feedback controller and the corresponding closed-loop dynamics are described by

$$\begin{bmatrix} u \\ y \end{bmatrix} = \begin{bmatrix} -\hat{Y} + MQ \\ \hat{X} + NQ \end{bmatrix} r_y, r_y = \hat{N}_d d. \tag{12.41}$$

It seems that the closed-loop is stable and nice disturbance rejection performance is achievable, when a good UIO is designed. Unfortunately, under existence of the multiplicative uncertainties $(\Delta A, \Delta B, \Delta C, \Delta D)$, d is a function of $\left(u, y, d_p, d_s\right)$. To examine the system stability and control performance, a careful system analysis is needful. To this end, the (left) coprime factor uncertainty is adopted for modelling the uncertainty system (12.37),

$$\begin{aligned}
y &= G_{u,\Delta}u + G_{d,\Delta}\bar{d}, \bar{d} = \begin{bmatrix} d_p \\ d_s \end{bmatrix}, \\
G_{u,\Delta} &= \left(\hat{M} + \Delta_{\hat{M}}\right)^{-1}\left(\hat{N} + \Delta_{\hat{N}}\right), G_{d,\Delta} = \left(\hat{M} + \Delta_{\hat{M}}\right)^{-1}\left(\hat{N}_{\bar{d}} + \Delta_{\hat{N}_{\bar{d}}}\right), \\
\hat{N}_{\bar{d}} &= \left(A - LC, \bar{E}_{\bar{d}} - L\bar{F}_{\bar{d}}, WC, W\bar{F}_{\bar{d}}\right), \bar{E}_{\bar{d}} = \left[\, I \ 0 \,\right], \bar{F}_{\bar{d}} = \left[\, 0 \ I \,\right].
\end{aligned}$$

The computation of $\Delta_{\hat{M}}, \Delta_{\hat{N}}, \Delta_{\hat{N}_{\bar{d}}}$ is attributed to the following factorisations,

$$\begin{aligned}
&\hat{M}_\Delta = (A + \Delta A - L(C + \Delta C), -L, W(C + \Delta C), W), \\
&\hat{N}_\Delta = \begin{pmatrix} A + \Delta A - L(C + \Delta C), (B + \Delta B) - L(D + \Delta D), \\ W(C + \Delta C), W(D + \Delta D) \end{pmatrix}, \\
&\hat{N}_{\bar{d},\Delta} = \left(A + \Delta A - L(C + \Delta C), \bar{E}_{\bar{d}} - L\bar{F}_{\bar{d}}, W(C + \Delta C), W\bar{F}_{\bar{d}}\right), \\
&\Delta_{\hat{M}} = \hat{M}_\Delta - \hat{M}, \Delta_{\hat{N}} = \hat{N}_\Delta - \hat{N}, \Delta_{\hat{N}_{\bar{d}}} = \hat{N}_{\bar{d},\Delta} - \hat{N}_{\bar{d}}.
\end{aligned}$$

It is straightforward that

$$r_y = \hat{N}_d d = \Delta_{\hat{N}} u - \Delta_{\hat{M}} y + \left(\hat{N}_{\bar{d}} + \Delta_{\hat{N}_{\bar{d}}}\right)\bar{d}.$$

Substituting r_y in (12.41) gives

$$\begin{bmatrix} u \\ y \end{bmatrix} = \begin{bmatrix} -\hat{Y} + MQ \\ \hat{X} + NQ \end{bmatrix} \left(\begin{bmatrix} \Delta_{\hat{N}} & -\Delta_{\hat{M}} \end{bmatrix} \begin{bmatrix} u \\ y \end{bmatrix} + \left(\hat{N}_{\bar{d}} + \Delta_{\hat{N}_{\bar{d}}}\right)\bar{d} \right) \Longrightarrow$$
$$\begin{bmatrix} u \\ y \end{bmatrix} = \left(I + \begin{bmatrix} -\hat{Y} + MQ \\ \hat{X} + NQ \end{bmatrix} \begin{bmatrix} \Delta_{\hat{N}} & -\Delta_{\hat{M}} \end{bmatrix} \right)^{-1} \begin{bmatrix} -\hat{Y} + MQ \\ \hat{X} + NQ \end{bmatrix} \left(\hat{N}_{\bar{d}} + \Delta_{\hat{N}_{\bar{d}}}\right)\bar{d}. \tag{12.42}$$

According to Small Gain Theorem, the system stability is guaranteed, when

$$\left\| \begin{bmatrix} -\hat{Y} + MQ \\ \hat{X} + NQ \end{bmatrix} \right\|_\infty \left\| \begin{bmatrix} \Delta_{\hat{N}} & -\Delta_{\hat{M}} \end{bmatrix} \right\|_\infty < 1.$$

For engineering applications, the concept of stability margin defined by

$$b := \left\| \begin{bmatrix} -\hat{Y} + MQ \\ \hat{X} + NQ \end{bmatrix} \right\|_\infty^{-1}$$

is a safety performance index that is used in fault-tolerant control as well. It is obvious that, in order to assure the required stability margin, the feedback controller should be designed so that

$$\left\| \begin{bmatrix} -\hat{Y} + MQ \\ \hat{X} + NQ \end{bmatrix} \right\|_\infty^{-1} > \left\| \begin{bmatrix} \Delta_{\hat{N}} & -\Delta_{\hat{M}} \end{bmatrix} \right\|_\infty \Longleftrightarrow$$
$$\left\| \begin{bmatrix} \Delta_{\hat{N}} & -\Delta_{\hat{M}} \end{bmatrix} \right\|_\infty < b.$$

In this regard, determining Q by solving the subsequent optimisation problem is helpful,

$$\min_{Q \in \mathcal{RH}_\infty} \left\| \begin{bmatrix} -\hat{Y} + MQ \\ \hat{X} + NQ \end{bmatrix} \right\|_\infty.$$

It is worth mentioning that the so-called high-gain observer scheme is a popular technique of designing an UIO. At the core of a high-gain observer is an observer-gain that is of the form

$$L = \frac{1}{\varepsilon} L_0, \varepsilon \to 0. \tag{12.43}$$

Recall that

$$\begin{bmatrix} -\hat{Y} \\ \hat{X} \end{bmatrix} = \begin{bmatrix} F \\ C + DF \end{bmatrix} (zI - A - BF)^{-1} L + \begin{bmatrix} 0 \\ I \end{bmatrix}.$$

The use of such a high-gain observer might considerably reduce the system stability margin. Hence, a careful stability examination is essential, when a high-gain observer is integrated into a DOBC system.

At the end of this section, we would like, in the form of an example, to provide a more general form of DOBC according to the conceptual scheme shown Fig. 12.1 and delineate that such DOBC design can be elegant solved in the unified framework.

Example 12.3 Consider the system model

$$y = G_u u + G_d d, G_u = (A, B, C, D).$$

Let

$$u = u_0 + u_{in}, u_{in} = K_{in} (G_u u_0 - y).$$

Here, u_{in} is the disturbance compensator in the inner loop, and K_{in} serves as the inner controller (compensator) and is a design parameter. Note that, if G_u is (left) invertible, setting

$$K_{in} = Q_1 G_u^-, G_u^- G_u = I$$

leads to $u_{in} = Q_1 \left(u_0 - G_u^- y\right)$, the same configuration of the inner loop sketched in Fig. 12.1. Routine computations yield

$$y = G_u u_0 + (I + K_{in} G_u)^{-1} G_d d =: G_u u_0 + y_{in}. \tag{12.44}$$

Before dealing with the design of u_0, the control input of the outer loop, we would like to have a close look at the dynamics of the inner loop. The inner loop is similar to the well-established internal model control (IMC) scheme shown in Fig. 3.4, whose dynamics are subject to

$$\begin{bmatrix} u_{in} \\ y_{in} \end{bmatrix} = \begin{bmatrix} -(I + K_{in} G_u)^{-1} K_{in} \\ (I + K_{in} G_u)^{-1} \end{bmatrix} G_d d.$$

The inner compensator K_{in} is designed to assure the inner loop stability and to dampen the influence of d. By means of the Youla parameterisation for the given plant model G_u, K_{in} can be written as

$$\begin{aligned} K_{in} &= \left(\hat{Y} - M Q_{in}\right) \left(\hat{X} + N Q_{in}\right)^{-1} \\ &= \left(X + Q_{in} \hat{N}\right)^{-1} \left(Y - Q_{in} \hat{M}\right), Q_{in} \in \mathcal{RH}_\infty. \end{aligned}$$

By routine calculations using the Bezout identity, it turns out

$$\begin{bmatrix} -(I + K_{in} G_u)^{-1} K_{in} \\ (I + K_{in} G_u)^{-1} \end{bmatrix} = \begin{bmatrix} -\hat{Y} + M Q_{in} \\ \hat{X} + N Q_{in} \end{bmatrix} \hat{M} \Longrightarrow$$

$$\begin{bmatrix} u_{in} \\ y_{in} \end{bmatrix} = \begin{bmatrix} -\hat{Y} + MQ_{in} \\ \hat{X} + NQ_{in} \end{bmatrix} \hat{M} G_d d. \tag{12.45}$$

Now, consider the outer loop controller,

$$u_0 = K_0 y,$$

which is parametrised by

$$\begin{aligned} K_0 &= -\left(\hat{Y} - MQ_0\right)\left(\hat{X} + NQ_0\right)^{-1} \\ &= -\left(X + Q_0\hat{N}\right)^{-1}\left(Y - Q_0\hat{M}\right), \, Q_0 \in \mathcal{RH}_\infty. \end{aligned}$$

The loop dynamics are governed by

$$\begin{bmatrix} u_0 \\ y \end{bmatrix} = \begin{bmatrix} -\hat{Y} + MQ_0 \\ \hat{X} + NQ_0 \end{bmatrix} \hat{M} y_{in} = \begin{bmatrix} -\hat{Y} + MQ_0 \\ \hat{X} + NQ_0 \end{bmatrix} \hat{M}\left(\hat{X} + NQ_{in}\right)\hat{M} G_d d.$$

Since, according to the Bezout identity

$$\begin{gathered} \hat{M}\left(\hat{X} + NQ_{in}\right) = I - \hat{N}\hat{Y} + \hat{N}MQ_{in} \Longrightarrow \\ \begin{bmatrix} -\hat{Y} \\ \hat{X} \end{bmatrix} \hat{M}\left(\hat{X} + NQ_{in}\right) = \begin{bmatrix} -\hat{Y} \\ \hat{X} \end{bmatrix} - \begin{bmatrix} -\hat{Y} \\ \hat{X} \end{bmatrix} \hat{N}\left(\hat{Y} - \hat{M}Q_{in}\right) \\ = \begin{bmatrix} -\hat{Y} \\ \hat{X} \end{bmatrix} + \begin{bmatrix} M \\ N \end{bmatrix} X\left(\hat{Y} - \hat{M}Q_{in}\right) + \begin{bmatrix} I \\ 0 \end{bmatrix}\left(\hat{Y} - \hat{M}Q_{in}\right), \end{gathered}$$

it holds

$$\begin{gathered} \begin{bmatrix} u_0 \\ y \end{bmatrix} = \begin{bmatrix} -\hat{Y} + MQ \\ \hat{X} + NQ \end{bmatrix} \hat{M} G_d d + \begin{bmatrix} I \\ 0 \end{bmatrix}\left(\hat{Y} - \hat{M}Q_{in}\right)\hat{M} G_d d, \\ Q = X\left(\hat{Y} - \hat{M}Q_{in}\right) + Q_0\hat{M}\left(\hat{X} + NQ_{in}\right). \end{gathered}$$

Taking into account of u_{in} in (12.45), the overall loop dynamics are described by

$$\begin{bmatrix} u \\ y \end{bmatrix} = \begin{bmatrix} u_0 + u_{in} \\ y \end{bmatrix} = \begin{bmatrix} -\hat{Y} + MQ \\ \hat{X} + NQ \end{bmatrix} \hat{M} G_d d. \tag{12.46}$$

The overall loop dynamics (12.46) mean that the aforementioned DOBC design with the inner and outer loops is a special case of the unified framework solution. In other words, control performance that is achievable using the aforementioned DOBC design scheme can be accomplished by a controller

$$\begin{aligned} u &= Ky, \\ K &= -\left(\hat{Y} - MQ\right)\left(\hat{X} + NQ\right)^{-1} \\ &= -\left(X + Q\hat{N}\right)^{-1}\left(Y - Q\hat{M}\right), Q \in \mathcal{RH}_\infty. \end{aligned}$$

Moreover, setting Q as

$$Q = \arg \min_{Q \in \mathcal{RH}_\infty} \left\| \begin{bmatrix} -\hat{Y} + MQ \\ \hat{X} + NQ \end{bmatrix} \hat{M} G_d \right\|_\infty$$

results in the optimal disturbance rejection performance.

12.3 Fault-Tolerant Observers: A Reinforcement Learning Approach

In this and the next sections, we are going to leverage our knowledge of normalised SKR/SIR, co-inner/inner and their relations to the optimal observer and controller to approach fault-tolerant control issues. To be specific, the following system operation scenario is patterned as a benchmark. Suppose that the closed-loop dynamic is governed by

$$\begin{cases} x(k+1) = Ax(k) + Bu(k) \\ y(k) = Cx(k), \end{cases} \tag{12.47}$$

$$\begin{cases} \hat{x}(k+1) = A\hat{x}(k) + Bu(k) + L_0\left(y(k) - \hat{y}(k)\right) \\ u(k) = F_0\hat{x}(k) + v(k). \end{cases} \tag{12.48}$$

Hereby, v is the reference signal, (F_0, L_0) are the state feedback control gain matrix and observer gain matrix, respectively. They are designed according to Theorem 2.1. In other words, the corresponding SKR and SIR of (12.47)–(12.48) are normalised SKR and SIR, respectively, which can be interpreted as an optimal observer/residual generator and optimal LQ controller. In the sequel, system operations subject to (12.47)–(12.48) are referred as the nominal dynamic. For the sake of simplicity, only a full-order state observer and an observer-based state feedback controller are considered. During system operations, uncertainties around the system, ageing of system components, even anomaly caused by faulty operation conditions might lead to variations in the system matrices (A, B, C, D) . And this in turn leads to system performance degradation. The main objective of this section is to formulate the problem of recovering optimal observer performance as an optimisation problem, and solve it using reinforcement learning (RL) technique.

12.3.1 Basic Idea and Problem Formulation

Application of RL-based algorithms to adaptive optimal (feedback) control is state of the art in control-theoretic research domain. Informally speaking, a RL-based control is implemented in a feedback loop. Different from a classic control loop, where process measurements are fed back, in the RL-loop, the value of the objective function is computed aiming to evaluate the system response to the (last) update of the control law (improvement of policy). The further update of the control law is then calculated toward optimising the objective function. This is a cause-and-effect iteration process, running consecutively in time. It is apparent that such RL-based control strategy cannot be directly transferred to RL-based observer optimisation. The elemental problem is that an observer does not directly act on the process. Consequently, there exists no cause-and-effect between update of observer and process response.

The objective of applying an observer is to optimally estimate, for instance, state variables in the system. This implies that estimation errors caused by uncertainties in the system should be minimised. Recall the projection-based estimation scheme introduced in Sect. 12.1.3, in which the projection system

$$\begin{bmatrix} \Delta u \\ \Delta y \end{bmatrix} = K_{\tilde{G}_0} K_{G_0} \begin{bmatrix} u \\ y \end{bmatrix}$$

delivers an optimal estimate (in the LS-sense) for the uncertainty in the process data (u, y). In particular, the co-state vector $\lambda(k)$ models the dynamics of the state estimation error. The configuration of the projection system, the functionalities of the two subsystems, K_{G_0} and $K_{\tilde{G}_0}$, and the control-theoretic characterisation of the system variables inspire the conception of the following RL-based fault-tolerant observer (FT-observer).

As showcased in Fig. 12.2, the essential RL working principle, the cause-and-effect iteration, is realised by the projection system serving both as "plant" and

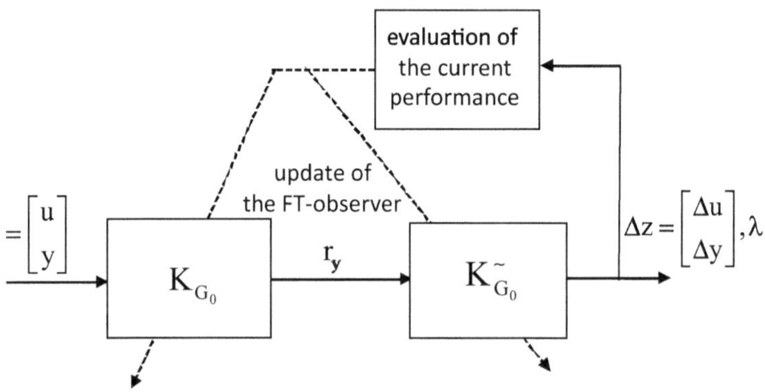

Fig. 12.2 A schematic description of RL-based FT-observer

"controller" (observer gain) that are to be updated, supervised by the (observer) performance evaluation and policy improvement algorithm. To be specific, given a batch of data, $\{u(k), y(k), k \in [k_0, k_0 + N]\}$, with a large number N,

- in the first step, the set of the residual signal $\left\{r_y(k), k \in [k_0, k_0 + N]\right\}$ is generated using the SKR K_{G_0} of the projection system,

$$K_{G_0}^{\sim} K_{G_0} : \begin{cases} \hat{x}(k+1) = A_K \hat{x}(k) + B_K \begin{bmatrix} u(k) \\ y(k) \end{bmatrix} \\ r_y(k) = C_K \hat{x}(k) + D_K \begin{bmatrix} u(k) \\ y(k) \end{bmatrix} \\ \lambda(k) = A_K^T \lambda(k+1) + C_K^T r_y(k) \\ \begin{bmatrix} \Delta u(k) \\ \Delta y(k) \end{bmatrix} = B_K^T \lambda(k+1) + D_K^T r_y(k), \end{cases} \tag{12.49}$$

$$A_K = A - L_0 C,\ B_K = \begin{bmatrix} B & L_0 \end{bmatrix},\ C_K = -W_0 C,\ D_K = \begin{bmatrix} 0 & W_0 \end{bmatrix},$$

- then, $K_{G_0}^{\sim}$ is computed, backward in time, that delivers

$$\{\Delta u(k), \Delta y(k), \lambda(k), k \in [k_0, k_0 + N]\}$$

  as "measurement data" set.
- For improving the observer performance, a target model

$$\begin{cases} \lambda(k) = A_{K,\Delta}^T \lambda(k+1) \\ \Delta z(k) = \begin{bmatrix} \Delta u(k) \\ \Delta y(k) \end{bmatrix} = B_{K,\Delta}^T \lambda(k+1), \end{cases} \tag{12.50}$$

  is adopted for computing the update of the observer gain L_0. Here, $\left(A_{K,\Delta}^T, B_{K,\Delta}^T\right)$ represent the corresponding system matrices during faulty operations, and the target model will be clarified in more detail in the subsequent section.
- The corresponding objective function, J_N, is defined as

$$J_N = \|\Delta z(k)\|_{2,(-\infty,N]}^2 = \sum_{k=-\infty}^{N} \lambda^T(k) B_K B_K^T \lambda(k). \tag{12.51}$$

Aiming at exploring the learning algorithm, needed control-theoretic preliminaries are firstly summarised in the next subsection. It is followed by delineating the learning algorithm.

12.3.2 Theoretic Preliminary

To begin with, the objective of applying RL-methods to learn FT-observer is specified. It is dedicated to recovering the system estimation performance which degrades due to variations in the system matrices (C, A) caused by faults. We assume that there exists no changes in the system observability structure. Specifically, the observability indices, $\{\mu_j, j = 1, \ldots, m\}$, are invariant. Denominating system matrices during faulty operations by (C_Δ, A_Δ), it will be firstly delineated that

$$A_\Delta = A + L_\Delta C_\Delta, C_\Delta = \Delta_C C \tag{12.52}$$

for some (unknown) matrices (C_Δ, L_Δ). Relations (12.52) are fundamental to realise RL-based FT observers.

Consider (C, A) in the , denoted by (C_o, A_o),

$$A_o = \begin{bmatrix} A_{o,11} & \cdots & A_{o,1m} \\ \vdots & \ddots & \vdots \\ A_{o,m1} & \cdots & A_{o,mm} \end{bmatrix}, A_{o,ij} = \begin{bmatrix} 0 \cdots 0 & \alpha_{1,j} \\ \vdots \cdots \vdots & \vdots \\ 0 \cdots 0 & \alpha_{\mu_i-1,j} \\ 0 \cdots 0 & \alpha_{\mu_i,j} \end{bmatrix} \in \mathbb{R}^{\mu_i \times \mu_j}, i \neq j,$$

$$A_{o,jj} = \begin{bmatrix} 0 & 0 & \cdots & 0 & \beta_{1,j} \\ 1 & 0 & \cdots & 0 & \beta_{2,j} \\ \vdots & \ddots & \ddots & \ddots & \vdots \\ 0 & \cdots & 1 & 0 & \beta_{\mu_j-1,j} \\ 0 & 0 & \cdots & 1 & \beta_{\mu_j,j} \end{bmatrix} \in \mathbb{R}^{\mu_j \times \mu_j}, j = 1, \ldots, m,$$

$$C_o = \begin{bmatrix} C_{o,1} & \cdots & C_{o,m} \end{bmatrix}, C_{o,i} = \begin{bmatrix} 0_{m \times (\mu_j - 1)} & c_i \end{bmatrix} \in \mathbb{R}^{m \times \mu_i}, i = 1, \ldots, m,$$

$$\Pi_C := \begin{bmatrix} c_1 & \cdots & c_m \end{bmatrix} = \begin{bmatrix} 1 & 0 & \cdots & \cdots & 0 \\ \gamma_{1,2} & 1 & 0 & \cdots & 0 \\ \vdots & \gamma_{2,3} & \ddots & \ddots & \vdots \\ \vdots & \cdots & \ddots & 1 & 0 \\ \gamma_{1,m} & \gamma_{2,m} & \cdots & \gamma_{m-1,m} & 1 \end{bmatrix} \in \mathbb{R}^{m \times m}, \tag{12.53}$$

where $\alpha_{l,j}, l = 1, \ldots, \mu_i, i, j = 1, \ldots, m, i \neq j,$ $\beta_{l,j}, l = 1, \ldots, \mu_j, j = 1, \ldots, m,$ $\gamma_{i,j}, j = i + 1, \ldots, m, i = 1, \ldots, m,$ are arbitrary real numbers. It is supposed that faults cause variations in $\alpha_{l,j}$ and $\beta_{l,j}$. Observe that

$$\Pi_C^{-1} C_o = \begin{bmatrix} \Pi_C^{-1} C_{o,1} & \cdots & \Pi_C^{-1} C_{o,m} \end{bmatrix}, \Pi_C^{-1} c_i = \begin{bmatrix} 0 \cdots 0 \, 1 \, 0 \cdots 0 \end{bmatrix}^T, \tag{12.54}$$

$i = 1, \ldots, m$. Thus, by routine computations, A_o can be decomposed into

$$A_o = A_{o,0} + L_\Delta C_o, A_{o,0} = diag\left(A_{o,1}, \cdots, A_{o,m}\right), \tag{12.55}$$

$$A_{o,j} = \begin{bmatrix} 0 & 0 & 0 & \cdots & 0 \\ 1 & 0 & 0 & \cdots & 0 \\ \vdots & \ddots & \ddots & \ddots & \vdots \\ 0 & \cdots & 1 & 0 & 0 \\ 0 & 0 & \cdots & 1 & 0 \end{bmatrix} \in \mathbb{R}^{\mu_j \times \mu_j}, j = 1, \ldots, m, L_\Delta = L_{\Delta_f} \Pi_C^{-1},$$

$$L_{\Delta_f} = \begin{bmatrix} \Delta_{f,1} \\ \vdots \\ \Delta_{f,m} \end{bmatrix}, \Delta_{f,i} = \begin{bmatrix} \alpha_{1,1} & \cdots & \alpha_{1,i-1} & \beta_{1,i} & \alpha_{1,i+1} & \cdots & \alpha_{1,m} \\ \vdots & \ldots & \vdots & \vdots & \vdots & \vdots & \vdots \\ \alpha_{\mu_i 1} & \cdots & \alpha_{\mu_i,i-1} & \beta_{\mu_i,i} & \alpha_{\mu_i,i+1} & \cdots & \alpha_{\mu_i,m} \end{bmatrix}.$$

That means, variations in $A = A_o$ are fully modelled by L_{Δ_f}. As a result, updating the observer gain matrix L_0 is sufficient to compensate the variations.

Next, consider an arbitrary matrix A and suppose that faults cause changes in A. Let T_{Δ_f} be a state transformation leading to

$$A_o = T_{\Delta_f} A_\Delta T_{\Delta_f}^{-1}, C_o = C T_{\Delta_f}^{-1}, B_o = T_{\Delta_f} B. \tag{12.56}$$

As well-known, system input-output dynamics are invariant to a state transformation. In our case, this is explained as follows. According to (12.67), after the state transformation the corresponding (optimal) observer gain matrix, denoted by L_{0,Δ_f}, is given by

$$L_{0,\Delta_f} = \left(A_o P_o C_o^T + B_o D^T\right)\left(C_o P_o C_o^T + I + DD^T\right)^{-1},$$

where P_o is the solution of the Riccati equation

$$P_o = A_o P_o A_o^T + B_o B_o^T - L_{0,\Delta_f}\left(I + DD^T + C_o P_o C_o^T\right) L_{0,\Delta_f}^T.$$

On account of (12.56), we obtain

$$P_o = T_{\Delta_f}^{-1} P T_{\Delta_f}^{-T} \Longrightarrow L_{0,\Delta_f} = T_{\Delta_f} L_0.$$

This implies, under the state transformation

$$\bar{x}(k) = T_{\Delta_f} \hat{x}(k), \bar{\lambda}(k) = T_{\Delta_f}^{-1} \lambda(k),$$

the projection-based system,

$$K_{\tilde{G}_0} K_{G_0} : \begin{cases} \bar{x}(k+1) = \bar{A}_K \bar{x}(k) + \bar{B}_K \begin{bmatrix} u(k) \\ y(k) \end{bmatrix} \\ r_y(k) = \bar{C}_K \bar{x}(k) + D_K \begin{bmatrix} u(k) \\ y(k) \end{bmatrix} \\ \bar{\lambda}(k) = \bar{A}_K^T \bar{\lambda}(k+1) + \bar{C}_K^T r_y(k) \\ \begin{bmatrix} \Delta u(k) \\ \Delta y(k) \end{bmatrix} = \bar{B}_K^T \bar{\lambda}(k+1) + D_K^T r_y(k), \end{cases} \tag{12.57}$$

with the corresponding system matrices

$$\begin{aligned} \bar{A}_K &= T_{\Delta_f} A_\Delta T_{\Delta_f}^{-1} - T_{\Delta_f} L_0 C T_{\Delta_f}^{-1} = A_o - L_{0,\Delta_f} C_o, \\ \bar{B}_K &= \begin{bmatrix} T_{\Delta_f} B - T_{\Delta_f} L_0 D & T_{\Delta_f} L_0 \end{bmatrix} = \begin{bmatrix} B_o - L_{0,\Delta_f} D & L_{0,\Delta_f} \end{bmatrix}, \\ \bar{C}_K &= -W_0 C T_{\Delta_f}^{-1} = -W_0 C_o \end{aligned}$$

possess the identical input-output dynamics like the projection system (12.49) with the state vector $\left(\hat{x}(k), \lambda(k)\right)$. It can be therefore concluded that tuning the observer gain matrix L_0 is sufficient to cope with the changes in A.

Finally, variations in system matrix C is specified. It follows from the aforementioned discussions that, by state transformations T_o, T_{Δ_f}, the pairs (C, A) and (C_Δ, A_Δ) can be respectively written as

$$(C, A) = \left(C_o T_o^{-1}, T_o A_o T_o^{-1}\right), (C_\Delta, A_\Delta) = \left(C_o T_{\Delta_f}^{-1}, T_{\Delta_f} A_o T_{\Delta_f}^{-1}\right),$$

and the pair (A_o, C_o) is subject to

$$A_o = A_{o,0} + L_\Delta C_o, C_o = \Pi_C \bar{C}_0, \bar{C}_0 = \begin{bmatrix} \bar{C}_{o,1} & \cdots & \bar{C}_{o,m} \end{bmatrix}.$$

Hereby, the (i, μ_i)-th entry of matrix $\bar{C}_{o,i}$ equals to one and all others are zero. Hence, it is apparent that $\bar{C}_0$ is independent of system parameters and their variations. Note that L_Δ and Π_C are of different values during fault-free and faulty operations. In order to distinguish between these two cases for Π_C, define

$$\Pi_C = \begin{cases} \Pi_{C,o}, \text{ fault-free} \\ \Pi_{C,f}, \text{ faulty} \end{cases}, \Delta_C = \Pi_{C,f} \Pi_{C,o}^{-1} \Longrightarrow \tag{12.58}$$

$$C_o = \begin{cases} C_{o,o}, \text{ fault-free} \\ \Delta_C C_{o,o}, \text{ faulty} \end{cases}, C_\Delta = \Delta_C C_{o,o} T_{\Delta_f}^{-1} = \Delta_C C.$$

Consequently, the variations in C caused by faults are characterised by $\left(\Delta_C, T_{\Delta_f}^{-1}\right)$.

On account of (12.52), the target model (12.50) is specified as follows,

$$\begin{cases} \lambda(k) = (A_\Delta - LC_\Delta)^T \lambda(k+1) \\ \quad\quad = (A + L_\Delta \Delta_C C)^T \lambda(k+1) - C^T \Delta_C^T L^T \lambda(k+1) \\ \Delta z(k) = \begin{bmatrix} B^T \\ L^T \end{bmatrix} \lambda(k+1), \end{cases} \tag{12.59}$$

where L represents the observer gain matrix to be "learnt". Defining

$$u_\lambda(k+1) := -L^T \lambda(k+1)$$

allows us to bring the target model (12.50) and the objective function (12.51) into the new form,

$$\begin{cases} \lambda(k) = A_\Delta^T \lambda(k+1) + C_\Delta^T u_\lambda(k+1) \\ \Delta z(k) = \begin{bmatrix} B^T \\ 0 \end{bmatrix} \lambda(k+1) + \begin{bmatrix} 0 \\ -I \end{bmatrix} u_\lambda(k+1), \end{cases} \tag{12.60}$$

$$\begin{aligned} J_N &= \sum_{k=-\infty}^{N} \left(\lambda^T(k) B B^T \lambda(k) + u_\lambda^T(k) u_\lambda(k) \right) \\ &= \sum_{k=-\infty}^{N} \left[\lambda^T(k) \; u_\lambda^T(k) \right] R \begin{bmatrix} \lambda(k) \\ u_\lambda(k) \end{bmatrix}, \; R = \begin{bmatrix} BB^T & 0 \\ 0 & I \end{bmatrix}. \end{aligned}$$

On this basis, the optimisation problem towards RL-based FT-observers is formulated as

$$\min_{\substack{u_\lambda(k) \\ k \in (-\infty, N]}} J_N = \min_{\substack{u_\lambda(k) \\ k \in (-\infty, N]}} \sum_{k=-\infty}^{N} \left[\lambda^T(k) \; u_\lambda^T(k) \right] R \begin{bmatrix} \lambda(k) \\ u_\lambda(k) \end{bmatrix} \tag{12.61}$$
$$\text{s.t. (12.60).}$$

Before solving the above optimisation problem, the meaning of the target model (12.60) and the objective function J_N are illustrated. During faulty operations, the system dynamics are governed by

$$\begin{cases} x(k+1) = A_\Delta x(k) + Bu(k) \\ y(k) = C_\Delta x(k), \end{cases}$$

and the corresponding SKR dynamic is subject to

$$K_{G_0} : \begin{cases} \hat{x}(k+1) = A_K \hat{x}(k) + B_K \begin{bmatrix} u(k) \\ y(k) \end{bmatrix} \\ r_y(k) = C_K \hat{x}(k) + D_K \begin{bmatrix} u(k) \\ y(k) \end{bmatrix} \end{cases}$$

$$
= \begin{cases} \tilde{x}(k+1) = A_K \tilde{x}(k) + B_K \begin{bmatrix} \Delta u(k) \\ \Delta y(k) \end{bmatrix} \\ r_y(k) = C_K \tilde{x}(k) + D_K \begin{bmatrix} \Delta u(k) \\ \Delta y(k) \end{bmatrix}. \end{cases}
$$

Observe that

$$
\begin{aligned}
&\tilde{x}(k+1) = A_K \tilde{x}(k) + B_K \begin{bmatrix} \Delta u(k) \\ \Delta y(k) \end{bmatrix} \Longleftrightarrow \\
&\tilde{x}(k+1) = A\tilde{x}(k) - L_0 C \tilde{x}(k) + B\Delta u(k) + L_0 \Delta y(k), \\
&\begin{bmatrix} \Delta u(k) \\ \Delta y(k) \end{bmatrix} = B_K^T \lambda(k+1) + D_K^T r_y(k) = \begin{bmatrix} B^T \\ L_0^T \end{bmatrix} \lambda(k+1) + \begin{bmatrix} 0 \\ W_0^T \end{bmatrix} r_y(k).
\end{aligned}
$$

Now, varying the observer gain L so that

$$
L = L_\Delta + \tilde{L}\Delta_C^{-1} \Longrightarrow LC_\Delta = L_\Delta C_\Delta + \tilde{L}C
$$

yields

$$
\begin{aligned}
&\hat{x}(k+1) = A_\Delta \hat{x}(k) + Bu(k) + L\left(y(k) - C_\Delta \hat{x}(k)\right) \Longrightarrow \\
&\begin{cases} \tilde{x}(k+1) = (A_\Delta - LC_\Delta)\,\tilde{x}(k) = \left(A - \tilde{L}C\right)\tilde{x}(k) \\ r_y(k) = -W_0\left(y(k) - C_\Delta \hat{x}(k)\right), \end{cases} \\
&\lambda(k) = \left(A - \tilde{L}C\right)^T \lambda(k+1) - C_\Delta^T W_0^T r_y(k) \\
&= A_\Delta^T \lambda(k+1) + C_\Delta^T \left(u_\lambda(k+1) + W_0^T r_y(k)\right).
\end{aligned} \tag{12.62}
$$

It is apparent that for $\left(A - \tilde{L}C\right)$ being Schur, $r_y(k) \to 0$, and correspondingly,

$$
\lambda(k) \to 0 \Longrightarrow \begin{bmatrix} \Delta u(k) \\ \Delta y(k) \end{bmatrix} \to 0.
$$

Hence, the target model (12.60) depicts the dynamics of the system (12.62). Accordingly, the optimisation problem (12.61) implies finding the optimal $\tilde{L}$ with respect to the objective function (12.61). In fact,

$$
\tilde{L} = L_0 \Longrightarrow L = L_\Delta + L_0 \Delta_C^{-1} \tag{12.63}
$$

is the optimal solution of the optimisation problem (12.61). In this case, the optimal estimation performance is recovered.

Dynamic programming (DP) is a backward recursive computation algorithm that widely applied to solve optimisation problems, in particular in RL-based methods. For our purpose, the dual form of the standard DP algorithm is proved.

Theorem 12.3 *Given the objective function (12.51), it holds*

$$\min_{\substack{u_\lambda(k)\\ k\in(-\infty,N]}} J_N = \min_{u_\lambda(N)} \left(J^*_{N-1} + \left[\lambda^T(N)\ u^T_\lambda(N) \right] R \begin{bmatrix} \lambda(N) \\ u_\lambda(N) \end{bmatrix} \right), \tag{12.64}$$

$$J^*_{N-1} = \min_{u_\lambda(k), k\in(-\infty,N-1]} J_{N-1}.$$

Proof Let $i = N - k$. It turns out

$$J_N = \bar{J}_0 =: \sum_{i=0}^{\infty} \left[\lambda^T(N-i)\ u^T_\lambda(N-i) \right] R \begin{bmatrix} \lambda(N-i) \\ u_\lambda(N-i) \end{bmatrix}.$$

According to Bellman's optimality principle,

$$\bar{J}^*_0 = \min_{\substack{u_\lambda(i)\\ i\in[0,\infty)}} \bar{J}_0 = \min_{u_\lambda(N-0)} \left(\left[\lambda^T(N)\ u^T_\lambda(N) \right] R \begin{bmatrix} \lambda(N) \\ u_\lambda(N) \end{bmatrix} + \bar{J}^*_1 \right),$$

$$\begin{aligned} \bar{J}^*_1 &= \min_{u_\lambda(i), i\in[1,\infty)} \sum_{i=1}^{\infty} \left[\lambda^T(N-i)\ u^T_\lambda(N-i) \right] R \begin{bmatrix} \lambda(N-i) \\ u_\lambda(N-i) \end{bmatrix} \\ &= \min_{u_\lambda(k), k\in(-\infty,N-1]} J_{N-1} = J^*_{N-1}, \end{aligned}$$

which proves (12.64).

In order to evaluate the observer performance, the value function model of the objective function (12.51) is to be established.

Theorem 12.4 *Given the reference model (12.50) and the objective function (12.51), then it holds*

$$J^*_N = \min_{u_\lambda(k), k\in(-\infty,N]} J_N = \lambda^T(N) P \lambda(N), \tag{12.65}$$

where $P > 0$ is the solution of Riccati equation (2.40) given in Theorem 2.1.

Proof Recall that the normalised SKR K_{G_0} is co-inner. Moreover, for A^T_K being a Schur matrix, it turns out for $\lambda(N) = 0$,

$$\lambda(-\infty) = 0.$$

As a result, it follows from Theorem 5.4 that

$$\|\Delta z(k)\|^2_{2,(-\infty,N]} = \lambda^T(N) P \lambda(N).$$

The theorem is proved.

On the basis of Theorems 12.3–12.4, the optimisation problem (12.64) can be solved by means of the subsequent computation steps,

$$\begin{aligned}
& \left[\lambda^T(N)\ u_\lambda^T(N)\right] R \begin{bmatrix} \lambda(N) \\ u_\lambda(N) \end{bmatrix} + \lambda^T(N-1) P \lambda(N-1) \\
&= \left[\lambda^T(N)\ u_\lambda^T(N)\right] R \begin{bmatrix} \lambda(N) \\ u_\lambda(N) \end{bmatrix} \\
&+ \left(A_\Delta^T \lambda(N) + C_\Delta^T u_\lambda(N)\right)^T P \left(A_\Delta^T \lambda(N) + C_\Delta^T u_\lambda(N)\right) \\
&= \left[\lambda^T(N)\ u_\lambda^T(N)\right] S \begin{bmatrix} \lambda(N) \\ u_\lambda(N) \end{bmatrix}, \qquad (12.66) \\
S &= \begin{bmatrix} A_\Delta P A_\Delta^T + B B^T & A_\Delta P C_\Delta^T \\ C_\Delta P A_\Delta^T & C_\Delta P C_\Delta^T + I \end{bmatrix}.
\end{aligned}$$

By routine calculations and the Riccati equation (2.40), we obtain

$$u_\lambda^*(N) = -\left(C_\Delta P C_\Delta^T + I\right)^{-1} C_\Delta P A_\Delta^T \lambda(N) \Longrightarrow \tag{12.67}$$

$$L^* = \left(C_\Delta P C_\Delta^T + I\right)^{-1} C_\Delta P A_\Delta^T, \tag{12.68}$$

which leads to the optimal performance value,

$$J_N^* = \lambda^T(N) P \lambda(N).$$

It is noteworthy that the optimal solution (12.67) is nothing other than the one given in Theorem 2.1. On the other hand, (12.66)–(12.67) are essential for our subsequent work on the realisation of a RL-based observer.

The last action of constructing the FT-observer is updating the system matrices of the observer as

$$\hat{x}(k+1) = A_\Delta \hat{x}(k) + Bu(k) + L\left(y(k) - C_\Delta \hat{x}(k)\right).$$

To this end, (L_Δ, Δ_C) are to be determined. Considering that, given S matrix, it holds

$$\begin{aligned}
C_\Delta P C_\Delta^T + I &= \Delta_C C P C^T \Delta_C^T + I = S_{22} \Longrightarrow \Delta_C C P C^T \Delta_C^T = S_{22} - I, \\
P &= A_\Delta P A_\Delta^T + B B^T - A_\Delta P C_\Delta^T \left(C_\Delta P C_\Delta^T + I\right)^{-1} C_\Delta P A_\Delta^T \\
&= S_{11} - S_{12} S_{22}^{-1} S_{21},
\end{aligned}$$

where S_{ij} is the (i, j)-block of matrix $S, i, j = 1, 2$. Sequentially, Δ_C can be determined by solving the matrix equation,

$$S_{22} - I = \Delta_C C P C^T \Delta_C^T,$$

for given P and C. Note that, according to (12.53) and (12.58), Δ_C is a lower triangular matrix with the diagonal entries equal to one. Furthermore, it follows from (12.63) that

$$L_\Delta = L - L_0 \Delta_C^{-1}$$

with $L = L^*$, L_0 being the setting of the nominal (optimal) observer gain matrix. Hence, (L_Δ, Δ_C) are finally determined.

12.3.3 A Q-learning Algorithm of RL-Based Fault-Tolerant Observers

Q-learning is a classic RL-algorithm. The endeavours in this subsection is to leverage Q-learning for a learning algorithm for a RL-based FT-observer. Note that differing from RL-based feedback control systems, updating the system matrices in $K_{\tilde{G}_0} K_{G_0}$ is necessary in the RL-based FT-observer (12.49).

At first, we introduce a Q-function,

$$Q\left(\lambda(N), u_\lambda(N)\right) = J_N = \sum_{k=-\infty}^{N} \gamma^{N-k} \left[\lambda^T(k) \; u_\lambda^T(k) \right] R \begin{bmatrix} \lambda(k) \\ u_\lambda(k) \end{bmatrix}, \tag{12.69}$$

the centerpiece of Q-learning scheme. According to (12.66), (12.69) is further written into

$$\begin{gathered} Q\left(\lambda(N), u_\lambda(N)\right) = \left[\lambda^T(N) \; u_\lambda^T(N) \right] S \begin{bmatrix} \lambda(N) \\ u_\lambda(N) \end{bmatrix}, \\ S = \begin{bmatrix} S_{11} & S_{12} \\ S_{12}^T & S_{22} \end{bmatrix}, \\ S_{11} = \gamma A_\Delta P A_\Delta^T + B B^T, S_{12} = \gamma A_\Delta P C_\Delta^T, S_{22} = \gamma C_\Delta P C_\Delta^T + I. \end{gathered}$$

Hereby, $0 < \gamma < 1$ is a discount weighting factor that reduces the weight of costs incurred in the past. Let $Q^*\left(\lambda(N), u_\lambda(N)\right)$ be the optimal Q-function,

$$Q^*\left(\lambda(N), u_\lambda(N)\right) = \min_{u_\lambda(k), k\in(-\infty,N]} Q\left(\lambda(N), u_\lambda(N)\right) = \min_{u_\lambda(k), k\in(-\infty,N]} J_N.$$

Attributed to Theorem 12.3, it turns out

$$\begin{gathered} Q^*\left(\lambda(N), u_\lambda(N)\right) \\ = \min_{u_\lambda(N)} \left(\left[\lambda^T(N) \; u_\lambda^T(N) \right] R \begin{bmatrix} \lambda(N) \\ u_\lambda(N) \end{bmatrix} + \gamma Q^*\left(\lambda(N-1), u_\lambda(N-1)\right) \right) \\ \Longrightarrow \left[\lambda^T(N) \; u_\lambda^{*T}(N) \right] S \begin{bmatrix} \lambda(N) \\ u_\lambda^*(N) \end{bmatrix} = \\ \left[\lambda^T(N) \; u_\lambda^{*T}(N) \right] R \begin{bmatrix} \lambda(N) \\ u_\lambda^*(N) \end{bmatrix} + \gamma \left[\lambda^T(N-1) \; u_\lambda^{*T}(N-1) \right] S \begin{bmatrix} \lambda(N-1) \\ u_\lambda^*(N-1) \end{bmatrix}. \end{gathered} \tag{12.70}$$

We are now in a position to introduce a Q-learning algorithm for RL-based FT-observers. The Q-learning algorithm consists of two major steps, (i) identification of Q-function, (ii) update of $u_\lambda(N)$ and the observer system. For our purpose, the well-established on-policy value iteration strategy is applied. Suppose that the j-th iteration is performed, corresponding matrix S and the update policy are

$$S_j = \begin{bmatrix} S_{11,j} & S_{12,j} \\ S_{21,j} & S_{22,j} \end{bmatrix},$$

and $u_{\lambda,j}(k) = -S_{22,j}^{-1} S_{21,j} \lambda(k)$, respectively. At the $(j+1)$-th iteration, equation

$$\begin{aligned} &\left[\lambda^T(N)\; u_{\lambda,j}^T(N) + \varepsilon^T(N)\right] S_{j+1} \begin{bmatrix} \lambda(N) \\ u_{\lambda,j}(N) + \varepsilon(N) \end{bmatrix} \\ &= \left[\lambda^T(N)\; u_{\lambda,j}^T(N)\right] R \begin{bmatrix} \lambda(N) \\ u_{\lambda,j}(N) \end{bmatrix} \\ &+\gamma \left[\lambda^T(N-1)\; u_{\lambda,j}^T(N-1)\right] S_j \begin{bmatrix} \lambda(N-1) \\ u_{\lambda,j}(N-1) \end{bmatrix} \end{aligned}$$

serves for the identification of S_{j+1} with $\varepsilon(N)$ being probing noise added to assure sufficient excitation (satisfying the persistence of excitation condition). Let

$$\begin{aligned} \varphi_{j+1} &= vec\left(S_{j+1}\right), \\ \phi_j^T(N) &= \left[\lambda^T(N)\; u_{\lambda,j}^T(N) + \varepsilon^T(N)\right] \otimes \left[\lambda^T(N)\; u_{\lambda,j}^T(N) + \varepsilon^T(N)\right], \end{aligned}$$

and write the above equation as

$$\begin{aligned} \phi_j^T(N)\,\varphi_{j+1} &= \theta_j(N), \\ \theta_j(N) &= \left[\lambda^T(N)\; u_{\lambda,j}^T(N)\right] R \begin{bmatrix} \lambda(N) \\ u_{\lambda,j}(N) \end{bmatrix} \\ &+\gamma \left[\lambda^T(N-1)\; u_{\lambda,j}^T(N-1)\right] S_j \begin{bmatrix} \lambda(N-1) \\ u_{\lambda,j}(N-1) \end{bmatrix}. \end{aligned}$$

Identification

- Form the data matrix:

$$\Phi_j^T(k) = \begin{bmatrix} \phi_j^T(k) \\ \vdots \\ \phi_j^T(k-M) \end{bmatrix}, \Theta_j(k) = \begin{bmatrix} \theta_j(k) \\ \vdots \\ \theta_j(k-M) \end{bmatrix}, k \in (-\infty, N],$$

  Hereby, M is sufficient large so that $\Phi_j^T(k)$ is of full column-rank,

- Form S_{j+1} using the identified φ_{j+1}, which is the LS solution of the following equation

$$\Phi_j^T(k)\,\varphi_{j+1} = \Theta_j(k).$$

Update

- Update $u_\lambda(k+1), k \in (-\infty, N]$,

$$\begin{aligned} u_{\lambda,j+1}(k+1) &= -S_{22,j+1}^{-1} S_{21,j+1} \lambda(k+1) \\ &\Longrightarrow L_{j+1} = S_{12,j+1} S_{22,j+1}^{-1}, \end{aligned} \tag{12.71}$$

- Update (L_Δ, Δ_C) :
  - calculate P_{j+1},

$$P_{j+1} = S_{11,j+1} - S_{12,j+1} S_{22,j+1}^{-1} S_{21,j+1},$$

  - solve

$$S_{22} - I = \Delta_{C,j+1} C P_{j+1} C^T \Delta_{C,j+1}^T,$$

  for $\Delta_{C,j+1}$
  - calculate

$$L_{\Delta,j+1} = L_{j+1} - L_0 \Delta_{C,j+1}^{-1};$$

- Update the fault-tolerant observer

$$\begin{cases} \hat{x}(k+1) = A_{K,j+1}\hat{x}(k) + B_{K,j+1}\begin{bmatrix} u(k) \\ y(k) \end{bmatrix} \\ r_y(k) = C_{K,j+1}\hat{x}(k) + D_{K,j+1}\begin{bmatrix} u(k) \\ y(k) \end{bmatrix} \\ \lambda(k) = A_{K,j+1}^T \lambda(k+1) + C_{K,j+1}^T r_y(k) \\ \Delta z(k) = B_{K,j+1}^T \lambda(k+1) + D_{K,j+1}^T r_y(k), \end{cases}$$

$$A_{K,j+1} = A + L_{\Delta,j+1}\Delta_{C,j+1} - L_{j+1}C,\ B_{K,j+1} = \begin{bmatrix} B & L_{j+1} \end{bmatrix},$$
$$C_{K,j+1} = -W_{0,j+1}\Delta_{C,j+1}C,\ D_K = \begin{bmatrix} 0 & W_{0,j+1} \end{bmatrix},\ W_{0,j+1} = S_{22,j+1}^{-1/2}.$$

At the end of this section, we would like to remark that as the dual form of the well-established RL-based LQ-optimal control, the convergence of the above algorithm can be proved along with the line of the proof of RL-based LQ-optimal control algorithms. To this end, a number of references are introduced at the end of this chapter.

12.4 A Fault-Tolerant Control Scheme

12.4.1 An Observer-Based Fault-Tolerant LQ Control: Problem Formulation

Consider the control system (12.47)–(12.48). Suppose that faults cause variations in the system matrices (A, B), and consequently result in control performance degradation. The objective of applying RL-methods to update the observer-based controller is to recover the system performance. It is assumed that the faults will not cause changes in the system controllability structure. Specifically, the controllability indices, $\{\mu_j, j = 1, \ldots, p\}$, are invariant. Observe that this is a dual problem addressed in the previous section about fault-tolerant observers. As such, it is clear that the faulty system matrices can be modelled by

$$A_\Delta = A + B_\Delta F_\Delta,\ B_\Delta = B\Delta_B,$$
$$\Delta_B = \begin{bmatrix} 1 & \gamma_{12} & \cdots & \cdots & \gamma_{1p} \\ 0 & 1 & \gamma_{23} & \cdots & \gamma_{2p} \\ \vdots & 0 & \ddots & \ddots & \vdots \\ \vdots & \cdots & \ddots & 1 & \gamma_{p-1p} \\ 0 & & \cdots & 0 & 1 \end{bmatrix} \in \mathbb{R}^{p\times p},$$

where $F_\Delta \in \mathbb{R}^{p\times n}$ is an unknown matrix, $\gamma_{ij}, i = 1, \ldots, p-1, i < j \le p$, are some unknown constants. Moreover, it is assumed that the faulty system $(C, A + B_\Delta F_\Delta)$ remains observable or detectable.

Thanks to the duality with the theoretical results introduced in Subsection 12.3.2, we will directly formulate our problem to be addressed in the subsequent study without highlighting the theoretic handling behind that. The reader is referred to Sect. 12.3.2 for the needed control-theoretic preliminary.

It is apparent that corresponding to the faulty plant dynamic

$$x(k+1) = A_\Delta x(k) + B_\Delta u(k) = (A + B_\Delta F_\Delta)\, x(k) + B\Delta_B u(k),$$

a state feedback controller,

$$u(k) = \left(-F_\Delta + \Delta_B^{-1} F_0\right) x(k), \tag{12.72}$$

would fully recover the system nominal (optimal) dynamic,

$$x(k+1) = Ax(k) + Bu(k) = (A + BF_0)\, x(k) \tag{12.73}$$

with F_0 as the nominal (optimal) feedback gain. However, both F_Δ and Δ_B are unknown, and $x(k)$ is not accessible. Nevertheless, it inspires us to impose the controller to be

$$u(k) = -F\hat{x}(k) + VF_0\hat{x}(k) = \begin{bmatrix} F & V \end{bmatrix} \begin{bmatrix} -I \\ F_0 \end{bmatrix} \hat{x}(k), \tag{12.74}$$

and use the existing technique to determine the controller parameter matrices (F, V) towards (12.72). To this end, the problem to be explored is formulated as follows.

Consider that the following closed-loop dynamic under faulty operation conditions,

$$\begin{cases} x(k+1) = (A + B_\Delta F_\Delta) x(k) + B_\Delta \left(-F\hat{x}(k) + Vu^*(k)\right) \\ u(k) = -F\hat{x}(k) + Vu^*(k) \\ \hat{x}(k+1) = A\hat{x}(k) + Bu^*(k) + L_0 \left(y(k) - C\hat{x}(k)\right) \\ y(k) = Cx(k). \end{cases} \tag{12.75}$$

Note that the observer is driven by $u^*(k)$ that represents the nominal system input, e.g. $u^*(k) = F_0\hat{x}(k) + v(k)$. Correspondingly, the dynamic of the estimation error, $e(k) = x(k) - \hat{x}(k)$, is governed by

$$\begin{aligned} e(k+1) &= (A - L_0C + B_\Delta F_\Delta) e(k) + B \begin{pmatrix} \Delta_B (F_\Delta - F) \hat{x}(k) \\ + (\Delta_B V - I) u^*(k) \end{pmatrix} \\ &= (A - L_0C) e(k) + B \begin{pmatrix} \Delta_B \left(F_\Delta x(k) - F\hat{x}(k)\right) \\ + (\Delta_B V - I) u^*(k) \end{pmatrix}, \end{aligned} \tag{12.76}$$

$$r(k) = Ce(k). \tag{12.77}$$

The problem to be studied is formulated as finding (F, V) given the system model (12.76), and tuning the observer gain matrix so that the system dynamic and the corresponding performance are recovered.

12.4.2 An Iterative Estimation Based Solution

Let

$$d(k) = \Delta_B \left(F_\Delta x(k) - F\hat{x}(k)\right) + (\Delta_B V - I) u^*(k). \tag{12.78}$$

The estimation error dynamic (12.76) can be written into

$$\begin{cases} e(k+1) = (A - L_0C) e(k) + Bd(k) \\ r(k) = Ce(k). \end{cases} \tag{12.79}$$

When the existing conditions for a UIO are satisfied, applying a UIO to the system (12.76) allows us to estimate $d(k)$ and $e(k)$, as described in Sect. 12.1.1. Otherwise, the LS unknown input estimation algorithm introduced in Sect. 12.1.2 is applied.

Once $d(k)$ and $e(k)$ have been estimated, F_Δ and Δ_B will be identified based on the relation

$$d(k) = \Delta_B \left(F_\Delta x(k) - F\hat{x}(k) \right) + \left(\Delta_B V - I \right) u^*(k), \tag{12.80}$$

$$x(k) = \hat{x}(k) + e(k). \tag{12.81}$$

To be specific, suppose that (F, V) is determined iteratively. For $F = F_i, V = V_i$, F_Δ, Δ_B can be, according to (12.80), determined using the model

$$\left[\Delta_B F_\Delta \ \Delta_B \right] \left[\begin{array}{c} x(k) \\ V_i u^*(k) - F_i \hat{x}(k) \end{array} \right] = d(k) + u^*(k) \tag{12.82}$$

and sufficient number of data $\left(x(k), \hat{x}(k), u^*(k) \right)$. Subsequently, F_{i+1}, V_{i+1} are set to be

$$\left(F_{i+1}, V_{i+1} \right) = \left(F_\Delta, \Delta_B^{-1} \right). \tag{12.83}$$

The realisation of the aforementioned solution is summarised in the following algorithm.

Algorithm 12.2 *Observer-based fault-tolerant LQ control*

Step 0: *Suppose that the i-th iteration delivers* L_i, F_i, V_i;
Step 1: *Apply* L_i, F_i, V_i *to controller* (12.74) *and run the observer* (12.75);
Step 2: *Run a UIO or an LS input estimator based on the model* (12.79) *that delivers the estimates* $d(k), e(k)$;
Step 3: *Collect sufficient number of data* $\left(x(k), u^*(k), \hat{x}(k) \right)$, *solve* (12.82) *and determine* F_{i+1}, V_{i+1} *based on the model* (12.83);
Step 4: *Check*

$$\| F_{i+1} - F_i \| \le \varepsilon_F, \| V_{i+1} - V_i \| \le \varepsilon_V,$$

where $(\varepsilon_F, \varepsilon_V)$*are pre-defined tolerance values. In the case of "No", go to Step 1, otherwise, stop.*

After (Δ_B, F_Δ) have been identified, re-constructing the observer-based residual generator as

$$\begin{aligned} \hat{x}(k+1) &= A\hat{x}(k) + Bu(k) + L_0 \left(y(k) - C\hat{x}(k) \right) \\ u(k) &= -F\hat{x}(k) + Vu^*(k), u^*(k) = F_0 \hat{x}(k) + v(k) \end{aligned}$$

enables fully recover the nominal system dynamics,

$$\begin{cases} x(k+1) = (A + BF_0) x(k) + Bv(k) - Be(k) \\ e(k+1) = (A - L_0 C) e(k) \\ y(k) = C\hat{x}(k) + W_0^{-1} r_y(k). \end{cases}$$

12.5 Notes and References

This chapter has focused on estimation and estimation-based control. Undoubtedly, estimation is an essential and challenging issue for successfully dealing with fault diagnosis and control. Our objective is twofold. On the one hand, the existing unknown input estimation and disturbance control methods are re-visited in our unified framework of control and detection. On the other hand, some methods developed in the unified framework in the previous chapters like projection technique, co-inner-based LS estimation, are leveraged to explore advanced estimation and control methods and algorithms.

For our first objective, UIO and DOBC schemes have been re-visited. The research on UIO technique began in 80s. To our best knowledge, the first work dedicated to UIO-based fault detection was reported by Wünnenberg and Frank at an international conference in 1986 and published in the conference proceedings [210]. Although the UIO and UIO-based fault detection techniques have undergone a dynamic development in the past decades and remain nowadays an active research area focusing on diverse variations of UIO-methods, the work reported by Wünnenberg in his dissertation [209] and by Hou and his co-authors in a series of publications [87, 88, 90] laid down the rigorous theoretic foundation for subsequent investigations in this field. Reviewing the state of the art publications, it can be observed that two issues have drawn the main research attention, namely state estimation decoupled from unknown inputs, and unknown inputs and (additive) fault estimation respectively. The latter is primarily devoted to fault estimation and disturbance estimation embedded in the DOBC framework aiming at disturbance rejection. In our examination of UIO schemes in Sect. 12.1.1, we have explored the aspect of UIOs as an estimator of unknown input vector. As illustrated in Example 12.1, the well-known UIOs are in fact an unknown input estimator as well. Hereby, a backward computation is embedded in the estimation algorithm. In fact, this observation has motivated our subsequent work on LS estimation issues, where backward computations build a necessary step. It can be noticed that in the fault diagnosis research area, it is supposed that, without specification, fault estimation is performed real-time and at each sampling time instant. In real engineering applications, technical specifications on fault estimations are considerably different. Firstly, fault estimation is a step after a fault is detected. Secondly, sufficient data (over a time interval and in the form of a data batch) are collected for fault estimation. These facts allow us to achieve fault estimation by means of backward computations. From the information viewpoint, backward computations enable gaining more information about the uncertainty to be estimated.

Our re-visit of DOBC schemes in Sect. 12.2 serves for the purpose of exploring DOBC schemes in the unified framework of control and detection. The review part is mainly based on the review paper [21] and the references therein. Hereby, attention has been paid to the initial idea of DOBC, namely generating the residual $u - \hat{u}$ for disturbance compensation (rejection). It has been shown that such residual generation is indeed a special case of the output residual generation embedded in the unified framework of control and detection. Moreover, in Theorem 12.2 and Corollary 12.1, general existence conditions to achieve a decoupling of closed-loop dynamics from the disturbance are provided. Concerning DOBC design for systems with uncertainties, the stability concern is closely examined. In the DOBC framework, it is a popular strategy to model uncertainties by means of (additive) unknown inputs. In other words, parametric or multiplicative uncertainties are summed up as additive disturbances and then handled (estimated) using high gain observers [21, 61]. It is well-known [6] that a high-gain observer is of the property given in (12.43). As illustrated in Sect. 12.2.3, the use of a high-gain observer might result in closed-loop instability.

LS estimation schemes of unknown inputs (together with state variables) and uncertainties corrupted in the process data build another focus of this chapter. In a certain sense, LS estimation can be understood as the counterpart of Kalman filter-based estimation for deterministic disturbances. On the other hand, we observe that research on LS estimation schemes is less popular, and receives remarkably less attention, for instance, in comparison with H_∞-optimal estimation. In our previous work [39], we noticed this situation and called for more attention to this issue. In Sect. 12.1.2, we have firstly introduced the definition of the LS estimation problem under consideration, i.e. Definition 12.1, along the line described in [208]. It is noteworthy that LS estimation problem has been comprehensively investigated by Willems [208]. It seems that, reviewing the literature, this gorgeous and elegant work did not draw deserved attention in the fault diagnosis research community. As described in Definition 12.1, the LS estimation is an optimal estimation problem expressed in the form of (12.15), by which the norms of the estimated unknown input vector and the (unknown) initial state vector are minimised. Importantly, the achieved estimates should satisfy the system equations. This feature distinguishes the LS estimation schemes from those H_∞-optimal ones, and gives us a convincing argument for recommending LS algorithms for fault estimation, rather than H_∞-methods. It is worth mentioning that LS estimation of unknown inputs in stochastic processes is comprehensively discussed in [236] (Chap. 5). The work in Sect. 12.1.3 on estimating uncertainties corrupted in process data can be applied, for instance, for data analysis, detection and monitoring of multiplicative faults in processes. To this end, the projection technique is leveraged for uncertainties estimation. It is remarkable that both aforementioned estimation schemes have one point in common that backward computations are an essential step of the estimation algorithms.

The successful exploration of the projection technique to uncertainty estimation has inspired our endeavour to learn fault-tolerant observers. The basic idea of our solution is to address variations in the system dynamics caused by faults by means of estimation and compensation in a learning iteration. Herewith, the term learning is understood in the regard of the so-called reinforcement learning-based control [107]. Nevertheless, the proposed RL-based optimal observer differs from the RL-based LQ control in the working principle. It is known that RL-based control is performed based on interactions between the plant and controller. Roughly speaking, each update of the control law is evaluated according to the objective function and the loop response to the controller update. The subsequent update will then be computed to optimise the objective function. Such a cause-and-effect iteration process runs consecutively (feedforward) in time. On the contrary, there exists no cause-and-effect between update of the observer and process response, because an observer does not directly act on the process. To solve this problem, $K_{G_0}^{\sim} K_{G_0}$ is applied to project the data (u, y) onto the system residual subspace. Recall that the co-state vector $\lambda(k)$ in $K_{G_0}^{\sim}$ models the dynamics of the unknown dynamics (caused by the faults). Hence, $K_{G_0}^{\sim}$ is interpreted as the process, while K_{G_0}, the observer, posses as the controller. In an iteration process, the controller (observer) will then be learnt so that the uncertainties are compensated. The system configuration and the learning process are schematically depicted in Fig. 12.2. It is of considerable interest to notice that our work on optimal fault-tolerant observers presented in Sect. 12.3 is dual to the RL-aided LQ optimal control. To be specific,

- the optimisation problem is formulated as an LS estimation, dual to the well-known LQ control, and
- it is solved with respect to the co-state variables and the conjugate system of the system SKR, dual to the state feedback and the system SIR,
- the learning algorithm is performed sequentially backwards in time, dual to feed-forward in time as usually done in RL technique (refer to [107]). Consequently, the DP is computed backwards.

Also for these reasons, we prefer call the proposed method RL-based fault-tolerant observers rather than optimal adaptive one.

A critic issue to learn an observer consists in the fact that any observer is a realisation of a model. In other words, a prerequisite is available model knowledge of the process under observation. In this context, it is assumed in our work that the observability indices are invariant with respect to the faulty operations. It is known in well-established linear system theory that under this assumption the variations in the system matrices (C, A) can be modelled by (12.52) [128]. It is noteworthy that such an assumption imposes no strong restriction on the applicability of the proposed scheme. For instance, when a single output system is observable, the observability index is constantly equal to n. For systems with multiple outputs, it is known [140] that if the system remains observable, then for a vector $\alpha \in \mathbb{R}^m$ generated randomly from a uniform distribution, it holds with probability one that

$$rank \begin{bmatrix} \alpha C \\ \alpha CA \\ \vdots \\ \alpha CA^{n-1} \end{bmatrix} = n.$$

Accordingly, the system (C, A) becomes an observable single output system with $(\bar{c}, A)$, $\bar{c} = \alpha C$. As a result, the RL-based algorithm proposed in Sect. 12.3.3 delivers a sub-optimal solution as long as the system is observable, and independent of the observability indices. We would like to call the reader's attention to the possibility of applying the proposed algorithm to learn an observer on the assumption that the observability indices, $\left\{\mu_j, j = 1, \ldots, m\right\}$, are known. In this case, the system matrices (C, A) can be modelled as

$$A = A_0 + LC_o, C = \Pi C_0, A_{o,0} = diag\left(A_{o,1}, \ldots, A_{o,m}\right),$$

$$A_{o,j} = \begin{bmatrix} 0 & 0 & 0 & \cdots & 0 \\ 1 & 0 & 0 & \cdots & 0 \\ \vdots & \ddots & \ddots & \ddots & \vdots \\ 0 & \cdots & 1 & 0 & 0 \\ 0 & 0 & \cdots & 1 & 0 \end{bmatrix} \in \mathbb{R}^{\mu_j \times \mu_j}, j = 1, \ldots, m,$$

$$C_o = \left[C_{o,1} \cdots C_{o,m} \right], C_{o,i} = \left[0_{m\times(\mu_j-1)} \; c_i \right] \in \mathbb{R}^{m\times\mu_i}, i = 1, \ldots, m,$$

$$c_i = \left[0 \cdots 0\; 1\; 0 \cdots 0\right]^T, \Pi = \begin{bmatrix} 1 & 0 & \cdots & \cdots & 0 \\ \gamma_{1,2} & 1 & 0 & \cdots & 0 \\ \vdots & \gamma_{2,3} & \ddots & \ddots & \vdots \\ \vdots & \cdots & \ddots & 1 & 0 \\ \gamma_{1,m} & \gamma_{2,m} & \cdots & \gamma_{m-1,m} & 1 \end{bmatrix} \in \mathbb{R}^{m\times m},$$

where (C_0, A_0) are known, since they are determined by the observability indices, and (L, Π) will be learnt.

The observer-based fault-tolerant LQ control scheme presented in Sect. 12.4 is indeed an example of applying LS estimation algorithms to fault-tolerant control and performance recovery. At the core of this scheme is the closed-loop model (12.75) that leads to the estimation error dynamics (12.76) with the unknown input given in (12.78). It is worth emphasising that the assumption on the invariance of the controllability indices can be released. As a dual result to the aforementioned observability discussion, on the assumption of controllability transforming a multiple input system to a single input system allows us to find a solution that enables a full recovery of the optimal system dynamic.

Chapter 13
Control and Detection Methods of Finite Samples System Representations

In this chapter, applications of finite samples system representations introduced in Sect. 7.3 to control and detection issues are explored. The intended explorations are twofold: firstly, an immediate use of the finite samples system representations, given commonly used analytical models in the time or frequency domains, to simplify computations for design and implementation of control and detection systems, secondly, use of the finite samples system representations in the data-driven fashion for the control and detection purposes. This means that the finite samples system representations are built by means of data rather than analytical models.

In our subsequent work, it is assumed that finite samples system representations under consideration are associated with

- the nominal state space model,

$$x(k+1) = Ax(k) + Bu(k), x(0) = x_0, \tag{13.1}$$
$$y(k) = Cx(k) + Du(k), \tag{13.2}$$

  where $x \in \mathbb{R}^n$ is the state vector and x_0 is the initial condition of the system, $u \in \mathbb{R}^p$ and $y \in \mathbb{R}^m$ are the plant input and output vectors, respectively,
- the SIR model,

$$\left\{ \begin{array}{l} x(k+1) = (A+BF)x(k) + BVv(k) \\ \begin{bmatrix} u(k) \\ y(k) \end{bmatrix} = \begin{bmatrix} F \\ C+DF \end{bmatrix} x(k) + \begin{bmatrix} V \\ DV \end{bmatrix} v(k), \end{array} \right. \tag{13.3}$$

$$\begin{bmatrix} u(z) \\ y(z) \end{bmatrix} = \begin{bmatrix} M(z) \\ N(z) \end{bmatrix} v(z), \tag{13.4}$$

S. X. Ding, *Unified Framework of Control and Detection*,
https://doi.org/10.1007/978-3-662-72417-0_13

and SKR model,

$$\begin{cases} \hat{x}(k+1) = A\hat{x}(k) + Bu(k) + L\left(y(k) - \hat{y}(k)\right) \\ r_y(k) = Wr_0(k), r_0(k) = y(k) - \hat{y}(k), \hat{y}(k) = C\hat{x}(k) + Du(k), \end{cases} \tag{13.5}$$

$$r_y(z) = \hat{M}(z)y(z) - \hat{N}(z)u(z). \tag{13.6}$$

13.1 Summary of Finite Samples System Representations

To begin with, we summarise the finite samples system representations and some associated results introduced in Sect. 7.3. Relevant supplements will then be added, including some new results.

13.1.1 Finite Samples System Representations Induced System Models

Corresponding to the SIRs and SKRs of the plant and controller, four finite samples system representations have been derived. These are

- the s-samples long data image representation of the plant (s-SIRP),

$$\begin{bmatrix} u_s(k) \\ y_s(k) \end{bmatrix} = I_{G,s} v_{s+n}(k-n) \in \mathbb{R}^{s(p+m)}, \ I_{G,s} = \begin{bmatrix} M_s \\ N_s \end{bmatrix} V_{s+n},$$

  where $V_{s+n} \in \mathbb{R}^{p(s+n) \times p(s+n)}$ is an invertible matrix and $v_{s+n}(k-n) \in \mathbb{R}^{p(s+n)}$ the latent variable,
- the $(s+n)$-samples long data kernel representation of the plant ($(s+n)$-SKRP),

$$r_s(k) = K_{G,s+n} \begin{bmatrix} u_{s+n}(k-n) \\ y_{s+n}(k-n) \end{bmatrix} = 0, \tag{13.7}$$

$$K_{G,s+n} = R_s \begin{bmatrix} -\hat{N}_{s+n} & \hat{M}_{s+n} \end{bmatrix} \in \mathbb{R}^{sm \times (m+p)(s+n)}, \tag{13.8}$$

  where $r_s(k) \in \mathbb{R}^{sm}$ is the residual vector and $R_s \in \mathbb{R}^{sm \times sm}$ is an invertible matrix,
- the s-samples long data image representation of the controller (s-SIRC),

$$\begin{bmatrix} u_s(k) \\ y_s(k) \end{bmatrix} = I_{K,s} r_{s+n}(k-n), \ I_{K,s+n} = \begin{bmatrix} -\hat{Y}_s & M_s \\ \hat{X}_s & N_s \end{bmatrix} \begin{bmatrix} I \\ \bar{R}_{s+n} \end{bmatrix},$$

  where the residual vector $r_{s+n}(k) \in \mathbb{R}^{(s+n)m}$ serves as the latent variable, $\bar{R}_{s+n} \in \mathbb{R}^{(s+n)p \times sm}$ is a matrix whose rank is smaller than sm,

- the $(s+n)$-samples long data kernel representation of the controller ($(s+n)$-SKRC),

$$v_s(k) = K_{K,s+n}\begin{bmatrix} u_{s+n}(k-n) \\ y_{s+n}(k-n) \end{bmatrix} = 0,\ K_{K,s+n} = \begin{bmatrix} I & \bar{V}_s \end{bmatrix}\begin{bmatrix} X_{s+n} & Y_{s+n} \\ -\hat{N}_{s+n} & \hat{M}_{s+n} \end{bmatrix},$$

  where $v_s(k) \in \mathbb{R}^{sp}$ is the input residual signal, $\bar{V}_s \in \mathbb{R}^{sp\times sm}$ is a matrix whose rank is smaller than sp.

Given an LTI system, any s-samples data $(u, y) \in \mathcal{L}_{[0,\infty)}^{s(m+p)}$ can be modelled by

$$\begin{aligned}\begin{bmatrix} u_s(k) \\ y_s(k) \end{bmatrix} &= I_{G,s}v_{s+n}(k-n) + I_{K,s}r_{s+n}(k-n) \\ &= \begin{bmatrix} M_s \\ N_s \end{bmatrix}\bar{v}_{s+n}(k-n) + \left(\begin{bmatrix} -\hat{Y}_s \\ \hat{X}_s \end{bmatrix} + \begin{bmatrix} M_s \\ N_s \end{bmatrix}\bar{R}_{s+n}\right) r_{s+n}(k-n) \\ &= \begin{bmatrix} M_s & -\hat{Y}_s \\ N_s & \hat{X}_s \end{bmatrix}\begin{bmatrix} I & \bar{R}_{s+n} \\ 0 & I \end{bmatrix}\begin{bmatrix} \bar{v}_{s+n}(k-n) \\ r_{s+n}(k-n) \end{bmatrix},\end{aligned} \tag{13.9}$$

$\bar{v}_{s+n}(k-n) = V_{s+n}v_{s+n}(k-n)$. The model (13.9) is also the s-samples representation of the closed-loop dynamics corresponding to

$$\begin{bmatrix} u \\ y \end{bmatrix} = \begin{bmatrix} M \\ N \end{bmatrix} v + \begin{bmatrix} -\hat{Y} \\ \hat{X} \end{bmatrix} r_y \in \mathcal{H}_2,$$

which is the SIR-SKR-model (7.55) as well. As its dual form, the $(s+n)$-samples kernel data representation is known as

$$\begin{aligned}\begin{bmatrix} v_s(k) \\ r_s(k) \end{bmatrix} &= \begin{bmatrix} K_{K,s+n} \\ K_{G,s+n} \end{bmatrix}\begin{bmatrix} u_{s+n}(k-n) \\ y_{s+n}(k-n) \end{bmatrix} \\ &= \begin{bmatrix} I & \bar{V}_s \\ 0 & I \end{bmatrix}\begin{bmatrix} X_{s+n} & Y_{s+n} \\ -\hat{N}_{s+n} & \hat{M}_{s+n} \end{bmatrix}\begin{bmatrix} u_{s+n}(k-n) \\ y_{s+n}(k-n) \end{bmatrix}.\end{aligned} \tag{13.10}$$

Equations (13.9) and (13.10) can be interpreted as the finite samples form of the Bezout identity.

13.1.2 Finite Samples Image and Residual Subspaces

Given $(s+n)$-SIRP and $(s+n)$-SKRP, finite samples image and kernel subspaces are defined by

$$\mathcal{I}_{G,s+n} = \left\{ \begin{bmatrix} u_{s+n}(k) \\ y_{s+n}(k) \end{bmatrix}, \begin{bmatrix} u_{s+n}(k) \\ y_{s+n}(k) \end{bmatrix} = \begin{bmatrix} M_{s+n} \\ N_{s+n} \end{bmatrix} v_{s+2n}(k-n), \atop v_{s+2n}(k-n) \in \mathcal{L}_{[0,\infty)} \right\}, \tag{13.11}$$

$$\mathcal{K}_{G,s+n} = \left\{ \begin{bmatrix} u_{s+n}(k-n) \\ y_{s+n}(k-n) \end{bmatrix} \in \mathcal{L}_{[0,\infty)}, K_{G,s+n} \begin{bmatrix} u_{s+n}(k-n) \\ y_{s+n}(k-n) \end{bmatrix} = 0 \right\}. \tag{13.12}$$

Often, we are interested in the complemented subspace of $\mathcal{I}_{G,s+n}$, $\mathcal{I}_{G,s+n}^{\perp}$, that consists of the process data generated under non-nominal operation conditions, namely, the data are corrupted by uncertainties and even faults. We call $\mathcal{I}_{G,s+n}^{\perp}$ residual subspace, denoted by $\mathcal{R}_{G,s+n}$. Remember the relation

$$K_{G,s+n} I_{G,s+n} = 0.$$

Apparently, the complementary subspace of $\mathcal{K}_G$ defined by

$$\mathcal{K}_{G,s+n}^{\perp} := \left\{ \begin{bmatrix} u_{s+n}(k-n) \\ y_{s+n}(k-n) \end{bmatrix}, \begin{bmatrix} u_{s+n}(k-n) \\ y_{s+n}(k-n) \end{bmatrix} = K_{G,s+n}^{T} r_s(k), r_s(k) \in \mathcal{L}_{[0,\infty)} \right\}$$

belongs to the complementary subspace of $\mathcal{I}_{G,s+n}$, i.e.

$$\forall \begin{bmatrix} u_{s+n}(k-n) \\ y_{s+n}(k-n) \end{bmatrix} \in \mathcal{I}_{G,s+n}, \begin{bmatrix} u_{s+n}(k-n) \\ y_{s+n}(k-n) \end{bmatrix}^{T} K_{G,s+n}^{T} = 0.$$

On the other hand, differing from the result $\mathcal{I}_G = \mathcal{K}_G$,

$$\mathcal{I}_{G,s+n} \subseteq \mathcal{K}_{G,s+n}.$$

In particular,

$$\dim\left(\mathcal{I}_{G,s+n}\right) = (s+n)\,p + n \leq \dim\left(\mathcal{K}_{G,s+n}\right) = (s+n)\,p + nm.$$

Hence,

$$\mathcal{K}_{G,s+n}^{\perp} \subseteq \mathcal{I}_{G,s+n}^{\perp} = \mathcal{R}_{G,s+n}.$$

An alternative way to analytically describe the residual subspace $\mathcal{R}_{G,s+n}$ is to apply orthogonal projections, which are related to the concepts of normalised SIR $I_{G,s+n}$ and SKR $K_{G,s+n}$.

13.1.3 Normalised SIR and SKR, and Residual Subspace

In the unified framework, the normalised SIR and SKR play an indisputable role. In this subsection, their finite samples counterparts, $I_{G_0,s}$ and $K_{G_0,s+n}$, are addressed.

Recall that

$$\begin{bmatrix} M_s \\ N_s \end{bmatrix} \in \mathbb{R}^{(p+m)s \times p(s+n)}, rank \begin{bmatrix} M_s \\ N_s \end{bmatrix} = sp + n,$$
$$\begin{bmatrix} -\hat{N}_{s+n} & \hat{M}_{s+n} \end{bmatrix} \in \mathbb{R}^{ms \times (s+n)(p+m)}, rank \begin{bmatrix} -\hat{N}_{s+n} & \hat{M}_{s+n} \end{bmatrix} = ms.$$

By means of SVD, it turns out

$$\begin{bmatrix} M_s \\ N_s \end{bmatrix} = \begin{bmatrix} U_{I_G,1} & U_{I_G,2} \end{bmatrix} \begin{bmatrix} \Sigma_{I_G} & 0 \\ 0 & 0 \end{bmatrix} \begin{bmatrix} V_{I_G,1}^T \\ V_{I_G,2}^T \end{bmatrix} = U_{I_G,1} \Sigma_{I_G} V_{I_G,1}^T,$$
$$U_{I_G,1} \in \mathbb{R}^{(p+m)s \times (sp+n)}, rank \left(U_{I_G,1} \right) = sp + n,$$
$$V_{I_G,1}^T \in \mathbb{R}^{(sp+n) \times p(s+n)}, rank \left(V_{I_G,1} \right) = sp + n,$$
$$\Sigma_{I_G} = diag \left(\sigma_1, \ldots, \sigma_{sp+n} \right), \sigma_i \neq 0, i = 1, \ldots, sp + n,$$
$$\begin{bmatrix} -\hat{N}_{s+n} & \hat{M}_{s+n} \end{bmatrix} = U_{K_G} \begin{bmatrix} \Sigma_{K_G} & 0 \end{bmatrix} \begin{bmatrix} V_{K_G,1}^T \\ V_{K_G,2}^T \end{bmatrix} = U_{K_G} \Sigma_{K_G} V_{K_G,1}^T,$$
$$U_{K_G} \in \mathbb{R}^{ms \times sm}, U_{K_G} U_{K_G}^T = I,$$
$$V_{K_G,1}^T \in \mathbb{R}^{sm \times (s+n)(p+m)}, rank \left(V_{K_G,1} \right) = sm,$$
$$\Sigma_{K_G} = diag \left(\sigma_1, \ldots, \sigma_{sm} \right), \sigma_i \neq 0, i = 1, \ldots, sm.$$

As a result,

$$I_{G_0,s} = \begin{bmatrix} M_s \\ N_s \end{bmatrix} V_o, V_o = V_{I_G,1} \Sigma_{I_G}^{-1}, \tag{13.13}$$
$$K_{G_0,s+n} = R_o \begin{bmatrix} -\hat{N}_{s+n} & \hat{M}_{s+n} \end{bmatrix}, R_o = \Sigma_{K_G}^{-1} U_{K_G}^T. \tag{13.14}$$

It is worth remarking that $I_{G_0,s}$ and $K_{G_0,s+n}$ are non-causal systems.

With the aid of the normalised image representation $I_{G_0,s}$, we are now in a position to give an analytical representation of the residual subspace. Let

$$\mathcal{P}_{\mathcal{I}_{G,s}} := I_{G_0,s} I_{G_0,s}^T$$

be an operator of the orthogonal projection onto $\mathcal{I}_{G,s}$. It is apparent that

$$\mathcal{P}_{\mathcal{R}_{G,s}} := I - I_{G_0,s} I_{G_0,s}^T$$

imposes an operator of the orthogonal projection onto $\mathcal{R}_{G,s}$. Because

$$rank \left(I - I_{G_0,s} I_{G_0,s}^T \right) = (p + m) s - ps - n = ms - n,$$

we obtain, by means of standard decomposition algorithms,

$$\begin{aligned} I - I_{G_0,s} I_{G_0,s}^T &= R_{G_0,s} R_{G_0,s}^T, \ R_{G_0,s} \in \mathbb{R}^{(p+m)s \times (ms-n)}, \\ rank\left(R_{G_0,s}\right) &= ms - n, \ R_{G_0,s}^T R_{G_0,s} = I. \end{aligned}$$

Consequently, the residual subspace $\mathcal{R}_{G,s}$ is defined by

$$\mathcal{R}_{G,s} = \left\{ \begin{bmatrix} u_s(k) \\ y_s(k) \end{bmatrix}, \begin{bmatrix} u_s(k) \\ y_s(k) \end{bmatrix} = R_{G_0,s} r, \ r \in \mathcal{L}_{[0,\infty)}^{ms-n} \right\}.$$

Note that $I_{G_0,s}$ and $R_{G_0,s}$ can be alternatively expressed by

$$I_{G_0,s} = U_{I_G,1}, \ R_{G_0,s} = U_{I_G,2},$$

respectively.

13.2 Detection and Control Schemes Based on Finite Samples System Representations

Finite samples system representations are characterised by their algebraic models that simplify design and online computations. On the assumption of available system models, we explore applications of finite samples system representations to fault detection and control.

13.2.1 *Finite Samples Kernel Model-Based Fault Detection*

The kernel model

$$r_s(k) = R_s \begin{bmatrix} -\hat{N}_{s+n} & \hat{M}_{s+n} \end{bmatrix} \begin{bmatrix} u_{s+n}(k-n) \\ y_{s+n}(k-n) \end{bmatrix} \tag{13.15}$$

is a residual generator. And, it is a finite impulse response (FIR) filter. From this point of view, it is similar to the so-called parity vector-based residual generator. On the other hand, its configuration is different to that of a parity vector-based residual generator. Alternatively, the residual vector $r_s(k)$ can be generated using an observer-based residual generator as well. To this end, suppose that the residual generator (13.5) delivers the residual vector r_0. Now, switching a post-filter $R(z)$,

$$R(z) = I - C\left(zI - A_{L_d}\right)^{-1}(L_d - L) \in \mathcal{RH}_\infty,$$

to r_0 yields a deadbeat observer-based residual generator

$$\begin{cases} \hat{x}(k+1) = (A - L_d C)\,\hat{x}(k) + (B - L_d D)\,u(k) + L_d y(k) \\ r_0(k) = y(k) - \hat{y}(k),\, \hat{y}(k) = C\hat{x}(k) + Du(k). \end{cases} \tag{13.16}$$

From $r_0(k)$, $r_s(k)$ is constructed as

$$r_s(k) = \begin{bmatrix} r_0(k+1) \\ \vdots \\ r_0(k+s) \end{bmatrix},$$

which is subject to (13.15) with $R_s = I$. It is worth noticing that by means of model (13.15) with

$$R_s = \begin{bmatrix} I & 0 & \cdots & 0 \\ C\,(L_d - L) & I & \ddots & \vdots \\ \vdots & \ddots & \ddots & 0 \\ CA_L^{s-2}\,(L_d - L) & \cdots & C\,(L_d - L) & I \end{bmatrix},$$

the kernel representation is given by

$$R_s \hat{N}_{s+n} = \begin{bmatrix} CA_L^{n-1} B_L & \cdots & D & 0 & \cdots & 0 \\ CA_L^{n} B_L & \cdots & CB_L & D & \ddots & \vdots \\ \vdots & \vdots & \vdots & \ddots & \ddots & 0 \\ CA_L^{n+s-2} B_L & \cdots & CA_L^{s-2} B_L & \cdots & CB_L & D \end{bmatrix},$$

$$R_s \hat{M}_{s+n} = \begin{bmatrix} -CA_L^{n-1} L & \cdots & I & 0 & \cdots & 0 \\ -CA_L^{n} L & \cdots & -CL & I & \ddots & \vdots \\ \vdots & \vdots & \vdots & \ddots & \ddots & 0 \\ -CA_L^{n+s-2} L & \cdots & -CA_L^{s-2} L & \cdots & -CL & I \end{bmatrix},$$

$$A_L = A - LC,\, A_{L_d} = A - L_d C,\, B_L = B - LD.$$

This implies,

$$r_s(k) = y_s(k) - \hat{y}_s(k) = R_s \begin{bmatrix} -\hat{N}_{s+n} & \hat{M}_{s+n} \end{bmatrix} \begin{bmatrix} u_{s+n}(k-n) \\ y_{s+n}(k-n) \end{bmatrix} \Longrightarrow$$

$$\hat{y}_s(k) = H_{u,s+n} u_{s+n}(k-n) + H_{y,s+n} y_{s+n}(k-n),$$

$$H_{u,s+n} = \begin{bmatrix} CA_L^{n-1} B_L & \cdots & D & 0 & \cdots & 0 \\ CA_L^{n} B_L & \cdots & CB_L & D & \ddots & \vdots \\ \vdots & \vdots & \vdots & \ddots & \ddots & 0 \\ CA_L^{n+s-2} B_L & \cdots & CA_L^{s-2} B_L & \cdots & CB_L & D \end{bmatrix},$$

$$H_{y,s+n} = \begin{bmatrix} CA_L^{n-1}L & \cdots & 0 & 0 & \cdots & 0 \\ CA_L^{n}L & \cdots & CL & 0 & \ddots & \vdots \\ \vdots & \vdots & \vdots & \ddots & \ddots & 0 \\ CA_L^{n+s-2}L & \cdots & CA_L^{s-2}L & \cdots & CL & 0 \end{bmatrix}.$$

It is interesting to observe that $R_s \hat{N}_{s+n}$ does not depend on L_d.

Next, a finite samples kernel model-based scheme is introduced for detecting faults in LTI systems with additive disturbances. For our purpose, the nominal model (13.1)–(13.2) is extended to include unknown input vector (disturbances),

$$\begin{cases} x(k+1) = Ax(k) + Bu(k) + E_d d(k),\ x(0) = x_0 \\ y(k) = Cx(k) + Du(k) + F_d d(k), \end{cases} \tag{13.17}$$

where $d \in \mathbb{R}^{k_d}$ is an unknown input vector bounded by

$$\frac{d_s^T(k) d_s(k)}{s} \leq \delta_d,$$

and matrices (E_d, F_d) are known and of appropriate dimensions. To deal with more critical operations in practice, it is assumed that

$$k_d \geq m. \tag{13.18}$$

Suppose that there exists no knowledge of faults or the subspace spanned by the potential faults covers the whole (output) measurement subspace. As known in the literature, under such conditions the optimal (detection) solution is the so-called unified solution. For finite samples residual signals, the corresponding optimal solution is computed using the subsequent algorithm.

Algorithm 13.1 *Optimal fault detection of LTI-systems with unknown inputs*

Step 1: *Offline: Construction of matrix* $H_{d,s+n}$,

$$H_{d,s+n} = \begin{bmatrix} CA^{n-1}E_d & \cdots & F_d & 0 & \cdots & 0 \\ CA^{n}E_d & \cdots & CE_d & F_d & \ddots & \vdots \\ \vdots & \vdots & \vdots & \ddots & \ddots & 0 \\ CA^{n+s-2}E_d & \cdots & CA^{s-2}E_d & \cdots & CE_d & F_d \end{bmatrix};$$

Step 2: *Computation of* Σ_d^{-1},

$$\Sigma_d = H_{d,s+n} H_{d,s+n}^T,$$

on the assumption that

$$rank\left(\begin{bmatrix} CA^{n-1}E_d & \cdots & F_d & 0 & \cdots & 0 \\ CA^{n}E_d & \cdots & CE_d & F_d & \ddots & \vdots \\ \vdots & \vdots & \vdots & \ddots & \ddots & 0 \\ CA^{n+s-2}E_d & \cdots & CA^{s-2}E_d & \cdots & CE_d & F_d \end{bmatrix}\right) = sm; \tag{13.19}$$

Step 3: *Construction of the optimal residual generator* (13.15), *where R_s is selected so that*

$$R_s\hat{N}_{s+n} = \begin{bmatrix} CA^{n-1}B & \cdots & D & 0 & \cdots & 0 \\ CA^{n}B & \cdots & CB & D & \ddots & \vdots \\ \vdots & \vdots & \vdots & \ddots & \ddots & 0 \\ CA^{n+s-2}B & \cdots & CA^{s-2}B & \cdots & CB & D \end{bmatrix},$$

$$R_s\hat{M}_{s+n} = \begin{bmatrix} 0 & \cdots & I & 0 & \cdots & 0 \\ 0 & \cdots & 0 & I & \ddots & \vdots \\ \vdots & \vdots & \vdots & \ddots & \ddots & 0 \\ 0 & \cdots & 0 & \cdots & 0 & I \end{bmatrix};$$

Step 4: *Online: collection of residual signal r_s*, and *computation of the evaluation function*

$$J = \frac{r_s^T(k)\Sigma_d^{-1}r_s(k)}{s},$$

where

$$r_s(k) = R_s H_{d,s+n} d_s(k);$$

Step 5: *Detection logic,*

$$\begin{cases} J \leq J_{th} \Longrightarrow \text{fault-free} \\ J > J_{th} \Longrightarrow \text{faulty}, \end{cases} \quad J_{th} := \delta_d.$$

It is remarkable to notice that the matrix $H_{d,s+n}$ is merely composed of the system matrices (A, C, E_d, F_d) and independent of the observer gain matrix. In other words, $H_{d,s+n}$ is independent of the construction of the residual generator. This result is attributed to the fact that for any observer gain matrix L

$$H_{d,s+n,L} = \hat{N}_{d,s+n} = \begin{bmatrix} CA_L^{n-1}E_{d,L} & \cdots & F_d & 0 & \cdots & 0 \\ CA_L^{n}E_{d,L} & \cdots & CE_{d,L} & F_d & \ddots & \vdots \\ \vdots & \vdots & \vdots & \ddots & \ddots & 0 \\ CA_L^{n+s-2}E_{d,L} & \cdots & CA_L^{s-2}E_{d,L} & \cdots & CE_{d,L} & F_{d,L} \end{bmatrix}$$

can be equivalently written as

$$H_{d,s+n,L} = R_s \begin{bmatrix} CA^{n-1}E_d & \cdots & F_d & 0 & \cdots & 0 \\ CA^n E_d & \cdots & CE_d & F_d & \ddots & \vdots \\ \vdots & \vdots & \vdots & \ddots & \ddots & 0 \\ CA^{n+s-2}E_d & \cdots & CA^{s-2}E_d & \cdots & CE_d & F_d \end{bmatrix} = R_s H_{d,s+n}$$

with the invertible matrix R_s. Consequently,

$$\left(H_{d,s+n,L} H_{d,s+n,L}^T\right)^{-1} = \left(R_s \Sigma_d R_s^T\right)^{-1} \Longrightarrow$$

$$J = \frac{r_s^T(k) R_s^T \left(H_{d,s+n,L} H_{d,s+n,L}^T\right)^{-1} R_s r_s(k)}{s} = \frac{r_s^T(k) \Sigma_d^{-1} r_s(k)}{s}.$$

In comparison with the existing (comparable) optimal fault detection systems, the advantages of the proposed solution are obvious. To be specific,

- the design and construction of the residual generator based on the kernel model (13.15) is considerably simpler than the design of a parity vector-based residual generator, where solution of the so-called parity equation is to be performed. In particular, concerning the design of optimal detection systems, for parity relation-based methods, an optimal matrix is to be found in the parity space, in order to achieve an optimal parity relation-based fault detection, while the above introduced optimal solution depends exclusively on the system matrices and is achieved without involved computation,
- as known, the unified solution or equivalently the H_2-observer-based detection system implies a system co-inner-outer factorisation, whose existence conditions require, amongst others,

$$rank\,(F_d) = m.$$

  In comparison, the existence conditions for assumption (13.19) are (significantly) less strict. For instance, if

$$rank \begin{bmatrix} CE_d & F_d \end{bmatrix} = m,$$

  condition (13.19) holds. Moreover, solution of a Riccati equation is needed for computing a co-inner-outer factorisation, which requires considerably more computations than the realisation of Algorithm 13.1. From the realisation point of view, the residual evaluation function adopted in the unified solution is

$$J = \left\| r_y(k) \right\|_2^2 = \sum_{k=0}^{\infty} r_y^T(k) r_y(k).$$

In engineering applications, it can only be approximated in a finite time interval, for instance,

$$J \approx \sum_{k=k_0}^{k_1} r_y^T(k) r_y(k).$$

In this regard, the proposed Algorithm 13.1 offers an easily realisable and optimal engineering solution.

13.2.2 Projection-Based Fault Detection

In this subsection, issues of fault detection of uncertain LTI-systems based on the finite samples representations are addressed. As discussed in Chap. 9, projection-based methods are capable of dealing with fault detection in dynamic systems with multiplicative (parametric) uncertainties. In the sequel, a projection-based method based on the finite samples image representation is introduced.

The first step to our aim is an orthogonal projection-based residual generation. Let

$$\begin{aligned}
\mathcal{P}_{\mathcal{I}_{G,s}} &= I_{G_0,s} I_{G_0,s}^T, \\
I_{G_0,s} &= \begin{bmatrix} M_s \\ N_s \end{bmatrix} V_o = U_{I_G,1}, \; V_o = V_{I_G,1} \Sigma_{I_G}^{-1}, \\
M_s &= \begin{bmatrix} CA^{n-1}B & \cdots & D & 0 & \cdots & 0 \\ CA^n B & \cdots & CB & D & \ddots & \vdots \\ \vdots & \vdots & \vdots & \ddots & \ddots & 0 \\ CA^{n+s-2}B & \cdots & CA^{s-2}B & \cdots & CB & D \end{bmatrix}, N_s = \begin{bmatrix} 0 & I \end{bmatrix},
\end{aligned}$$

where $\left(V_{I_G,1}, \Sigma_{I_G}\right)$ are found by an SVD of $I_{G,s}$,

$$I_{G,s} = \begin{bmatrix} M_s \\ N_s \end{bmatrix} = \begin{bmatrix} U_{I_G,1} & U_{I_G,2} \end{bmatrix} \begin{bmatrix} \Sigma_{I_G} & 0 \\ 0 & 0 \end{bmatrix} \begin{bmatrix} V_{I_G,1}^T \\ V_{I_G,2}^T \end{bmatrix},$$
$$U_{I_G,1} U_{I_G,1}^T + U_{I_G,2} U_{I_G,2}^T = I,$$

as described in Sect. 13.1.3. Again, notice that (M_s, N_s) are constructed independently of the state feedback gain matrix F (F_d). This result is attributed to the Eq. (7.79) proved in Theorem 7.4. Subsequently, the projection-based residual is generated by

$$r_{\mathcal{P}_{\mathcal{I}_{G,s}},s}(k) = \left(\mathcal{I} - \mathcal{P}_{\mathcal{I}_{G,s}}\right) \begin{bmatrix} u_s(k) \\ y_s(k) \end{bmatrix} = U_{I_G,2} U_{I_G,2}^T \begin{bmatrix} u_s(k) \\ y_s(k) \end{bmatrix} \in \mathbb{R}^{s(m+p)}.$$

The corresponding residual evaluation function is defined by

$$J_{rel} = \frac{\left\| r_{\mathcal{P}_{\mathcal{I}_{G,s}},s}(k) \right\|}{\left\| \begin{bmatrix} u_s(k) \\ y_s(k) \end{bmatrix} \right\|} = \frac{\left\| U_{I_G,2}^T \begin{bmatrix} u_s(k) \\ y_s(k) \end{bmatrix} \right\|}{\left\| \begin{bmatrix} u_s(k) \\ y_s(k) \end{bmatrix} \right\|}.$$

Now, let $\left(\Delta_{M_s}, \Delta_{N_s}\right)$ denote the variations induced by parametric uncertainties, leading to

$$\begin{bmatrix} u_s(k) \\ y_s(k) \end{bmatrix} = \begin{bmatrix} M_s + \Delta_{M_s} \\ N_s + \Delta_{N_s} \end{bmatrix} v_{\Delta,s+n}(k-n).$$

It is assumed that $\left(\Delta_{M_s}, \Delta_{N_s}\right)$ is bounded by

$$\left\| \Delta_{I_G} \right\|_2 \le \delta_{\Delta_{I_G}} < 1, \ \Delta_{I_G} = \begin{bmatrix} \Delta_{M_s} \\ \Delta_{N_s} \end{bmatrix}.$$

For our detection purpose, define the subspace

$$\mathcal{I}_{G_\Delta,s} = \left\{ \begin{bmatrix} u_{\Delta,s}(k) \\ y_{\Delta,s}(k) \end{bmatrix}, \begin{array}{c} \begin{bmatrix} u_{\Delta,s}(k) \\ y_{\Delta,s}(k) \end{bmatrix} = \begin{bmatrix} M_s + \Delta_{M_s} \\ N_s + \Delta_{N_s} \end{bmatrix} v_{\Delta,s+n}(k-n), \\ v_{\Delta,s+n}(k-n) \in \mathcal{L}_{[0,\infty)} \end{array} \right\}$$

and the orthogonal projection operator

$$\mathcal{P}_{\mathcal{I}_{G_\Delta,s}} = I_{G_{\Delta,0},s} I_{G_{\Delta,0},s}^T, \ I_{G_{\Delta,0},s}^T I_{G_{\Delta,0},s} = I.$$

On this basis, gap metric $\delta\left(\mathcal{I}_{G_\Delta,s}, \mathcal{I}_{G,s}\right)$ is introduced to measure the similarity between $\mathcal{I}_{G,s}$ and $\mathcal{I}_{G_\Delta,s}$ aiming to determine the threshold. Recall that

$$\delta\left(\mathcal{I}_{G_\Delta,s}, \mathcal{I}_{G,s}\right) = \sup_{\begin{bmatrix} u_{\Delta,s}(k) \\ y_{\Delta,s}(k) \end{bmatrix} \in \mathcal{I}_{G_\Delta,s}} \ \inf_{\begin{bmatrix} u_s(k) \\ y_s(k) \end{bmatrix} \in \mathcal{I}_{G,s}} \frac{\left\| \begin{bmatrix} u_{\Delta,s}(k) \\ y_{\Delta,s}(k) \end{bmatrix} - \begin{bmatrix} u_s(k) \\ y_s(k) \end{bmatrix} \right\|}{\left\| \begin{bmatrix} u_{\Delta,s}(k) \\ y_{\Delta,s}(k) \end{bmatrix} \right\|}.$$

Let

$$\begin{bmatrix} u_{\Delta,s}(k) \\ y_{\Delta,s}(k) \end{bmatrix} = \begin{bmatrix} M_s + \Delta_{M_s} \\ N_s + \Delta_{N_s} \end{bmatrix} v_{s+n}(k-n), \begin{bmatrix} u_s(k) \\ y_s(k) \end{bmatrix} = \begin{bmatrix} M_s \\ N_s \end{bmatrix} v_{s+n}(k-n).$$

It holds

$$\delta\left(\mathcal{I}_{G_\Delta,s},\mathcal{I}_{G,s}\right)=\sup_{\begin{bmatrix}u_{\Delta,s}(k)\\ y_{\Delta,s}(k)\end{bmatrix}\in\mathcal{I}_{G_\Delta,s}}\frac{\left\|U_{I_G,2}^T\begin{bmatrix}\Delta_{M_s}\\ \Delta_{N_s}\end{bmatrix}v_{s+n}(k-n)\right\|}{\left\|I_{G_{\Delta,0},s}v_{s+n}(k-n)\right\|},$$

$$\left\|I_{G_{\Delta,0},s}v_{s+n}(k-n)\right\|=\left\|v_{s+n}(k-n)\right\|\Longrightarrow$$

$$\delta\left(\mathcal{I}_{G_\Delta,s},\mathcal{I}_{G,s}\right)=\sup_{\begin{bmatrix}\Delta_{M_s}\\ \Delta_{N_s}\end{bmatrix}}\left\|U_{I_G,2}^T\begin{bmatrix}\Delta_{M_s}\\ \Delta_{N_s}\end{bmatrix}\right\|_2\leq\delta_{\Delta_{I_G}}<1.$$

As a result, the threshold is set to be

$$J_{th,rel}=\sup_{\begin{bmatrix}\Delta_{M_s}\\ \Delta_{N_s}\end{bmatrix}}J_{rel}=\sup_{\begin{bmatrix}\Delta_{M_s}\\ \Delta_{N_s}\end{bmatrix}}\frac{\left\|r_{\mathcal{P}_{\mathcal{I}_{G,s}},s}(k)\right\|}{\left\|\begin{bmatrix}u_{\Delta,s}(k)\\ y_{\Delta,s}(k)\end{bmatrix}\right\|}=\delta_{\Delta_{I_G}}.$$

13.2.3 Fault Detection in Feedback Control Systems

Fault detection in feedback control systems with (parametric) uncertainties is a challenging issue. In the framework of model-based fault detection, threshold settings in $\mathcal{H}_2$-subspace might be too conservative, leading to lower detection rate. This motivates us to leverage the finite samples system representations to deal with this issue.

Suppose that the control system under consideration is given by

$$\begin{cases}\hat{x}(k+1)=(A+BF)\hat{x}(k)+Bv(k)+Lr(k),\ \hat{x}(0)=0\\ \begin{bmatrix}u(k)\\ y(k)\end{bmatrix}=\begin{bmatrix}F\\ C+DF\end{bmatrix}\hat{x}(k)+\begin{bmatrix}I\\ D\end{bmatrix}v(k)+\begin{bmatrix}0\\ I\end{bmatrix}r(k)\\ r(k)=y(k)-C\hat{x}(k)-Du(k),\end{cases}\tag{13.20}$$

and the plant SIR is modelled by

$$I_G=I_{G,0}+\Delta_{I_G}=\begin{bmatrix}M\\ N\end{bmatrix},\ I_{G,0}=\begin{bmatrix}M_0\\ N_0\end{bmatrix},\ \Delta_{I_G}=\begin{bmatrix}\Delta_M\\ \Delta_N\end{bmatrix}\in\mathcal{RH}_\infty,$$

where $I_{G,0}$ represents the nominal SIR and Δ_{I_G} the uncertainty. It follows from Theorem 4.6 that

$$\begin{bmatrix} u \\ y \end{bmatrix} = \begin{bmatrix} M_0 \\ N_0 \end{bmatrix} v + \begin{bmatrix} -\hat{Y}_0 \\ \hat{X}_0 \end{bmatrix} \delta_{I_K} \left(I + \delta_{I_G}\right)^{-1} v, \tag{13.21}$$

$$\hat{X}_0 = (A + BF, L, C + DF, I), \hat{Y}_0 = (A + BF, L, -F, 0),$$

where $\left(\delta_{I_K}, \delta_{I_G}\right)$ are defined by

$$\Delta_{I_G} = \begin{bmatrix} M_0 \\ N_0 \end{bmatrix} \delta_{I_G} + \begin{bmatrix} -\hat{Y}_0 \\ \hat{X}_0 \end{bmatrix} \delta_{I_K}.$$

Now, we construct a residual generator as

$$r = \begin{bmatrix} u \\ y \end{bmatrix} - \begin{bmatrix} M_0 \\ N_0 \end{bmatrix} v = \begin{bmatrix} -\hat{Y}_0 \\ \hat{X}_0 \end{bmatrix} \delta_{I_K} \left(I + \delta_{I_G}\right)^{-1} v. \tag{13.22}$$

The established procedure to manage residual evaluation is to use $\mathcal{L}_2$-norm of r as the evaluation function,

$$J = \|r\|_2,$$

and the corresponding threshold is

$$J_{th} = \sup_{\left(\delta_{I_K}, \delta_{I_G}\right)} \left\| \begin{bmatrix} -\hat{Y}_0 \\ \hat{X}_0 \end{bmatrix} \delta_{I_K} \left(I + \delta_{I_G}\right)^{-1} \right\|_\infty \|v\|_2.$$

It is apparent that the related upper-bound of the $\mathcal{H}_\infty$-norm may lead to (considerably) conservative threshold setting, and for practical applications, the $\mathcal{L}_2$-norms of the related signals are to be approximated by the Euclidean norms of the signals over a finite time interval. These motivate us to explore a solution using the associated finite samples representations.

Along the lines of the discussion in Sect. 7.3.5, let us firstly introduce notations,

$$\bar{v}(z) = V_1(z)v(z), V_1 = I + (F - F_d)(zI - A_F)^{-1} B,$$
$$\bar{R} = (F_d - F)\left(zI - A_{F_d}\right)^{-1} L, A_{F_d} = A + BF_d.$$

The closed-loop dynamic is then rewritten as

$$\begin{bmatrix} u \\ y \end{bmatrix} = \begin{bmatrix} M_0 \\ N_0 \end{bmatrix} \bar{v} + \left(\begin{bmatrix} -\hat{Y}_0 \\ \hat{X}_0 \end{bmatrix} + \begin{bmatrix} M_0 \\ N_0 \end{bmatrix} \bar{R} \right) \delta_{I_K} \left(I + \delta_{I_G}\right)^{-1} \bar{v},$$

where

$$u = F_d \hat{x}(k) + \bar{v}(k).$$

The corresponding finite samples model is described by

$$\begin{bmatrix} u_s(k) \\ y_s(k) \end{bmatrix} = \begin{bmatrix} M_s \\ N_s \end{bmatrix} \bar{v}_{s+n}(k-n) + \begin{bmatrix} -\hat{Y}_s & M_s \\ \hat{X}_s & N_s \end{bmatrix} \bar{v}_{\delta,s+n}(k-n),$$

$$\bar{v}_\delta = \begin{bmatrix} I \\ \bar{R} \end{bmatrix} \delta_{I_K} \left(I + \delta_{I_G}\right)^{-1} \bar{v},\ \bar{v}_{\delta,s+n}(k-n) = \begin{bmatrix} \bar{v}_\delta(k-n+1) \\ \vdots \\ \bar{v}_\delta(k+s) \end{bmatrix}.$$

Analogous to the residual generator (13.22), the residual generator is built by

$$r_s(k) = \begin{bmatrix} u_s(k) \\ y_s(k) \end{bmatrix} - \begin{bmatrix} M_s \\ N_s \end{bmatrix} \bar{v}_{s+n}(k-n),$$

whose dynamic is subject to

$$r_s(k) = \begin{bmatrix} -\hat{Y}_s & M_s \\ \hat{X}_s & N_s \end{bmatrix} \bar{v}_{\delta,s+n}(k-n).$$

Now, define an optimal residual evaluation function

$$J = \frac{1}{s} r_s^T(k) \Sigma_s^{-1} r_s(k),$$

$$\Sigma_s = \begin{bmatrix} -\hat{Y}_s & M_s \\ \hat{X}_s & N_s \end{bmatrix} \begin{bmatrix} -\hat{Y}_s & M_s \\ \hat{X}_s & N_s \end{bmatrix}^T.$$

The threshold is

$$\begin{aligned} J_{th} &= \sup_{(\delta_{I_K}, \delta_{I_G})} \frac{\bar{v}_{\delta,s+n}^T(k-n)\bar{v}_{\delta,s+n}(k-n)}{s} \\ &= \sup_{(\delta_{I_K}, \delta_{I_G})} \left\| \begin{bmatrix} I \\ \bar{R} \end{bmatrix} \delta_{I_K} \left(I + \delta_{I_G}\right)^{-1} \right\|_\infty^2 \frac{\bar{v}_{s+n}^T(k-n)\bar{v}_{s+n}(k-n)}{s}. \end{aligned}$$

Note that on the assumption

$$\begin{bmatrix} -\hat{N}_0 & \hat{M}_0 \\ X_0 & Y_0 \end{bmatrix} \begin{bmatrix} \Delta_M \\ \Delta_N \end{bmatrix} = \begin{bmatrix} \delta_{IK} \\ \delta_{IG} \end{bmatrix}, \left\| \begin{bmatrix} \delta_{IK} \\ \delta_{IG} \end{bmatrix} \right\|_\infty \leq \delta_{I_G},$$

we obtain

$$\sup_{(\delta_{I_K}, \delta_{I_G})} \left\| \begin{bmatrix} I \\ \bar{R} \end{bmatrix} \delta_{I_K} \left(I + \delta_{I_G}\right)^{-1} \right\|_\infty^2 = \left(1 + \|\bar{R}\|_\infty^2\right) \frac{\delta_{I_G}^2}{1 - \delta_{I_G}^2}.$$

Remark 13.1 In fact, the uncertainty

$$\begin{bmatrix} \delta_{I_K} \\ \delta_{IG} \end{bmatrix} = \begin{bmatrix} -\hat{N}_0 & \hat{M}_0 \\ X_0 & Y_0 \end{bmatrix} \begin{bmatrix} \Delta_M \\ \Delta_N \end{bmatrix}$$

can be approximately expressed in terms of a finite samples representation, which enables a tighter estimate of the upper-bound of $\left(\delta_{I_K}, \delta_{I_G}\right)$.

13.2.4 Finite Samples Representation-Based Predictive Control

In its original form, MPC is a time-varying state feedback controller. It is state of the art that a state observer is applied, delivering a state estimate, when the state vector is not measured. We now introduce an MPC system that is designed based on a finite samples model and hence without explicit use of a state observer. The problem setup is as follows.

Suppose that a feedback control system builds the inner loop, whose dynamic is governed by

$$\begin{cases} x(k+1) = Ax(k) + Bu(k), x(0) = x_0 \\ y(k) = Cx(k) + Du(k) \\ u(z) = K(z)y(z) + v(z). \end{cases} \tag{13.23}$$

The output feedback control law, $K(z)$, is embedded in the loop to assure the loop stability and required robustness. The signal vector v is the output of an MPC algorithm, performing a predictive (optimal) tracking control. To be specific, consider a reference model presented in the form of an SIR,

$$\begin{bmatrix} u_{ref} \\ y_{ref} \end{bmatrix} = \begin{bmatrix} M_r \\ N_r \end{bmatrix} v_{ref} =: I_{G_r} v_{ref}. \tag{13.24}$$

The objective of predictive tracking control is to minimise the predicted tracking error over the time interval $[k+1, k+s]$, $s \geq n$,

$$\sum_{i=1}^{s} \left(\begin{bmatrix} u_{ref}(k+i) \\ y_{ref}(k+i) \end{bmatrix} - \begin{bmatrix} u(k+i\,|k) \\ y(k+i\,|k) \end{bmatrix} \right)^T \left(\begin{bmatrix} u_{ref}(k+i) \\ y_{ref}(k+i) \end{bmatrix} - \begin{bmatrix} u(k+i\,|k) \\ y(k+i\,|k) \end{bmatrix} \right).$$

Hereby, $(u(k+i\,|k), y(k+i\,|k))$, $i = 1, \ldots, s$, denote the predicted input and output variables given the data up to k.

For our purpose, the output feedback controller is equivalently written into its observer-based parameterisation form,

$$u(z) = F\hat{x} + Qr_y + \left(X + Q\hat{N}\right) v,$$

with the (output) residual vector r_y and (stable) parameter system Q. Correspondingly, the dynamics of the inner loop are expressed by

$$\begin{bmatrix} u \\ y \end{bmatrix} = \begin{bmatrix} M \\ N \end{bmatrix} \left(X + Q\hat{N} \right) v + \begin{bmatrix} -\hat{Y} + MQ \\ \hat{X} + NQ \end{bmatrix} r_y. \tag{13.25}$$

Recall that the second term in the above expression represents the influence of uncertainties, expressed in terms of the residual r_y and suppressed by setting Q. In other words, to optimise the tracking performance, the latent variable

$$\bar{v} = \left(X + Q\hat{N} \right) v$$

is to be appropriately designed. To this end, we begin with the finite samples representations of the reference model (13.24) and the inner loop dynamics (13.25). Suppose that $x_r \in \mathbb{R}^n$ denotes the hidden state vector of the reference model (13.24) whose state space representation is described by

$$\begin{cases} x_r(k+1) = (A_r + B_r F_r) x_r(k) + B_f v_{ref}(k) \\ y_{ref}(k) = (C_r + D_r F_r) x_r(k) + D_r v_{ref}(k) \\ u_{ref}(k) = F_r x_r(k) + v_{ref}(k) \end{cases} \tag{13.26}$$

Rewrite it into

$$\begin{bmatrix} u_{ref} \\ y_{ref} \end{bmatrix} = \begin{bmatrix} M_{r,d} \\ N_{r,d} \end{bmatrix} \bar{v}_{ref},$$
$$\bar{v}_{ref}(z) = V_1(z)v(z), \; V_1 = I + (F_r - F_d)\left(zI - A_{r,F}\right)^{-1} B_r,$$

where $A_{r,F} = A_r + B_r F_r$, $\left(M_{r,d}, N_{r,d}\right)$ is the SIR of the reference model subject to (13.26) with F_r being replaced by F_d. Hence, we have the finite samples representation of the reference model (13.24),

$$\begin{bmatrix} u_{ref,s}(k) \\ y_{ref,s}(k) \end{bmatrix} = \begin{bmatrix} M_{r,s} \\ N_{r,s} \end{bmatrix} \bar{v}_{ref,s+n}(k-n). \tag{13.27}$$

Similarly, the finite samples representation of the inner loop dynamics (13.25) is expressed by

$$\begin{bmatrix} u_s(k) \\ y_s(k) \end{bmatrix} = \begin{bmatrix} M_s \\ N_s \end{bmatrix} \bar{v}_{s+n}(k-n) + \begin{bmatrix} -\hat{Y}_s & M_s \\ \hat{X}_s & N_s \end{bmatrix} \begin{bmatrix} r_{y,s+n}(k-n) \\ r_{y,Q,s+n}(k-n) \end{bmatrix}, \tag{13.28}$$
$$\bar{v} := \left(I + (F - F_d)(zI - A_F)^{-1} B \right) \left(X + Q\hat{N} \right) v,$$
$$r_{y,Q} = \left(Q + \bar{R} \right) r_y, \; \bar{R} = (F_d - F)\left(zI - A_{F_d}\right)^{-1} L,$$

$$\bar{v}_{s+n}(k-n)=\begin{bmatrix}\bar{v}(k-n+1)\\ \vdots\\ \bar{v}(k+s)\end{bmatrix}, r_{y,Q,s+n}(k-n)=\begin{bmatrix}r_{y,Q}(k-n+1)\\ \vdots\\ r_{y,Q}(k+s)\end{bmatrix}.$$

For a predictive control, only $\bar{v}(k+i), i=1,\ldots,s,$ are arbitrarily selectable. Therefore, $\begin{bmatrix}M_s\\ N_s\end{bmatrix}\bar{v}_{s+n}(k-n)$ is further split into

$$\begin{bmatrix}M_s\\ N_s\end{bmatrix}\bar{v}_{s+n}(k-n)=\begin{bmatrix}\bar{M}_0\\ \bar{N}_0\end{bmatrix}\bar{v}_n(k-n)+\begin{bmatrix}\bar{M}_s\\ \bar{N}_s\end{bmatrix}\bar{v}_s(k), \tag{13.29}$$

$$\begin{bmatrix}\bar{M}_0\\ \bar{N}_0\end{bmatrix}=\begin{bmatrix}H_{u,x,F}H_{x,v,F}\\ H_{y,x,F}H_{x,v,F}\end{bmatrix}, \begin{bmatrix}\bar{M}_s\\ \bar{N}_s\end{bmatrix}=\begin{bmatrix}H_{u,v,F}\\ H_{y,v,F}\end{bmatrix}.$$

Considering that no observer is explicitly integrated in the inner loop, a finite samples SKR of the plant,

$$r_s(k)=\begin{bmatrix}-\hat{N}_s & \hat{M}_s\end{bmatrix}\begin{bmatrix}u_{s+n}(k-n)\\ y_{s+n}(k-n)\end{bmatrix}, r_s(k)=\begin{bmatrix}r(k-n+1)\\ \vdots\\ r(k+s)\end{bmatrix}, \tag{13.30}$$

is constructed and used for generating r_y and $r_{y,Q}$ by

$$r_y=Rr, r_{y,Q}=\left(Q+\bar{R}\right)Rr, R=I-C\left(zI-A_L\right)^{-1}\left(L-L_d\right).$$

Starting from the finite samples representations (13.27) and (13.28), we are in a position to specify the optimal predictive tracking control problem. Given the predictive reference model

$$\begin{bmatrix}u_{ref,s}(k)\\ y_{ref,s}(k)\end{bmatrix}=\begin{bmatrix}M_{r,s}\\ N_{r,s}\end{bmatrix}\bar{v}_{ref,s+n}(k-n), \tag{13.31}$$

and predictive inner loop model,

$$\begin{bmatrix}u_s(k\,|k)\\ y_s(k\,|k)\end{bmatrix}=\begin{bmatrix}M_s\\ N_s\end{bmatrix}\bar{v}_{s+n}(k-n)+\begin{bmatrix}-\hat{Y}_s & M_s\\ \hat{X}_s & N_s\end{bmatrix}\begin{bmatrix}r_{y|k,s+n}(k-n)\\ r_{y|k,Q,s+n}(k-n)\end{bmatrix}, \tag{13.32}$$

$$r_{y|k,s+n}(k-n)=\begin{bmatrix}r_{y|k}(k-n+1)\\ \vdots\\ r_{y|k}(k+s)\end{bmatrix},$$

$$r_{y|k,Q,s+n}(k-n)=\begin{bmatrix}r_{y|k,Q}(k-n+1)\\ \vdots\\ r_{y|k,Q}(k+s)\end{bmatrix},$$

where $r_{y|k,s+n}(k-n)$ and $r_{y|k,Q,s+n}(k-n)$ represent the influences of the residuals in the past (up to the time instant k) on the inner loop dynamics over the time interval $[k+1, k+s]$, and are defined by

$$\begin{bmatrix} u_{s+n}(k-n\,|k) \\ y_{s+n}(k-n\,|k) \end{bmatrix} = \begin{cases} \begin{bmatrix} u_{s+n}(k-n+i) \\ y_{s+n}(k-n+i) \end{bmatrix}, i = 1, \ldots, n \\ 0, i = n+1, \ldots, n+s, \end{cases}$$

$$r_{s,|k}(k) = \begin{bmatrix} r_{|k}(k+1) \\ \vdots \\ r_{|k}(k+s) \end{bmatrix} = \begin{bmatrix} -\hat{N}_s & \hat{M}_s \end{bmatrix} \begin{bmatrix} u_{s+n}(k-n\,|k) \\ y_{s+n}(k-n\,|k) \end{bmatrix}$$

$$= \begin{bmatrix} -H_{\hat{x},L} H_{\hat{x},u,L} & -H_{\hat{x},L} H_{\hat{x},y,L} \end{bmatrix} \begin{bmatrix} u_n(k-n) \\ y_n(k-n) \end{bmatrix},$$

$$\begin{bmatrix} r_{y|k} \\ r_{y|k,Q} \end{bmatrix} = \begin{bmatrix} I \\ (Q+\bar{R}) \end{bmatrix} R r_{|k},$$

build the predicted tracking error,

$$\begin{bmatrix} u_{ref,s}(k) \\ y_{ref,s}(k) \end{bmatrix} - \begin{bmatrix} u_s(k\,|k) \\ y_s(k\,|k) \end{bmatrix} = \begin{bmatrix} M_{r,s} \\ N_{r,s} \end{bmatrix} \bar{v}_{ref,s+n}(k+s-n) - \begin{bmatrix} \bar{M}_0 \\ \bar{N}_0 \end{bmatrix} \bar{v}_n(k-n)$$
$$- \begin{bmatrix} \bar{M}_s \\ \bar{N}_s \end{bmatrix} \bar{v}_s(k) - \begin{bmatrix} -\hat{Y}_s & M_s \\ \hat{X}_s & N_s \end{bmatrix} \begin{bmatrix} r_{y|k,s+n}(k-n) \\ r_{y|k,Q,s+n}(k-n) \end{bmatrix},$$

and find $\bar{v}_s(k)$ that minimises the objective function

$$J = \left\| \begin{bmatrix} u_{ref,s}(k) \\ y_{ref,s}(k) \end{bmatrix} - \begin{bmatrix} u_s(k\,|k) \\ y_s(k\,|k) \end{bmatrix} \right\|.$$

The fact that the predictive model (13.32) comprises the influence of uncertainties distinguishes the proposed finite samples predictive model from those generally applied in the existing MPC schemes.

By a routine SVD computation,

$$\begin{bmatrix} \bar{M}_s \\ \bar{N}_s \end{bmatrix} = \begin{bmatrix} \bar{U}_{I_G,1} & \bar{U}_{I_G,2} \end{bmatrix} \begin{bmatrix} \bar{\Sigma}_{I_G} & 0 \\ 0 & 0 \end{bmatrix} \begin{bmatrix} \bar{V}_{I_G,1}^T \\ \bar{V}_{I_G,2}^T \end{bmatrix} = \bar{U}_{I_G,1} \bar{\Sigma}_{I_G} \bar{V}_{I_G,1}^T, \tag{13.33}$$

it is apparent that the optimal solution $\bar{v}_s(k)$ solves the equation

$$\bar{\Sigma}_{I_G} \bar{V}_{I_G,1}^T \bar{v}_s(k) = \bar{U}_{I_G,1}^T \left(\begin{array}{l} \begin{bmatrix} M_{r,s} \\ N_{r,s} \end{bmatrix} \bar{v}_{ref,s+n}(k+s-n) - \begin{bmatrix} \bar{M}_0 \\ \bar{N}_0 \end{bmatrix} \bar{v}_n(k-n) \\ - \begin{bmatrix} -\hat{Y}_s & M_s \\ \hat{X}_s & N_s \end{bmatrix} \begin{bmatrix} r_{y|k,s+n}(k-n) \\ r_{y|k,Q,s+n}(k-n) \end{bmatrix} \end{array} \right).$$

The corresponding LS solution $\bar{\upsilon}_s^*(k)$ is then given by

$$\bar{\upsilon}_s^*(k) = \bar{V}_{I_G,1}\bar{\Sigma}_{I_G}^{-1}\bar{U}_{I_G,1}^T \left(\begin{array}{c} \begin{bmatrix} M_{r,s} \\ N_{r,s} \end{bmatrix} \bar{\upsilon}_{ref,s+n}(k+s-n) - \begin{bmatrix} \bar{M}_0 \\ \bar{N}_0 \end{bmatrix} \bar{\upsilon}_n(k-n) \\ - \begin{bmatrix} -\hat{Y}_s & M_s \\ \hat{X}_s & N_s \end{bmatrix} \begin{bmatrix} r_{y|k,s+n}(k-n) \\ r_{y|k,Q,s+n}(k-n) \end{bmatrix} \end{array} \right). \tag{13.34}$$

Finally, the input $\bar{\upsilon}^*(k+1)$ is applied to

$$\upsilon = \left(\left(I + (F - F_d)(zI - A_F)^{-1} B \right) \left(X + Q\hat{N} \right) \right)^{-1} \bar{\upsilon}^*, \tag{13.35}$$

which delivers the reference signal $\upsilon(k+1)$ to the inner loop (13.23).

The main design and implementation steps are summarised in the following algorithm.

Algorithm 13.2 *Predictive tracking control*

Step 1: *Offline: Build the finite samples representation* (13.27) *of the reference model* (13.24);
Step 2: *Build the residual generator* (13.30);
Step 3: *Build the finite samples representation of the predictive inner loop model* (13.25) *in two steps*:

- *Build*

$$\begin{bmatrix} M_s \\ N_s \end{bmatrix} \bar{\upsilon}_{s+n}(k-n) = \begin{bmatrix} \bar{M}_0 \\ \bar{N}_0 \end{bmatrix} \bar{\upsilon}_n(k-n) + \begin{bmatrix} \bar{M}_s \\ \bar{N}_s \end{bmatrix} \bar{\upsilon}_s(k),$$

- *Build*

$$\begin{bmatrix} -\hat{Y}_s & M_s \\ \hat{X}_s & N_s \end{bmatrix} \begin{bmatrix} r_{y|k,s+n}(k-n) \\ r_{y|k,Q,s+n}(k-n) \end{bmatrix};$$

Step 4: *Perform an SVD* (13.33);
Step 5: *Online: Compute the optimal solution*$\bar{\upsilon}_s^*(k)$*according to* (13.34);
Step 6: *Compute* $\upsilon(k+1)$*according to* (13.35) *and output it to the inner loop.*

Concerning the control performance, let us examine the tracking error,

$$\begin{bmatrix} u_{ref,s}(k) \\ y_{ref,s}(k) \end{bmatrix} - \begin{bmatrix} u_s(k|k) \\ y_s(k|k) \end{bmatrix}$$
$$= \left(I - \bar{U}_{I_G,1}\bar{U}_{I_G,1}^T\right) \left(\begin{array}{c} \begin{bmatrix} M_{r,s} \\ N_{r,s} \end{bmatrix} \bar{\upsilon}_{ref,s+n}(k+s-n) - \begin{bmatrix} \bar{M}_0 \\ \bar{N}_0 \end{bmatrix} \bar{\upsilon}_n(k-n) \\ - \begin{bmatrix} -\hat{Y}_s & M_s \\ \hat{X}_s & N_s \end{bmatrix} \begin{bmatrix} r_{y|k,s+n}(k-n) \\ r_{y|k,Q,s+n}(k-n) \end{bmatrix} \end{array} \right)$$

$$= \bar{U}_{I_G,2}\bar{U}_{I_G,2}^T \left(\begin{array}{c} \begin{bmatrix} M_{r,s} \\ N_{r,s} \end{bmatrix} \bar{v}_{ref,s+n}(k+s-n) - \begin{bmatrix} \bar{M}_0 \\ \bar{N}_0 \end{bmatrix} \bar{v}_n(k-n) \\ - \begin{bmatrix} -\hat{Y}_s & M_s \\ \hat{X}_s & N_s \end{bmatrix} \begin{bmatrix} r_{y|k,s+n}(k-n) \\ r_{y|k,Q,s+n}(k-n) \end{bmatrix} \end{array} \right) \Longrightarrow$$

$$J = \left\| \begin{bmatrix} u_{ref,s}(k) \\ y_{ref,s}(k) \end{bmatrix} - \begin{bmatrix} u_s(k|k) \\ y_s(k|k) \end{bmatrix} \right\|$$

$$= \left\| \left(\begin{array}{c} \begin{bmatrix} M_{r,s} \\ N_{r,s} \end{bmatrix} \bar{v}_{ref,s+n}(k+s-n) - \begin{bmatrix} \bar{M}_0 \\ \bar{N}_0 \end{bmatrix} \bar{v}_n(k-n) \\ - \begin{bmatrix} -\hat{Y}_s & M_s \\ \hat{X}_s & N_s \end{bmatrix} \begin{bmatrix} r_{y|k,s+n}(k-n) \\ r_{y|k,Q,s+n}(k-n) \end{bmatrix} \end{array} \right) \right\|_{\bar{U}_{I_G,2}\bar{U}_{I_G,2}^T} .$$

At the end of this study, we would like to emphasise that an optimal feedback control with respect to the parameter system Q, for instance,

$$\min_{Q \in \mathcal{RH}_\infty} \left\| \begin{bmatrix} -\hat{Y} + MQ \\ \hat{X} + NQ \end{bmatrix} \right\|_\infty ,$$

may improve the predictive control performance as well, because it reduces the influence of uncertainties on the predictive tracking error via the term

$$\begin{bmatrix} -\hat{Y}_s & M_s \\ \hat{X}_s & N_s \end{bmatrix} \begin{bmatrix} r_{y|k,s+n}(k-n) \\ r_{y|k,Q,s+n}(k-n) \end{bmatrix} .$$

13.3 Data-Driven Unified Framework of Control and Detection

We now remove the assumption of the available plant model and study detection and control issues in the data-driven fashion, namely, detection and control systems will be designed, analysed and implemented model-free and on the basis of collected raw process input and output data. To this aim, data-driven version and interpretations of some essential concepts and core results of the model-based unified framework are explored.

13.3.1 A Review of State of the Art Work and Open Issues

We begin with a short review of state of the art research in the data-driven control domain, and identify open issues to motivate the subsequent work.

Willems's fundamental lemma is at the core of the data-driven control technique that has become popular in the recent decade. Informally speaking, given a process input and output data set, $\begin{bmatrix} U_s \\ Y_s \end{bmatrix}$, the fundamental lemma claims that any system trajectory $\begin{bmatrix} u_s(k) \\ y_s(k) \end{bmatrix}$ can be described by

$$\begin{bmatrix} u_s(k) \\ y_s(k) \end{bmatrix} = \begin{bmatrix} U_s \\ Y_s \end{bmatrix} v_l,$$

for some latent vector v_l, and for any given v_l, $\begin{bmatrix} U_s \\ Y_s \end{bmatrix} v_l$ builds a trajectory subject to the system dynamic. Hereby, the input data set U_s should be persistently exciting, i.e. the rank condition

$$rank \begin{bmatrix} U_s \\ Y_s \end{bmatrix} = sp + n \tag{13.36}$$

is satisfied. In Sect. 7.3, it has been illustrated that the data set (U_s, Y_s) can be generated by

$$V_{s+n,M} = \left[v_{s+n}(k-n) \cdots v_{s+n}(k-n+M-1) \right],$$

the data set of the latent variable $v_{s+n}(k-n+i), i = 0, \ldots, M-1$, satisfying

$$rank \left(V_{s+n,M} \right) = (s+n)\, p,$$

so that condition (13.36) holds. In this regard, the input signal is called persistently exciting.

In a mathematically and control theoretically rigorous survey paper, Markovsky and Dörfler have systematically reviewed data-driven analysis and control methods in the framework of behavioural systems theory. Concerning control issues, two open-loop (feedforward) control schemes have been presented. One is the so-called data-driven linear quadratic (LQ) tracking control, which is formulated as, given the reference vector,

$$\begin{bmatrix} u_{r,N}(k) \\ y_{r,N}(k) \end{bmatrix} = \begin{bmatrix} u(k+1) \\ \vdots \\ u(k+N) \\ y(k+1) \\ \vdots \\ y(k+N) \end{bmatrix},$$

with $N \geq 1$ being the length (sample number) of the future time interval, and the data set

$$\begin{bmatrix} U_s \\ Y_s \end{bmatrix}, s = k_p + N,$$

that satisfies the persistency of excitation condition with k_p representing the needed number of samples for re-constructing the initial state vector, solve the following optimisation problem,

$$\min_{u_s(k), y_s(k), v_l} \left\| y_{r,N}(k) - y_N(k) \right\|_Q^2 + \left\| u_{r,N}(k) - u_N(k) \right\|_R^2 \tag{13.37}$$

$$\text{s.t.} \begin{bmatrix} u_{k_p}(k-k_p) \\ y_{k_p}(k-k_p) \\ u_N(k) \\ y_N(k) \end{bmatrix} = \begin{bmatrix} U_{k_p+N} \\ Y_{k_p+N} \end{bmatrix} v_l,$$

where (Q, R) are weighting matrices. Another is data-driven predictive control. Among a number of schemes, the so-called data-enabled predictive control (DeePC) is the popular one. Analogous to the LQ tracking control, DeePC scheme is formulated as

$$\min_{u_s(k), y_s(k), v_l} \left\| y_{r,N}(k) - y_N(k) \right\|_Q^2 + \left\| u_{r,N}(k) - u_N(k) \right\|_R^2 \tag{13.38}$$

$$\text{s.t.} \begin{bmatrix} u_{k_p}(k-k_p) \\ y_{k_p}(k-k_p) \\ u_N(k) \\ y_N(k) \end{bmatrix} = \begin{bmatrix} U_{k_p+N} \\ Y_{k_p+N} \end{bmatrix} v_l,$$

$$u_N(k) \in \mathcal{U}, \; y_N(k) \in \mathcal{Y},$$

where $(u_N(k), y_N(k))$ are N-step ahead predicted input and output vector, and $(\mathcal{U}, \mathcal{Y})$ represent the constraint sets of the input and output vectors. Recall that

$$rank \begin{bmatrix} U_{k_p+N} \\ Y_{k_p+N} \end{bmatrix} = \left(k_p + N\right) p + n, \; v_l \in \mathbb{R}^M,$$

and generally $M > \left(k_p + N\right) p + n$. The additional degree of design freedom is often used to enhance the system robustness. To this aim, it has been proposed to add a regularisation term given by

$$\left\| (I - \Pi) v_l \right\|^2, \; \Pi = \begin{bmatrix} U_{k_p} \\ Y_{k_p} \\ U_N \end{bmatrix}^{\dagger} \begin{bmatrix} U_{k_p} \\ Y_{k_p} \\ U_N \end{bmatrix}, \tag{13.39}$$

$$U_{k_p} = \left[u_{k_p}(j) \cdots u_{k_p}(j+M-1) \right] \in \mathbb{R}^{pk_p \times M},$$
$$Y_{k_p} = \left[y_{k_p}(j) \cdots y_{k_p}(j+M-1) \right] \in \mathbb{R}^{mk_p \times M},$$
$$U_N = \left[u_N(j+k_p) \cdots u_N(j+k_p+M-1) \right] \in \mathbb{R}^{pN \times M},$$

in the optimisation problem (13.38), where $[\cdot]^{\dagger}$ denotes the Moore–Penrose inverse. The idea behind (13.39) is to dampen the influence of uncertainties, since the matrix

$I - \Pi$ spans the (output) residual subspace. To clarify this aspect, consider the SKR-based system model,

$$\begin{cases} \hat{x}(k+1) = A\hat{x}(k) + Bu(k) + Lr_0(k) \\ y(k) = C\hat{x}(k) + Du(k) + r_0(k), \end{cases} \tag{13.40}$$
$$r_0(k) = y(k) - C\hat{x}(k) - Du(k),$$

from which a finite samples input-output model is built as follows,

$$y_s(k) = H_{\hat{x}}\hat{x}(k+1) + H_u u_s(k) + H_r r_{0,s}(k), \tag{13.41}$$
$$H_{\hat{x}} = \begin{bmatrix} C \\ CA \\ \vdots \\ CA^{s-1} \end{bmatrix} \in \mathbb{R}^{sm\times n},\ H_r = \begin{bmatrix} I & 0 & \cdots & 0 \\ CL & I & \ddots & \vdots \\ \vdots & \ddots & \ddots & 0 \\ CA^{s-2}L & \cdots & CL & I \end{bmatrix} \in \mathbb{R}^{sm\times sm},$$
$$H_u = \begin{bmatrix} D & 0 & \cdots & 0 \\ CB & D & \ddots & \vdots \\ \vdots & \ddots & \ddots & 0 \\ CA^{s-2}B & \cdots & CB & D \end{bmatrix} \in \mathbb{R}^{sm\times sp},\ r_{0,s}(k) = \begin{bmatrix} r_0(k+1) \\ \vdots \\ r_0(k+s) \end{bmatrix}.$$

Now, substituting $\hat{x}(k+1)$ by input and output vectors,

$$\hat{x}(k+1) = A_{L_0}\hat{x}(k) + B_{L_0}u(k) + L_0 y(k)$$
$$= \begin{bmatrix} H_{\hat{x},u} & H_{\hat{x},r} \end{bmatrix} \begin{bmatrix} u_n(k-n) \\ y_n(k-n) \end{bmatrix},\ A_{L_0} = A - L_0C,\ B_{L_0} = B - L_0D,$$
$$H_{\hat{x},u} = \begin{bmatrix} A_{L_0}^{n-1}B_{L_0} & \cdots & B_{L_0} \end{bmatrix},\ H_{\hat{x},y} = \begin{bmatrix} A_{L_0}^{n-1}L_0 & \cdots & L_0 \end{bmatrix}$$

with deadbeat observer gain matrix L_0, yields

$$y_s(k) = \begin{bmatrix} H_{\hat{x}}H_{\hat{x},u} & H_{\hat{x}}H_{\hat{x},y} & H_u \end{bmatrix} \begin{bmatrix} u_n(k-n) \\ y_n(k-n) \\ u_s(k) \end{bmatrix} + H_r r_{0,s}(k). \tag{13.42}$$

The data-driven (matrix) input-output model follows immediately from (13.42),

$$Y_N = \begin{bmatrix} H_{\hat{x}}H_{\hat{x},u} & H_{\hat{x}}H_{\hat{x},y} & H_u \end{bmatrix} \begin{bmatrix} U_{k_p} \\ Y_{k_p} \\ U_N \end{bmatrix} + H_r R_{0,N}, \tag{13.43}$$
$$R_{0,N} = \begin{bmatrix} r_{0,N}(j+k_p) & \cdots & r_{0,N}(j+k_p+M-1) \end{bmatrix} \in \mathbb{R}^{mN\times M},$$
$$Y_N = \begin{bmatrix} y_N(j+k_p) & \cdots & y_N(j+k_p+M-1) \end{bmatrix} \in \mathbb{R}^{mN\times M}.$$

Hence, the matrix

$$Y_N \begin{bmatrix} U_{k_p} \\ Y_{k_p} \\ U_N \end{bmatrix}^{\dagger} =: \left[H_{\hat{x}} H_{\hat{x},u} \;\; H_{\hat{x}} H_{\hat{x},y} \;\; H_u \right]$$

delivers an LS estimate of $\left[H_{\hat{x}} H_{\hat{x},u} \;\; H_{\hat{x}} H_{\hat{x},y} \;\; H_u \right]$, and correspondingly,

$$\hat{y}_N(k) := Y_N \begin{bmatrix} U_{k_p} \\ Y_{k_p} \\ U_N \end{bmatrix}^{\dagger} \begin{bmatrix} u_{k_p}(k-k_p) \\ y_{k_p}(k-k_p) \\ u_N(k) \end{bmatrix}$$

is an LS estimate of $y_N(k)$. Therefore,

$$y_N(k) - \hat{y}_N(k) = (I - \Pi) \begin{bmatrix} u_{k_p}(k-k_p) \\ y_{k_p}(k-k_p) \\ u_N(k) \end{bmatrix} = H_r r_{0,s}(k)$$

builds a residual vector, and the matrix $I - \Pi$ spans the residual subspace.

For details about the aforementioned methods, the reader is referred to the references given at the end of this chapter.

Although robustness issues have received certain attention in the current research of data-driven control, no specified endeavour dedicated to uncertainties in the data set (U_s, Y_s) has been reported. In light of the unified control and detection framework, state of the art research efforts focus on the system image subspace, namely

$$\begin{bmatrix} u_s(k) \\ y_s(k) \end{bmatrix} = \begin{bmatrix} U_s \\ Y_s \end{bmatrix} v_l,$$

which represents the system nominal dynamics, while the residual subspace has been rarely examined. It is known that the residual subspace is indeed the core of fault detection and feedback control. This notice inspires our subsequent endeavours to leverage model-based unified framework to address data-driven detection and control issues.

13.3.2 Data-Driven Image and Residual Subspaces

In order to extend the model-based image, kernel representations as well the associated image and residual subspaces to the data-driven framework, following definitions and notations are introduced.

Definition 13.1 Given data sets collected from the process (13.1)–(13.2) during nominal operations,

$$U_s = \begin{bmatrix} u_s(j) \cdots u_s(j+M-1) \end{bmatrix} \in \mathbb{R}^{ps \times M},$$
$$Y_s = \begin{bmatrix} y_s(j) \cdots y_s(j+M-1) \end{bmatrix} \in \mathbb{R}^{ms \times M},$$

with $j \in \mathbb{N} := \{1, 2, \ldots\}$, $s > n$,

$$M \geq (s+n)\,p, rank \begin{bmatrix} U_s \\ Y_s \end{bmatrix} = sp + n, \tag{13.44}$$

the data matrix

$$I_{DD,s} := \begin{bmatrix} U_s \\ Y_s \end{bmatrix} \in \mathbb{R}^{(p+m)s \times M} \tag{13.45}$$

is called data-driven image representation of order s. Correspondingly, the subspace

$$\mathcal{I}_{DD,s} = \left\{ \begin{bmatrix} u_s(k) \\ y_s(k) \end{bmatrix}, \begin{bmatrix} u_s(k) \\ y_s(k) \end{bmatrix} = I_{DD,s} v_l,\, v_l \in \mathcal{L}^M_{[0,\infty)} \right\}, \tag{13.46}$$

$\forall k \in \mathbb{N}$, is called data-driven image subspace of order s. The vector $v_l \in \mathbb{R}^M$ is called latent variable.

Differing from the model-based framework, data-driven kernel representation will be defined in terms of the null-subspace of the image representation. To this end, perform an SVD of $I_{DD,s}$,

$$I_{DD,s} = \begin{bmatrix} U_{I_{DD},1} \; U_{I_{DD},2} \end{bmatrix} \begin{bmatrix} \Sigma_{I_{DD}} & 0 \\ 0 & 0 \end{bmatrix} \begin{bmatrix} V^T_{I_{DD},1} \\ V^T_{I_{DD},2} \end{bmatrix} = U_{I_{DD},1} \Sigma_{I_{DD}} V^T_{I_{DD},1}, \tag{13.47}$$

$$\Sigma_{I_{DD}} = diag\left(\sigma_1, \ldots, \sigma_\gamma\right) \in \mathbb{R}^{\gamma \times \gamma}, \gamma = rank\left(I_{DD,s}\right) = sp + n. \tag{13.48}$$

By means of $U_{I_{DD},2} \in \mathbb{R}^{(p+m)s \times \eta}$,

$$\eta = (p+m)\,s - \gamma = ms - n,$$

data-driven kernel representation is now defined.

Definition 13.2 Given the data set $I_{DD,s}$ whose SVD is given by (13.47), the matrix

$$K_{DD,s} := U^T_{I_{DD},2} \in \mathbb{R}^{\eta \times (p+m)s}$$

is called data-driven kernel representation of order s.

It is clear that

$$\forall \begin{bmatrix} u_s(k) \\ y_s(k) \end{bmatrix} \in \mathcal{I}_{DD,s}, \, K_{DD,s} \begin{bmatrix} u_s(k) \\ y_s(k) \end{bmatrix} = U_{I_{DD},2}^T \begin{bmatrix} u_s(k) \\ y_s(k) \end{bmatrix} = 0.$$

That means,

$$r_{K,\eta}(k) = K_{DD,s} \begin{bmatrix} u_s(k) \\ y_s(k) \end{bmatrix} \in \mathbb{R}^{\eta}, \tag{13.49}$$

is a kernel-based residual generator with the residual vector $r_{K,\eta}(k)$ as the output. Keeping it in mind, the data-driven residual subspace, as the orthogonal complement of $\mathcal{I}_{DD,s}$, is defined.

Definition 13.3 The subspace

$$\mathcal{R}_{DD,s} = \left\{ \begin{bmatrix} u_s(k) \\ y_s(k) \end{bmatrix}, \begin{bmatrix} u_s(k) \\ y_s(k) \end{bmatrix} = K_{DD,s}^T r_{K,\eta}(k), \, r_{K,\gamma}(k) \in \mathcal{L}_{[0,\infty)} \right\}, \tag{13.50}$$

$\forall k \in \mathbb{N}$, is called data-driven residual subspace of order s.

Observe that

$$\forall \begin{bmatrix} u_s(k) \\ y_s(k) \end{bmatrix} \in \mathcal{R}_{DD,s}, \begin{bmatrix} u_s(k) \\ y_s(k) \end{bmatrix} = K_{DD,s}^T r_{K,\eta}(k) = K_{DD,s}^T K_{DD,s} \begin{bmatrix} u_s(k) \\ y_s(k) \end{bmatrix},$$
$$K_{DD,s}^T K_{DD,s} = U_{I_{DD},2} U_{I_{DD},2}^T, \, U_{I_{DD},2}^T U_{I_{DD},2} = I.$$

Hence, $K_{DD,s}^T K_{DD,s}$ imposes an orthogonal projection onto the residual subspace of the process data. Similarly,

$$U_{I_{DD},1} U_{I_{DD},1}^T = I_{DD,s} \Sigma_{I_{DD}}^{-2} I_{DD,s}^T, \, U_{I_{DD},1} \in \mathbb{R}^{(p+m)s \times \gamma}$$

defines an orthogonal projection onto the data-driven image subspace, i.e.

$$U_{I_{DD},1} U_{I_{DD},1}^T \begin{bmatrix} u_s(k) \\ y_s(k) \end{bmatrix} \in \mathcal{I}_{DD,s},$$
$$\forall \begin{bmatrix} u_s(k) \\ y_s(k) \end{bmatrix} \in \mathcal{I}_{DD,s}, \, U_{I_{DD},1} U_{I_{DD},1}^T \begin{bmatrix} u_s(k) \\ y_s(k) \end{bmatrix} = \begin{bmatrix} u_s(k) \\ y_s(k) \end{bmatrix}.$$

Attributed to the SVD (13.47), a further useful and essential relation holds

$$K_{DD,s}^T K_{DD,s} + I_{DD,s} \Sigma_{I_{DD}}^{-2} I_{DD,s}^T = U_{I_{DD},1} U_{I_{DD},1}^T + U_{I_{DD},2} U_{I_{DD},2}^T = I.$$

Against the aforementioned background, we are now in a position to address data-driven image and kernel representations given a sufficient number of real process data, which are corrupted with noises and uncertainties. Let

$$\begin{bmatrix} U_s \\ Y_s \end{bmatrix} = \begin{bmatrix} u_s(j) \cdots u_s(j+M-1) \\ y_s(j) \cdots y_s(j+M-1) \end{bmatrix} \in \mathbb{R}^{(p+m)s \times M},$$

for some $j \in \mathbb{N}$, and sufficient large $s (> n)$. Hereby, M and U_s satisfy the condition (13.44). Do an SVD on the data set (U_s, Y_s)

$$\begin{bmatrix} U_s \\ Y_s \end{bmatrix} = U \Sigma V^T. \tag{13.51}$$

Note that, due to the existence of noises and uncertainties, it is assumed, without loss of generality, that

$$rank \begin{bmatrix} U_s \\ Y_s \end{bmatrix} = \min\{(p+m)s, M\} =: \xi.$$

Accordingly,

$$\Sigma = \begin{cases} \begin{bmatrix} \Sigma_\xi \ 0 \end{bmatrix}, \xi = (p+m)s \\ \begin{bmatrix} \Sigma_\xi \\ 0 \end{bmatrix}, \xi = M \neq (p+m)s \quad \Sigma_\xi = diag\left(\sigma_1, \ldots, \sigma_\xi\right). \\ \Sigma_\xi, \xi = M = (p+m)s, \end{cases}$$

In the sequel, it is assumed that M is sufficient large and

$$rank \begin{bmatrix} U_s \\ Y_s \end{bmatrix} = \xi = (p+m)s.$$

In engineering applications, it is a reasonable and convincing assumption that the influence of noises and uncertainties on the system dynamics is considerably weaker than the impact of the targeted (control) input signal. Under this assumption, the singular values of Σ_ξ, $\sigma_i, i = 1, \ldots, \xi$, are divided into two groups,

$$\sigma_1, \ldots, \sigma_\gamma >> \sigma_{\gamma+1}, \ldots, \sigma_\xi,$$

for some integer $\gamma > s$, which stands for the rank of the image representation $\mathcal{I}_{DD,s}$. To determine γ, the well-known AIC/BIC-based Model Selection methods for noises or Cross-validation and Rank estimation methods for (deterministic) uncertainties can be applied, respectively.

Once γ is determined, the SVD (13.51) is further written into

$$\begin{bmatrix} U_s \\ Y_s \end{bmatrix} = \begin{bmatrix} U_{I_{DD}} \ U_{K_{DD}} \end{bmatrix} \begin{bmatrix} \Sigma_{I_{DD}} & 0 & 0 \\ 0 & \Sigma_{K_{DD}} & 0 \end{bmatrix} \begin{bmatrix} V_{I_{DD}}^T \\ V_{K_{DD}}^T \\ V_{K_{DD},0}^T \end{bmatrix},$$

$$\Sigma_{I_{DD}} = diag\left(\sigma_1, \ldots, \sigma_\gamma\right), \Sigma_{K_{DD}} = diag\left(\sigma_{\gamma+1}, \ldots, \sigma_\xi\right),$$

based on which the data-driven image and kernel representations are set as

$$I_{DD,s} := U_{I_{DD}} \Sigma_{I_{DD}}, \; K_{DD,s} := \Sigma_{K_{DD}} U_{K_{DD}}^T \tag{13.52}$$

$$\Longrightarrow \begin{bmatrix} U_s \\ Y_s \end{bmatrix} = \begin{bmatrix} I_{DD,s} & K_{DD,s}^T \end{bmatrix} \begin{bmatrix} V_{I_{DD}}^T \\ V_{K_{DD}}^T \end{bmatrix}. \tag{13.53}$$

Accordingly, the orthogonal projections onto the image and residual subspaces are respectively defined by

$$\mathcal{P}_{\mathcal{I}_{DD,s}} := U_{I_{DD}} U_{I_{DD}}^T = I_{DD,s} \Sigma_{I_{DD}}^{-2} I_{DD,s}^T, \tag{13.54}$$

$$\mathcal{P}_{\mathcal{R}_{DD,s}} := U_{K_{DD}} U_{K_{DD}}^T = K_{DD,s}^T \Sigma_{K_{DD}}^{-2} K_{DD,s}, \tag{13.55}$$

and it holds

$$U_{I_{DD}} U_{I_{DD}}^T + U_{K_{DD}} U_{K_{DD}}^T = I_{DD,s} \Sigma_{I_{DD}}^{-2} I_{DD,s}^T + K_{DD,s}^T \Sigma_{K_{DD}}^{-2} K_{DD,s} = I. \tag{13.56}$$

Note that, by means of the above orthogonal projections, the data set can be written into

$$\begin{bmatrix} U_s \\ Y_s \end{bmatrix} = \mathcal{P}_{\mathcal{I}_{DD,s}} \begin{bmatrix} U_s \\ Y_s \end{bmatrix} + \mathcal{P}_{\mathcal{R}_{DD,s}} \begin{bmatrix} U_s \\ Y_s \end{bmatrix}$$

$$= U_{I_{DD}} U_{I_{DD}}^T \begin{bmatrix} U_s \\ Y_s \end{bmatrix} + U_{K_{DD}} U_{K_{DD}}^T \begin{bmatrix} U_s \\ Y_s \end{bmatrix}.$$

Hence, we have

$$\Sigma_{I_{DD}} V_{I_{DD}}^T = U_{I_{DD}}^T \begin{bmatrix} U_s \\ Y_s \end{bmatrix}, \; \Sigma_{K_{DD}} V_{K_{DD}}^T = U_{K_{DD}}^T \begin{bmatrix} U_s \\ Y_s \end{bmatrix}. \tag{13.57}$$

13.3.3 Data-Driven System Models, Bezout Identity and Unified Framework

In Sect. 7.1, a data structure and the associated SIR-SKR-based system model have been established, whose core is the Bezout identity and the projection method. On the other hand, it has been claimed in Sect. 7.3 that in the framework of the finite samples system representations, the Bezout identity does not exist in its original form. In this subsection, on the basis of the image and kernel representations (13.52), the projection operators (13.54)–(13.55) and the relation (13.56), data-driven system models of finite samples and a corresponding data-driven version of the Bezout identity are put forward, which are of the similar forms like the comparable ones known in the unified framework.

Given process data $(u_s(k), y_s(k))$, by means of orthogonal projections we decompose them into

$$\begin{bmatrix} u_s(k) \\ y_s(k) \end{bmatrix} = \mathcal{P}_{\mathcal{I}_{DD,s}} \begin{bmatrix} u_s(k) \\ y_s(k) \end{bmatrix} + \mathcal{P}_{\mathcal{R}_{DD,s}} \begin{bmatrix} u_s(k) \\ y_s(k) \end{bmatrix}$$
$$= I_{DD,s} \Sigma_{I_{DD}}^{-2} I_{DD,s}^T \begin{bmatrix} u_s(k) \\ y_s(k) \end{bmatrix} + K_{DD,s}^T \Sigma_{K_{DD}}^{-2} K_{DD,s} \begin{bmatrix} u_s(k) \\ y_s(k) \end{bmatrix},$$

which is further written as

$$\begin{bmatrix} u_s(k) \\ y_s(k) \end{bmatrix} = \bar{I}_{DD,s} v_{\bar{I},\gamma}(k) + \bar{K}_{DD,s}^T r_{\bar{K},\eta}(k), \tag{13.58}$$

where $\bar{I}_{DD,s}$, $\bar{K}_{DD,s}$ are normalised image and kernel representations,

$$\bar{I}_{DD,s} = I_{DD,s} \Sigma_{I_{DD}}^{-1} = U_{I_{DD}}, \ \bar{K}_{DD,s}^T = K_{DD,s}^T \Sigma_{K_{DD}}^{-1} = U_{K_{DD}},$$

and $\left(v_{\bar{I},\gamma}(k), r_{\bar{K},\eta}(k)\right)$ are the residual pair,

$$v_{\bar{I},\gamma}(k) = \bar{I}_{DD,s}^T \begin{bmatrix} u_s(k) \\ y_s(k) \end{bmatrix} = U_{I_{DD}}^T \begin{bmatrix} u_s(k) \\ y_s(k) \end{bmatrix}, \tag{13.59}$$

$$r_{\bar{K},\eta}(k) = \bar{K}_{DD,s} \begin{bmatrix} u_s(k) \\ y_s(k) \end{bmatrix} = U_{K_{DD}}^T \begin{bmatrix} u_s(k) \\ y_s(k) \end{bmatrix}. \tag{13.60}$$

Equations (13.59)–(13.60) are understood as the residual generation models. Notice that models (13.58) and (13.59)–(13.60) give a one-to-one mapping between the process data $(u_s(k), y_s(k))$ and the residual data $\left(v_{\bar{I},\gamma}(k), r_{\bar{K},\eta}(k)\right)$, the spirit of the unified framework of control and detection. Mathematically, this one-to-one mapping is expressed by

$$\begin{bmatrix} \bar{I}_{DD,s} & \bar{K}_{DD,s}^T \end{bmatrix} \begin{bmatrix} \bar{I}_{DD,s}^T \\ \bar{K}_{DD,s} \end{bmatrix} = \begin{bmatrix} \bar{I}_{DD,s}^T \\ \bar{K}_{DD,s} \end{bmatrix} \begin{bmatrix} \bar{I}_{DD,s} & \bar{K}_{DD,s}^T \end{bmatrix} = \begin{bmatrix} I & 0 \\ 0 & I \end{bmatrix}, \tag{13.61}$$

called data-driven Bezout identity.

Remember that $\bar{K}_{DD,s} \bar{I}_{DD,s} = 0$. Hence, for some $Q \in \mathbb{R}^{\gamma \times \eta}$,

$$\begin{bmatrix} \bar{I}_{DD,s}^T - Q\bar{K}_{DD,s} \\ \bar{K}_{DD,s} \end{bmatrix} \begin{bmatrix} \bar{I}_{DD,s} & \bar{K}_{DD,s}^T + \bar{I}_{DD,s} Q \end{bmatrix} = \begin{bmatrix} I & 0 \\ 0 & I \end{bmatrix}, \tag{13.62}$$
$$\begin{bmatrix} \bar{I}_{DD,s} & \bar{K}_{DD,s}^T + \bar{I}_{DD,s} Q \end{bmatrix} \begin{bmatrix} \bar{I}_{DD,s}^T - Q\bar{K}_{DD,s} \\ \bar{K}_{DD,s} \end{bmatrix} = \begin{bmatrix} I & 0 \\ 0 & I \end{bmatrix},$$

which is the extended data-driven Bezout identity. On this basis, we finally have a new data-driven system model,

$$\begin{bmatrix} u_s(k) \\ y_s(k) \end{bmatrix} = \bar{I}_{DD,s} v_\gamma(k) + \left(\bar{K}_{DD,s}^T + \bar{I}_{DD,s} Q\right) r_{\bar{K},\eta}(k) \tag{13.63}$$

$$= \bar{I}_{DD,s}\left(v_\gamma(k) + Q r_{\bar{K},\eta}(k)\right) + \bar{K}_{DD,s}^T r_{\bar{K},\eta}(k), \tag{13.64}$$

$$v_{\bar{I},\gamma}(k) = v_\gamma(k) + Q r_{\bar{K},\eta}(k), \forall v_\gamma(k) \in \mathbb{R}^\gamma,$$

that is parametrised by matrix Q. Comparing with the data structure (7.8) in the unified framework, the following interesting points are ascertained.

- Similar to the data structure (7.8), any uncertainties in the process are reflected by the residual $r_{\bar{K},\eta}(k)$, and
- The variable and parameters being usable for manipulating the process dynamics are the latent variable $v_{\bar{I},\gamma}(k)$ and parameter Q.
- Informally speaking, $v_{\bar{I},\gamma}(k)$, as the input of the image system that models the nominal system dynamics, navigates the system to move along a pre-defined target trajectory, while turning Q is to attenuate the effect of uncertainty so that the system dynamic tracks the target trajectory as close as possible.
- In the aforementioned regard, the results presented in these two subsections, amongst others, the data-driven image and kernel representations, the image and residual subspaces, the Bezout identity (13.62) as well as the data-driven model (13.63), are understood as the data-driven unified framework of control and detection.
- Note that, differing from the (model-based) unified framework, in the data-driven model (13.63), the residual $r_{\bar{K},\eta}(k)$ affects the system dynamics via $\bar{K}_{DD,s}^T$, rather than via $\left(\hat{X}, \hat{Y}\right)$ which is in fact an observer, as known in the (model-based) unified framework.

13.3.4 Data-Driven Image Factor Uncertainty

In the model-based framework, coprime factor uncertainties are widely adopted to model uncertainties in a process. Analogous to it, data-driven image factor uncertainty is introduced below. Consider the data set

$$\begin{bmatrix} U_s \\ Y_s \end{bmatrix} = \begin{bmatrix} I_{DD,s} & K_{DD,s}^T \end{bmatrix} \begin{bmatrix} V_{I_{DD}}^T \\ V_{K_{DD}}^T \end{bmatrix} \tag{13.65}$$

that is corrupted with uncertainties, and define it as the image representation of the uncertain system, denoted by

$$I_{\Delta,DD,s} := \begin{bmatrix} U_s \\ Y_s \end{bmatrix} \Longrightarrow \begin{bmatrix} u_s(k) \\ y_s(k) \end{bmatrix} = \begin{bmatrix} U_s \\ Y_s \end{bmatrix} v_{l,\Delta}.$$

Now, let us define data-driven image factor uncertainty as

$$\Delta_{I_{DD,s}} := I_{\Delta,DD,s} - I_{DD,s} V_{I_{DD}}^T.$$

Hence, the data set can be alternatively expressed by

$$\begin{bmatrix} U_s \\ Y_s \end{bmatrix} = I_{DD,s} V_{I_{DD}}^T + \Delta_{I_{DD,s}}.$$

It should be noticed that, differing from the model-based formulation, $I_{\Delta,DD,s}$ and $I_{DD,s}$ have different dimension, namely,

$$I_{\Delta,DD,s} \in \mathbb{R}^{(p+m)s \times M}, \; I_{DD,s} \in \mathbb{R}^{(p+m)s \times \gamma}.$$

Furthermore, it is apparent that $\Delta_{I_{DD,s}}$ equals $K_{DD,s}^T V_{K_{DD}}^T$. That means, in the data-driven framework, the image factor uncertainty models uncertainties belonging to the system residual subspace exclusively. This feature distinguishes it from the model-based image factor uncertainty. In other words, those uncertainties belonging to the system image subspace are represented by $V_{I_{DD}}^T$.

13.4 Data-Driven Fault Detection Schemes

13.4.1 Data-Driven Fault Detection in Systems with Unknown Inputs

An immediate application of the data-driven unified framework is data-driven fault detection. To begin with, one-class fault detection issues for systems with unknown inputs are considered.

Suppose that sufficient process data are collected from a process whose dynamics are governed by (13.17) and recast in the form of a data matrix expressed by

$$\begin{bmatrix} U_s \\ Y_s \end{bmatrix} = \begin{bmatrix} I_{DD,s} & K_{DD,s}^T \end{bmatrix} \begin{bmatrix} V_{I_{DD}}^T \\ V_{K_{DD}}^T \end{bmatrix}. \tag{13.66}$$

It follows from the analysis in Sect. 13.2.1 that the dynamic of the residual generator (13.60) is subject to

$$r_{\bar{K},\eta}(k) = R_\eta \bar{K}_{DD,s} \begin{bmatrix} u_s(k) \\ y_s(k) \end{bmatrix} = R_\eta U_{K_{DD}}^T \begin{bmatrix} 0 \\ H_{d,s+n} \end{bmatrix} d_{s+n}(k-n), \tag{13.67}$$

$$H_{d,s+n} = \begin{bmatrix} CA^{n-1}E_d & \cdots & F_d & 0 & \cdots & 0 \\ CA^n E_d & \cdots & CE_d & F_d & \ddots & \vdots \\ \vdots & \vdots & \vdots & \ddots & \ddots & 0 \\ CA^{n+s-2}E_d & \cdots & CA^{s-2}E_d & \cdots & CE_d & F_d \end{bmatrix},$$

where $R_\eta \in \mathbb{R}^{\eta \times \eta}$ is a post-filter, $\eta = ms - n$. The Eq. (13.67) is attributed to the relations

$$\begin{bmatrix} u_s(k) \\ y_s(k) \end{bmatrix} = \begin{bmatrix} M_s \\ N_s \end{bmatrix} v_{s+n}(k-n) + \begin{bmatrix} 0 \\ H_{d,s+n} \end{bmatrix} d_{s+n}(k-n),$$

$$U_{K_{DD}}^T \begin{bmatrix} M_s \\ N_s \end{bmatrix} = 0.$$

Next, under the use of the residual evaluation function

$$J = \frac{\left\| r_{\bar{K},\eta}(k) \right\|^2}{\eta},$$

a threshold is to be determined. It follows from (13.67) that

$$U_{K_{DD}} \Sigma_{K_{DD}} V_{K_{DD}}^T = \begin{bmatrix} 0 \\ H_{d,s+n} \end{bmatrix} D_{s+n}, \tag{13.68}$$

$$D_{s+n} = \left[d_{s+n}(j-n) \cdots d_{s+n}(j-n+M-1) \right],$$

for some integer $j \geq n$. Notice that, in the data-driven framework, we gain access to information about unknown input variable generally through the output data. Subsequently, differing from model-based detection schemes, the threshold setting is explored on the basis of (13.68). Consider that

$$\left\| \begin{bmatrix} 0 \\ H_{d,s+n} \end{bmatrix} D_{s+n} \right\|_F^2 = \left\| H_{d,s+n} D_{s+n} \right\|_F^2$$

$$= \sum_{i=0}^{M-1} \left\| H_{d,s+n} d_{s+n}(j-n+i) \right\|^2 = \left\| \Sigma_{K_{DD}} V_{K_{DD}}^T \right\|_F^2 = \sum_{i=1}^{\eta} \sigma_{\gamma+i}^2.$$

The last equation is understood as the average energy contained in the unknown input. On this account, we introduce the assumption that

$$\sup \left\| H_{d,s+n} d_{s+n}(k-n) \right\|^2 \leq \delta_d, \tag{13.69}$$

$$\delta_d := \frac{\eta}{M} \sigma_{\gamma+1}^2, \sigma_{\gamma+1} = \max \left\{ \sigma_{\gamma+i}, i = 1, \ldots, \eta \right\}. \tag{13.70}$$

and define the threshold as

$$J_{th} := \sup_{\substack{d_{s+n}(k-n) \\ \text{s.t. (13.69)}}} J = \frac{1}{\eta} \sup_{\substack{d_{s+n}(k-n) \\ \text{s.t. (13.69)}}} \left\| r_{\bar{K},\eta}(k) \right\|^2 . \tag{13.71}$$

By routine computations, it turns out

$$\begin{aligned} J &= \frac{\left\| r_{\bar{K},\eta}(k) \right\|^2}{\eta} = \frac{\left\| R_\eta U_{K_{DD}}^T \begin{bmatrix} 0 \\ H_{d,s+n} \end{bmatrix} d_{s+n}(k-n) \right\|^2}{\eta} \\ &= \frac{\left\| R_\eta U_{K_{DD},2}^T H_{d,s+n} d_{s+n}(k-n) \right\|^2}{\eta}, \end{aligned}$$

where $U_{K_{DD},2}^T$ is a sub-matrix of $U_{K_{DD}}^T$ subject to

$$U_{K_{DD}}^T = \begin{bmatrix} U_{K_{DD},1}^T & U_{K_{DD},2}^T \end{bmatrix}, U_{K_{DD},2}^T \in \mathbb{R}^{\eta \times sm}.$$

Set R_η being a pseudo inverse of $U_{K_{DD},2}^T$ and thus

$$\left\| R_\eta U_{K_{DD},2}^T \right\|_2 = 1.$$

We have

$$J \le \frac{\left\| H_{d,s+n} d_{s+n}(k-n) \right\|^2}{\eta} \le \frac{\sigma_{\gamma+1}^2}{M},$$

based on which the threshold is finally defined as,

$$J_{th} = \sup_{\substack{d_{s+n}(k-n) \\ \text{s.t. (13.69)}}} J = \frac{\sigma_{\gamma+1}^2}{M}. \tag{13.72}$$

In an analogous manner, we now address fault detection in a stochastic process. We consider again process dynamics governed by (13.17). Hereby, the unknown input d is a stochastic process subject to

$$d(k) = \begin{bmatrix} w(k) \\ v(k) \end{bmatrix} \in \mathbb{R}^{k_w+m}, w(k) \sim \mathcal{N}(0, \Sigma_w), v(k) \sim \mathcal{N}(0, \Sigma_v),$$
$$\mathbb{E}\left(\begin{bmatrix} w(i) \\ v(i) \end{bmatrix} \begin{bmatrix} w(j) \\ v(j) \end{bmatrix}^T \right) = \begin{bmatrix} \Sigma_w & S_{wv} \\ S_{wv}^T & \Sigma_v \end{bmatrix} \delta_{ij}, \delta_{ij} = \begin{cases} 1, i = j \\ 0, i \neq j, \end{cases}$$

and the corresponding system matrices E_d and F_d are

$$E_d = \begin{bmatrix} E & 0 \end{bmatrix}, F_d = \begin{bmatrix} 0 & I \end{bmatrix}.$$

It follows from (13.68) that

$$U_{K_{DD}}^T \begin{bmatrix} 0 \\ H_{d,s+n} \end{bmatrix} D_{s+n} \left(U_{K_{DD}}^T \begin{bmatrix} 0 \\ H_{d,s+n} \end{bmatrix} D_{s+n} \right)^T = \Sigma_{K_{DD}}^2. \tag{13.73}$$

For sufficiently large M, matrix

$$\frac{1}{M+1} U_{K_{DD}}^T \begin{bmatrix} 0 \\ H_{d,s+n} \end{bmatrix} D_{s+n} \left(U_{K_{DD}}^T \begin{bmatrix} 0 \\ H_{d,s+n} \end{bmatrix} D_{s+n} \right)^T$$

gives an approximation of the covariance matrix of the residual vector,

$$\mathbb{E}\left(r_{K,\eta}(k) r_{K,\eta}^T(k)\right) =: \Sigma_r \in \mathbb{R}^{\eta \times \eta}, \, r_{K,\eta}(k) = K_{DD,s} \begin{bmatrix} u_s(k) \\ y_s(k) \end{bmatrix}.$$

Since

$$r_{K,\eta}(k) \sim \mathcal{N}\left(0, \Sigma_r\right),$$

the T^2 test statistic,

$$J_{T^2} := r_{K,\eta}^T(k) \Sigma_r^{-1} r_{K,\eta}(k),$$

is a powerful test statistic, and satisfies

$$J_{T^2} \sim \chi^2\left(\eta\right).$$

Accordingly, the threshold is set for a given pre-defined FAR.

13.4.2 Data-Driven Projection-Based Fault Detection

This subsection is devoted to exploring a data-driven version of the projection-based fault detection method presented in Sect. 13.2.2. We present two approaches. In the first one, consider the data matrix

$$\begin{bmatrix} U_s \\ Y_s \end{bmatrix} = I_{DD,s} V_{I_{DD}}^T + K_{DD,s}^T V_{K_{DD}}^T,$$

and suppose that the uncertainty in the system is modelled by $K_{DD,s}^T V_{K_{DD}}^T$. Attributed to the identity (13.56), the orthogonal projection $\mathcal{P}_{\mathcal{R}_{DD,s}}$ onto the residual subspace $\mathcal{R}_{DD,s}$ is applied for the residual generation purpose,

$$r_{\mathcal{R}_{DD,s},\xi}(k) = \mathcal{P}_{\mathcal{R}_{DD,s}} \begin{bmatrix} u_s(k) \\ y_s(k) \end{bmatrix} = U_{K_{DD}} U_{K_{DD}}^T \begin{bmatrix} u_s(k) \\ y_s(k) \end{bmatrix} \in \mathbb{R}^{\xi}, \tag{13.74}$$

$\xi = (p+m)\,s$. The corresponding residual evaluation function is defined by

$$J_{rel} = \frac{\left\| r_{\mathcal{R}_{DD,s},\xi}(k) \right\|}{\left\| \begin{bmatrix} u_s(k) \\ y_s(k) \end{bmatrix} \right\|} = \frac{\left(\sum_{i=1}^{\xi} \left[u_s^T(k)\; y_s^T(k) \right] U_{K_{DD}} U_{K_{DD}}^T \begin{bmatrix} u_s(k) \\ y_s(k) \end{bmatrix} \right)^{1/2}}{\left\| \begin{bmatrix} u_s(k) \\ y_s(k) \end{bmatrix} \right\|}. \tag{13.75}$$

Concerning the threshold setting, projecting the data matrix to the residual subspace,

$$U_{K_{DD}} U_{K_{DD}}^T \begin{bmatrix} U_s \\ Y_s \end{bmatrix} = U_{K_{DD}} \Sigma_{K_{DD}} V_{K_{DD}}^T,$$

yields

$$\left\| U_{K_{DD}} \Sigma_{K_{DD}} V_{K_{DD}}^T \right\|_2 = \sigma_{\gamma+1}.$$

As a result, the threshold is set to be

$$J_{rel,th} = \sup J_{rel} = \sigma_{\gamma+1}.$$

Next, consider the fault detection problem, when the system dynamic is under influence of possible process uncertainties during operations and thus subject to

$$\begin{bmatrix} u_s(k) \\ y_s(k) \end{bmatrix} = \begin{bmatrix} U_{\Delta,s} \\ Y_{\Delta,s} \end{bmatrix} v_{I,\Delta}, \tag{13.76}$$

$$rank \begin{bmatrix} U_{\Delta,s} \\ Y_{\Delta,s} \end{bmatrix} = s\,(m+p). \tag{13.77}$$

The corresponding image subspace is defined by

$$\mathcal{I}_{\Delta,DD,s} = \left\{ \begin{bmatrix} u_s(k) \\ y_s(k) \end{bmatrix}, \begin{bmatrix} u_s(k) \\ y_s(k) \end{bmatrix} = \begin{bmatrix} U_{\Delta,s} \\ Y_{\Delta,s} \end{bmatrix} v_{I,\Delta}, v_{I,\Delta} \in \mathcal{L}_{[0,\infty)} \right\}. \tag{13.78}$$

For our subsequent work, the normalised image representation of the uncertain system $\bar{I}_{\Delta,DD,s}$ is introduced as follows

$$\begin{bmatrix} U_{\Delta,s} \\ Y_{\Delta,s} \end{bmatrix} = U_{I_\Delta} \left[\Sigma_\Delta \; 0 \right] \begin{bmatrix} V_{I_\Delta,1} \\ V_{I_\Delta,2} \end{bmatrix} = U_{I_\Delta} \Sigma_\Delta V_{I_\Delta,1} \tag{13.79}$$

$$\bar{I}_{\Delta,DD,s} := U_{I_\Delta}. \tag{13.80}$$

Hence,

$$\forall \begin{bmatrix} u_{\Delta,s}(k) \\ y_{\Delta,s}(k) \end{bmatrix} \in \mathcal{I}_{\Delta,DD,s}, \exists \bar{v}_{I,\Delta}, \text{ s.t. } \begin{bmatrix} u_{\Delta,s}(k) \\ y_{\Delta,s}(k) \end{bmatrix} = \bar{I}_{\Delta,DD,s} \bar{v}_{I,\Delta}.$$

Let us decompose the normalised image representation $\bar{I}_{\Delta,DD,s}$ into

$$\bar{I}_{\Delta,DD,s} = U_{I_{DD}} U_{I_{DD}}^T \bar{I}_{\Delta,DD,s} + U_{K_{DD}} U_{K_{DD}}^T \bar{I}_{\Delta,DD,s}.$$

Now, on the assumption that the value $\delta_{\Delta I_{DD}}$,

$$\left\| U_{K_{DD}} U_{K_{DD}}^T \bar{I}_{\Delta,DD,s} \right\|_2 = \left\| U_{K_{DD}}^T \bar{I}_{\Delta,DD,s} \right\|_2 \leq \delta_{\Delta I_{DD}} < 1$$

is the margin of tolerable uncertainties, a projection-based detection scheme is proposed. Again, the residual generator (13.74) and evaluation function (13.75) are applied. To find a threshold, the gap metric between $\mathcal{I}_{\Delta,DD,s}$ and $\mathcal{I}_{DD,s}$, i.e.

$$\delta\left(\mathcal{I}_{\Delta,DD,s}, \mathcal{I}_{DD,s}\right) = \sup_{\begin{bmatrix} u_{\Delta,s}(k) \\ y_{\Delta,s}(k) \end{bmatrix}} \inf_{\begin{bmatrix} u_s(k) \\ y_s(k) \end{bmatrix}} \frac{\left\| \begin{bmatrix} u_{\Delta,s}(k) \\ y_{\Delta,s}(k) \end{bmatrix} - \begin{bmatrix} u_s(k) \\ y_s(k) \end{bmatrix} \right\|}{\left\| \begin{bmatrix} u_{\Delta,s}(k) \\ y_{\Delta,s}(k) \end{bmatrix} \right\|}, \tag{13.81}$$

$$\begin{bmatrix} u_{\Delta,s}(k) \\ y_{\Delta,s}(k) \end{bmatrix} \in \mathcal{I}_{\Delta,DD,s}, \begin{bmatrix} u_s(k) \\ y_s(k) \end{bmatrix} \in \mathcal{I}_{DD,s}$$

is determined at first. Because,

$$dist\left(\begin{bmatrix} u_{\Delta,s}(k) \\ y_{\Delta,s}(k) \end{bmatrix}, \mathcal{I}_{DD,s} \right) = \left\| U_{K_{DD}} U_{K_{DD}}^T \begin{bmatrix} u_{\Delta,s}(k) \\ y_{\Delta,s}(k) \end{bmatrix} \right\| = \left\| U_{K_{DD}}^T \begin{bmatrix} u_{\Delta,s}(k) \\ y_{\Delta,s}(k) \end{bmatrix} \right\|,$$

it holds

$$\begin{aligned} \delta\left(\mathcal{I}_{\Delta,DD,s}, \mathcal{I}_{DD,s}\right) &= \sup_{\begin{bmatrix} u_{\Delta,s}(k) \\ y_{\Delta,s}(k) \end{bmatrix} \in \mathcal{I}_{\Delta,DD,s}} \frac{\left\| U_{K_{DD}}^T \bar{I}_{\Delta,DD,s} \bar{v}_{I,\Delta} \right\|}{\left\| \bar{I}_{\Delta,DD,s} \bar{v}_{I,\Delta} \right\|} \\ &= \sup_{\bar{v}_{I,\Delta}} \frac{\left\| U_{K_{DD}}^T \bar{I}_{\Delta,DD,s} \bar{v}_{I,\Delta} \right\|}{\left\| \bar{v}_{I,\Delta} \right\|} = \left\| U_{K_{DD}}^T \bar{I}_{\Delta,DD,s} \right\|_2. \end{aligned} \tag{13.82}$$

That is, the gap metric $\delta\left(\mathcal{I}_{\Delta,DD,s}, \mathcal{I}_{DD,s}\right)$ equals the 2-norm of the uncertainty matrix $U_{K_{DD}}^T \bar{I}_{\Delta,DD,s}$. We summarise this result as a theorem.

Theorem 13.1 *Given the image subspaces $\mathcal{I}_{DD,s}$ (13.46) and $\mathcal{I}_{\Delta,DD,s}$ (13.78) with the normalised kernel $U_{K_{DD}}^T$ and image representation $\bar{I}_{\Delta,DD,s}$ given in (13.78), respectively, then the gap metric is given by*

$$\delta\left(\mathcal{I}_{\Delta,DD,s}, \mathcal{I}_{DD,s}\right) = \left\| U_{K_{DD}}^T \bar{I}_{\Delta,DD,s} \right\|_2. \tag{13.83}$$

It is of interest to note that the equation in (13.83) is a data-driven version of the (model-based) gap metric computation given in Theorem 9.1.

The immediate application of the above theorem is the threshold setting, namely,

$$J_{rel,th} = \sup_{\begin{bmatrix} u_{\Delta,s}(k) \\ y_{\Delta,s}(k) \end{bmatrix} \in \mathcal{I}_{\Delta,DD,s} \left\| U_{K_{DD}}^T \bar{I}_{\Delta,DD,s} \right\|_2 \leq \delta_{\Delta I_{DD}}} J_{rel} = \delta_{\Delta I_{DD}}.$$

13.5 Data-Driven Control Schemes

As reviewed in Sect. 13.3.1, investigations on data-driven MPC build the mainstream in this research domain. In contrast, publications on data-driven feedback control are rare in the literature. In this section, we endeavour to study data-driven feedback control issues.

It can be observed that the commonly used data-driven model

$$\begin{bmatrix} u_s(k) \\ y_s(k) \end{bmatrix} = \begin{bmatrix} U_s \\ Y_s \end{bmatrix} v_l \tag{13.84}$$

is a finite impulse response (FIR) system, while any feedback systems are generally infinite impulse response (IIR) systems. Furthermore, both MPC and residual generation schemes presented in the previous sections are to be implemented in the moving window pattern. These two facts motivate us to recast the data-driven model (13.84) in the form of an IIR system and, based on it, to leverage the existing feedback control schemes to realise data-driven feedback control.

13.5.1 Alternative Data-Driven System Representations

Proceeding from the data decomposition,

$$\begin{aligned} \begin{bmatrix} U_s \\ Y_s \end{bmatrix} &= \begin{bmatrix} U_{I_{DD}} & U_{K_{DD}} \end{bmatrix} \begin{bmatrix} \Sigma_{I_{DD}} & 0 \\ 0 & \Sigma_{K_{DD}} \end{bmatrix} \begin{bmatrix} V_{I_{DD}}^T \\ V_{K_{DD}}^T \end{bmatrix} \\ &= \begin{bmatrix} I_{DD,s} & K_{DD,s}^T \end{bmatrix} \begin{bmatrix} V_{I_{DD}}^T \\ V_{K_{DD}}^T \end{bmatrix}, \end{aligned}$$

alternative data-driven image representations are firstly established in the sequel, whose realisations match the moving window pattern.

Let us denote by

$$\begin{bmatrix} u_{I,s}(k) \\ y_{I,s}(k) \end{bmatrix} = I_{DD,s} V_{I_{DD}}^T v_l$$

the FIR image representation. For our purpose, re-order data vector

$$\begin{bmatrix} u_{I,s}(k) \\ y_{I,s}(k) \end{bmatrix} = \begin{bmatrix} u_I(k+1) \\ \vdots \\ u_I(k+s) \\ y_I(k+1) \\ \vdots \\ y_I(k+s) \end{bmatrix}$$

into

$$z_{I,s}(k) = \begin{bmatrix} z_I(k+1) \\ \vdots \\ z_I(k+s) \end{bmatrix}, z_I(k+i) = \begin{bmatrix} u_I(k+i) \\ y_I(k+i) \end{bmatrix}, i = 1, \ldots, s.$$

It is well-known that

$$z_{I,s}(k) = P \begin{bmatrix} u_{I,s}(k) \\ y_{I,s}(k) \end{bmatrix}$$

where P is a permutation matrix. Below, we consider the data model of the form

$$z_{I,s}(k) = \hat{I}_{DD,s} V_{I_{DD}}^T v_l, \; \hat{I}_{DD,s} = P I_{DD,s}. \tag{13.85}$$

Expressing $z_I(k+s)$ in terms of

$$z_I(k+s) = B_z V_{I_{DD}}^T v_l,$$
$$B_z := \hat{I}_{DD,s}\left((p+m)(s-1)+1 : (p+m)s\right) \in \mathbb{R}^{(p+m)\times\gamma},$$

we are able to write $z_{I,s}(k+1)$ as

$$z_{I,s}(k+1) = A_{bus} z_{I,s}(k) + \begin{bmatrix} 0 \\ B_z \end{bmatrix} V_{I_{DD}}^T v_l, \tag{13.86}$$

where $A_{bus} \in \mathbb{R}^{(p+m)s\times(p+m)s}$ is a blockwise upper shift matrix, i.e.

$$A_{bus} = \begin{bmatrix} 0 & I & 0 & \cdots & 0 \\ 0 & 0 & I & \cdots & 0 \\ \vdots & \vdots & \ddots & \ddots & \vdots \\ 0 & 0 & \cdots & 0 & I \\ 0 & 0 & \cdots & 0 & 0 \end{bmatrix}.$$

Model (13.86) is an IIR system and describes the system data evolution in the moving window pattern. It can be interpreted as a state space representation as well.

Bearing in mind that the required response of the system output is usually the control target, the model (13.86) is further simplified by removing the input variable $u(k)$. Specifically, consider

$$y_{I,s}(k) = \begin{bmatrix} y_I(k+1) \\ \vdots \\ y_I(k+s) \end{bmatrix},$$

and $y_I(k+s)$ expressed by

$$y_I(k+s) = B_y v_l.$$

Hereby, B_y is constructed as follows. Let

$$\begin{aligned} I_s &:= I_{DD,s} V_{I_{DD}}^T \in \mathbb{R}^{(p+m)s \times M}, \\ B_y &= I_s\left(ps + m(s-1) + 1 : (p+m)s, :\right) \in \mathbb{R}^{m \times M}. \end{aligned}$$

Analogous to (13.86), $y_{I,s}(k+1)$ is recast into

$$y_{I,s}(k+1) = A_{bus,y} y_{I,s}(k) + \begin{bmatrix} 0 \\ B_y \end{bmatrix} v_l(k), \tag{13.87}$$

where $A_{bus,y}$ is an $(ms \times ms)$ blockwise upper shift matrix. System (13.87) can be interpreted as a nominal closed-loop input-output dynamic model, and serves for our subsequent work on feedback controller design. Note that the latent variable v_l serves as the input (control) signal and is updated at each sampling time. Subsequently, it is denoted by $v_l(k)$ in the sequel.

Similarly, $u_{I,s}(k+1)$ can be written into

$$\begin{aligned} u_{I,s}(k+1) &= A_{bus,u} u_{I,s}(k) + \begin{bmatrix} 0 \\ B_u \end{bmatrix} V_{I_{DD}}^T v_l(k), \\ B_u &= I_s\left(p(s-1) + 1 : ps, :\right) \in \mathbb{R}^{p \times M}, \end{aligned} \tag{13.88}$$

with $A_{bus,u} \in \mathbb{R}^{ps \times ps}$ being a blockwise upper shift matrix. The model (13.87)–(13.88) is called recursive image representation of the system modelled by (13.84).

Aiming at designing a feedback controller, the influence of the uncertainties in the system is now taken into account. Let

$$\begin{aligned} \begin{bmatrix} u_s(k) \\ y_s(k) \end{bmatrix} &= \mathcal{P}_{\mathcal{I}_{DD,s}} \begin{bmatrix} u_s(k) \\ y_s(k) \end{bmatrix} + \mathcal{P}_{\mathcal{R}_{DD,s}} \begin{bmatrix} u_s(k) \\ y_s(k) \end{bmatrix} \\ &= I_{DD,s} \Sigma_{I_{DD}}^{-2} I_{DD,s}^T \begin{bmatrix} u_s(k) \\ y_s(k) \end{bmatrix} + K_{DD,s}^T \Sigma_{K_{DD}}^{-2} K_{DD,s} \begin{bmatrix} u_s(k) \\ y_s(k) \end{bmatrix}, \end{aligned}$$

be the data-driven model of the system with

$$r_{\mathcal{R}_{DD,s},\xi}(k) = \mathcal{P}_{\mathcal{R}_{DD,s}} \begin{bmatrix} u_s(k) \\ y_s(k) \end{bmatrix},$$

representing the uncertainties corrupted in the data. By means of the orthogonal projection $\mathcal{P}_{\mathcal{R}_{DD,s}}$, the residual vector is built, and the process data vector (in the open-loop configuration) is given by

$$\begin{bmatrix} u_s(k) \\ y_s(k) \end{bmatrix} = \begin{bmatrix} u_{I,s}(k) \\ y_{I,s}(k) \end{bmatrix} + r_{\mathcal{R}_{DD,s},\xi}(k)$$
$$= I_{DD,s} V_{I_{DD}}^T v_l(k) + U_{K_{DD}} r_{\bar{K},\eta}(k). \tag{13.89}$$

Denoting by

$$E_y = U_{K_{DD}} (ps + m(s-1) + 1 : (p+m)s, :) \in \mathbb{R}^{m \times \eta},$$
$$E_u = U_{K_{DD}} (p(s-1) + 1 : ps, :) \in \mathbb{R}^{p \times \eta},$$

the influence of the residual signal (uncertainties) on the process data, we have

$$y_s(k+1) = A_{bus,y} y_s(k) + \begin{bmatrix} 0 \\ B_y \end{bmatrix} v_l(k) + \begin{bmatrix} 0 \\ E_y \end{bmatrix} r_{\bar{K},\eta}(k), \tag{13.90}$$

$$u_s(k+1) = A_{bus,u} u_s(k) + \begin{bmatrix} 0 \\ B_u \end{bmatrix} v_l(k) + \begin{bmatrix} 0 \\ E_u \end{bmatrix} r_{\bar{K},\eta}(k). \tag{13.91}$$

These two models are the data-driven process model in the open-loop configuration, that lays the basis for designing data-driven feedback controllers.

13.5.2 Residual-Driven Feedback Control Schemes

That $A_{bus,y}$ is an $(ms \times ms)$ blockwise upper shift matrix assures the stabilisability of system (13.90). Hence, when there exists no (multiplicative) model uncertainty, setting

$$v_l(k) = F_y y_s(k) \tag{13.92}$$

yields

$$y_s(k+1) = \left(A_{bus,y} + \begin{bmatrix} 0 \\ B_y F_y \end{bmatrix} \right) y_s(k) + \begin{bmatrix} 0 \\ E_y \end{bmatrix} r_{\bar{K},\eta}(k),$$

$$A_{bus,y}+\begin{bmatrix}0\\B_yF_y\end{bmatrix}=\begin{bmatrix}0&I&0&\cdots&0\\0&0&I&\cdots&0\\\vdots&\vdots&\ddots&\ddots&\vdots\\0&0&\cdots&0&I\\B_yF_{y,1}&B_yF_{y,2}&\cdots&B_yF_{y,s-1}&B_yF_{y,s}\end{bmatrix},$$

$$F_y=\begin{bmatrix}F_{y,1}\cdots F_{y,s}\end{bmatrix},$$

where $F_{y,i}\in\mathbb{R}^{M\times m}$, $i=1,\ldots,s$, are design parameter matrices and serve as (state) feedback gain matrices. For instance, by means of pole assignment, the system eigen-dynamics can be regulated. Note that, when the system (13.90) is not fully controllable, those uncontrollable poles are equal to zeros.

In order to enhance the system robustness against unknown inputs reflected by the residual vector $r_{\bar{K},\eta}(k)$, the feedback controller (13.90) is extended to

$$v_l(k)=F_yy_s(k)+F_Er_{\bar{K},\eta}(k). \tag{13.93}$$

Hereby, the feedback gain matrix $F_E\in\mathbb{R}^{M\times\eta}$ is the optimum solution of the problem

$$F_E=\arg\min_{F_E}\left\|B_yF_E+E_y\right\|_2=B_y^{\dagger}E_y. \tag{13.94}$$

In (13.94), matrix B_y^+ is the pseudo-inverse (Moore-Penrose inverse) of B_y. The resulting closed-loop dynamic is governed by

$$y_s(k+1)=\left(A_{bus,y}+\begin{bmatrix}0\\B_yF_y\end{bmatrix}\right)y_s(k)+\begin{bmatrix}0\\B_yF_E+E_y\end{bmatrix}r_{\bar{K},\eta}(k).$$

In the presence of multiplicative model uncertainties, the stability issue should be explored. To this end, consider

$$\begin{aligned}r_{\bar{K},\eta}(k)&=\bar{K}_{DD,s}\begin{bmatrix}u_s(k)\\y_s(k)\end{bmatrix}=\bar{K}_{DD,s}\left(I_{DD,s}V_{I_{DD}}^T+K_{DD,s}^TV_{K_{DD}}^T\right)v_l(k)\\&=\Sigma_{K_{DD}}V_{K_{DD}}^Tv_l(k)=\Sigma_{K_{DD}}V_{K_{DD}}^T\left(F_yy_s(k)+F_Er_{\bar{K},\eta}(k)\right).\end{aligned}$$

on the assumption that the model uncertainties are modelled by $K_{DD,s}^TV_{K_{DD}}^T$.

Remark 13.2 In real applications, the matrix $V_{K_{DD}}^T$ is generally unknown. In that case, the upper-bound of $\left\|V_{K_{DD}}^T\right\|_2$ can be substituted used.

It yields, on the assumption of regular $\left(I-\Sigma_{K_{DD}}V_{K_{DD}}^TF_E\right)$,

$$\begin{aligned}y_s(k+1)=&\left(A_{bus,y}+\begin{bmatrix}0\\B_yF_y\end{bmatrix}\right)y_s(k)\\&+\begin{bmatrix}0\\B_yF_E+E_y\end{bmatrix}\left(I-\Sigma_{K_{DD}}V_{K_{DD}}^TF_E\right)^{-1}\Sigma_{K_{DD}}V_{K_{DD}}^TF_yy_s(k).\end{aligned} \tag{13.95}$$

Introducing the notation

$$G_y(z) = \left(zI - \left(A_{bus,y} + \begin{bmatrix} 0 \\ B_y F_y \end{bmatrix}\right)\right)^{-1} \begin{bmatrix} 0 \\ B_y F_E + E_y \end{bmatrix},$$

then the overall system (13.95) is stable, if

$$\left\|G_y(z)\right\|_\infty \left\|\left(I - \Sigma_{K_{DD}} V_{K_{DD}}^T F_E\right)^{-1} \Sigma_{K_{DD}} V_{K_{DD}}^T F_y\right\|_2 < 1. \tag{13.96}$$

We now analyse the condition (13.96) thoroughly with regard of controller design.

- First of all, the condition (13.96) is attributed to the Small Gain theorem, since the system (13.95) is equivalent to the loop

$$\begin{cases} y_s(z) = G_y(z)u(z) \\ u(z) = \left(I - \Sigma_{K_{DD}} V_{K_{DD}}^T F_E\right)^{-1} \Sigma_{K_{DD}} V_{K_{DD}}^T F_y y_s(z). \end{cases}$$

- With respect to the condition (13.96), the feedback gain matrix F_y is to be designed to assure the stability of $G_y(z)$, i.e. $A_{bus,y} + \begin{bmatrix} 0 \\ B_y F_y \end{bmatrix}$ is a Schur matrix, and to mitigate $\left\|G_y(z)\right\|_\infty$.
- It is also necessary to require the invertibility of $\left(I - \Sigma_{K_{DD}} V_{K_{DD}}^T F_E\right)$. A sufficient condition to this end is

$$\rho\left(\Sigma_{K_{DD}} V_{K_{DD}}^T F_E\right) < 1,$$

  where $\rho\left(\Sigma_{K_{DD}} V_{K_{DD}}^T F_E\right)$ is the spectral radius of $\Sigma_{K_{DD}} V_{K_{DD}}^T F_E$. A somewhat more conservative condition is

$$\left\|\Sigma_{K_{DD}} V_{K_{DD}}^T F_E\right\|_2 < 1. \tag{13.97}$$

  As illustrated in Sect. 13.3.4,

$$\begin{aligned} K_{DD,s}^T V_{K_{DD}}^T &= \bar{K}_{DD,s}^T \Sigma_{K_{DD}} V_{K_{DD}}^T = \Delta_{I_{DD,s}} \Longrightarrow \\ \bar{K}_{DD,s} K_{DD,s}^T V_{K_{DD}}^T &= \Sigma_{K_{DD}} V_{K_{DD}}^T = \bar{K}_{DD,s} \Delta_{I_{DD,s}}, \end{aligned}$$

  namely, $\Sigma_{K_{DD}} V_{K_{DD}}^T$ represents the uncertainty in the residual. While F_E is the residual feedback gain matrix, (13.97) poses a constraint on the optimisation problem (13.94).

It is worth emphasising that similar results are achievable for the case that $V_{K_{DD}}^T$ is unknown and its norm-based upper-bound is alternatively used.

Finally, we give the FIR system form of the closed-loop dynamics. Differing from the data-driven model (13.84), the latent variable in the closed-loop is updated at each sampling time. To model such effect, the data-driven model (13.84) is extended to

$$\begin{bmatrix} u_s(k) \\ y_s(k) \end{bmatrix} = P^{-1} \begin{bmatrix} Z_1 v_l(k+1) \\ \vdots \\ Z_s v_l(k+s) \end{bmatrix}, Z = \begin{bmatrix} Z_1 \\ \vdots \\ Z_s \end{bmatrix} = P \begin{bmatrix} U_s \\ Y_s \end{bmatrix}, \tag{13.98}$$

$$U_s = \begin{bmatrix} U_{s,1} \\ \vdots \\ U_{s,s} \end{bmatrix}, U_{s,i} \in \mathbb{R}^{p \times M}, Y_s = \begin{bmatrix} Y_{s,1} \\ \vdots \\ Y_{s,s} \end{bmatrix}, Y_{s,i} \in \mathbb{R}^{m \times M}, i = 1, \ldots, s,$$

where P is a permutation matrix that maps $\begin{bmatrix} u_s(k) \\ y_s(k) \end{bmatrix}$ to $z_s(k)$,

$$z_s(k) = \begin{bmatrix} z(k+1) \\ \vdots \\ z(k+s) \end{bmatrix}, z(k+i) = \begin{bmatrix} u(k+i) \\ y(k+i) \end{bmatrix}, i = 1, \ldots, s.$$

13.5.3 A Data-Driven Predictive Control Scheme with Embedded Feedback Control

This subsection is devoted to a data-driven predictive control scheme. Proceeding from the data-driven closed-loop model (13.98) with the controller

$$v_l(k+i) = F_y y_s(k-s+i) + F_E r_{\bar{K},\eta}(k-s+i) + v_r, i = 1, \ldots, s, \tag{13.99}$$

where $F_y y_s(k-s+i) + F_E r_{\bar{K},\eta}(k-s+i)$ is the feedback controller (13.93) and v_r is a constant vector to be determined for the predictive control purpose, the data-driven system model with $v_l(k+i)$ as the input and $(u_s(k), y_s(k))$ as the output is given by

$$\begin{bmatrix} u_s(k) \\ y_s(k) \end{bmatrix} = P^{-1} \begin{bmatrix} Z_1 \left(F_y y_s(k-s+1) + F_E r_{\bar{K},\eta}(k-s+1) \right) \\ \vdots \\ Z_s \left(F_y y_s(k) + F_E r_{\bar{K},\eta}(k) \right) \end{bmatrix} + \begin{bmatrix} U_s \\ Y_s \end{bmatrix} v_r. \tag{13.100}$$

In the sequel, the model (13.100) will serve as the prediction model. It is remarkable that due to the embedded feedback controller (13.99), the predicted output, $(u_s(k), y_s(k))$, will depend on the system past dynamics. To be specific, suppose that given online data up to the current time instant k, $(u(k+i), y(k+i))$, $i = 1, \ldots, N$, are to be predicted. In this regard, the model (13.100) is, for $s = k_p + N$, written into

$$\begin{bmatrix} u_{k_p+N}(k-k_p) \\ y_{k_p+N}(k-k_p) \end{bmatrix} = P^{-1} \begin{bmatrix} Z_1 \begin{pmatrix} F_y y_{k_p+N}(k-(2k_p+N)+1) \\ +F_E r_{\bar{K},\eta}(k-(2k_p+N)+1) \end{pmatrix} \\ \vdots \\ Z_{k_p+N}\left(F_y y_{k_p+N}(k-k_p) + F_E r_{\bar{K},\eta}(k-k_p)\right) \end{bmatrix} + \begin{bmatrix} U_{k_p+N} \\ Y_{k_p+N} \end{bmatrix} v_r, \tag{13.101}$$

for some integer $k_p \geq 1$, $N > k_p$. In order to distinguish the measurement data from the predicted variables, the following notations are introduced,

$$\begin{bmatrix} u(i) \\ y(i) \end{bmatrix} = \begin{cases} \begin{bmatrix} u(i) \\ y(i) \end{bmatrix}, i \leq k \\ \begin{bmatrix} \hat{u}(i) \\ \hat{y}(i) \end{bmatrix}, i > k \end{cases} \tag{13.102}$$

with $(\hat{u}(i), \hat{y}(i))$ denoting the predicted $(u(i), y(i))$. In this regard, the first term in the right-hand side of (13.101) is split into

$$\begin{bmatrix} Z_1 \begin{pmatrix} F_y y_{k_p+N}(k-(2k_p+N)+1) \\ +F_E r_{\bar{K},\eta}(k-(2k_p+N)+1) \end{pmatrix} \\ \vdots \\ Z_{k_p+N}\left(F_y y_{k_p+N}(k-k_p) + F_E r_{\bar{K},\eta}(k-k_p)\right) \end{bmatrix} = \alpha_1 + \alpha_2 + \alpha_3,$$

where vectors α_1 and α_2 are of the following forms,

$$\alpha_1 = diag\left(\Pi_1, \ldots, \Pi_{k_p+N}\right) \begin{bmatrix} y_{k_p+N}(k-(2k_p+N)+1) \\ \vdots \\ y_{k_p+N}(k-N-k_p) \\ y_{k_p+N}(k-N-k_p) \\ \vdots \\ y_{k_p+N}(k-N-k_p) \end{bmatrix}, \alpha_2 = \begin{bmatrix} 0 \\ \Phi \end{bmatrix} \hat{y}_N(k)$$

with some matrices $\Pi_i, i = 1, \ldots, k_p + N$,and Φ, which are calculated by means of the known matrices $Z_i, i = 1, \ldots, k_p + N$, and F_y. The vector α_3 consists of the residual vector, where, corresponding to

$$r_{\bar{K},\eta}(k) = \bar{K}_{DD,s} \begin{bmatrix} u_s(k) \\ y_s(k) \end{bmatrix},$$

it is supposed that

$$r_{\bar{K},\eta}(k-(2k_p+N)+i)=\begin{cases}\bar{K}_{DD,k_p+N}\begin{bmatrix}u_{k_p+N}(k-(2k_p+N)+i)\\ y_{k_p+N}(k-(2k_p+N)+i)\end{bmatrix}, i\leq k_p\\ 0, i>k_p,\end{cases}$$

leading to

$$\alpha_3=\begin{bmatrix}Z_1F_Er_{\bar{K},\eta}(k-(2k_p+N)+1)\\ \vdots\\ Z_{k_p}F_Er_{\bar{K},\eta}(k-(k_p+N))\\ 0\end{bmatrix}.$$

Consequently, the model (13.101) becomes

$$z_{k_p+N}(k-k_p)=\alpha_1+\alpha_2+\alpha_3+Zv_r, \tag{13.103}$$

$$Z=\begin{bmatrix}U_{k_p+N}\\ Y_{k_p+N}\end{bmatrix}, z_{k_p+N}(k-k_p)=\begin{bmatrix}z(k-k_p+1)\\ \vdots\\ z(k)\\ \hat{z}(k+1)\\ \vdots\\ \hat{z}(k+N)\end{bmatrix}, \hat{z}(k+i)=\begin{bmatrix}\hat{u}(k+i)\\ \hat{y}(k+i)\end{bmatrix},$$

$i=1,\ldots,N$. Finally, the predictive control problem is formulated as: given the reference vector

$$z_{ref,N}(k)=\begin{bmatrix}u_{ref}(k+1)\\ y_{ref}(k+1)\\ \vdots\\ u_{ref}(k+N)\\ y_{ref}(k+N)\end{bmatrix},$$

solve the optimisation problem

$$\min_{v_r}\left\|z_{ref,N}(k)-\hat{z}_N(k)\right\|, \hat{z}_N(k)=\begin{bmatrix}\hat{z}(k+1)\\ \vdots\\ \hat{z}(k+N)\end{bmatrix} \tag{13.104}$$
$$\text{s.t. } \hat{z}_N(k) \text{ is subject to model (13.103).}$$

13.6 Notes and References

Serving as a data-driven model, finite samples system representations are widely applied to data-driven control of dynamic systems. The activities in this thematic area, as reported in major journals and conferences, build the mainstream in the research domain of data-driven control. It can be observed that no research efforts have been reported on the use of finite samples system representations in the model-based framework. In the first part of this chapter, the potential of model-based finite samples system representations for fault detection and predictive control has been comprehensively examined, making the initiation of a new subject. The first results described in Sect. 13.2 exemplify numerous benefits, amongst others,

- significant reduction of computation efforts, both for offline design and, particularly, for online implementation,
  Specifically, some complicated computation algorithms for optimal detection and control system design in Hilbert space are simplified by solutions in Euclidean space, like gap metric computation in Sect. 13.2.2, dealing with projection-based fault detection, upper-bound estimation towards threshold setting discussed in Sect. 13.2.3 or MMP solution applied in predictive control in Sect. 13.2.4.
- engineering-equivalent realisation,
  The fault detection system design, from residual generation to residual evaluation and threshold setting, is performed directly on the basis of the finite samples system representations without any approximation.
- release of existence conditions for optimal solutions.
  As is known, design of optimal fault detection systems should satisfy certain existence conditions. Design based on the finite samples system representations might release some hard existence conditions. For instance, in Sect. 13.2.1 it has been demonstrated that the existence condition for an H_2-observer-based detection system becomes significantly less strict.

Undoubtedly, the theoretic fundament of the popular data-driven control technique is the fundamental lemma [207], namely, any system trajectory $\begin{bmatrix} u_s(k) \\ y_s(k) \end{bmatrix}$ can be described by

$$\begin{bmatrix} u_s(k) \\ y_s(k) \end{bmatrix} = \begin{bmatrix} U_s \\ Y_s \end{bmatrix} v_l,$$

for some latent vector v_l, and for any given v_l, $\begin{bmatrix} U_s \\ Y_s \end{bmatrix} v_l$ builds a trajectory subject to the system dynamic. Hereby, (U_s, Y_s) are input and output data matrices and U_s is persistently exciting. In their survey paper, Markovsky and Dörfler [138] have systematically reviewed data-driven analysis and control methods in the framework of behavioral systems theory. Thereby, two control schemes have motivated and inspired our work. The one is the LQ tracking control, as formulated in (13.37), which has been firstly proposed in [137]. In the review paper [138], various solutions of the LQ tracking control problem have been introduced. Another one is data-driven predictive control, in particular, DeePC scheme. DeePC scheme was proposed by Coulson et al. [26], whose formulation is analogous to the LQ tracking control and given in (13.38). Making use of the available degree of design freedom, Dörfler et al. [53] have extended DeePC scheme by adding the regularisation term (13.39) into the objective function (13.38) to enhance the system robustness.

Although efforts to enhance system robustness have been made in the current research of data-driven control, issues about uncertainties in the data set (U_s, Y_s) have drawn limited attention. Inspired by the notice that the residual subspace is at the core of fault detection and feedback control, which has been rarely explored in state of the art research on data-driven control, data-driven image, kernel representations, the associated image and residual subspaces, and some related issues have been addressed in Sects. 13.3.2–13.3.4.

A key step to separate uncertainties in the data set is performed by an SVD,

$$\begin{bmatrix} U_s \\ Y_s \end{bmatrix} = U \Sigma V^T.$$

On the assumption that the influence of noises and uncertainties on the system dynamics is considerably weaker than the impact of the targeted control input signal that forms the nominal system dynamic, the singular values received by the SVD are divided into two groups,

$$\sigma_1, \ldots, \sigma_\gamma >> \sigma_{\gamma+1}, \ldots, \sigma_{(m+p)s}.$$

Hereby, the integer γ plays a central role, which defines the dimension of the image representation and subspace. There are well-established methods to determine γ, for instance, the AIC/BIC-based Model Selection methods for stochastic systems or Cross-validation and Rank estimation methods for systems with deterministic uncertainties [12, 29, 126, 132, 195]. It is remarkable that in the context of low-rank approximation of the data matrix, use of an SVD algorithm to approximate (identify) the image subspace, the low-rank approximation has been described in [138]. Thereby, the kernel representation, the matrix corresponding to the group of those smaller singular values, has not been taken into account. In other words, the residual subspace, as described in our work, has not been investigated.

Putting it in a nutshell, the significant results achieved in this part of the work are

- the normalised image and kernel representations,
- the data-driven Bezout identity (13.61) and its extended form (13.62) and based on it,
- the relation between the process input-output data and the residual signals,

$$\begin{bmatrix} u_s(k) \\ y_s(k) \end{bmatrix} = \bar{I}_{DD,s} v_{\bar{I},\gamma}(k) + \bar{K}_{DD,s}^T r_{\bar{K},\eta}(k),$$

  which serves as a data model as well, and

- description of uncertainties in the data set (U_s, Y_s).

They lay the foundation for the subsequent work on data-driven fault detection and control and can be understood as the data-driven version of the unified framework of control and detection.

The data-driven fault detection schemes are presented in Sect. 13.4, including

- data-driven fault detection in systems with unknown inputs, and
- data-driven projection-based fault detection.

Comparing with their model-based counterpart, these detection schemes, particular the projection-based scheme, are remarkably advantageous in offline design and online computations.

Section 13.5 deals with data-driven control issues. Noticing that existing data-driven control schemes are mainly implemented in the moving window pattern, an alternative data-driven image representation has been developed, which models the system dynamics in the moving window pattern. From the system-theoretic viewpoint, the raw data as an FIR system is recast as an IIR system. On this basis, a data-driven feedback control scheme has been presented. The resulted control system is fully residual-driven and fits the moving window operating pattern. A robustness analysis with respect to multiplicative uncertainties has been performed as well.

The data-driven predictive control with embedded feedback control is the last issue addressed in this chapter. This control system is characterised by an embedded feedback controller, which differs from the state of the art data-driven MPC systems. In fact, this control scheme is a closed-loop MPC that is of high robustness against uncertainties in the process under consideration.

Part III
Secure Industrial Cyber-Physical Control Systems

Chapter 14
Analysis of Industrial Cyber-Physical Systems Under Cyber-Attacks

In the era of industry 4.0, automatic control systems as the centrepiece of industrial cyber-physical systems (CPSs) are fully equipped with intelligent sensors, actuators and an excellent information infrastructure. It is a logical consequence of ever increasing demands for system performance and production efficiency that today's automatic control systems are of an extremely high degree of integration, automation and complexity. Maintaining reliable and safe operations of automatic control systems is of elemental importance for optimally managing industrial CPSs over the whole operation life-cycle. As an indispensable maintenance functionality, real-time monitoring, diagnosis and tolerant control are widely integrated in automatic control systems and run parallel to the embedded control systems.

In a traditional automatic control system, monitoring, diagnosis and tolerant control are mainly dedicated to maintaining functionalities of sensors and actuators as the key components embedded in the system. As a response to widely networking in modern automatic control systems, monitoring, diagnosis and tolerant control of networked control systems as a whole have received considerable attention as well in recent years. Over the past three decades, innumerable capable diagnosis and fault-tolerant control schemes have been developed with various specifications. Recently, a new type of malfunctions, the so-called cyber-attacks on automatic control systems, have drawn attention to the urgent need for developing new monitoring, diagnosis and tolerant control strategies. Cyber-attacks can not only considerably affect functionalities of sensors and actuators, but also impair communications among the system components and sub-systems, which may cause immense damages during system operations. In addition, different from technical faults, cyber-attacks are artificially created and could be designed by attackers in such a way that they cannot be detected using the existing diagnosis techniques. Such cyber-attacks are called stealthy. A further type of cyber-attacks are the so-called eavesdropping attacks. Although such attacks do not cause changes in system dynamics and performance degradation, they enable an adversary to gain system knowledge which can then be used to design, for instance, stealthy attacks. With rapidly increasing threats of

S. X. Ding, *Unified Framework of Control and Detection*,
https://doi.org/10.1007/978-3-662-72417-0_14

cyber-crime and the resulted damages to industrial CPSs, cyber-security imposes new challenging issues in research of monitoring, diagnosis and tolerant control, which should be managed with high reliability and in an integrated fashion with the existing control and monitoring algorithms and facilities.

In this and next chapters, we are going to address cyber-security issues arising in industrial CPSs. Our goal is to deal with those relevant cyber-security issues in a systematic and reliable manner in the unified framework of control and detection. The major focus will be on attack detection and resilient control. To begin with, we will in this chapter

- specify the configuration and models of industrial CPSs under consideration,
- study the dynamics of CPSs under defined types of cyber-attacks, and analyse the possible impairment of system dynamics,
- investigate the potential of the existing control and detection methods to manage the system performance impairment due to the cyber-attacks, and
- formulate cyber-attack detection and resilient control issues to be solved in the subsequent chapters.

14.1 Models of CPSs Under Consideration

Figure 14.1 is a sketch of the basic configuration of industrial cyber-physical systems considered in our subsequent investigation. It is state of the art in industrial practice that an industrial CPS consists of two major parts, one is located at the production process (plant) with embedded real-time control and monitoring sub-systems, and another part is a control station, in which advanced control, (production) optimisation as well as process monitoring and diagnosis algorithms run. Via communication

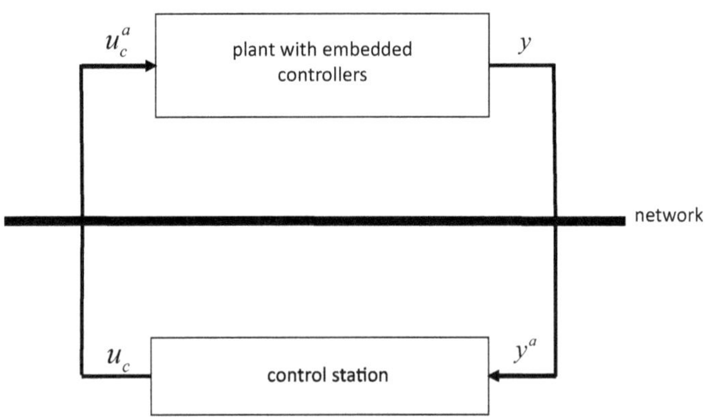

Fig. 14.1 CPS set-up under consideration

networks that connect the both system parts, the process operation data are transmitted to the control station, and the control station sends control commands to the embedded real-time systems at the plant side. Hence, such a CPS is closed-loop configured. Here, the data/signals transmitted from the plant to the control station and from the control station to the plant are schematically denoted by y and u_c, respectively. In our study, it is supposed that cyber-attacks on CPS are injected signals and data over the communication channels. Accordingly, the transmitted variables y and u_c are denoted by y^a and u_c^a on the receiver side to indicate that the received data are corrupted by attack signals/data.

It is supposed that the nominal plant dynamic is modelled by the discrete-time I/O-model (2.1) whose minimal state space realisation is described by (2.2)–(2.3). In order to model all possible uncertainties in the plant dynamics, the kernel-based system I/O-model,

$$\begin{cases} \hat{x}(k+1) = A\hat{x}(k) + Bu(k) + LW^{-1}r_y(k) \\ r_y(k) = W\left(y(k) - \hat{y}(k)\right), \hat{y}(k) = C\hat{x}(k) + Du(k) \\ y(k) = \hat{y}(k) + W^{-1}r_y(k), \end{cases} \tag{14.1}$$

is adopted for modelling plant dynamics. Observe that this model can be equivalently written into

$$y(z) = G(z)u(z) + \left(I + (zI - A)^{-1} L\right) W^{-1} r_y(z),$$
$$G(z) = D + C\left(zI - A\right)^{-1} B = \hat{M}^{-1}(z)\hat{N}(z).$$

It is of interest to notice that

$$\left(I + (zI - A)^{-1} L\right) W^{-1} = \hat{M}^{-1}(z)$$
$$\Longrightarrow y(z) = G(z)u(z) + \hat{M}^{-1}(z)r_y(z), \tag{14.2}$$
$$\Longleftrightarrow \hat{M}(z)y(z) - \hat{N}(z)u(z) = r_y(z). \tag{14.3}$$

That is, the kernel-based system I/O-model is in fact equivalent to the observer-based residual generator. Accordingly, in our subsequent work, the kernel-based system I/O-model (14.1), (14.2) and the residual generator model (14.3) will be alternately used.

Although the kernel-based system I/O-model (14.1) or (14.2) describes the system dynamics under all types of possible uncertainties, for the purpose of simplifying our investigation on the impact of cyber-attacks on the system performance, we make the following assumption in this chapter:

Assumption: Only $\mathcal{L}_2$-bounded additive disturbances, including unknown inputs, noises, and unknown initial (state) conditions, are considered in the plant (dynamic) model.

In other words, the residual vector r_y can be modelled by d, an unknown input vector that is independent of (u, y).

In the CPS configuration sketched in Fig. 14.1, the plant control input u is composed of u_c $\left(u_c^a\right)$ and u_e,

$$u(k) = u_c(k) + u_e(k),$$

where u_e represents the input vector of the embedded controller. In the sequel, it is assumed that the CPS closed-loop is stable. Moreover, to meet real industrial requirements on system security and safety, it is further assumed that the local embedded control system stabilises the control-loop when the communication between the plant and control station fails. In other words, the closed-loop,

$$\begin{cases} y(z) = G(z)u(z) + \hat{M}^{-1}(z)r_y(z) \\ u(z) = u_e(z) = K_e(z)y(z), \end{cases} \tag{14.4}$$

is stabilised by K_e, where K_e denotes the embedded controller.

Recall that, according to Theorem 3.1, the stabilising controller $u_e = K_e y$ can be written as

$$\begin{cases} \hat{x}(k+1) = A\hat{x}(k) + Bu_e(k) + LW^{-1}r_y(k) \\ r_y(k) = W\left(y(k) - \hat{y}(k)\right), \hat{y}(k) = C\hat{x}(k) + Du_e(k), \end{cases} \tag{14.5}$$

$$u_e(z) = F_e\hat{x}(z) + Q_e(z)r_y(z), Q_e(z) \in \mathcal{RH}_\infty. \tag{14.6}$$

Here, $F_e\hat{x}(z)$ is an observer-based state feedback controller with the feedback gain F_e and the observer gain L that assure $A + BF_e$ and $A - LC$ being Schur, respectively. Note that, following the discussion in Chap. 3, the LCF of the controller K_e is given by

$$K_e(z) = -\left(X(z) + Q_e(z)\hat{N}(z)\right)^{-1}\left(Y(z) - Q_e(z)\hat{M}(z)\right), \tag{14.7}$$

where $(X, Y) \in \mathcal{RH}_\infty$, $\left(\hat{M}, \hat{N}\right) \in \mathcal{RH}_\infty$ are LCPs and defined by (2.26) and (2.5), respectively. In fact, the equivalence becomes apparent, when the following results are considered,

$$u_e = K_e y \Longrightarrow \left(X + Q_e\hat{N}\right)u_e + \left(Y - Q_e\hat{M}\right)y = 0,$$

$$Xu_e + Yy = u_e - F_e\hat{x}, \hat{M}y - \hat{N}u_e = r_y \Longrightarrow u_e = F_e\hat{x} + Q_e r_y.$$

On the other hand, embedded in the overall CPS closed-loop shown in Fig. 14.1, these two different configurations, the one given by (14.5)–(14.6) in a closed-loop and another one (14.7) in an open-loop, may result in different system responses to the reference signal and, in particular, to attack signals. We would like to call the reader's attention to this fact, which will be delineated in detail in the following and in the next sections.

Since the CPS in the closed-loop is stable, it is known from Theorem 3.1 that u_c can be parametrised by

$$u_c(z) = Q_c(z)r_y^a(z) + \bar{v}(z), \bar{v}(z) = V(z)v(z), Q_c(z) \in \mathcal{RH}_\infty. \tag{14.8}$$

Here, r_y^a represents the residual signal generated at the control station using the transmitted plant output y^a that may be corrupted by the cyber-attack a_y, Q_c is a parameterisation system, and $\bar{v}$ is the output of a dynamic system V (as a feedforward controller) driven by the reference signal v, whose setting may depend on the configuration of the embedded closed-loop. We would like to illustrate the essential role of the feedback controller $Q_c r_y^a$ implemented at the control station. It is twofold:

- $Q_c r_y^a$, as an additional feedback signal, for enhancing the system performance; for instance, when faults in the plant are detected, updating $Q_c r_y^a$ could recover the control performance,
- $Q_c r_y^a$, as a residual signal, serves for system monitoring.

For the residual generation purpose, the following observer-based system is applied,

$$\begin{cases} \hat{x}^a(k+1) = A\hat{x}^a(k) + Bu_{est}(k) + LW^{-1}r_y^a(k) \\ r_y^a(k) = W\left(y^a(k) - \hat{y}^a(k)\right), \hat{y}^a(k) = C\hat{x}^a(k) + Du_{ec}(k) \end{cases}$$
$$\Longrightarrow Q_c r_y^a = Q_c\left(\hat{M}y^a - \hat{N}u_{est}\right),$$

where u_{est} represents the “estimate” of the control signal u with y^a $\left(r_y^a\right)$ instead of y $\left(r_y\right)$,

$$u_{est}(z) = F_e\hat{x}^a(z) + Q_e r_y^a(z) + Q_c r_y^a(z) + \bar{v}(z).$$

Note that, due to possible cyber-attacks, u_{est} calculated at the control station and u implemented on the plant can be different.

Finally, the overall controller of the CPS is examined with respect to the two different configurations of the embedded closed-loop and without considering cyber-attacks:

- the embedded closed-loop given by (14.5)–(14.6),

$$\begin{gathered} u(z) = F_e\hat{x}(z) + Q_e(z)r_y(z) + Q_c(z)r_y(z) + \bar{v}(z) \Longrightarrow \\ X(z)u(z) + Y(z)y(z) = Q(z)\left(\hat{M}(z)y(z) - \hat{N}(z)u(z)\right) + \bar{v}(z) \Longrightarrow \\ u(z) = V^{-1}(z)U(z)y(z) + V^{-1}(z)\bar{v}(z), \\ V(z) = X(z) + Q(z)\hat{N}(z), U(z) = -Y(z) + Q(z)\hat{M}(z) \in \mathcal{RH}_\infty, \\ Q(z) = Q_e(z) + Q_c(z) \in \mathcal{RH}_\infty. \end{gathered}$$

Setting

$$\bar{v}(z) = V(z)v(z)$$

with v as the reference signal leads to

$$u(z) = V^{-1}(z)U(z)\,y(z) + v(z),$$

- the embedded closed-loop given by (14.7),

$$\begin{aligned} u &= K_e y + Q_c r_y + \bar{v} \Longrightarrow \\ \left(X + Q_e\hat{N}\right)u &= -\left(Y - Q_e\hat{M}\right)y + \left(X + Q_e\hat{N}\right)\left(Q_c\left(\hat{M}y - \hat{N}u\right) + \bar{v}\right). \end{aligned}$$

Setting

$$Q = Q_e + \left(X + Q_e\hat{N}\right)Q_c, \tag{14.9}$$

$$\bar{v} = \left(X + Q_e\hat{N}\right)^{-1}\left(X + Q\hat{N}\right)v = \left(I + \left(X + Q_e\hat{N}\right)^{-1}Q_c\hat{N}\right)v \tag{14.10}$$

gives

$$\begin{aligned} &u(z) = V^{-1}(z)U(z)\,y(z) + v(z), \\ &V(z) = X(z) + Q(z)\hat{N}(z), U(z) = -Y(z) + Q(z)\hat{M}(z) \in \mathcal{RH}_\infty. \end{aligned}$$

In summary, the overall controller can be equivalently written as

$$\begin{aligned} &u(z) = K(z)y(z) + v(z) = F_e\hat{x}(z) + Q(z)r_y(z) + \bar{v}(z), \\ &K(z) = V^{-1}(z)\,U(z)\,. \end{aligned} \tag{14.11}$$

It is noteworthy that the aforementioned CPS configuration is equivalent to the standard one mostly adopted in the literature and modelled by

$$\begin{cases} y(z) = G(z)u(z) + \hat{M}^{-1}(z)r_y(z) \\ u(z) = K(z)y(z) + v(z). \end{cases} \tag{14.12}$$

Different from our study, the reported (state of the art) investigations assume that the controller u is implemented on the control station. In order to demonstrate the equivalence, let us write the controller in (14.12) into

$$\begin{aligned} u &= F_0\hat{x} + Q_0 r_{y,0} + \bar{v}_0, r_{y,0} = \hat{M}_0 y - \hat{N}_0 u, \\ \bar{v}_0 &= \left(X_0 + Q_0\hat{N}_0\right)v, \end{aligned}$$

where, according to the Youla parameterisation (refer to Theorem 3.1),

$$K = -\left(X_0 + Q_0\hat{N}_0\right)^{-1}\left(Y_0 - Q_0\hat{M}_0\right),$$

and (X_0, Y_0), $\left(\hat{M}_0, \hat{N}_0\right)$ are the LCP of the controller and the plant with F_0 as the state feedback gain matrix. It is known from Theorem 3.4, Corollary 3.1 and the discussion in Sect. 3.2.2 that u can be equivalently written as

$$u = F_e\hat{x} + Qr_y + \bar{v},$$

which can be further implemented in the CPS configuration considered in our work. In fact, for the same reason, the assumption that the equivalent controller is an observer-based state feedback controller with the feedback gain matrix F_e, as described by (14.11), suffers no loss of generality, because by setting Q an observer-based state feedback controller with an arbitrary state feedback gain, say F_0, can be realised.

14.2 Cyber-Attacks and Impairment of System Dynamics

14.2.1 Models of Cyber-Attacks Under Consideration

From the system control-theoretic viewpoint, cyber-attacks can be grouped into

- additive attacks modelled by

$$y^a(k) = y(k) + a_y(k), \tag{14.13}$$
$$u_c^a(k) = u_c(k) + a_{u_c}(k), \tag{14.14}$$

  where a_y and a_{u_c} represent the (additive) cyber-attacks injected into the output and input communication channels, respectively, and
- multiplicative attacks modelled by

$$y^a(z) = \Delta_y y(z), \Delta_y \in \mathcal{RH}_\infty, \tag{14.15}$$
$$u_c^a(z) = \Delta_u u_c(z), \Delta_u \in \mathcal{RH}_\infty. \tag{14.16}$$

  Here, Δ_y and Δ_u denote stable unknown and uncertain systems. The above model can also be equivalently written into

$$y^a(z) = y(z) + \delta_y y(z), \delta_y = \Delta_y - I \in \mathcal{RH}_\infty, \tag{14.17}$$
$$u_c^a(z) = u_c(z) + \delta_u u_c(z), \delta_u = \Delta_u - I \in \mathcal{RH}_\infty. \tag{14.18}$$

Numerous real cases of Denial of Service (DoS) and replay attacks leading to enormous impairment of industrial CPSs have been reported in the past decades. They are identified as critical cyber-threats to the IT-infrastructure of CPSs. Both attacks are multiplicative cyber-attacks. While DoS attacks make communication (network) resource unavailable or strongly limited available for the automatic control system by temporarily or indefinitely disrupting (data transmission) service, replay attacks "replay" data from a different context into the original ones during data transmission.

Additive cyber-attacks are often studied in the academic community. It is an active research field, in which the construction of cyber-attacks is simulated. Considering that cyber-attacks are often designed by the perpetrator (attacker) in such a way that they are undetectable for the monitoring system and therefore called stealthy attacks, investigations on stealthy attacks have received considerable research attention. The so-called zero dynamics, covert and kernel attacks are classical types of such stealthy attacks. The reader is referred to the references given at the end of this chapter for more detailed descriptions of the aforementioned cyber-attacks and state of the art of research in this area.

14.2.2 Closed-Loop Dynamics of the Plant with the Embedded Controller

We are going to analyse the impairment of CPS dynamics caused by cyber-attacks in two steps. Viewing the CPS as a cascade control system, the closed-loop dynamics of the plant with the embedded controller are firstly modelled as the inner loop. The control station builds the outer loop, based on which the impact of cyber-attacks on the CPS dynamics is then analysed. In this subsection, we model the closed-loop dynamics of the inner loop in the control and detection unified framework.

With u_c^a and y as the input and output variables, the control system at the plant side is described by

$$\begin{cases} y(z) = G(z)u(z) + \hat{M}^{-1}(z)r_y(z) \\ u(z) = u_e(z) + u_c^a(z). \end{cases}$$

$$u_e = \begin{cases} K_e y, & \text{controller is an open-loop subject to (14.7)} \\ F_e\hat{x} + Q_e r_y, & \text{controller is a closed-loop subject to (14.5)–(14.6).} \end{cases}$$

Let

$$K_e = -\left(X + Q_e\hat{N}\right)^{-1}\left(Y - Q_e\hat{M}\right) = -\left(\hat{Y} - MQ_e\right)\left(\hat{X} + NQ_e\right)^{-1}$$

be the Youla parameterisation form of K_e. According to Theorem 3.1, it holds

$$\begin{bmatrix} u \\ y \end{bmatrix} = \begin{bmatrix} M \\ N \end{bmatrix} \bar{u}_c^a + \begin{bmatrix} -\hat{Y} + MQ_e \\ \hat{X} + NQ_e \end{bmatrix} r_y, \tag{14.19}$$

$$\bar{u}_c^a = \begin{cases} \left(X + Q_e\hat{N}\right)u_c^a, & \text{if } u_e = K_e y \\ u_c^a, & \text{if } u_e = F_e\hat{x} + Q_e r_y. \end{cases} \tag{14.20}$$

Equations (14.19)–(14.20) model the dynamics of the inner loop consisting of the plant and the embedded controller.

14.2.3 Impairment of System Dynamics Caused by Additive Cyber-Attacks

We now analyse the possible impairment of the system dynamics under additive cyber-attacks based on (14.19)–(14.20). At first, the system dynamic with the embedded control loop modelled by (14.7) is examined.

Recall that, at the control station, the control signal u_c is computed,

$$u_c = Q_c r_y^a + \bar{v},$$

where the residual signal r_y^a is generated using the available data y^a and an estimate of the plant input u, designated by u_{est},

$$\begin{aligned} r_y^a &= \hat{M} y^a - \hat{N} u_{est}, \\ u_{est} &= K_e y^a + u_c = Q_c r_y^a + \bar{v}. \end{aligned}$$

Hence, for the additive attacks modelled by (14.13)–(14.14), it turns out

$$\begin{aligned} r_y^a &= r_y + \hat{M} a_y - \hat{N} K_e a_y + \hat{N} a_{u_c}, \\ u_c^a &= Q_c r_y^a + \bar{v} + a_{u_c} = Q_c r_y + \bar{v} + Q_c \left(\hat{M} - \hat{N} K_e\right) a_y + \left(I + Q_c \hat{N}\right) a_{u_c}. \end{aligned}$$

Observe that, by means of the Youla parameterisation of K_e and the Bezout identity, we have

$$\begin{aligned} \hat{M} - \hat{N} K_e = \left(\hat{X} + N Q_e\right)^{-1} &\Longrightarrow \hat{M} a_y - \hat{N} K_e a_y = \left(\hat{X} + N Q_e\right)^{-1} a_y \\ &\Longrightarrow r_y^a = r_y + \left(\hat{X} + N Q_e\right)^{-1} a_y + \hat{N} a_{u_c}, \qquad (14.21) \\ u_c^a &= Q_c r_y + \bar{v} + Q_c \left(\hat{X} + N Q_e\right)^{-1} a_y + \left(I + Q_c \hat{N}\right) a_{u_c}. \qquad (14.22) \end{aligned}$$

Substituting u_c^a into the inner loop model (14.19) yields

$$\begin{aligned} \begin{bmatrix} u \\ y \end{bmatrix} = \begin{bmatrix} M \\ N \end{bmatrix} \left(X + Q_e \hat{N}\right) \bar{v} + \begin{bmatrix} -\hat{Y} + M \left(Q_e + \left(X + Q_e \hat{N}\right) Q_c\right) \\ \hat{X} + N \left(Q_e + \left(X + Q_e \hat{N}\right) Q_c\right) \end{bmatrix} r_y \\ + \begin{bmatrix} M \\ N \end{bmatrix} \left(X + Q_e \hat{N}\right) \left(Q_c \left(\hat{X} + N Q_e\right)^{-1} a_y + \left(I + Q_c \hat{N}\right) a_{u_c}\right). \end{aligned}$$

Let us set

$$d_{a,o} := \left(X + Q_e\hat{N}\right)\left(Q_c\left(\hat{X} + NQ_e\right)^{-1}a_y + \left(I + Q_c\hat{N}\right)a_{u_c}\right),$$
$$Q := Q_e + \left(X + Q_e\hat{N}\right)Q_c,$$

and recall the setting (14.10) for $\bar{v}$. We finally have

$$\begin{bmatrix} u \\ y \end{bmatrix} = \begin{bmatrix} M \\ N \end{bmatrix}\left(v + d_{a,o}\right) + \begin{bmatrix} M \\ N \end{bmatrix}\begin{bmatrix} -\hat{Y} + MQ \\ \hat{X} + NQ \end{bmatrix} r_y. \tag{14.23}$$

Here, $d_{a,o}$ represents the unknown input induced by the (additive) cyber-attacks. Defining

$$Q_c = \bar{Q}_c\left(\hat{X} + NQ_e\right), \bar{Q}_c \in \mathcal{RH}_\infty,$$

$d_{a,o}$ can also be written as

$$d_{a,o} = \left(X + Q_e\hat{N}\right)\left(\bar{Q}_c a_y + \left(I + \bar{Q}_c\left(\hat{X} + NQ_e\right)\hat{N}\right)a_{u_c}\right), \tag{14.24}$$

and thus is $\mathcal{L}_2$-bounded. It is of interest to notice that the overall controller can be equivalently written as

$$u = V^{-1}Uy + v + V^{-1}d_{a,o}, \tag{14.25}$$
$$V = X + Q\hat{N}, U = -Y + Q\hat{M}. \tag{14.26}$$

Next, the closed-loop dynamic is examined, when the embedded control system is observer-based, as described by (14.5)–(14.6). In this configuration, the residual r_y^a is generated at the control station as follows,

$$\begin{cases} \hat{x}^a(k+1) = A\hat{x}^a(k) + Bu_{est}(k) + LW^{-1}r_y^a(k) \\ r_y^a(k) = W\left(y^a(k) - \hat{y}^a(k)\right), \hat{y}^a(k) = C\hat{x}^a(k) + Du_{est}(k) \end{cases}$$
$$\Longrightarrow r_y^a(z) = \hat{M}(z)y^a(z) - \hat{N}(z)u_{est}(z),$$

and the control estimate is computed according to

$$\begin{aligned} u_{est} &= F_e\hat{x}^a + Q_e r_y^a + Q_c r_y^a + \bar{v}(z) \\ &= -\left(\hat{Y} - M\left(Q_e + Q_c\right)\right)r_y^a + M\bar{v}. \end{aligned}$$

Observe that

$$\begin{aligned} r_y^a &= r_y + \hat{M} a_y - \hat{N}\left(u_{est} - u\right) \\ &= r_y + \hat{M} a_y - \hat{N}\left(\left(-\hat{Y} + M Q_e\right)\left(r_y^a - r_y\right) - a_{u_c}\right) \Longrightarrow \\ r_y^a &= r_y + \left(\hat{X} + N Q_e\right)^{-1} \hat{M}^{-1}\left(\hat{M} a_y + \hat{N} a_{u_c}\right), \end{aligned} \tag{14.27}$$

which leads to

$$u_c^a = Q_c r_y + \bar{v} + Q_c\left(\hat{X} + N Q_e\right)^{-1} \hat{M}^{-1}\left(\hat{M} a_y + \hat{N} a_{u_c}\right) + a_{u_c}.$$

Substituting u_c^a into the inner loop model (14.19) results in

$$\begin{bmatrix} u \\ y \end{bmatrix} = \begin{bmatrix} M \\ N \end{bmatrix} \bar{v} + \begin{bmatrix} -\hat{Y} + M Q \\ \hat{X} + N Q \end{bmatrix} r_y + \begin{bmatrix} M \\ N \end{bmatrix} d_{a,c}, \tag{14.28}$$

$$\begin{aligned} Q &:= Q_e + Q_c, \\ d_{a,c} &:= Q_c\left(\hat{X} + N Q_e\right)^{-1} \hat{M}^{-1}\left(\hat{M} a_y + \hat{N} a_{u_c}\right) + a_{u_c} \\ &= \hat{Q}_c\left(\hat{M} a_y + \hat{N} a_{u_c}\right) + a_{u_c}, \text{ for } Q_c = \hat{Q}_c \hat{M}\left(\hat{X} + N Q_e\right). \end{aligned}$$

Note that

$$u = F_e \hat{x} + Q_e r_y + Q_c r_y + \bar{v} + d_{a,c}.$$

Thus, u can be written, analogously to (14.25), as

$$u = V^{-1} U y + v + V^{-1} d_{a,c}. \tag{14.29}$$

Summarising (14.23) and (14.28) as well as (14.25) and (14.29) gives the following theorem that uniformly describes the dynamics of the closed-loop CPS under additive cyber-attacks.

Theorem 14.1 *Given CPS modelled by (14.2) and (14.11) with the embedded control system (14.5)–(14.6) or (14.7), then the system dynamics under additive attacks (14.13)–(14.14) are governed by*

$$\begin{bmatrix} u \\ y \end{bmatrix} = \begin{bmatrix} M \\ N \end{bmatrix} \left(\bar{v} + d_a\right) + \begin{bmatrix} -\hat{Y} + M Q \\ \hat{X} + N Q \end{bmatrix} r_y, \tag{14.30}$$

where $\bar{v} = \left(X + Q\hat{N}\right) v$,

$$d_a = \begin{cases} d_{a,o}, \text{ if } u_e = K_e y \\ d_{a,c}, \text{ if } u_e = F_e \hat{x} + Q_e r_y \end{cases}$$

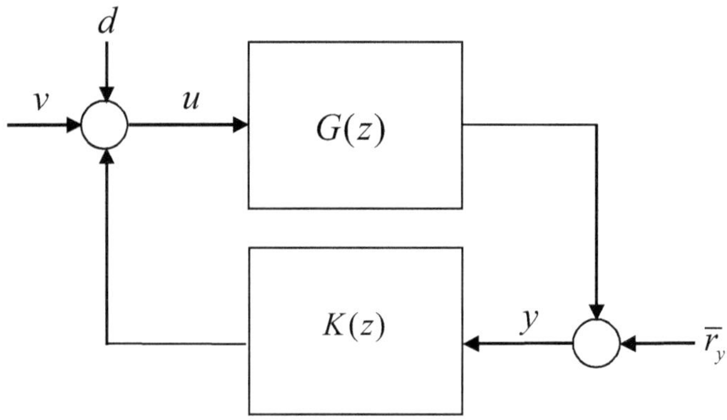

Fig. 14.2 The equivalent CPS closed-loop under additive cyber-attacks

is an $\mathcal{L}_2$-bounded signal, (linearly) depends on the cyber-attacks a_y and a_{u_c}, and r_y is independent of a_y and a_{u_c}. Equivalently, the CPS dynamics can also be uniformly described by

$$\begin{cases} y(z) = G(z)u(z) + \hat{M}^{-1}(z)r_y(z) \\ u(z) = K(z)y(z) + v(z) + d(z), \end{cases} \tag{14.31}$$

$$K = -\left(X + Q\hat{N}\right)^{-1}\left(Y - Q\hat{M}\right),$$

$$d = \left(X + Q\hat{N}\right)^{-1} d_a.$$

Proof Equations (14.30) and (14.31) are the results of (14.23) and (14.28) as well as (14.25) and (14.29), respectively. Moreover, note that

$$r_y = \hat{M}y - \hat{N}u, \begin{bmatrix} -\hat{N} & \hat{M} \end{bmatrix} \begin{bmatrix} M \\ N \end{bmatrix} = 0.$$

Consequently, r_y is independent of a_y and a_{u_c}.

According to (14.31), the equivalent configuration of the CPS under additive cyber-attacks is depicted in Fig. 14.2 with $\bar{r}_y = \hat{M}^{-1}r_y$. Since r_y is independent of a_y and a_{u_c}, it is apparent that the cyber-attacks a_y and a_{u_c} affect the system response to the reference signal and thus lead to variations in the system image subspace.

Example 14.1 As an example, it is shown that the CPS response (14.30) can be derived from the closed-loop dynamics (14.31). In fact, it follows from (14.31) that

$$\begin{bmatrix} u \\ y \end{bmatrix} = \begin{bmatrix} I & -K \\ -G & I \end{bmatrix}^{-1} \begin{bmatrix} v + d \\ \hat{M}^{-1}r_y \end{bmatrix}$$

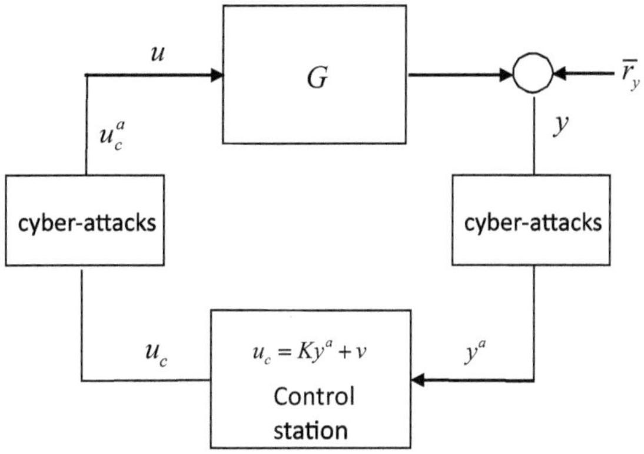

Fig. 14.3 A CPS configuration considered in Examples 14.2 and 14.3

$$= \begin{bmatrix} X + Q\hat{N} & Y - Q\hat{M} \\ -\hat{N} & \hat{M} \end{bmatrix}^{-1} \begin{bmatrix} \left(X + Q\hat{N}\right)(v + d) \\ r_y \end{bmatrix}$$

$$= \begin{bmatrix} M & -\hat{Y} + MQ \\ N & \hat{X} + NQ \end{bmatrix} \begin{bmatrix} \left(X + Q\hat{N}\right)(v + d) \\ r_y \end{bmatrix}$$

$$= \begin{bmatrix} M \\ N \end{bmatrix} (\bar{v} + d_a) + \begin{bmatrix} -\hat{Y} + MQ \\ \hat{X} + NQ \end{bmatrix} r_y.$$

Example 14.2 In this example, we would like to demonstrate that, under the additive cyber-attacks, the response of a CPS, whose configuration is shown in Fig. 14.3 and typically adopted in the literature, is of the identical model form given by (14.30). To this end, notice that the dynamic of the CPS is subject to (14.31) with

$$u = u_c^a = K\left(y + a_y\right) + v + a_{u_c} \Longrightarrow d = K a_y + a_{u_c},$$

where $K = -\left(X + Q\hat{N}\right)^{-1}\left(Y - Q\hat{M}\right)$ is the Youla parameterisation of the controller K. It follows from the proof of Theorem 14.1 that

$$\begin{bmatrix} u \\ y \end{bmatrix} = \begin{bmatrix} M \\ N \end{bmatrix} (\bar{v} + d_a) + \begin{bmatrix} -\hat{Y} + MQ \\ \hat{X} + NQ \end{bmatrix} r_y,$$

$$d_a = \left(X + Q\hat{N}\right) d = -\left(Y - Q\hat{M}\right) a_y + \left(X + Q\hat{N}\right) a_{u_c}.$$

14.2.4 Impairment of System Dynamics Caused by Multiplicative Cyber-Attacks

Instinctively, the closed-loop dynamics of the CPS under multiplicative cyber-attacks would be considerably impaired by the cyber-attacks and could eventually become unstable. Below, we address this issue and begin with the embedded control loop modelled by (14.7). Bearing in mind the attack model (14.15)–(14.16), we obtain,

- on the plant side,

$$u = u_e + u_c^a = K_e y + \Delta_u u_c, \tag{14.32}$$

$$r_y = \hat{M} y - \hat{N} u = \hat{M} y - \hat{N} K_e y - \hat{N} \Delta_u u_c, \tag{14.33}$$

  where u_c is the (control) signal sent by the control station,
- at the control station,

$$r_y^a = \hat{M} y^a - \hat{N} u_{est}, u_{est} = K_e y^a + u_c,$$
$$u_c = Q_c r_y^a + \bar{v}, \bar{v} = \left(X + Q_e \hat{N}\right)^{-1} \left(X + Q \hat{N}\right) v, y^a = \Delta_y y.$$

Since

$$u_{est} = K_e y^a + Q_c r_y^a + \bar{v}, u = K_e y + \Delta_u \left(Q_c r_y^a + \bar{v}\right) \Longrightarrow$$
$$u_{est} = u + K_e \delta_y y - \delta_u \left(Q_c r_y^a + \bar{v}\right),$$

it turns out

$$r_y^a = \hat{M} y^a - \hat{N} u_{est} = r_y + \hat{M} \delta_y y - \hat{N} \left(K_e \delta_y y - \delta_u \left(Q_c r_y^a + \bar{v}\right)\right)$$
$$\Longrightarrow r_y^a = \left(I - \hat{N} \delta_u Q_c\right)^{-1} \left(r_y + \left(\hat{M} - \hat{N} K_e\right) \delta_y y + \hat{N} \delta_u \bar{v}\right).$$

Observing that $\hat{M} - \hat{N} K_e = \left(\hat{X} + N Q_e\right)^{-1}$, it holds

$$r_y^a = \left(I - \hat{N} \delta_u Q_c\right)^{-1} \left(r_y + \left(\hat{X} + N Q_e\right)^{-1} \delta_y y + \hat{N} \delta_u \bar{v}\right),$$

which yields

$$u_c^a = \Delta_u \left(Q_c r_y^a + \bar{v}\right)$$
$$= \Delta_u Q_c \left(I - \hat{N} \delta_u Q_c\right)^{-1} \left(r_y + \left(\hat{X} + N Q_e\right)^{-1} \delta_y y + \hat{N} \delta_u \bar{v}\right) + \Delta_u \bar{v}.$$

Substituting u_c^a into (14.19) leads to

$$\begin{bmatrix} u \\ y \end{bmatrix} = \begin{bmatrix} M \\ N \end{bmatrix} \left(X + Q_e \hat{N}\right) \Delta_u \bar{v} + \begin{bmatrix} -\hat{Y} + M Q_e \\ \hat{X} + N Q_e \end{bmatrix} r_y$$

$$+\begin{bmatrix} M \\ N \end{bmatrix} \left(X + Q_e \hat{N}\right) \Delta_u Q_c \left(I - \hat{N}\delta_u Q_c\right)^{-1} \left(\begin{array}{r} r_y + \\ \left(\hat{X} + N Q_e\right)^{-1} \delta_y y + \hat{N}\delta_u \bar{v} \end{array} \right).$$

Let

$$Q_c = \hat{Q}_c \left(\hat{X} + N Q_e\right), \hat{Q}_c \in \mathcal{RH}_\infty,$$
$$\Psi = \left(X + Q_e \hat{N}\right) \Delta_u \left(I - Q_c \hat{N} \delta_u\right)^{-1} \hat{Q}_c \delta_y.$$

Note that

$$\hat{N} \left(X + Q_e \hat{N}\right)^{-1} = \left(\hat{X} + N Q_e\right)^{-1} N \Longrightarrow Q_c \hat{N} = \hat{Q}_c N \left(X + Q_e \hat{N}\right).$$

Matrix Ψ can also be written as

$$\Psi = \Delta \bar{u} \left(I - \hat{Q}_c N \bar{\delta}_u\right)^{-1} \hat{Q}_c \delta_y,$$
$$\Delta \bar{u} = \left(X + Q_e \hat{N}\right) \Delta_u, \bar{\delta}_u = \left(X + Q_e \hat{N}\right) \delta_u.$$

After some routine calculations, it turns out

$$\begin{aligned}
\begin{bmatrix} u \\ y \end{bmatrix} &= \begin{bmatrix} M \\ N \end{bmatrix} (I - \Psi N)^{-1} \Delta \bar{u} \left(I - \hat{Q}_c N \bar{\delta}_u\right)^{-1} \bar{v} \\
&+ \begin{bmatrix} -\hat{Y} \\ \hat{X} \end{bmatrix} r_y + \begin{bmatrix} M \\ N \end{bmatrix} (I - \Psi N)^{-1} \left(\Psi \hat{X} + Q_e\right) r_y \\
&+ \begin{bmatrix} M \\ N \end{bmatrix} (I - \Psi N)^{-1} \left(X + Q_e \hat{N}\right) \Delta_u \left(I - Q_c \hat{N} \delta_u\right)^{-1} Q_c r_y \\
&= \begin{bmatrix} M \\ N \end{bmatrix} (I - \Psi N)^{-1} \Delta \bar{u} \left(I - \hat{Q}_c N \bar{\delta}_u\right)^{-1} \bar{v} + \begin{bmatrix} -\hat{Y} + M Q_e \\ \hat{X} + N Q_e \end{bmatrix} r_y \\
&+ \begin{bmatrix} M \\ N \end{bmatrix} (I - \Psi N)^{-1} \Delta \bar{u} \left(I - \hat{Q}_c N \bar{\delta}_u\right)^{-1} \hat{Q}_c \Delta_y \left(\hat{X} + N Q_e\right) r_y.
\end{aligned} \tag{14.34}$$

A further useful relation is

$$\begin{aligned}
u &= K_e y + \Delta_u \left(Q_c r_y^a + \bar{v}\right) \Longrightarrow \\
\left(X + Q_e \hat{N}\right) u + \left(Y - Q_e \hat{M}\right) y &= \left(X + Q_e \hat{N}\right) \Delta_u \left(Q_c r_y^a + \bar{v}\right) \\
\left(X + Q_e \hat{N}\right) u + \left(Y - Q_e \hat{M}\right) y - \Psi y &= \left(X + Q_e \hat{N}\right) \Delta_u Q_c \left(I - \hat{N} \delta_u Q_c\right)^{-1} r_y \\
&+ \left(X + Q_e \hat{N}\right) \Delta_u \left(I - Q_c \hat{N} \delta_u\right)^{-1} \bar{v}.
\end{aligned} \tag{14.35}$$

Now, we consider the observer-based embedded control loop modelled by (14.5)–(14.6). It holds,

- on the plant side,

$$\begin{aligned}u &= u_e + u_c^a = F_e\hat{x} + Q_e r_y + \Delta_u\left(Q_c r_y^a + \bar{v}\right) \Longrightarrow \\ &\left(X + Q_e\hat{N}\right)u + \left(Y - Q_e\hat{M}\right)y = \Delta_u\left(Q_c r_y^a + \bar{v}\right),\end{aligned} \tag{14.36}$$

- at the control station,

$$\begin{aligned}&\begin{cases}\hat{x}^a(k+1) = A\hat{x}^a(k) + Bu_{est}(k) + LW^{-1}r_y^a(k) \\ r_y^a(k) = W\left(y^a(k) - \hat{y}^a(k)\right), \hat{y}^a(k) = C\hat{x}^a(k) + Du_{est}(k),\end{cases} \\ u_{est} &= F_e\hat{x}^a + Q_e r_y^a + u_c, u_c = Q_c r_y^a + \bar{v} \\ &\Longrightarrow \left(X + Q_e\hat{N}\right)u_{est} + \left(Y - Q_e\hat{M}\right)y^a = Q_c r_y^a + \bar{v}, \\ r_y^a &= \hat{M}y^a - \hat{N}u_{est}.\end{aligned} \tag{14.37}$$

Observe that

$$\begin{aligned}u_{est} &= K_e y^a + \left(X + Q_e\hat{N}\right)^{-1}\left(Q_c r_y^a + \bar{v}\right) \\ &= u + K_e\delta_y y - \left(X + Q_e\hat{N}\right)^{-1}\delta_u\left(Q_c r_y^a + \bar{v}\right) \Longrightarrow \\ r_y^a &= \hat{M}y^a - \hat{N}u_{est} = r_y + \left(\hat{M} - \hat{N}K_e\right)\delta_y y + \hat{N}\left(X + Q_e\hat{N}\right)^{-1}\delta_u\left(Q_c r_y^a + \bar{v}\right) \\ &= r_y + \left(\hat{X} + NQ_e\right)^{-1}\left(\delta_y y + N\delta_u\left(Q_c r_y^a + \bar{v}\right)\right) \Longrightarrow \\ r_y^a &= \left(I - \left(\hat{X} + NQ_e\right)^{-1}N\delta_u Q_c\right)^{-1}\left(r_y + \left(\hat{X} + NQ_e\right)^{-1}\left(\delta_y y + N\delta_u\bar{v}\right)\right).\end{aligned} \tag{14.38}$$

In the above calculation, relations

$$\hat{M} - \hat{N}K_e = \left(\hat{X} + NQ_e\right)^{-1}, \hat{N}\left(X + Q_e\hat{N}\right)^{-1} = \left(\hat{X} + NQ_e\right)^{-1}N$$

have been used. Accordingly,

$$\begin{aligned}u_c^a &= \Delta_u\left(Q_c r_y^a + \bar{v}\right) \\ &= \Delta_u\left(I - Q_c\left(\hat{X} + NQ_e\right)^{-1}N\delta_u\right)^{-1}Q_c\left(r_y + \left(\hat{X} + NQ_e\right)^{-1}\delta_y y\right) \\ &+ \Delta_u\left(I - Q_c\left(\hat{X} + NQ_e\right)^{-1}N\delta_u\right)^{-1}\bar{v}.\end{aligned}$$

Substituting u_c^a into (14.19) yields

$$\begin{bmatrix} u \\ y \end{bmatrix} = \begin{bmatrix} M \\ N \end{bmatrix} \Delta_u \left(I - Q_c \left(\hat{X} + N Q_e \right)^{-1} N \delta_u \right)^{-1} \bar{v} + \begin{bmatrix} -\hat{Y} + M Q_e \\ \hat{X} + N Q_e \end{bmatrix} r_y$$
$$+ \begin{bmatrix} M \\ N \end{bmatrix} \Delta_u \left(I - Q_c \left(\hat{X} + N Q_e \right)^{-1} N \delta_u \right)^{-1} Q_c \left(r_y + \left(\hat{X} + N Q_e \right)^{-1} \delta_y y \right).$$

Setting

$$Q_c = \hat{Q}_c \left(\hat{X} + N Q_e \right), \hat{Q}_c \in \mathcal{RH}_\infty,$$
$$\Psi = \Delta_u \left(I - \hat{Q}_c N \delta_u \right)^{-1} \hat{Q}_c \delta_y,$$

and by some routine calculations, it turns out

$$\begin{bmatrix} u \\ y \end{bmatrix} = \begin{bmatrix} M \\ N \end{bmatrix} (I - \Psi N)^{-1} \Delta_u \left(I - \hat{Q}_c N \delta_u \right)^{-1} \bar{v} + \begin{bmatrix} -\hat{Y} + M Q_e \\ \hat{X} + N Q_e \end{bmatrix} r_y$$
$$+ \begin{bmatrix} M \\ N \end{bmatrix} (I - \Psi N)^{-1} \Delta_u \left(I - \hat{Q}_c N \delta_u \right)^{-1} \hat{Q}_c \Delta_y \left(\hat{X} + N Q_e \right) r_y. \tag{14.39}$$

Furthermore, it follows from (14.36) and (14.38) that

$$\left(X + Q_e \hat{N} \right) u + \left(Y - Q_e \hat{M} \right) y - \Psi y = \Delta_u \left(I - \hat{Q}_c N \delta_u \right) \bar{v}$$
$$+ \Delta_u \left(I - \hat{Q}_c N \delta_u \right)^{-1} \hat{Q}_c \left(\hat{X} + N Q_e \right) r_y. \tag{14.40}$$

Analogously to Theorems 14.1, 14.2 and the associated corollaries summarise the aforementioned results about the CPS dynamics under multiplicative cyber-attackers.

Theorem 14.2 *Given CPS modelled by (14.2) and (14.11) with the embedded control system (14.5)–(14.6) or (14.7), then the system dynamics under multiplicative attacks (14.15)–(14.16) are governed by*

$$\begin{bmatrix} u \\ y \end{bmatrix} = \begin{bmatrix} M \\ N \end{bmatrix} (I - \Psi N)^{-1} \Delta \bar{u} \left(I - \hat{Q}_c N \bar{\delta}_u \right)^{-1} \bar{v} + \begin{bmatrix} -\hat{Y} + M Q_e \\ \hat{X} + N Q_e \end{bmatrix} r_y$$
$$+ \begin{bmatrix} M \\ N \end{bmatrix} (I - \Psi N)^{-1} \Delta \bar{u} \left(I - \hat{Q}_c N \bar{\delta}_u \right)^{-1} \hat{Q}_c \Delta_y \left(\hat{X} + N Q_e \right) r_y, \tag{14.41}$$

where the transfer function matrix Ψ and signals $\bar{\delta}_u$, $\Delta \bar{u}$, $\bar{v}$ are respectively defined by

$$\Psi = \Delta \bar{u} \left(I - \hat{Q}_c N \bar{\delta}_u \right)^{-1} \hat{Q}_c \delta_y, \ Q_c = \hat{Q}_c \left(\hat{X} + N Q_e \right), \hat{Q}_c \in \mathcal{RH}_\infty,$$

$$\bar{\delta}_u = \begin{cases} \left(X + Q_e\hat{N}\right)\delta_u, \ if\, u_e = K_e y \\ \delta_u, \ if\, u_e = F_e\hat{x} + Q_e r_y, \end{cases}$$

$$\Delta\bar{u} = \begin{cases} \left(X + Q_e\hat{N}\right)\Delta u, \ if\, u_e = K_e y \\ \Delta u, \ if\, u_e = F_e\hat{x} + Q_e r_y, \end{cases}$$

$$\bar{v} = \begin{cases} \left(I + \left(X + Q_e\hat{N}\right)^{-1} Q_c\hat{N}\right)v, \ if\, u_e = K_e y \\ \left(X + Q_e\hat{N} + Q_c\hat{N}\right)v, \ if\, u_e = F_e\hat{x} + Q_e r_y. \end{cases}$$

Proof The proof follows immediately from (14.34) and (14.39).

Theorem 14.2 describes the CPS dynamics with respect to the multiplicative cyber-attacks. In expression (14.41), the response of the plant signals (u, y) to the reference signal and the unknown inputs in the plant, modelled by r_y, is given. We would like to place a greater emphasis on the fact that r_y is independent of the cyber-attacks, as far as no uncertainty in the plant model is under consideration. For the analysis of the impact of the cyber-attacks, it is of special interest to derive the subsequent equivalent expression of (14.41).

Corollary 14.1 *Given CPS modelled by (14.2) and (14.11) with the embedded control system (14.5)–(14.6) or (14.7), then the system dynamics under multiplicative attacks (14.15)–(14.16) can be equivalently expressed by*

$$\begin{bmatrix} u \\ y \end{bmatrix} = \begin{bmatrix} M \\ N \end{bmatrix}\left(X + Q\hat{N}\right)v + \begin{bmatrix} -\hat{Y} + MQ \\ \hat{X} + NQ \end{bmatrix} r_y + \begin{bmatrix} M \\ N \end{bmatrix} Q_\Delta \bar{v} + \begin{bmatrix} M \\ N \end{bmatrix} \Delta Q r_y, \tag{14.42}$$

where the transfer function matrices Q, Q_Δ and ΔQ are defined by

$$Q = \begin{cases} Q_e + \left(X + Q_e\hat{N}\right)\hat{Q}_c\left(\hat{X} + NQ_e\right), \ if\, u_e = K_e y \\ Q_e + \hat{Q}_c\left(\hat{X} + NQ_e\right), \ if\, u_e = F_e\hat{x} + Q_e r_y, \end{cases} \tag{14.43}$$

$$Q_\Delta = \begin{cases} \left(\left(X + Q_e\hat{N}\right)\hat{Q}_c\left(\Delta_y N\bar{\delta}_u + \delta_y N\right) + \bar{\delta}_u\right)\Theta, \ if\, u_e = K_e y \\ \left(\hat{Q}_c\left(\Delta_y N\bar{\delta}_u + \delta_y N\right) + \bar{\delta}_u\right)\Theta, \ if\, u_e = F_e\hat{x} + Q_e r_y, \end{cases} \tag{14.44}$$

$$\Theta = \left(I - \hat{Q}_c\left(\Delta_y N\bar{\delta}_u + \delta_y N\right)\right)^{-1},$$

$$\Delta Q = \begin{cases} \left(\left(X + Q_e\hat{N}\right)\hat{Q}_c\delta_y + Q_\Delta\hat{Q}_c\Delta_y\right)\left(\hat{X} + NQ_e\right), \ if\, u_e = K_e y \\ \left(\hat{Q}_c\delta_y + Q_\Delta\hat{Q}_c\Delta_y\right)\left(\hat{X} + NQ_e\right), \ if\, u_e = F_e\hat{x} + Q_e r_y. \end{cases} \tag{14.45}$$

Proof To prove the result, we firstly rewrite $(I - \Psi N)^{-1}\Delta\bar{u}\left(I - \hat{Q}_c N\bar{\delta}_u\right)^{-1}$. For $u_e = K_e y$,

$$
\begin{aligned}
&(I-\Psi N)^{-1}\Delta\bar{u}\left(I-\hat{Q}_cN\bar{\delta}_u\right)^{-1} \\
&=\left(X+Q_e\hat{N}\right)(I+\delta_u)\left(I-\left(I-\hat{Q}_cN\bar{\delta}_u\right)^{-1}\hat{Q}_c\delta_yN\Delta\bar{u}\right)^{-1}\left(I-\hat{Q}_cN\bar{\delta}_u\right)^{-1} \\
&=\left(X+Q_e\hat{N}\right)(I+\delta_u)\left(I-\hat{Q}_c\Delta_yN\bar{\delta}_u-\hat{Q}_c\delta_yN\right)^{-1} \\
&=\left(X+Q_e\hat{N}\right)+\left(X+Q_e\hat{N}\right)\hat{Q}_c\left(\Delta_yN\bar{\delta}_u+\delta_yN\right)\left(I-\hat{Q}_c\left(\Delta_yN\bar{\delta}_u+\delta_yN\right)\right)^{-1} \\
&\quad+\bar{\delta}_u\left(I-\hat{Q}_c\left(\Delta_yN\bar{\delta}_u+\delta_yN\right)\right)^{-1}.
\end{aligned}
$$

Set

$$
Q_\Delta=\left(\left(X+Q_e\hat{N}\right)\hat{Q}_c\left(\Delta_yN\bar{\delta}_u+\delta_yN\right)+\bar{\delta}_u\right)\left(I-\hat{Q}_c\left(\Delta_yN\bar{\delta}_u+\delta_yN\right)\right)^{-1}.
$$

Consequently,

$$
(I-\Psi N)^{-1}\Delta\bar{u}\left(I-\hat{Q}_cN\bar{\delta}_u\right)^{-1}\bar{v}=\left(X+Q_e\hat{N}\right)v+Q_\Delta\bar{v} \tag{14.46}
$$

$$
\begin{aligned}
&(I-\Psi N)^{-1}\Delta\bar{u}\left(I-\hat{Q}_cN\bar{\delta}_u\right)^{-1}\hat{Q}_c\Delta_y\left(\hat{X}+NQ_e\right)r_y \\
&=\left(X+Q_e\hat{N}\right)\hat{Q}_c\left(\hat{X}+NQ_e\right)r_y+\left(X+Q_e\hat{N}\right)\hat{Q}_c\delta_y\left(\hat{X}+NQ_e\right)r_y \\
&\quad+Q_\Delta\hat{Q}_c\Delta_y\left(\hat{X}+NQ_e\right)r_y.
\end{aligned} \tag{14.47}
$$

Similarly, for $u_e=F_e\hat{x}+Q_er_y$, we have

$$
\begin{aligned}
&(I-\Psi N)^{-1}\Delta\bar{u}\left(I-\hat{Q}_cN\bar{\delta}_u\right)^{-1}\bar{v}=\left(X+Q_e\hat{N}\right)v+Q_\Delta\bar{v}, \\
&(I-\Psi N)^{-1}\Delta\bar{u}\left(I-\hat{Q}_cN\bar{\delta}_u\right)^{-1}\hat{Q}_c\Delta_y\left(\hat{X}+NQ_e\right)r_y \\
&=\hat{Q}_c\left(\hat{X}+NQ_e\right)r_y+\left(\hat{Q}_c\delta_y+Q_\Delta\hat{Q}_c\Delta_y\right)\left(\hat{X}+NQ_e\right)r_y, \qquad (14.48) \\
&Q_\Delta=\left(\hat{Q}_c\left(\Delta_yN\bar{\delta}_u+\delta_yN\right)+\bar{\delta}_u\right)\left(I-\hat{Q}_c\left(\Delta_yN\bar{\delta}_u+\delta_yN\right)\right)^{-1}.
\end{aligned}
$$

Summarising the above results and noting the definitions of Q, Q_Δ and ΔQ defined (14.43)–(14.45) finally prove (14.42).

In (14.42), the first two terms give the nominal system response (attack-free dynamic), while the third and forth terms describe the deviations, caused by the cyber-attacks, in the responses to the reference signal and to the plant residual (unknown input), respectively.

It is obvious that the multiplicative cyber-attacks might affect the CPS stability. The following corollary provides us with the conditions of setting the parameter system Q_c to assure the CPS stability.

Corollary 14.2 *Given CPS modelled by (14.2) and (14.8) and (equivalently (14.11)), let*

$$Q_c = \hat{Q}_c\left(\hat{X} + NQ_e\right), \hat{Q}_c \in \mathcal{RH}_\infty,$$

and suppose that $\hat{Q}_c$ stabilises

$$\left(I - \hat{Q}_c N \bar{\delta}_u\right)^{-1} \hat{Q}_c = \hat{Q}_c\left(I - N\bar{\delta}_u \hat{Q}_c\right)^{-1}. \tag{14.49}$$

Then, $\Psi \in \mathcal{RH}_\infty$, signal

$$\Delta v := \Delta\bar{u}\left(I - \hat{Q}_c N\bar{\delta}_u\right)^{-1} \bar{v}$$

is $\mathcal{L}_2$-bounded, and the system dynamic is stable if

$$\left\|\Delta\bar{u}\left(I - \hat{Q}_c N\bar{\delta}_u\right)^{-1} \hat{Q}_c \delta_y N\right\|_\infty < 1. \tag{14.50}$$

Proof It is apparent that Δv is $\mathcal{L}_2$-bounded and $\Psi \in RH_\infty$, if condition (14.49) is satisfied. Furthermore, according to the Small Gain Theorem, it can be concluded that

$$\|\Psi N\|_\infty < 1 \Longleftrightarrow \left\|\Delta\bar{u}\left(I - \hat{Q}_c N\bar{\delta}_u\right)^{-1} \hat{Q}_c \delta_y N\right\|_\infty < 1$$

results in

$$(I - \Psi N)^{-1} \in \mathcal{RH}_\infty,$$

which implies, together with the conditions of $\mathcal{L}_2$-bounded Δv and stable Ψ, that the system dynamic is stable.

Next, note that the cyber-attacks induced uncertainties can be interpreted as the uncertainties in the controller. To be specific, consider the relations (14.35) and (14.40), which are uniformly written as

$$\begin{aligned}&\left(X + Q_e\hat{N}\right)u + \left(Y - Q_e\hat{M}\right)y - \Psi y\\ &= \Delta v + \Delta\bar{u}\left(I - \hat{Q}_c N\bar{\delta}_u\right)^{-1} \hat{Q}_c\left(\hat{X} + NQ_e\right)r_y.\end{aligned} \tag{14.51}$$

Observe that

$$\Delta\bar{u}\left(I - \hat{Q}_c N\bar{\delta}_u\right)^{-1} \hat{Q}_c\left(\hat{X} + NQ_e\right) = \bar{Q}_c + Q_\delta, \tag{14.52}$$

$$\bar{Q}_c = \begin{cases} \left(X + Q_e\hat{N}\right)\hat{Q}_c\left(\hat{X} + NQ_e\right), \text{ if } u_e = K_e y \\ \hat{Q}_c\left(\hat{X} + NQ_e\right), \text{ if } u_e = F_e\hat{x} + Q_e r_y, \end{cases}$$

$$Q_\delta = \begin{cases} \left(I + \left(X + Q_e\hat{N}\right)\hat{Q}_c N\right)\Gamma, \text{ if } u_e = K_e y \\ \left(I + \hat{Q}_c N\right)\Gamma, \text{ if } u_e = F_e\hat{x} + Q_e r_y, \end{cases}$$

$$\Gamma := \left(I - \bar{\delta}_u\hat{Q}_c N\right)^{-1}\bar{\delta}_u\hat{Q}_c\left(\hat{X} + NQ_e\right).$$

We have

$$\left(X + (Q + Q_\delta)\hat{N}\right)u + \left(Y - (Q + Q_\delta)\hat{M} - \Psi\right)y = \Delta v. \tag{14.53}$$

Here, Q is given in (14.43). Let

$$\Delta K_u = Q_\delta\hat{N}, \Delta K_y = -Q_\delta\hat{M} - \Psi.$$

The overall controller is equivalent to

$$u = K_\Delta y + \left(X + Q\hat{N} + \Delta K_u\right)^{-1}\Delta v, \tag{14.54}$$
$$K_\Delta = \left(X + Q\hat{N} + \Delta K_u\right)^{-1}\left(Y - Q\hat{M} + \Delta K_y\right).$$

Summarising the plant dynamic (14.2) and the overall controller (14.55) results in an equivalent model of the closed-loop dynamics of CPSs under the multiplicative cyber-attacks.

Corollary 14.3 *Given CPS modelled by (14.2) and (14.11) with the embedded control system (14.5)–(14.6) or (14.7), then the system dynamics under multiplicative attacks (14.15)–(14.16) can be equivalently described by*

$$\begin{cases} y(z) = G(z)u(z) + \hat{M}^{-1}(z)r_y(z) \\ u(z) = K_\Delta(z)y(z) + \left(X(z) + Q(z)\hat{N}(z) + \Delta K_u\right)^{-1}\Delta v. \end{cases} \tag{14.55}$$

Figure 14.4 sketches an equivalent configuration of the CPS closed-loop dynamic under multiplicative cyber-attacks, where

$$\Delta\bar{v} = \left(X + Q\hat{N} + \Delta K_u\right)^{-1}\Delta v.$$

Comparing with the nominal controller (14.11), the uncertainties induced by Δ_u and Δ_y, ΔK_u and ΔK_y, are a part of the controller K_Δ. Remember that the system responses with respect to (plant) model uncertainties have been comprehensively investigated in Chap. 4. In fact, robust control and fault-tolerant control theories are

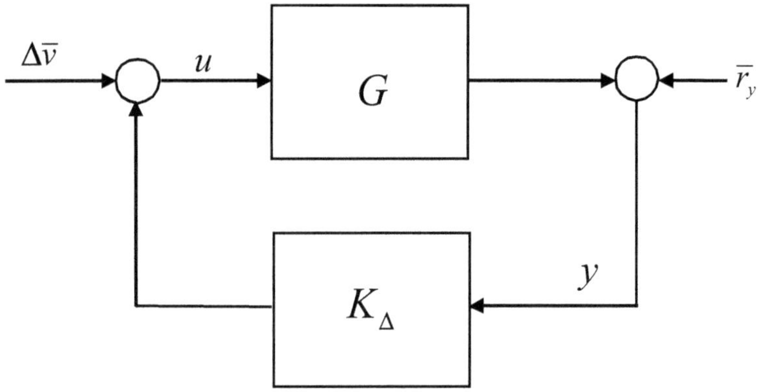

Fig. 14.4 An equivalent configuration of CPS closed-loop under multiplicative cyber-attacks

dedicated to uncertainties and faults in the plant. To our best knowledge, no result has been reported on controller design towards enhancing system robustness against uncertainties in the controller.

Example 14.3 As an exercise, this example gives an alternative proof of (14.42) using the equivalent closed-loop model of CPSs in Corollary 14.3. To this end, write (14.55) into

$$\begin{bmatrix} u \\ y \end{bmatrix} = \begin{bmatrix} X + (Q + Q_\delta)\hat{N} + & Y - (Q + Q_\delta)\hat{M} - \Psi \\ -\hat{N} & \hat{M} \end{bmatrix}^{-1} \begin{bmatrix} \Delta v \\ r_y \end{bmatrix}.$$

It follows from the Bezout identity that

$$\begin{bmatrix} X + (Q + Q_\delta)\hat{N} + & Y - (Q + Q_\delta)\hat{M} - \Psi \\ -\hat{N} & \hat{M} \end{bmatrix}^{-1}$$

$$= \begin{bmatrix} M & -\hat{Y} + M(Q + Q_\delta) \\ N & \hat{X} + N(Q + Q_\delta) \end{bmatrix} \begin{bmatrix} I - \Psi N & -\Psi\left(\hat{X} + N(Q + Q_\delta)\right) \\ 0 & I \end{bmatrix}^{-1}$$

$$= \begin{bmatrix} M(I - \Psi N)^{-1} & M(I - \Psi N)^{-1}\Psi\left(\hat{X} + N(Q + Q_\delta)\right) - \hat{Y} + M(Q + Q_\delta) \\ N(I - \Psi N)^{-1} & N(I - \Psi N)^{-1}\Psi\left(\hat{X} + N(Q + Q_\delta)\right) + \hat{X} + N(Q + Q_\delta) \end{bmatrix}.$$

Accordingly, it holds

$$\begin{bmatrix} u \\ y \end{bmatrix} = \begin{bmatrix} M \\ N \end{bmatrix}(I - \Psi N)^{-1}\Delta v + \begin{bmatrix} -\hat{Y} + MQ \\ \hat{X} + NQ \end{bmatrix} r_y$$
$$+ \begin{bmatrix} M \\ N \end{bmatrix}\left((I - \Psi N)^{-1}\Psi\left(\hat{X} + N(Q + Q_\delta)\right) + Q_\delta\right) r_y.$$

Observe that

$$
\begin{gathered}
(I-\Psi N)^{-1}\Psi\left(\hat{X}+N\left(Q+Q_\delta\right)\right)+Q_\delta= \\
(I-\Psi N)^{-1}\Psi\left(\hat{X}+NQ_e\right)+(I-\Psi N)^{-1}\Psi N\bar{Q}_c+(I-\Psi N)^{-1}Q_\delta .
\end{gathered}
$$

Recall (14.52) that leads to

$$
\begin{gathered}
Q_\delta=\Delta\bar{u}\left(I-\hat{Q}_cN\bar{\delta}_u\right)^{-1}\hat{Q}_c\left(\hat{X}+NQ_e\right)-\bar{Q}_c\Longrightarrow \\
(I-\Psi N)^{-1}\Psi N\bar{Q}_c+(I-\Psi N)^{-1}Q_\delta= \\
=(I-\Psi N)^{-1}\Delta\bar{u}\left(I-\hat{Q}_cN\bar{\delta}_u\right)^{-1}\hat{Q}_c\left(\hat{X}+NQ_e\right)-\bar{Q}_c, \\
(I-\Psi N)^{-1}\Psi+(I-\Psi N)^{-1}\Delta\bar{u}\left(I-\hat{Q}_cN\bar{\delta}_u\right)^{-1}\hat{Q}_c \\
=(I-\Psi N)^{-1}\Delta\bar{u}\left(I-\hat{Q}_cN\bar{\delta}_u\right)^{-1}\hat{Q}_c\Delta_y\Longrightarrow \\
(I-\Psi N)^{-1}\Psi\left(\hat{X}+N\left(Q+Q_\delta\right)\right)+Q_\delta \\
=(I-\Psi N)^{-1}\Delta\bar{u}\left(I-\hat{Q}_cN\bar{\delta}_u\right)^{-1}\hat{Q}_c\Delta_y\left(\hat{X}+NQ_e\right)-\bar{Q}_c .
\end{gathered}
$$

According to (14.46), (14.47) and (14.48), we finally obtain

$$
\begin{gathered}
(I-\Psi N)^{-1}\Delta v=\left(X+Q\hat{N}\right)v+Q_\Delta\bar{v}, \\
(I-\Psi N)^{-1}\Psi\left(\hat{X}+N\left(Q+Q_\delta\right)\right)+Q_\delta=\Delta Q .
\end{gathered}
$$

Thus, (14.42) is proved in terms of the equivalent CPS model (14.55).

Example 14.4 In this example, we delineate the response of a CPS with the configuration shown in Fig. 14.3 and under the multiplicative cyber-attacks. The CPS dynamic is modelled by

$$
\left\{\begin{array}{l}
y(z)=G(z)u(z)+\hat{M}^{-1}(z)r_y(z) \\
u(z)=u_c^a(z),\ u_c^a(z)=\left(I+\delta_u\right)\left(K(z)y^a(z)+v(z)\right) \\
y^a(z)=\Delta_y y(z)=\left(I+\delta_y\right)y(z),
\end{array}\right. \tag{14.56}
$$

$$
K=-\left(X+Q\hat{N}\right)^{-1}\left(Y-Q\hat{M}\right).
$$

Write u_c^a into

$$
\begin{gathered}
u_c^a(z)=\left(I+\delta_u\right)\left(K(z)\left(I+\delta_y\right)y(z)+v(z)\right) \\
=K(z)y(z)+v(z)+K(z)\delta_y y(z)+\delta_u\left(K(z)\left(I+\delta_y\right)y(z)+v(z)\right).
\end{gathered}
$$

It turns out

$$\left(X + Q\hat{N}\right) u + \left(Y - Q\hat{M}\right) y = \left(\begin{array}{c} \left(X + Q\hat{N}\right) \delta_u K(z) \left(I + \delta_y\right) \\ -\left(Y - Q\hat{M}\right) \delta_y \end{array}\right) y$$
$$+ \left(X + Q\hat{N}\right) \left(\delta_u + I\right) v \Longrightarrow$$
$$\begin{bmatrix} X + Q\hat{N} & Y - Q\hat{M} \\ -\hat{N} & \hat{M} \end{bmatrix} \begin{bmatrix} u \\ y \end{bmatrix} = \begin{bmatrix} \bar{v} + \Delta v + \Delta K_y y \\ r_y \end{bmatrix},$$
$$\Delta v = \left(X + Q\hat{N}\right) \delta_u v,$$
$$\Delta K_y = -\left(X + Q\hat{N}\right) \delta_u K \left(I + \delta_y\right) + \left(Y - Q\hat{M}\right) \delta_y.$$

By means of the Bezout identify, we have

$$\begin{bmatrix} u \\ y \end{bmatrix} = \begin{bmatrix} X + Q\hat{N} & Y - Q\hat{M} + \Delta K_y \\ -\hat{N} & \hat{M} \end{bmatrix}^{-1} \begin{bmatrix} \bar{v} + \Delta v \\ r_y \end{bmatrix}$$
$$\begin{bmatrix} u \\ y \end{bmatrix} = \begin{bmatrix} M & -\hat{Y} + MQ \\ N & \hat{X} + NQ \end{bmatrix} \begin{bmatrix} \left(I + \Delta K_y \hat{N}\right)^{-1} \left(\bar{v} + \Delta v - \Delta K_y \hat{M} r_y\right) \\ r_y \end{bmatrix}$$
$$= \begin{bmatrix} M \\ N \end{bmatrix} \left(I + \Delta K_y \hat{N}\right)^{-1} \left(\bar{v} + \Delta v - \Delta K_y \hat{M} r_y\right) + \begin{bmatrix} -\hat{Y} + MQ \\ \hat{X} + NQ \end{bmatrix} r_y,$$

which is analogous to (14.41) in Theorem 14.2. Note that the controller (14.56) can be equivalently written as

$$u(z) = K_{\Delta}(z) y(z) + \Delta_u v(z),$$
$$K_{\Delta} = -\left(X + Q\hat{N}\right)^{-1} \left(Y - Q\hat{M} + \Delta K_y\right).$$

To put it in a nutshell, this example demonstrates as well that

- the cyber-attacks attack the image subspace of the plant,
- their impact on the CPS dynamics can be equivalently modelled as uncertainties in the feedback controller, and
- this conclusion is independent of the configuration of the feedback controller embedded in the CPS under consideration.

14.3 Analysis of System Dynamics Under Cyber-Attacks

In this section, we will shortly analyse the results achieved in the previous section. It builds the basis for formulating the problems to be solved in dealing with diagnosis and control of CPS, which will be addressed in the subsequent chapters.

To begin with, we recall some basic results in the control and detection framework.

- The system signal (data) space $\begin{bmatrix} u \\ y \end{bmatrix}$ consists of the image and residual subspaces. While the image subspace is modelled by the SIR of the plant, $\begin{bmatrix} M \\ N \end{bmatrix}$, the residual subspace can be presented by the SIR of the controller, $\begin{bmatrix} -\hat{Y} \\ \hat{X} \end{bmatrix}$;
- the image subspace is the set of the system response (u, y) to the reference signal v $(\bar{v})$ and thus called nominal dynamic, and the residual subspace is composed of the data as the responses to the uncertainties in the plant;
- any uncertainties in the system, denoted by Δ, can be modelled by

$$\Delta = \begin{bmatrix} M \\ N \end{bmatrix} \Delta_{im} + \begin{bmatrix} -\hat{Y} \\ \hat{X} \end{bmatrix} \Delta_{res}.$$

  Here, Δ_{im} and Δ_{res} are called uncertainties in the image and residual subspaces, respectively;
- the observer-based residual generator,

$$r_y = \hat{M} y - \hat{N} u,$$

  only reflects the uncertainties in the residual subspace, since

$$\begin{bmatrix} -\hat{N} & \hat{M} \end{bmatrix} \begin{bmatrix} M \\ N \end{bmatrix} = 0, \begin{bmatrix} -\hat{N} & \hat{M} \end{bmatrix} \begin{bmatrix} -\hat{Y} \\ \hat{X} \end{bmatrix} = I;$$

- consequently, feedback control, according to Theorem 3.1, is the feedback of the uncertainties in the residual subspace.

We now analyse the major conclusions in Theorems 14.1–14.2 about the impairment of system dynamics induced by cyber-attacks. As the consequences of the aforementioned properties, it can be claimed that

- the cyber-attacks, both the additive and multiplicative ones considered in our work, lead to variations in the system image subspace. That means, the system response to the reference signal can considerably deviate from the nominal value. In typical control engineering applications like tracking control, model reference adaptive control, leader-follower formation control, such attacks result in grave control performance degradation;

- on the other hand, an observer-based process monitoring and fault diagnosis system embedded in the control system on the plant side can well operate independently of the cyber-attacks, since those faults and the corresponding control performance degradations in the plant solely cause variations in the system residual subspace. We would like to place a greater emphasis on this conclusion, which illustrates that the discussion in the literature about the challenge of isolating (distinguishing) cyber-attacks and (physical) faults makes less practical sense;
- equivalently, it is apparent that the residual generator,

$$r_y = \hat{M}y - \hat{N}u,$$

embedded and implemented on the plant side cannot detect cyber-attacks;
- the traditional feedback control strategies are less effective in enhancing the CPS robustness against cyber-attacks.

The last two points motivate a close loop at the residual r_y^a, generated at the control station and serving as an alternative. We will focus on examining the role of r_y^a in detecting the cyber-attacks and improving the system robustness.

We firstly consider the residual dynamics under the additive cyber-attacks. According to (14.21) and (14.27), the dynamics of the residual generator $Q_c r_y^a$ are governed by,

- in case of the embedded control system (14.7),

$$Q_c r_y^a = \bar{Q}_c \left(\hat{X} + NQ_e\right) r_y + \bar{Q}_c \left(a_y + \hat{X}\hat{N}a_{u_c} + NQ_e\hat{N}a_{u_c}\right) \tag{14.57}$$

$$= \bar{Q}_c \left(\hat{X} + NQ_e\right) r_y + \bar{Q}_c \left(a_y + N\left(X + Q_e\hat{N}\right) a_{u_c}\right), \tag{14.58}$$

with $\bar{Q}_c \in \mathcal{RH}_\infty$,
- in case of the embedded control system (14.5)–(14.6),

$$Q_c r_y^a = \bar{Q}_c \left(I + \hat{N}Q_c\right) r_y + \bar{Q}_c \left(\hat{M}a_y + \hat{N}a_{u_c}\right), \bar{Q}_c \in \mathcal{RH}_\infty. \tag{14.59}$$

On the assumption that the plant is operating fault-free, r_y merely contains uncertainties in the plant, like noises, disturbances and parameter variations. Hence, the cyber-attacks a_y and a_{u_c} are structurally detectable by means of $Q_c r_y^a$ on the assumption that $\left(a_{u_c}, a_y\right)$ does not lie in the null-space of

$$\begin{bmatrix} N\left(X + Q_e\hat{N}\right) & I \end{bmatrix}, \text{ for the dynamic subject to (14.57)}$$
$$\text{or } \begin{bmatrix} \hat{N} & \hat{M} \end{bmatrix}, \text{ for the dynamic subject to (14.59).}$$

Concerning the multiplicative cyber-attacks, a comprehensive analysis is necessary. To this end, we examine the dynamics of $Q_c r_y^a$ with the aid of Theorem 14.2

and its proof. It is apparent that the detection of $\Delta_y\left(\delta_y\right)$ and $\Delta_u\left(\delta_y\right)$ by means of $Q_c r_y^a$ depends on $\bar{v}$ and r_y, and is generally possible as far as $\bar{v}$ and r_y are not too small so that there exist sufficient excitations.

Next, we reveal the role of $Q_c r_y^a$ in the control signal in the context of the observer-based Youla parameterisation. This view is motivated by the fact that $Q_c r_y^a$ is indeed the control signal (online) calculated at the control station to optimise or maintain the overall CPS performance.

When additive cyber-attacks are injected into the system, it follows from Theorem 14.1 that

$$\begin{bmatrix} u \\ y \end{bmatrix} = \begin{bmatrix} M \\ N \end{bmatrix} (\bar{v} + d_a) + \begin{bmatrix} -\hat{Y} + MQ \\ \hat{X} + NQ \end{bmatrix} r_y,$$

$$d_a = \begin{cases} d_{a,o}, \text{ if } u_e = K_e y \\ d_{a,c}, \text{ if } u_e = F_e \hat{x} + Q_e r_y. \end{cases}$$

Consequently,

- tuning Q_c can reduce

$$\left\| \begin{bmatrix} -\hat{Y} + MQ \\ \hat{X} + NQ \end{bmatrix} \right\|$$

  aiming at recovering the system performance suffered from degradation caused by the unknown inputs in the plant, represented by r_y, and
- tuning $\bar{Q}_c$, $\hat{Q}_c$ is able to reduce

$$\begin{cases} \left\| \begin{bmatrix} I + \bar{Q}_c \left(\hat{X} + N Q_e \right) \hat{N} & \bar{Q}_c \end{bmatrix} \right\|, \text{ if } u_e = K_e y \\ \left\| \begin{bmatrix} I + \hat{Q}_c \hat{N} & \hat{Q}_c \hat{M} \end{bmatrix} \right\|, \text{ if } u_e = F_e \hat{x} + Q_e r_y, \end{cases}$$

  in order to reject $\left(a_{u_c}, a_y\right)$.

Unfortunately, there exists no sufficient degree of freedom for the realisation of the above two tuning objectives, since $\hat{Q}_c$ is a linear function of Q_c. In other words, once Q_c is fixed for achieving an optimal control performance (during attack-free operations), there is no further possibility to modify $\hat{Q}_c$ to enhance the robustness against attacks.

Concerning the multiplicative cyber-attacks, the capability of $Q_c r_y^a$ feedback to enhance the robustness against attacks is considerably limited. According to (14.42) in Corollary 14.1, the initial task of feeding back $Q_c r_y^a$ is to reject r_y representing unknown inputs in the plant, as indicated by the second term in (14.42), i.e. minimising

$$\left\| \begin{bmatrix} -\hat{Y} + MQ \\ \hat{X} + NQ \end{bmatrix} \right\|.$$

The impact of the multiplicative cyber-attacks on the system response is described by the third and fourth terms in (14.42). Apparently, $\hat{Q}_c$ as a linear function of Q_c cannot efficiently mitigate the influence of the attacks on the system dynamics. In this case, a reasonable setting of $\hat{Q}_c$ is to guarantee the stability of $\left(I - \hat{Q}_c N\bar{\delta}_u\right)^{-1} \hat{Q}_c$ and make it as small as possible so that

$$\left\| \Delta\bar{u} \left(I - \hat{Q}_c N\bar{\delta}_u\right)^{-1} \hat{Q}_c \delta_y N \right\|_\infty < 1.$$

As a result, the closed-loop dynamic, according to Corollary 14.2, is stable.

14.4 Problem Formulation

On the basis of the analysis results of CPS dynamics in response to the cyber-attacks, presented in the previous sections, the following issues will be addressed in the subsequent chapters.

- Detection of stealthy attacks
  Stealthy attacks are a class of cyber-attacks which cannot be detected timely and reliably, but often remarkably degrade the system performance. The fact that cyber-attacks are artificially created by attackers makes their detection considerably difficult. We will deal with this problem from two different aspects:
  - investigating existence conditions of stealthy attacks, and
  - extending the existing model-based detection and diagnosis schemes towards timely and reliable detection of this type of cyber-attacks.
- Resilient control of industrial CPSs
  The two essential conclusions from our analysis in the previous two sections,
  - cyber-attacks exclusively impair the system image subspace, and
  - the (plant) residual signal r_y is independent of (decoupled from) cyber-attacks,

  enforce us, on the one hand, to search alternative (feedback) control solutions, inspire us, on the other hand, to apply the feedback of a residual signal generated on the controller side and implemented in the control station. Analysis and synthesis of such type of feedback control systems are the major efforts of this work.

The foundation of our study is the unified control and detection framework with its capable and efficient methods for system analysis and synthesis.

14.5 Notes and References

Automatic control systems are essential system parts of many industrial CPSs and their flawless operations are of elemental importance for optimal system operation and high product quality. That cyber-attacks can cause immense damages on industrial CPSs is a common cognition. It is therefore not surprising that automatic control systems are often immediate targets of cyber-attacks on industrial CPSs. Driven by the rapidly increasing industrial demands for higher cyber-security, detection and control of cyber-attacks on automatic control systems has drawn incredible research attention in the current decade. Excellent reviews of state of the art of research in this thematic area can be found in the surveys published in [31, 32, 72, 155, 187, 214, 223, 241] and the references therein.

Among various types of cyber-attacks, integrity attacks are specially directed to automatic control systems [31, 74, 155]. By injecting attack signals into system input and output channels, e.g. via I/O and network interfaces, integrity attacks can lead to remarkable system performance degradations and even catastrophic damages. An early and reliable detection of integrity attacks is becoming a vital requirement on cyber-security of industrial CPSs, for instance, for power control systems [143]. In the literature, a great number of results have been reported about detecting the so-called replay, zero dynamics, covert attacks, false data injection attacks or kernel attacks, which are stealthy integrity attacks and cannot, structurally, be detected by means of a standard observer-based fault detection system [31, 45, 74, 187] or fault detection methods for NCSs [48, 204, 205, 218]. The zero dynamics, covert attacks, false data injection attacks or kernel attacks as the general form of such type of cyber-attacks, are additive attacks, as modelled by (14.13)–(14.14). Differently, replay attacks are a special type of multiplicative attacks [45]. This type of attacks impose a realistic cyber-threat on industrial CPSs, since their realisation does not require knowledge of the plant and embedded control systems. A further class of multiplicative cyber-attacks are DoS attacks, which, due to their easy implementation and various forms, are one of the most frequently performed cyber-attacks on IT-infrastructures. In their pioneering work, Persis and Tesi have reported a comprehensive study on modelling and control issues of DoS attacks on automatic control systems [160].

In the recent decade, a trend of investigating control-theoretically aided design of cyber-attacks on control performance of automatic control systems can be observed. Most of such design schemes result in additive cyber-attacks [22, 170, 228], and only few of them lead to multiplicative cyber-attacks [182].

Our works in this chapter are not in the trend of the current research on detection and control of cyber-attacks issues. We observe that most of the results reported in the literature are focused on impact analysis of cyber-attacks on (a) test and evaluation functions in dealing with detection issues, (b) control performance when control issues are under consideration, and (c) both of them, when the attack design schemes are involved. As a result, most of the reported schemes have been developed on the assumption of a given (attack) detection scheme or/and system control laws under consideration. A systematic, control and detection scheme independent analysis is apparently missing. Furthermore, the CPSs under consideration are rarely analysed with regard to their closed-loop configurations. An immediate consequence is that the proposed design algorithms for attack detection and control are limited to updating the existing methods, schemes and restricted in their application. Motivated by these observations, in particular inspired by the analysis results in the control and detection unified framework, our efforts in this chapter have been made mainly on

- analysis of closed-loop dynamics of CPSs under defined types of cyber-attacks and the attack-induced impairment of system dynamics, and
- analysis of the potential of the existing control and detection methods to manage the system performance impairment due to the cyber-attacks.

They are the basis for formulating the tasks of studying cyber-attack detection and resilient control issues and methods in the subsequent chapters. At the beginning of our work, the configuration and models of industrial CPSs under consideration have been specified. It is worth emphasising that the CPS model, (14.2) and (14.11), adopted in our work is formally different from the existing model form mostly used in the literature (refer to Fig. 14.3). At the end of Sect. 14.1, we have delineated the equivalence between these two different model forms and, in the unified control and detection framework, explained that our CPS model imposes no restriction on the generality of the achieved analysis results and is in fact of a general form of feedback control systems. This conclusion has also been validated in Examples 14.2 and 14.4, which demonstrate that the analysis results of attack's impacts on the CPS dynamics are independent of the configuration of the feedback controller embedded in the CPS under consideration. On the other hand, and more importantly, the CPS configuration adopted in our work is supposed to meet real industrial requirements on system security and safety. For instance, when the communication between the plant and control station fails, the local embedded control system should guarantee the loop stability. It is also noteworthy that, differing from state of the art research, the attack models considered in our work, (14.15)–(14.16) and (14.13)–(14.14), are solely characterised and classified by their influences on system dynamics. To put it in a nutshell, the introduced CPS models, including cyber-attack models, enable a systematic and general analysis of CPS system dynamics under cyber-attacks, which builds a solid fundament for the research work in the subsequent chapters towards developing reliable and capable attack detection and resilient control schemes.

In Sect. 4.2, the closed-loop dynamics of CPSs under cyber-attacks have been examined. The major results are summarised in Theorems 14.1 and 14.2, in which the responses of (u, y) to the reference signal and unknown inputs in the plant, and above all, to the cyber-attacks, additive and multiplicative ones respectively, are described. Among a number of results presented in the subsequent discussions, the most important and distinguishing conclusion is that the cyber-attacks exclusively affect the plant image subspace. In other words, and as apparently sketched in Figs. 14.2 and 14.3, the impacts of the cyber-attacks on the closed-loop dynamics are equivalently modelled (i) as an unknown input added to the reference signal, when the additive attacks act on the CPS, and (ii) as uncertainty in the feedback controller in the case of the multiplicative attacks. A comparison with the traditional feedback control loops with unknown inputs and (plant) model uncertainties as well as faults immediately shows the difference. It is remarkably surprising that this essential feature has not been reported in the literature and among innumerable publications on cyber-security issues of automatic control systems. This result is so far of considerable importance, because the immediate consequences/claims follow,

- detection and control of CPSs under cyber-attacks and dynamic processes suffering uncertainties or faults are two different but dual problems. This conclusion illustrates that the common view of considering distinguishing/isolating cyber-attacks and physical faults as a challenging issue, because of their similar forms, is less convincing,
- instead, the real challenges are the development of novel detection and (feedback) control schemes that are capable to detect and control the uncertainties in the plant image subspace sketched in Figs. 14.2 and 14.3. It is known that the traditional detection and control methods are dedicated to handling of uncertainties and faults in the plant. Since those uncertainties, including unknown inputs and faults, are indeed uncertainties in the (plant) residual subspace, which is a complement of the image subspace, the traditional methods are less (or even not) capable to detect and control of cyber-attacks, and
- concretely, methods should be developed to detect and control uncertainties in the (plant) image subspace. To this end, the analysis results on the unified framework of control and detection, presented in the first part of this book will serve as a useful and capable tool.

As a direct result of Theorem 14.2, the existence conditions of the loop stability, when multiplicative cyber-attacks are injected in the system, have been derived and given in Corollary 14.1.

Based on the aforementioned results, some elemental system properties of CPSs under cyber-attacks have been analysed in Sect. 14.3. The focus has been on the role of the residual signal r_y^a generated in the control station in the context of cyber-attacks and the "controller" $u_c = Q_c r_y^a$ for control and detection purposes. It has been delineated that their capability of detecting and suppressing cyber-attacks is strongly limited. All these results have motivated the problem formulation given in Sect. 14.4 as tasks for our work in the subsequent chapters.

Chapter 15
Detection of Cyber-Attacks and Faults, and Resilient CPS

Among various types of cyber-attacks, integrity attacks are specially directed to automatic control systems. By injecting attack signals into system input and output channels, e.g. via I/O and network interfaces, integrity attacks lead to remarkable system performance degradations and even catastrophic damages. An early and reliable detection of integrity attacks is becoming a vital requirement on cyber-security of industrial CPSs. Thanks to its well-established control-theoretic framework in the past three decades, observer-based fault detection technique is widely accepted as an efficient method, among numerous ones, to deal with detection of integrity attacks on automatic control systems. Unfortunately, different from technical faults, cyber-attacks are artificially created and can be designed and generated by an adversary. It is particularly insidious, when cyber-attacks are constructed in such a way that they are stealthy and cannot be detected using the existing detection techniques. This observation and some real cases with stealthy cyber-attacks on industrial automatic control systems have strongly motivated researchers to improve the existing detection schemes and develop alternative solutions. In this regard, a great number of results have been reported about detecting the so-called replay, zero dynamics, covert attacks and false data injection, which are stealthy integrity attacks and cannot be detected by means of the standard observer-based fault detection technique without modifications on the applied algorithms. Representative solutions are the watermark detection scheme, the moving target method and the auxiliary system aided detection scheme.

Reviewing the related publications in the literature, we observe that

- stealthiness of cyber-attacks is defined with respect to the detection system assumed to be implemented for detecting cyber-attacks. In other words, for different detection systems or algorithms, the definitions of stealthy attacks could be different.
- On the other hand, the currently popular research area of designing cyber-attacks leads to creation of various types of cyber-attacks, which are stealthy with regard to a specified detection scheme, an unusual procedure.

S. X. Ding, *Unified Framework of Control and Detection*,
https://doi.org/10.1007/978-3-662-72417-0_15

- Moreover, the trend in the current research on detecting stealthy attacks is integrating the classical cyber-attack detection methods known in computer and communication engineering and the traditional model-, in particular, observer-based detection methods. Little attention is paid to the specific properties of cyber-attacks in feedback control systems, as described in the previous chapter.

In this chapter, we will approach the above issues from different aspects. The main objectives of our work are, in the unified framework,

- to model cyber-attacks with respect to their impact on the system dynamics,
- to specify cyber-attacks in the unified framework of control and detection and introduce the so-called kernel attacks, and on this basis,
- to analyse existence conditions of kernel attacks in industrial CPSs with various configurations,
- to develop advanced schemes for cyber-attack detection,
- to examine simultaneous detection of cyber-attacks and faults, and finally
- to explore configuration of resilient CPSs and possible solutions for resilient CPSs.

At first, a uniform model form of cyber-attacks on automatic control systems is introduced. Specifically, not only attacks that are injected into the system as unknown inputs, but also attacks that build a loop with the system variables are under consideration. The latter type of attacks are particularly critical, because they may cause remarkable changes in the system eigen-dynamic.

Against the background of the current research on stealthy attacks, kernel attacks are introduced. Differing from typical stealthy attacks addressed in the literature, kernel attacks are defined in terms of system input and output signals in CPSs, which lie in the system kernel subspace. Consequently, the existence of kernel attacks is independent of specifications of a detection system applied for detection purpose. On this basis, relations of kernel attacks and typical stealthy attacks addressed in the literature are explored, and existence conditions of kernel attacks are examined with respect to various configurations of CPSs.

In the last part of this chapter, capable detection schemes of cyber-attacks, in particular, kernel attacks, are proposed. Distinctly differing from the existing detection schemes, the core of the proposed schemes is the generation of the input residual and fusion of the input and output residuals. Proceeding from these detection schemes, issues of simultaneous detection of cyber-attacks and faults and configuration of resilient CPSs are addressed.

15.1 System Models and Configurations

In our subsequent work, CPSs schematically sketched in Fig. 15.1 are under consideration. Aiming at dealing with various types of CPSs, the configurations of the plant and monitoring and control (MC) station may be differently modelled. In this

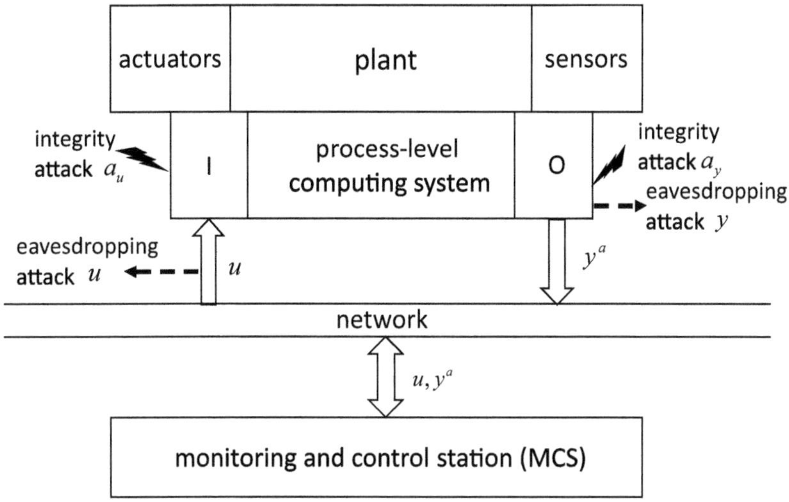

Fig. 15.1 CPS set-up under consideration

section, we briefly summarise the models and configurations of CPSs and cyber-attacks considered in our work.

15.1.1 CPS Models

As shown in Fig. 15.1, a CPS consists of two main parts, the plant with the embedded sensors and actuators as well as a computing unit, which may include embedded control algorithms (controller), and the monitoring and control station, at which CPS monitoring and control algorithms are implemented. We introduce the models in each of these two parts separately.

- On the plant side:
  Two types of plant models are considered, the open-loop and the closed-loop, as described below:
  - open-loop configuration

$$\begin{cases} y(z) = G(z)u(z) + \hat{M}^{-1}(z)r_y(z) \\ u(z) = u_{MC}^a(z), \end{cases} \tag{15.1}$$

    where u_{MC}^a is the control signal received from the MC-station. In other words, the dynamic system running at the plant side is open-loop configured;
  - closed-loop configuration

$$y(z) = G(z)u(z) + \hat{M}^{-1}(z)r_y(z), \tag{15.2}$$
$$u(z) = u_e(z) + u_{MC}^a(z), \tag{15.3}$$

where u_e is the output of the local controller embedded in the dynamic system on the plant side, which might be of two different forms,

$$u_e(z) = K_e(z)y(z), \tag{15.4}$$
$$u_e(z) = F_e\hat{x}(z) + Q_e(z)r_y(z), Q_e(z) \in \mathcal{RH}_\infty, \tag{15.5}$$
$$\begin{cases} \hat{x}(k+1) = A\hat{x}(k) + Bu_e(k) + LW^{-1}r_y(k) \\ r_y(k) = W\left(y(k) - \hat{y}(k)\right), \hat{y}(k) = C\hat{x}(k) + Du_e(k), \end{cases} \tag{15.6}$$

as introduced in Chap. 14 with the Youla parameterisation of K_e

$$K_e = -\left(X + Q_e\hat{N}\right)^{-1}\left(Y - Q_e\hat{M}\right),$$

Hence, the local system is closed-loop configured.

On demand, an observer-based residual generator,

$$r_y(z) = \hat{M}(z)y(z) - \hat{N}(z)u(z), \tag{15.7}$$
$$\begin{cases} \hat{x}(k+1) = A\hat{x}(k) + Bu(k) + LW^{-1}r_y(k) \\ r_y(k) = W\left(y(k) - \hat{y}(k)\right), \hat{y}(k) = C\hat{x}(k) + Du(k), \end{cases} \tag{15.8}$$

can be installed for detecting faults in the plant asset.

- At the MC-station:
  With respect to the various system configurations on the plant side, two types of dynamic (control) systems are considered:
  - corresponding to the open-loop configuration (15.1),

    $$u_{MC}(z) = K(z)\, y^a(z) + v(z) \tag{15.9}$$

    serves as the feedback controller. The CPS configuration modelled by (15.1) and (15.9) is mostly adopted in the literature;
  - in the case of the closed-loop configuration (15.2) with the two possible control laws, (15.4) and (15.5),

    $$u_{MC}(z) = Q_{MC}(z)\, r_y^a(z) + \bar{v}(z) \tag{15.10}$$

    will be implemented. Here, r_y^a is the residual signal corrupted with the possible cyber-attacks, $\bar{v}$ is the reference signal with a pre-filter defined by (refer to Chap. 14)

$$\bar{v} = \begin{cases} \left(X + Q_e \hat{N}\right)^{-1} \left(X + Q\hat{N}\right) v, \text{ corresponding to (15.4)} \\ \left(X + Q\hat{N}\right) v, \text{ corresponding to (15.5)}, \end{cases}$$

$$Q = \begin{cases} Q_e + \left(X + Q_e \hat{N}\right) Q_{MC}, \text{ corresponding to (15.4)} \\ Q_e + Q_{MC}, \text{ corresponding to (15.5)}. \end{cases} \tag{15.11}$$

Independent of the system configuration of u_{MC}, a fault/attack detection system is to be installed at the MC-station, whose core is an observer-based residual generator,

$$r_y^a(z) = \hat{M}(z) y^a(z) - \hat{N}(z) u_{est}(z), \tag{15.12}$$

$$\begin{cases} \hat{x}(k+1) = A\hat{x}(k) + Bu_{est}(k) + LW^{-1} r_y^a(k) \\ r_y^a(k) = W\left(y^a(k) - \hat{y}(k)\right), \hat{y}(k) = C\hat{x}(k) + Du_{est}(k), \end{cases} \tag{15.13}$$

where u_{est} designates the estimate of the input u acting on the plant system and is computed as follows:

– in the case of the open-loop configuration (15.1),

$$u_{est}(z) = u_{MC}(z) = K(z)\, y^a(z) + v(z), \tag{15.14}$$

– for the closed-loop configuration (15.2),

$$u_{est} = \begin{cases} K_e y^a + Q_{MC} r_y^a + \bar{v}, \text{ corresponding to (15.4)} \\ F_e \hat{x} + (Q_e + Q_{MC})\, r_y^a + \bar{v}, \text{ corresponding to (15.5)}, \end{cases} \tag{15.15}$$
$$\hat{x}(k+1) = A\hat{x}(k) + Bu_{est}(k) + LW^{-1} r_y^a(k).$$

To simplify our subsequent work, (15.15) is uniformly written into

$$u_{est} = K_e y^a + u_{MC}, \tag{15.16}$$

$$u_{MC} := \begin{cases} Q_{MC} r_y^a + \bar{v}, \text{ corresponding to (15.4)} \\ \left(X + Q_e \hat{N}\right)^{-1} \left(Q_{MC} r_y^a + \bar{v}\right) \text{corresponding to (15.5)}. \end{cases} \tag{15.17}$$

15.1.2 Integrity Attack Models Under Consideration

The attack models introduced in Chap. 14, (14.13)–(14.14) and (14.15)–(14.16), are extended to take account of those cyber-attacks artificially designed and created by attackers. To simplify the notation, we write them in a uniform form

$$y^a(k) = y(k) + a_y(k), \tag{15.18}$$

$$u^a_{MC}(k) = u_{MC}(k) + a_{u_{MC}}(k). \tag{15.19}$$

Here, $a_{u_{MC}}$ and a_y are modelled by

$$\begin{bmatrix} a_{u_{MC}}(z) \\ a_y(z) \end{bmatrix} = \Pi^a(z) \begin{bmatrix} u_{MC}(z) \\ y(z) \end{bmatrix} + \begin{bmatrix} \varepsilon_u(z) \\ \varepsilon_y(z) \end{bmatrix}, \Pi^a(z) \in \mathcal{RH}_\infty. \tag{15.20}$$

Signals ε_y and ε_u are $\mathcal{L}_2$-bounded, unknown, and assumed to be independent of u and y. Thus, they represent additive attacks. In order to model those cyber-attacks which are designed and artificially generated by attackers, a (stable) dynamic system Π^a with (u_{MC}, y) as its input variables is introduced. The idea behind this system is to model possible cyber-attacks artificially designed on the assumption that attackers are in possession of data (u_{MC}, y). $\Pi^a \begin{bmatrix} u_{MC} \\ y \end{bmatrix}$ are multiplicative attacks.

In the literature, cyber-attacks on actuators and sensors are called false-data injection attacks as well. Our major attention is paid to those false-data injection attacks that cause (considerable) changes in the system dynamics, but cannot be detected by a standard observer-based detector. They are called stealthy. The so-called zero dynamics, covert and replay attacks are the typical stealthy attacks addressed in the literature. Below is a short description of these attack types. Note that these three types of cyber-attacks are defined with respect to the CPS configuration given by (15.1) and (15.9).

15.1.2.1 Zero Dynamics Attacks

Roughly speaking, a zero dynamics attack is referred to an additive attack a_u on the actuators, which causes no response at the system output over the detection time interval $[k_0, k_0 + N]$ and thus cannot be detected. The corresponding attack model is

$$x(k+1) = Ax(k) + B(u(k) + a_u(k)), \tag{15.21}$$

$$y^a(k) = Cx(k) + D(u(k) + a_u(k)) \tag{15.22}$$

with $y^a(k)$ satisfying the condition

$$\forall k \in [k_0, k_0 + N], y^a(k) = y(k). \tag{15.23}$$

It is obvious that the existence condition of zero dynamics attacks can be expressed by means of the LCF of the plant as

$$\hat{N}(z)\, a_u(z) = 0. \tag{15.24}$$

15.1.2.2 Covert Attacks

Introduced by Smith 2015, covert attacks are modelled by

$$x(k+1) = Ax(k) + B\left(u(k) + a_u(k)\right), \tag{15.25}$$

$$y^a(k) = Cx(k) + D\left(u(k) + a_u(k)\right) + a_y(k) \tag{15.26}$$

with a_u and a_y denoting additive attacks on the actuators and sensors, respectively, and satisfying

$$a_y(z) + G(z)a_u(z) = 0 \Longrightarrow \forall k \in [k_0, k_0 + N], y^a(k) = y(k). \tag{15.27}$$

It is straightforward that the existence condition for covert attacks is

$$\hat{M}(z)a_y(z) + \hat{N}(z)\,a_u(z) = 0. \tag{15.28}$$

15.1.2.3 Replay Attacks

Replay attacks are performed on the assumption that the plant is operating in the steady state, which yields

$$y(k) \approx y(k-i), i = 1, \cdots.$$

Consequently, the attacker can "replay", e.g. over the time interval $[k, k+M]$, the sensor data collected in the past, for instance, by means of an eavesdropping attack, and simultaneously inject signals in the actuators. Denote by $y\left(k_0+i\right), k_0+M < k, i = 0, 1, \cdots, M$, the data collected and saved by the attacker. Replay attacks can be modelled by

$$\begin{aligned} x(j+1) &= Ax(j) + B\left(u(j) + a_u(j)\right), j \in [k, k+M], \\ y^a(j) &= Cx(j) + D\left(u(j) + a_u(j)\right) + a_y(j), \\ a_y(j) &= y(k_0 + j - k) - \left(Cx(j) + D\left(u(j) + a_u(j)\right)\right). \end{aligned}$$

As a result,

$$\forall j \in [k, k+M], y^a(j) = y(j - k + k_0) \approx y(j) = Cx(j) + Du(j). \tag{15.29}$$

Notice that the attack signal $a_y(j)$ depends on the plant state vector $x(j)$ and is, therefore, not a pure additive attack signal.

In summary, it can be seen that the above three types of attacks have one thing in common that they do not cause changes in the measurement output and hence cannot be traced in terms of the output variables. As a result, these attacks cannot be detected using an observer-based detection scheme.

15.2 Kernel Attacks: A General form of Stealthy Integrity Attacks

In this section, we firstly investigate the structural existence conditions for stealthy attacks. Here, we place a greater emphasis on the structure conditions. That means, the stealthiness of an integrity attack is examined solely in terms of available system knowledge, i.e. the plant model, and the available on-line information (data), and it is independent of the schemes used for the detection purpose. On the basis of these conditions, we will derive a general form of stealthy attacks. It is called kernel attacks, and generalises different types of stealthy integrity attacks, including the three types of integrity attacks introduced in the previous section.

15.2.1 Structurally Stealthy Integrity Attacks

To deal with structural system properties, we consider the plant model

$$y(z) = G(z)u(z)$$

without taking into account the plant uncertainty modelled by the residual r_y. Its kernel representation form is

$$\hat{M}(z)y(z) - \hat{N}(z)u(z) = 0.$$

The available on-line information at the CM-station, where the attack detection is performed, is (u_{est}, y^a). It is worth emphasising that u_{est} is an estimate of u at the MC-station and equal to u if the CPS is free of cyber-attacks. On account of this set-up, we have

Definition 15.1 An integrity cyber-attack is called structurally stealthy if

$$\forall u_{MC}, \hat{M}(z)y^a(z) - \hat{N}(z)u_{est}(z) = 0. \tag{15.30}$$

The meaning of this definition is apparent. It implies that a structurally stealthy cyber-attack should not yield any variation in the nominal system dynamic, which is represented by the kernel representation of the plant.

We now study the generalisation of structurally stealthy attacks and their existence conditions. Recall that $\left(\hat{M}, \hat{N}\right)$ is the LCP of the plant model G, and

$$\mathcal{K}_G = \left\{\begin{bmatrix} u \\ y \end{bmatrix} : \begin{bmatrix} -\hat{N} & \hat{M} \end{bmatrix}\begin{bmatrix} u \\ y \end{bmatrix} = 0, \begin{bmatrix} u \\ y \end{bmatrix} \in \mathcal{H}_2\right\} \tag{15.31}$$

is the kernel subspace of the plant. Accordingly, we have the following theorem.

Theorem 15.1 *An integrity attack is structurally stealthy if and only if*

$$\begin{bmatrix} u_{est}(z) \\ y^a(z) \end{bmatrix} \in \mathcal{K}_G. \tag{15.32}$$

Proof The proof follows immediately from Definition 15.1 and (15.31).

The property (15.32) motivates the definition of kernel attacks.

Definition 15.2 A structurally stealthy attack is called kernel attack.

In the next example, it is demonstrated that the zero dynamics, covert and replay attacks are kernel attacks.

Example 15.1 At first, we check a zero dynamics attack. Given the CPS configuration (15.1) and (15.9),

$$\begin{aligned} \begin{bmatrix} -\hat{N}(z) & \hat{M}(z) \end{bmatrix} \begin{bmatrix} u_{est}(z) \\ y^a(z) \end{bmatrix} &= \begin{bmatrix} -\hat{N}(z) & \hat{M}(z) \end{bmatrix} \begin{bmatrix} u(z) - a_u(z) \\ y(z) \end{bmatrix} \\ &= -\hat{N}(z)a_u(z). \end{aligned}$$

It follows from (15.24) that

$$\hat{N}(z)a_u(z) = 0 \Longrightarrow \begin{bmatrix} -\hat{N}(z) & \hat{M}(z) \end{bmatrix} \begin{bmatrix} u(z) \\ y^a(z) \end{bmatrix} = 0,$$

i.e. the zero dynamics attack is a kernel attack.

Next, consider the dynamic (15.30) under a covert attack,

$$\begin{bmatrix} -\hat{N}(z) & \hat{M}(z) \end{bmatrix} \begin{bmatrix} u_{est}(z) \\ y^a(z) \end{bmatrix} = \hat{N}(z)a_u(z) + \hat{M}(z)a_y(z).$$

According to (15.24), it holds

$$\hat{N}(z)a_u(z) + \hat{M}(z)a_y(z) = 0 \Longrightarrow \begin{bmatrix} -\hat{N}(z) & \hat{M}(z) \end{bmatrix} \begin{bmatrix} u_{est}(z) \\ y^a(z) \end{bmatrix} = 0.$$

Therefore, the covert attack is obviously a kernel attack.

Concerning replay attacks, recall the relation (15.29). It turns out, for $j \in [k, k+M]$,

$$\begin{aligned} &\hat{M}(z)y^a(z) - \hat{N}(z)\,u_{est}(z) = y^a(j) - \hat{y}^a(j) \\ &\approx Cx(j) + Du(j) - \left(C\hat{x}(j) + Du(j)\right) = 0 \Longrightarrow \begin{bmatrix} u_{est} \\ y^a \end{bmatrix} \in \mathcal{K}_G. \end{aligned}$$

Thus, the replay attack is a kernel attack.

15.2.2 Existence Conditions of Additive Kernel Attacks

We now study existence conditions of additive kernel attacks, i.e. additive structurally stealthy cyber-attacks. We begin with CPSs whose configuration is subject to (15.1) and (15.9).

Theorem 15.2 *Given a CPS subject to (15.1) and (15.9), suppose that the CPS is under additive cyber-attacks modelled by (15.20) with $\Pi^a = 0$, then the cyber-attacks are kernel attacks, if and only if*

$$\hat{N} a_{u_{MC}} + \hat{M} a_y = 0. \tag{15.33}$$

Proof As delineated in Example 14.2, the closed-loop dynamic of the CPS is governed by

$$\begin{bmatrix} u \\ y \end{bmatrix} = \begin{bmatrix} M \\ N \end{bmatrix} (\bar{v} + d_a) + \begin{bmatrix} -\hat{Y} + MQ \\ \hat{X} + NQ \end{bmatrix} r_y,$$
$$d_a = \left(X + Q\hat{N}\right) d = -\left(Y - Q\hat{M}\right) a_y + \left(X + Q\hat{N}\right) a_{u_c}.$$

According to the attack model (15.18)–(15.19), it turns out

$$\begin{bmatrix} u_{est} \\ y^a \end{bmatrix} = \begin{bmatrix} u - a_{u_{MC}} \\ y + a_y \end{bmatrix} = \begin{bmatrix} M \\ N \end{bmatrix} (\bar{v} + d_a) + \begin{bmatrix} -\hat{Y} + MQ \\ \hat{X} + NQ \end{bmatrix} r_y + \begin{bmatrix} -a_{u_{MC}} \\ a_y \end{bmatrix}.$$

Consequently, for $r_y = 0$,

$$\begin{bmatrix} -\hat{N} & \hat{M} \end{bmatrix} \begin{bmatrix} u_{est} \\ y^a \end{bmatrix} = \begin{bmatrix} -\hat{N} & \hat{M} \end{bmatrix} \begin{bmatrix} -a_{u_{MC}} \\ a_y \end{bmatrix},$$

from which it becomes clear that (15.33) is a necessary and sufficient condition for $\left(a_{u_{MC}}, a_y\right)$ being kernel attacks.

Next, the closed-loop configurations of CPSs given by (15.2) and (15.10) are considered. The major result is summarised in the following theorem.

Theorem 15.3 *Given a CPS subject to (15.2) and (15.10), suppose that the CPS is under additive cyber-attacks modelled by (15.20) with $\Pi^a = 0$, then the cyber-attacks are kernel attacks, if and only if*

- *for $u_e = K_e y$,*

$$a_y + \left(\hat{X} + NQ_e\right) \hat{N} a_{u_{MC}} = a_y + N\left(X + Q_e \hat{N}\right) a_{u_{MC}} = 0, \tag{15.34}$$

- *for $u_e = F_e \hat{x} + Q_e r_y$,*

$$\hat{N} a_{u_{MC}} + \hat{M} a_y = 0. \tag{15.35}$$

Proof Recall the main result in Theorem 14.1,

$$\begin{bmatrix} u \\ y \end{bmatrix} = \begin{bmatrix} M \\ N \end{bmatrix} (\bar{v} + d_a) + \begin{bmatrix} -\hat{Y} + MQ \\ \hat{X} + NQ \end{bmatrix} r_y.$$

Here, d_a is given in Theorem 14.1, but not relevant for the subsequent investigation. Since

$$u = u_e + u_{MC}^a, u_{MC} = Q_{MC} r_y^a + \bar{v},$$
$$u_e = \begin{cases} K_e y, \text{ in the case of (15.4)} \\ F_e \hat{x} + Q_e r_y = -\left(\hat{Y} - MQ_e\right) r_y, \text{ in the case of (15.5),} \end{cases}$$

and the estimate of u at the CM-station, u_{est}, is given by

$$u_{est} = \begin{cases} K_e y^a + u_{MC} \\ F_e \hat{x} + Q_e r_y^a + u_{MC} \end{cases} = \begin{cases} u + K_e \left(y^a - y\right) + u_{MC} - u_{MC}^a \\ u - \left(\hat{Y} - MQ_e\right)\left(r_y^a - r_y\right) + u_{MC} - u_{MC}^a \end{cases}$$
$$= \begin{cases} u + K_e a_y - a_{u_{MC}} \\ u - \left(\hat{Y} - MQ_e\right)\left(r_y^a - r_y\right) - a_{u_{MC}}. \end{cases}$$

Using (14.27), the last equation is written as

$$u_{est} = u - \left(\hat{Y} - MQ_e\right)\left(\hat{X} + NQ_e\right)^{-1} \hat{M}^{-1}\left(\hat{M}a_y + \hat{N}a_{u_{MC}}\right) - a_{u_{MC}}$$
$$= u + K_e \hat{M}^{-1}\left(\hat{M}a_y + \hat{N}a_{u_{MC}}\right) - a_{u_{MC}}.$$

Consequently, for $r_y = 0$, $u_{est} = K_e y^a + u_{MC}$,

$$\begin{bmatrix} u_{est} \\ y^a \end{bmatrix} = \begin{bmatrix} u + K_e a_y - a_{u_{MC}} \\ y + a_y \end{bmatrix} = \begin{bmatrix} M \\ N \end{bmatrix} (\bar{v} + d_a) + \begin{bmatrix} K_e a_y - a_{u_{MC}} \\ a_y \end{bmatrix}$$
$$\Longrightarrow \begin{bmatrix} -\hat{N} & \hat{M} \end{bmatrix} \begin{bmatrix} u_{est} \\ y^a \end{bmatrix} = \begin{bmatrix} -\hat{N} & \hat{M} \end{bmatrix} \begin{bmatrix} K_e a_y - a_{u_{MC}} \\ a_y \end{bmatrix}$$
$$= \left(\hat{M} - \hat{N} K_e\right) a_y + \hat{N} a_{u_{MC}}.$$

Using the Youla parameterisation of K_e,

$$K_e = -\left(\hat{Y} - MQ_e\right)\left(\hat{X} + NQ_e\right)^{-1},$$

and the Bezout identity, the last expression is written as

$$\begin{bmatrix} -\hat{N} & \hat{M} \end{bmatrix} \begin{bmatrix} u_{est} \\ y^a \end{bmatrix} = \left(\hat{M} - \hat{N} K_e \right) a_y + \hat{N} a_{u_{MC}} =$$

$$\left(\hat{X} + N Q_e \right)^{-1} a_y + \hat{N} a_{u_{MC}} = \left(\hat{X} + N Q_e \right)^{-1} \begin{bmatrix} I & \left(\hat{X} + N Q_e \right) \hat{N} \end{bmatrix} \begin{bmatrix} a_{u_{MC}} \\ a_y \end{bmatrix}.$$

Furthermore, for $r_y = 0$, $u_{est} = F_e \hat{x} + Q_e r_y^a + u_{MC}$,

$$\begin{aligned} \begin{bmatrix} u_{est} \\ y^a \end{bmatrix} &= \begin{bmatrix} u + K_e \hat{M}^{-1} \left(\hat{M} a_y + \hat{N} a_{u_{MC}} \right) - a_{u_{MC}} \\ y + a_y \end{bmatrix} \\ &= \begin{bmatrix} M \\ N \end{bmatrix} (\bar{v} + d_a) + \begin{bmatrix} K_e \hat{M}^{-1} \left(\hat{M} a_y + \hat{N} a_{u_{MC}} \right) - a_{u_{MC}} \\ a_y \end{bmatrix} \\ \Longrightarrow \begin{bmatrix} -\hat{N} & \hat{M} \end{bmatrix} \begin{bmatrix} u_{est} \\ y^a \end{bmatrix} &= \begin{bmatrix} -\hat{N} & \hat{M} \end{bmatrix} \begin{bmatrix} K_e \hat{M}^{-1} \left(\hat{M} a_y + \hat{N} a_{u_{MC}} \right) - a_{u_{MC}} \\ a_y \end{bmatrix} \\ &= \left(I - \hat{N} K_e \hat{M}^{-1} \right) \left(\hat{M} a_y + \hat{N} a_{u_{MC}} \right). \end{aligned}$$

As a result, we finally have (15.34) and (15.35).

15.2.3 Existence Conditions of Multiplicative Kernel Attacks

We begin with a relatively simple CPS configuration, CPSs subject to (15.1) and (15.9) and suffered from the multiplicative cyber-attacks modelled by (15.20). In this case,

$$u_{est} = u_{MC}.$$

For the sake of simplicity, it is firstly assumed that $\varepsilon_u = 0$, $\varepsilon_y = 0$.

Taking account of the attack model (15.20), it is evident that

$$\begin{aligned} \begin{bmatrix} u_{MC} \\ y^a \end{bmatrix} &= \begin{bmatrix} u \\ y \end{bmatrix} + \begin{bmatrix} -I & 0 \\ 0 & I \end{bmatrix} \Pi^a \begin{bmatrix} u_{MC} \\ y \end{bmatrix} \Longrightarrow \\ \begin{bmatrix} u_{MC} \\ y^a \end{bmatrix} &= \left(I + \begin{bmatrix} -I & 0 \\ 0 & I \end{bmatrix} \Pi^a \begin{bmatrix} I & 0 \\ 0 & 0 \end{bmatrix} \right)^{-1} \left(I + \begin{bmatrix} -I & 0 \\ 0 & I \end{bmatrix} \Pi^a \begin{bmatrix} 0 & 0 \\ 0 & I \end{bmatrix} \right) \begin{bmatrix} u \\ y \end{bmatrix} \\ &= \begin{bmatrix} u \\ y \end{bmatrix} + \begin{bmatrix} -I & 0 \\ 0 & I \end{bmatrix} \left(I + \Pi^a \begin{bmatrix} -I & 0 \\ 0 & 0 \end{bmatrix} \right)^{-1} \Pi^a \begin{bmatrix} -I & 0 \\ 0 & I \end{bmatrix} \begin{bmatrix} u \\ y \end{bmatrix}. \end{aligned} \quad (15.36)$$

Here, without loss of generality, it is assumed that $I + \begin{bmatrix} I & 0 \\ 0 & 0 \end{bmatrix} \Pi^a$ is invertible. Prior to finding the existence conditions of kernel attacks, it is useful to examine the closed-loop dynamic of the CPS under consideration. To this end, the following theorem is proved.

Theorem 15.4 *Given a CPS subject to (15.1) and (15.9), suppose that the CPS is under multiplicative cyber-attacks modelled by (15.20) with $\varepsilon_u = 0$, $\varepsilon_y = 0$, then the closed-loop dynamic is governed by*

$$\begin{aligned} \begin{bmatrix} u \\ y \end{bmatrix} &= \begin{bmatrix} M \\ N \end{bmatrix} \bar{v} + \begin{bmatrix} -\hat{Y} + MQ \\ \hat{X} + NQ \end{bmatrix} r_y \\ &- \begin{bmatrix} M \\ N \end{bmatrix} \left(I + \Phi \begin{bmatrix} M \\ N \end{bmatrix} \right)^{-1} \Phi \begin{bmatrix} M \\ N \end{bmatrix} \bar{v} \\ &- \begin{bmatrix} M \\ N \end{bmatrix} \left(I + \Phi \begin{bmatrix} M \\ N \end{bmatrix} \right)^{-1} \Phi \begin{bmatrix} -\hat{Y} + MQ \\ \hat{X} + NQ \end{bmatrix} r_y, \\ \Phi &= - \begin{bmatrix} X + Q\hat{N} & -Y + Q\hat{M} \end{bmatrix} \Pi^a \left(I + \begin{bmatrix} I & 0 \\ 0 & 0 \end{bmatrix} \Pi^a \right)^{-1}. \end{aligned} \tag{15.37}$$

Moreover, the CPS dynamic can be equivalently expressed by

$$\begin{cases} y = Gu + \hat{M}^{-1} r_y \\ u = K_\Delta y + \left(X + Q\hat{N} + \Delta K_u \right)^{-1} (\bar{v} + \Delta_v), \end{cases} \tag{15.38}$$

$$\Delta_v = - \left(I + \Phi \begin{bmatrix} M \\ N \end{bmatrix} \right)^{-1} \Phi \begin{bmatrix} M \\ N \end{bmatrix} \bar{v}, \tag{15.39}$$

$$K_\Delta = - \left(X + Q\hat{N} + \Delta K_u \right)^{-1} \left(Y - Q\hat{M} + \Delta K_y \right), \tag{15.40}$$

$$\Delta K_u = - \left(I + \Phi \begin{bmatrix} M \\ N \end{bmatrix} \right)^{-1} \Phi \begin{bmatrix} -\hat{Y} + MQ \\ \hat{X} + NQ \end{bmatrix} \hat{N}, \tag{15.41}$$

$$\Delta K_y = - \left(I + \Phi \begin{bmatrix} M \\ N \end{bmatrix} \right)^{-1} \Phi \begin{bmatrix} -\hat{Y} + MQ \\ \hat{X} + NQ \end{bmatrix} \hat{M}. \tag{15.42}$$

Proof Consider the closed-loop dynamic

$$\begin{aligned} \begin{bmatrix} u \\ y \end{bmatrix} &= \begin{bmatrix} I & -K \\ -G & I \end{bmatrix}^{-1} \begin{bmatrix} Ka_y + v + a_{u_{MC}} \\ \hat{M}^{-1} r_y \end{bmatrix} \\ &= \begin{bmatrix} X + Q\hat{N} & Y - Q\hat{M} \\ -\hat{N} & \hat{M} \end{bmatrix}^{-1} \begin{bmatrix} \left(X + Q\hat{N} \right) \left(Ka_y + v + a_{u_{MC}} \right) \\ r_y \end{bmatrix} \\ &= \begin{bmatrix} M & -\hat{Y} + MQ \\ N & \hat{X} + NQ \end{bmatrix} \begin{bmatrix} -\left(Y - Q\hat{M} \right) a_y + \bar{v} + \left(X + Q\hat{N} \right) a_{u_{MC}} \\ r_y \end{bmatrix} \end{aligned}$$

$$
\begin{aligned}
&= \begin{bmatrix} M \\ N \end{bmatrix} \left(\bar{v} + \begin{bmatrix} X + Q\hat{N} & -Y + Q\hat{M} \end{bmatrix} \Pi^a \begin{bmatrix} u_{MC} \\ y \end{bmatrix} \right) \\
&+ \begin{bmatrix} -\hat{Y} + MQ \\ \hat{X} + NQ \end{bmatrix} r_y.
\end{aligned}
$$

Observe that

$$
\begin{bmatrix} u_{MC} \\ y \end{bmatrix} = \begin{bmatrix} u \\ y \end{bmatrix} + \begin{bmatrix} -I & 0 \\ 0 & 0 \end{bmatrix} \Pi^a \begin{bmatrix} u_{MC} \\ y \end{bmatrix} = \left(I + \begin{bmatrix} I & 0 \\ 0 & 0 \end{bmatrix} \Pi^a \right)^{-1} \begin{bmatrix} u \\ y \end{bmatrix},
$$

which yields

$$
\begin{aligned}
&\begin{bmatrix} u \\ y \end{bmatrix} = \begin{bmatrix} M \\ N \end{bmatrix} \left(\bar{v} - \Phi \begin{bmatrix} u \\ y \end{bmatrix} \right) + \begin{bmatrix} -\hat{Y} + MQ \\ \hat{X} + NQ \end{bmatrix} r_y \Longrightarrow \\
&\begin{bmatrix} u \\ y \end{bmatrix} = \left(I + \begin{bmatrix} M \\ N \end{bmatrix} \Phi \right)^{-1} \left(\begin{bmatrix} M \\ N \end{bmatrix} \bar{v} + \begin{bmatrix} -\hat{Y} + MQ \\ \hat{X} + NQ \end{bmatrix} r_y \right).
\end{aligned}
$$

After some routine calculations using the relation

$$
\left(I + \begin{bmatrix} M \\ N \end{bmatrix} \Phi \right)^{-1} = I - \begin{bmatrix} M \\ N \end{bmatrix} \left(I + \Phi \begin{bmatrix} M \\ N \end{bmatrix} \right)^{-1} \Phi,
$$

it turns out

$$
\begin{aligned}
\begin{bmatrix} u \\ y \end{bmatrix} &= \begin{bmatrix} M \\ N \end{bmatrix} \bar{v} + \begin{bmatrix} -\hat{Y} + MQ \\ \hat{X} + NQ \end{bmatrix} r_y \\
&- \begin{bmatrix} M \\ N \end{bmatrix} \left(I + \Phi \begin{bmatrix} M \\ N \end{bmatrix} \right)^{-1} \Phi \begin{bmatrix} M \\ N \end{bmatrix} \bar{v} \\
&- \begin{bmatrix} M \\ N \end{bmatrix} \left(I + \Phi \begin{bmatrix} M \\ N \end{bmatrix} \right)^{-1} \Phi \begin{bmatrix} -\hat{Y} + MQ \\ \hat{X} + NQ \end{bmatrix} r_y.
\end{aligned} \tag{15.43}
$$

To prove (15.38), multiplying the both sides of (15.43) by

$$
\begin{bmatrix} M & -\hat{Y} + MQ \\ N & \hat{X} + NQ \end{bmatrix}^{-1} = \begin{bmatrix} X + Q\hat{N} & Y - Q\hat{M} \\ -\hat{N} & \hat{M} \end{bmatrix}
$$

leads to

$$
\begin{aligned}
&\begin{bmatrix} X + Q\hat{N} & Y - Q\hat{M} \\ -\hat{N} & \hat{M} \end{bmatrix} \begin{bmatrix} u \\ y \end{bmatrix} = \begin{bmatrix} \bar{v} + \Delta_v + \Delta_{r_y} \\ r_y \end{bmatrix}, \\
&\Delta_v = -\left(I + \Phi \begin{bmatrix} M \\ N \end{bmatrix} \right)^{-1} \Phi \begin{bmatrix} M \\ N \end{bmatrix} \bar{v},
\end{aligned}
$$

$$\Delta_{r_y} = -\left(I + \Phi \begin{bmatrix} M \\ N \end{bmatrix}\right)^{-1} \Phi \begin{bmatrix} -\hat{Y} + MQ \\ \hat{X} + NQ \end{bmatrix} r_y.$$

We finally have

$$\begin{aligned} y &= Gu + \hat{M}^{-1} r_y, \\ u &= K_\Delta y + \left(X + Q\hat{N} + \Delta K_u\right)^{-1} (\bar{v} + \Delta_v), \\ K_\Delta &= -\left(X + Q\hat{N} + \Delta K_u\right)^{-1} \left(Y - Q\hat{M} + \Delta K_y\right). \end{aligned}$$

The theorem is proved.

Remark 15.1 Theorem 15.4 demonstrates again that the cyber-attacks exclusively act on the plant image subspace, and the corresponding impact can be equivalently modelled by the uncertainties in the controller, ΔK_u, ΔK_y, as well as by the variation of the reference signal, Δ_v.

Applying the result in Theorem 15.4 and setting $r_y = 0$, we have immediately the subsequent lemma that is crucial for achieving the main results of this section.

Lemma 15.1 *Given a CPS subject to (15.1) and (15.9), suppose that the CPS is under multiplicative cyber-attacks modelled by*

$$\begin{bmatrix} a_{u_{MC}} \\ a_y \end{bmatrix} = \Pi^a \begin{bmatrix} u_{MC} \\ y \end{bmatrix}.$$

Then, $\left(a_{u_{MC}}, a_y\right)$ *are kernel attacks, if and only if*

$$\forall \alpha, \begin{bmatrix} -I & 0 \\ 0 & I \end{bmatrix} \left(I + \Pi^a \begin{bmatrix} -I & 0 \\ 0 & 0 \end{bmatrix}\right)^{-1} \Pi^a \begin{bmatrix} -M \\ N \end{bmatrix} \alpha \in \mathcal{I}_G. \tag{15.44}$$

Proof It follows from (15.36) and Theorem 15.4 that

$$\begin{aligned} &\begin{bmatrix} -\hat{N} & \hat{M} \end{bmatrix} \begin{bmatrix} u_{est} \\ y^a \end{bmatrix} = \begin{bmatrix} -\hat{N} & \hat{M} \end{bmatrix} \begin{bmatrix} u_{MC} \\ y^a \end{bmatrix} \\ &= \begin{bmatrix} -\hat{N} & \hat{M} \end{bmatrix} \begin{bmatrix} -I & 0 \\ 0 & I \end{bmatrix} \left(I + \Pi^a \begin{bmatrix} -I & 0 \\ 0 & 0 \end{bmatrix}\right)^{-1} \Pi^a \begin{bmatrix} -I & 0 \\ 0 & I \end{bmatrix} \begin{bmatrix} u \\ y \end{bmatrix} \\ &= \begin{bmatrix} -\hat{N} & \hat{M} \end{bmatrix} \begin{bmatrix} -I & 0 \\ 0 & I \end{bmatrix} \left(I + \Pi^a \begin{bmatrix} -I & 0 \\ 0 & 0 \end{bmatrix}\right)^{-1} \Pi^a \begin{bmatrix} -M \\ N \end{bmatrix} \left(I + \Phi \begin{bmatrix} M \\ N \end{bmatrix}\right)^{-1} \bar{v}. \end{aligned}$$

According Theorem 15.1, it holds

$$\begin{bmatrix} u_{est} \\ y^a \end{bmatrix} \in \mathcal{K}_G \Longrightarrow \begin{bmatrix} u_{est} \\ y^a \end{bmatrix} \in \mathcal{I}_G,$$

because $\mathcal{I}_G = \mathcal{K}_G$, from which it becomes clear that

$$\forall \alpha, \begin{bmatrix} -I & 0 \\ 0 & I \end{bmatrix} \left(I + \Pi^a \begin{bmatrix} -I & 0 \\ 0 & 0 \end{bmatrix} \right)^{-1} \Pi^a \begin{bmatrix} -M \\ N \end{bmatrix} \alpha \in \mathcal{I}_G.$$

is a necessary and sufficient condition for $\left(a_{u_{MC}}, a_y\right)$ being kernel attacks.

We are in a position to present the main results of this section.

Theorem 15.5 *Given a CPS subject to (15.1) and (15.9), suppose that the CPS is under multiplicative cyber-attacks modelled by (15.20) with $\varepsilon_u = 0, \varepsilon_y = 0$, then $\left(a_{u_{MC}}, a_y\right)$ given by*

$$\begin{bmatrix} a_{u_{MC}} \\ a_y \end{bmatrix} = \Pi^a \begin{bmatrix} u_{MC} \\ y \end{bmatrix} \Longrightarrow \Pi^a = \begin{bmatrix} -M \\ N \end{bmatrix} R \tag{15.45}$$

are kernel attacks, where $R \in \mathcal{RH}_\infty$ is a stable paratermisation system.

Proof Note that

$$\forall \alpha, \begin{bmatrix} -I & 0 \\ 0 & I \end{bmatrix} \left(I + \Pi^a \begin{bmatrix} -I & 0 \\ 0 & I \end{bmatrix} \right)^{-1} \Pi^a \begin{bmatrix} -M \\ N \end{bmatrix} \alpha \in \mathcal{I}_G \Longleftrightarrow$$
$$\exists\, Q \in \mathcal{RH}_\infty, \text{s.t.} \left(I + \Pi^a \begin{bmatrix} -I & 0 \\ 0 & 0 \end{bmatrix} \right)^{-1} \Pi^a \begin{bmatrix} -M \\ N \end{bmatrix} = \begin{bmatrix} -M \\ N \end{bmatrix} Q.$$

Consequently, setting Π^a as

$$\Pi^a = \begin{bmatrix} -M \\ N \end{bmatrix} R,\ R \in \mathcal{RH}_\infty$$

yields

$$\begin{aligned} &\left(I + \Pi^a \begin{bmatrix} -I & 0 \\ 0 & 0 \end{bmatrix} \right)^{-1} \Pi^a \begin{bmatrix} -M \\ N \end{bmatrix} \\ &= \left(I + \begin{bmatrix} -M \\ N \end{bmatrix} R \begin{bmatrix} -I & 0 \\ 0 & 0 \end{bmatrix} \right)^{-1} \begin{bmatrix} -M \\ N \end{bmatrix} R \begin{bmatrix} -M \\ N \end{bmatrix} \\ &= \begin{bmatrix} -M \\ N \end{bmatrix} \left(I + R \begin{bmatrix} -I & 0 \\ 0 & 0 \end{bmatrix} \begin{bmatrix} -M \\ N \end{bmatrix} \right)^{-1} R \begin{bmatrix} -M \\ N \end{bmatrix} \Longrightarrow \\ &Q = \left(I + R \begin{bmatrix} -I & 0 \\ 0 & 0 \end{bmatrix} \begin{bmatrix} -M \\ N \end{bmatrix} \right)^{-1} R \begin{bmatrix} -M \\ N \end{bmatrix}. \end{aligned}$$

As a result, it is proved that

$$\begin{bmatrix} a_{u_{MC}} \\ a_y \end{bmatrix} = \begin{bmatrix} -M \\ N \end{bmatrix} R \begin{bmatrix} u_{MC} \\ y \end{bmatrix}$$

are kernel attacks and parametrised by R.

The subsequent theorem is an extension of the results in Theorems 15.2 and 15.4–15.5 and gives a general description of the CPS dynamic under cyber-attacks (15.20) and the necessary and sufficient condition of kernel cyber-attacks.

Theorem 15.6 *Given a CPS subject to (15.1) and (15.9) and under cyber-attacks modelled by (15.20), the dynamic of the CPS is governed by*

$$\begin{aligned}\begin{bmatrix} u \\ y \end{bmatrix} &= \begin{bmatrix} M \\ N \end{bmatrix} \bar{v} + \begin{bmatrix} -\hat{Y} + MQ \\ \hat{X} + NQ \end{bmatrix} r_y \\ &- \begin{bmatrix} M \\ N \end{bmatrix} \left(I + \Phi \begin{bmatrix} M \\ N \end{bmatrix} \right)^{-1} \Phi \begin{bmatrix} M \\ N \end{bmatrix} \bar{v} \\ &+ \begin{bmatrix} M \\ N \end{bmatrix} \left(I + \Phi \begin{bmatrix} M \\ N \end{bmatrix} \right)^{-1} \begin{bmatrix} X + Q\hat{N} & -Y + Q\hat{M} \end{bmatrix} \begin{bmatrix} \varepsilon_u \\ \varepsilon_y \end{bmatrix} \\ &- \begin{bmatrix} M \\ N \end{bmatrix} \left(I + \Phi \begin{bmatrix} M \\ N \end{bmatrix} \right)^{-1} \Phi \begin{bmatrix} -\hat{Y} + MQ \\ \hat{X} + NQ \end{bmatrix} r_y. \end{aligned} \tag{15.46}$$

The attacks are kernel attacks if and only if $(\varepsilon_u, \varepsilon_y)$ *is modelled by*

$$\begin{bmatrix} \varepsilon_u \\ \varepsilon_y \end{bmatrix} = \begin{bmatrix} -M \\ N \end{bmatrix} \xi, \xi \in \mathcal{H}_2, \tag{15.47}$$

and Π^a *satisfies (15.45). In this case,*

$$r_y^a = \begin{bmatrix} -\hat{N} & \hat{M} \end{bmatrix} \begin{bmatrix} u_{MC} \\ y^a \end{bmatrix} = r_y. \tag{15.48}$$

Proof Analogous to the proof of Theorem 15.4, it can be proved that

$$\begin{aligned}\begin{bmatrix} u \\ y \end{bmatrix} &= \begin{bmatrix} I & -K \\ -G & I \end{bmatrix}^{-1} \begin{bmatrix} Ka_y + v + a_{u_{MC}} \\ \hat{M}^{-1} r_y \end{bmatrix} \\ &= \begin{bmatrix} M \\ N \end{bmatrix} \left(\bar{v} + \begin{bmatrix} X + Q\hat{N} & -Y + Q\hat{M} \end{bmatrix} \left(\Pi^a \begin{bmatrix} u_{MC} \\ y \end{bmatrix} + \begin{bmatrix} \varepsilon_u \\ \varepsilon_y \end{bmatrix} \right) \right) \\ &+ \begin{bmatrix} -\hat{Y} + MQ \\ \hat{X} + NQ \end{bmatrix} r_y, \end{aligned}$$

from which (15.46) follows. Next, note that

$$\begin{bmatrix} u_{MC} \\ y^a \end{bmatrix} = \begin{bmatrix} u \\ y \end{bmatrix} + \begin{bmatrix} -I & 0 \\ 0 & I \end{bmatrix} \Pi^a \begin{bmatrix} u_{MC} \\ y \end{bmatrix} + \begin{bmatrix} -\varepsilon_u \\ \varepsilon_y \end{bmatrix} \Longrightarrow$$

$$\begin{bmatrix} u_{MC} \\ y^a \end{bmatrix} = \begin{bmatrix} u \\ y \end{bmatrix} + \begin{bmatrix} -I & 0 \\ 0 & I \end{bmatrix} \left(I + \Pi^a \begin{bmatrix} -I & 0 \\ 0 & 0 \end{bmatrix} \right)^{-1} \Pi^a \begin{bmatrix} -I & 0 \\ 0 & I \end{bmatrix} \begin{bmatrix} u \\ y \end{bmatrix}$$
$$+ \begin{bmatrix} -I & 0 \\ 0 & I \end{bmatrix} \left(I + \Pi^a \begin{bmatrix} -I & 0 \\ 0 & 0 \end{bmatrix} \right)^{-1} \begin{bmatrix} \varepsilon_u \\ \varepsilon_y \end{bmatrix}.$$

It yields

$$\begin{aligned}
\begin{bmatrix} -\hat{N} & \hat{M} \end{bmatrix} \begin{bmatrix} u_{MC} \\ y^a \end{bmatrix} &= \begin{bmatrix} -\hat{N} & \hat{M} \end{bmatrix} \begin{bmatrix} -I & 0 \\ 0 & I \end{bmatrix} \left(I + \Pi^a \begin{bmatrix} -I & 0 \\ 0 & 0 \end{bmatrix} \right)^{-1} \Pi^a \begin{bmatrix} -I & 0 \\ 0 & I \end{bmatrix} \begin{bmatrix} u \\ y \end{bmatrix} \\
&\quad + \begin{bmatrix} -\hat{N} & \hat{M} \end{bmatrix} \begin{bmatrix} -I & 0 \\ 0 & I \end{bmatrix} \left(I + \Pi^a \begin{bmatrix} -I & 0 \\ 0 & 0 \end{bmatrix} \right)^{-1} \begin{bmatrix} \varepsilon_u \\ \varepsilon_y \end{bmatrix} \\
&= \begin{bmatrix} -\hat{N} & \hat{M} \end{bmatrix} \begin{bmatrix} -I & 0 \\ 0 & I \end{bmatrix} \left(I + \Pi^a \begin{bmatrix} -I & 0 \\ 0 & 0 \end{bmatrix} \right)^{-1} \Pi^a \begin{bmatrix} -M \\ N \end{bmatrix} \left(I + \Phi \begin{bmatrix} M \\ N \end{bmatrix} \right)^{-1} (\bar{v} + \eta) \\
&\quad - \begin{bmatrix} -\hat{N} & \hat{M} \end{bmatrix} \begin{bmatrix} -I & 0 \\ 0 & I \end{bmatrix} \left(I + \Pi^a \begin{bmatrix} -I & 0 \\ 0 & 0 \end{bmatrix} \right)^{-1} \Pi^a \begin{bmatrix} -M \\ N \end{bmatrix} \times \\
&\quad \left(I + \Phi \begin{bmatrix} M \\ N \end{bmatrix} \right)^{-1} \Phi \begin{bmatrix} -\hat{Y} + MQ \\ \hat{X} + NQ \end{bmatrix} r_y + r_y \\
&\quad + \begin{bmatrix} -\hat{N} & \hat{M} \end{bmatrix} \begin{bmatrix} -I & 0 \\ 0 & I \end{bmatrix} \left(I + \Pi^a \begin{bmatrix} -I & 0 \\ 0 & 0 \end{bmatrix} \right)^{-1} \begin{bmatrix} \varepsilon_u \\ \varepsilon_y \end{bmatrix}, \\
\eta &= \begin{bmatrix} X + Q\hat{N} & -Y + Q\hat{M} \end{bmatrix} \begin{bmatrix} \varepsilon_u \\ \varepsilon_y \end{bmatrix}.
\end{aligned}$$

Since

$$\begin{bmatrix} \varepsilon_u \\ \varepsilon_y \end{bmatrix} \text{ and } \Pi^a \begin{bmatrix} u_{MC} \\ y \end{bmatrix}$$

are independent, according to Theorem 15.5, kernel attacks Π^a satisfying (15.45) lead to

$$\begin{bmatrix} -\hat{N} & \hat{M} \end{bmatrix} \begin{bmatrix} -I & 0 \\ 0 & I \end{bmatrix} \left(I + \Pi^a \begin{bmatrix} -I & 0 \\ 0 & 0 \end{bmatrix} \right)^{-1} \Pi^a \begin{bmatrix} -M \\ N \end{bmatrix} = 0.$$

Hence, it holds

$$\begin{bmatrix} -\hat{N} & \hat{M} \end{bmatrix} \begin{bmatrix} u_{MC} \\ y^a \end{bmatrix} = r_y + \begin{bmatrix} -\hat{N} & \hat{M} \end{bmatrix} \begin{bmatrix} -I & 0 \\ 0 & I \end{bmatrix} \left(I + \Pi^a \begin{bmatrix} -I & 0 \\ 0 & 0 \end{bmatrix} \right)^{-1} \begin{bmatrix} \varepsilon_u \\ \varepsilon_y \end{bmatrix},$$

from which it is apparent that $(a_{u_{MC}}, a_y)$ are kernel attacks, if and only if (15.45) is satisfied and

$$\begin{bmatrix} -I & 0 \\ 0 & I \end{bmatrix} \left(I + \Pi^a \begin{bmatrix} -I & 0 \\ 0 & 0 \end{bmatrix} \right)^{-1} \begin{bmatrix} \varepsilon_u \\ \varepsilon_y \end{bmatrix} \in \mathcal{I}_G \Longleftrightarrow$$

$$\text{for some } \zeta \in \mathcal{H}_2, \left(I + \Pi^a \begin{bmatrix} -I & 0 \\ 0 & 0 \end{bmatrix} \right)^{-1} \begin{bmatrix} \varepsilon_u \\ \varepsilon_y \end{bmatrix} = \begin{bmatrix} -M \\ N \end{bmatrix} \zeta$$

$$\Longleftrightarrow \begin{bmatrix} \varepsilon_u \\ \varepsilon_y \end{bmatrix} = \left(I + \Pi^a \begin{bmatrix} -I & 0 \\ 0 & 0 \end{bmatrix} \right) \begin{bmatrix} -M \\ N \end{bmatrix} \zeta.$$

Recall that

$$\Pi^a = \begin{bmatrix} -M \\ N \end{bmatrix} R \Longrightarrow$$

$$\left(I + \Pi^a \begin{bmatrix} -I & 0 \\ 0 & 0 \end{bmatrix} \right) \begin{bmatrix} -M \\ N \end{bmatrix} = \begin{bmatrix} -M \\ N \end{bmatrix} \left(I + R \begin{bmatrix} -I & 0 \\ 0 & 0 \end{bmatrix} \right). \tag{15.49}$$

As a result, (15.47) is proved. Finally, we obtain

$$\begin{bmatrix} -\hat{N} & \hat{M} \end{bmatrix} \begin{bmatrix} u_{MC} \\ y^a \end{bmatrix} = r_y.$$

The consequence of Theorem 15.6 is that such kernel attacks will cause no variation in the residual r_y^a and thus cannot be detected using observer-based detection schemes.

Corollary 15.1 *Given a CPS subject to (15.1) and (15.9) and under cyber-attacks modelled by (15.20), suppose that* $(a_{u_{MC}}, a_y)$ *are kernel attacks, then they can be parametrised by*

$$\begin{bmatrix} a_{u_{MC}} \\ a_y \end{bmatrix} = \begin{bmatrix} -M \\ N \end{bmatrix} \left(R \begin{bmatrix} u_{MC} \\ y \end{bmatrix} + \xi \right). \tag{15.50}$$

Proof The proof follows directly from (15.45) and (15.49).

We now shortly analyse how the kernel attacks affect the CPS dynamic. To this end, recall that

$$\begin{bmatrix} u \\ y \end{bmatrix} = \begin{bmatrix} M \\ N \end{bmatrix} \left(\bar{v} + \begin{bmatrix} X + Q\hat{N} & -Y + Q\hat{M} \end{bmatrix} \begin{bmatrix} a_{u_{MC}} \\ a_y \end{bmatrix} \right) + \begin{bmatrix} -\hat{Y} + MQ \\ \hat{X} + NQ \end{bmatrix} r_y.$$

Substituting $(a_{u_{MC}}, a_y)$ in (15.50) into the above equation gives

$$\begin{aligned} \begin{bmatrix} u \\ y \end{bmatrix} &= \begin{bmatrix} M \\ N \end{bmatrix} \left(\bar{v} - \left(R \begin{bmatrix} u_{MC} \\ y \end{bmatrix} + \xi \right) \right) + \begin{bmatrix} -\hat{Y} + MQ \\ \hat{X} + NQ \end{bmatrix} r_y \\ &= \begin{bmatrix} M \\ N \end{bmatrix} (\bar{v} - \xi) - \begin{bmatrix} M \\ N \end{bmatrix} R \begin{bmatrix} u_{MC} \\ y \end{bmatrix} + \begin{bmatrix} -\hat{Y} + MQ \\ \hat{X} + NQ \end{bmatrix} r_y. \end{aligned}$$

In other words,

$$-R\begin{bmatrix} u_{MC} \\ y \end{bmatrix} - \xi$$

describes the changes in the system image subspace caused by the kernel attack. Notice that the term $R\begin{bmatrix} u_{MC} \\ y \end{bmatrix}$ might cause stability concerning.

15.2.4 Existence Conditions of Multiplicative Kernel Attacks in CPS with Embedded Plant Controller

We now address issues related to CPSs subject to (15.1) and (15.2) that suffer from the multiplicative cyber-attacks modelled by (15.20), and investigate existence conditions of kernel attacks. Remember that, different from our study in the previous subsection, (15.16) holds for the closed-loop configuration (15.2). Corresponding to it, we have

$$u = K_e y + \left(X + Q_e \hat{N}\right)^{-1} u_{MC}^a,$$

$$u_{MC}^a := \begin{cases} \left(X + Q_e \hat{N}\right)\left(Q_{MC} r_y^a + \bar{v} + a_{u_{MC}}\right), \text{ corresponding to (15.4)} \\ Q_{MC} r_y^a + \bar{v} + a_{u_{MC}}, \text{ corresponding to (15.5).} \end{cases}$$

Accordingly,

$$\begin{bmatrix} u_{est} \\ y^a \end{bmatrix} = \begin{bmatrix} u \\ y \end{bmatrix} + \begin{bmatrix} -Z\left(X + Q_e \hat{N}\right)^{-1} & K_e \\ 0 & I \end{bmatrix} \begin{bmatrix} a_{u_{MC}} \\ a_y \end{bmatrix},$$

$$r_y^a = \hat{M} y^a - \hat{N} u_{est} = r_y + \begin{bmatrix} -\hat{N} & \hat{M} \end{bmatrix} \begin{bmatrix} -Z\left(X + Q_e \hat{N}\right)^{-1} & K_e \\ 0 & I \end{bmatrix} \begin{bmatrix} a_{u_{MC}} \\ a_y \end{bmatrix},$$

$$Z := \begin{cases} X + Q_e \hat{N}, \text{ corresponding to (15.4)} \\ I, \text{ corresponding to (15.5).} \end{cases}$$

It leads to, after some routine calculations,

$$u = Ky + v + \left(X + Q\hat{N}\right)^{-1} \bar{\Phi} \begin{bmatrix} a_{u_{MC}} \\ a_y \end{bmatrix}, \tag{15.51}$$

$$K = -\left(X + Q\hat{N}\right)^{-1}\left(Y - Q\hat{M}\right),$$

$$Q = \begin{cases} Q_e + \left(X + Q_e \hat{N}\right) Q_{MC}, \text{ corresponding to (15.4)} \\ Q_e + Q_{MC}, \text{ corresponding to (15.5),} \end{cases} \tag{15.52}$$

$$\bar{\Phi} = Q_{MC} \begin{bmatrix} -\hat{N} & \hat{M} \end{bmatrix} \begin{bmatrix} -Z\left(X + Q_e\hat{N}\right)^{-1} & K_e \\ 0 & I \end{bmatrix} + \begin{bmatrix} Z & 0 \end{bmatrix}. \tag{15.53}$$

Observe that

$$\begin{aligned} \begin{bmatrix} u_{MC} \\ y \end{bmatrix} &= \begin{bmatrix} I & -K_e \\ 0 & I \end{bmatrix}\begin{bmatrix} u \\ y \end{bmatrix} + \begin{bmatrix} -Z\left(X + Q_e\hat{N}\right)^{-1} a_{u_{MC}} \\ 0 \end{bmatrix} \\ &= \begin{bmatrix} I & -K_e \\ 0 & I \end{bmatrix}\begin{bmatrix} u \\ y \end{bmatrix} - \begin{bmatrix} Z\left(X + Q_e\hat{N}\right)^{-1} & 0 \\ 0 & 0 \end{bmatrix}\left(\Pi^a \begin{bmatrix} u_{MC} \\ y \end{bmatrix} + \begin{bmatrix} \varepsilon_u \\ \varepsilon_y \end{bmatrix}\right) \\ \Longrightarrow \begin{bmatrix} u_{MC} \\ y \end{bmatrix} &= \left(I + \begin{bmatrix} Z\left(X + Q_e\hat{N}\right)^{-1} & 0 \\ 0 & 0 \end{bmatrix}\Pi^a\right)^{-1} \times \\ &\left(\begin{bmatrix} I & -K_e \\ 0 & I \end{bmatrix}\begin{bmatrix} u \\ y \end{bmatrix} - \begin{bmatrix} Z\left(X + Q_e\hat{N}\right)^{-1} & 0 \\ 0 & 0 \end{bmatrix}\begin{bmatrix} \varepsilon_u \\ \varepsilon_y \end{bmatrix}\right). \end{aligned} \tag{15.54}$$

Proceeding from (15.51) and (15.54), the closed-loop dynamic is derived along the lines of Theorem 15.4.

Theorem 15.7 *Given a CPS subject to (15.1) and (15.10), and multiplicative cyber-attacks modelled by (15.20), then the closed-loop dynamic is governed by*

$$\begin{aligned} \begin{bmatrix} u \\ y \end{bmatrix} &= \begin{bmatrix} M \\ N \end{bmatrix}\bar{v} + \begin{bmatrix} -\hat{Y} + MQ \\ \hat{X} + NQ \end{bmatrix} r_y \\ &- \begin{bmatrix} M \\ N \end{bmatrix}\left(I + \Phi\begin{bmatrix} M \\ N \end{bmatrix}\right)^{-1} \Phi \begin{bmatrix} M \\ N \end{bmatrix}\bar{v} \\ &- \begin{bmatrix} M \\ N \end{bmatrix}\left(I + \Phi\begin{bmatrix} M \\ N \end{bmatrix}\right)^{-1} \Phi \begin{bmatrix} -\hat{Y} + MQ \\ \hat{X} + NQ \end{bmatrix} r_y \end{aligned} \tag{15.55}$$

$$+ \begin{bmatrix} M \\ N \end{bmatrix}\left(I + \Phi\begin{bmatrix} M \\ N \end{bmatrix}\right)^{-1} \Psi \begin{bmatrix} \varepsilon_u \\ \varepsilon_y \end{bmatrix}, \tag{15.56}$$

$$\Phi = -\bar{\Phi}\Pi^a\left(I + \begin{bmatrix} Z\left(X + Q_e\hat{N}\right)^{-1} & 0 \\ 0 & 0 \end{bmatrix}\Pi^a\right)^{-1}\begin{bmatrix} I & -K_e \\ 0 & I \end{bmatrix}, \tag{15.57}$$

$$\Psi = \bar{\Phi} - \left(I + \begin{bmatrix} Z\left(X + Q_e\hat{N}\right)^{-1} & 0 \\ 0 & 0 \end{bmatrix}\Pi^a\right)^{-1}\begin{bmatrix} Z\left(X + Q_e\hat{N}\right)^{-1} & 0 \\ 0 & 0 \end{bmatrix}. \tag{15.58}$$

Moreover, the CPS dynamic can be equivalently expressed by

$$\begin{cases} y = Gu + \hat{M}^{-1} r_y \\ u = K_\Delta y + \left(X + Q\hat{N} + \Delta K_u\right)^{-1} \eta\left(\bar{v}, a_{u_{MC}}, a_y\right), \end{cases} \tag{15.59}$$

$$\eta\left(\bar{v}, a_{u_{MC}}, a_y\right) = \left(I + \Phi \begin{bmatrix} M \\ N \end{bmatrix}\right)^{-1} \left(-\Phi \begin{bmatrix} M \\ N \end{bmatrix} \bar{v} + \Psi \begin{bmatrix} \varepsilon_u \\ \varepsilon_y \end{bmatrix}\right), \tag{15.60}$$

$$K_\Delta = -\left(X + Q\hat{N} + \Delta K_u\right)^{-1} \left(Y - Q\hat{M} + \Delta K_y\right), \tag{15.61}$$

$$\Delta K_u = -\left(I + \Phi \begin{bmatrix} M \\ N \end{bmatrix}\right)^{-1} \Phi \begin{bmatrix} -\hat{Y} + MQ \\ \hat{X} + NQ \end{bmatrix} \hat{N}, \tag{15.62}$$

$$\Delta K_y = -\left(I + \Phi \begin{bmatrix} M \\ N \end{bmatrix}\right)^{-1} \Phi \begin{bmatrix} -\hat{Y} + MQ \\ \hat{X} + NQ \end{bmatrix} \hat{M}. \tag{15.63}$$

Proof According to (15.51), the closed-loop dynamic is governed by

$$\begin{aligned} \begin{bmatrix} u \\ y \end{bmatrix} &= \begin{bmatrix} I & -K \\ -G & I \end{bmatrix}^{-1} \begin{bmatrix} v + \left(X + Q\hat{N}\right)^{-1} \bar{\Phi} \begin{bmatrix} a_{u_{MC}} \\ a_y \end{bmatrix} \\ \hat{M}^{-1} r_y \end{bmatrix} \\ &= \begin{bmatrix} M & -\hat{Y} + MQ \\ N & \hat{X} + NQ \end{bmatrix} \begin{bmatrix} \bar{v} + \bar{\Phi} \begin{bmatrix} a_{u_{MC}} \\ a_y \end{bmatrix} \\ r_y \end{bmatrix} \\ &= \begin{bmatrix} M \\ N \end{bmatrix} \bar{v} + \begin{bmatrix} -\hat{Y} + MQ \\ \hat{X} + NQ \end{bmatrix} r_y + \begin{bmatrix} M \\ N \end{bmatrix} \bar{\Phi} \left(\Pi^a \begin{bmatrix} u_{MC} \\ y \end{bmatrix} + \begin{bmatrix} \varepsilon_u \\ \varepsilon_y \end{bmatrix}\right). \end{aligned}$$

Substituting (u_{MC}, y) by the relation given in (15.54) yields

$$\begin{bmatrix} u \\ y \end{bmatrix} = \left(I + \begin{bmatrix} M \\ N \end{bmatrix} \Phi\right)^{-1} \left(\begin{bmatrix} M \\ N \end{bmatrix} \bar{v} + \begin{bmatrix} -\hat{Y} + MQ \\ \hat{X} + NQ \end{bmatrix} r_y + \begin{bmatrix} M \\ N \end{bmatrix} \Psi \begin{bmatrix} \varepsilon_u \\ \varepsilon_y \end{bmatrix}\right).$$

The remaining steps of the proof can be performed exactly along the lines of the proof of Theorem 15.4, and hence are omitted.

Next, we explore existence conditions for kernel attacks. To begin with, consider the residual r_y^a, for $r_y = 0$,

$$r_y^a = \begin{bmatrix} -\hat{N} & \hat{M} \end{bmatrix} \begin{bmatrix} -Z\left(X + Q_e\hat{N}\right)^{-1} & K_e \\ 0 & I \end{bmatrix} \left(\Pi^a \begin{bmatrix} u_{MC} \\ y \end{bmatrix} + \begin{bmatrix} \varepsilon_u \\ \varepsilon_y \end{bmatrix}\right).$$

Substituting (u_{MC}, y) subject to (15.54) into it gives

$$r_y^a = \begin{bmatrix} -\hat{N} & \hat{M} \end{bmatrix} \begin{bmatrix} -Z\left(X + Q_e\hat{N}\right)^{-1} & K_e \\ 0 & I \end{bmatrix} \left(\theta(u, y) + \varphi\left(\varepsilon_u, \varepsilon_y\right)\right),$$

$$\theta(u, y) = \Pi^a \left(I + \begin{bmatrix} Z\left(X + Q_e\hat{N}\right)^{-1} & 0 \\ 0 & 0 \end{bmatrix} \Pi^a\right)^{-1} \begin{bmatrix} I & -K_e \\ 0 & I \end{bmatrix} \begin{bmatrix} u \\ y \end{bmatrix},$$

$$\varphi\left(\varepsilon_u, \varepsilon_y\right) = \begin{pmatrix} I - \Pi^a \left(I + \begin{bmatrix} Z\left(X + Q_e\hat{N}\right)^{-1} & 0 \\ 0 & 0 \end{bmatrix} \Pi^a\right)^{-1} \times \\ \begin{bmatrix} Z\left(X + Q_e\hat{N}\right)^{-1} & 0 \\ 0 & 0 \end{bmatrix} \end{pmatrix} \begin{bmatrix} \varepsilon_u \\ \varepsilon_y \end{bmatrix}.$$

Using Theorem 15.7, $\theta(u, y)$ is further written as

$$\theta(u, y) = \Pi^a \left(I + \begin{bmatrix} Z\left(X + Q_e\hat{N}\right)^{-1} & 0 \\ 0 & 0 \end{bmatrix} \Pi^a\right)^{-1} \begin{bmatrix} I & -K_e \\ 0 & I \end{bmatrix} \begin{bmatrix} M \\ N \end{bmatrix}$$
$$\times \left(I + \Phi \begin{bmatrix} M \\ N \end{bmatrix}\right)^{-1} \left(\bar{v} + \Psi \begin{bmatrix} \varepsilon_u \\ \varepsilon_y \end{bmatrix}\right).$$

From this result, the existence conditions for kernel attacks become obvious.

Theorem 15.8 *Given a CPS subject to (15.1) and (15.10), suppose that the CPS is under multiplicative cyber-attacks modelled by (15.20), then* $\left(a_{u_{MC}}, a_y\right)$ *are kernel attacks if*

$$\begin{bmatrix} -\hat{N} & \hat{M} \end{bmatrix} \begin{bmatrix} -Z\left(X + Q_e\hat{N}\right)^{-1} & K_e \\ 0 & I \end{bmatrix} \Pi^a = 0, \tag{15.64}$$

$$\begin{bmatrix} -\hat{N} & \hat{M} \end{bmatrix} \begin{bmatrix} -Z\left(X + Q_e\hat{N}\right)^{-1} & K_e \\ 0 & I \end{bmatrix} \begin{bmatrix} \varepsilon_u \\ \varepsilon_y \end{bmatrix} = 0. \tag{15.65}$$

Proof The proof is straightforward. It is apparent that (15.64) leads to

$$\begin{bmatrix} -\hat{N} & \hat{M} \end{bmatrix} \begin{bmatrix} -Z\left(X + Q_e\hat{N}\right)^{-1} & K_e \\ 0 & I \end{bmatrix} \theta(u, y) = 0$$

$$\begin{bmatrix} -\hat{N} & \hat{M} \end{bmatrix} \begin{bmatrix} -Z\left(X + Q_e\hat{N}\right)^{-1} & K_e \\ 0 & I \end{bmatrix} \varphi\left(\varepsilon_u, \varepsilon_y\right)$$
$$= \begin{bmatrix} -\hat{N} & \hat{M} \end{bmatrix} \begin{bmatrix} -Z\left(X + Q_e\hat{N}\right)^{-1} & K_e \\ 0 & I \end{bmatrix} \begin{bmatrix} \varepsilon_u \\ \varepsilon_y \end{bmatrix}.$$

Consequently, (15.65) results in

$$\begin{bmatrix} -\hat{N} & \hat{M} \end{bmatrix} \begin{bmatrix} -Z\left(X + Q_e\hat{N}\right)^{-1} K_e \\ 0 & I \end{bmatrix} \varphi\left(\varepsilon_u, \varepsilon_y\right) = 0.$$

On account of (15.64) and (15.65), it can be claimed that the construction of kernel attacks depends on the configuration and design of the controller embedded at the plant side.

15.3 Detection of Cyber-Attacks

We now begin with addressing issues of detecting cyber-attacks. It is supposed that the detection is to be performed in the MC-station using the available online data set, which may be different with regard to the CPS configuration. The main attention will be paid to detection of kernel attacks.

15.3.1 An Observer-Based Detection Scheme Under the Existing CPS Configurations

The discussions in Chap. 14 and in the previous section make it plain that the observer-based residual signal r_y^a,

$$r_y^a = \hat{M} y^a - \hat{N} u_{est},$$

is a natural information provider about the cyber-attacks $\left(a_y, a_{u_{CM}}\right)$. Here, u_{est} is the estimate (calculated value) of the process input at the MC-station that takes different forms depending on the CPS configurations. To be specific, consider firstly the CPS configuration subject to (15.1) and (15.9), for which

$$u_{est} = u_{MC}, \begin{bmatrix} u_{est} \\ y^a \end{bmatrix} = \begin{bmatrix} u \\ y \end{bmatrix} + \begin{bmatrix} -a_{u_{MC}} \\ a_y \end{bmatrix}$$
$$\Longrightarrow r_y^a = \hat{M} y^a - \hat{N} u_{est} = r_y + \hat{M} a_y + \hat{N} a_{u_{MC}}.$$

For a CPS subject to (15.1) and (15.10), it holds

$$\begin{bmatrix} u_{est} \\ y^a \end{bmatrix} = \begin{bmatrix} u \\ y \end{bmatrix} + \begin{bmatrix} -Z\left(X + Q_e\hat{N}\right)^{-1} K_e \\ 0 & I \end{bmatrix} \begin{bmatrix} a_{u_{MC}} \\ a_y \end{bmatrix}$$

$$\Longrightarrow r_y^a = \hat{M} y^a - \hat{N} u_{est}$$
$$= r_y + \hat{M} a_y + \hat{N} Z \left(X + Q_e \hat{N} \right)^{-1} a_{u_{MC}} + \left(\hat{M} - \hat{N} K_e \right) a_y.$$

Notice that in both cases, the effects of r_y and $\left(a_{u_{MC}}, a_y\right)$ are coupled that makes a simultaneous detection of faults and cyber-attacks difficult. During fault-free operations, the cyber-attack detection is a standard robust detection problem, where the residual r_y is handled as uncertainties. Aiming at finding an alternative and more efficient solution, we now examine the dynamic of (u_{est}, y^a) in the CPS-loop. It is straightforward that

- in the case of CPS subject to (15.1) and (15.9),

$$\begin{bmatrix} u_{est} \\ y^a \end{bmatrix} = \begin{bmatrix} I & -K \\ -G & I \end{bmatrix}^{-1} \begin{bmatrix} v + K a_y + a_{u_{MC}} \\ \hat{M} r_y \end{bmatrix} + \begin{bmatrix} -a_{u_{MC}} \\ a_y \end{bmatrix}$$
$$= \begin{bmatrix} X + Q\hat{N} & Y - Q\hat{M} \\ -\hat{N} & \hat{M} \end{bmatrix}^{-1} \begin{bmatrix} \bar{v} - \left(Y - Q\hat{M} \right) a_y + \left(X + Q\hat{N} \right) a_{u_{MC}} \\ r_y \end{bmatrix}$$
$$+ \begin{bmatrix} -a_{u_{MC}} \\ a_y \end{bmatrix},$$

  which can be further written into

$$\begin{bmatrix} u_{est} \\ y^a \end{bmatrix} = \begin{bmatrix} X + Q\hat{N} & Y - Q\hat{M} \\ -\hat{N} & \hat{M} \end{bmatrix}^{-1} \begin{bmatrix} \bar{v} - \left(Y - Q\hat{M} \right) a_y + \left(X + Q\hat{N} \right) a_{u_{MC}} \\ r_y \end{bmatrix}$$
$$+ \begin{bmatrix} X + Q\hat{N} & Y - Q\hat{M} \\ -\hat{N} & \hat{M} \end{bmatrix}^{-1} \begin{bmatrix} \left(Y - Q\hat{M} \right) a_y - \left(X + Q\hat{N} \right) a_{u_{MC}} \\ \hat{N} a_{u_{MC}} + \hat{M} a_y \end{bmatrix}$$
$$= \begin{bmatrix} X + Q\hat{N} & Y - Q\hat{M} \\ -\hat{N} & \hat{M} \end{bmatrix}^{-1} \begin{bmatrix} \bar{v} \\ r_y + \hat{N} a_{u_{MC}} + \hat{M} a_y \end{bmatrix}$$
$$= \begin{bmatrix} M \\ N \end{bmatrix} \bar{v} + \left(\begin{bmatrix} -\hat{Y} \\ \hat{X} \end{bmatrix} + \begin{bmatrix} M \\ N \end{bmatrix} Q \right) r_y^a, \tag{15.66}$$

- for a CPS subject to (15.1) and (15.10),

$$\begin{bmatrix} u_{est} \\ y^a \end{bmatrix} = \begin{bmatrix} X + Q_e\hat{N} & Y - Q_e\hat{M} \\ -\hat{N} & \hat{M} \end{bmatrix}^{-1} \begin{bmatrix} \bar{v} + Z \left(Q_{MC} r_y^a + a_{u_{MC}} \right) \\ r_y \end{bmatrix}$$
$$+ \begin{bmatrix} -Z \left(X + Q_e\hat{N} \right)^{-1} & K_e \\ 0 & I \end{bmatrix} \begin{bmatrix} a_{u_{MC}} \\ a_y \end{bmatrix},$$

which is then written into

$$\begin{aligned}
\begin{bmatrix} u_{est} \\ y^a \end{bmatrix} &= \begin{bmatrix} X + Q_e\hat{N} & Y - Q_e\hat{M} \\ -\hat{N} & \hat{M} \end{bmatrix}^{-1} \begin{bmatrix} \bar{v} + Z\left(Q_{MC}r_y^a + a_{u_{MC}}\right) \\ r_y \end{bmatrix} \\
&+ \begin{bmatrix} X + Q_e\hat{N} & Y - Q_e\hat{M} \\ -\hat{N} & \hat{M} \end{bmatrix}^{-1} \begin{bmatrix} -\left(X + Q_e\hat{N}\right)Z\left(X + Q_e\hat{N}\right)^{-1} a_{u_{MC}} \\ \hat{N}Z\left(X + Q_e\hat{N}\right)^{-1} a_{u_{MC}} + \left(\hat{M} - \hat{N}K_e\right)a_y \end{bmatrix} \\
&= \begin{bmatrix} X + Q_e\hat{N} & Y - Q_e\hat{M} \\ -\hat{N} & \hat{M} \end{bmatrix}^{-1} \begin{bmatrix} \bar{v} + ZQ_{MC}r_y^a \\ r_y + \hat{N}Z\left(X + Q_e\hat{N}\right)^{-1} a_{u_{MC}} \\ +\left(\hat{M} - \hat{N}K_e\right)a_y \end{bmatrix} \\
&= \begin{bmatrix} X + Q_e\hat{N} & Y - Q_e\hat{M} \\ -\hat{N} & \hat{M} \end{bmatrix}^{-1} \left(\begin{bmatrix} \bar{v} \\ 0 \end{bmatrix} + \begin{bmatrix} ZQ_{MC} \\ I \end{bmatrix} r_y^a\right) \\
&= \begin{bmatrix} M \\ N \end{bmatrix}\bar{v} + \left(\begin{bmatrix} -\hat{Y} \\ \hat{X} \end{bmatrix} ZQ_{MC} + \begin{bmatrix} M \\ N \end{bmatrix} Q_e\right) r_y^a. \qquad (15.67)
\end{aligned}$$

From (15.66) and (15.67) it becomes apparent that

- r_y and $\left(a_{u_{MC}}, a_y\right)$ affect the identical subspace in the data space (u_{est}, y^a), namely, the image subspace of the controller, and it is impossible to distinguish r_y and $\left(a_{u_{MC}}, a_y\right)$,
- more crucially, it is standard in robust control theory that

$$\left\| \begin{bmatrix} -\hat{Y} \\ \hat{X} \end{bmatrix} + \begin{bmatrix} M \\ N \end{bmatrix} Q \right\|_\infty$$

  is minimised to enhance the robustness against disturbance. Although such a controller design cannot decrease the impact of cyber-attacks on the system dynamic, as illustrated in Chap. 14, it can remarkably impair the capability of detecting cyber-attacks,
- in the case of kernel attacks, that is

$$r_y^a = r_y,$$

  and r_y^a is independent of $\left(a_{u_{MC}}, a_y\right)$, it is impossible to detect the attacks in the data space (u_{est}, y^a).

These observations give us reasons to develop alternative detection methods.

15.3.2 A Preliminary Detection Scheme with a Modified CPS Configuration

The fact that $\left(a_{u_{MC}}, a_y\right)$ and r_y are indistinguishable in the data space (u_{est}, y^a) implies that additional information about the CPS dynamics is needed. Intuitively, information related to the input variable u on the plant side is an immediate candidate. Before proposing an alternative detection scheme, we would like to examine the necessary information for the detection and control purposes and propose a modified CPS configuration.

15.3.2.1 Residuals are the Essential Information for Detection and Control

As described in the first part of the book, the basis of the unified framework of control and detection is the one-to-one mapping between the process data (u, y) and the residual signals $\left(r_u, r_y\right)$ given by

$$\begin{bmatrix} r_u \\ r_y \end{bmatrix} = \begin{bmatrix} X & Y \\ -\hat{N} & \hat{M} \end{bmatrix} \begin{bmatrix} u \\ y \end{bmatrix} \Longleftrightarrow \begin{bmatrix} u \\ y \end{bmatrix} = \begin{bmatrix} M & -\hat{Y} \\ N & \hat{X} \end{bmatrix} \begin{bmatrix} r_u \\ r_y \end{bmatrix}.$$

This core relation enables control and detection using $\left(r_u, r_y\right)$ and the reference signal v.

15.3.2.2 Controller and Detector Can be Realised by an Observer-Based Residual Generation System

According to the state space representations of $\left(\hat{M}, \hat{N}\right)$ and (X, Y), (2.5) and (2.26), the residual generators and the controller can be realised as follows

$$\begin{cases} \hat{x}(k+1) = (A - LC)\,\hat{x}(k) + (B - LD)\,u(k) + Ly(k) \\ r_u(k) = V^{-1}\left(u(k) - F\hat{x}(k)\right) \\ r_y(k) = W\left(y(k) - \hat{y}(k)\right) = W\left(y(k) - C\hat{x}(k) - Du(k)\right) \\ u(z) = F\hat{x}(z) + Q\,(z)\,r_y(z) + \bar{v}(z),\, \bar{v}(z) = \left(X(z) + Q(z)\hat{N}(z)\right) v(z). \end{cases} \tag{15.68}$$

On account of (15.68), the subsequent CPS configuration is proposed, which is similar to the one with the embedded closed-loop given by (14.5)–(14.6), but with different data transmissions:

- on the plant side, the received signal transmitted by the MC-station is u_{MC}^a,

$$u_{MC}^a = u_{MC} + a_{u_{MC}},$$

the performed (online) computation is

$$\begin{cases} \hat{x}(k+1) = (A - LC)\,\hat{x}(k) + (B - LD)\,u(k) + Ly(k) \\ Vr_u(k) = u(k) - F\hat{x}(k) \\ r_y(k) = W\left(y(k) - \hat{y}(k)\right) = W\left(y(k) - C\hat{x}(k) - Du(k)\right) \\ u(z) = F\hat{x}(z) + u^a_{MC}(z), \end{cases} \tag{15.69}$$

and $\left(r_u, r_y\right)$ are sent to the MC-station,

- at the MC-station, the received signals are

$$\begin{bmatrix} r^a_u \\ r^a_y \end{bmatrix} = \begin{bmatrix} r_u \\ r_y \end{bmatrix} + \begin{bmatrix} a_{r_u} \\ a_{r_y} \end{bmatrix},$$

the running online computations are

– for the control purpose

$$u_{MC} = Qr^a_y + \bar{v} = Qr_y + \bar{v} + Qa_{r_y},$$

– for attack and fault detection,

$$\bar{r}_u = Vr^a_u - u_{MC} = u - F\hat{x}(z) - Qr_y - \bar{v} - Qa_{r_y} + Va_{r_u}$$

as well as fusion of $\bar{r}_u$ and r_y towards detecting cyber-attacks and faults,

and u_{MC} is sent to the plant.

The crucial arguments for the above CPS configuration are

- cyber-security: with the embedded controller at the plant side, (15.69), the system operation is assured for the case that the communications fail,
- with $\left(r^a_u, r^a_y\right)$ sufficient information is available at the MC-station, and
- the transmission of the residuals $\left(r_u, r_y\right)$, instead of the process data y, increases the cyber-security with respect to the data privacy.

Next, the closed-loop dynamic of the CPS system with the plant model (15.1) is analysed. We begin with (u, y), which showcases the impact of the cyber-attacks on the plant dynamic. It is straightforward that

$$\begin{aligned} \begin{bmatrix} u \\ y \end{bmatrix} &= \begin{bmatrix} X + Q\hat{N} & Y - Q\hat{M} \\ -\hat{N} & \hat{M} \end{bmatrix}^{-1} \begin{bmatrix} \bar{v} + Qa_{r_y} + a_{u_{MC}} \\ r_y \end{bmatrix} \\ &= \begin{bmatrix} M \\ N \end{bmatrix} \left(\bar{v} + Qa_{r_y} + a_{u_{MC}}\right) + \begin{bmatrix} -\hat{Y} + MQ \\ \hat{X} + NQ \end{bmatrix} r_y. \end{aligned} \tag{15.70}$$

As expected, this result conforms with the claim that the cyber-attacks exclusively affect the image subspace of the plant. Concerning the residuals $\left(\bar{r}_u, r_y^a\right)$ for the detection purpose, it turns out, after some routine calculations,

$$\begin{aligned}\begin{bmatrix} \bar{r}_u \\ r_y^a \end{bmatrix} &= \begin{bmatrix} u - F\hat{x} - Qr_y - \bar{v} - Qa_{r_y} + Va_{r_u} \\ r_y + a_{r_y} \end{bmatrix} \\ &= \begin{bmatrix} X & Y \\ -\hat{N} & \hat{M} \end{bmatrix} \begin{bmatrix} u \\ y \end{bmatrix} + \begin{bmatrix} -Qr_y - \bar{v} - Qa_{r_y} + Va_{r_u} \\ a_{r_y} \end{bmatrix} \\ &= \begin{bmatrix} a_{uMC} + Va_{r_u} \\ a_{r_y} + r_y \end{bmatrix}. \end{aligned} \tag{15.71}$$

The last equation is attributed to the Bezout identity. The response of the residuals $\left(\bar{r}_u, r_y^a\right)$ described by (15.71) tellingly demonstrates that

- the attacks on the control related signals, $\left(a_{uMC}, a_{r_u}\right)$, is decoupled from the uncertainty (including faults) in the plant, and can be detected by means of $\bar{r}_u$,
- a_{r_y} can be detected, when there exists no fault in the plant, and in other words,
- a reliable fault detection is possible, when the CPS is not under cyber-attack a_{r_y}.

On the other hand, it follows from (15.71) that it is facile for the attacker to design $\left(a_{uMC}, a_{r_u}\right)$ so that

$$a_{uMC} + Va_{r_u} = 0$$

and consequently they become undetectable. Moreover, maybe more crucially, the proposed CPS configuration with a fixed state feedback gain F described in the controller (15.68) is inflexible and often cannot meet requirements for high resilience on a CPS. These observations motivate and drive the subsequent work.

15.4 A Cyber-Attack and Fault Detection Scheme, and a Resilient CPS Configuration

We now extend the CPS configuration proposed in the previous section aiming at the realisation of the feedback control system (15.68) and, more importantly, a resilient configuration so that the CPS system is capable for (i) cyber-attack and fault detection, (ii) online adaptation of the control law to recover control performance degradation.

15.4.1 A Resilient CPS Configuration

We begin with a detailed description of the proposed configuration.

- On the plant side, in addition to the implementation of the observer-based controller and residual generators (15.69),

$$u = F_0\hat{x} + u^a_{MC},$$

a standard residual-driven fault detection algorithm will run. Here, F_0 is the default setting of the embedded controller. Let J and J_{th} denote the residual evaluation function and the corresponding threshold, respectively. The adopted detection logic is,

$$\begin{cases} J > J_{th} \Longrightarrow \text{faulty} \\ J \leq J_{th} \Longrightarrow \text{faulty-free.} \end{cases} \tag{15.72}$$

The data transmitted to the MC-station are

$$\begin{bmatrix} r_u + \Gamma r_y \\ r_y \end{bmatrix} = \begin{bmatrix} X_1 & Y_1 \\ -\hat{N} & \hat{M} \end{bmatrix} \begin{bmatrix} u \\ y \end{bmatrix} + \begin{bmatrix} \Gamma r_y \\ 0 \end{bmatrix}.$$

Hereby, Γ is a design parameter, and adding r_y to r_u serves for twofold purposes, enhancing the data privacy and simplifying attack detection. The latter will be illustrated in the next subsection. Moreover, the MC-station will be informed if a fault is detected. Here, we would like to call the reader's attention that LCP (X_1, Y_1),

$$X_1 = (A_L, -B_L, F_1, I), Y_1 = (A_L, -L, F_1, 0),$$

is different from

$$X_0 = (A_L, -B_L, F_0, I), Y_1 = (A_L, -L, F_0, 0)$$

due to the different state feedback gain matrices F_1, F_0. In the subsequent subsection, the setting of (X_1, Y_1) will be explained.

- At the MC-station, using the received signals

$$\begin{bmatrix} r^a_u \\ r^a_y \end{bmatrix} = \begin{bmatrix} r_u + \Gamma r_y \\ r_y \end{bmatrix} + \begin{bmatrix} a_{r_u} \\ a_{r_y} \end{bmatrix},$$

the controller and the cyber-attack detector are implemented:

  - the controller u_{MC}

$$u_{MC} = Q r^a_y + \bar{v},$$

$$Q = \begin{cases} Q_0, J \leq J_{th} \\ Q_1, J > J_{th}, \end{cases}, \bar{v} = \begin{cases} \bar{v}_0 = \left(X_0 + Q_0\hat{N}\right) v, J \leq J_{th} \\ \bar{v}_1 = \left(X_0 + Q_1\hat{N}\right) v, J > J_{th}, \end{cases} \tag{15.73}$$

  where the notations Q_0, Q_1, $\bar{v}_0$ and $\bar{v}_1$ will be explained in the subsequent subsection;
  - the cyber-attack detector r^a_y and

$$\bar{r}_u = \begin{cases} r_u^a - V_{10}\left(\bar{v} + Qr_y^a\right), \, J \leq J_{th} \\ r_u^a - V_{10}\left(\bar{v} + Qr_y^a\right) - \bar{V}_{10} r_y^a, \, J > J_{th}, \end{cases} \tag{15.74}$$

where

$$V_{10} = I + (F_1 - F_0)\left(zI - A_{F_1}\right)^{-1} B \in \mathcal{RH}_\infty, \tag{15.75}$$

$$\bar{V}_{10} = \Gamma + (F_0 - F_1)\left(zI - A_{F_0}\right)^{-1} L \in \mathcal{RH}_\infty, \tag{15.76}$$

$$\hat{X}_1 = \left(A_{F_1}, L, C_{F_1}, I\right), \hat{Y}_1 = \left(A_{F_1}, -L, F_1, 0\right). \tag{15.77}$$

15.4.2 Basic Working Principle of the Resilient CPS

The major difference to the CPS proposed in the last section is that the controller and detector in the resilient CPS are set demand-driven. To be specific, when no fault is detected, the default controller is in operation, which is equivalent to

$$u = -\left(X_0 + Q_0\hat{N}\right)^{-1}\left(Y_0 - Q_0\hat{M}\right) y + \bar{v}_0, \tag{15.78}$$

$$\bar{v}_0 = \left(X_0 + Q_0\hat{N}\right) v.$$

As illustrated in the previous section (refer to (15.71)), if (X_0, Y_0) is applied as the generator of r_u, it holds

$$\bar{r}_u = a_{u_{MC}} + Va_{r_u}.$$

Considering that the attacker is able to construct stealthy attacks by setting

$$a_{u_{MC}} + Va_{r_u} = 0,$$

the detection of $\left(a_{u_{MC}}, a_{r_u}\right)$ can become very difficult. Moreover, in terms of u and $X_0 u + Y_0 y$, the key variable of the CPS, $F_0\hat{x}$, can be directly identified. In order to detect $\left(a_{u_{MC}}, a_{r_u}\right)$ reliably and assure the data privacy, transmission of

$$r_u + \Gamma r_y = X_1 u + Y_1 y + \Gamma r_y = u - F_1\hat{x} + \Gamma r_y$$

is proposed, where $F_1 \neq F_0$. Hence,

$$r_u^a = X_1 u + Y_1 y + \Gamma r_y + a_{r_u}.$$

The following theorem describes a basic relation used to construct a reliable cyber-attack detector.

Theorem 15.9 *Given the CPS model*

$$\begin{cases} y = Gu + \hat{M}^{-1} r_y \\ u = F_0 \hat{x} + u^a_{MC}, \\ u^a_{MC} = \bar{v} + Q r^a_y + a_{u_{MC}}, \end{cases} \tag{15.79}$$

it holds

$$r^a_u = X_1 u + Y_1 y + \Gamma r_y = V_{10} \left(\bar{v} + Q r^a_y + a_{u_{MC}} \right) + \bar{V}_{10} r_y + a_{r_u}, \tag{15.80}$$

where V_{10} and $\bar{V}_{10}$ are given in (15.75) and (15.76), respectively.

Proof Observe that

$$u = F_0 \hat{x} + u^a_{MC} \Longrightarrow X_0 u + Y_0 y = \bar{v} + Q r^a_y + a_{u_{MC}}.$$

By means of Theorem 3.4 and the result in Example 3.4, $X_1 u + Y_1 y$ can then be written into

$$X_1 u + Y_1 y = V_{10} \left(X_0 u + Y_0 y \right) + \begin{bmatrix} X_0 & Y_0 \end{bmatrix} \begin{bmatrix} -\hat{Y}_1 \\ \hat{X}_1 \end{bmatrix} r_y,$$

$$\begin{bmatrix} X_0 & Y_0 \end{bmatrix} \begin{bmatrix} -\hat{Y}_1 \\ \hat{X}_1 \end{bmatrix} = \left(F_0 - F_1 \right) \left(zI - A_{F_0} \right)^{-1} L.$$

Consequently,

$$\begin{aligned} r^a_u &= V_{10} \left(X_0 u + Y_0 y \right) + \bar{V}_{10} r_y + a_{r_u} \\ &= V_{10} \left(\bar{v} + Q r^a_y + a_{u_{MC}} \right) + \bar{V}_{10} r_y + a_{r_u}. \end{aligned}$$

This completes the proof.

It is obvious that the design of the cyber-attack detector (15.74) is the use of the relation (15.80). It is apparent that during fault-free operations,

$$\begin{bmatrix} \bar{r}_u \\ r^a_y \end{bmatrix} = \begin{bmatrix} V_{10} a_{u_{MC}} + a_{r_u} \\ a_{r_y} \end{bmatrix} + \begin{bmatrix} \bar{V}_{10} \\ I \end{bmatrix} r_y. \tag{15.81}$$

Since r_y represents the uncertainties in the process, e.g. noises or disturbances, and is independent of the cyber-attacks, both a_{r_y} and $V_{10} a_{u_{MC}} + a_{r_u}$ can be well detected by means of a standard fault detection algorithm.

When a fault is detected, both the controller and cyber-attack detector are re-set, as described in (15.73) and (15.74). The controller is equivalent to

$$u = -\left(X_0 + Q_1 \hat{N} \right)^{-1} \left(Y_0 - Q_1 \hat{M} \right) y + \bar{v}_1,$$

and the detector is subject to

$$\bar{r}_u = V_{10} a_{u_{MC}} + a_{r_u} - \bar{V}_{10} a_{r_y}. \tag{15.82}$$

Note that in this case, $\bar{r}_u$ is independent of r_y and thus will not be affected by the fault.

15.4.3 Analysis and Design of Controllers and Detectors

15.4.3.1 On Controller Design

We firstly examine the CPS dynamic and discuss the controller design issues. As delineated in the last section, the CPS dynamic is governed by

$$\begin{aligned}\begin{bmatrix} u \\ y \end{bmatrix} &= \begin{bmatrix} X_0 + Q\hat{N} & Y_0 - Q\hat{M} \\ -\hat{N} & \hat{M} \end{bmatrix}^{-1} \begin{bmatrix} \bar{v} + Qa_y + a_{u_{MC}} \\ r_y \end{bmatrix} \\ &= \begin{bmatrix} M_0 \\ N_0 \end{bmatrix} \left(\bar{v} + Qa_{r_y} + a_{u_{MC}} \right) + \left(\begin{bmatrix} -\hat{Y}_0 \\ \hat{X}_0 \end{bmatrix} + \begin{bmatrix} M_0 \\ N_0 \end{bmatrix} Q \right) r_y. \end{aligned} \tag{15.83}$$

In the case that a fault is detected, Q is re-set as Q_1 aiming at recovering the control performance suffered from degradation caused by the fault. To this end, there exist a number of established methods, for instance, solving the optimisation problem

$$Q_1 = \arg \min_{Q \in \mathcal{RH}_\infty} \left\| \begin{bmatrix} -\hat{Y}_0 \\ \hat{X}_0 \end{bmatrix} + \begin{bmatrix} M_0 \\ N_0 \end{bmatrix} Q \right\|.$$

It is worth mentioning that, according to Corollary 3.1, the setting of a feedback controller $u = Ky$, parametrised by

$$u = -\left(X + Q\hat{N} \right)^{-1} \left(Y - Q\hat{M} \right) y,$$

is invariant with respect to the state feedback gain F (adopted in (X_0, Y_0), $\left(\hat{X}_0, \hat{Y}_0\right)$ and (M_0, N_0)). Hence, re-setting the parameterisation matrix Q is equivalent to redesigning the controller K.

Although the feedback controller is inefficient to suppress the impact of the cyber-attacks $\left(a_y, a_{u_{MC}}\right)$ on the system dynamic, the cognition that $\left(a_y, a_{u_{MC}}\right)$ exclusively attack the system image subspace inspires us to make use of the existing redundancy in the feedback controller design to deal with cyber-attacks. Below is a short description of the basic idea. Consider the nominal controller

$$u = -\left(X_0 + Q_0\hat{N}\right)^{-1}\left(Y_0 - Q_0\hat{M}\right)y + v \Longleftrightarrow$$
$$u = F_0\hat{x} + Q_0 r_y^a + \bar{v}_0 \Longleftrightarrow X_0 u + Y_0 y = Q_0 r_y^a + \bar{v}_0.$$

Now, set

$$u = F_0\hat{x} + Q_u r_u + u_{MC}^a, u_{MC} = Q r_y^a + Q_v v,$$
$$r_u = X_0 u + Y_0 y, Q_u, Q_v \in \mathcal{RH}_\infty.$$

Hereby, Q_u, Q_v and Q are parameter systems to be designed. Write the controller into

$$u = F_0\hat{x} + Q r_y^a + Q_u r_u + Q_v v + a_{u_{MC}} \Longleftrightarrow$$
$$\bar{Q}_u^{-1} X_0 u + \bar{Q}_u^{-1} Y_0 y = Q r_y + Q_v v + Q a_{r_y} + a_{u_{MC}},$$

where $\bar{Q}_u^{-1} = I - Q_u \in \mathcal{RH}_\infty$ and Q_u is selected so that $\bar{Q}_u$ is stable. Accordingly, the feedback controller is given by

$$u = -\left(\bar{Q}_u^{-1} X_0 + Q\hat{N}\right)^{-1}\left(\left(\bar{Q}_u^{-1} Y_0 - Q\hat{M}\right)y + Q_v v + Q a_{r_y} + a_{u_{MC}}\right) =$$
$$-\left(X_0 + \bar{Q}\hat{N}\right)^{-1}\left(Y_0 - \bar{Q}\hat{M}\right)y + \left(X_0 + \bar{Q}\hat{N}\right)^{-1}\bar{Q}_u\left(Q_v v + Q a_{r_y} + a_{u_{MC}}\right),$$
$$\bar{Q} = \bar{Q}_u Q \in \mathcal{RH}_\infty,$$

and the closed-loop dynamic is subject to

$$\begin{bmatrix} u \\ y \end{bmatrix} = \begin{bmatrix} M_0 \\ N_0 \end{bmatrix}\bar{Q}_u\left(Q_v v + Q a_{r_y} + a_{u_{MC}}\right) + \left(\begin{bmatrix} -\hat{Y}_0 \\ \hat{X}_0 \end{bmatrix} + \begin{bmatrix} M_0 \\ N_0 \end{bmatrix}\bar{Q}\right) r_y. \quad (15.84)$$

Setting $\bar{Q}_u$ and Q_v so that

$$\left\|\bar{Q}_u\right\|_\infty << 1, \quad (15.85)$$
$$Q_v = (I - Q_u)\left(X_0 + Q_0\hat{N}\right) \quad (15.86)$$

yields

$$\begin{bmatrix} u \\ y \end{bmatrix} = \begin{bmatrix} M_0 \\ N_0 \end{bmatrix}(\bar{v}_0 + a) + \left(\begin{bmatrix} -\hat{Y}_0 \\ \hat{X}_0 \end{bmatrix} + \begin{bmatrix} M_0 \\ N_0 \end{bmatrix}\bar{Q}\right) r_y, \quad (15.87)$$
$$a = \bar{Q}_u\left(Q a_{r_y} + a_{u_{MC}}\right).$$

Moreover, in order to assure the identical feedback control performance, define

$$\bar{Q} = \bar{Q}_u Q = Q_0 \Longrightarrow Q = (I - Q_u) Q_0.$$

As a result, the resulted system dynamic is of the nice properties that

- the feedback control performance (response to r_y) and the response to the reference signal v are fully recovered, and
- the impact of $(a_y, a_{u_{MC}})$ on the system dynamic is suppressed to

$$\|a\|_2 \leq \left\|\bar{Q}_u\right\|_\infty \left\|Qa_{r_y} + a_{u_{MC}}\right\|_2 << \left\|Qa_{r_y} + a_{u_{MC}}\right\|_2.$$

The following algorithm is proposed for the online realisation of the above idea, when cyber-attacks $(a_y, a_{u_{MC}})$ are detected.

Algorithm: *Resilient control algorithm in case of detected cyber-attacks* $(a_y, a_{u_{MC}})$

Step 0: Define $0 < \gamma << 1$ and $\bar{Q}_u$ satisfying

$$\left\|\bar{Q}_u\right\|_\infty < \gamma, \bar{Q}_u^{-1} \in \mathcal{RH}_\infty;$$

Step 1: The MC-station sends the message "detected cyber-attacks" and signal

$$u_{MC} = (I - Q_u) Q_0 r_y^a + (I - Q_u) \left(X_0 + Q_0 \hat{N}\right) v$$

to the plant;
Step 2: On the plant side, u is set to be

$$u = F_0 \hat{x} + \left(I - \bar{Q}_u^{-1}\right) r_u + u_{MC}^a.$$

It is noteworthy that the capability of the controller in the present form is considerably limited in dealing with cyber-attacks. To be specific, the feedback controller has no impact on the dynamics induced by the cyber-attacks, as illustrated by (15.84). The utilisation of the redundancy in the controller configuration, equivalent with use of the residual signal r_u, is restricted to a class of cyber-attacks. The control-theoretic explanation for this restriction is the way how r_u is used. As shown in (15.84), the use of $Q_u r_u$ serves as a feedforward controller in the loop and, consequently, is inefficient. In the next chapter, a resilient control scheme will be proposed, which is based on a feedback of the residual signal r_u in the loop.

15.4.3.2 On Detector Design

Next, the detection issues are addressed. Concerning fault detection, it is suggested to apply

- the unified solution or Kalman filter presented in Chap. 8 as efficient solutions if additive faults are to be detected in a plant with unknown input disturbances (noises), and
- the projection-based method introduced in Chap. 9, when multiplicative faults and model uncertainties are under consideration. In order to save online computations at the plant side, the method of projecting the data onto $\mathcal{L}_2$ subspace is preferred.

We would like to call the reader's attention to the fact that the projection-based detection algorithm should be realised in the open-loop configuration, although the closed-loop feedback control system is addressed. The reason is that the control system at the plant side only receives the merged signal $u_{MC} = Qr_y^a + \bar{v}$ instead of a direct access to $\bar{v}$. To be specific, the following algorithm is applied on the assumption that the plant is modelled by

$$y = Gu, G = NM^{-1} = (N_0 + \Delta_N)(M_0 + \Delta_M)^{-1},$$
$$\left\| \Delta_{I_G} \right\|_\infty \leq \delta_{\Delta_{I_G}} < 1, \Delta_{I_G} = \begin{bmatrix} \Delta_M \\ \Delta_N \end{bmatrix},$$

where the RCP (M_0, N_0) represents the nominal dynamic and (Δ_M, Δ_N) the right-coprime factor (RCF) uncertainty.

Algorithm: *Fault detection in the case of existence of multiplicative model uncertainties*

Step 1: Implementation of (open-loop) residual generation

$$r_y = \begin{bmatrix} -\hat{N}_0 & \hat{M}_0 \end{bmatrix} \begin{bmatrix} u \\ y \end{bmatrix},$$

where $\left(\hat{M}_0, \hat{N}_0\right)$ denotes the normalised LCP,
Step 2: Calculation of the residual evaluation function

$$J_{rel} = \frac{\left\| r_{\mathcal{I}_G^{\mathcal{L}_2}} \right\|_2}{\left\| \begin{bmatrix} u \\ y \end{bmatrix} \right\|_2} = \frac{\left\| r_y \right\|_2}{\left\| \begin{bmatrix} u \\ y \end{bmatrix} \right\|_2}, \tag{15.88}$$

Step 3: Detection of faults by means of the detection logic

$$\text{Detection logic}: \begin{cases} J_{rel} > J_{rel,th} \Longrightarrow \text{a fault is detected} \\ J_{rel} \leq J_{rel,th} \Longrightarrow \text{fault-free}, \end{cases}$$

$$J_{rel,th} = \delta_{\Delta_{I_G}} \sqrt{\frac{1 - \left\| r_y \right\|_2^2}{1 - \delta_{\Delta_{I_G}}^2}}. \tag{15.89}$$

The reader is referred to Sect. 9.4 for details. It is of interest to notice that both the residual evaluation function J_{rel} (15.88) and the threshold $J_{rel,th}$ (15.89) are independent of the cyber-attacks, because r_y is decoupled from a_{r_y}, $a_{u_{MC}}$ and a_{r_u}.

We now devote our attention to detecting cyber-attacks, the main focus of our work. At first, detection of a_{r_y}, $a_{u_{MC}}$ and a_{r_u} is considered on the assumption that no fault is detected. The fact that the residual r_y is independent of the cyber-attacks $\left(a_{r_y}, a_{u_{MC}}, a_{r_u}\right)$ allows us to model the system dynamic with respect to the cyber-attacks and r_y (as uncertainties) uniformly as

$$r = G_{r_y} r_y + f^a, \tag{15.90}$$
$$G_{r_y} = \begin{cases} I, \text{ if } r = r_y^a \\ \bar{V}_{10}, \text{ if } r = \bar{r}_u, \end{cases} \quad f^a = \begin{cases} a_{r_y}, \text{ if } r = r_y^a \\ V_{10} a_{u_{MC}} + a_{r_u}, \text{ if } r = \bar{r}_u \end{cases}$$

with the additive vector f^a representing the cyber-attacks. We examine the following two cases.

- The system uncertainty is noises and unknown inputs. To simplify our study, it is assumed that zero-mean white noises or $\mathcal{L}_2$-bounded disturbances are under consideration so that r_y delivered by the residual generator at the plant side is either a white noise series or $\left\|r_y\right\|_2 \leq \delta_d$, where δ_d denotes the $\mathcal{L}_2$-upper-bound of the disturbance vector d. It enables us to detect
  - a_{r_y} by means of

$$J_{a_{r_y}} = \begin{cases} r_y^T \Sigma_{r_y}^{-1} r_y, \text{ if } r_y \text{ is white noise series} \\ \left\|r_y\right\|_2, \text{ if } r_y \text{ is } \mathcal{L}_2\text{-bounded disturbance,} \end{cases}$$
$$J_{a_{r_y},th} = \begin{cases} \chi_\alpha^2, \text{ if } r_y \text{ is white noise series} \\ \delta_d, \text{ if } r_y \text{ is } \mathcal{L}_2\text{-bounded disturbance,} \end{cases}$$
$$\text{Detection logic}: \begin{cases} J_{a_{r_y}} > J_{a_{r_y},th} \Longrightarrow a_{r_y} \text{ is detected} \\ J_{a_{r_y}} \leq J_{a_{r_y},th} \Longrightarrow \text{no cyber-attack } a_{r_y}, \end{cases}$$

    where α is the (pre-defined) false alarm rate,
  - $V_{10} a_{u_{MC}} + a_{r_u} =: b$ by (i) designing a Kalman filter or an H_2 observer-based residual generator (the unified solution) based on the model (15.90),

$$\bar{r} = Rr = R\bar{V}_{10} r_y$$

    so that $\bar{r} \sim \mathcal{N}(0, \Sigma_{\bar{r}})$ or $\|\bar{r}\|_2 \leq \sup \left\|r_y\right\|_2 = \delta_d$,(ii)

$$J_b = \begin{cases} \bar{r}^T \Sigma_{\bar{r}}^{-1} \bar{r}, \text{ if } \bar{r} \text{ is white noise series} \\ \|\bar{r}\|_2, \text{ if } \bar{r} \text{ is } \mathcal{L}_2\text{-bounded disturbance,} \end{cases}$$
$$J_{b,th} = \begin{cases} \chi_\alpha^2, \text{ if } \bar{r} \text{ is white noise series} \\ \delta_d, \text{ if } \bar{r} \text{ is } \mathcal{L}_2\text{-bounded disturbance,} \end{cases}$$

$$\text{Detection logic} : \begin{cases} J_b > J_{b,th} \Longrightarrow V_{10} a_{u_{MC}} + a_{r_u} \text{ is detected} \\ J_b \leq J_{b,th} \Longrightarrow \text{no cyber-attack } V_{10} a_{u_{MC}} + a_{r_u}. \end{cases}$$

Note that the design parameter Γ in

$$\bar{V}_{10} = \Gamma + (F_0 - F_1)\left(zI - A_{F_0}\right)^{-1} L$$

is selected so that the existence conditions for a Kalman filter or an H_2 observer-based residual generator are satisfied.

- The system is modelled by

$$G = NM^{-1} = (N_0 + \Delta_N)(M_0 + \Delta_M)^{-1} \tag{15.91}$$

with the nominal dynamic represented by the SIR (M_0, N_0) and right-coprime factor uncertainty (Δ_M, Δ_N). It follows from Lemma 4.2 that

$$\begin{aligned} r_y &= \begin{bmatrix} -\hat{N}_0 & \hat{M}_0 \end{bmatrix} \begin{bmatrix} u \\ y \end{bmatrix} \\ &= \begin{bmatrix} -\hat{N}_0 & \hat{M}_0 \end{bmatrix} \begin{bmatrix} \Delta_M \\ \Delta_N \end{bmatrix} \left(I + \begin{bmatrix} V & -U \end{bmatrix} \begin{bmatrix} \Delta_M \\ \Delta_N \end{bmatrix} \right)^{-1} \bar{v}, \\ V &= \left(X_0 + Q_0 \hat{N} \right), U = \left(Y_0 - Q_0 \hat{M} \right), \bar{v} = V v, \end{aligned}$$

where $\left(\hat{M}_0, \hat{N}_0\right)$ denotes the normalised LCP applied for the projection-based residual generation on the plant side, and

$$K = -\left(X_0 + Q_0 \hat{N}\right)^{-1} \left(Y_0 - Q_0 \hat{M}\right)$$

is the equivalent feedback gain. On the assumption

$$\left\| \begin{bmatrix} V & -U \\ -\hat{N}_0 & \hat{M}_0 \end{bmatrix} \begin{bmatrix} \Delta_M \\ \Delta_N \end{bmatrix} \right\|_\infty = c < 1, \tag{15.92}$$

the following detection strategy is proposed:

– for detecting a_{r_y}

$$\begin{aligned} J_{a_{r_y}} &= \left\| r_y \right\|_2, \\ J_{a_{r_y},th} &= \frac{c}{\sqrt{1-c^2}} \left\| \bar{v} \right\|_2, \end{aligned} \tag{15.93}$$

$$\text{Detection logic} : \begin{cases} J_{a_{r_y}} > J_{a_{r_y},th} \Longrightarrow a_{r_y} \text{ is detected} \\ J_{a_{r_y}} \leq J_{a_{r_y},th} \Longrightarrow \text{no cyber-attack } a_{r_y}, \end{cases}$$

– for detecting $b := V_{10}a_{u_{MC}} + a_{r_u}$, (i) do a co-inner-outer factorisation of $\bar{V}_{10}$,

$$\bar{V}_{10} = \bar{V}_{10,co}\bar{V}_{10,ci},$$

with co-outer $\bar{V}_{10,co}$ and co-inner $\bar{V}_{10,ci}$, (ii) set

$$R = \bar{V}_{10,co}^{-1} \Longrightarrow \bar{r} = R\bar{V}_{10}r_y = \bar{V}_{10,ci}r_y,$$

(iii) online compute

$$\begin{aligned} J_b &= \|\bar{r}\|_2, \\ J_{b,th} &= \frac{c}{\sqrt{1-c^2}}\|\bar{v}\|_2, \\ \text{Detection logic}: &\begin{cases} J_b > J_{b,th} \Longrightarrow & V_{10}a_{u_{MC}} + a_{r_u} \text{ is detected} \\ J_b \leq J_{b,th} \Longrightarrow & \text{no cyber-attack } V_{10}a_{u_{MC}} + a_{r_u}. \end{cases} \end{aligned} \tag{15.94}$$

We now shortly highlight the above detection strategy. Firstly, the adoption of the plant model (15.91) loses no generality. As delineated in Sect. 4.7, there exists one-to-one mapping between the right-coprime factor and left-coprime factor uncertainty with respect to the closed-loop dynamics. The assumption (15.92) is a sufficient condition for the closed-loop stability and thus considered by the controller design. The threshold setting (15.93) follows from a known result that leads to

$$\left\| \begin{bmatrix} -\hat{N}_0 & \hat{M}_0 \end{bmatrix} \begin{bmatrix} \Delta_M \\ \Delta_N \end{bmatrix} \left(I + \begin{bmatrix} V & -U \end{bmatrix} \begin{bmatrix} \Delta_M \\ \Delta_N \end{bmatrix} \right)^{-1} \right\|_\infty \leq \frac{c}{\sqrt{1-c^2}}, \tag{15.95}$$

when the inequality (15.92) holds. The corresponding reference is given at the end of this chapter. Based on the same result and the fact that

$$\left\| \bar{V}_{10,ci}r_y \right\|_2 = \left\| r_y \right\|_2,$$

we have the threshold setting (15.94). Here, the design parameter Γ in $\bar{V}_{10}$ is selected to assure the existence of a co-inner-outer factorisation. Roughly speaking, the design of a residual generator based on a co-inner-outer factorisation is equivalent to the design of a Kalman filter or an H_2 observer-based residual generator, as discussed in Chap. 8. In this context, the selection of Γ is equivalent to the previous case.

It is noteworthy that, while a_{r_y} can be well detected, it is possible that attackers construct $a_{u_{MC}}$ and a_{r_u} so that they are undetectable, namely they satisfy the condition

$$V_{10}a_{u_{MC}} + a_{r_u} = 0.$$

To this end, attackers should gain knowledge of V_{10} subject to (15.75). On the other hand, the matrix F_1 can be selected arbitrarily and randomly, and changed from time to time so that it is hard for attackers to identify V_{10}. There are numerous strategies for this purpose. The reader is referred to the references given at the end of this chapter.

At the end of this section, we address attack detection in the case of a detected fault. Consider that the residual r_y is now corrupted with the fault, the aforementioned detection schemes based on $\left(r_y^a, \bar{r}_u\right)$ are no more reliable. As an alternative solution, the residual signal $\bar{r}_u$,

$$\bar{r}_u = r_u^a - V_{10}\left(\bar{v} + Q r_y^a\right) - \bar{V}_{10} r_y^a,$$

given in (15.74) is built. A routine calculation yields

$$\bar{r}_u = V_{10} a_{u_{MC}} + a_{r_u} - \bar{V}_{10} a_{r_y}. \tag{15.96}$$

Even though $\bar{r}_u$ is independent of r_y and hence not affected by the fault, as a linear combination of the three cyber-attacks $\left(a_{r_y}, a_{u_{MC}}, a_{r_u}\right)$, the probability that $\bar{r}_u = 0$ sharply increases, and the channel noises induced by transmitting r_y, u_{MC} and r_u result in raised false alarm rate.

15.5 Notes and References

Among various types of cyber-attacks, integrity attacks are specially directed to automatic control systems [31, 74]. By injecting attack signals into system input and output channels, e.g. via I/O and network interfaces, integrity attacks can lead to remarkable system performance degradations and even catastrophic damages. An early and reliable detection of integrity attacks is becoming a vital requirement on cyber-security of industrial CPSs, for instance, for power control systems [143]. Thanks to its well-established theoretical framework in the past three decades, observer-based fault detection technique [36] is widely accepted as an efficient method, among numerous ones, to deal with detection of integrity attacks on control systems [31, 74, 187]. Unfortunately, differing from technical faults, cyber-attacks are artificially created and can be designed and generated by an adversary. It is particularly insidious when cyber-attacks are generated in such a way that they cannot be detected using the known detection techniques. Such cyber-attacks are called stealthy. This observation and some real examples with stealthy cyber-attacks have strongly motivated researchers to improve the existing detection schemes and develop alternative solutions. In this regard, a great number of results have been reported about detecting the so-called replay, zero dynamics and covert attacks, which are stealthy integrity attacks as the standard observer-based fault detection technique cannot detect them without modifications on the applied algorithms [31, 74, 187]. Representative solutions are the watermark detection scheme [142], the moving target method [206] and the auxiliary system aided detection scheme [176], just citing the initial works on these methods. Motivated by the above observation and in particular driven by those questions, amongst others, what is the general form of stealthy integrity attacks? What are the existence conditions for such stealthy integrity attacks? Is it possible to develop a general observer-based scheme applied to detecting integrity attacks in automatic control systems? Satisfactory answers to these questions help us (i) to reveal possible weaknesses of observer-based detection technique by dealing with integrity cyber-attacks, and thus (ii) to prevent new variations of stealthy integrity attacks, and (iii) to develop new detection schemes, in particular such ones that are able to detect major types of integrity attacks. The main objective of our work is to investigate possible answers to the above questions. Different from the reported studies, our work studies the issues of stealthy integrity attacks in the unified framework of control and detection.

At first, a uniform model form of cyber-attacks on automatic control systems is introduced. Specifically, not only attacks that are injected into the system as unknown inputs, but also attacks that build a loop with the system variables are under consideration. The latter type of attacks are particularly critical, because they may cause remarkable changes in the system eigen-dynamic.

Against the background of the current research on stealthy attacks, kernel attacks are introduced. Distinctly differing from typical stealthy attacks addressed in the literature, kernel attacks are defined in terms of system input and output signals in CPSs, which lie in the system kernel subspace. Consequently, the existence of kernel attacks is independent of specifications of a detection system applied for detection purpose. On this basis, relations of kernel attacks and typical stealthy attacks addressed in the literature are explored, and existence conditions of kernel attacks are examined with respect to various configurations of CPSs. To be specific,

- the existence conditions for additive kernel attacks have been proved in Theorem 15.3,
- Theorem 15.5 provides the existence conditions of multiplicative kernel attacks and their parameterisation form, while Corollary 15.1 gives the general form of additive and multiplicative kernel attacks, when the CPS is subject to (15.1) and (15.9).
- In the case that the CPS is subject to (15.1) and (15.10), the existence conditions of multiplicative kernel attacks have been proved in Theorem 15.8.

It is noteworthy that the aforementioned explorations on the kernel attacks are based on a comprehensive analysis of the system dynamics under cyber-attacks in the unified framework of control and detection. The results in Theorems 15.2, 15.4, 15.6–15.7 and their proofs have reassured that

- the additive cyber-attacks act on the system image subspace, and
- the multiplicative cyber-attacks affect both the image and residual subspaces.

In Sect. 15.3, cyber-attack detection methods have been investigated. At the beginning, application of standard observer-based detection systems running on at the MC-station to attack detection is examined. It has been showcased that it is impossible to (i) distinguish r_y from $\left(a_{u_{MC}}, a_y\right)$ that makes simultaneous detection of faults and cyber-attacks pretty difficult, and (ii) detect the attacks in the data space (u_{est}, y^a) at the MC-station side. These observations motivate our subsequent work on alternative detection methods. The first alternative approach is based on the Bezout identity and the associated one-to-one mapping from the process input and output data (u, y) to the residual pair $\left(r_u, r_y\right)$. It has been demonstrated that the use of $\left(r_u, r_y\right)$ transmitted from the plant side to the MC-station (i) enables a decoupling of attacks $\left(a_{u_{MC}}, a_{r_u}\right)$ from faults in the plant, and (ii) allows a reliable fault detection, when the CPS is not under cyber-attack a_{r_y}. It is remarkable that the use of the residual r_u for attack detection has been firstly proposed in [45]. On the other hand, it has been showcased that the proposed CPS configuration with a fixed state feedback gain is inflexible and often cannot meet requirements for high resilience on CPSs. Moreover, it is facile for the attacker to design kernel attacks. To deal with these issues, a cyber-attack and fault detection scheme and a resilient CPS configuration have been presented in Sect. 15.4. The basic idea of the proposed schemes is to (i) encode the residual signal r_u by

$$r_u = X_1 u + Y_1 y + \Gamma r_y$$

with the LCP (X_1, Y_1) and matrix Γ that are to be designed, and (ii) embed r_u into the controller. It is worth mentioning that the latter measure is to enhance the resilience to cyber-attacks and has no impact on the feedback control performance. At the end of this section, algorithms are described dedicated to the design of detectors and controllers on the basis of the proposed CPS configurations.

Chapter 16
Resilient and Fault-Tolerant Control of CPSs Under Cyber-Attacks

Reviewing the literature on resilient control of CPSs leaves an impression that there exists, amongst a number of research topics, a strong focus on

- attack detection,
- state estimation of CPSs under attacks, and
- event-triggered control of CPSs under attacks.

Closely related to the first two topics, control-theory aided design of cyber-attacks seems to be drawing incredible attention. In this regard, issues related to stealthy attacks are of particular research interest. Resilient control and fault-tolerant control (FTC) are common in their line of action, consisting of two main functional units, detection and countermeasure as reaction. On the other hand, distinguishing differences between them can be well recognised. While the integrated design of fault detection and tolerant control is state of the art of advanced FTC technique, most of studies on resilient control of CPSs address detection, estimation and control issues separately. A further evident difference is that feedback mechanism is the most commonly used control strategy in FTC to enhance the system tolerance, which distinguishes, in its working principle, from event-triggered control schemes whose core is resource management.

An interesting observation of the research works in the domain of resilient control of CPSs is that resilient control schemes reported in the literature have been often developed on the basis of the well-established classic feedback control system configuration. Our study on the unified framework of control and detection in Part I and on the use of the residual signal of controller r_u naturally triggers our inquisitiveness for alternative solutions and modifications on system configuration. To be specific, embedding the residual signal r_u both in detection and control mechanisms is the centrepiece of our effort towards an integrated design of a CPS control system that is capable of detecting faults and cyber-attacks and, simultaneously enables resilient and fault-tolerant control of the CPS under attacks or faults. It is the objective of this chapter.

S. X. Ding, *Unified Framework of Control and Detection*,
https://doi.org/10.1007/978-3-662-72417-0_16

Besides detection and control functional considerations in the design of CPS configuration, specifications concerning safe and secure operation of CPSs will be taken into account. These include,

- "fail-safe" cyber-security,
- the data privacy, and
- online computation and communication effort.

Since fault diagnosis and FTC methods are comprehensively presented in the previous parts, and most of the reported methods are technically mature, the main work in this chapter is dedicated to (i) attack detection, including detection of stealthy attacks, (ii) resilient control of CPSs under attacks, and (iii) interactions of fault detection, FTC and attack detection and resilient control, as well as the associated interfaces.

This chapter comprises two parts. The first part is devoted to potential control schemes that aim to repress control performance degradation caused by cyber-attacks using the residual signal r_u. The proposed control scheme with feedback of r_u will then be integrated in the design of a resilient and fault-tolerant control system for CPSs. The main efforts in this part are dedicated to detection schemes of cyber-attacks, including some types of stealthy attacks.

16.1 Basic Resilient Control Mechanisms

16.1.1 Motivation

Consider the closed-loop dynamic of a CPS in attack-free operation,

$$\begin{bmatrix} u \\ y \end{bmatrix} = \begin{bmatrix} M \\ N \end{bmatrix} V v + \left(\begin{bmatrix} -\hat{Y} \\ \hat{X} \end{bmatrix} + \begin{bmatrix} M \\ N \end{bmatrix} Q \right) r_y,$$

and remember the essential result that any LTI control schemes are well parametrised by the feedforward controller V and feedback system Q. Moreover, recall the insightful conclusions drawn from our study in the previous chapter that

- the residual signal r_y is fully decoupled from any cyber-attacks,
- the closed-loop dynamic under any cyber-attacks is fully modelled by

$$\begin{aligned} \begin{bmatrix} u \\ y \end{bmatrix} &= \begin{bmatrix} M \\ N \end{bmatrix} \left(v + \vartheta_a + \Pi_a \begin{bmatrix} v \\ r_y \end{bmatrix} \right) + \left(\begin{bmatrix} -\hat{Y} \\ \hat{X} \end{bmatrix} + \begin{bmatrix} M \\ N \end{bmatrix} Q \right) r_y & (16.1) \\ &= \begin{bmatrix} M \\ N \end{bmatrix} \left(v + \vartheta_a + \Pi_{a,v} v \right) + \begin{bmatrix} M \\ N \end{bmatrix} \left(\Pi_{a,r_y} + Q \right) r_y + \begin{bmatrix} -\hat{Y} \\ \hat{X} \end{bmatrix} r_y & (16.2) \end{aligned}$$

with ϑ_a and $\Pi_a = \begin{bmatrix} \Pi_{a,v} & \Pi_{a,r_y} \end{bmatrix}$ parametrising additive and multiplicative types of attacks schematically.

From the information point of view, system/process faults and attacks are uncertainties that enter into the CPS closed-loop through two independent residual signal channels, the former one via the channel of the residual r_y, and the latter via the r_u channel. The response description (16.1) suggests that an effective control mechanism to suppress the impairment of system performance provoked by cyber-attacks should act on the r_u channel directly. When attacks are exclusively targeted at the feedback control dynamic, attacks will be injected into the system in form of a multiplicative attack, as indicated in the second term of the expression (16.2).

Feedforward and feedback are two essential control mechanisms. In the previous chapter, application of the residual r_u in the form of a feedforward control to enhance resilience against attacks in CPSs has been demonstrated. In this section, we delineate two basic control mechanisms in a compact form that will serve for the design of resilient CPSs discussed in the subsequent sections.

16.1.2 Models and Closed-Loop Dynamics

We begin with a short description of the system model forms under consideration and the analysis of the associated closed-loop dynamics. Suppose that the plant is modelled by

$$y(z) = G(z)u(z) + \hat{M}^{-1}(z)r_y(z) \tag{16.3}$$

with the state space representation

$$\begin{cases} \hat{x}(k+1) = A\hat{x}(k) + Bu(k) + LW^{-1}r_y(k) \\ r_y(k) = W\left(y(k) - \hat{y}(k)\right), \hat{y}(k) = C\hat{x}(k) + Du(k) \\ y(k) = \hat{y}(k) + W^{-1}r_y(k). \end{cases} \tag{16.4}$$

The preliminary controller is given in the observer-based form of Youla parameterisation subject to

$$\begin{cases} u(k) = F\hat{x}(k) + r_{y,Q}(k) + \eta(k) \\ r_{y,Q}(z) = Q(z)\left[-\hat{N}(z)\ \hat{M}(z)\right]\begin{bmatrix} u(z) \\ y(z) \end{bmatrix}, Q(z) \in \mathcal{RH}_\infty, \end{cases} \tag{16.5}$$

where η is a vector and, from the viewpoint of the plant side, is handled as an unknown input. Viewed from the MC-station, η consists of two components, represented by

$$\eta = Q_{\eta_0}\eta_0 + Q_{\eta_\delta}\eta_\delta,$$

where η_0 denotes the known input like the reference signal or/and control feedback signal, while η_δ represents an unknown input that can be cyber-attacks. In a conventional CPS control system, it is natural that Q_{η_0}, Q_{η_δ} are not specified and thus equivalent to a unit matrix. Our endeavour to develop a resilient control scheme is formulated as increasing the ratio R_{η_0/η_δ},

$$R_{\eta_0/\eta_\delta} = \frac{\left\| Q_{\eta_\delta} \eta_0 \right\|}{\left\| Q_{\eta_0} \eta_\delta \right\|},$$

by designing Q_{η_0} and Q_{η_δ}. Note that, without considering model uncertainties, a feedback controller is incapable of suppressing η. In other words, the design of the cyber-attack resilient controller and the feedback controller (for system control performance) can be addressed separately.

It is known that (16.5) leads to

$$\begin{aligned} \left(X + V^{-1} Q \hat{N}\right) u + \left(Y - V^{-1} Q \hat{M}\right) y = V^{-1} \eta \Longrightarrow \\ u = -\left(X + V^{-1} Q \hat{N}\right)^{-1} \left(Y - V^{-1} Q \hat{M}\right) y + \left(X + V^{-1} Q \hat{N}\right)^{-1} V^{-1} \eta, \\ r_u = X u + Y y = V^{-1} \left(r_{y,Q} + \eta\right). \end{aligned}$$

Here, V is the constant matrix given in (2.25)–(2.26), an algebraic pre-filter.

Remark 16.1 Without loss of generality, it is assumed $V = I$, $W = I$ in our subsequent study.

It follows from (16.5) that

$$r_{y,Q} + \eta = u - F \hat{x} = r_u.$$

Subsequently, the closed-loop dynamic can be equivalently modelled by

$$\begin{cases} \hat{x}(k+1) = (A + BF) \hat{x}(k) + B r_u(k) + L r_y(k) \\ \begin{bmatrix} u(k) \\ y(k) \end{bmatrix} = \begin{bmatrix} F \\ C + DF \end{bmatrix} \hat{x}(k) + \begin{bmatrix} I \\ D \end{bmatrix} r_u(k) + \begin{bmatrix} 0 \\ I \end{bmatrix} r_y(k). \end{cases} \tag{16.6}$$

That means, the system is driven by the both residual signals, r_u and r_y, which can be further written as

$$\begin{aligned} \begin{bmatrix} u \\ y \end{bmatrix} &= \begin{bmatrix} X + Q\hat{N} & Y - Q\hat{M} \\ -\hat{N} & \hat{M} \end{bmatrix}^{-1} \begin{bmatrix} \eta \\ r_y \end{bmatrix} \\ &= \begin{bmatrix} M \\ N \end{bmatrix} \left(X + Q\hat{N}\right) \eta + \begin{bmatrix} -\hat{Y} + MQ \\ \hat{X} + NQ \end{bmatrix} r_y, \end{aligned}$$

when r_y and η are unknown inputs and independent of (u, y).

In the case that model uncertainties are concerned, for instance, modelled by

$$y = (N + \Delta_N)(M + \Delta_M)^{-1} u + d, \tag{16.7}$$

where d is an unknown input vector, $(\Delta_M, \Delta_N) \in \mathcal{RH}_\infty$ are model uncertainties, the residual signal r_y in the plant model (16.3) is subject to

$$r_y = \begin{bmatrix} -\hat{N} & \hat{M} \end{bmatrix} \begin{bmatrix} u \\ y \end{bmatrix} = \begin{bmatrix} -\hat{N} & \hat{M} \end{bmatrix} \begin{bmatrix} \Delta_M \\ \Delta_N \end{bmatrix} \zeta + \hat{M} d,$$
$$\zeta = (M + \Delta_M)^{-1} u \Longleftrightarrow u = (M + \Delta_M) \zeta \Longrightarrow$$
$$y = \hat{M}^{-1} \hat{N} u + \hat{M}^{-1} \begin{bmatrix} -\hat{N} & \hat{M} \end{bmatrix} \begin{bmatrix} \Delta_M \\ \Delta_N \end{bmatrix} \zeta + d.$$

Here, ζ is some $\mathcal{L}_2$-bounded vector and serves as a hidden variable. Moreover, it holds

$$\begin{bmatrix} u \\ y \end{bmatrix} = \left(\begin{bmatrix} M \\ N \end{bmatrix} + \begin{bmatrix} \Delta_M \\ \Delta_N \end{bmatrix} \right) \zeta + \begin{bmatrix} 0 \\ d \end{bmatrix},$$
$$\left(X + Q\hat{N} \right) u + \left(Y - Q\hat{M} \right) y = \eta \Longrightarrow$$
$$\zeta = \left(I + \begin{bmatrix} X + Q\hat{N} & Y - Q\hat{M} \end{bmatrix} \begin{bmatrix} \Delta_M \\ \Delta_N \end{bmatrix} \right)^{-1} \left(\eta - \left(Y - Q\hat{M} \right) d \right).$$

By the following routine calculations,

$$\begin{bmatrix} u \\ y \end{bmatrix} = \begin{bmatrix} I & -K \\ -G & I \end{bmatrix}^{-1} \begin{bmatrix} \left(X + Q\hat{N} \right)^{-1} \eta \\ \hat{M}^{-1} r_y \end{bmatrix}$$
$$= \begin{bmatrix} X + Q\hat{N} & Y - Q\hat{M} \\ -\hat{N} & \hat{M} \end{bmatrix}^{-1} \begin{bmatrix} \eta \\ \begin{bmatrix} -\hat{N} & \hat{M} \end{bmatrix} \begin{bmatrix} \Delta_M \\ \Delta_N \end{bmatrix} \zeta + \hat{M} d \end{bmatrix}$$
$$= \begin{bmatrix} M & -\hat{Y} + MQ \\ N & \hat{X} + NQ \end{bmatrix} \begin{bmatrix} \eta \\ \Psi_2 (I + \Psi_1)^{-1} \left(\eta - \left(Y - Q\hat{M} \right) d \right) + \hat{M} d \end{bmatrix},$$
$$\Psi = \begin{bmatrix} \Psi_1 \\ \Psi_2 \end{bmatrix} = \begin{bmatrix} X + Q\hat{N} & Y - Q\hat{M} \\ -\hat{N} & \hat{M} \end{bmatrix} \begin{bmatrix} \Delta_M \\ \Delta_N \end{bmatrix},$$

we have, as shown in Sect. 4.7 as well,

$$\begin{bmatrix} u \\ y \end{bmatrix} = \begin{bmatrix} M \\ N \end{bmatrix} \eta + \begin{bmatrix} -\hat{Y} + MQ \\ \hat{X} + NQ \end{bmatrix} \Psi_2 (I + \Psi_1)^{-1} \eta$$
$$+ \begin{bmatrix} -\hat{Y} + MQ \\ \hat{X} + NQ \end{bmatrix} \left(\hat{M} - \Psi_2 (I + \Psi_1)^{-1} \left(Y - Q\hat{M} \right) \right) d.$$

It is obvious that

- tuning the parameterisation system Q can efficiently reduce

$$\left\| \begin{bmatrix} -\hat{Y} + MQ \\ \hat{X} + NQ \end{bmatrix} \right\|_\infty$$

  and thus suppress the impact of d and the model uncertainty on the nominal closed-loop dynamic. It is, on the other hand, rarely effective for handling the dynamic with respect to η due to the first term in the above description of the closed-loop dynamic,
- the closed-loop is stable, when

$$\|\Psi_1\|_\infty < 1.$$

  In our subsequent investigation, as done in most of our works in the previous chapters, it is assumed that

$$\|\Psi\|_\infty < 1,$$

  which is a sufficient condition for $\|\Psi_1\|_\infty < 1$.

16.1.3 A Resilient Control Mechanism

The aforementioned discussion adumbrates that feedback of the residual signal r_u is capable of dealing with unknown inputs affecting the plant image subspace. To this end, extending the controller by adding $\bar{Q}_u r_u$ to

$$\begin{aligned} u &= F\hat{x} + r_{y,Q} + \eta + \bar{Q}_u r_u, \ \bar{Q}_u \in \mathcal{RH}_\infty, \\ r_u &= Xu + Yy = u - F\hat{x}, \end{aligned} \tag{16.8}$$

yields

$$\begin{aligned} \left(I - \bar{Q}_u\right) r_u = r_{y,Q} + \eta &\Longrightarrow r_u = \left(I - \bar{Q}_u\right)^{-1} \left(r_{y,Q} + \eta\right) \\ &\Longrightarrow u = F\hat{x} + \left(I - \bar{Q}_u\right)^{-1} \left(r_{y,Q} + \eta\right). \end{aligned}$$

Setting

$$Q_u^{-1} = I - \bar{Q}_u \in \mathcal{RH}_\infty$$

results in finally

$$u = F\hat{x} + Q_u \left(r_{y,Q} + \eta\right). \tag{16.9}$$

The control law (16.9) gives the interpretation of this control mechanism in the context of a feedforward controller Q_u, realised by using the residual signal r_u. On the other hand, the control law (16.8) can be written into

$$\begin{aligned}
\left(I - \bar{Q}_u\right)(Xu + Yy) &= r_{y,Q} + \eta \Longrightarrow \\
\left(Q_u^{-1}X + Q\hat{N}\right)u + \left(Q_u^{-1}Y - Q\hat{M}\right)y &= \eta \Longrightarrow \\
u &= Ky + \left(Q_u^{-1}X + Q\hat{N}\right)^{-1}\eta, \\
K &= -\left(Q_u^{-1}X + Q\hat{N}\right)^{-1}\left(Q_u^{-1}Y - Q\hat{M}\right),
\end{aligned}$$

an output feedback controller. Since

$$\begin{aligned}
K &= -\left(Q_u^{-1}X + Q\hat{N}\right)^{-1}\left(Q_u^{-1}Y - Q\hat{M}\right) \\
&= -\left(X + Q_y\hat{N}\right)^{-1}\left(Y - Q_y\hat{M}\right),
\end{aligned}$$

with $Q_y = Q_u Q$, this is the Youla parameterisation of the feedback controller. In other words, Q_u can be viewed as a redundant system. We firstly discuss this aspect.

As a result of the Youla parameterisation, the closed-loop dynamic is governed by

$$\begin{aligned}
\begin{bmatrix} u \\ y \end{bmatrix} &= \begin{bmatrix} X + Q_y\hat{N} & Y - Q_y\hat{M} \\ -\hat{N} & \hat{M} \end{bmatrix}^{-1}\begin{bmatrix} Q_u\eta \\ r_y \end{bmatrix} \\
&= \begin{bmatrix} M \\ N \end{bmatrix} Q_u\eta + \begin{bmatrix} -\hat{Y} + MQ_y \\ \hat{X} + NQ_y \end{bmatrix} r_y.
\end{aligned}$$

It is apparent that tuning Q_y and Q_u can repress the impact of r_y and η on the loop dynamic, respectively. More interestingly, Q_y and Q_u can be designed separately.

Now, we devote ourselves to the closed-loop dynamic under the existence of the right coprime factor uncertainty described in (16.7). It turns out

$$\begin{aligned}
u &= F\hat{x} + Q_u\left(r_{y,Q} + \eta\right) \Longrightarrow \\
\left(X + Q_y\hat{N}\right)u &+ \left(Y - Q_y\hat{M}\right)y = Q_u\eta \Longrightarrow \\
\zeta &= \left(I + \begin{bmatrix} X + Q_y\hat{N} & Y - Q_y\hat{M} \end{bmatrix}\begin{bmatrix} \Delta_M \\ \Delta_N \end{bmatrix}\right)^{-1}\left(Q_u\eta - \left(Y - Q_y\hat{M}\right)d\right), \\
r_y &= \begin{bmatrix} -\hat{N} & \hat{M} \end{bmatrix}\begin{bmatrix} \Delta_M \\ \Delta_N \end{bmatrix}\zeta + \hat{M}d \\
&= \Psi_2\left(I + \Psi_1\right)^{-1}\left(Q_u\eta - \left(Y - Q_y\hat{M}\right)d\right) + \hat{M}d,
\end{aligned}$$

$$\Psi = \begin{bmatrix} \Psi_1 \\ \Psi_2 \end{bmatrix} = \begin{bmatrix} X + Q_y \hat{N} & Y - Q_y \hat{M} \\ -\hat{N} & \hat{M} \end{bmatrix} \begin{bmatrix} \Delta_M \\ \Delta_N \end{bmatrix}.$$

Accordingly, the closed-loop dynamic is as follows:

$$\begin{aligned} \begin{bmatrix} u \\ y \end{bmatrix} &= \begin{bmatrix} X + Q_y \hat{N} & Y - Q_y \hat{M} \\ -\hat{N} & \hat{M} \end{bmatrix}^{-1} \begin{bmatrix} Q_u \eta \\ r_y \end{bmatrix} \\ &= \begin{bmatrix} M \\ N \end{bmatrix} Q_u \eta + \begin{bmatrix} -\hat{Y} + M Q_y \\ \hat{X} + N Q_y \end{bmatrix} \Psi_2 (I + \Psi_1)^{-1} Q_u \eta \\ &+ \begin{bmatrix} -\hat{Y} + M Q_y \\ \hat{X} + N Q_y \end{bmatrix} \left(\hat{M} - \Psi_2 (I + \Psi_1)^{-1} \left(Y - Q_y \hat{M} \right) \right) d. \end{aligned} \tag{16.10}$$

As expected, the feedback of the residual signal r_u is useful to handle the influence of η. It is also interesting to observe that tuning Q_y to reduce

$$\left\| \begin{bmatrix} -\hat{Y} + M Q_y \\ \hat{X} + N Q_y \end{bmatrix} \right\|_\infty$$

increases the CPS robustness against both d and η.

Example 16.1 The aforementioned resilient control scheme is illustrated in this example. Under consideration is the uncertainty model (16.7), where the model uncertainty (Δ_M, Δ_N) is specified by

$$\begin{bmatrix} \Delta_M \\ \Delta_N \end{bmatrix} = \begin{bmatrix} -\hat{Y} \\ \hat{X} \end{bmatrix} S, \; S \in \mathcal{RH}_\infty.$$

As is known, given the default controller (16.5), the closed-loop dynamic is governed by

$$\begin{aligned} \begin{bmatrix} u \\ y \end{bmatrix} &= \begin{bmatrix} M \\ N \end{bmatrix} \eta + \begin{bmatrix} -\hat{Y} + M Q \\ \hat{X} + N Q \end{bmatrix} S (I - QS)^{-1} \eta \\ &+ \begin{bmatrix} -\hat{Y} + M Q \\ \hat{X} + N Q \end{bmatrix} \left(\hat{M} - S (I - QS)^{-1} \left(Y - Q \hat{M} \right) \right) d. \end{aligned}$$

Now, applying the resilient controller (16.9) results in

$$\begin{aligned} \begin{bmatrix} u \\ y \end{bmatrix} &= \begin{bmatrix} M \\ N \end{bmatrix} Q_u \eta + \begin{bmatrix} -\hat{Y} + M Q_y \\ \hat{X} + N Q_y \end{bmatrix} S \left(I - Q_y S \right)^{-1} Q_u \eta \\ &+ \begin{bmatrix} -\hat{Y} + M Q_y \\ \hat{X} + N Q_y \end{bmatrix} \left(\hat{M} - S \left(I - Q_y S \right)^{-1} \left(Y - Q_y \hat{M} \right) \right) d. \end{aligned} \tag{16.11}$$

Hence, the design problem can be finally formulated as: given $\alpha > 0$, $\beta > 0$,

- find Q_y such that

$$\left\| \begin{bmatrix} -\hat{Y} + MQ_y \\ \hat{X} + NQ_y \end{bmatrix} \right\|_\infty \leq \alpha, \left(I - Q_y S\right)^{-1} \in \mathcal{RH}_\infty,$$

- find Q_u so that

$$\left\| \left(I - Q_y S\right)^{-1} Q_u \right\|_\infty \leq \beta.$$

16.1.4 A Resilient Control Mechanism by Means of a Feedback over r_u Channel

A clear disadvantage of the aforementioned control mechanism is that it cannot significantly increase the R_{η_0/η_δ} ratio,

$$R_{\eta_0/\eta_\delta} = \frac{\left\| Q_{\eta_\delta} \eta_0 \right\|}{\left\| Q_{\eta_0} \eta_\delta \right\|} = \frac{\left\| Q_u \eta_0 \right\|}{\left\| Q_u \eta_\delta \right\|} \approx \frac{\left\| \eta_0 \right\|}{\left\| \eta_\delta \right\|}.$$

Differing from it, the main objective of the control scheme proposed below is to cushion the unknown input term η_δ. The basic idea of this scheme is to integrate the information relevant to η_δ into the feedback signal and transmit it over the network.

We begin with the CPS configuration equipped with the controller (16.8), whose dynamic is subject to

$$\begin{bmatrix} u \\ y \end{bmatrix} = \begin{bmatrix} M \\ N \end{bmatrix} Q_u \left(\eta_0 + \eta_\delta\right) + \begin{bmatrix} -\hat{Y} + M\bar{Q} \\ \hat{X} + N\bar{Q} \end{bmatrix} r_y.$$

Inspired by the idea of feeding back the residual r_u, construct the residual signal $r_{y,u}$,

$$r_{y,u} := Q_{r,1} r_y + Q_{r,2} r_u. \tag{16.12}$$

on the plant side, and feed it back to the MC-station. Note that $r_{y,u}$ comprises information about possible faults (in r_y), the unknown input η_δ and the signal η_0 that is known on the side of the MC-station, where $\left(Q_{r,1}, Q_{r,2}\right) \in \mathcal{RH}_\infty$ are design parameters to be determined. For our purpose, designate r_u by

$$r_u = Q r_y + Q_{\eta_0} \eta_0 + Q_{\eta_\delta} \eta_\delta,$$

for some $\left(Q, Q_{\eta_0}, Q_{\eta_\delta}\right) \in \mathcal{RH}_\infty$. Now, set the control input u_{MC} (at the MC-station) as

$$\begin{aligned} u_{MC} &= r_{y,u} - Q_{r,2} Q_{\eta_0} \eta_0 = Q_{r,1} r_y + Q_{r,2} r_u - Q_{r,2} Q_{\eta_0} \eta_0 \\ &= Q_{r,1} r_y + Q_{r,2} \left(u_{MC} + \eta \right) - Q_{r,2} Q_{\eta_0} \eta_0 \\ \Longrightarrow u_{MC} &= \left(I - Q_{r,2} \right)^{-1} \left(Q_{r,1} r_y + Q_{r,2} \eta - Q_{r,2} Q_{\eta_0} \eta_0 \right). \end{aligned}$$

Accordingly, the control signal u is subject to

$$\begin{aligned} u &= F\hat{x} + u_{MC} + \eta \\ &= F\hat{x} + \left(I - Q_{r,2} \right)^{-1} \left(Q_{r,1} r_y + Q_{r,2} \eta - Q_{r,2} Q_{\eta_0} \eta_0 \right) + \eta \Longrightarrow \\ r_u &= \left(I - Q_{r,2} \right)^{-1} Q_{r,1} r_y + \left(I + \left(I - Q_{r,2} \right)^{-1} Q_{r,2} \right) \eta_\delta \\ &+ \left(I + \left(I - Q_{r,2} \right)^{-1} Q_{r,2} \left(I - Q_{\eta_0} \right) \right) \eta_0, \end{aligned}$$

which implies

$$\begin{aligned} Q &= \left(I - Q_{r,2} \right)^{-1} Q_{r,1}, \ Q_{\eta_0} = I, \ Q_{\eta_\delta} = \left(I - Q_{r,2} \right)^{-1} \Longrightarrow \\ r_u &= Q r_y + Q_{\eta_0} \eta_0 + Q_{\eta_\delta} \eta_\delta = Q r_y + \eta_0 + \left(I - Q_{r,2} \right)^{-1} \eta_\delta. \end{aligned}$$

As a result, the closed-loop dynamic is governed by

$$\begin{bmatrix} u \\ y \end{bmatrix} = \begin{bmatrix} M \\ N \end{bmatrix} \left(\eta_0 + \left(I - Q_{r,2} \right)^{-1} \eta_\delta \right) + \begin{bmatrix} -\hat{Y} + MQ \\ \hat{X} + NQ \end{bmatrix} r_y. \tag{16.13}$$

Hence, the design problem is formulated as:

- selecting $Q_{\eta_\delta} \in \mathcal{RH}_\infty$ so that

  $$Q_{\eta_\delta}^{-1} \in \mathcal{RH}_\infty, \ \left\| Q_{\eta_\delta} \right\|_\infty \leq \gamma, \gamma << 1,$$

  and setting

  $$Q_{r,2} = I - Q_{\eta_\delta}^{-1} \in \mathcal{RH}_\infty,$$

- setting

  $$Q_{r,1} = \left(I - Q_{r,2} \right) Q$$

  so that the optimal r_y rejection is achieved.

It is apparent that the proposed feedback mechanism assures a significant increase of the ratio

$$\frac{\|\eta_0\|_2}{\left\| \left(I - Q_{r,2} \right)^{-1} \eta_\delta \right\|_2} \geq \frac{1}{\gamma} \frac{\|\eta_0\|_2}{\|\eta_\delta\|_2} >> \frac{\|\eta_0\|_2}{\|\eta_\delta\|_2}, \tag{16.14}$$

without a change of the rejection feedback performance.

The following algorithm summarises the major steps for the online implementation of the above resilient feedback control scheme.

Algorithm: *Resilient feedback control over the r_u channel*
Step 1: Generate the residuals $\left(r_y, r_u\right)$ on the plant side,

$$\begin{aligned} r_y &= \hat{M} y - \hat{N} u, \\ r_u &= u - F\hat{x} = Xu + Yy; \end{aligned}$$

Step 2: Build the residual $r_{y,u}$ on the plant side,

$$r_{y,u} = Q_{\eta_\delta}^{-1} Q r_y + \left(I - Q_{\eta_\delta}^{-1}\right) r_u;$$

Step 3: Construct u_{MC} at the MC-station,

$$u_{MC} = r_{y,u} - \left(I - Q_{\eta_\delta}^{-1}\right) \eta_0.$$

16.2 A Design Scheme for Resilient Control of CPSs

This section is devoted to the design of a resilient control system for CPSs with the configuration sketched in Fig. 15.1. The main design goals are

- high resilience of CPSs under cyber-attacks,
- high fault-tolerance (resilience).

Besides these, further aspects should be taken into account, including

- "fail-safe" cyber-security,
  That means, the control system is capable of working in a pre-defined safe operating state, when the network fails.
- the data privacy,
  The process data (u, y, v) should be prevented from possible eavesdropping.
- online computation and communication effort.

To this end, the detection and control schemes and methods introduced in the unified framework of control and detection and in the previous section serve as the methodological basis.

16.2.1 Control System Configuration and Information Pattern

On the plant side, it is supposed that the process model of the plant is described by (16.3). The control and detection system on this side is composed of (i) an observer of the form (16.4) that builds the system core, and based on it, (ii) an observer-based state feedback controller $F\hat{x}$ and two residual generators,

$$\begin{aligned} r_y &= \hat{M}y - \hat{N}u, \\ r_u &= u - F\hat{x} = Xu + Yy, \\ r_{y,u} &= Q_{r,1}r_y + Q_{r,2}r_u, \end{aligned} \tag{16.15}$$

where $\left(Q_{r,1}, Q_{r,2}\right) \in \mathcal{RH}_\infty$ are design parameters. The controller $F\hat{x}$ is a part of the overall control system, and assures the stable process operation, when the network communication fails. In this way, the required "fail-safe" cyber-security is guaranteed. The residual generators serve for various control and detection functionalities. The generated residual signals r_y and $r_{y,u}$ are used for detecting faults and cyber-attacks, respectively, and simultaneously, act as information providers for the feedback control of the plant and cyber-attacks. It is noteworthy that to preserve the data privacy, $r_{y,u}$ is transmitted over the communication network, instead of the process data (u, y). In this context, the residual generators pose as an encoder.

Concerning the information pattern, the available process data are (u, y) with

$$u = F\hat{x} + u_{MC}^a.$$

Here, u_{MC}^a is the signal received from the MC-station corrupted by cyber-attack $a_{u_{MC}}$. The residual signal $r_{y,u}$ is sent to the MC-station over the communication network. Moreover, when a fault is detected, the corresponding message is sent to the MC-station.

To put it in a nutshell, the main tasks of the control and detection system at the plant side are

- fault detection,
- implementation of an observer-based state feedback control that works independently of the communication network, and
- encoding the information needed for cyber-attack detection and controller u_{MC} as the residual signal $r_{y,u}$.

The control and detection system at the MC-station consists of control law u_{MC},

$$u_{MC} = Q_{u_{MC}} r_{y,u}^a + v, \tag{16.16}$$

where $Q_{u_{MC}} \in \mathcal{RH}_\infty$ is the design parameter, and

$$v = Q_v v_0, \ Q_v \in \mathcal{RH}_\infty, \tag{16.17}$$

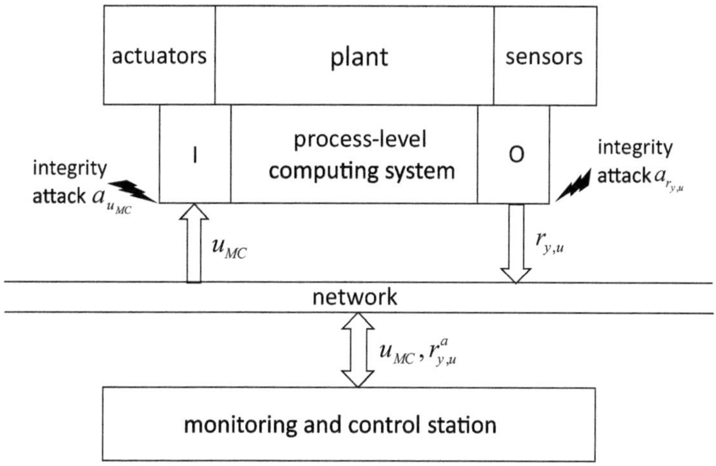

Fig. 16.1 A resilient CPS control system

with v_0 as the target reference signal and Q_v a pre-filter. The available data are $\left(v_0, r^a_{y,u}\right)$, where $r^a_{y,u}$ is received from the plant and corrupted by the cyber-attack $a_{r_{y,u}}$. The main tasks of the MC-station consist of

- detection of cyber-attacks based on the available data $\left(v_0, r^a_{y,u}\right)$,
- computation of Q and $Q_{u_{MC}}$ with regard to the various CPS operating states, amongst others, (i) control performance optimisation during fault- and attack-free operations, (ii) performance degradation recovery (PDR) in case of faulty-operation, and (iii) resilient control, when the CPS is under cyber-attacks. Switching between the control laws is triggered by the MC-station with respect to the system operation conditions.

The structural configuration of the resilient CPS is schematically sketched in Fig. 16.1.

Remark 16.2 Before beginning with system design, we would like to emphasise the essential role of the reference signal v considered in our CPS configuration sketched in Fig. 16.1, where v is included in u_{MC}. In fact, this is an elementary difference between the CPS configuration and the one mostly adopted in the literature on the resilient control of CPSs under attacks, where an observer or a Kalman filter-based state feedback controller

$$u(k) = F\hat{x}(k)$$

is under consideration. While such a control law with $v = 0$ is a prevalent practice in control theory dealing with feedback control loops, it is doubtful in CPS applications. In this regard, we would like to call the reader's attention to the following three aspects.

- Engineering practice: In an industrial CPS, the MC-station supervises and manages the CPS operation. The reference signal v is an essential signal that commands the CPS to perform a defined operation. In particular, such a reference signal depends on events and tasks and will often be online computed as a part of a decision made on the MC-station.
- Information pattern: As described before, the MC-station and the plant controller (at the plant side) have different information patterns. On account of the aforementioned real-time character of the reference signal, it is consequential that the plant controller possesses no real-time knowledge of v. The assumption $u(k) = F\hat{x}(k)$ implicitly means that the plant controller knows the reference signal online, i.e. $v = 0$. In fact, under such an information pattern specification, handling of resilient control of CPSs, both for attack detection and control, becomes simplified, but unrealistic and less engineering relevant.
- Control-theoretic handling: An immediate consequence of the assumption $u(k) = F\hat{x}(k)$ is that the closed-loop dynamic during attack-free operations is governed by

$$\begin{bmatrix} u \\ y \end{bmatrix} = \begin{bmatrix} -\hat{Y} \\ \hat{X} \end{bmatrix} r_y. \tag{16.18}$$

Considering that in the literature, the state space model

$$\begin{cases} x(k+1) = Ax(k) + Bu(k) + E_d d(k), x(0) = x_0 \\ y(k) = Cx(k) + Du(k) + F_d d(k), \end{cases}$$

is state of the art, where d represents (additive) disturbances, either process and measurement noises or $\mathcal{L}_2$-bounded unknown input vector, the closed loop-dynamic is described in the state space representation as

$$\begin{cases} x(k+1) = A_F x(k) + E_d d(k) - BFe(k), x(0) = x_0 \\ e(k+1) = A_L e(k) + (E_d - LF_d) d(k), e(0) = x_0 \\ y(k) = (C + DF) x(k) + DFe(k) + F_d d(k) \\ u(k) = F\hat{x}(k), \end{cases}$$
$$A_F = A + BF, A_L = A - LC,$$

where $\hat{x}(k)$ is the state estimate delivered by a full-order state observer, whose initial value is set, without loss of generality, equal to zero. Note that the above dynamic model is equivalent to

$$\begin{cases} \hat{x}(k+1) = A_F \hat{x}(k) + Lr_y(k), \hat{x}(0) = 0 \\ e(k+1) = A_L e(k) + (E_d - LF_d) d(k), e(0) = x_0 \\ r_y(k) = Ce(k) + F_d d(k) \\ y(k) = (C + DF) \hat{x}(k) + r_y(k) \\ u(k) = F\hat{x}(k), \end{cases}$$

which is the state space realisation of (16.18). In this regard, the nominal CPS dynamic (for $d = 0$) is often formulated as the system response to the initial state value, as reported in the literature. Note that in this case,

$$e(k+1) = A_L e(k), e(0) = x_0,$$

and thus the overall system dynamic can be equivalently described by

$$G_0(x_0) : \begin{cases} \begin{bmatrix} \hat{x}(k+1) \\ e(k+1) \end{bmatrix} = \begin{bmatrix} A_F & LC \\ 0 & A_L \end{bmatrix} \begin{bmatrix} \hat{x}(k) \\ e(k) \end{bmatrix} + \begin{bmatrix} 0 \\ A_L \end{bmatrix} x_0 \delta(k) \\ \begin{bmatrix} u_0(k) \\ y_0(k) \end{bmatrix} = \begin{bmatrix} F & 0 \\ C + DF & C \end{bmatrix} \begin{bmatrix} \hat{x}(k) \\ e(k) \end{bmatrix} \\ \delta(k) = \begin{cases} 1, k = 0 \\ 0, k \neq 0, \end{cases} e(0) = 0, \hat{x}(k) = 0, \end{cases}$$

$$\mathcal{R}_{x(0)} := \left\{ \begin{bmatrix} u_0 \\ y_0 \end{bmatrix}, \begin{bmatrix} u_0 \\ y_0 \end{bmatrix} = G_0(x_0), x_0 \in \mathbb{R}^n \right\}$$

The subspace $\mathcal{R}_{x(0)}$ is apparently different from the nominal system dynamic described in our framework in terms of the plant image subspace,

$$\left\{ \begin{bmatrix} u \\ y \end{bmatrix}, \begin{bmatrix} u \\ y \end{bmatrix} = \begin{bmatrix} M \\ N \end{bmatrix} v, v \in \mathcal{L}_2 \right\}.$$

One consequence is, for instance, that existence conditions have been reported in the literature on the stealthy attacks, which are different from ones presented in the last two chapters. Herewith, we would like to emphasise that in the unified framework of control and detection, initial conditions of state variables are addressed as uncertainty due to their unknown feature, which is apprehended by the residual signal r_y. This is indeed a prevalent practice in control theory, for instance in the Kalman filter theory. It is noteworthy as well that in the model setup under consideration, the response of (u, y) to the (unknown) state initial value x_0 can be neglected after the system operates under nominal conditions over a certain time interval. This phase is called calibration in engineering practice and is an essential phase of process operations.

16.2.2 System Design

We now begin with the centrepiece of our work, the design of detection and control sub-systems. Four sub-systems are to be designed, dedicated to the functionalities: (i) fault detection, (ii) cyber-attack detection, (iii) performance optimisation and recovery, and (iv) resilient control of CPS under cyber-attacks. The two issues (ii) and (iv) are the main focus of our study.

16.2.2.1 Fault Detection System Design

On account of the fact that the residual signal r_y is perfectly decoupled from cyber-attacks, an observer-based residual generation system running at the plant side is designed for detecting faults. Among the numerous methods introduced in the previous parts, following observer-based fault detection systems are proposed for our purpose,

- a Kalman-filter based fault detection system, described in Sect. 8.2.1, when noises are dominant in r_y,
- an $\mathcal{H}_2$ observer-based fault detection system, the so-called unified solution, which is well described in Sect. 8.2.2, when an unknown input vector d is under consideration, and
- a normalised SKR-based fault detection system that implies an $\mathcal{L}_2$ orthogonal projection-based residual generation, as elaborately described in Chap. 9, in the case of dominant model uncertainties.

The reader's attention is drawn here especially to the fact that these three methods not only deliver optimal fault detection performance, but also result in an orthogonal decomposition of the process data, as delineated in Chap. 9. To be specific,

- in the stochastic case, y is decomposed as

$$y = \hat{y} + r_y,$$

  where $\hat{y}$ is the output estimate delivered by the Kalman-filter, and the residual r_y is innovation,
- in the case of the unknown input d,

$$y = \hat{y} + r_y$$

  with $\hat{y}$ representing the output estimate given by the $\mathcal{H}_2$ observer, and
- for the plant with model uncertainties, the projection-based residual generation leads to

$$\begin{bmatrix} u \\ y \end{bmatrix} = \begin{bmatrix} \hat{u} \\ \hat{y} \end{bmatrix} + r_{\mathcal{I}_G}, \begin{bmatrix} \hat{u} \\ \hat{y} \end{bmatrix} = \mathcal{P}_{\mathcal{I}_G} \begin{bmatrix} u \\ y \end{bmatrix},$$
$$\left\| r_{\mathcal{I}_G} \right\|_2 = \left\| r_y \right\|_2,$$

  where $(\hat{u}, \hat{y})$ is the projection of the process data (u, y) onto the plant image subspace.

Recall that the estimates contain full information of the process model and the associated nominal system dynamic, and the generated residuals are of the property that

$$\begin{cases} \mathbb{E}\left(\hat{y}^T r_y\right) = 0, \text{ in the case of noises} \\ \left\langle \hat{y}, r_y \right\rangle = 0, \text{ in the case of unknown input} \\ \left\langle \begin{bmatrix} \hat{u} \\ \hat{y} \end{bmatrix}, r_{\mathcal{I}_G} \right\rangle = 0, \text{ in the case of model uncertainties.} \end{cases}$$

In other words, the residual signals are perpendicular to the estimates and thus do not comprise process information. That means, transmission of residual signals preserves the data privacy efficiently.

16.2.2.2 Cyber-Attack Detection

Cyber-attack detection is performed at the MC-station under the condition that no fault is detected (by the fault detection system at the plant side). To formulate the detection problem to be solved, the impact of cyber-attacks on $r^a_{y,u}$ is firstly examined. It follows from the relations,

$$\begin{aligned} r^a_{y,u} &= Q_{r,1} r_y + Q_{r,2} r_u + a_{r_{y,u}}, \\ r_u &= u - F\hat{x} = u_{MC} + a_{u_{MC}} \\ u_{MC} &= Q_{u_{MC}}\left(r^a_{y,u} - Q_{r,2} v\right) + v = Q_{u_{MC}} r^a_{y,u} + \left(I - Q_{u_{MC}} Q_{r,2}\right) v, \end{aligned}$$

and by some routine calculations that

$$r^a_{y,u} = \left(I - Q_{r,2} Q_{u_{MC}}\right)^{-1}\left(Q_{r,1} r_y + Q_{r,2} a_{u_{MC}} + a_{r_{y,u}}\right) + Q_{r,2} v, \tag{16.19}$$

from which it becomes apparent that the residual $r^a_{y,u}$ contains the necessary information for detecting the cyber-attacks $\left(a_{u_{MC}}, a_{r_{y,u}}\right)$. Let

$$\eta_a = Q_{r,2} a_{u_{MC}} + a_{r_{y,u}}, \tag{16.20}$$

$$r_{\eta_a} = \left(I - Q_{r,2} Q_{u_{MC}}\right)\left(r^a_{y,u} - Q_{r,2} v\right), \tag{16.21}$$

and assume either $r_y \sim \mathcal{N}(0, \Sigma_r)$ or $r_y = G_d d$, $G_d \in \mathcal{RH}_\infty$ is co-inner so that

$$\left\| r_y \right\|_2 \leq \delta_d,$$

where d is an unknown input vector with an $\mathcal{L}_2$ upper-bound δ_d. On account of the above definitions and assumptions, the cyber-attack detection problem is formulated as: given the residual model

$$r_{\eta_a} = Q_{r,1} r_y + \eta_a, \tag{16.22}$$

and a test statistic or a residual evaluation function

$$J = \left\| R r_{\eta_a} \right\|_2,$$

find the post-filter $R \in \mathcal{RH}_\infty$, a threshold J_{th}, as well as the conditions for $Q_{r,1}$, $Q_{r,2}$ so that the cyber-attack η_a can be detected using the detection logic

$$\text{detection logic}: \begin{cases} J - J_{th} > 0 \Longrightarrow \text{cyber-attacks} \\ J - J_{th} \leq 0 \Longrightarrow \text{no cyber-attack.} \end{cases} \tag{16.23}$$

Remark 16.3 Both $Q_{r,1}$, $Q_{r,2}$ are, together with $Q_{u_{MC}}$, design parameters for the controller design. In other words, the conditions determined by solving the above-defined detection problem are to be considered, when necessary, as the controller design is addressed. Moreover, although $Q_{r,2}$ is not explicitly included in the model (16.22), it is a part of the residual generator (16.15) for $r_{y,u}$, on which the residual r_{η_a} is built.

We would like to call the reader's attention to an essential difference between the two problems of fault detection and cyber-attack detection. Differing from typical technical faults caused by ageing or damage of system components, cyber-attacks are artificially generated signals injected into the system over a certain time interval. It is similar to the so-called intermittent faults. In this regard, the detection task comprises not only detecting injection of cyber-attacks as timely and reliable as possible, but also detecting switching-off of cyber-attacks. The latter is of a special importance for the reason that the cyber-attack resilient control law is to be switched to the nominal controller, once the injected cyber-attacks disappear. On account of these basic requirements, the task of detecting cyber-attacks is split into three sub-tasks,

- detecting cyber-attacks during system attack-free operations,
- monitoring the detected cyber-attacks, when the CPS is under cyber-attacks, and
- making a decision of the existence of cyber-attacks.

Recall that transmitting the residual signal r_u to the MC-station makes it easier to cope with detection of cyber-attacks. On the other hand, it implies increasing data transmissions. Furthermore, transmitting raw data r_u increases the risk of eavesdropping attacks that enable an easy generation of stealthy cyber-attacks. On account of these concerns, the residual signal $r_{y,u}$ as a fusion of r_y and r_u, as given in (16.15), is generated and transmitted over the network. In order to reduce the stress of data transmission, the dimension of $r_{y,u}$, and thus the demanded channel capacity, is limited by

$$\dim\left(r_{y,u}\right) = m.$$

In the sequel, the aforementioned three sub-tasks are addressed.

Detection of cyber-attacks during fault-free operations
Consider $r^a_{y,u}$, r_{η_a} and relation (16.19) that yields

$$r_{\eta_a} = \left(I - Q_{r,2}Q_{u_{MC}}\right)\left(r^a_{y,u} - Q_{r,2}v\right) = Q_{r,1}r_y + Q_{r,2}a_{u_{MC}} + a_{r_{y,u}}.$$

That is, r_{η_a} solely contains redundant information about the plant uncertainties comprised in r_y, when there exists no cyber-attack. Consequently, the detection scheme is designed depending on the specifications of the plant uncertainties. We begin with the case $r_y \sim \mathcal{N}\left(0, \Sigma_{r_y}\right)$ and suppose that the system r_{η_a} is operating in a stationary state for some $Q_{r,1}$, and $r_{\eta_a} \sim \mathcal{N}\left(0, \Sigma_{r_{\eta a}}\right)$. Here, $\Sigma_{r_{\eta a}}$ is a positive definite constant matrix. As described in Chap. 8, a standard detection algorithm for detecting cyber-attack η_a on the model assumption (16.22) is the χ^2 test statistic,

$$\chi^2(m) = r_{\eta_a}^T \Sigma_{r_{\eta a}}^{-1} r_{\eta_a} \underset{\mathcal{H}_1}{\overset{\mathcal{H}_0}{\lesseqgtr}} J_{th,\chi_\alpha^2} \tag{16.24}$$

with the null hypothesis $\mathcal{H}_0$ for attack-free, the alternative hypothesis $\mathcal{H}_1$ for attacks, and the threshold J_{th,χ_α^2} for the given FAR upper-bound α.

In general, a (steady state) Kalman filter is designed on the assumption that

$$Q_{r,1} = \left(A_{r,1}, B_{r,1}, C_{r,1}, D_{r,1}\right), \tag{16.25}$$

and $D_{r,1}$ is of full row rank. Recall that any observer (and so Kalman filter) can be equivalently realised by a post-filter. It follows from Lemma 3.1 that setting

$$R = \left(A_{r,1} - LC_{r,1}, -L, \Sigma_{r,1}^{-1} C_{r,1}, \Sigma_{r,1}^{-1}\right) \tag{16.26}$$

results in

$$\begin{aligned} RQ_{r,1} &= \left(A_{r,1} - LC_{r,1}, B_{r,1} - LD_{r,1}, \Sigma_{r,1}^{-1} C_{r,1}, \Sigma_{r,1}^{-1} D_{r,1}\right), \\ \bar{r}_{\eta_a} &:= RQ_{r,1} r_y \sim \mathcal{N}(0, I), \end{aligned}$$

where L is the Kalman filter gain satisfying

$$\begin{aligned} L &= \left(A_{r,1} P C_{r,1}^T + B_{r,1} \Sigma_{r_y} D_{r,1}^T\right) \Sigma_{r,1}^{-1}, \\ \Sigma_{r,1} &= C_{r,1} P C_{r,1}^T + D_{r,1} \Sigma_{r_y} D_{r,1}^T, \end{aligned}$$

and P is the solution of the following Riccati equation,

$$P - A_{r,1} P A_{r,1}^T - B_{r,1} \Sigma_{r_y} B_{r,1}^T + L \Sigma_{r,1} L^T = 0.$$

Consequently, the χ^2 test statistic

$$\chi^2(m) = \bar{r}_{\eta_a}^T \bar{r}_{\eta_a} \underset{\mathcal{H}_1}{\overset{\mathcal{H}_0}{\lesseqgtr}} J_{th,\chi_\alpha^2},$$

is adopted for the attack detection.

For the case,

$$r_y = G_d d, \left\|r_y\right\|_2 \leq \delta_d,$$

the analogous design procedure, namely the unified solution, is followed. Let

$$Q_{r,1} = Q_{r,1}^{o} Q_{r,1}^{i}$$

be a co-inner-outer factorisation of $Q_{r,1}$. Then,

$$R = \left(Q_{r,1}^{o}\right)^{-1} \Longrightarrow \bar{r}_{\eta_a} = R Q_{r,1} r_y = Q_{r,1}^{i} r_y \Longrightarrow$$
$$J = \left\|\bar{r}_{\eta_a}\right\|_2 \leq d \Longrightarrow J = \left\|\bar{r}_{\eta_a}\right\|_2 \underset{\mathcal{H}_1}{\overset{\mathcal{H}_0}{\lessgtr}} d.$$

Remember that

$$\begin{aligned} \eta_a &= Q_{r,2} a_{u_{MC}} + a_{r_{y,u}}, \\ r_{y,u}^{a} &= Q_{r,1} r_y + Q_{r,2} r_u. \end{aligned}$$

It is not to obviate that attackers may create stealthy attacks, i.e. the attack pair $\left(a_{u_{MC}}, a_{r_{y,u}}\right)$ is designed so that

$$Q_{r,2} a_{u_{MC}} + a_{r_{y,u}} = 0.$$

Note that in this case attackers should know $Q_{r,2}$. At the end of this chapter, we are going to address issues concerning stealthy attacks, their generation and detection.

Monitoring of cyber-attacks and detecting disappearance of cyber-attacks Given the residual model (16.22), our detection problem is formulated as: finding a post-filter R, a test statistic J as well as a threshold so that switching off of cyber-attacks is optimally detected. To simplify our study, the following assumptions are made,

- $r_y \sim \mathcal{N}\left(0, \Sigma_{r_y}\right)$,
- $Q_{r,1}$ is subject to (16.25),
- residual data $r_y\left(k_0+1\right), \ldots, r_y\left(k_0+s\right)$ are available for the detection purpose, where s is an integer and serves as a hyperparameter.

On account of our resilient control strategy, "weak attacks" can be well tolerated. In this context, it is said

$$\begin{cases} \left\|\eta_a\right\|_{RMS,[k+1,k+\tau]} \leq L_{\eta_a,l} \Longrightarrow \text{attack-free} \\ \left\|\eta_a\right\|_{RMS,[k+1,k+\tau]} \geq L_{\eta_a,u} \Longrightarrow \text{cyber-attack}. \end{cases} \tag{16.27}$$

Here, $\left\|\eta_a\right\|_{RMS,[k+1,k+\tau]}$ is the RMS of η_a defined by

$$\left\|\eta_a\right\|_{RMS,[k+1,k+\tau]} := \sqrt{\frac{1}{\tau} \sum_{i=k+1}^{k+\tau} \eta_a^T(i)\, \eta_a(i)}, \tau > s,$$

and $L_{\eta_a,l}$, $L_{\eta_a,u}$ are the lower- and upper-bounds, $L_{\eta_a,l} < L_{\eta_a,u}$.

Now, we begin with the formulation of the detection problem to be addressed. Consider the model

$$\bar{r}_{\eta_a} = RQ_{r,1}r_y + \bar{\eta}_a \sim \mathcal{N}(\bar{\eta}_a, I), \bar{\eta}_a = R\eta_a \in \mathbb{C}^m, \eta_a \in \mathbb{C}^m,$$

and write it into

$$\bar{r}_{\eta_a,s}(k_O) = \varepsilon_{\eta_a,s}(k_O) + \bar{\eta}_{a,s}(k_O), \bar{r}_{\eta_a,s}(k_O) = \begin{bmatrix} \bar{r}_{\eta_a}(k_O+1) \\ \vdots \\ \bar{r}_{\eta_a}(k_O+s) \end{bmatrix}, \tag{16.28}$$

$$\varepsilon_{\eta_a,s}(k_O) = \begin{bmatrix} \varepsilon_{\eta_a}(k_O+1) \\ \vdots \\ \varepsilon_{\eta_a}(k_O+s) \end{bmatrix} \sim \mathcal{N}(0, I), \bar{\eta}_{a,s}(k_O) = \begin{bmatrix} \bar{\eta}_a(k_O+1) \\ \vdots \\ \bar{\eta}_a(k_O+s) \end{bmatrix}.$$

By means of the state space realisation (16.26) of R with $x_R(k)$ denoting its state vector, $\bar{\eta}_{a,s}(k_O)$ is given by

$$\begin{aligned}
\bar{\eta}_{a,s}(k_O) &= H_{o,s}A_{r,L}^{\gamma}x_R(k_O+1-\gamma) + H_{\bar{\eta}_{a,s,\gamma}}\eta_{a,s+\gamma}(k_O-\gamma) \in \mathbb{R}^{ms} \\
A_{R,L} &= A_{r,1} - LC_{r,1}B_R = -L, C_R = \Sigma_{r,1}^{-1}C_{r,1}, D_R = \Sigma_{r,1}^{-1}, \\
H_{o,s} &= \begin{bmatrix} C_R \\ C_R A_{R,L} \\ \vdots \\ C_R^{s-1}A_{R,L} \end{bmatrix}, \eta_{a,s+\gamma}(k_O-\gamma) = \begin{bmatrix} \eta_a(k_O-\gamma+1) \\ \vdots \\ \eta_a(k_O+s) \end{bmatrix} \in \mathbb{R}^{m(s+\gamma)}, \\
H_{\bar{\eta}_{a,s,\gamma}} &= \begin{bmatrix} H_{\bar{\eta}_{a,s,\gamma},1} & H_{\bar{\eta}_{a,s,\gamma},2} \end{bmatrix}, H_{\bar{\eta}_{a,s,\gamma},1} = H_{o,s}\begin{bmatrix} A_{r,L}^{\gamma-1} & \cdots & A_{r,L} \end{bmatrix}, \\
H_{\bar{\eta}_{a,s,\gamma},2} &= \begin{bmatrix} D_R & 0 & \cdots & 0 \\ C_R B_R & D_R & \ddots & \vdots \\ \vdots & \ddots & \ddots & 0 \\ C_R A_{R,L}^{s-2}B_R & \cdots & C_R B_R & D_R \end{bmatrix} \in \mathbb{R}^{ms\times ms}.
\end{aligned}$$

On the assumption that the integer γ is sufficiently large so that $A_{r,L}^{\gamma} \approx 0$, $\bar{\eta}_{a,s}(k_O)$ is well approximated by

$$\bar{\eta}_{a,s}(k_O) = H_{\bar{\eta}_{a,s,\gamma}}\eta_{a,s+\gamma}(k_O-\gamma).$$

Note that $rank\left(H_{\bar{\eta}_{a,s,\gamma}}\right) \leq ms$. In the sequel, without loss of generality, it is assumed that

$$rank\left(H_{\bar{\eta}_{a,s,\gamma}}\right) = ms.$$

Moreover, the detection logic (16.27) is re-formulated as

$$\begin{cases} \left\| \bar{\eta}_{a,s}\left(k_0\right) \right\|^2 \leq \tau \lambda_{\min}\left(H_{\bar{\eta}_{a,s,\gamma}} H_{\bar{\eta}_{a,s,\gamma}}^T\right) L_0^2 \Longrightarrow \text{attack-free} \\ \left\| \bar{\eta}_{a,s}\left(k_0\right) \right\|^2 \geq \tau \lambda_{\max}\left(H_{\bar{\eta}_{a,s,\gamma}} H_{\bar{\eta}_{a,s,\gamma}}^T\right) L_0^2 \Longrightarrow \text{cyber-attack} \end{cases} \tag{16.29}$$

for $\tau = s + \gamma$ and some constant $L_0 > 0$. Here, $\lambda_{\max}\left(H_{\bar{\eta}_{a,s,\gamma}} H_{\bar{\eta}_{a,s,\gamma}}^T\right)$ and with $\lambda_{\min}\left(H_{\bar{\eta}_{a,s,\gamma}} H_{\bar{\eta}_{a,s,\gamma}}^T\right)$ are the maximum and minimum eigenvalues of $H_{\bar{\eta}_{a,s,\gamma}} H_{\bar{\eta}_{a,s,\gamma}}^T$, respectively. The detection logic (16.29) is interpreted as follows. Suppose that $\eta_{a,s+\gamma}\left(k_0 - \gamma\right)$ doesn't belong to the null space of $H_{\bar{\eta}_{a,s,\gamma}}$, i.e.

$$H_{\bar{\eta}_{a,s,\gamma}} \eta_{a,s+\gamma}\left(k_0 - \gamma\right) \neq 0.$$

Then,

$$\begin{aligned} &\left\| \bar{\eta}_{a,s}\left(k_0\right) \right\|^2 = \eta_{a,s+\gamma}^T\left(k_0 - \gamma\right) H_{\bar{\eta}_{a,s,\gamma}}^T H_{\bar{\eta}_{a,s,\gamma}} \eta_{a,s+\gamma}\left(k_0 - \gamma\right) \Longrightarrow \\ &\left\| \bar{\eta}_{a,s}\left(k_0\right) \right\|^2 \geq \lambda_{\min}\left(H_{\bar{\eta}_{a,s,\gamma}} H_{\bar{\eta}_{a,s,\gamma}}^T\right) \eta_{a,s+\gamma}^T\left(k_0 - \gamma\right) \eta_{a,s+\gamma}\left(k_0 - \gamma\right), \\ &\left\| \bar{\eta}_{a,s}\left(k_0\right) \right\|^2 \leq \lambda_{\max}\left(H_{\bar{\eta}_{a,s,\gamma}} H_{\bar{\eta}_{a,s,\gamma}}^T\right) \eta_{a,s+\gamma}^T\left(k_0 - \gamma\right) \eta_{a,s+\gamma}\left(k_0 - \gamma\right). \end{aligned}$$

Consequently, it leads to the detection logic (16.27),

$$\left\| \eta_{a,s+\gamma}\left(k_0 - \gamma\right) \right\|^2 \leq \tau L_0^2 \Longleftrightarrow \sqrt{\frac{1}{\tau} \sum_{i=k_0-\gamma+1}^{k_0+s} \eta_a^T(i)\, \eta_a(i)} \leq L_0,$$

$$\left\| \eta_{a,s+\gamma}\left(k_0 - \gamma\right) \right\|^2 \geq \tau L_0^2 \Longleftrightarrow \sqrt{\frac{1}{\tau} \sum_{i=k_0-\gamma+1}^{k_0+s} \eta_a^T(i)\, \eta_a(i)} \geq L_0.$$

Now, our detection problem is formulated as: given the model (16.28) and the log-likelihood ratio

$$\begin{gathered} \ln \frac{p_{\left\| \bar{\eta}_{a,s}(k_0) \right\|^2 \leq L_l}\left(\bar{r}_{\eta_a,s}\left(k_0\right)\right)}{p_{\left\| \bar{\eta}_{a,s}(k_0) \right\|^2 \geq L_u}\left(\bar{r}_{\eta_a,s}\left(k_0\right)\right)}, \\ L_l^2 := \tau \lambda_{\min}\left(H_{\bar{\eta}_{a,s,\gamma}} H_{\bar{\eta}_{a,s,\gamma}}^T\right) L_0^2,\ L_u^2 := \tau \lambda_{\max}\left(H_{\bar{\eta}_{a,s,\gamma}} H_{\bar{\eta}_{a,s,\gamma}}^T\right) L_0^2. \end{gathered}$$

find the GLR

$$J = \ln \frac{\sup\limits_{\bar{\eta}_{a,s}(k_0)} p_{\left\| \bar{\eta}_{a,s}(k_0) \right\|^2 \leq L_l^2}\left(\bar{r}_{\eta_a,s}\left(k_0\right)\right)}{\sup\limits_{\bar{\eta}_{a,s}(k_0)} p_{\left\| \bar{\eta}_{a,s}(k_0) \right\|^2 \geq L_u^2}\left(\bar{r}_{\eta_a,s}\left(k_0\right)\right)} \tag{16.30}$$

as the test statistic and the corresponding threshold so that

$$J \underset{\mathcal{H}_1}{\overset{\mathcal{H}_0}{\lessgtr}} J_{th},$$

where the null hypothesis $\mathcal{H}_0$ is for attacks, i.e. $\|\eta_a\|_{RMS,[k_0-\gamma+1,k_0-\gamma+\tau]} \geq L_0$, and the alternative hypothesis is for attack-free, $\|\eta_a\|_{RMS,[k+1,k+\tau]} < L_0$. Below, the solution is outlined.

Next, the optimisation problem (16.30) is addressed. To begin with, J is written into

$$\begin{aligned} J = & \frac{1}{2} \sup_{\substack{\bar{\eta}_{a,s}(k_0) \\ \|\bar{\eta}_{a,s}(k_0)\|^2 \leq L_l^2}} - \left\| \bar{r}_{\eta_a,s}(k_0) - \bar{\eta}_{a,s}(k_0) \right\|^2 \\ & - \frac{1}{2} \sup_{\substack{\bar{\eta}_{a,s}(k_0) \\ \|\bar{\eta}_{a,s}(k_0)\|^2 \geq L_u^2}} - \left\| \bar{r}_{\eta_a,s}(k_0) - \bar{\eta}_{a,s}(k_0) \right\|^2, \end{aligned}$$

which can be solved equivalently by means of the following two optimisation problems,

$$\begin{aligned} & \frac{1}{2} \inf_{\substack{\bar{\eta}_{a,s}(k_0) \\ \|\bar{\eta}_{a,s}(k_0)\|^2 \leq L_l^2}} \left\| \bar{r}_{\eta_a,s}(k_0) - \bar{\eta}_{a,s}(k_0) \right\|^2, \\ & \frac{1}{2} \inf_{\substack{\bar{\eta}_{a,s}(k_0) \\ \|\bar{\eta}_{a,s}(k_0)\|^2 \geq L_u^2}} \left\| \bar{r}_{\eta_a,s}(k_0) - \bar{\eta}_{a,s}(k_0) \right\|^2. \end{aligned}$$

Accordingly, it turns out, for $\left\| \bar{\eta}_{a,s}(k_0) \right\|^2 \leq L_l^2$,

$$\begin{aligned} J &= \frac{1}{2} \left\| \bar{r}_{\eta_a,s}(k_0) - \hat{\eta}_{a,s}^u(k_0) \right\|^2, \\ \hat{\eta}_{a,s}^u(k_0) &= \frac{\bar{r}_{\eta_a,s}(k_0) L_u}{\sqrt{\bar{r}_{\eta_a,s}^T(k_0) \bar{r}_{\eta_a,s}(k_0)}}, \end{aligned}$$

for $\left\| \bar{\eta}_{a,s}(k_0) \right\|^2 \geq L_u^2$,

$$\begin{aligned} J &= -\frac{1}{2} \left\| \bar{r}_{\eta_a,s}(k_0) - \hat{\eta}_{a,s}^l(k_0) \right\|^2, \\ \hat{\eta}_{a,s}^l(k_0) &= \frac{\bar{r}_{\eta_a,s}(k_0) L_l}{\sqrt{\bar{r}_{\eta_a,s}^T(k_0) \bar{r}_{\eta_a,s}(k_0)}}, \end{aligned}$$

and for $L_l^2 \leq \left\| \bar{\eta}_{a,s}(k_0) \right\|^2 \leq L_u^2$,

$$J = \frac{1}{2} \left\| \bar{r}_{\eta_a,s}(k_0) - \hat{\eta}_{a,s}^u(k_0) \right\|^2 - \frac{1}{2} \left\| \bar{r}_{\eta_a,s}(k_0) - \hat{\eta}_{a,s}^l(k_0) \right\|^2.$$

Some routine calculations give

$$J = \begin{cases} \frac{1}{2} \left(\left\| \bar{r}_{\eta_a,s}(k_0) \right\| - L_u \right)^2, \left\| \bar{\eta}_{a,s}(k_0) \right\|^2 \leq L_l^2 \\ \frac{1}{2} \begin{pmatrix} \left(\left\| \bar{r}_{\eta_a,s}(k_0) \right\| - L_u \right)^2 \\ - \left(\left\| \bar{r}_{\eta_a,s}(k_0) \right\| - L_l \right)^2 \end{pmatrix}, L_l^2 \leq \left\| \bar{\eta}_{a,s}(k_0) \right\|^2 \leq L_u^2 \\ -\frac{1}{2} \left(\left\| \bar{r}_{\eta_a,s}(k_0) \right\| - L_l \right)^2, \left\| \bar{\eta}_{a,s}(k_0) \right\|^2 \geq L_u^2. \end{cases}$$

Next, a threshold J_{th} is determined under the condition of a (maximum) FAR equal to α. That means

$$\sup_{\left\| \bar{\eta}_{a,s}(k_0) \right\|^2 \geq L_u^2} \Pr \left(J > J_{th} \left| \bar{r}_{\eta_a,s}(k_0) \sim \mathcal{N} \left(\bar{\eta}_{a,s}(k_0), I \right) \right. \right) = \alpha.$$

Observe that the core of the log-likelihood ratio J is $\left(\left\| \bar{r}_{\eta_a,s}(k_0) \right\| - L_\mu \right)^2, \mu = l, u$. Let

$$\left(\left\| \bar{r}_{\eta_a,s}(k_0) \right\| - L_\mu \right)^2 = J_\mu \left(\bar{r}_{\eta_a,s}(k_0) \right).$$

Then,

$$\left(J_\mu^{1/2} \left(\bar{r}_{\eta_a,s}(k_0) \right) + L_\mu \right)^2 = \left\| \bar{r}_{\eta_a,s}(k_0) \right\|^2.$$

Recall that $\left\| \bar{r}_{\eta_a,s}(k_0) \right\|^2$ is subject to a non-central χ^2 distribution with ms degrees of freedom and the non-centrality parameter $\left\| \bar{\eta}_{a,s}(k_0) \right\|^2$. Since $\bar{\eta}_{a,s}(k_0)$ is unknown and no a prior information of $\bar{\eta}_{a,s}(k_0)$ is available, it is suggested to calculate the FAR with respect to

$$\left\| \bar{\eta}_{a,s}(k_0) \right\|^2 = L_u^2.$$

This means, we check the FAR corresponding to the lower-bound of the $\left\| \bar{\eta}_{a,s}(k_0) \right\|^2$, when attacks exist. In a certain sense, this is the worst case. As a result, we finally have

$$\left(J_\mu^{1/2} \left(\bar{r}_{\eta_a,s}(k_0) \right) + L_\mu \right)^2 = \left\| \bar{r}_{\eta_a,s}(k_0) \right\|^2 \sim \chi^2 \left(sm, L_u^2 \right),$$

based on it, the threshold J_{th} can be calculated.

16.2.2.3 Performance Optimisation and Recovery

When faults are detected on the plant side, the MC-station is informed and corresponding algorithms will run at the MC-station to recover the system performance suffered from the degradation caused by the faults. To be specific, the parameter system in the controller (16.5) will be online tuned. There exist numerous performance recovery methods, some of which were introduced in the first part of this book. Further references for the existing methods are provided at the end of this chapter.

16.2.2.4 Resilient Control of CPSs Under Cyber-Attacks

We now address the design issues of a resilient controller, which is switched on under the condition that cyber-attacks are detected and the plant works fault-free. To be specific, consider the process model (16.3) of the plant, the controller and cyber-attack models

$$u = F\hat{x} + u_{MC}^{a}, \tag{16.31}$$

$$u_{MC}^{a} = Q_{u_{MC}}\left(r_{y,u}^{a} - Q_{r,2}v\right) + v + a_{u_{MC}}, \tag{16.32}$$

$$r_{y,u}^{a} = Q_{r,1}r_{y} + Q_{r,2}r_{u} + a_{r_{y,u}} \tag{16.33}$$

$$= \left(I - Q_{r,2}Q_{u_{MC}}\right)^{-1}\left(Q_{r,1}r_{y} + Q_{r,2}a_{u_{MC}} + a_{r_{y,u}}\right) + Q_{r,2}v \tag{16.34}$$

$$\Longrightarrow u_{MC}^{a} = \left(I - Q_{u_{MC}}Q_{r,2}\right)^{-1}Q_{u_{MC}}\left(Q_{r,1}r_{y} + a_{r_{y,u}}\right) + \left(I - Q_{u_{MC}}Q_{r,2}\right)^{-1}a_{u_{MC}} + v. \tag{16.35}$$

Since

$$r_{y} = \hat{M}y - \hat{N}u, r_{u} = u - F\hat{x} = Xu + Yy,$$

the closed-loop dynamic is described by

$$\begin{aligned}
\begin{bmatrix} u \\ y \end{bmatrix} &= \begin{bmatrix} X & Y \\ -\hat{N} & \hat{M} \end{bmatrix}^{-1}\begin{bmatrix} u_{MC} + a_{u_{MC}} \\ r_{y} \end{bmatrix} \\
&= \begin{bmatrix} M \\ N \end{bmatrix}\left(v + \left(I - Q_{u_{MC}}Q_{r,2}\right)^{-1}\left(Q_{u_{MC}}a_{r_{y,u}} + a_{u_{MC}}\right)\right) \\
&\quad + \left(\begin{bmatrix} -\hat{Y} \\ \hat{X} \end{bmatrix} + \begin{bmatrix} M \\ N \end{bmatrix}\left(I - Q_{u_{MC}}Q_{r,2}\right)^{-1}Q_{u_{MC}}Q_{r,1}\right)r_{y} \\
&= \begin{bmatrix} M \\ N \end{bmatrix}\left(v + \left(I - Q_{u_{MC}}Q_{r,2}\right)^{-1}\vartheta_{a}\right) + \left(\begin{bmatrix} -\hat{Y} \\ \hat{X} \end{bmatrix} + \begin{bmatrix} M \\ N \end{bmatrix}Q_{u_{MC}}\bar{Q}_{r,1}\right)r_{y},
\end{aligned} \tag{16.36}$$

$$\vartheta_{a} := Q_{u_{MC}}a_{r_{y,u}} + a_{u_{MC}}, \bar{Q}_{r,1} := \left(I - Q_{r,2}Q_{u_{MC}}\right)^{-1}Q_{r,1}.$$

Based on it, the resilient controller design is formulated as: given $\gamma_{r_y} > 0$ and $\gamma_{\vartheta_a} > 0$ that represent the resilient control performance, find $Q_{r,1}$, $Q_{r,2}$ and $Q_{u_{MC}}$ such that $\left(I - Q_{u_{MC}}Q_{r,2}\right)^{-1}$ is stable, and

$$\left\| \left(I - Q_{u_{MC}} Q_{r,2}\right)^{-1} \right\|_{\infty} \leq \gamma_{\vartheta_a}, \tag{16.37}$$

$$\left\| \begin{bmatrix} -\hat{Y} \\ \hat{X} \end{bmatrix} + \begin{bmatrix} M \\ N \end{bmatrix} Q \right\|_{\infty} \leq \gamma_{r_y}, \tag{16.38}$$

$$Q := Q_{u_{MC}} \bar{Q}_{r,1}.$$

It is plain that the condition (16.38) describes the system robustness against system uncertainties presented by the residual r_y, while the condition (16.37) imposes the resilient requirement dealing with the cyber-attack ϑ_a. Observe that for the resilient controller subject to (16.37)–(16.38), $Q_{r,1}$, $Q_{r,2}$ and $Q_{u_{MC}}$ are redundant. In fact, $Q_{r,2}$ can be used as an encoder to complicate the construction of a stealthy attack vector η_a defined in (16.20). In this case, $Q_{u_{MC}}$ is firstly determined given $Q_{r,2}$, and then $Q_{r,1}$ is found for given $Q_{u_{MC}}$.

It is noteworthy that the capability of the aforementioned design scheme is limited to deal with plants with additive disturbances. Below, we consider plants with (multiplicative) model uncertainties and additive disturbances. It is supposed that the model uncertainty is presented in the form of (16.7) with

$$\begin{bmatrix} \Delta_M \\ \Delta_N \end{bmatrix} = \begin{bmatrix} -\hat{Y} \\ \hat{X} \end{bmatrix} S \Longrightarrow$$
$$\begin{bmatrix} u \\ y \end{bmatrix} = \left(\begin{bmatrix} M \\ N \end{bmatrix} + \begin{bmatrix} -\hat{Y} \\ \hat{X} \end{bmatrix} S \right) \left(u_{MC} + a_{u_{MC}}\right) + \begin{bmatrix} 0 \\ d \end{bmatrix}.$$

Here, the unknown system $S \in \mathcal{RH}_{\infty}$, $\|S\|_{\infty} \leq \gamma_s < 1$, parametrises the model uncertainty and d is the unknown input vector. As is known,

$$\begin{aligned} r_y &= \begin{bmatrix} -\hat{N} & \hat{M} \end{bmatrix} \begin{bmatrix} u \\ y \end{bmatrix} = S\left(u_{MC} + a_{u_{MC}}\right) + \hat{M} d \\ &= S\left(v + Q_{u_{MC}}\left(Q_{r,1} r_y + Q_{r,2}\left(r_u - v\right)\right) + Q_{u_{MC}} a_{r_{y,u}} + a_{u_{MC}}\right) + \hat{M} d \Longrightarrow \\ r_y &= \left(I - S Q_{u_{MC}} Q_{r,1}\right)^{-1} \left(S\left(\left(I - Q_{u_{MC}} Q_{r,2}\right) v + Q_{u_{MC}} Q_{r,2} r_u + \vartheta_a\right) + \hat{M} d\right). \end{aligned}$$

It yields

$$\begin{bmatrix} r_u \\ r_y \end{bmatrix} = \begin{bmatrix} X & Y \\ -\hat{N} & \hat{M} \end{bmatrix} \begin{bmatrix} u \\ y \end{bmatrix} =$$
$$\begin{bmatrix} \left(I - Q_{u_{MC}} Q_{r,2}\right) v + Q_{u_{MC}}\left(Q_{r,1} r_y + Q_{r,2} r_u\right) + \vartheta_a \\ \left(I - S Q_{u_{MC}} Q_{r,1}\right)^{-1} \left(S\left(\left(I - Q_{u_{MC}} Q_{r,2}\right) v + Q_{u_{MC}} Q_{r,2} r_u + \vartheta_a\right) + \hat{M} d\right) \end{bmatrix}$$
$$\Longrightarrow \begin{bmatrix} r_u \\ r_y \end{bmatrix} = (I - \Pi)^{-1} \begin{bmatrix} I & 0 \\ \bar{Q}_1 S & \bar{Q}_1 \hat{M} \end{bmatrix} \begin{bmatrix} \left(I - Q_{u_{MC}} Q_{r,2}\right) v + \vartheta_a \\ d \end{bmatrix},$$
$$\bar{Q}_1 = \left(I - S Q_{u_{MC}} Q_{r,1}\right)^{-1}, \Pi = \begin{bmatrix} Q_{u_{MC}} Q_{r,2} & Q_{u_{MC}} Q_{r,1} \\ \bar{Q}_1 S Q_{u_{MC}} Q_{r,2} & 0 \end{bmatrix}.$$

Accordingly, the closed-loop dynamic is governed by

$$\begin{aligned}\begin{bmatrix} u \\ y \end{bmatrix} &= \begin{bmatrix} X & Y \\ -\hat{N} & \hat{M} \end{bmatrix}^{-1} \begin{bmatrix} r_u \\ r_y \end{bmatrix} \\ &= \begin{bmatrix} M & -\hat{Y} \\ N & \hat{X} \end{bmatrix} (I - \Pi)^{-1} \begin{bmatrix} I & 0 \\ \bar{Q}_1 S & \bar{Q}_1 \hat{M} \end{bmatrix} \begin{bmatrix} \left(I - Q_{u_{MC}} Q_{r,2}\right) v + \vartheta_a \\ d \end{bmatrix}.\end{aligned}$$

By means of the following routine calculations,

$$\begin{aligned}&(I - \Pi)^{-1} = \\ &\begin{bmatrix} \bar{Q}_2 (I - Q_1 S Q_2)^{-1} & 0 \\ 0 & \bar{Q}_1 (I - S Q_2 Q_1)^{-1} \end{bmatrix} \begin{bmatrix} I & Q_{u_{MC}} Q_{r,1} \\ S Q_2 & I - S Q_{u_{MC}} Q_{r,1} \end{bmatrix}, \\ &(I - \Pi)^{-1} \begin{bmatrix} I & 0 \\ \bar{Q}_1 S & \bar{Q}_1 \hat{M} \end{bmatrix} = \\ &\begin{bmatrix} \bar{Q}_2 (I - Q_1 S Q_2)^{-1} & 0 \\ 0 & \bar{Q}_1 (I - S Q_2 Q_1)^{-1} \end{bmatrix} \begin{bmatrix} I + Q_1 S & Q_1 \hat{M} \\ S (I + Q_2) & \hat{M} \end{bmatrix}, \\ &\bar{Q}_2 = \left(I - Q_{u_{MC}} Q_{r,2}\right)^{-1}, Q_1 = Q_{u_{MC}} Q_{r,1} \bar{Q}_1, Q_2 = Q_{u_{MC}} Q_{r,2} \bar{Q}_2,\end{aligned}$$

the system responses (u, y) to the reference v, the cyber-attack vector ϑ_a and the unknown input vector d are described respectively as,

$$\begin{bmatrix} u \\ y \end{bmatrix} = \begin{bmatrix} u_v \\ y_v \end{bmatrix} + \begin{bmatrix} u_{\vartheta_a} \\ y_{\vartheta_a} \end{bmatrix} + \begin{bmatrix} u_d \\ y_d \end{bmatrix},$$

$$\begin{aligned}\begin{bmatrix} u_v \\ y_v \end{bmatrix} &= \begin{bmatrix} M \\ N \end{bmatrix} \bar{Q}_2 (I - Q_1 S Q_2)^{-1} (I + Q_1 S) \bar{Q}_2^{-1} v \\ &+ \begin{bmatrix} -\hat{Y} \\ \hat{X} \end{bmatrix} \bar{Q}_1 (I - Q_2 Q_1 S)^{-1} S (I + Q_2) \bar{Q}_2^{-1} v,\end{aligned} \tag{16.39}$$

$$\begin{aligned}\begin{bmatrix} u_{\vartheta_a} \\ y_{\vartheta_a} \end{bmatrix} &= \begin{bmatrix} M \\ N \end{bmatrix} \bar{Q}_2 (I - Q_1 S Q_2)^{-1} (I + Q_1 S) \vartheta_a \\ &+ \begin{bmatrix} -\hat{Y} \\ \hat{X} \end{bmatrix} \bar{Q}_1 (I - Q_2 Q_1 S)^{-1} S (I + Q_2) \vartheta_a,\end{aligned} \tag{16.40}$$

$$\begin{aligned}\begin{bmatrix} u_d \\ y_d \end{bmatrix} &= \begin{bmatrix} M \\ N \end{bmatrix} \bar{Q}_2 (I - Q_1 S Q_2)^{-1} Q_1 \hat{M} d \\ &+ \begin{bmatrix} -\hat{Y} \\ \hat{X} \end{bmatrix} \bar{Q}_1 (I - S Q_2 Q_1)^{-1} \hat{M} d.\end{aligned} \tag{16.41}$$

Based on (16.39)–(16.41), the demands for the resilient controller design are formulated as follows:

- the selection of $Q_{r,1}$, $Q_{r,2}$ and $Q_{u_{MC}}$ should assure the stability of

$$\bar{Q}_1 = \left(I - SQ_{u_{MC}}Q_{r,1}\right)^{-1}, \bar{Q}_2 = \left(I - Q_{u_{MC}}Q_{r,2}\right)^{-1},$$

- determine $Q_{r,1}$, $Q_{r,2}$ and $Q_{u_{MC}}$ so that

$$Q_1 = Q_{u_{MC}}Q_{r,1}\left(I - SQ_{u_{MC}}Q_{r,1}\right)^{-1}, Q_2 = Q_{u_{MC}}Q_{r,2}\left(I - Q_{u_{MC}}Q_{r,2}\right)^{-1}$$

  are stable and

$$\|Q_1 S Q_2\|_\infty < 1, \|SQ_2Q_1\|_\infty < 1, \|Q_1 S\|_\infty < 1$$

  which assures that $(I - Q_1 S Q_2)^{-1}, (I - SQ_2Q_1)^{-1}$ are stable and their $\mathcal{H}_\infty$ norm is close to 1, and
- set $Q_{r,1}$, $Q_{r,2}$ and $Q_{u_{MC}}$ for the possibly small $\mathcal{H}_\infty$ norms of $\bar{Q}_1$, $\bar{Q}$, Q_1 and Q_2.

As a result, the subsequent design algorithm is proposed,

- given $Q_{r,2}$, set $Q_{u_{MC}}$ so that

$$\inf_{Q_{u_{MC}} \in \mathcal{RH}_\infty} \left\|Q_{u_{MC}}Q_{r,2}\left(I - Q_{u_{MC}}Q_{r,2}\right)^{-1}\right\|_\infty$$
$$\text{subject to } \left\|\left(I - Q_{u_{MC}}Q_{r,2}\right)^{-1}\right\|_\infty \leq \gamma_1 << 1,$$

- given $Q_{u_{MC}}$, Q_2, set $Q_{r,1}$ so that

$$\inf_{\substack{Q_{r,1} \in \mathcal{RH}_\infty \\ \|S\|_\infty \leq \gamma_s}} \left\|\left(I - SQ_{u_{MC}}Q_{r,1}\right)^{-1}\right\|_\infty$$
$$\text{subject to } \left\|Q_{u_{MC}}Q_{r,1}S\left(I - Q_{u_{MC}}Q_{r,1}S\right)^{-1}Q_2\right\|_\infty \leq \gamma_3 < 1.$$

Here, γ_1, γ_2 and γ_3 are given performance indices.

16.3 Stealthy Attacks and Performance-Based Detection

The resilient control scheme proposed in the previous section presupposes that the cyber-attacks $\left(a_{r_{y,u}}, a_{u_{MC}}\right)$ can be well detected. Examining the proposed detection scheme makes it clear that, when

$$\eta_a = Q_{r,2}a_{u_{MC}} + a_{r_{y,u}} = 0,$$

a detection of $\left(a_{r_{y,u}}, a_{u_{MC}}\right)$ becomes impossible. In fact, this is the case that $\left(a_{r_{y,u}}, a_{u_{MC}}\right)$ are designed as stealthy cyber-attacks. This observation and a pretty impressive amount of publications on attack design issues raise our interest on a close analysis of stealthy cyber-attacks on the control performance of the CPSs under consideration and, more importantly, on the development of a detection scheme of stealthy cyber-attacks.

16.3.1 On Stealthy Attacks

Recall that our cyber-attack detection scheme is based on the signal models (16.19)–(16.22). To be specific, if $r_y \sim \mathcal{N}\left(0, \Sigma_{r_y}\right)$ is a stochastic process, χ^2 test statistic is adopted, while in the case of an $\mathcal{L}_2$-bounded r_y, the unified solution delivers the optimal detection. Stealthy attacks are such attacks that will not cause detectable changes in the detection test statistic or the residual evaluation function. On the other hand, the true intention of an attacker is to degrade the system control performance. It follows from the closed-loop dynamic (16.36), i.e.

$$\begin{bmatrix} u \\ y \end{bmatrix} = \begin{bmatrix} M \\ N \end{bmatrix}\left(v + \left(I - Q_{u_{MC}} Q_{r,2}\right)^{-1} \vartheta_a\right) + \left(\begin{bmatrix} -\hat{Y} \\ \hat{X} \end{bmatrix} + \begin{bmatrix} M \\ N \end{bmatrix} Q_{u_{MC}} \bar{Q}_{r,1}\right) r_y,$$

that the cyber-attack $\vartheta_a = Q_{u_{MC}} a_{r_{y,u}} + a_{u_{MC}}$ degrades the system tracking behaviour, while a change of the feedback mechanism of r_y will lead to feedback performance degradation. Hence, from the attacker's viewpoint, the design of cyber-attacks is an optimisation problem aiming to maximise the control performance degradation as the optimisation objective and attack stealthiness as a constraint. This motivates us to examine the relations between closed-loop dynamic and the residual dynamic given by

$$r_{y,u}^a = \left(I - Q_{r,2} Q_{u_{MC}}\right)^{-1}\left(Q_{r,1} r_y + \eta_a\right) + Q_{r,2} v. \tag{16.42}$$

It is apparent that, to attack the system tracking behaviour stealthily, the attack design is to be formulated as maximising the tracking error,

$$\begin{bmatrix} \Delta u \\ \Delta y \end{bmatrix} := \begin{bmatrix} u \\ y \end{bmatrix} - \begin{bmatrix} M \\ N \end{bmatrix} v = \begin{bmatrix} M \\ N \end{bmatrix}\left(I - Q_{u_{MC}} Q_{r,2}\right)^{-1} \vartheta_a,$$

by selecting

$$\vartheta_a = Q_{u_{MC}} a_{r_{y,u}} + a_{u_{MC}}$$

that subject to

$$\eta_a = Q_{r,2} a_{u_{MC}} + a_{r_{y,u}} = 0. \tag{16.43}$$

The stealthiness condition (16.43) poses the following relations,

$$\vartheta_a = \left(I - Q_{u_{MC}} Q_{r,2}\right) a_{u_{MC}} \Longrightarrow \begin{bmatrix} \Delta u \\ \Delta y \end{bmatrix} = \begin{bmatrix} M \\ N \end{bmatrix} a_{u_{MC}}. \tag{16.44}$$

Remember that cyber-attacks cause no change in the residual r_y. This implies that degrading the system feedback dynamic can only be achieved by changing the feedback mechanism

$$\begin{bmatrix} -\hat{Y} \\ \hat{X} \end{bmatrix} + \begin{bmatrix} M \\ N \end{bmatrix} Q_{u_{MC}} \bar{Q}_{r,1}.$$

Because the nominal setting for a robust control dynamic is the result of such an optimisation problem like

$$\inf_{Q \in \mathcal{RH}_\infty} \left\| \begin{bmatrix} -\hat{Y} \\ \hat{X} \end{bmatrix} + \begin{bmatrix} M \\ N \end{bmatrix} Q \right\|_\infty,$$
$$Q = Q_{u_{MC}} \bar{Q}_{r,1},$$

a degradation is possible when Q is so modified that the above norm becomes larger. Observing the residual $r^a_{y,u}$, such degradation is realisable by constructing the attack as follows,

$$\begin{aligned} a_{r_{y,u}} &= a^1_{r_{y,u}} + a^2_{r_{y,u}}, \\ a^2_{r_{y,u}} &= Q_a r^a_{y,u},\ Q_a \in \mathcal{RH}_\infty, \end{aligned}$$

where Q_a is a design parameter. That is, $a_{r_{y,u}}$ consists of an additive and a multiplicative attacks. Accordingly, the residual is governed by

$$\begin{aligned} r^a_{y,u} &= \left(I - Q_{r,2} Q_{u_{MC}} - Q_a\right)^{-1} \left(Q_{r,1} r_y + \bar{\eta}_a\right) + \bar{Q}_{r,2} v, \\ \bar{Q}_{r,2} &= \left(I - Q_{r,2} Q_{u_{MC}} - Q_a\right)^{-1} \left(I - Q_{r,2} Q_{u_{MC}}\right), \\ \bar{\eta}_a &= Q_{r,2} a_{u_{MC}} + a^1_{r_{y,u}}, \end{aligned} \tag{16.45}$$

and the closed-loop dynamic is subject to

$$\begin{aligned} u^a_{MC} &= Q_{u_{MC}} \left(r^a_{y,u} - Q_{r,2} v\right) + v + a_{u_{MC}} \\ &= Q_{r,a} \left(Q_{r,1} r_y + \bar{\eta}_a\right) + \left(I + Q_{v,a}\right) v + a_{u_{MC}}, \\ \begin{bmatrix} u \\ y \end{bmatrix} &= \begin{bmatrix} X & Y \\ -\hat{N} & \hat{M} \end{bmatrix}^{-1} \begin{bmatrix} u^a_{MC} \\ r_y \end{bmatrix} \\ &= \begin{bmatrix} M \\ N \end{bmatrix} \left(\left(I + Q_{v,a}\right) v + \hat{\vartheta}_a\right) + \left(\begin{bmatrix} -\hat{Y} \\ \hat{X} \end{bmatrix} + \begin{bmatrix} M \\ N \end{bmatrix} Q_{r,a} Q_{r,1}\right) r_y, \\ Q_{v,a} &= Q_{u_{MC}} \left(\bar{Q}_{r,2} - Q_{r,2}\right), \end{aligned}$$

$$Q_{r,a} = Q_{u_{MC}} \left(I - Q_{r,2} Q_{u_{MC}} - Q_a \right)^{-1},$$
$$\hat{\vartheta}_a = Q_{r,a} a^1_{r_{y,u}} + \left(I + Q_{r,a} Q_{r,2} \right) a_{u_{MC}}.$$

It is clear that tuning Q_a may degrade the feedback control performance

$$\left(\begin{bmatrix} -\hat{Y} \\ \hat{X} \end{bmatrix} + \begin{bmatrix} M \\ N \end{bmatrix} Q_{r,a} Q_{r,1} \right) r_y.$$

Note that any change in the feedback behaviour will also cause variation in the tracking behaviour.

In order to find the existence conditions for stealthy attacks on the feedback control loop, consider the residual dynamic (16.45) under the conditions, (i) $r_y \sim \mathcal{N}\left(0, \Sigma_{r_y}\right)$, and (ii) $\mathcal{L}_2$-bounded unknown input r_y. In the case (i), $r^a_{y,u}$ is subject to $\mathcal{N}\left(\mathbb{E} r^a_{y,u}, \Sigma_{r^a_{y,u}}\left(Q_a\right)\right)$,

$$\mathbb{E} r^a_{y,u} = \left(I - Q_{r,2} Q_{u_{MC}} - Q_a \right)^{-1} \bar{\eta}_a + \bar{Q}_{r,2} v,$$
$$\Sigma_{r^a_{y,u}}\left(Q_a\right) = \mathbb{E}\left(r^a_{y,u} - \mathbb{E} r^a_{y,u} \right) \left(r^a_{y,u} - \mathbb{E} r^a_{y,u} \right)^T.$$

Here, $\Sigma_{r^a_{y,u}}\left(Q_a\right)$ denotes the covariance matrix as a function of attack $a^2_{r_{y,u}} = Q_a r^a_{y,u}$. On account of the fact that a stochastic process with Gaussian distribution is fully characterised by its expectation value and covariance matrix, the following conditions guarantee the stealthiness of the cyber-attacks under consideration,

$$a_{u_{MC}} = 0, \left(I - Q_{r,2} Q_{u_{MC}} - Q_a \right)^{-1} a^1_{r_{y,u}} + \bar{Q}_{r,2} v = 0$$
$$\Longrightarrow a^1_{r_{y,u}} + \left(I - Q_{r,2} Q_{u_{MC}} \right) v = 0, \tag{16.46}$$
$$\Sigma_{r^a_{y,u}}\left(Q_a\right) = \Sigma_{r^a_{y,u}}(0). \tag{16.47}$$

Remark 16.4 In the literature, conditions like (16.46)–(16.47) are often called strict stealthiness conditions. Such conditions can be relaxed by

$$\left\| a^1_{r_{y,u}} + \left(I - Q_{r,2} Q_{u_{MC}} \right) v \right\| \leq \gamma_E, \tag{16.48}$$
$$\left\| \Sigma^{-1/2}_{r^a_{y,u}}\left(Q_a\right) \Sigma_{r^a_{y,u}}(0)\, \Sigma^{-1/2}_{r^a_{y,u}}\left(Q_a\right) \right\| \leq \gamma_\Sigma, \tag{16.49}$$

for some signal and matrix norms whose selection depends on the test statistic used for the detection purpose. Here, γ_E is a small constant, and γ_Σ can be smaller than but close to one.

Design of stealthy attacks is a trend topic in recent research on CPS security. It can be observed that a great number of reported design schemes are based on the assumption that attackers have access to information like system models and control laws, and are in the position to be able to design attacks by means of advanced control-theoretic methods. On the other hand, it is rare that such design is devoted to attacking feedback control performance. Below, we delineate a more realistic and practical scenario for constructing stealthy cyber-attacks that degrade the feedback control performance.

Firstly, it is assumed that the CPS system under consideration is operating in the steady state with a constant reference vector before $\left(a_{y,u}, a_{u_{MC}}\right)$ are added so that the residual vector $r_{y,u}$ is a weakly stationary stochastic process. That is

$$\begin{aligned}\mathbb{E}r_{y,u} &= \lim_{z\to 1} Q_{r,2}v =: \zeta,\\ \Sigma_{r_{y,u}} &= \mathbb{E}\left(r_{y,u}-\zeta\right)\left(r_{y,u}-\zeta\right)^T,\\ r_{y,u}-\zeta &= \left(I - Q_{r,2}Q_{u_{MC}}\right)^{-1} Q_{r,1}r_y,\end{aligned}$$

ζ and $\Sigma_{r_{y,u}}$ are a constant vector and matrix, respectively. Suppose that attackers collect (sufficient) N number of data of $r_{y,u}$ over the time interval $[k_1, k_N]$ and calculate the maximal likelihood estimates of ζ and $\Sigma_{r_{y,u}}$,

$$\hat{\zeta} = \frac{1}{N}\sum_{i=1}^{N}\zeta\left(k_i\right), \hat{\Sigma}_{r_{y,u}} = \frac{1}{N}\sum_{i=1}^{N}\left(r_{y,u}\left(k_i\right)-\hat{\zeta}\right)\left(r_{y,u}\left(k_i\right)-\hat{\zeta}\right)^T.$$

Under such conditions, stealthy attacks can be constructed as follows:

$$r_{y,u}^a = \Xi\Pi\left(r_{y,u}-\zeta\right)+\zeta.$$

Here, Π is a Kalman filter satisfying

$$\begin{aligned}\hat{x}_\Pi(k+1\,|k) &= A_\Pi\hat{x}_\Pi(k\,|k-1) + K(k)\Delta r_{y,u}(k),\\ \Delta r_{y,u}(k) &= r_{y,u}(k) - \zeta - C_\Pi\hat{x}_\Pi(k\,|k-1), \hat{x}_\Pi(0) = 0,\\ K(k) &= A_\Pi P(k\,|k-1)C_\Pi^T\Sigma_{\Delta r_{y,u}}^{-1}(k),\\ P(k+1\,|k) &= A_\Pi P(k\,|k-1)A_\Pi^T - K(k)\Sigma_{\Delta r_{y,u}}(k)K^T(k),\\ \Sigma_{\Delta r_{y,u}}(k) &= C_\Pi P(k\,|k-1)C_\Pi^T + \hat{\Sigma}_{r_{y,u}},\end{aligned}$$

and Ξ is a matrix

$$\Xi = U\hat{\Sigma}_{r_{y,u}}^{1/2}\Sigma_{\Delta r_{y,u}}^{-1/2}(k), UU^T = I.$$

It is easy to examine that

$$\begin{aligned}\mathbb{E}r_{y,u}^a &= \mathbb{E}r_{y,u} = \zeta,\\ \Sigma_{r_{y,u}^a} &= \mathbb{E}\left(r_{y,u}^a-\zeta\right)\left(r_{y,u}^a-\zeta\right)^T = \Sigma_{r_{y,u}},\end{aligned}$$

so that the added cyber-attack is (strictly) stealthy. Moreover, the closed-loop dynamic is subject to

$$\begin{bmatrix} u \\ y \end{bmatrix} = \begin{bmatrix} X & Y \\ -\hat{N} & \hat{M} \end{bmatrix}^{-1} \begin{bmatrix} u_{MC} \\ r_y \end{bmatrix} =$$
$$\begin{bmatrix} M \\ N \end{bmatrix} v + \left(\begin{bmatrix} -\hat{Y} \\ \hat{X} \end{bmatrix} + \begin{bmatrix} M \\ N \end{bmatrix} Q_{u_{MC}} \Xi \Pi \bar{Q}_{r,1} \right) r_y.$$

It is interesting to notice that, to achieve an optimal robustness against uncertainty, it is a prevalent design strategy that $Q = Q_{u_{MC}} \bar{Q}_{r,1}$ is set as

$$Q = \arg \inf_{Q \in \mathcal{RH}_\infty} \left\| \begin{bmatrix} -\hat{Y} \\ \hat{X} \end{bmatrix} + \begin{bmatrix} M \\ N \end{bmatrix} Q \right\|_\infty .$$

Consequently, setting $\Xi \Pi = -I$ could considerably degrade the feedback performance. In other words, without any process model knowledge, constructing the cyber-attack simply as

$$r_{y,u}^a = -\left(r_{y,u} - \zeta\right) + \zeta = -r_{y,u} + 2\zeta \Longrightarrow$$
$$\begin{bmatrix} u \\ y \end{bmatrix} = \begin{bmatrix} M \\ N \end{bmatrix} v + \left(\begin{bmatrix} -\hat{Y} \\ \hat{X} \end{bmatrix} - \begin{bmatrix} M \\ N \end{bmatrix} Q_{u_{MC}} \bar{Q}_{r,1} \right) r_y$$

can result in a remarkable degradation of the system robustness, and the attack is stealthy.

Next, we consider more general system operation conditions with $\mathcal{L}_2$-bounded r_y. It is of interest to notice that, to attack the system feedback dynamic, injecting attack signals in the communication channel from the plant to the MC-station is an efficient scheme. Thus, in the case that $\bar{Q}_{r,2} v$ is sufficiently small or $v = 0$, as assumed in most of the results reported in the literature, stealthy attacks on the system feedback dynamic can be realised as follows,

$$\begin{aligned} r_{y,u}^a &= \Pi_a r_{y,u} \simeq \Pi_a \left(I - Q_{r,2} Q_{u_{MC}}\right)^{-1} Q_{r,1} r_y, \\ \left\| r_{y,u}^a \right\|_2 &\leq \left\| r_{y,u} \right\|_2 . \end{aligned} \quad (16.50)$$

Here, the attack detection scheme introduced in the last subsection with $\left\| r_{y,u}^a \right\|_2$ as the evaluation function is assumed to be used. Analogous to the aforementioned scheme for the stochastic process $r_{y,u}$, Π_a parametrising the cyber-attack is designed as a co-inner, which guarantees the inequality (16.50). Accordingly, the closed-loop dynamic is governed by

$$\begin{bmatrix} u \\ y \end{bmatrix} = \begin{bmatrix} X & Y \\ -\hat{N} & \hat{M} \end{bmatrix}^{-1} \begin{bmatrix} u_{MC} \\ r_y \end{bmatrix}$$

$$= \begin{bmatrix} M \\ N \end{bmatrix} v + \left(\begin{bmatrix} -\hat{Y} \\ \hat{X} \end{bmatrix} + \begin{bmatrix} M \\ N \end{bmatrix} Q_{u_{MC}} \Pi_a \bar{Q}_{r,1} \right) r_y.$$

Consequently, by tuning Π_a the system feedback dynamic is attacked.

16.3.2 Performance Degradation Monitoring-Based Detection of Stealthy Attacks

Recall that stealthy attacks are those attacks that cannot be detected by means of a residual-based detection algorithm. That means, on the other hand, it is not excluded that this class of attacks can be detected using other detection schemes. In this subsection, we schematically delineate the idea of control performance monitoring-based detection of stealthy attacks.

We firstly summarise some key facts and call the reader's attention to them.

- Cyber-attacks don't affect r_y, and hence degradations of feedback control performance can only be caused by attacking the feedback mechanism of the form

$$\begin{bmatrix} -\hat{Y} \\ \hat{X} \end{bmatrix} + \begin{bmatrix} M \\ N \end{bmatrix} Q.$$

- That means, on the other hand, signal r_{PD} generated by

$$r_{PD} = \Psi \left(\begin{bmatrix} -\hat{Y} \\ \hat{X} \end{bmatrix} + \begin{bmatrix} M \\ N \end{bmatrix} Q \right) r_y \tag{16.51}$$

  is the information carrier of degradation in feedback control performance that is capable of detecting performance degradation (PD) caused by cyber-attacks. Here, $\Psi \in \mathcal{RH}_\infty$ is an arbitrary stable dynamic system.
- The existence condition for (additive) stealthy cyber-attacks is

$$Q_{r,2} a_{u_{MC}} + a_{r_{y,u}} = 0,$$

  while the degradation of tracking (control) performance is activated by

$$\vartheta_a = Q_{u_{MC}} a_{r_{y,u}} + a_{u_{MC}}.$$

- It yields

$$\begin{bmatrix} I & Q_{u_{MC}} \\ Q_{r,2} & I \end{bmatrix} \begin{bmatrix} a_{u_{MC}} \\ a_{r_{y,u}} \end{bmatrix} = \begin{bmatrix} \vartheta_a \\ 0 \end{bmatrix},$$

which implies that, if

$$rank \begin{bmatrix} I & Q_{u_{MC}} \\ Q_{r,2} & I \end{bmatrix} > rank \begin{bmatrix} Q_{r,2} \; I \end{bmatrix},$$

attackers are able to construct stealthy cyber-attacks that degrade the system tracking performance. In fact, according to (16.44), it holds,

$$\begin{bmatrix} \Delta u \\ \Delta y \end{bmatrix} = \begin{bmatrix} M \\ N \end{bmatrix} \left(I - Q_{u_{MC}} Q_{r,2} \right)^{-1} \vartheta_a = \begin{bmatrix} M \\ N \end{bmatrix} a_{u_{MC}}, \tag{16.52}$$

with $(\Delta u, \Delta y)$ denoting the response of (u, y) to ϑ_a.

- Accordingly, stealthy attacks subject to the existence condition $Q_{r,2} a_{u_{MC}} + a_{r_{y,u}} = 0$ degrade the system tracking performance, as far as $a_{u_{MC}} \neq 0$. They can be detected, when the residual signal $r_{y,u}$ is constructed differently, for instance as

$$r_{y,u} = Q_{r,1} r_y + \bar{Q}_{r,2} r_u, \left(\bar{Q}_{r,2} - Q_{r,2} \right) a_{u_{MC}} \neq 0.$$

- Hereby, it is noteworthy that such stealthy attacks make the proposed feedback mechanism ineffective, as indicated by (16.52). This fact reveals the damage potential of such cyber-attacks and poses the importance of their detection.

Now, let us introduce the idea of performance degradation monitoring-based detection of stealthy attacks. Herewith, stealthy attacks are defined as those cyber-attacks satisfying (16.43) and/or

$$\Sigma_{r_{y,u}^a} = \Sigma_{r_{y,u}}.$$

For our purpose, we firstly address stealthy attacks on the system feedback and tracking performance separately. Without loss of generality, it is assumed that the stealthy attacks on the system feedback result in

$$\begin{bmatrix} u_f \\ y_f \end{bmatrix} = \left(\begin{bmatrix} -\hat{Y} \\ \hat{X} \end{bmatrix} + \begin{bmatrix} M \\ N \end{bmatrix} Q_{u_{MC}} \Pi_a \bar{Q}_{r,1} \right) r_y$$

with $\left(u_f, y_f \right)$ representing the response of (u, y) to the stealthy attacks. Here, Π_a parametrises the cyber-attacks with

$$\Pi_a = \begin{cases} I, & \text{attack-free} \\ \neq I, & \text{attacks.} \end{cases} \tag{16.53}$$

It is reasonable to assume that, during the nominal system operation, (u, y) is considerably robust against uncertainties, which is, for instance, achieved by

$$\left\| \begin{bmatrix} -\hat{Y} \\ \hat{X} \end{bmatrix} + \begin{bmatrix} M \\ N \end{bmatrix} Q \right\|_\infty \leq \gamma.$$

On the other hand, attackers will design the attacks in such a way that

$$\left\| \begin{bmatrix} -\hat{Y} \\ \hat{X} \end{bmatrix} + \begin{bmatrix} M \\ N \end{bmatrix} Q_{u_{MC}} \Pi_a \bar{Q}_{r,1} \right\|_\infty > \gamma.$$

Motivated by our aforementioned discussion and the residual (16.51), we now focus on detecting attacks on the feedback mechanism and propose to construct $r_{y,u}$ as follows,

$$r_{y,u} = Q_{r,1} r_y + \bar{r}_{PD}, \tag{16.54}$$

where $\bar{r}_{PD}$ is generated on the plant side by

$$\bar{r}_{PD} = \Psi \begin{bmatrix} u \\ y \end{bmatrix}$$

which results in, during attack-free operations,

$$r_{PD} = \bar{r}_{PD} - \Psi \begin{bmatrix} M \\ N \end{bmatrix} v = \Psi \left(\begin{bmatrix} -\hat{Y} \\ \hat{X} \end{bmatrix} + \begin{bmatrix} M \\ N \end{bmatrix} Q \right) r_y.$$

Here, the design parameter Ψ can be, for instance, simply selected equal to

$$\Psi = \begin{bmatrix} 0 & P \end{bmatrix} \Longrightarrow r_{PD} = P \left(\hat{X} + NQ \right) r_y$$

for some $P \in \mathcal{RH}_\infty$. Note that the feedback of r_u is not included in $r_{y,u}$, since our attention is paid to the multiplicative attack Π_a and such a handling simplifies the problem solution. The subsequent work is dedicated to the analysis of system dynamics under such cyber-attacks and, based on it, the development of a performance degradation detection scheme.

Corresponding to the nominal controller law (16.31)–(16.33), the controller is now set as

$$\begin{aligned} u &= F\hat{x} + u_{MC} - Q_{u_{MC}} \bar{r}_{PD}, \\ u_{MC} &= Q_{u_{MC}} r^a_{y,u} + v. \end{aligned}$$

Let

$$\begin{aligned} r^a_{y,u} &= \Pi_a \left(r_{y,u} - Q_v v \right) + Q_v v, \\ Q_v &:= \Psi \begin{bmatrix} M \\ N \end{bmatrix} \end{aligned}$$

with Π_a as the cyber-attack system, and assume that attackers are in possession of knowledge of $Q_v v$. We now examine the dynamics of the residual generator and the closed-loop. It follows from

$$\begin{aligned}
u &= F\hat{x} + Q_{u_{MC}}\left(\Pi_a\left(Q_{r,1}r_y + \Psi\begin{bmatrix} u \\ y \end{bmatrix} - Q_v v\right) + Q_v v\right) + v - Q_{u_{MC}}\Psi\begin{bmatrix} u \\ y \end{bmatrix} \\
&= F\hat{x} + Q_{u_{MC}}\Pi_a Q_{r,1}r_y + Q_{u_{MC}}\left(\Pi_a - I\right)\Psi\left(\begin{bmatrix} u \\ y \end{bmatrix} - \begin{bmatrix} M \\ N \end{bmatrix} v\right) + v
\end{aligned}$$

that

$$\begin{aligned}
&\begin{bmatrix} u \\ y \end{bmatrix} = \begin{bmatrix} M \\ N \end{bmatrix} v + \left(\begin{bmatrix} -\hat{Y} \\ \hat{X} \end{bmatrix} + \begin{bmatrix} M \\ N \end{bmatrix} Q_{u_{MC}}\Pi_a Q_{r,1}\right) r_y \\
&\quad + \begin{bmatrix} M \\ N \end{bmatrix} Q_{u_{MC}}\left(\Pi_a - I\right)\Psi\left(\begin{bmatrix} u \\ y \end{bmatrix} - \begin{bmatrix} M \\ N \end{bmatrix} v\right) \Longrightarrow \\
&\begin{bmatrix} u \\ y \end{bmatrix} = \begin{bmatrix} M \\ N \end{bmatrix} v + \left(\begin{bmatrix} -\hat{Y} \\ \hat{X} \end{bmatrix} + \begin{bmatrix} M \\ N \end{bmatrix} \Phi\right) r_y, \qquad (16.55) \\
&\Phi = Q_{u_{MC}}\left(I - \left(\Pi_a - I\right)\Psi\begin{bmatrix} M \\ N \end{bmatrix} Q_{u_{MC}}\right)^{-1}\left(\Pi_a Q_{r,1} + \left(\Pi_a - I\right)\Psi\begin{bmatrix} -\hat{Y} \\ \hat{X} \end{bmatrix}\right).
\end{aligned}$$

From (16.55) it can be seen clearly that during attack-free operations, i.e. $\Pi_a = I$,

$$\begin{bmatrix} u \\ y \end{bmatrix} = \begin{bmatrix} M \\ N \end{bmatrix} v + \left(\begin{bmatrix} -\hat{Y} \\ \hat{X} \end{bmatrix} + \begin{bmatrix} M \\ N \end{bmatrix} Q_{u_{MC}} Q_{r,1}\right) r_y$$

which is exactly the nominal closed-loop dynamic, and the feedback performance degradation caused by $\Pi_a \neq I$ is modelled by

$$\begin{bmatrix} u_f \\ y_f \end{bmatrix} = \left(\begin{bmatrix} -\hat{Y} \\ \hat{X} \end{bmatrix} + \begin{bmatrix} M \\ N \end{bmatrix} \Phi\right) r_y.$$

Consequently, it holds

$$\begin{aligned}
r_{y,u}^a &= \Pi_a\left(r_{y,u} - Q_v v\right) + Q_v v, \\
&= \Pi_a\left(Q_{r,1}r_y + \Psi\begin{bmatrix} u \\ y \end{bmatrix} - \Psi\begin{bmatrix} M \\ N \end{bmatrix} v\right) + Q_v v \\
&= \Pi_a Q_{r,1}r_y + \Pi_a\Psi\left(\begin{bmatrix} -\hat{Y} \\ \hat{X} \end{bmatrix} + \begin{bmatrix} M \\ N \end{bmatrix} Q_{u_{MC}}\Pi_a Q_{r,1}\right) r_y + Q_v v.
\end{aligned}$$

Hence, residual signal

$$\begin{aligned}
r^a_{PDD} &:= r^a_{y,u} - Q_v v \\
&= \Pi_a \left(Q_{r,1} + \Psi \left(\begin{bmatrix} -\hat{Y} \\ \hat{X} \end{bmatrix} + \begin{bmatrix} M \\ N \end{bmatrix} Q_{u_{MC}} \Pi_a Q_{r,1} \right) \right) r_y
\end{aligned}$$

contains the full information about the feedback performance degradation and is used for the detection purpose. To this end, design Ψ so that

$$\left\| \Psi \left(\begin{bmatrix} -\hat{Y} \\ \hat{X} \end{bmatrix} + \begin{bmatrix} M \\ N \end{bmatrix} Q_{u_{MC}} Q_{r,1} \right) \right\|_2 >> \left\| Q_{r,1} \right\|_2 ,$$

where $\|\cdot\|_2$ denotes $\mathcal{H}_2$-norm of a transfer function matrix. On account of the fact that the cyber-attacks target at reducing the system robustness leading to a larger

$$\left\| \Psi \left(\begin{bmatrix} -\hat{Y} \\ \hat{X} \end{bmatrix} + \begin{bmatrix} M \\ N \end{bmatrix} \Phi \right) \right\|_2 ,$$

it is expected that

$$\begin{aligned}
&\left\| \Psi \left(\begin{bmatrix} -\hat{Y} \\ \hat{X} \end{bmatrix} + \begin{bmatrix} M \\ N \end{bmatrix} \Phi \right) \right\|_2 > \\
&\left\| \Psi \left(\begin{bmatrix} -\hat{Y} \\ \hat{X} \end{bmatrix} + \begin{bmatrix} M \\ N \end{bmatrix} Q_{u_{MC}} Q_{r,1} \right) \right\|_2 >> \left\| Q_{r,1} \right\|_2 .
\end{aligned}$$

Considering moreover that Π_a should be an inner, the condition of attack stealthiness, the standard detection schemes, for (i) $r_y \sim \mathcal{N}\left(0, \Sigma_{r_y}\right)$ and (ii) $\mathcal{L}_2$-bounded unknown input r_y, can be applied to detect the attacks. For the former case, it is a problem of detecting increasing covariance matrix of r_{PDD}, while for the latter, the detection can be achieved by examining the $\mathcal{L}_2$-norm of r_{PDD}. For instance, for case (i), during attack-free operation, the covariance matrix of r_{PDD} can be calculated based on the model

$$r_{PDD} = \left(Q_{r,1} + \Psi \left(\begin{bmatrix} -\hat{Y} \\ \hat{X} \end{bmatrix} + \begin{bmatrix} M \\ N \end{bmatrix} Q_{u_{MC}} Q_{r,1} \right) \right) r_y .$$

To simplify the online implementation of the detection algorithm, a post-filter $R(z)$ is added so that

$$\bar{r}_{PDD}(z) = R(z) r_{PDD}(z) \sim \mathcal{N}\left(0, \Sigma_{\bar{r}_{PDD}}\right).$$

Using the generalised likelihood ratio detection method, we have the test statistic,

$$\begin{aligned} J &= \frac{N}{2} \ln \frac{\det \left(\Sigma_{\bar{r}_{PDD}} \right)}{\det \left(\hat{\Sigma}_{\bar{r}^a_{PDD}} \right)} \\ &+ \frac{1}{2} \sum_{i=1}^{N} \left(\left(\bar{r}^a_{PDD} (k+i) \right)^T \left(\Sigma^{-1}_{\bar{r}_{PDD}} - \hat{\Sigma}^{-1}_{\bar{r}^a_{PDD}} \right) \bar{r}^a_{PDD} (k+i) \right), \\ \hat{\Sigma}_{\bar{r}^a_{PDD}} &= \frac{1}{N} \sum_{i=1}^{N} \bar{r}^a_{PDD} (k+i) \left(\bar{r}^a_{PDD} (k+i) \right)^T, \end{aligned}$$

and determine the threshold accordingly. Here, it is supposed that N data, $\bar{r}^a_{PDD} (k+i), i = 1, \ldots, N$, have been collected. For case (ii), similar scheme can be followed. Firstly, design the post-filter using the unified solution so that

$$\|\bar{r}_{PDD}\|_{2,[k+1,k+N]} \leq \delta_d,$$

where δ_d is the upper-bound of the unknown input d. Thus, by means of the evaluation function J and the threshold J_{th},

$$J = \left\| \bar{r}^a_{PDD} \right\|_{2,[k+1,k+N]}, \; J_{th} = \delta_d,$$

the detection is performed. The reader is referred to the literature given at the end of this chapter for such standard detection schemes.

We now give ourselves over to the (additive) stealthy attacks subject to (16.43). Similar to the construction of the residual (16.54), signal $\bar{r}_{PD}$ is added, leading to

$$\begin{aligned} r_{y,u} &= Q_{r,1} r_y + Q_{r,2} r_u + \bar{r}_{PD}, \\ \bar{r}_{PD} &= \Psi \begin{bmatrix} u \\ y \end{bmatrix}. \end{aligned} \tag{16.56}$$

The controller is set to be

$$\begin{aligned} u &= F\hat{x} + u^a_{MC} - Q_{u_{MC}} \bar{r}_{PD}, \\ u_{MC} &= Q_{u_{MC}} \left(r^a_{y,u} - Q_{r,2} v \right) + v. \end{aligned}$$

Let

$$r^a_{y,u} = r_{y,u} + a_{r_{y,u}}, \; u^a_{MC} = u_{MC} + a_{u_{MC}}.$$

Next, we examine the system dynamics. Since

$$\begin{aligned} r_u = u - F\hat{x} = Q_{u_{MC}} &\left(Q_{r,1} r_y + Q_{r,2} r_u - Q_{r,2} v + \bar{r}_{PD} + a_{r_{y,u}} \right) \\ &+ v + a_{u_{MC}} - Q_{u_{MC}} \bar{r}_{PD} \Longrightarrow \end{aligned}$$

$$
\begin{aligned}
&r_u = \left(I - Q_{u_{MC}} Q_{r,2}\right)^{-1} \left(Q_{u_{MC}} \left(Q_{r,1} r_y - Q_{r,2} v + a_{r_{y,u}}\right) + v + a_{u_{MC}}\right),\\
&Q_{r,2} a_{u_{MC}} + a_{r_{y,u}} = 0 \Longrightarrow Q_{u_{MC}} a_{r_{y,u}} + a_{u_{MC}} = \left(I - Q_{u_{MC}} Q_{r,2}\right) a_{u_{MC}}\\
&\qquad \Longrightarrow u = F\hat{x} + v + a_{u_{MC}} + \left(I - Q_{u_{MC}} Q_{r,2}\right)^{-1} Q_{u_{MC}} Q_{r,1} r_y,
\end{aligned}
$$

the closed-loop dynamic is governed by

$$
\begin{bmatrix} u \\ y \end{bmatrix} = \begin{bmatrix} M \\ N \end{bmatrix} \left(v + a_{u_{MC}}\right) + \left(\begin{bmatrix} -\hat{Y} \\ \hat{X} \end{bmatrix} + \begin{bmatrix} M \\ N \end{bmatrix} Q_{u_{MC}} \bar{Q}_{r,1}\right) r_y,
$$

which results in

$$
\begin{aligned}
r_{y,u}^a &= \left(I - Q_{u_{MC}} Q_{r,2}\right)^{-1} r_y + Q_{r,2} v + \bar{r}_{PD}\\
&= \left(I - Q_{u_{MC}} Q_{r,2}\right)^{-1} r_y + Q_{r,2} v\\
&+ \Psi \left(\begin{bmatrix} M \\ N \end{bmatrix} \left(v + a_{u_{MC}}\right) + \left(\begin{bmatrix} -\hat{Y} \\ \hat{X} \end{bmatrix} + \begin{bmatrix} M \\ N \end{bmatrix} Q_{u_{MC}} \bar{Q}_{r,1}\right) r_y\right).
\end{aligned}
$$

Thus, building residual signal

$$
r_{PDD} := r_{y,u}^a - \left(Q_{r,2} + Q_v\right) v
$$

gives

$$
\begin{aligned}
r_{PDD} &= Q_v a_{u_{MC}} + \Gamma r_y, \\
\Gamma &= \Psi \left(\begin{bmatrix} -\hat{Y} \\ \hat{X} \end{bmatrix} + \begin{bmatrix} M \\ N \end{bmatrix} Q_{u_{MC}} \bar{Q}_{r,1}\right) + \left(I - Q_{u_{MC}} Q_{r,2}\right)^{-1}.
\end{aligned} \tag{16.57}
$$

Detecting $a_{u_{MC}}$ on the basis of model (16.57) is a standard fault (change) detection problem, both for (i) $r_y \sim \mathcal{N}\left(0, \Sigma_{r_y}\right)$ and (ii) $\mathcal{L}_2$-bounded unknown input r_y. The reader can refer to the fault detection methods introduced in Chap. 8 and the literature given at the end of this chapter.

16.3.3 On Implementation

In the previous and this sections, three different forms of the residual signal $r_{y,u}$ have been proposed, (16.15), (16.54) and (16.56). While the first one is applied to the detection of cyber-attacks in general, the latter two are dedicated to detecting stealthy attacks. It is noteworthy that, independent of which of the three residual forms is adopted, the overall closed-loop dynamics under attack-free operations are identical with the nominal system dynamic. For a reliable detection of cyber-attacks with high resilience, a schedule mechanism is to be designed that manages switchings between the three detection algorithms. For instance, run the detection algorithm based the residual $r_{y,u}$ (16.15) as the base detection scheme, and switch to an algorithm with residual $r_{y,u}$ (16.54) or (16.56) over a short time interval, and then return to the base detection scheme. The schedule mechanism can be triggered by a time function or, more reliably, by a random generator. A meaningful side-effect of such a schedule mechanism is to enhance the resilience against eavesdropping attacks, which is a typical strategy followed by a cyber adversary to design stealthy attacks.

16.4 Notes and References

Resilient control of CPSs under cyber-attacks is the objective of this chapter. In the control community, resilient control of CPSs is mostly understood and handled analogously to a fault-tolerant control (FTC) system, which mainly consists of detection and control as the two essential functional units [31, 33, 74]. It is interesting to observe how researchers from computer science assess the efforts made by the control community towards resilient control of CPSs. In their survey paper [179], the authors describe the contributions made by the control community with the following text, "*although these approaches are called resilient by control theorists, from a cyber-security standpoint, resilient control is still dependent on a detection and reaction paradigm. In other words, although resilient control incorporates in the traditional fault-tolerant control new strategies to face cyber-security breaches, it still aims at determining how a controller can detect, correctly estimate the system state, and recalculate the required commands despite malicious data. It also aims at responding to the attacks with appropriate countermeasures to achieve stability and graceful degradation while the system is under attack*". Without doubt, this analysis appropriately summarises the state of the art of research on resilient control of CPSs in the control community. It can be observed that observer/Kalman filter based state estimation of systems under cyber-attacks has been a dominant thematic area of resilient control of CPSs [33]. Such handling is popular and common in control theory, considering that there exist rich robust and fault-tolerant control methods to deal with control performance degradation caused by the cyber-attacks, as far as the system state variables are accurately reconstructed. As appropriate control-theoretic countermeasures, the so-called resource-aware secure control methods [33], like event-triggered control algorithms or switching control mechanisms, are prevalent resilient control schemes. Roughly speaking, these methods make the CPS highly resilient against attacks by means of optimal management of data communications among subsystems in the CPS.

In this chapter, we have handled resilient control issues from different aspects and on account of specifications, which demand not only high resilience of CPSs under cyber-attacks and high fault-tolerance against process faults, but also

- "fail-safe" cyber-security,
- the data privacy, and
- online computation and communication effort.

16.4.1 Summary

For our purpose, in the first part of this chapter, alternative control strategies aiming at suppressing the influence of cyber-attacks have been addressed. Having the closed-loop dynamic under cyber-attacks in mind,

$$\begin{bmatrix} u \\ y \end{bmatrix} = \begin{bmatrix} M \\ N \end{bmatrix} \left(v + \vartheta_a + \Pi_a \begin{bmatrix} v \\ r_y \end{bmatrix} \right) + \left(\begin{bmatrix} -\hat{Y} \\ \hat{X} \end{bmatrix} + \begin{bmatrix} M \\ N \end{bmatrix} Q \right) r_y,$$

where ϑ_a and Π_a parametrise additive and multiplicative types of attacks, respectively, it is natural that our study has focused on the use of the residual signal r_u given by

$$r_u = \begin{bmatrix} X & Y \end{bmatrix} \begin{bmatrix} u \\ y \end{bmatrix} = v + \vartheta_a + \Pi_a \begin{bmatrix} v \\ r_y \end{bmatrix} + Q r_y.$$

To this end, two schemes have been proposed in Sect. 16.1. The first one is to add r_u into the input signal u,

$$u = F\hat{x} + Q r_y + \bar{Q}_u r_u + v,$$

which leads to

$$\begin{bmatrix} u \\ y \end{bmatrix} = \begin{bmatrix} M \\ N \end{bmatrix} Q_u \left(v + \vartheta_a + \Pi_a \begin{bmatrix} v \\ r_y \end{bmatrix} + Q r_y \right) + \begin{bmatrix} -\hat{Y} \\ \hat{X} \end{bmatrix} r_y,$$
$$Q_u^{-1} = I - \bar{Q}_u.$$

The last expression is equivalent to the controller setting

$$u = F\hat{x} + Q_u \left(Q r_y + v \right).$$

In this regard, this control mechanism is called feedforward control. Note that by the online implementation of this control scheme, the MC-station is not able to tune the controller parameter Q_u due to the lack of information about r_u.

With the transmission of r_u from the plant side to the MC-station, the second control mechanism differs from the first one. In other words, it realises a feedback control via the channel of r_u. In comparison with the first scheme, the feedback of r_u is of the advantages that

- information contained in r_u about attacks can be used for the detection purpose, that is, it allows an integrated design of control and detection,
- it enables tuning of controller parameter on the MC-station, and
- it can enhance the system robustness against model uncertainties, a typical capability of feedback control.

To be specific, a new type of residual signal is defined,

$$r_{y,u} = Q_{r,1} r_y + Q_{r,2} r_u,$$

and the controller is set to be

$$u = F\hat{x} + Q_{u_{MC}} \left(r_{y,u} - Q_{r,2} v \right) + v$$

with the design/tuning parameters $Q_{u_{MC}}$, $Q_{r,1}$ and $Q_{r,2}$ for controller and detector.

Both control mechanisms evidently differ from state of the art methods for resilient control in CPSs.

In the second part of this chapter, a comprehensive design scheme towards a resilient and fault-tolerant control system of CPSs has been proposed, which consists of four sub-systems, (i) fault detection, (ii) cyber-attack detection, (iii) performance optimisation and recovery, and (iv) resilient control of CPS under cyber-attacks, whereby issues (ii) and (iv) are the main study focus.

The work begins in Sect. 16.2 with the control system configuration and information pattern. Shown in Fig. 16.1, the proposed control system configuration with an embedded controller directly placed at the plant side and transmission of the residual signal meets the requirements for "fail-safe" cyber-security and the data privacy. It is worth emphasising that any output feedback controllers, including observer-based state feedback controllers typically adopted in the literature on resilient control of CPSs, can be equivalently built in the configuration given in Fig. 16.1. On the other hand, differing from the state of the art research, in the CPS configuration of our work the reference signal is included. In Remark 16.2, we have outlined the importance of the reference signal, in particular in dealing with secure CPSs, and clearly pointed out the deficits, when the reference signal is not considered.

According to the conclusion that the residual signal r_y is decoupled from cyber-attacks, in the proposed CPS control system, fault detection is performed directly on the plant side and based on r_y. This enables a reliable and timely fault detection. Once a fault is detected, a message is sent to the MC-station. For the detection end, well-established fault detection methods [36, 39, 218] can be applied, among them Kalman filter- and $\mathcal{H}_2$ observer-based fault detection systems for processes corrupted with additive disturbances, noises or $\mathcal{L}_2$-bounded unknown inputs, or the projection-based method proposed in Chap. 9 for processes with uncertainties are recommended. Besides their optimal detection performance, the residual signals generated in such detection systems do not comprise process information. That means, transmission of residual signals prevents the data privacy efficiently.

Cyber-attack detection is performed at the MC-station under the condition that no fault is detected. On the basis of the attack and residual models, the detection model (16.22) is achieved, i.e.

$$r_{\eta_a} = Q_{r,1} r_y + \eta_a,$$

with η_a consisting of $a_{u_{MC}}$ and a_{r_y}, the attacks on the channels of r_u and r_y. On the assumption of additive attacks, such a detection problem is trivial. Analogous to the fault detection issues, well-established fault detection methods [36, 39] can be applied for this purpose as well, for instance, the Kalman filter solution when r_y is white noise, and $\mathcal{H}_2$ observer-based solution, when r_y is $\mathcal{L}_2$-bounded unknown input. Nevertheless, differing from the general fault detection problem, cyber-attack detection demands for detecting both attack injection and disappearance, in order to achieve an efficient and reliable resilient control. In this regard, a hypothesis detection problem,

$$J \underset{\mathcal{H}_1}{\overset{\mathcal{H}_0}{\lesseqgtr}} J_{th},$$

is formulated, where the null hypothesis $\mathcal{H}_0$ is for attacks, and the alternative hypothesis $\mathcal{H}_1$ is for attack-free. The reader is referred to [8] for a detailed description of the detection scheme.

The application of the control mechanism of feeding back the residual signal r_u enables us to optimise the resilient performance by tuning $Q_{u_{MC}}$ and $Q_{r,1}$. To be specific, given the closed-loop dynamic (16.36) with additive attacks ϑ_a and known $Q_{r,2}$,

$$\begin{aligned}\begin{bmatrix} u \\ y \end{bmatrix} &= \begin{bmatrix} M \\ N \end{bmatrix}\left(v + \left(I - Q_{u_{MC}} Q_{r,2}\right)^{-1} \vartheta_a\right) \\ &+ \left(\begin{bmatrix} -\hat{Y} \\ \hat{X} \end{bmatrix} + \begin{bmatrix} M \\ N \end{bmatrix} Q_{u_{MC}} \left(I - Q_{r,2} Q_{u_{MC}}\right)^{-1} Q_{r,1}\right) r_y,\end{aligned}$$

find $Q_{u_{MC}}, Q_{r,1} \in \mathcal{RH}_\infty$ so that

$$\begin{aligned}\left\| \left(I - Q_{u_{MC}} Q_{r,2}\right)^{-1} \right\|_\infty &\le \gamma_{\vartheta_a}, \\ \left\| \begin{bmatrix} -\hat{Y} \\ \hat{X} \end{bmatrix} + \begin{bmatrix} M \\ N \end{bmatrix} Q_{u_{MC}} \left(I - Q_{r,2} Q_{u_{MC}}\right)^{-1} Q_{r,1} \right\|_\infty &\le \gamma_{r_y},\end{aligned}$$

a trivial $\mathcal{H}_\infty$-optimisation problem. In case of existence of model uncertainties, the control system design becomes considerably involved. A trade-off design of a sub-optimal resilient control has been suggested at the end of Sect. 16.2.

The capacity of the aforementioned attack detection methods is so far limited, since there exist attacks that cannot be detected using the residual signal $r_{y,u}$. They are those attacks subject to (16.43),

$$Q_{r,2}a_{u_{MC}} + a_{r_{y,u}} = 0.$$

Observe that this stealthiness condition implies that the residual signal $r^a_{y,u}$ is fully decoupled from $\left(a_{u_{MC}}, a_{r_{y,u}}\right)$, and thus such class of attacks is stealthy, independent of the system dynamics and applied detection algorithms. It is interesting to notice that the existence condition (16.43) is similar to the so-called covert attacks [185]. The difference between them lies in the transfer function $Q_{r,2}$, which is the transfer function of the plant in case of a covert attack. Specifically, $Q_{r,2}$ is a design parameter and can also be online updated during system operations, while the transfer function of the plant is fixed. This fact makes the realisation of such stealthy attacks more difficult.

There exists a further class of stealthy attacks, whose generation depends on the system dynamics and the detection algorithms. In fact, detection and construction (design) of this class of stealthy attacks are the recent trending topic. Technically speaking, the reported methods can be schematically formulated as: given a model

$$r \sim \mathcal{N}(0, \Sigma_r), r \in \mathbb{R}^m, \tag{16.58}$$

representing the nominal (attack-free) residual signal, find a linear mapping of r,

$$\bar{r} = f(r) \in \mathbb{R}^m$$

so that

- the existing attack detector cannot detect changes in the signal r, e.g. $\bar{r}$ is of the similar statistic like r, and
- the control performance is considerably degraded.

Before introducing the basic idea to approach the problem solution, we would like to exemplify the background of the above formulation. On the assumption of LTI model with white process and measurement noises, the use of a (steady) Kalman filter delivers an optimal state estimation $\hat{x}(k)$ (in the sense of minimum covariance of the state estimation error) and an innovation series as residual (refer to Sect. 8.2.1). The latter can be uniformly modelled by (16.58). When the mapping function f is a linear operator, $\bar{r}$ should be a normally distributed random vector as well. Consequently, a (linear) mapping f leading to

$$\bar{r} \sim \mathcal{N}(0, \Sigma_{\bar{r}}), \Sigma_{\bar{r}} = \Sigma_r, \tag{16.59}$$

guarantees the strictly stealthy attacks. Notice that this conclusion is independent of the test statistic implemented on the attack detection, and hence holds also for Kullback-Leibler divergence (KLD) based test statistic [229]. In the literature, a considerable number of design methods have been reported. Guo et al. [76] proposed the idea of a manipulating r by

$$\bar{r} = \Pi_a r$$

with a unitary matrix (or operator) Π_a so that the condition (16.59) or similar conditions are satisfied. This idea has been followed by numerous subsequent works, see, for instance, [97, 122, 171, 229, 242]. It is noteworthy that differing from above-mentioned works, [203] has addressed the issue of stealthy attack design for CPSs with $\mathcal{L}_2$ -bounded unknown inputs, where the residual evaluation function is the $\mathcal{L}_2$-norm of the residual signal over a time interval. A cyber-attack is then (strictly) stealthy, if it holds

$$\|\bar{r}\|_{2,[k_0,k_1]} \leq \|r\|_{2,[k_0,k_1]} . \tag{16.60}$$

It is noteworthy that availability of process model knowledge or collection of sufficient amount of data, by which process model can be identified, are the widely accepted and adopted assumptions, on which the developed design methods for stealthy attacks are based. In fact, if the stealthiness aspect is under consideration exclusively, stealthy attacks can be constructed more practically without rich process model knowledge. As schematically described in the previous section, in case of model (16.58), Σ_r can be well estimated by means of sufficient number of data $r(k)$. Then, constructing an LTI or even time-varying co-inner (Kalman filter) with $r(k)$ as its input delivers $\bar{r}(k)$ as output corrupted with cyber-attacks. When the $\mathcal{L}_2$-norm of the residual signal serves as evaluation function, a co-inner with $r(k)$ as its input guarantees the stealthiness condition (16.60). It should be emphasised that in both cases no knowledge of the process model is necessary. Note that, under the CPS configuration studied in our work, the corresponding (design) scenario of this type of stealthy attacks is

$$r_{y,u}^a = \Pi_a \left(r_{y,u} - Q_v v \right) + Q_v v.$$

Naturally, constructing stealthy attacks is not the true target of attackers. Degrading system performance is the attacker's intention. In the literature, attacks on the estimation performance of the embedded observer are the focus of the existing research, both for control and (remote) monitoring of CPSs [33, 179]. Roughly speaking, there are two mostly studied scenarios of attacks on $\hat{x}(k)$, (i) increasing the covariance matrix $e(k) = x(k) - \hat{x}(k)$, (ii) increasing $\|e(k)\|$. Let us examine the influence of such attacks on the CPS closed-loop dynamic

$$\begin{bmatrix} u \\ y \end{bmatrix} = \begin{bmatrix} M \\ N \end{bmatrix} v + \left(\begin{bmatrix} -\hat{Y} \\ \hat{X} \end{bmatrix} + \begin{bmatrix} M \\ N \end{bmatrix} Q \right) r_y,$$

whose state space representation is

$$\begin{cases} \hat{x}(k+1) = (A+BF)\,\hat{x}(k) + B\left(r_{y,Q}(k)+v(k)\right) + Lr_y(k), \hat{x}\,(0) = 0 \\ u(k) = F\hat{x}\,(k) + r_{y,Q}(k) + v(k) \\ y(k) = (C+DF)\,\hat{x}(k) + D\left(r_{y,Q}(k)+v(k)\right) + r_y(k) \\ r_y(k) = y(k) - \hat{y}(k) = y(k) - (C+DF)\,\hat{x}(k) - D\left(r_{y,Q}(k)+v(k)\right) \\ r_{y,Q}(z) = Q(z)r_y(z). \end{cases}$$

Note that in the CPS configuration we addressed, r_y is decoupled from the attacks, and the attacks can be equivalently modelled by attacks on $r_{y,Q}$, that is

$$r^a_{y,Q}(z) = \begin{cases} Q(z)\Pi_a(z)r_y(z), & \text{case (i)} \\ Q(z)r_y(z) + \vartheta_a(z), & \text{case (ii)}. \end{cases}$$

It yields

$$\begin{cases} \hat{x}(k+1) = (A+BF)\,\hat{x}(k) + B\left(r^a_{y,Q}(k)+v(k)\right) + Lr_y(k), \hat{x}\,(0) = 0 \\ u(k) = F\hat{x}\,(k) + r^a_{y,Q}(k) + v(k) \\ y(k) = (C+DF)\,\hat{x}(k) + D\left(r^a_{y,Q}(k)+v(k)\right) + r_y(k). \end{cases}$$

Accordingly, such attacks result in changes in the closed-loop dynamic equivalently modelled by

$$\begin{bmatrix} u \\ y \end{bmatrix} = \begin{bmatrix} M \\ N \end{bmatrix}(v+\vartheta_a) + \left(\begin{bmatrix} -\hat{Y} \\ \hat{X} \end{bmatrix} + \begin{bmatrix} M \\ N \end{bmatrix} Q\Pi_a\right) r_y.$$

This result has motivated us to detect attacks satisfying the stealthy conditions

$$Q_{r,2}a_{u_{MC}} + a_{r_{y,u}} = 0 \text{ or } r^a_{y,u} = \Pi_a\left(r_{y,u} - Q_v v\right) + Q_v v$$

by means of performance degradation monitoring. To this end, a performance indicator,

$$\bar{r}_{PD} = \Psi \begin{bmatrix} u \\ y \end{bmatrix},$$

is embedded in the residual signal $r_{y,u}$,

$$r_{y,u} = Q_{r,1}r_y + Q_{r,2}r_u + \bar{r}_{PD},$$

based on which two detection schemes have been proposed. The first one is dedicated to detecting stealthy attacks subject to $r^a_{y,u} = \Pi_a\left(r_{y,u} - Q_v v\right) + Q_v v$, which leads to detecting changes in the covariance matrix of signal r_{PDD},

$$r_{PDD} = \left(Q_{r,1} + \Psi\left(\begin{bmatrix} -\hat{Y} \\ \hat{X} \end{bmatrix} + \begin{bmatrix} M \\ N \end{bmatrix} Q_{u_{MC}} Q_{r,1}\right)\right) r_y,$$
$$r_y \sim \mathcal{N}\left(0, \Sigma_{r_y}\right),$$

caused by the attacks. The solution consists of two steps: (I) design a post-filter $R(z)$ so that

$$\bar{r}_{PDD}(z) = R(z) r_{PDD}(z), \bar{r}_{PDD}(k) \sim \mathcal{N}\left(0, \Sigma_{\bar{r}_{PDD}}\right),$$

i.e. $\bar{r}_{PDD}(k)$ is a white noise series with (constant) covariance matrix $\Sigma_{\bar{r}_{PDD}}$, (II) detection of changes in $\Sigma_{\bar{r}_{PDD}}$. The first step can be achieved by a co-inner-outer factorisation (equivalent to a Kalman filter). The reader is referred to Chap. 8, [36, 39], while the second step is a trivial detection problem and can be well solved using GLR- and KLD-methods, [8, 39], or using Riemannian distance of Symmetric Positive Definite (SPD) Matrices [39].

The second scheme deals with detecting stealthy attacks subject to $Q_{r,2} a_{u_{MC}} + a_{r_{y,u}} = 0$. To this end, the detection model (16.57), i.e.

$$r_{PDD} = Q_v a_{u_{MC}} + \Gamma r_y,$$
$$\Gamma = \Psi\left(\begin{bmatrix} -\hat{Y} \\ \hat{X} \end{bmatrix} + \begin{bmatrix} M \\ N \end{bmatrix} Q_{u_{MC}} \bar{Q}_{r,1}\right) + \left(I - Q_{u_{MC}} Q_{r,2}\right)^{-1},$$

has been established. When

$$rank\left(Q_v\right) = p = \dim\left(u\right) < m,$$

a co-inner-outer factorisation of Q_v,

$$Q_v = Q_v^{co} Q_v^{ci},$$

is performed firstly. The residual signal r_{PDD} is further transferred to

$$\hat{r}_{PDD} = \left(Q_v^{co}\right)^{-1} r_{PDD} = \hat{a}_{u_{MC}} + \hat{\Gamma} r_y$$
$$\hat{a}_{u_{MC}} = Q_v^{ci} a_{u_{MC}}, \hat{\Gamma} = \left(Q_v^{co}\right)^{-1} \Gamma.$$

Consequently, we write detection model in a uniform form,

$$\hat{r}_{PDD} = R_{PDD} r_{PDD} = \hat{a}_{u_{MC}} + \hat{\Gamma} r_y,$$
$$R_{PDD} = \begin{cases} \left(Q_v^{co}\right)^{-1}, rank\left(Q_v\right) = p < m \\ I, rank\left(Q_v\right) = m, \end{cases}$$
$$\hat{\Gamma} = \begin{cases} \left(Q_v^{co}\right)^{-1} \Gamma, rank\left(Q_v\right) = p < m \\ \Gamma, rank\left(Q_v\right) = m. \end{cases}$$

$$\hat{a}_{u_{MC}} = \begin{cases} Q_v^{ci} a_{u_{MC}}, rank\left(Q_v\right) = p < m \\ Q_v a_{u_{MC}}, rank\left(Q_v\right) = m. \end{cases}$$

It is apparent that detecting $\hat{a}_{u_{MC}}$ is a standard fault (change) detection problem, and thus can be solved analogously to the detection scheme proposed in the first case. That is, (I) design a post filter R by means of a co-inner-outer factorisation of $\hat{\Gamma} = \hat{\Gamma}^{co}\hat{\Gamma}^{ci}$,

$$R = \left(\hat{\Gamma}^{co}\right)^{-1},$$

(II) detection of $\tilde{a}_{u_{MC}} = \left(\hat{\Gamma}^{co}\right)^{-1} \hat{a}_{u_{MC}}$. Note that in the case of $r_y \sim \mathcal{N}\left(0, \Sigma_{r_y}\right)$, detection of $\tilde{a}_{u_{MC}}$ is equivalent to detecting changes in the expectation value of $\left(\hat{\Gamma}^{co}\right)^{-1} \hat{r}_{PDD}$. As a result, the GLR method leads to the χ^2 test statistic. For $\mathcal{L}_2$-bounded unknown input r_y and on the assumption that r_y is generated by the unified solution ($\mathcal{H}_2$-observer-based residual generator) [36, 39], detection is achieved as: (i) defining the evaluation function J,

$$J = \left\| \left(\hat{\Gamma}^{co}\right)^{-1} \hat{r}_{PDD} \right\|_{2,[k_0,k_1]},$$

and (ii) setting threshold J_{th},

$$J_{th} = \delta_d,$$

where δ_d is the upper-bound of the $\mathcal{L}_2$-norm of the unknown input in the system [36, 39].

We would like to emphasise that we have introduced the both schemes of performance-based detection of stealthy attacks separately in order to simplify the problem formulation and the corresponding solution. For the practical application, it is recommended to design the detector in an integrated form with the uniform residual

$$r_{y,u} = Q_{r,1} r_y + Q_{r,2} r_u + \bar{r}_{PD},$$

and, based on it, the uniform detection model.

It is of interest to observe that research secure issues of CPSs under stealthy cyber-attacks on a part of the communication (data transmission) channels seem to draw recent research attention. For instance, in [141], the so-called stealthy local covert attacks have been considered, and the authors in [125] have studied this issue in the context of stealthy attacks with energy constraints. In the regard of unified framework of control and detection, such class of attacks can be modelled by $\left(A_{r_{y,u}} a_{r_{y,u}}, A_{u_{MC}} a_{u_{MC}}\right)$

$$
\begin{aligned}
r_{y,u}^{a} &= r_{y,u} + A_{r_{y,u}} a_{r_{y,u}}, \; u_{MC}^{a} = u_{MC} + A_{u_{MC}} a_{u_{MC}}, \\
A_{r_{y,u}} &= diag\left(\delta_1, \ldots, \delta_m\right), \delta_i = \begin{cases} 1, & \text{attack on the } i\text{-th channel of } r_{y,u} \\ 0, & \text{no attack on the } i\text{-th channel of } r_{y,u}, \end{cases} \\
A_{u_{MC}} &= diag\left(\delta_1, \ldots, \delta_p\right), \delta_j = \begin{cases} 1, & \text{attack on the } j\text{-th channel of } u_{MC} \\ 0, & \text{no attack on the } j\text{-th channel of } u_{MC}, \end{cases} \\
i &= 1, \ldots, m, , j = 1, \ldots, p.
\end{aligned}
$$

Accordingly, the attack detection residual is subject to,

$$
\begin{aligned}
r_{y,u}^{a} &= r_{y,u} + A_{r_{y,u}} a_{r_{y,u}} = Q_{r,1} r_y + Q_{r,2} r_u + A_{r_{y,u}} a_{r_{y,u}} \Longrightarrow \\
r_{\eta_a} &= \left(I - Q_{r,2} Q_{u_{MC}}\right)\left(r_{y,u}^{a} - Q_{r,2} v\right) \\
&= Q_{r,1} r_y + Q_{r,2} A_{u_{MC}} a_{u_{MC}} + A_{r_{y,u}} a_{r_{y,u}},
\end{aligned}
$$

and the closed-loop dynamic is governed by

$$
\begin{aligned}
\begin{bmatrix} u \\ y \end{bmatrix} &= \begin{bmatrix} M \\ N \end{bmatrix} \left(v + \left(I - Q_{u_{MC}} Q_{r,2}\right)^{-1} \left(Q_{u_{MC}} A_{r_{y,u}} a_{r_{y,u}} + A_{u_{MC}} a_{u_{MC}}\right)\right) \\
&+ \left(\begin{bmatrix} -\hat{Y} \\ \hat{X} \end{bmatrix} + \begin{bmatrix} M \\ N \end{bmatrix} Q_{u_{MC}} \left(I - Q_{r,2} Q_{u_{MC}}\right)^{-1} Q_{r,1}\right) r_y.
\end{aligned}
$$

It immediately becomes apparent that the stealthiness condition is

$$
Q_{r,2} A_{u_{MC}} a_{u_{MC}} + A_{r_{y,u}} a_{r_{y,u}} = 0,
$$

and in case of such stealthy attacks, the closed-loop dynamic is governed by

$$
\begin{aligned}
\begin{bmatrix} u \\ y \end{bmatrix} &= \begin{bmatrix} M \\ N \end{bmatrix} \left(v + A_{u_{MC}} a_{u_{MC}}\right) \\
&+ \left(\begin{bmatrix} -\hat{Y} \\ \hat{X} \end{bmatrix} + \begin{bmatrix} M \\ N \end{bmatrix} Q_{u_{MC}} \left(I - Q_{r,2} Q_{u_{MC}}\right)^{-1} Q_{r,1}\right) r_y.
\end{aligned}
$$

If multiplicative attacks on a part of $r_{y,u}$ are considered, for instance,

$$
r_{y,u} = \begin{bmatrix} r_{y,u,a} \\ r_{y,u,o} \end{bmatrix}, \begin{cases} r_{y,u,a}: & \text{on which attacks are injected} \\ r_{y,u,o}: & \text{on which no attacks are added,} \end{cases}
$$

then stealthy attacks on the feedback mechanism are described by

$$
r_{y,u}^{a} = \begin{bmatrix} \Pi_a \left(r_{y,u,a} - G_{v,a} v\right) + G_{v,a} v \\ r_{y,u,o} \end{bmatrix},
$$

where $G_{v,a}v$ denotes the transfer function from the reference v to the residual $r_{y,u,a}$. Corresponding to the residual signal $r_{y,u} = Q_{r,1}r_y + Q_{r,2}r_u$, the closed-loop dynamic is given by

$$
\begin{aligned}
u_{MC} &= Q_{u_{MC}}\left(r_{y,u}^{a} - Q_{r,2}v\right) + v \Longrightarrow \\
u_{MC} &= \left(I - Q_{u_{MC}}\bar{\Pi}_a Q_{r,2}\right)^{-1} Q_{u_{MC}}\bar{\Pi}_a Q_{r,1} r_y, \\
\begin{bmatrix} u \\ y \end{bmatrix} &= \begin{bmatrix} M \\ N \end{bmatrix} v + \left(\begin{bmatrix} -\hat{Y} \\ \hat{X} \end{bmatrix} + \begin{bmatrix} M \\ N \end{bmatrix} Q_{u_{MC}}\left(I - \bar{\Pi}_a Q_{r,2} Q_{u_{MC}}\right)^{-1} \bar{\Pi}_a Q_{r,1}\right) r_y, \\
\bar{\Pi}_a &= \begin{bmatrix} \Pi_a & 0 \\ 0 & I \end{bmatrix}.
\end{aligned}
$$

16.4.2 Concluding Remarks

Apparently, our work on secure CPSs in this and previous two chapters differs from state of the art research in this thematic area. The main differences can be summarised from the following four aspects,

- system integrity,
- system configuration,
- the transmitted signals, and
- system analysis and synthesis.

Aspect system integrity: The CPS control is designed to meet the requirements on cyber-security, resilience and fault-tolerance. While cyber-security demands for fail-safe system operation and data privacy, resilient and fault-tolerant control requires reliable and timely detection of cyber-attacks and technical anomalies (faults), and performing efficient countermeasures.

Aspect system configuration: The proposed system configuration shown in Fig. 16.1 and the distributed control and detection functionalities guarantee a secure system operation also in case of broken communication, and enable real-time updating the control and detection algorithms by tuning the system parameters in the MC-station. To be specific, the embedded observer-based state feedback controller and fault detector at the plant side control ensure the overall system stability and timely detection of system faults. The control and detection algorithms implemented on the MC-station considerably enhance the capability of attack-resilience and fault-tolerance by tuning the system parameters $\left(Q_{u_{MC}}, Q_{r,1}, Q_{r,2}\right)$.

Aspect transmitted signals: Signal feedback is the centrepiece of any control and detection systems. Without exaggeration, the spirit of the proposed CPS control and detection system is the feedback of the residual signal

$$r_{y,u} = Q_{r,1}r_y + Q_{r,2}r_u,$$

that throughout reflects the spirit of system integrity. The residual $r_{y,u}$ integrates all necessary information needed for feedback control and attack detection. The residual signal r_y, as a part of the control feedback information, serves to prevent adversary to gain system knowledge by means of eavesdropping attacks as well. This function can be interpreted as a natural realisation of differential privacy for the control system data [103] and thus increases the data privacy. Note that the control command u_{MC} sent by the MC-station to the plant side contains r_y as well.

Aspect system analysis and synthesis: The aforementioned system configurations, functionalities and algorithms are designed on the control-theoretic basis and the unified framework of control and detection. The core of the proposed analysis and design methods and algorithms consists in the Bezout identity and, based on it, the one-to-one mapping between the system data (u, y) and the residual signals (r_u, r_y). Amongst a number of essential results, the following two,

- system faults and cyber-attacks act on two (perpendicular) signal subspaces, i.e. faults on the residual subspace, while cyber-attacks on the image subspace,
- a full parameterisation of CPS closed-loop dynamic,

$$\begin{bmatrix} u \\ y \end{bmatrix} = \begin{bmatrix} M \\ N \end{bmatrix} Q_u r_u + \left(\begin{bmatrix} -\hat{Y} \\ \hat{X} \end{bmatrix} + \begin{bmatrix} M \\ N \end{bmatrix} Q_r \right) r_y,$$

  where $(Q_u, Q_r) \in \mathcal{RH}_\infty$ are the parameter systems,

are of elementary importance. They significantly facilitate the design and solutions of the associated problems, which are otherwise hard and considerably complicated to be handled in the state space and using the input-output system dynamics.

Chapter 17
A Survey of the Unified Framework of Control and Detection

In the previous chapters, the main concepts, schemes and methods of the unified framework have been introduced problem- and application oriented, respectively. The main focus of this chapter is on surveying the unified framework as a whole and in a three-layer structure, as sketched in Fig. 17.1.

17.1 Bezout Identity: The Core of the Unified Framework

Bezout identity is a famous result in the field of algebra. In the context of coprime factorisations of transfer functions, the Bezout identity of the form

$$\begin{bmatrix} M & -\hat{Y} \\ N & \hat{X} \end{bmatrix} \begin{bmatrix} X & Y \\ -\hat{N} & \hat{M} \end{bmatrix} = \begin{bmatrix} I & 0 \\ 0 & I \end{bmatrix}, \tag{17.1}$$

together with its extension

$$\begin{bmatrix} M & -\hat{Y} + MQ \\ N & \hat{X} + NQ \end{bmatrix} \begin{bmatrix} X + Q\hat{N} & Y - Q\hat{M} \\ -\hat{N} & \hat{M} \end{bmatrix} = \begin{bmatrix} I & 0 \\ 0 & I \end{bmatrix}, \ Q \in \mathcal{RH}_\infty, \tag{17.2}$$

is crucial for controller synthesis and robustness analysis. Specifically, it is the mathematical basis for the parameterisation of all stabilising controllers and stability analysis of feedback control systems. At the core of the unified framework, the role of the Bezout identity is multifaceted.

S. X. Ding, *Unified Framework of Control and Detection*,
https://doi.org/10.1007/978-3-662-72417-0_17

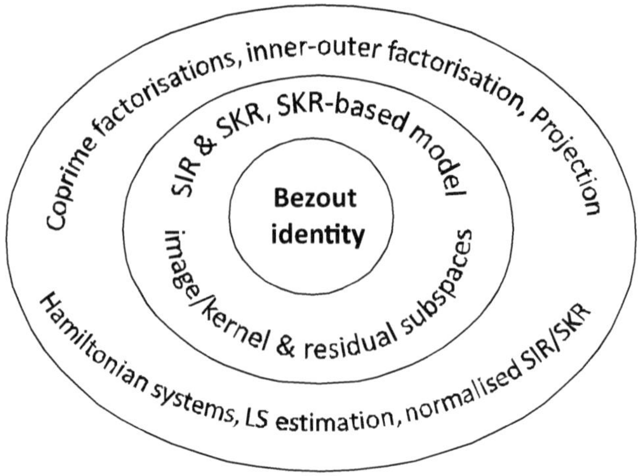

Fig. 17.1 The unified framework of control and detection

17.1.1 Perspective Significance

Informally speaking, control theory is based on input and output system dynamics, either described by a transfer function or by a state space representation. Schematically, a control loop is described by

$$\begin{cases} y = Gu \\ u = Ky. \end{cases} \tag{17.3}$$

By means of the Bezout identity, any dynamic systems, both open and closed-loop configurated, can be modelled by

$$\begin{bmatrix} u \\ y \end{bmatrix} = \begin{bmatrix} M & -\hat{Y} + MQ \\ N & \hat{X} + NQ \end{bmatrix} \begin{bmatrix} v \\ r_y \end{bmatrix} = \begin{bmatrix} M \\ N \end{bmatrix} v + \begin{bmatrix} -\hat{Y} + MQ \\ \hat{X} + NQ \end{bmatrix} r_y, \tag{17.4}$$

where (v, r_y) are two signal vectors and generated by

$$\begin{bmatrix} v \\ r_y \end{bmatrix} = \begin{bmatrix} X + Q\hat{N} & Y - Q\hat{M} \\ -\hat{N} & \hat{M} \end{bmatrix} \begin{bmatrix} u \\ y \end{bmatrix}. \tag{17.5}$$

The system (17.4) is the basis model form and underlies the investigations in the unified framework. It is of practical interest to notice the state space representation of (17.4),

$$\begin{cases} \hat{x}(k+1) = A_F \hat{x}(k) + B\left(v(k) + r_Q(k)\right) + LW^{-1} r_y(k) \\ \begin{bmatrix} u(k) \\ y(k) \end{bmatrix} = \begin{bmatrix} F \\ C_F \end{bmatrix} \hat{x}(k) + \begin{bmatrix} I \\ D \end{bmatrix} r_Q(k) + \begin{bmatrix} 0 \\ I \end{bmatrix} r_y(k), \\ r_Q(z) = Q(z) r_y(z), \end{cases}$$

an observer-based system driven by the residual r_y and reference v. On the assumption that there exist no (multiplicative) model uncertainties in the system,

- $\left(v, r_y\right)$ are independent of (u, y), when v is not set as a function of y, and
- v and r_y are independent, i.e. there exists no root-cause relation between them.

These two properties are of elemental significance. From the viewpoint of system configuration, the loop (17.3) is broken, which simplifies stability analysis and assures the well-posedness for the realisation. The system dynamics, both in open and closed-loop configurations, are structurally decomposed into two parts. The one represents the nominal system dynamic, and another one comprises the influence of uncertainties in the system. In particular, the nominal dynamic is uniquely determined by the latent variable (reference) v rather than the transfer function from u to y. In this regard, the feedback control and feedforward control can be handled separately, a general form of separation principle.

The dynamic model form (17.4) enables us to deal with control and detection issues in process input and output signal subspaces. To be specific, the SIR $\begin{bmatrix} M \\ N \end{bmatrix}$ spans the image subspace in Hilbert space and fully models the system nominal dynamics, while its (orthogonal) complement represents the system dynamics induced by uncertainties and forms the residual subspace. The residual subspace is a capable model form that fully characterises any uncertainties inside and around the system, and enables the use of firmly established methods like the projection technique to deal with fault diagnosis and feedback control problems capably and efficiently. Mathematically, the residual subspace can be constructed by $\begin{bmatrix} -\hat{Y} + MQ^* \\ \hat{X} + NQ^* \end{bmatrix}$ with Q^* as the solution of the following MMP,

$$Q^* = \arg \min_{Q \in \mathcal{RH}_\infty} \left\| \begin{bmatrix} -\hat{Y} + MQ \\ \hat{X} + NQ \end{bmatrix} \right\|_\infty .$$

An important and promising perspective to address detection and control issues in the system signal subspaces is the remarkable potential to transfer the established model-based methods to data-driven algorithms and implementations, including machine learning-based solutions, for instance, our endeavours in Chap. 13.

17.1.2 Variations of Bezout Identity

The Bezout identity (17.1) obviously applies only to LTI systems. Below is a summary of the variations of Bezout identity for different classes of dynamic systems addressed in our previous study.

- LTV systems: given the operators, $(\mathcal{M}, \mathcal{N})$ and $\left(\hat{\mathcal{M}}, \hat{\mathcal{N}}\right)$ as well as $(\mathcal{X}, \mathcal{Y})$ and $\left(\hat{\mathcal{X}}, \hat{\mathcal{Y}}\right)$, whose state space representations are respectively

$$\begin{gathered}
\begin{bmatrix} u \\ y \end{bmatrix} = \mathcal{I}_G v, \mathcal{I}_G = \begin{bmatrix} \mathcal{M} \\ \mathcal{N} \end{bmatrix}, \\
\mathcal{I}_G : \begin{cases} x(k+1) = (A(k) + B(k)F(k))\, x(k) + B(k)V(k)v(k) \\ \begin{bmatrix} u(k) \\ y(k) \end{bmatrix} = \begin{bmatrix} F(k) \\ C(k) + D(k)F(k) \end{bmatrix} x(k) + \begin{bmatrix} V(k) \\ D(k)V(k) \end{bmatrix} v(k), \end{cases}
\end{gathered} \tag{17.6}$$

$$\begin{gathered}
r_y = \mathcal{K}_G \begin{bmatrix} u \\ y \end{bmatrix} = 0, \mathcal{K}_G = \begin{bmatrix} -\hat{\mathcal{N}} & \hat{\mathcal{M}} \end{bmatrix}, \\
\mathcal{K}_G : \begin{cases} \hat{x}(k+1) = (A(k) - L(k)C(k))\, \hat{x}(k) + \begin{pmatrix} (B(k) - L(k)D(k))\, u(k)) \\ +L(k)y(k) \end{pmatrix} \\ r_y(k) = W(k) \left(-C(k)\hat{x}(k) + \begin{bmatrix} -D(k) & I \end{bmatrix} \begin{bmatrix} u(k) \\ y(k) \end{bmatrix} \right), \end{cases}
\end{gathered} \tag{17.7}$$

as well as

$$\begin{gathered}
\begin{bmatrix} u \\ y \end{bmatrix} = \mathcal{I}_K r_y, \mathcal{I}_K = \begin{bmatrix} -\hat{\mathcal{Y}} \\ \hat{\mathcal{X}} \end{bmatrix}, \\
\mathcal{I}_K : \begin{cases} \hat{x}(k+1) = (A(k) + B(k)F(k))\, \hat{x}(k) - L(k)W^{-1}(k)r_y(k) \\ \begin{bmatrix} u(k) \\ y(k) \end{bmatrix} = \begin{bmatrix} F(k) \\ C(k) + D(k)F(k) \end{bmatrix} \hat{x}(k) + \begin{bmatrix} 0 \\ W^{-1}(k) \end{bmatrix} r_y(k), \end{cases} \\
v = \mathcal{K}_K \begin{bmatrix} u \\ y \end{bmatrix}, \mathcal{K}_K = \begin{bmatrix} \mathcal{X} & \mathcal{Y} \end{bmatrix}, \\
\mathcal{K}_K : \begin{cases} \hat{x}(k+1) = (A(k) - L(k)C(k))\, \hat{x}(k) - \begin{pmatrix} (B(k) - L(k)D(k))\, u(k)) \\ -L(k)y(k) \end{pmatrix} \\ v(k) = V^{-1}(k) \left(F(k)\hat{x}(k) + \begin{bmatrix} I & 0 \end{bmatrix} \begin{bmatrix} u(k) \\ y(k) \end{bmatrix} \right), \end{cases}
\end{gathered}$$

the corresponding Bezout identity comprises

$$\begin{gathered}
\hat{\mathcal{M}} \circ \mathcal{N} = \hat{\mathcal{N}} \circ \mathcal{M}, \mathcal{Y} \circ \hat{\mathcal{X}} = \mathcal{X} \circ \hat{\mathcal{Y}}, \\
\mathcal{X} \circ \mathcal{M} + \mathcal{Y} \circ \mathcal{N} = \mathcal{I}, \hat{\mathcal{N}} \circ \hat{\mathcal{Y}} + \hat{\mathcal{M}} \circ \hat{\mathcal{X}} = \mathcal{I} \Longrightarrow \\
\begin{bmatrix} \mathcal{K}_K \circ \mathcal{I}_G & \mathcal{K}_G \circ \mathcal{I}_G \\ \mathcal{K}_K \circ \mathcal{I}_K & \mathcal{K}_G \circ \mathcal{I}_K \end{bmatrix} = \begin{bmatrix} \mathcal{I} & 0 \\ 0 & \mathcal{I} \end{bmatrix}.
\end{gathered} \tag{17.8}$$

- finite samples representations: given the data matrix,

$$\begin{bmatrix} U_s \\ Y_s \end{bmatrix} \in \mathbb{R}^{(p+m)s \times M}, \; M \geq (p+m)\, s,$$

satisfying

$$rank \begin{bmatrix} U_s \\ Y_s \end{bmatrix} = (p+m)\, s,$$

whose SVD is expressed by

$$\begin{aligned} \begin{bmatrix} U_s \\ Y_s \end{bmatrix} &= \begin{bmatrix} U_{I_{DD}} \; U_{K_{DD}} \end{bmatrix} \begin{bmatrix} \Sigma_{I_{DD}} & 0 & 0 \\ 0 & \Sigma_{K_{DD}} & 0 \end{bmatrix} \begin{bmatrix} V_{I_{DD}}^T \\ V_{K_{DD}}^T \\ V_{K_{DD},0}^T \end{bmatrix} \\ &= U_{I_{DD}} \Sigma_{I_{DD}} V_{I_{DD}}^T + U_{K_{DD}} \Sigma_{K_{DD}} V_{K_{DD}}^T, \\ \Sigma_{I_{DD}} &= diag\left(\sigma_1, \cdots, \sigma_\gamma\right), \, \Sigma_{K_{DD}} = diag\left(\sigma_{\gamma+1}, \cdots, \sigma_\xi\right), \end{aligned} \tag{17.9}$$

and suppose that

$$\sigma_1, \cdots, \sigma_\gamma >> \sigma_{\gamma+1}, \cdots, \sigma_\xi, \tag{17.10}$$

for some integer $\gamma > s$, then the matrices

$$I_{DD,s} = U_{I_{DD}} \Sigma_{I_{DD}}, \, K_{DD,s} = \Sigma_{K_{DD}} U_{K_{DD}}^T$$

are called data-driven image representation and kernel representations, respectively. Normalising $I_{DD,s}, K_{DD,s}$ by

$$\bar{I}_{DD,s} = I_{DD,s} \Sigma_{I_{DD}}^{-1} = U_{I_{DD}}, \, \bar{K}_{DD,s} = \Sigma_{K_{DD}}^{-1} K_{DD,s} = U_{K_{DD}}^T$$

yields

$$\begin{bmatrix} \bar{I}_{DD,s} \; \bar{K}_{DD,s}^T \end{bmatrix} \begin{bmatrix} \bar{I}_{DD,s}^T \\ \bar{K}_{DD,s} \end{bmatrix} = \begin{bmatrix} \bar{I}_{DD,s}^T \\ \bar{K}_{DD,s} \end{bmatrix} \begin{bmatrix} \bar{I}_{DD,s} \; \bar{K}_{DD,s}^T \end{bmatrix} = \begin{bmatrix} I & 0 \\ 0 & I \end{bmatrix}. \tag{17.11}$$

Although (17.11) seems remarkably different from the Bezout identity (17.1), it is called data-driven Bezout identity. This denomination is attributed to the central role of the Bezout identity in the unified framework. As described by (17.4) and (17.5), the Bezout identity defines a one-to-one map between the process data (u, y) and the two signal (residual) vectors $\left(v, r_y\right)$, the core of the unified framework. In this context, with the aid of (17.11), data $(u_s(k), y_s(k))$ can be written as

$$\begin{bmatrix} u_s(k) \\ y_s(k) \end{bmatrix} = \bar{I}_{DD,s}\bar{I}_{DD,s}^T \begin{bmatrix} u_s(k) \\ y_s(k) \end{bmatrix} + \bar{K}_{DD,s}^T \bar{K}_{DD,s} \begin{bmatrix} u_s(k) \\ y_s(k) \end{bmatrix}.$$

Because

$$v_{\bar{I},\gamma}(k) = \bar{I}_{DD,s}^T \begin{bmatrix} u_s(k) \\ y_s(k) \end{bmatrix}, r_{\bar{K},\eta}(k) = \bar{K}_{DD,s} \begin{bmatrix} u_s(k) \\ y_s(k) \end{bmatrix}$$

represent two signal (residual) vectors that serve as the latent variables of two orthogonal subspaces, respectively defined by $\bar{I}_{DD,s}$ and $K_{DD,s}^T$, it becomes clear that (17.11) defines a one-to-one mapping between $(u_s(k), y_s(k))$ and $\left(v_{\bar{I},\gamma}(k), r_{\bar{K},\eta}(k)\right)$ as well.

- nonlinear (affine) systems: given system Σ,

$$\Sigma : \begin{cases} \dot{x} = a(x) + B(x)u, x(0) = x_0 \\ y = c(x) + D(x)u, \end{cases} \tag{17.12}$$

whose SIR and SKR are

$$\Sigma_{\mathcal{I}} : \begin{cases} \dot{x} = a_I(x) + B_I(x)v \\ \begin{bmatrix} u \\ y \end{bmatrix} = c_I(x) + D_I(x)v, \end{cases} \tag{17.13}$$

$$a_I(x) = a(x) + B(x)g(x), B_I(x) = B(x)V(x),$$

$$c_I(x) = \begin{bmatrix} g(x) \\ c(x) + D(x)g(x) \end{bmatrix}, D_I(x) = \begin{bmatrix} V(x) \\ D(x)V(x) \end{bmatrix},$$

$$\Sigma_{\mathcal{K}} : \begin{cases} \dot{\hat{x}} = a(\hat{x}) + B(\hat{x})u + L(\hat{x})W^{-1}(\hat{x})r_y \\ = a_K(\hat{x}) + B_K(\hat{x}) \begin{bmatrix} u \\ y \end{bmatrix} \\ r_y = c_K(\hat{x}) + D_K(\hat{x}) \begin{bmatrix} u \\ y \end{bmatrix}, \end{cases} \tag{17.14}$$

$$a_K(\hat{x}) = a(\hat{x}) - L(\hat{x})c\left(\hat{x}\right), B_K(\hat{x}) = \left[\, B(\hat{x}) - L(\hat{x})D\left(\hat{x}\right) \; L(\hat{x}) \,\right],$$

$$c_K(\hat{x}) = -W(\hat{x})c\left(\hat{x}\right), D_K(\hat{x}) = \left[\, -W(\hat{x})D\left(\hat{x}\right) \; W(\hat{x}) \,\right],$$

and suppose that $\Sigma_{\mathcal{I}}$ and $\Sigma_{\mathcal{K}}$ are coprime and thus there exist a stable left inverse of $\Sigma_{\mathcal{I}}$ and right inverse of $\Sigma_{\mathcal{K}}$, i.e.

$$\Sigma_{\mathcal{I}}^- \circ \Sigma_{\mathcal{I}} = I, \Sigma_{\mathcal{K}} \circ \Sigma_{\mathcal{K}}^- = I.$$

Constructing $\left(\Sigma_{\mathcal{I}}^-, \Sigma_{\mathcal{K}}^-\right)$ based on a state observer, as described in Chap. 6, it holds,

$$\Sigma_{\mathcal{I}}^- \circ \Sigma_{\mathcal{K}}^- = 0.$$

Moreover, remember

$$\Sigma_{\mathcal{K}} \circ \Sigma_{\mathcal{I}} = 0.$$

We have finally the following identity

$$\begin{bmatrix} \Sigma_{\mathcal{I}}^{-} \circ \Sigma_{\mathcal{I}} & \Sigma_{\mathcal{K}} \circ \Sigma_{\mathcal{I}} \\ \Sigma_{\mathcal{I}}^{-} \circ \Sigma_{\mathcal{K}}^{-} & \Sigma_{\mathcal{K}} \circ \Sigma_{\mathcal{K}}^{-} \end{bmatrix} = \begin{bmatrix} I & 0 \\ 0 & I \end{bmatrix}, \tag{17.15}$$

which can be interpreted as the Bezout identity . Unfortunately, it cannot be directly applied to the analysis of closed-loop dynamics. To this end, residual signals, (v, r_y) , have been introduced in Chap. 6 with the aid of $\Sigma_{\mathcal{I}}^{-}$ and $\Sigma_{\mathcal{K}}$,

$$\begin{bmatrix} v \\ r_y \end{bmatrix} = \begin{bmatrix} \Sigma_{\mathcal{I}}^{-} \begin{pmatrix} u \\ y \end{pmatrix} \\ \Sigma_{\mathcal{K}} \begin{pmatrix} u \\ y \end{pmatrix} \end{bmatrix} =: \mathcal{K}_{\Sigma,C} \begin{pmatrix} u \\ y \end{pmatrix}. \tag{17.16}$$

It is observer-based. Suppose that $\mathcal{K}_{\Sigma,C}$ has a stable right inverse $\mathcal{K}_{\Sigma,C}^{-}$ described by

$$\mathcal{K}_{\Sigma,C}^{-} = \begin{cases} \dot{\hat{x}} = a_I(\hat{x}) + \begin{bmatrix} B_I(\hat{x}) & L(\hat{x}) W^{-1}(\hat{x}) \end{bmatrix} \begin{bmatrix} v \\ r_y \end{bmatrix} \\ \begin{bmatrix} u \\ y \end{bmatrix} = \begin{bmatrix} g(\hat{x}) \\ -\left(c(\hat{x}) + D(\hat{x}) g(\hat{x})\right) \end{bmatrix} + \begin{bmatrix} V(\hat{x}) & 0 \\ -D(\hat{x}) V(\hat{x}) & W^{-1}(\hat{x}) \end{bmatrix} \begin{bmatrix} v \\ r_y \end{bmatrix}, \end{cases}$$

then the loop dynamics can be expressed as

$$\begin{bmatrix} u \\ y \end{bmatrix} = \mathcal{K}_{\Sigma,C}^{-} \left(\begin{bmatrix} v \\ r_y \end{bmatrix} \right)$$

with the residuals (v, r_y) as the loop input.

17.2 System Models and Subspaces: The Cornerstones of the Unified Framework

Distinctly different to traditional analysis and synthesis methods in control theory, SIR and SKR models as well as their various extensions build the cornerstones of the unified framework. Given the system model

$$G = \hat{M}^{-1} \hat{N} = N M^{-1} = (A, B, C, D), \, y = G u,$$

its SIR and SKR are, respectively,

$$\begin{bmatrix} u \\ y \end{bmatrix} = I_G v, \, I_G = \begin{bmatrix} M \\ N \end{bmatrix} v, \, r_y = K_G \begin{bmatrix} u \\ y \end{bmatrix}, \, K_G = \begin{bmatrix} -\hat{N} & \hat{M} \end{bmatrix}$$

with the state space representations,

$$I_G : \begin{cases} x(k+1) = A_F x(k) + BVv(k) \\ \begin{bmatrix} u(k) \\ y(k) \end{bmatrix} = \begin{bmatrix} F \\ C_F \end{bmatrix} x(k) + \begin{bmatrix} I \\ D \end{bmatrix} Vv(k), \end{cases}$$

$$K_G : \begin{cases} \hat{x}(k+1) = A_L \hat{x}(k) + B_L u(k) + LW^{-1} r_y(k) \\ r_y(k) = W\left(y(k) - \hat{y}(k)\right), \hat{y}(k) = C\hat{x}(k) + Du(k). \end{cases}$$

The significance of the SIR and SKR lies in

- their duality, modelling the system dynamics from different aspects,
- modelling the subspace of the process input and output variables (u, y) belonging to $\mathcal{H}_2$, which conforms to engineering practice,
- mapping possible uncertainties, including model uncertainties, faults or disturbances, into the residual signal generated by the SKR, and
- substituting the causality between the input u and the output y by the causality between the latent variable v (as input) and (u, y) (as output).

These properties enable us to extract new aspects in dealing with analysis and synthesis issues, offer alternative, often more efficient solutions, and support us to bridge different research domains. Doubtlessly, an immediate and meaningful exploration is the introduction of system image/kernel and residual subspaces.

The system image subspace,

$$\mathcal{I}_G = \left\{ \begin{bmatrix} u \\ y \end{bmatrix} : \begin{bmatrix} u \\ y \end{bmatrix} = \begin{bmatrix} M \\ N \end{bmatrix} v, v \in \mathcal{H}_2 \right\} \subset \mathcal{H}_2, \tag{17.17}$$

represents the system nominal dynamics in $\mathcal{H}_2$. Its dual form is the kernel subspace,

$$\mathcal{K}_G = \left\{ \begin{bmatrix} u \\ y \end{bmatrix} \in \mathcal{H}_2 : \begin{bmatrix} -\hat{N} & \hat{M} \end{bmatrix} \begin{bmatrix} u \\ y \end{bmatrix} = 0 \right\}, \tag{17.18}$$

$$\mathcal{K}_G = \mathcal{I}_G, \tag{17.19}$$

which is seldom applied to system analysis and synthesis directly. In contrast, the residual subspace,

$$\mathcal{R}_G = \mathcal{I}_G^{\perp} = \mathcal{K}_G^{\perp},$$

as the orthogonal complement of $\mathcal{I}_G$, plays a central role in the unified framework. Residual subspace is an abused term and might be confusing. Note that $\mathcal{R}_G$ is composed of (u, y) rather than residual signals. The residual subspace $\mathcal{R}_G$ is characterised by the feature that

$$\forall \begin{bmatrix} u \\ y \end{bmatrix} \in \mathcal{R}_G, r_y = \begin{bmatrix} -\hat{N} & \hat{M} \end{bmatrix} \begin{bmatrix} u \\ y \end{bmatrix} \neq 0. \tag{17.20}$$

This implies that uncertainty and fault detection and estimation will be consummated in the residual subspace $\mathcal{R}_G$ using the residual signal r_y. The definition of the system residual subspace,

$$\mathcal{R}_G = \left\{ \begin{bmatrix} u \\ y \end{bmatrix} : \begin{bmatrix} u \\ y \end{bmatrix} = \mathcal{P}_{\mathcal{H}_2} \begin{bmatrix} -\hat{N}^{\sim} \\ \hat{M}^{\sim} \end{bmatrix} r_y, r_y \in \mathcal{H}_2 \right\},$$

is rarely applied in the unified framework. Hereby, $\mathcal{P}_{\mathcal{H}_2} : \mathcal{L}_2 \rightarrow \mathcal{H}_2$ is an orthogonal operator representing a projection onto $\mathcal{H}_2$. It is worth remarking that the notation r_y as the latent variable is adopted to draw the reader's attention to the application context of the residual subspace. It is interesting to recall that the system residual subspace $\mathcal{R}_G$ can be constructed by means of the following algorithm,

- given the normalised SIR and SKR, (M_0, N_0) and $\left(\hat{M}_0, \hat{N}_0\right)$, as well as the corresponding RCP $\left(\hat{X}_0, \hat{Y}_0\right)$, normalising $\left(\hat{X}_0, \hat{Y}_0\right)$ by

$$\begin{bmatrix} -\hat{Y}_{0,N} \\ \hat{X}_{0,N} \end{bmatrix} = \begin{bmatrix} -\hat{Y}_0 \\ \hat{X}_0 \end{bmatrix} V_0^{-1},$$

  where V_0^{-1} is given by a spectral decomposition,

$$V_0^{\sim} V_0 = \begin{bmatrix} -\hat{Y}_0^{\sim} & \hat{X}_0^{\sim} \end{bmatrix} \begin{bmatrix} -\hat{Y}_0 \\ \hat{X}_0 \end{bmatrix}, V_0, V_0^{-1} \in \mathcal{RH}_{\infty},$$

- solving the MMP

$$Q^* = \inf_{Q \in \mathcal{RH}_{\infty}} \left\| \begin{bmatrix} -\hat{Y}_{0,N} \\ \hat{X}_{0,N} \end{bmatrix} - \begin{bmatrix} M_0 \\ N_0 \end{bmatrix} Q \right\|$$

  for Q^*, and finally
- computing

$$Q_{\mathcal{H}_2} = Q^* V_0$$

  and defining

$$\mathcal{R}_G = \left\{ \begin{bmatrix} u \\ y \end{bmatrix}, \begin{bmatrix} u \\ y \end{bmatrix} = \left(\begin{bmatrix} -\hat{Y}_0 \\ \hat{X}_0 \end{bmatrix} - \begin{bmatrix} M_0 \\ N_0 \end{bmatrix} Q_{\mathcal{H}_2} \right) r_y, r_y \in \mathcal{H}_2 \right\}. \tag{17.21}$$

Definition (17.21) and the associated algorithm are useful for analysis and synthesis of closed-loop dynamics.

In addition to $\mathcal{I}_G$ and $\mathcal{R}_G$ in $\mathcal{H}_2$ subspace, $\mathcal{I}_G^{\mathcal{L}_2}$ and $\mathcal{R}_G^{\mathcal{L}_2}$ in $\mathcal{L}_2$ subspace play an important role in dealing with detection and estimation issues as well. $\mathcal{I}_G^{\mathcal{L}_2}$ and $\mathcal{R}_G^{\mathcal{L}_2}$ are defined by

$$\mathcal{I}_G^{\mathcal{L}_2} = \left\{ \begin{bmatrix} u \\ y \end{bmatrix} : \begin{bmatrix} u \\ y \end{bmatrix} = \begin{bmatrix} M \\ N \end{bmatrix} v, v \in \mathcal{L}_2 \right\}, \tag{17.22}$$

$$\mathcal{R}_G^{\mathcal{L}_2} = \left\{ \begin{bmatrix} u \\ y \end{bmatrix} : \begin{bmatrix} u \\ y \end{bmatrix} = \begin{bmatrix} -\hat{N}^{\sim} \\ \hat{M}^{\sim} \end{bmatrix} r_y, r_y \in \mathcal{L}_2 \right\}, \tag{17.23}$$

respectively. In $\mathcal{L}_2$ subspace, we are able to address non-causal systems and the related computations, which are often useful for optimal detection and estimation. In our subsequent discussion on projection-based methods, we will address this issue in more detail.

Nowadays, data-driven techniques become popular. The data-driven unified framework offers a counterpart of the model-based one. At the core of the data-driven unified framework is a data matrix (model) of the form

$$\begin{bmatrix} u_s \\ y_s \end{bmatrix} = \begin{bmatrix} U_s \\ Y_s \end{bmatrix} v_l, \begin{bmatrix} U_s \\ Y_s \end{bmatrix} \in \mathbb{R}^{(p+m)s \times M}, \tag{17.24}$$

where (U_s, Y_s) are input and output Hankel matrices, respectively, $v_l \in \mathbb{R}^M$ is the latent variable and M is a sufficiently large integer. In state of the art research on data-driven control, the rank condition

$$rank \begin{bmatrix} U_s \\ Y_s \end{bmatrix} = sp + n$$

is a common assumption. In this context, the model (17.24) represents the process nominal dynamic behaviour, and $M \geq (s+n)p$. In the data-driven framework introduced in the previous chapters, the assumption

$$rank \begin{bmatrix} U_s \\ Y_s \end{bmatrix} = (p+m)s, s \geq n,$$

is adopted in order to match engineering practice. That is, uncertainties in the system and data are taken into account. In this context, data-driven image and residual subspaces are respectively defined as

$$\mathcal{I}_{G,s} = \left\{ \begin{bmatrix} u_s \\ y_s \end{bmatrix}, \begin{bmatrix} u_s \\ y_s \end{bmatrix} = I_{DD,s} V_{I_{DD}}^T v_l, v_l \in \mathbb{R}^M \right\},$$
$$\mathcal{R}_{G,s} = \left\{ \begin{bmatrix} u_s \\ y_s \end{bmatrix}, \begin{bmatrix} u_s \\ y_s \end{bmatrix} = K_{DD,s}^T V_{K_{DD}}^T v_l, v_l \in \mathbb{R}^M \right\},$$
$$rank\left(I_{DD,s}\right) = \gamma = sp + n, rank\left(K_{DD,s}\right) = ms - n.$$

Hereby, the matrices $I_{DD,s}, V_{I_{DD}}^T, K_{DD,s}^T, V_{K_{DD}}^T$ and integer γ are given in (17.9)–(17.10). It is noteworthy that in the data-driven study, system causality is not taken into account.

In Chap. 7, the model (17.24) has been examined from a different aspect. With the aid of model-based SIR, the model

$$\begin{bmatrix} u_s(k) \\ y_s(k) \end{bmatrix} = \begin{bmatrix} M_s \\ N_s \end{bmatrix} v_{s+n}(k-n),\ v_{s+n}(k-n) \in \mathcal{L}_{[0,\infty)} \tag{17.25}$$

has been derived, where $(u_s(k), y_s(k))$ are samples of the signals $(u, y) \in \mathcal{H}_2$ over a (finite) time interval $[k+1, k+s]$, generated by

$$I_G : \begin{cases} x(i+1) = A_{F_d}x(i) + Bv(i) \\ \begin{bmatrix} u(i) \\ y(i) \end{bmatrix} = \begin{bmatrix} F_d \\ C_F \end{bmatrix} x(i) + \begin{bmatrix} I \\ D \end{bmatrix} v(i) \end{cases}$$

with a deadbeat state feedback controller,

$$u(i) = F_d x(i) + v(i), i \in [k-n, k+s].$$

Matrices (M_s, N_s) are finite samples SIR and satisfy

$$rank \begin{bmatrix} M_s \\ N_s \end{bmatrix} = sp + n.$$

In this context, the model (17.25) is a causal FIF and can be interpreted as a finite samples form of the model-based SIR. Accordingly, the image and residual subspaces are defined by

$$\begin{aligned} \mathcal{I}_{G,s} &= \left\{ \begin{bmatrix} u_s(k) \\ y_s(k) \end{bmatrix}, \begin{bmatrix} u_s(k) \\ y_s(k) \end{bmatrix} = \begin{bmatrix} M_s \\ N_s \end{bmatrix} v_{s+n}(k-n),\ v_{s+n}(k-n) \in \mathcal{L}_{[0,\infty)} \right\}, \\ \mathcal{R}_{G,s} &= \left\{ \begin{bmatrix} u_s(k) \\ y_s(k) \end{bmatrix}, \begin{bmatrix} u_s(k) \\ y_s(k) \end{bmatrix} = R_{G_0,s} r_l,\ r_l \in \mathbb{R}^{(ms-n)} \right\}, \end{aligned}$$

where $R_{G_0,s}$ is given by

$$\begin{aligned} I - I_{G_0,s} I_{G_0,s}^T &= R_{G_0,s} R_{G_0,s}^T,\ R_{G_0,s} \in \mathbb{R}^{(p+m)s \times (ms-n)}, \\ I_{G_0,s} &= \begin{bmatrix} M_s \\ N_s \end{bmatrix} V_o,\ I_{G_0,s}^T I_{G_0,s} = I \in \mathbb{R}^{(ps+n)\times(ps+n)}. \end{aligned}$$

The model (17.25) not only reveals a new aspect of the famous fundamental lemma, but also offers us computationally efficient engineering solutions in dealing with fault detection and estimation issues.

The issues concerning the image and residual subspaces for LTV and nonlinear systems will be addressed in the subsequent section.

A further class of system models is the so-called SKR-based input-output model, which has been widely applied to investigate various analysis and synthesis issues of systems with uncertainties, particularly those types, which are hardly modelled. On the basis of the observer-based unified system model presented in Sect. 4.6, the SKR-based system model,

$$\begin{cases} \hat{x}(k+1) = A\hat{x}(k) + Bu(k) + Lr_0(k), \hat{x}(0) = 0 \\ y(k) = C\hat{x}(k) + Du(k) + r_0(k), \end{cases} \tag{17.26}$$
$$r_0(k) = y(k) - \left(C\hat{x}(k) + Du(k)\right),$$

has been proposed, which fully describes the input-output dynamics of an open-loop system (plant), including all possible types of uncertainties, although they are not exclusively presented in the model. A frequency-domain form of the model (17.26) is given by

$$y(z) = G(z)u(z) + \hat{M}^{-1}r_0(z), G = (A, B, C, D). \tag{17.27}$$

In particular dealing with analysis of closed-loop dynamics, the model (17.27) might considerably simplify the related computations.

An extension of the model (17.26) to nonlinear systems is straightforward and described by, given the nonlinear system model (17.12),

$$\Sigma : \begin{cases} \dot{\hat{x}} = a(\hat{x}) + B(\hat{x})u + Lr_0, \hat{x}(0) = 0 \\ y = c(\hat{x}) + D(\hat{x})u + r_0, \end{cases} \tag{17.28}$$
$$r_0 = y - c(\hat{x}) - D(\hat{x})u.$$

The following properties distinguish the SKR-based model from the conventional model forms widely applied in control theory and engineering.

- The SKR-based model gives a uniform process model representation. It holds for all possible types of uncertainties.
- Information about uncertainties in the system like unknown inputs and parameter variations is fully embedded in the model and represented by the residual vector r_0.
- The SKR-based model is capable of online implementation. That means, the SKR-based model can run online and in parallel with the process under consideration. Thus, it can serve as an online simulation model.

We would like to place a greater emphasis on the subsequent two points:

- the residual signal r_0 is indeed generated by a system whose dynamic is governed by the uncertainties in the system and different from the one given in (17.26). In other words, the residual signal can be understood as an external input from an uncertain system, and

- the influence of the unknown initial value of the state vector, $x(0) = x_0$, is incorporated by the residual vector.

17.3 Basic Theoretic Methods

The methods summarised in this section are those, which are known in control theory and essential for our study on various detection and control problems in the unified framework. They often help us to gain deep insights into the addressed issues. They lay the foundation for solving control and detection problems specified in our work.

17.3.1 Normalised Coprime Factorisations

Normalised SIR and SKR have been applied throughout our work in this book. Denoted by (M_0, N_0) and $\left(\hat{M}_0, \hat{N}_0\right)$ the normalised SIR and SKR of LTI systems, they are characterised in the frequency domain by the condition

$$\begin{bmatrix} M_0^{\sim}(z) & N_0^{\sim}(z) \end{bmatrix} \begin{bmatrix} M_0(z) \\ N_0(z) \end{bmatrix} = M_0^{\sim}(z) M_0(z) + N_0^{\sim}(z) N_0(z) = I, \tag{17.29}$$

$$\begin{bmatrix} \hat{N}_0(z) & \hat{M}_0(z) \end{bmatrix} \begin{bmatrix} \hat{N}_0^{\sim}(z) \\ \hat{M}_0^{\sim}(z) \end{bmatrix} = \hat{N}_0(z) \hat{N}_0^{\sim}(z) + \hat{M}_0(z) \hat{M}_0^{\sim}(z) = I. \tag{17.30}$$

For LTV systems $(A(k), B(k), C(k)D(k))$, the normalised SIR and SKR given in (17.6)–(17.7) are defined in terms of the property that

$$\mathcal{I}_G^* \circ \mathcal{I}_G = \mathcal{I}, \mathcal{K}_G \circ \mathcal{K}_G^* = \mathcal{I},$$

with $\mathcal{I}_G^*$ and $\mathcal{K}_G^*$ as the adjoints of $\mathcal{I}_G$ and $\mathcal{K}_G$, respectively. With v and $\hat{v}$ as the input and output variables, respectively, the condition (17.29) is equivalently written into

$$\hat{v} = I_{G_0}^{\sim} I_{G_0} v = v, \ I_{G_0} = \begin{bmatrix} M_0(z) \\ N_0(z) \end{bmatrix}. \tag{17.31}$$

Similarly, it holds

$$\hat{v} = \mathcal{I}_G^* \circ \mathcal{I}_G(v) = v. \tag{17.32}$$

Systems satisfying (17.31) or (17.32) are called inner. The dual form of an inner is co-inner defined by

$$\hat{r}_y = K_{G_0} K_{G_0}^{\sim} r_y = r_y, \ K_{G_0} = \begin{bmatrix} \hat{N}_0 & \hat{M}_0 \end{bmatrix}, \tag{17.33}$$

$$\hat{r}_y = \mathcal{K}_G \circ \mathcal{K}_G^* \left(r_y\right) = r_y. \tag{17.34}$$

Given nonlinear system (17.12), the associated SIR (17.13) and SKR (17.14), build the Hamiltonian systems

$$(D\Sigma_{\mathcal{I}})^T \circ \Sigma_{\mathcal{I}} : \begin{cases} \dot{x} = \frac{\partial H}{\partial \lambda} (x, \lambda, \upsilon) \\ \dot{\lambda} = -\frac{\partial H}{\partial x} (x, \lambda, u) \\ \hat{\upsilon} = \frac{\partial H}{\partial u} (x, \lambda, \upsilon) , \end{cases} \tag{17.35}$$

$$H (x, \lambda, \upsilon) = \frac{1}{2} (c_I(x) + D_I(x)\upsilon)^T (c_I(x) + D_I(x)\upsilon) + \lambda^T (a_I(x) + B_I(x)\upsilon) ,$$

as well as

$$\Sigma_{\mathcal{K}} \circ (D\Sigma_{\mathcal{K}})^T : \begin{cases} \dot{x}_\Delta = \frac{\partial H(x_\Delta, \lambda, r_y)}{\partial \lambda} \\ \dot{\lambda} = -\frac{\partial H(x_\Delta, \lambda, r_y)}{\partial \hat{x}} \\ \hat{r}_y = \frac{\partial H(x_\Delta, \lambda, r_y)}{\partial r_y}, \end{cases} \tag{17.36}$$

$$H \left(x_\Delta, \lambda, r_y\right) = \frac{1}{2} r_a^T D(x_\Delta) D^T (x_\Delta) r_a + c_K^T (x_\Delta) r_y$$
$$+\lambda^T \left(a_K(x_\Delta) + \frac{1}{2} B_K(x_\Delta) B_K^T(x_\Delta)\lambda + B_K(x_\Delta) D_K^T(x_\Delta) r_y \right).$$

Systems (17.13) and (17.14) are normalised SIR and SKR, when

$$\hat{\upsilon} = \upsilon, \hat{r}_y = r_y,$$

respectively.

The computations of normalised SIR and SKR are described in Theorem 2.1 for LTI systems, Theorem 5.1 for LTV systems, and Theorems 6.1 and 6.2 for the nonlinear system (17.12). The core of these computations is the solution of Riccati equations (recursions) and HJEs for nonlinear systems. It is noteworthy that the conjugate systems, $I_{G_0}^{\sim}$ and $K_{G_0}^{\sim}$, the adjoints, $\mathcal{I}_G^*$ and $\mathcal{K}_G^*$, and systems $(D\Sigma_{\mathcal{I}})^T$, $(D\Sigma_{\mathcal{K}})^T$, are anti-stable systems. They are to be computed backwards in time. For instance,

$$\mathcal{I}_G^* : \begin{cases} \lambda(k) = A_I^T(k)\lambda(k+1) + C_I^T(k) \begin{bmatrix} u(k+1) \\ y(k+1) \end{bmatrix} \\ \hat{\upsilon}(k+1) = B_I^T(k)\lambda(k+1) + D_I^T(k) \begin{bmatrix} u(k+1) \\ y(k+1) \end{bmatrix}, \end{cases}$$

$$A_I(k) = A(k) + B(k)F(k), \; B_I(k) = B(k)V(k),$$

$$C_I(k) = \begin{bmatrix} F(k) \\ C(k) + D(k)F(k) \end{bmatrix}, \; D_I(k) = \begin{bmatrix} V(k) \\ D(k)V(k) \end{bmatrix}.$$

Doubtlessly, the so-called lossless property of the normalised SIR and SKR is essential for dealing with system analysis and syntheses, and hence is of elemental importance. For the normalised SIR, there exists a storage function $V(x) \geq 0$, $V(0) = 0$ so that

$$V(x(t_2)) - V(x(t_1)) = \frac{1}{2}\int_{t_1}^{t_2}\left(v^T v - \begin{bmatrix} u^T & y^T \end{bmatrix}\begin{bmatrix} u \\ y \end{bmatrix}\right) d\tau. \quad (17.37)$$

The storage function $V(x)$ is interpreted as the storage energy in the system. In this context, the relation (17.37) implies no loss of energy. Hence, the SIR is lossless. The dual form for the normalised SKR is given

$$V(x_\Delta(t_2)) - V(x_\Delta(t_1)) = \frac{1}{2}\int_{t_1}^{t_2}\left(\begin{bmatrix} \Delta u^T & \Delta y^T \end{bmatrix}\begin{bmatrix} \Delta u \\ \Delta y \end{bmatrix} - r^T r\right) d\tau, \quad (17.38)$$

for some storage function $V(x_\Delta) \geq 0$, $V(0) = 0$. Hereby, we place a greater emphasis on the meaning of (17.38). Taking an LTI SKR as an example, we have the state space representation of $K_{G_0}K_{G_0}^{\sim}$,

$$K_{G_0}K_{G_0}^{\sim} : \begin{cases} \lambda(k) = A_K^T \lambda(k+1) + C_K^T r_y(k+1) \\ \begin{bmatrix} \Delta u(k+1) \\ \Delta y(k+1) \end{bmatrix} = B_K^T \lambda(k+1) + D_K^T r_y(k+1) \\ x_\Delta(k+1) = A_K x_\Delta(k) + B_K \begin{bmatrix} \Delta u(k) \\ \Delta y(k) \end{bmatrix} \\ \hat{r}_y(k) = C_K x_\Delta(k) + D_K \begin{bmatrix} \Delta u(k) \\ \Delta y(k) \end{bmatrix}, \end{cases}$$

$$A_K = A - L_0 C,\ B_K = \begin{bmatrix} B - L_0 D & L_0 \end{bmatrix},$$
$$C_K = -W_0 C,\ D_K = \begin{bmatrix} -W_0 D & W_0 \end{bmatrix}.$$

The output of $K_{G_0}^{\sim}$, $(\Delta u, \Delta y)$ represents the uncertainty in the process data (u, y). As a response to it, the state variable x_Δ of K_{G_0} models the uncertain dynamic of the residual generation system. In this context, (17.38) implies that the residual signal r comprises full information about the uncertainty in the process data (u, y) and the residual generator (the normalised SKR) loses no information.

A further important property of normalised SIR and SKR is the relations between the state variable pair (x, λ) and the storage function $V(x)$. To be specific, for an SIR,

$$\lambda = \frac{\partial V(x)}{\partial x},$$

which is equal to, in the case of an LTI SIR,

$$V(x) = \frac{1}{2}x^T P x \Longrightarrow \lambda = \frac{\partial V(x)}{\partial x} = Px,$$

and for an SKR,

$$\begin{aligned} x &= \frac{\partial V(\lambda)}{\partial \lambda}, V(\lambda) = \frac{1}{2}\lambda^T P\lambda, x = P\lambda \Longleftrightarrow \\ \lambda &= \frac{\partial V(x)}{\partial x}, V(x) = \frac{1}{2}x^T P^{-1}x, \lambda = P^{-1}x. \end{aligned}$$

Against the background that

$$u(k) = Fx(k) + Vv(k),$$

an SIR is interpreted as a state feedback controller. In this regard, the normalised SIR is an LQ optimal controller. As the dual system, an SKR represents a residual generator,

$$r(k) = W\left(y(k) - \hat{y}(k)\right),$$

and the conjugate (adjoint) of the normalised SKR delivers an LS estimate of the uncertainty corrupted in the process data (u, y),

$$\begin{bmatrix} \Delta u \\ \Delta y \end{bmatrix} = K_{G_0}^{\sim} r_y.$$

Normalising SIR and SKR is equivalent to the so-called inner-outer factorisation (IOF) and co-inner-outer factorisation (CIOF), respectively. This aspect helps us to better understand computation of normalised SIR and SKR. For the sake of simplicity, we only consider SIR I_G, whose IOF is given by

$$I_G(z) = I_G^i(z)I_G^o(z),$$

with the inner system I_G^i and the outer I_G^o. Subsequently, the normalised SIR I_{G_0} can be determined by

$$I_{G_0}(z) = I_G^i(z)I_G^o(z)\left(I_G^o(z)\right)^{-1} = I_G^i(z),$$

which is equivalent to

$$I_{G_0}(z) = I_G(z)V(z), V(z) = \left(I_G^o(z)\right)^{-1}.$$

In other words, the normalisation of an SIR can be realised by adding a pre-filter. The reader is referred to the references given at the end of this chapter for the existence conditions and the associated computation algorithms of an IOF (as well as CIOF), from which it can be seen that there exist IOF and CIOF for SIR and SKR, respectively.

17.3.2 Projection Methods

In this subsection, orthogonal projections in Hilbert space are summarised. Orthogonal projection is a standard method dealing with classification problems. Roughly speaking, an orthogonal projection enables a fair measurement of the distance from a signal (vector) to a subspace or between two subspaces. In the unified framework, system dynamics are represented by signal and data subspaces like image and residual subspaces. As subspaces in Hilbert space, they are vector spaces endowed with an inner product $\langle \cdot, \cdot \rangle$. As a logical consequence, orthogonal projections are an efficient tool that is capable of solving detection and control problems in the unified framework.

The mathematical basis of projection methods is the definition of an inner product of two signal vectors. In the unified framework, three different inner products are considered for operations in different signal subspaces. They are,

- $\mathcal{L}_2$ subspace, $\mathcal{L}_2 = \mathcal{L}_2 (-\infty, 0] \oplus \mathcal{L}_2 [0, \infty)$:

$$\langle x, y \rangle = \sum_{k=-\infty}^{\infty} x^T(k) y(k), x, y \in \mathcal{L}_2, \tag{17.39}$$

- $\mathcal{H}_2$ subspace:

$$\langle x, y \rangle = \sum_{k=0}^{\infty} x^T(k) y(k), x, y \in \mathcal{H}_2, \tag{17.40}$$

- $\mathcal{L}_{2,[k_1,k_2]}$ subspace, $k_2 > k_1 \geq 0$:

$$\langle x, y \rangle = \sum_{k=k_1}^{k_2} x^T(k) y(k), x, y \in \mathcal{L}_{2,[k_1,k_2]}. \tag{17.41}$$

Correspondingly, following operators that project the process data (u, y) onto image and residual subspaces are useful in the unified framework,

- orthogonal projections onto $\mathcal{H}_2$ image and residual subspaces $\mathcal{I}_G$ and $\mathcal{R}_G$,

$$\mathcal{P}_{\mathcal{I}_G} = \mathcal{L}_{I_{G0}} \mathcal{P}_{\mathcal{H}_2} \mathcal{L}_{I_{\tilde{G}0}} : \mathcal{H}_2 \to \mathcal{H}_2, \tag{17.42}$$

$$\mathcal{P}_{\mathcal{R}_G} = \mathcal{P}_{\mathcal{H}_2} \mathcal{L}_{K_{\tilde{G}0}} \left(\mathcal{P}_{\mathcal{H}_2} \mathcal{L}_{K_{\tilde{G}0}} \right)^* : \mathcal{H}_2 \to \mathcal{H}_2. \tag{17.43}$$

While the projection $\mathcal{P}_{\mathcal{I}_G}$ onto image subspace is widely applied in robust control and also addressed in our study on fault detection, the projection onto the residual subspace $\mathcal{R}_G$ receives little attention. From system-theoretic viewpoint, these projections from $\mathcal{H}_2$ onto $\mathcal{H}_2$ assure the system causality, a basic requirement for real control. On the other hand, this property imposes restrictions on information processing and utilisation. In addition, the involved computations are challenging.

- orthogonal projections onto $\mathcal{L}_2$ image and residual subspaces $\mathcal{I}_G^{\mathcal{L}_2}$ and $\mathcal{R}_G^{\mathcal{L}_2}$,

$$\mathcal{P}_{\mathcal{I}_G^{\mathcal{L}_2}} = \mathcal{L}_{I_{G_0}} \mathcal{L}_{I_{G_0}^{\sim}} : \mathcal{L}_2 \to \mathcal{L}_2, \tag{17.44}$$

$$\mathcal{P}_{\mathcal{R}_G^{\mathcal{L}_2}} = \mathcal{L}_{K_{G_0}^{\sim}} \mathcal{L}_{K_{G_0}} : \mathcal{L}_2 \to \mathcal{L}_2. \tag{17.45}$$

From the point of view of detection and estimation, the projections $\mathcal{P}_{\mathcal{I}_G^{\mathcal{L}_2}}$ and $\mathcal{P}_{\mathcal{R}_G^{\mathcal{L}_2}}$ are of a number of advantages. For instance, it holds

$$\left\| \left(\mathcal{I} - \mathcal{P}_{\mathcal{I}_G^{\mathcal{L}_2}} \right) \begin{bmatrix} u \\ y \end{bmatrix} \right\|_2 = \left\| \mathcal{P}_{\mathcal{R}_G^{\mathcal{L}_2}} \begin{bmatrix} u \\ y \end{bmatrix} \right\|_2 = \left\| K_{G_0} \begin{bmatrix} u \\ y \end{bmatrix} \right\|_2 = \|r_y\|_2 .$$

That is, the $\mathcal{L}_2$-norm of the projection-based residual signal is equal to the $\mathcal{L}_2$-norm of the output residual r_y generated by a normalised SKR. Observe the state space realisation of $\mathcal{P}_{\mathcal{R}_G^{\mathcal{L}_2}}$ is equivalent to

$$\mathcal{P}_{\mathcal{R}_G^{\mathcal{L}_2}} : \begin{cases} x_\Delta(k+1) = A_K x_\Delta(k) + B_K \begin{bmatrix} \Delta u(k) \\ \Delta y(k) \end{bmatrix} \\ r_y(k) = C_K x_\Delta(k) + D_K \begin{bmatrix} \Delta u(k) \\ \Delta y(k) \end{bmatrix} \\ \lambda(k) = A_K^T \lambda(k+1) + C_K^T r_y(k) \\ \begin{bmatrix} \Delta u(k) \\ \Delta y(k) \end{bmatrix} = B_K^T \lambda(k+1) + D_K^T r_y(k). \end{cases}$$

Hereby, the vector

$$\begin{bmatrix} \Delta u(k) \\ \Delta y(k) \end{bmatrix} \in \mathcal{R}_G^{\mathcal{L}_2}$$

represents the uncertainty in the data $(u(k), y(k))$ and its estimate is delivered by the output of the projection system. In other words, the system $K_{G_0}^{\sim}$ serves as an indicator of the uncertainty in the data (u, y) by processing the residual signal r_y backwards in time. Such information is of essential importance for fault diagnosis and process monitoring.

- orthogonal projections onto $\mathcal{L}_{2,[k_1,k_2]}$ image and residual subspaces,

$$\mathcal{I}_G^{\mathcal{L}_{2,[k_1,k_2]}} = \left\{ \begin{bmatrix} u \\ y \end{bmatrix} : \begin{bmatrix} u \\ y \end{bmatrix} = \begin{bmatrix} M \\ N \end{bmatrix} v, v \in \mathcal{L}_{2,[k_1,k_2]} \right\},$$

$$\mathcal{R}_G^{\mathcal{L}_{2,[k_1,k_2]}} = \left\{ \begin{bmatrix} u \\ y \end{bmatrix} : \begin{bmatrix} u \\ y \end{bmatrix} = \begin{bmatrix} -\hat{N}^{\sim} \\ \hat{M}^{\sim} \end{bmatrix} r_y, r_y \in \mathcal{L}_{2,[k_1,k_2]} \right\},$$

respectively, given by

$$\mathcal{P}_{\mathcal{I}_G^{\mathcal{L}_{2,[k_1,k_2]}}} : \begin{cases} x(k+1) = A_I(k)x(k) + B_I(k)v(k) \\ \begin{bmatrix} u_a(k) \\ y_a(k) \end{bmatrix} = C_I(k)x(k) + D_I(k)v(k) \\ \lambda(k) = A_I^T(k)\lambda(k+1) + C_I^T(k) \begin{bmatrix} u(k+1) \\ y(k+1) \end{bmatrix} \\ v_a(k+1) = B_I^T(k)\lambda(k+1) + D_I^T(k) \begin{bmatrix} u(k+1) \\ y(k+1) \end{bmatrix}, \end{cases} \tag{17.46}$$

$$A_I(k) = A + BF(k),\ B_I(k) = BV(k),$$

$$C_I(k) = \begin{bmatrix} F(k) \\ C + DF(k) \end{bmatrix}, D_I(k) = \begin{bmatrix} V(k) \\ DV(k) \end{bmatrix}.$$

as well as

$$\mathcal{P}_{\mathcal{R}_G^{\mathcal{L}_2}} : \begin{cases} \hat{x}(k+1) = A_K(k)\hat{x}(k) + B_K(k) \begin{bmatrix} u(k) \\ y(k) \end{bmatrix} \\ r_y(k) = C_K\hat{x}(k) + D_K(k) \begin{bmatrix} u(k) \\ y(k) \end{bmatrix} \\ \lambda(k) = A_K^T(k)\lambda(k+1) + C_K^T(k)r_y(k) \\ \begin{bmatrix} \Delta u(k) \\ \Delta y(k) \end{bmatrix} = B_K^T(k)\lambda(k+1) + D_K^T(k)r_y(k), \end{cases} \tag{17.47}$$

$$A_K(k) = A - L(k)C,\ B_K(k) = \begin{bmatrix} B - L(k)D & L(k) \end{bmatrix},$$
$$C_K(k) = -W(k)C,\ D_K(k) = \begin{bmatrix} -W(k)D & W(k) \end{bmatrix}.$$

Notice that, although the system under consideration is time-invariant, the feedback gain and observer gain matrices $(F(k), L(k))$ as well as the pre- and post-filters $(V(k), W(k))$ are time-varying so that the corresponding projection systems are time-varying. The matrices $(F(k), L(k))$ and $(V(k), W(k))$ are given in Theorem 5.1. These two projections are of practical importance for the engineering applications. In particular, data processing in dealing with fault diagnosis and process monitoring should be generally conducted over a finite time interval, and the process data are often recorded and formatted batchwise. Also in the case of moving window mode of data processing are these projections useful.

It is noteworthy that the state space realisations of the adjoint systems (17.46) and (17.47) are expressed in slightly different forms concerning the system input signals. As known, the adjoint system of a given system is found using the duality identity. The adjoint system (17.46) is derived according to the duality identity,

$$\sum_{k=k_1}^{k_2-1} v^T(k)\, v_a(k+1) + x^T(k_1)\lambda(k_1) = \sum_{k=k_1}^{k_2-1} z_a^T(k)\, z(k+1) + x^T(k_2)\lambda(k_2),$$

$$z_a(k) = \begin{bmatrix} u_a(k) \\ y_a(k) \end{bmatrix}, z(k+1) = \begin{bmatrix} u(k+1) \\ y(k+1) \end{bmatrix},$$

while the adjoint system (17.47) is found according to the alternative duality identity form

$$\sum_{k=k_1}^{k_2} z^T(k)\, z_\Delta(k) + x^T(k_1)\lambda(k_1) = \sum_{k=k_1}^{k_2} r_y^T(k)\, r_y(k) + x^T(k_2+1)\lambda(k_2+1),$$

$$z_\Delta(k) = \begin{bmatrix} u_\Delta(k) \\ y_\Delta(k) \end{bmatrix}.$$

The reader is referred to Remark 5.2.

An orthogonal operator is idempotent and self-adjoint. The following properties of an orthogonal projection are useful for studies in the unified framework: given $y \in \mathcal{H}$, the operator $\mathcal{P}_\mathcal{V}$ that projects y onto $\mathcal{V} \in \mathcal{H}$, then it holds

$$y = \mathcal{P}_\mathcal{V} y + \mathcal{P}_{\mathcal{V}^\perp} y, \tag{17.48}$$

$$\|y\|_2^2 = \|\mathcal{P}_\mathcal{V} y\|^2 + \|\mathcal{P}_{\mathcal{V}^\perp} y\|^2, \tag{17.49}$$

$$dist\,(y, \mathcal{V}) = \inf_{x\in\mathcal{V}} \|y - x\|_2, \tag{17.50}$$

where $\mathcal{V}^\perp$ is the orthogonal complement of $\mathcal{V}$. Equation (17.48) gives a decomposition of variable y and the decomposition satisfies (17.49), which is the well-known Pythagorean equation. The Pythagorean equation tells us, $\forall x \in \mathcal{V}$,

$$\|y - x\|_2 \geq \|y - \mathcal{P}_\mathcal{V} y\|_2. \tag{17.51}$$

Consequently, the distance from y to the subspace $\mathcal{V}$ is defined by (17.50). On this basis, the concept of gap metric is established, which measures the similarity of two (closed) subspaces in Hilbert space $\mathcal{H}$. Given two closed subspaces $\mathcal{V}, \mathcal{U} \in \mathcal{H}$, the gap metric between them $\delta(\mathcal{V}, \mathcal{U})$ is defined by

$$\delta(\mathcal{V},\mathcal{U}) = \max\{\boldsymbol{\delta}(\mathcal{V},\mathcal{U}), \boldsymbol{\delta}(\mathcal{U},\mathcal{V})\},$$

$$\boldsymbol{\delta}(\mathcal{V},\mathcal{U}) = \sup_{\substack{x\in\mathcal{V} \\ \|x\|_2=1}} dist\,(x,\mathcal{U}) = \|(\mathcal{I} - \mathcal{P}_\mathcal{U})\mathcal{P}_\mathcal{V}\| = \sup_{x\in\mathcal{V}} \inf_{y\in\mathcal{U}} \frac{\|x-y\|_2}{\|x\|_2},$$

$$\boldsymbol{\delta}(\mathcal{U},\mathcal{V}) = \sup_{\substack{y\in\mathcal{U} \\ \|y\|_2=1}} dist\,(y,\mathcal{V}) = \|(\mathcal{I} - \mathcal{P}_\mathcal{V})\mathcal{P}_\mathcal{U}\| = \sup_{y\in\mathcal{U}} \inf_{x\in\mathcal{V}} \frac{\|y-x\|_2}{\|y\|_2}.$$

It is evident that $\delta(\mathcal{V},\mathcal{U}) \in [0, 1]$, and for $0 < \delta(\mathcal{V},\mathcal{U}) < 1$, $\boldsymbol{\delta}(\mathcal{V},\mathcal{U}) = \boldsymbol{\delta}(\mathcal{U},\mathcal{V}) = \delta(\mathcal{V},\mathcal{U})$. In the unified framework, computing the gap metric of two image or residual subspaces is generally performed with the aid of the system SIR and SKR. For instance, given two systems with SIRs, $I_{G_i}, i = 1, 2$, and SKRs, $K_{G_i}, i = 1, 2$, on the assumption of $0 < \delta\left(\mathcal{I}_{I_{G_1}}, \mathcal{I}_{I_{G_2}}\right) < 1$, the gap metric is calculated as follows,

$$\delta\left(\mathcal{I}_{I_{G_1}}^{\mathcal{L}_2}, \mathcal{I}_{I_{G_2}}^{\mathcal{L}_2}\right) = \left\| K_{G_{1,0}} I_{G_{2,0}} \right\|_\infty = \left\| K_{G_{2,0}} I_{G_{1,0}} \right\|_\infty,$$
$$\delta\left(\mathcal{I}_{I_{G_1}}^{\mathcal{H}_2}, \mathcal{I}_{I_{G_2}}^{\mathcal{H}_2}\right) = \min_{Q \in \mathcal{H}_\infty} \left\| I_{G_{1,0}} - I_{G_{2,0}} Q \right\|_\infty = \min_{Q \in \mathcal{H}_\infty} \left\| I_{G_{2,0}} - I_{G_{1,0}} Q \right\|_\infty.$$

Note that $\delta\left(\mathcal{I}_{I_{G_1}}^{\mathcal{L}_2}, \mathcal{I}_{I_{G_2}}^{\mathcal{L}_2}\right)$ can be equivalently written as

$$\delta\left(\mathcal{I}_{I_{G_1}}^{\mathcal{L}_2}, \mathcal{I}_{I_{G_2}}^{\mathcal{L}_2}\right) = \min_{Q \in \mathcal{L}_\infty} \left\| I_{G_{1,0}} - I_{G_{2,0}} Q \right\|_\infty = \min_{Q \in \mathcal{L}_\infty} \left\| I_{G_{2,0}} - I_{G_{1,0}} Q \right\|_\infty.$$

Hence, we obtain the relation

$$\delta\left(\mathcal{I}_{I_{G_1}}^{\mathcal{H}_2}, \mathcal{I}_{I_{G_2}}^{\mathcal{H}_2}\right) \geq \delta\left(\mathcal{I}_{I_{G_1}}^{\mathcal{L}_2}, \mathcal{I}_{I_{G_2}}^{\mathcal{L}_2}\right).$$

Given the process model,

$$G : \begin{cases} x(k+1) = Ax(k) + Bu(k) + E_d d(k), x(0) = x_0 \\ y(k) = Cx(k) + Du(k) + F_d d(k), \end{cases} \tag{17.52}$$

with known matrices E_d, F_d and F_d satisfying

$$rank\left(F_d\right) = m < k_d, \tag{17.53}$$

the optimal estimation of $d(k) \in \mathbb{R}^{k_d}$, an $\mathcal{L}_{2,[k_1,k_2]}$-bounded unknown input vector, is a challenging issue frequently met in engineering fault diagnosis and process monitoring practice. In Chap. 12, Algorithm 12.1 is provided to compute the LS estimate of $d(k)$. Blow, a projection-based solution is presented.

Suppose that a batch of data $(u(k), y(k))$, $k \in [k_1, k_2]$, is available for the estimation purpose. Design a time-varying SKR residual generator,

$$\mathcal{K}_G : \begin{cases} \hat{x}(k+1) = (A - L(k)C)\,\hat{x}(k) + \begin{bmatrix} B - L(k)D & L(k) \end{bmatrix} \begin{bmatrix} u(k) \\ y(k) \end{bmatrix} \\ r_y(k) = -W(k)C\hat{x}(k) + \begin{bmatrix} -W(k)D & W(k) \end{bmatrix} \begin{bmatrix} u(k) \\ y(k) \end{bmatrix}, \end{cases}$$

whose dynamic is governed by

$$\hat{\mathcal{N}}_d : \begin{cases} e(k+1) = (A - L(k)C)\,e(k) + (E_d - L(k)F_d)\,d(k) \\ r_y(k) = W(k)\left(Ce(k) + F_d d(k)\right). \end{cases} \tag{17.54}$$

Now, the estimation problem can be formulated as an orthogonal projection onto the subspace

$$\mathcal{R}_{\hat{\mathcal{N}}_d}^{\mathcal{L}_{2,[k_1,k_2]}} = \left\{\hat{d} : \hat{d} = \hat{\mathcal{N}}_d^* r_y, r_y \in \mathcal{L}_{2,[k_1,k_2]}\right\}.$$

To this end, let

$$\mathcal{P}_{\mathcal{R}_{\hat{\mathcal{N}}_d}^{\mathcal{L}_{2,[k_1,k_2]}}} = \hat{\mathcal{N}}_d^* \circ \hat{\mathcal{N}}_d$$

be the projection operator. The estimate $\hat{d}(k)$ is given by

$$\hat{d}(k) = \mathcal{P}_{\mathcal{R}_{\hat{\mathcal{N}}_d}^{\mathcal{L}_{2,[k_1,k_2]}}} \left(d(k)\right),$$

$$\mathcal{P}_{\mathcal{R}_{\hat{\mathcal{N}}_d}^{\mathcal{L}_{2,[k_1,k_2]}}} : \begin{cases} e(k+1) = (A - L(k)C)\, e(k) + (E_d - L(k)F_d)\, d(k) \\ r_y(k) = W(k)\,(Ce(k) + F_d d(k)) \\ \lambda(k) = (A - L(k)C)^T\, \lambda(k+1) + (W(k)C)^T\, r_y(k) \\ \hat{d}(k) = (E_d - L(k)F_d)^T\, \lambda(k+1) + (W(k)D)^T\, r_y(k) \end{cases}$$

for $k \in [k_1, k_2]$ and a given $\lambda(k_2+1)$. Hereby, $(L(k), W(k))$ are recursively computed according to Theorem 5.1 so that the time-varying system $\hat{\mathcal{N}}_d$ (17.54) is co-inner.

17.3.3 Hamiltonian Systems

Even though Hamiltonian systems have been addressed in the previous chapters, as issues of control and detection of nonlinear systems have been under consideration, Hamiltonian systems theory is a powerful mathematical framework for analysis and synthesis of linear dynamic systems as well. Characterised by a Hamiltonian function and the associated system dynamic questions, Hamiltonian systems theory reveals insightful aspects of dynamic systems, which are inspirational for exploring new ideas and concepts. In the control and detection framework, Hamiltonian systems enable us to deal with detection and control issues in the signal subspaces in a unified and elegant way. In particular, based on the Hamiltonian function, design of control and detection systems can be equivalently formulated as optimisation problems and solved in the ML-framework. In a certain sense, Hamiltonian systems are a firm bridge connecting model- and ML-based detection and control methods, which will play a central role in future work on detection and control.

Representatively, the nonlinear affine system model (17.12), its SIR and SKR (17.13)–(17.14) are under consideration in this subsection. The Hamiltonian extensions of $\Sigma_{\mathcal{I}}$ and $\Sigma_{\mathcal{K}}$ are

$$\left(\Sigma_{\mathcal{I}}, (D\Sigma_{\mathcal{I}})^T\right) : \begin{cases} \dot{x} = a_I(x) + B_I(x)v \\ \dot{\lambda} = -\left(\frac{\partial a_I(x)}{\partial x} + \frac{\partial B_I(x)}{\partial x}v\right)^T \lambda - \left(\frac{\partial c_I(x)}{\partial x} + \frac{\partial D_I(x)}{\partial x}v\right)^T v_a \\ z = c_I(x) + D_I(x)v \\ z_a = B_I^T(x)\lambda + D_I^T(x)v_a,\, z_a \in \mathbb{R}^p,\, v_a \in \mathbb{R}^{m+p}, \end{cases}$$

$$\left(\Sigma_{\mathcal{K}}, (D\Sigma_{\mathcal{K}})^T\right) : \begin{cases} \dot{\hat{x}} = a_K(\hat{x}) + B_K(\hat{x})z \\ \dot{\lambda} = -\left(\frac{\partial a_K(\hat{x})}{\partial \hat{x}} + \frac{\partial B_K(\hat{x})}{\partial \hat{x}}z\right)^T \lambda - \left(\frac{\partial c_K(\hat{x})}{\partial \hat{x}} + \frac{\partial D_K(\hat{x})}{\partial \hat{x}}z\right)^T z_a \\ r_y = c_K(\hat{x}) + D_K(\hat{x})z \\ r_{y,a} = B_K^T(\hat{x})\lambda + D_K^T(\hat{x})z_a, z_a \in \mathbb{R}^m, r_{y,a} \in \mathbb{R}^{m+p}, \end{cases}$$

respectively, where $z = \begin{bmatrix} u^T & y^T \end{bmatrix}^T$. In different configurations of the above Hamiltonian extensions, we are able to construct various Hamiltonian systems for diverse applications. In the unified framework, the subsequent four Hamiltonian systems are serviceable:

- Hamiltonian system configured by connecting z and v_a, $z = v_a$,

$$z_a = (D\Sigma_{\mathcal{I}})^T \circ \Sigma_{\mathcal{I}}(v), \tag{17.55}$$

$$H(x, \lambda, v) = \frac{1}{2}(c_I(x) + D_I(x)v)^T (c_I(x) + D_I(x)v) + \lambda^T (a_I(x) + B_I(x)v), \tag{17.56}$$

$$(D\Sigma_{\mathcal{I}})^T \circ \Sigma_{\mathcal{I}} : \begin{cases} \dot{x} = \frac{\partial H}{\partial \lambda}(x, \lambda, v) \\ \dot{\lambda} = -\frac{\partial H}{\partial x}(x, \lambda, v) \\ z_a = \frac{\partial H}{\partial v}(x, \lambda, v), \end{cases} \tag{17.57}$$

- Hamiltonian system configured by connecting z_a and v, $z_a = v$, $v_a := z$,

$$\hat{z} := \Sigma_{\mathcal{I}} \circ (D\Sigma_{\mathcal{I}})^T (z), \tag{17.58}$$

$$\begin{aligned} H(x, \lambda, z) &= \frac{1}{2}z^T D_I(x)D_I^T(x)z + c_I^T(x)z \\ &+ \lambda^T \left(a_I(x) + \frac{1}{2}B_I(x)B_I^T(x)\lambda + B_I(x)D_I^T(x)z\right), \end{aligned} \tag{17.59}$$

$$\Sigma_{\mathcal{I}} \circ (D\Sigma_{\mathcal{I}})^T : \begin{cases} \dot{x} = \frac{\partial H}{\partial \lambda}(x, \lambda, z) \\ \dot{\lambda} = -\frac{\partial H}{\partial x}(x, \lambda, z) \\ \hat{z} = \frac{\partial H}{\partial z}(x, \lambda, z), \end{cases} \tag{17.60}$$

  and the output $\hat{z}$ serves as an estimate of the input data z,
- Hamiltonian system configured by connecting z and $r_{y,a}$, $z = r_{y,a}$,

$$r_y = \Sigma_{\mathcal{K}} \circ (D\Sigma_{\mathcal{K}})^T (z_a), \tag{17.61}$$

$$\begin{aligned} H\left(\hat{x}, \lambda, z_a\right) &= \frac{1}{2}z_a^T D_K(\hat{x})D_K^T(\hat{x})z_a + c_K^T(\hat{x})z_a \\ &+ \lambda^T \left(a_K(\hat{x}) + \frac{1}{2}B_K(\hat{x})B_K^T(\hat{x})\lambda + B_K(\hat{x})D_K^T(\hat{x})z_a\right), \end{aligned} \tag{17.62}$$

$$\Sigma_{\mathcal{K}} \circ (D\Sigma_{\mathcal{K}})^T : \begin{cases} \dot{\hat{x}} = \frac{\partial H}{\partial \lambda}\left(\hat{x}, \lambda, z_a\right) \\ \dot{\lambda} = -\frac{\partial H}{\partial \hat{x}}\left(x, \lambda, z_a\right) \\ r_y = \frac{\partial H}{\partial z_a}\left(x, \lambda, z_a\right), \end{cases} \tag{17.63}$$

- Hamiltonian system configured by connecting z_a and r_y, $z_a = r_{y,a}$,

$$\Delta z = (D\Sigma_{\mathcal{K}})^T \circ \Sigma_{\mathcal{K}}(z), \Delta z = \begin{bmatrix} \Delta u \\ \Delta y \end{bmatrix} = r_{y,a}, \tag{17.64}$$

$$H\left(\hat{x}, \lambda, z\right) = \frac{1}{2}\left(c_K(\hat{x}) + D_K(\hat{x})z\right)^T \left(c_K(\hat{x}) + D_K(\hat{x})z\right) \\ +\lambda^T \left(a_K(\hat{x}) + B_K(\hat{x})z\right), \tag{17.65}$$

$$(D\Sigma_{\mathcal{K}})^T \circ \Sigma_{\mathcal{K}} : \begin{cases} \dot{\hat{x}} = \frac{\partial H}{\partial \lambda}\left(\hat{x}, \lambda, z\right) \\ \dot{\lambda} = -\frac{\partial H}{\partial \hat{x}}\left(x, \lambda, z\right) \\ \Delta z = \frac{\partial H}{\partial z}\left(x, \lambda, z\right), \end{cases} \tag{17.66}$$

and the output Δz is an estimate of the uncertainty in z.

The systems (17.55) and (17.61) are typical model forms applied for optimal control and estimation. Observe that Hamiltonian function (17.56) is equivalent to

$$H\left(x, \lambda, v\right) = \frac{1}{2}z^T z + \lambda^T \left(a_I(x) + B_I(x)v\right), \tag{17.67}$$

while the Legendre dual of Hamiltonian function $H\left(\hat{x}, \lambda, z_a\right)$ in (17.62)

$$H^{\times}\left(\hat{x}, \lambda, r_y\right) = r_y^T z_a - H\left(\hat{x}, \lambda, z_a\right)$$

can be equivalently written as

$$H^{\times}\left(\hat{x}, \lambda, r_y\right) = \frac{1}{2}r_{y,a}^T r_{y,a} - \lambda^T \left(a_K(\hat{x}) + B_K(\hat{x})r_{y,a}\right) \tag{17.68}$$

with $r_{y,a}$ as the output of $(D\Sigma_{\mathcal{K}})^T$, which yields

$$\Sigma_{\mathcal{K}} \circ (D\Sigma_{\mathcal{K}})^T : \begin{cases} \dot{\hat{x}} = -\frac{\partial H^{\times}(\hat{x}, \lambda, r_y)}{\partial \lambda} \\ \dot{\lambda} = \frac{\partial H^{\times}(\hat{x}, \lambda, r_y)}{\partial \hat{x}} \\ z_a = \frac{\partial H^{\times}(\hat{x}, \lambda, r_y)}{\partial r_y}. \end{cases} \tag{17.69}$$

In this context, the Hamiltonian functions (17.67) and (17.68) serve as the objective functions for optimal control and estimation of the systems (17.57) and (17.69), respectively. In this manner, the normalised SIR and SKR can be computed as well.

Similarly, the projection-based estimation systems (17.58) and (17.64) and the associated Hamiltonian functions (17.59) and (17.65) can be respectively rewritten as

$$H^{\times}\left(x, \lambda, \hat{z}\right) = \hat{z}^T z - H\left(x, \lambda, z\right) = \frac{1}{2} z_a^T z_a - \lambda^T \left(a_I(x) + B_I(x) z_a\right) \tag{17.70}$$

$$\Sigma_{\mathcal{I}} \circ \left(D\Sigma_{\mathcal{I}}\right)^T : \begin{cases} \dot{x} = -\frac{\partial H^{\times}(x,\lambda,\hat{z})}{\partial \lambda} \\ \dot{\lambda} = \frac{\partial H^{\times}(x,\lambda,\hat{z})}{\partial x} \\ z = \frac{\partial H^{\times}(x,\lambda,\hat{z})}{\partial \hat{z}}, \end{cases} \tag{17.71}$$

as well as

$$H\left(\hat{x}, \lambda, z\right) = \frac{1}{2} r_y^T r_y + \lambda^T \left(a_K(\hat{x}) + B_K(\hat{x}) z\right). \tag{17.72}$$

On the basis of (17.70)–(17.72), we are able to compute the projection-based estimates $\hat{z}$ and Δz by means of two optimisation problems.

17.4 Notes and References

It is not an exaggeration to state that Youla parameterisation of all stabilising controllers is a popular application of the Bezout identity in control theory. The claim that the Bezout identity is at the core of the unified framework of control and detection is however based on the fact that it establishes the one-to-one mapping between the process data (u, y) and the residuals $\left(r_u, r_y\right)$. It helps us to gain keen insights into complex control and detection issues and to develop alternative solutions, and builds the basis to decompose the process data space into the subspaces of the nominal and uncertain dynamics. In this regard, we would like to place a greater emphasis on the important role of the residuals $\left(r_u, r_y\right)$ in dealing with control and detection problems. While the output residual r_y is widely used for detection and estimation purposes, little attention has been paid to the input residual signal r_u. Our endeavours in analysis of closed-loop dynamics and impacts of uncertainties and faults on system performance, in the projection-based diagnosis and estimation applications have showcased the capability of $\left(r_u, r_y\right)$ to solve complex control and detection problems. A particular example hereby is the application to secure CPSs. As demonstrated in previous chapters and in [45, 109], the use of r_u is necessary to detect cyber-attacks and enhance CPS resilience, and the integrated use of $\left(r_u, r_y\right)$ enables simultaneous detection of faults and cyber-attacks and further fault-tolerant and cyber-attack resilient control and CPSs.

The system descriptions SIR and SKR seem to be a natural extension of conventional input-output transfer functions or state space representations. Specifically formulated, SIR and SKR are indeed the system representation forms of modelling the system dynamics in $\mathcal{H}_2$ or $\mathcal{L}_2$ subspace. Since in most of engineering applications the system signals or data belong to $\mathcal{L}_2$ subspace, SIR and SKR are capable model forms for system analysis and synthesis. In particular, on the basis of SIR and SKR, the system image and residual subspaces are defined. Introduction of image and residual subspaces to represent the system nominal and uncertain/faulty dynamics, and analysis of system behaviour in its image and residual subspaces are characteristic features of the unified framework. They not only serve as a capable tool to address control and detection issues, but also establish a bridge connecting model-based, data-driven and ML-based frameworks. It is worth mentioning a special application of the SKR to system modelling, the SKR-based system input-output model (17.26), which gives a uniform process model representation, is valid under all operating conditions, including faulty operations. Moreover, this model is capable for online realisation and can serve as an online simulation model. This model has been successfully applied to model and monitor nano-particle generation processes recently [28].

Computation algorithms of normalised SIR and SKR of plants are well-established and can be found in the literature for various types of dynamic systems. In our study on the unified framework of control and detection, they serve both as a tool for the design and analysis of projection-based systems and as benchmark solutions for control and detection problems. Normalised SIR and SKR can be achieved by means of IOF, CIOF and spectral factorisation methods [243].

Doubtlessly, orthogonal projection-based detection and system analysis methods serve as a key technique of the unified framework. It is a major tool to analyse system dynamics in the data subspaces and, based on it, to address detection, estimation and control issues. In robust control, projections onto $\mathcal{H}_2$ image subspaces have been comprehensively investigated, where the major focus is on stability analysis and uncertainty issues. Projections onto $\mathcal{H}_2$ residual subspaces have been rarely explored. The orthogonal projection onto $\mathcal{H}_2$ residual subspace using the operator (17.43) is given by [152]. For the detection and estimation aim, projections onto $\mathcal{L}_2$ image and kernel subspaces are especially useful. Here, the backward computation of the involved adjoint system enables us to gain information about uncertainties or faults and even estimate unknown inputs or uncertainties corrupted in the process data. It is noteworthy that the projection-based detection and control systems are of a natural configuration of an autoencoder widely applied in ML-based algorithms. An autoencoder consists of an encoder that compresses the process data into the latent variable, and a decoder that is driven by the latent variable and reconstructs the process variables reflecting nominal process operations [73, 84]. This promises autoencoder-based realisation of data-driven projection systems [112]. It is of special interest to notice the role of latent variables in an autoencoder. In the information theoretic framework, [68, 192] have proposed to generate the latent variable as the so-called minimal sufficient statistic. In the projection-based systems, the residual signals r_u or r_y acts as a latent variable. For instance, the residual r_y generated by the normalised SKR can be interpreted as minimum sufficient information for re-constructing uncertainties corrupted in the data or unknown inputs.

Hamiltonian systems theory unifies system dynamics, Hamiltonian function as an energy function that can serve as an objective function for optimisation as well, and the dual system structure in a framework. In this framework, normalised SIR and SKR, projection-based systems, and optimal control and detection as well as estimation can be unifiedly addressed. More importantly, a natural integration of Hamiltonian systems and Information geometry [3], as delineated in Chap. 11, promises a new research subject towards control-theory-informed machine learning (CTIML) aided control and detection in the unified framework.

References

1. I. Ahmed, T. Galoppo, X. Hu, Y. Ding, Graph regularized autoencoder and its application in unsupervised anomaly detection. IEEE Trans. Pattern Anal. Mach. Intell. **44**(8), 4110–4124 (2022)
2. E. Alcorta-Garcia, P. Frank, Deterministic nonlinear observer based approaches to fault diagnosis: a survey. Control. Eng. Pract. **5**(5), 663–670 (1997)
3. S. Amari, *Information Geometry and its Applications* (Springer, Japan, 2016)
4. B.D.O. Anderson, From youla-kucera to identification, adaptive and nonlinear control. Automatica **34**, 1485–1506 (1998)
5. A. Armeni, A. Casavola, E. Mosca, Robust fault detection and isolation for LPV systems under a sensitivity constraint. Int. J. Adapt. Control Signal Process. **23**, 55–72 (2009)
6. A. Atassi, H. Khalil, A separation principle for the stabilization of a class of nonlinear systems. IEEE Trans. Autom. Control **44**(9), 1672–1687 (1999)
7. J.A. Ball, A.J. van der Schaft, J-inner-outer factorization, j-spectral factorization, and robust control for nonlinear systems. IEEE Trans. Automatic Contr. **41**, 379–392 (1996)
8. M. Basseville, I. Nikiforov, *Detection of Abrupt Changes - Theory and Application* (Prentice-Hall, New Jersey, 1993)
9. R. Beard, *Failure Accomondation in Linear Systems Through Self-Reorganization*. Ph.D. Dissertation, MIT (1971)
10. J. Bendtsen, K. Trangbaek, J. Stoustrup, Plug-and-play control-modifying control systems online. IEEE Trans. Control Syst. Technol. **21**(1), 79–93 (2013)
11. Y. Bengio, A. Courville, P. Vincent, Representation learning: a review and new perspectives. IEEE Trans. Pattern Anal. Mach. Intell. **35**(8), 1798–1828 (2013)
12. P. Benner, M. Ohlberger, A. Cohen, K. Willcox, *Model Reduction and Approximation: Theory and Algorithms* (SIAM, Philadelphia, 2017)
13. M. Blanke, M. Kinnaert, J. Lunze, M. Staroswiecki, *Diagnosis and Fault-Tolerant Control*, 2nd edn. (Springer, Berlin Heidelberg, 2006)
14. J. Bokor, G. Balas, Detection filter design for LPV systems - a geometric approach. Automatica **40**, 511–518 (2004)
15. S. Boyd, L. Vandenberghe, *Convex Optimization* (Cambridge University Press, 2004)
16. L. Bregman, The relaxation method of finding the common point of convex sets and its application to the solution of problems in convex programming. USSR Comput. Math. Math. Phys. **7**(3), 200–217 (1967)
17. F.M. Callier, C.A. Desoer, *Linear System Theory* (Springer, New York, 1991)
18. M. Chadli, A. Abdo, S.X. Ding, H_-/H_∞ fault detection filter design for discrete-time takagi-sugeno fuzzy system. Automatica **49**, 1996–2005 (2013)

S. X. Ding, *Unified Framework of Control and Detection*,
https://doi.org/10.1007/978-3-662-72417-0

19. G. Chen, G. Chen, S.-H. Hsu, *Linear Stochastic Control Systems* (CRC Press, 1995)
20. J. Chen, R.J. Patton, *Robust Model-Based Fault Diagnosis for Dynamic Systems* (Kluwer Academic Publishers, Boston, 1999)
21. W.-H. Chen, J. Yang, L. Guo, S. Li, Disturbance-observer-based control and related methods - an overview. IEEE Trans. Industr. Electron. **63**(2), 1083–1095 (2016)
22. Y. Chen, S. Kar, J.M.F. Moura, Cyber-physical attacks with control objectives. IEEE Trans. Autom. Control **63**(5), 1418–1425 (2018)
23. L.H. Chiang, E.L. Russell, R.D. Braatz, *Fault Detection and Diagnosis in Industrial Systems* (Springer, London, 2001)
24. E.Y. Chow, A.S. Willsky, Analytical redundancy and the design of robust failure detection systems. IEEE Trans. Autom. Control **29**, 603–614 (1984)
25. C.-C. Chu, On discrete inner-outer and spectral factorizations. American Control Conference **1988**, 1699–1700 (1988)
26. J. Coulson, J. Lygeros, F. Doerfler, Data-enabled predictive control: in the shallows of the deepc, in *2019 18th European Control Conference (ECC)* (2019), pp. 307–312
27. P. Crouch, A. van der Schaft, *Variational and Hamiltonian Control Systems* (Springer, Basel, 1987)
28. D. Cuturic, J.V. Weidemann, S.X. Ding, E. Kruis, M.S. Obergfell, C. Louen, A control theory informed machine learning-based modeling and monitoring of nanoparticle generation processes. IEEE Trans. Ind. Inf. **20**(10), 7455–7465 (2025)
29. B.N. Datta, *Numerical Methods for Linear Control Systems* (Academic, 2004)
30. M.A. Davenport, R.G. Baraniuk, C.D. Scott, Tuning support vector machines for minimax and neyman-pearson classification. IEEE Trans. Pattern Anal. Mach. Intell. **32**, 1888–1898 (2010)
31. S.M. Dibaji, M. Pirani, D.B. Flamholz, A.M. Annaswamy, K.H. Johansson, A. Chakrabortty, A systems and control perspective of CPS security. Annu. Rev. Control. **47**, 394–411 (2019)
32. D. Ding, Q.-L. Han, Y. Xiang, X. Ge, X.-M. Zhang, A survey on security control and attack detection for industrial cyber-physical systems. Neurocomputing **275**, 1674–1683 (2018)
33. D. Ding, Q.-L. Han, X. Ge, J. Wang, Secure state estimation and control of cyber-physical systems: a survey. IEEE Trans. Syst. Man Cybern.: Syst. **51**(1), 176–190 (2021)
34. S.X. Ding, *Model-Based Fault Diagnosis Techniques - Design Schemes, Algorithms, and Tools* (Springer, 2008)
35. S.X. Ding, Integrated design of feedback controllers and fault detectors. Annu. Rev. Control. **33**, 124–135 (2009)
36. S.X. Ding, *Model-Based Fault Diagnosis Techniques - Design Schemes, Algorithms and Tools*, 2nd edn. (Springer, London, 2013)
37. S.X. Ding, *Data-Driven Design of Fault Diagnosis and Fault-Tolerant Control Systems* (Springer, London, 2014)
38. S.X. Ding, Data-driven design of monitoring and diagnosis systems for dynamic processes: a review of subspace technique based schemes and some recent results. J. Process Control **24**, 431–449 (2014)
39. S.X. Ding, *Advanced Methods for Fault Diagnosis and Fault-tolerant Control* (Springer, Berlin, 2020)
40. S.X. Ding, A note on diagnosis and performance degradation detection in automatic control systems towards functional safety and cyber security. Secur. Safety **1**, 2022004 (2022)
41. S.X. Ding, T. Jeinsch, P.M. Frank, E.L. Ding, A unified approach to the optimization of fault detection systems. Int. J. Adapt. Control Signal Process. **14**, 725–745 (2000)
42. S.X. Ding, L. Li, Control performance monitoring and degradation recovery in automatic control systems: a review, some new results, and future perspectives. Control. Eng. Pract. **111**, 104790 (2021)
43. S.X. Ding, L. Li, M. Kruger, Application of radomized algorithms to assessment and design of observer-based fault detection systems. Automatica **107**, 175–182 (2019)
44. S.X. Ding, L. Li, T. Liu, An alternative paradigm of fault diagnosis in dynamic systems: orthogonal projection-based methods. Automatica **183**, 112637 (2026)

45. S.X. Ding, L. Li, D. Zhao, C. Louen, T. Liu, Application of the unified control and detection framework to detecting stealthy integrity cyber-attacks on feedback control systems. Automatica **142**, 110352 (2022)
46. S.X. Ding, Y. Yang, Y. Zhang, L. Li, Data-driven realization of kernel and image representations and their application to fault detection and control system design. Automatica **50**, 2615–2623 (2014)
47. S.X. Ding, P. Zhang, A. Naik, E. Ding, B. Huang, Subspace method aided data-driven design of fault detection and isolation systems. J. Process Control **19**, 1496–1510 (2009)
48. S.X. Ding, P. Zhang, S. Yin, E.L. Ding, An integrated design framework of fault-tolerant wireless networked control systems for industrial automatic control applications. IEEE Trans. Industr. Inf. **9**(1), 462–471 (2013)
49. X. Ding, P.M. Frank, Frequency domain approach and threshold selector for robust model-based fault detection and isolation, in *Proceedings of the 1st IFAC Symposium on SAFEPROCESS* (1991)
50. X. Ding, L. Guo, P.M. Frank, A frequency domain approach to fault detection of uncertain dynamic systems, in *Proceedings of the 32nd Conference on Decision and Control*, Texas, USA (1993), pp. 1722–1727
51. X. Ding, L. Guo, P.M. Frank, Parametrization of linear observers and its application to observer design. IEEE Trans. on Autom. Contr. **39**, 1648–1652 (1994)
52. X. Ding, L. Guo, T. Jeinsch, A characterization of parity space and its application to robust fault detection. IEEE Trans. Autom. Control **44**(2), 337–343 (1999)
53. F. Doerfler, J. Coulson, I. Markovsky, Bridging direct and indirect data-driven control formulations via regularizations and relaxations. IEEE Trans. Autom. Control **68**(2), 883–897 (2023)
54. A. Feintuch, The time-varying gap and coprime factor perturbations. Math. Control Signal Syst. **8**, 352–374 (1995)
55. A. Feintuch, *Robust Control Theory in Hilbert Space* (Springer, New York, 1998)
56. T. Floquet, J.P. Barbot, W. Perruquetti, M. Djemai, On the robust fault detection via sliding mode disturbance observer. Int. J. Control **77**, 622–629 (2004)
57. B.A. Francis, *A Course in H-Infinity Control Theory* (Springer, Berlin - New York, 1987)
58. P.M. Frank, Fault diagnosis in dynamic systems using analytical and knowledge-based redundancy - a survey. Automatica **26**, 459–474 (1990)
59. P.M. Frank, X. Ding, Frequency domain approach to optimally robust residual generation and evaluation for model-based fault diagnosis. Automatica **30**, 789–904 (1994)
60. P.M. Frank, X. Ding, Survey of robust residual generation and evaluation methods in observer-based fault detection systems. J. Process Control **7**(6), 403–424 (1997)
61. L. Freidovich, H. Khalil, Performance recovery of feedback-linearization-based designs. IEEE Trans. Autom. Control **53**(10), 2324–2334 (2008)
62. K. Fujimoto, T. Sugie, State-space characterization of youla parametrization for nonlinear systems based on input-to-state stability, in *Proceedings of the 37th IEEE CDC*, vol. 4 (1998), pp. 2479–2484
63. K. Fujimoto, T. Sugie, Characterization of all nonlinear stabilizing controllers via observer-based kernel representations. Automatica **36**(8), 1123–1135 (2000)
64. Z.W. Gao, C. Cecati, S.X. Ding, A survey of fault diagnosis and fault-tolerant techniques, part i: fault diagnosis with model-based and signal-based approaches. IEEE Trans. Ind. Electr. **62**, 3757–3767 (2015)
65. Z.W. Gao, C. Cecati, S.X. Ding, A survey of fault diagnosis and fault-tolerant techniques, part ii: fault diagnosis with knowledge-based and hybrid/active approaches. IEEE Trans. Ind. Electr. **62**, 3768–3774 (2015)
66. Z.Q. Ge, Z.H. Song, *Multivariate Statistic Process Control* (Springer, London, 2013)
67. Z.Q. Ge, Z.H. Song, F.R. Gao, Review of recent research on data-based process monitoring. Ind. Eng. Chem. Res. **52**, 1446–1455 (2013)
68. B.C. Geiger, On information plane analyses of neural network classifiers–a review. IEEE Trans. Neural Netw. Learn. Syst. 1–13 (2021)

69. T.T. Georgiou, On the computation of the gap metric. Syst. Contr. Lett. **11**, 253–257 (1988)
70. T.T. Georgiou, M.C. Smith, Optimal robustness in the gap metric. IEEE Trans. Autom. Control **35**, 673–686 (1990)
71. J.J. Gertler, *Fault Detection and Diagnosis in Engineering Systems* (Marcel Dekker, New York Basel Hong Kong, 1998)
72. J. Giraldo, D. Urbina, A. Cardenas, J. Valente, M. Faisal, J. Ruths, N.O. Tippenhauer, H. Sandberg, R. Candell, A survey of physics-based attack detection in cyber-physical systems. ACM Comput. Surv. 51 (2018)
73. I. Goodfellow, Y. Bengio, A. Courville, *Deep Learning* (MIT Press, 2016)
74. P. Griffioen, S. Weerakkody, B. Sinopoli, O. Ozel, Y. Mo, A tutorial on detecting security attacks on cyber-physical systems, in *2019 18th European Control Conference (ECC)* (2019), pp. 979–984
75. G. Gu, L. Qiu, Connection of multiplicative/relative perturbation in coprime factors and gap metric uncertainty. Automatica **34**, 603–607 (1998)
76. Z. Guo, D. Shi, K.H. Johansson, L. Shi, Optimal linear cyber-attack on remote state estimation. IEEE Trans. Control Netw. Syst. **4**(1), 4–13 (2017)
77. F. Gustafsson, *Adaptive Filtering and Change Detection* (John Wiley and Sons, LTD, Chichester, 2000)
78. A. Halanay, V. Ionescu, *Time-Varying Discrete Linear Systems* (Springer, 2012)
79. H. Hammouri, M. Kinnaert, E.E. Yaagoubi, Observer-based approach to fault detection and isolation for nonlinear systems. IEEE Trans. Autom. Control **44**(10), 1879–1884 (1999)
80. H. Han, Y. Yang, L. Li, S.X. Ding, Performance-based fault detection and fault-tolerant control for nonlinear systems with t-s fuzzy implementation. IEEE Trans. Cybern. **51**, 801–814 (2021)
81. F. Hansen, G. Franklin, R. Kosut, Closed-loop identification via the fractional representation: experiment design. American Control Conference **1989**, 1422–1427 (1989)
82. X. He, Z. Wang, D. Zhou, Robust hinf filtering for time-delay systems with probabilistic sensor faults. IEEE Signal Process. Lett. **16**, 442–445 (2009)
83. D. Henry, A. Zolghadri, Design and analysis of robust residual generators for systems under feedback control. Automatica **41**, 251–264 (2005)
84. G.E. Hinton, R.R. Salakhutdinov, Reducing the dimensionality of data with neural networks. Science **313**(5786), 504–507 (2006)
85. D. Hoang, H. Kang, A survey on deep learning based bearing fault diagnosis. Neurocomputing **335**, 327–335 (2019)
86. J.W. Hoffmann, Normalized coprime factorizations in continuous and discrete time - a joint state-space approach. IMA J. Math. Control. Inf. **13**(4), 359–384 (1996)
87. M. Hou, P.C. Mueller, Disturbance decoupled observer design: a unified viewpoint. IEEE Trans. Autom. Control **39**(6), 1338–1341 (1994)
88. M. Hou, P. Mueller, Design of observers for linear systems with unknown inputs. IEEE Trans. Autom. Control **37**(6), 871–875 (1992)
89. M. Hou, R.J. Patton, An LMI approach to infinity fault detection observers, in *Proceedings of the UKACC International Conference on Control* (1996), pp. 305–310
90. M. Hou, R.J. Patton, Input observability and input reconstruction. Automatica **34**, 789–794 (1998)
91. C.-W. Hsu, C.-J. Lin, A comparison of methods for multiclass support vector machines. IEEE Trans. Neural Netw. **13**, 415–425 (2002)
92. Z. Hu, H. Zhao, J. Peng, Low-rank reconstruction-based autoencoder for robust fault detection. Control. Eng. Pract. **123**, 105156 (2022)
93. I. Hwang, S. Kim, Y. Kim, C.E. Seah, A survey of fault detection, isolation, and reconfiguration methods. IEEE Trans. Contr. Syst. Tech. **18**, 636–653 (2010)
94. R. Isermann, *Fault Diagnosis Systems* (Springer, Berlin Heidelberg, 2006)
95. W. Jia, J. Wang, Partial-nodes-based distributed fault detection and isolation for second-order multiagent systems with exogenous disturbances. IEEE Trans. Cybern. **52**(4), 2518–2530 (2022)

96. L. Jiang, Z. Ge, Z. Song, Semi-supervised fault classification based on dynamic sparse stacked auto-encoders model. Chemom. Intell. Lab. Syst. **168**, 72–83 (2017)
97. K. Jin, D. Ye, Optimal innovation-based stealthy attacks in networked lqg systems with attack cost. IEEE Trans. Cybern. **54**(2), 787–796 (2024)
98. I. Jolliffe, *Principal Component Analysis* (Springer, New York, Berlin, 1986)
99. H. Jones, *Failure Detection in Linear Systems*. Ph.D. dissertation, MIT (1973)
100. P. Kabore, H. Wang, Design of fault diagnosis filters and fault-tolerant control for a class of nonlinear systems. IEEE Trans. Autom. Control **46**, 1805–1810 (2001)
101. G. Karniadakis, I. Kevrekidis, L. Lu, P. Perdikaris, S. Wang, L. Yang, Physics-informed machine learning. Nat. Rev. Phys. **3**, 422–440 (2021)
102. T. Kato, *Perturbation Theory for Linear Operators* (Springer, Berlin, 1995)
103. Y. Kawano, M. Cao, Design of privacy-preserving dynamic controllers. IEEE Trans. Autom. Control **65**(9), 3863–3878 (2020)
104. K. Koenings, M. Krueger, H. Lou, S.X. Ding, A data-driven computation method for the gap metric and the optimal stability margin. IEEE Trans. Autom. Control **63**, 805–810 (2018)
105. V. Kucera, *Analysis and Design of Discrete Linear Control Systems* (Prentice Hall, 1991)
106. V. Kucera, Stability of discrete linear feedback systems. IFAC Proc. Vol. **8**(1), Part 1, 573–578 (1975), 6th IFAC World Congress (IFAC 1975) - Part 1: Theory, Boston/Cambridge, MA, USA, August 24–30 1975
107. F.L. Lewis, D. Vrabie, K.G. Vamvoudakis, Reinforcement learning and feedback control. IEEE Control Syst. Mag. 76–105 (2012)
108. G. Li, S.J. Qin, D. Zhou, Geometric properties of partial least squares for process monitoring. Automatica **46**, 204–210 (2010)
109. L. Li, S.X. Ding, M. Zhong, K. Peng, *Orthogonal projection-based fault detection for linear discrete-time varying systems* IEEE Trans. Autom, Control **70**(5), 3478–3485 (2025)
110. L. Li, S.X. Ding, Gap metric techniques and their application to fault detection performance analysis and fault isolation schemes. Automatica **118**, 109029 (2020)
111. L. Li, S.X. Ding, Optimal detection schemes for multiplicative faults in uncertain systems with application to rolling mill processes. IEEE Trans. Control Syst. Technol. **28**(6), 2432–2444 (2020)
112. L. Li, S.X. Ding, K. Liang, Z. Chen, T. Xue, Control theoretically explainable application of autoencoder methods to fault detection in nonlinear dynamic systems (2022). arXiv:2208.01921v1
113. L. Li, S.X. Ding, X. Peng, Optimal observer-based fault detection and estimation approaches for T-S fuzzy systems. IEEE Trans. Fuzzy Syst. **30**(2), 579–590 (2022)
114. L. Li, S.X. Ding, J. Qiu, Y. Yang, Real-time fault detection approach for nonlinear systems and its asynchronous T-S fuzzy observer-based implementation. IEEE Trans. Cybern. **47**, 283–294 (2017)
115. L. Li, S.X. Ding, J. Qui, Y. Yang, D. Xu, Fuzzy observer-based fault detection design approach for nonlinear processes. IEEE Trans. Syst. Man Cybern.: Syst. **47**, 1941–1952 (2017)
116. L. Li, S.X. Ding, J. Qui, Y. Yang, Y. Zhang, Weighted fuzzy observer-based fault detection approach for discrete-time nonlinear systems via piecewise-fuzzy lyapunov functions. IEEE Trans. Fuzzy Syst. **24**, 1320–1333 (2016)
117. L. Li, S.X. Ding, Y. Yang, K. Peng, J. Qiu, A fault detection approach for nonlinear systems based on data-driven realizations of fuzzy kernel representations. IEEE Trans. Fuzzy Syst. **26**, 1800–1812 (2018)
118. L. Li, Y. Yang, S.X. Ding, Y. Zhang, S. Zhai, On fault-tolerant control configurations for a class of nonlinear systems. J. Franklin Inst. **352**(4), 1397–1416 (2015)
119. Z. Li, Z. Duan, *Cooperative Control of Multi-Agent Systems: A Consensus Region Approach* (CRC Press, 2015)
120. Z. Li, E. Mazars, Z. Zhang, I.M. Jaimoukha, State-space solution to the h-/hinf fault detection problem. Int. J. Robst. Nonlinear Control **22**, 282–299 (2012)
121. Z. Lin, B. Chen, A. Saberi, Y. Shamash, Inner-outer factorization of discrete-time transfer function matrices. IEEE Trans. Circuits Syst. I: Fund. Theory Appl. **43**, 941–945 (1996)

122. H. Liu, Y. Ni, L. Xie, K.H. Johansson, How vulnerable is innovation-based remote state estimation: fundamental limits under linear attacks. Automatica **136**, 110079 (2022)
123. J. Liu, J.L. Wang, G.H. Yang, An LMI approach to minimum sensitivity analysis with application to fault detection. Automatica **41**, 1995–2004 (2005)
124. X. Liu, J. Yu, Y. Ye, Residual attention convolutional autoencoder for feature learning and fault detection in nonlinear industrial processes. Neural Comput. & Appl. **33**(19), 12 737–12 753 (2021)
125. X. Liu, G.-H. Yang, Optimal encryption strategy for cyber-physical systems against stealthy attacks with energy constraints: a stackelberg game approach. Inf. Sci. **610**, 674–693 (2022)
126. L. Ljung, *System Identification: Theory for the User* (Prentice Hall PTR, 1999)
127. W.-M. Lu, A state-space approach to parameterization of stabilizing controllers for nonlinear systems. IEEE Trans. Autom. Control **40**, 1576–1588 (1995)
128. D.G. Luenberger, Canonical forms for linear multivariable systems. IEEE Trans. Automatic Control 290–293 (1967)
129. H. Luo, *Plug-and-Play Monitoring and Performance Optimization for Industrial Automation Processes* (Springer Fachmedien Wiesbaden, 2016)
130. H. Luo, Z. Chen, L. Li, *Process Monitoring, Fault Diagnosis, and Tolerant Control for Complex Industrial Systems* (Springer Singapore, 2025)
131. H. Luo, X. Yang, M. Kruger, S.X. Ding, K. Peng, A plug-and play monitoring and control architecture for disturbance compensation in rolling mills. IEEE-ASME Trans. Mechatr. **23**, 200–210 (2018)
132. J.M. Maciejowski, *Multivariable Feedback Design* (Addison Welsley, 1989)
133. M. Mahmoud, J. Jiang, Y. Zhang, *Active Fault Tolerant Control Systems* (Springer, London, 2003)
134. I. Mahtout, F. Navas, V. Milanes, F. Nashashibi, Advances in Youla-Kucera parametrization: a review. Annu. Rev. Control. **49**, 81–94 (2020)
135. R. Mangoubi, *Robust Estimation and Failure Detection* (Springer, London, 1998)
136. R. Mangoubi, M. Desai, A. Edelmayer, P. Sammak, Robust detection and estimation in dynamic systems and statistical signal processing: intersection, parallel paths and applications. Eur. J. Control. **15**, 348–369 (2009)
137. I. Markovsky, P. Rapisarda, Data-driven simulation and control. Int. J. Control **81**, 1946–1959 (2008)
138. I. Markovsky, F. Doerfler, Behavioral systems theory in data-driven analysis, signal processing, and control. Annu. Rev. Control. **52**, 42–64 (2021)
139. P.P. Menon, C. Edwards, Robust fault estimation using relative information in linear multi-agent networks. IEEE Trans. Autom. Control **59**(2), 477–482 (2014)
140. G. Mercere, L. Bako, Parameterization and identification of multivariable state-space systems: a canonical approach. Automatica **47**, 1547–1555 (2011)
141. D. Mikhaylenko, P. Zhang, Stealthy local covert attacks on cyber-physical systems. IEEE Trans. Autom. Control **67**(12), 6778–6785 (2022)
142. Y. Mo, S. Weerakkody, B. Sinopoli, Physical authentication of control systems: designing watermarked control inputs to detect counterfeit sensor outputs. IEEE Control Syst. Mag. **35**, 93–109 (2015)
143. A.M. Mohan, N. Meskin, H. Mehrjerdi, A comprehensive review of the cyber-attacks and cyber-security on load frequency control of power systems. Energies 13 (2020)
144. M. Mueller, M. Cantoni, Normalized coprime representations for time-varying linear systems, in *49th IEEE Conference on Decision and Control (CDC)* (2010), pp. 7718–7723
145. H. Nagaoka, S. Amari, Differential geometry of smooth families of probability distributions. University of Tokyo, Technical Report Technical Report METR 82-7 (1982)
146. S.K. Nguang, P. Shi, S. Ding, Fault detection for uncertain fuzzy systems: an LMI approach. IEEE Trans. Fuzzy Syst. **15**, 1251–1262 (2007)
147. F. Nielsen, An elementary introduction to information geometry. Entropy **22**(10), 1100 (2020)
148. H. Niemann, J. Stoustrup, An architecture for fault tolerant controllers. Int. J. Control **78**(14), 1091–1110 (2005)

149. H. Niemann, J. Stoustrup, Passive fault tolerant control of a double inverted pendulum - a case study. Control. Eng. Pract. **13**(8), 1047–1059 (2005)
150. H. Noura, D. Theilliol, J. Ponsart, A. Chamseddine, *Faul-Tolerant Control Systems: Design and Practical Applications* (Springer, New York, NY, USA, 2009)
151. C. Oara, A. Varga, Computation of general inner-outer and spectral factorizations. IEEE Trans. Autom. Control **45**, 2307–2325 (2000)
152. R. Ober, J. Sefton, Stability of control systems and graphs of linear systems. Syst. & Control Lett. **17**(4), 265–280 (1991)
153. A. Paice, A. van der Schaft, The class of stabilizing nonlinear plant controller pairs. IEEE Trans. Automat. Control **41**, 634–645 (1994)
154. A. Paice, A. van der Schaft, Stable kernel representations as nonlinear left coprime factorizations, in *Proceedings of the 33rd IEEE CDC* (1994), pp. 2786–2791
155. F. Pasqualetti, F. Doerfler, F. Bullo, Attack detection and identification in cyber-physical systems. IEEE Trans. Autom. Control **58**(11), 2715–2729 (2013)
156. R.J. Patton, J. Chen, Optimal unknown input distribution matrix selection in robust fault diagnosis. Automatica **29**, 837–841 (1993)
157. R.J. Patton, P.M. Frank, R.N. Clark, *Fault Diagnosis in Dynamic Systems, Theory and Applications* (Prentice-Hall, Englewood Cliffs, NJ, 1989)
158. R.J. Patton, P.M. Frank, R.N. C (Eds.), *Issues of Fault Diagnosis for Dynamic Systems* (Springer, London, 2000)
159. C.D. Persis, A. Isidori, A geometric approach to nonlinear fault detection and isolation. IEEE Trans. Autom. Control **46**, 853–865 (2001)
160. C.D. Persis, P. Tesi, Input-to-state stabilizing control under denial-of-service. IEEE Trans. Autom. Control **60**(11), 2930–2944 (2015)
161. A.M. Pertew, H.J. Marquez, Q. Zhao, LMI-based sensor fault diagnosis for nonlinear lipschitz systems. Automatica **43**, 1464–1469 (2007)
162. M.A. Petersen, A.V. der Schaft, Nonsquare spectral factorization for nonlinear control systems. IEEE Trans. Autom. Control **50**, 286–298 (2005)
163. J.W. Polderman, J.C. Willens, *Introduction to Mathematical Systems Theory: A Behavioral Approach* (Springer, 1998)
164. S.J. Qin, Survey on data-driven industrial process monitoring and diagnosis. Annu. Rev. Control. **36**, 220–234 (2012)
165. S. Qin, Statistical process monitoring: basics and beyond. J. Chemom. **17**, 480–502 (2003)
166. Z. Qiu, J.J. Gertler, Robust FDI systems and h-infinity optimization. Proc. ACC **93**, 1710–1715 (1993)
167. F. Rambeaux, F. Hamelin, D. Sauter, Robust residual generation via LMI, in *Proceedings of the 14th IFAC World Congress*, Beijing, China (1999), pp. 240–246
168. M.L. Rank, H. Niemann, Norm based design of fault detectors. Int. J. Control **72**(9), 773–783 (1999)
169. R. Ravi, A. Pascoal, P. Khargonekar, Normalized coprime factorizations for linear time-varying systems. Syst. & Control Lett. **18**(6), 455–465 (1992)
170. X.-X. Ren, G.H. Yang, Kullback-leibler divergence-based optimal stealthy sensor attack against networked linear quadratic gaussian systems. IEEE Trans. Cybern. **52**(11), 11 539–11 548 (2022)
171. X.-X. Ren, G.-H. Yang, X.-G. Zhang, Optimal stealthy attack with historical data on cyber-physical systems. Automatica **151**, 110895 (2023)
172. Z. Ren, W. Zhang, Z. Zhang, A deep nonnegative matrix factorization approach via autoencoder for nonlinear fault detection. IEEE Trans. Ind. Inform. **16**(8), 5042–5052 (2020)
173. S.T. Roweis, L.K. Saul, Nonlinear dimensionality reduction by locally linear embedding. Science **290**, 2323–2326 (2000)
174. E.L. Russell, L. Chiang, R.D. Braatz, *Data-Driven Techniques for Fault Detection and Diagnosis in Chemical Processes* (Springer, London, 2000)
175. D. Sauter, F. Hamelin, Frequency-domain optimization for robust fault detection and isolation in dynamic systems. IEEE Trans. Autom. Control **44**, 878–882 (1999)

176. C. Schellenberger, P. Zhang, Detection of covert attacks on cyber-physical systems by extending the system dynamics with an auxiliary system, in *2017 IEEE 56th Annual Conference on Decision and Control (CDC)* (2017), pp. 1374–1379
177. J.M.A. Scherpen, A. van der Schaft, Normalized coprime factorization and balancing for unstable nonlinear systems. Int. J. Control **60**, 1193–1222 (1994)
178. B. Schoelkopf, J. Platt, J. Shawe-Taylor, A. Smola, R. Williamson, Estimating the support of a high-dimensional distribution. Neural Comput. **13**, 1443–1471 (2001)
179. M. Segovia-Ferreira, J. Rubio-Hernan, A. Cavalli, J. Garcia-Alfaro, A survey on cyber-resilience approaches for cyber-physical systems **56**(8) (2024). https://doi.org/10.1145/3652953
180. R. Seliger, P. Frank, Fault diagnosis by disturbance decoupled nonlinear observers, in *Proceedings of the CDC91*, Brighton, England (1991), pp. 2248–2253
181. I. Shames, A.M. Teixeira, H. Sandberg, K.H. Johansson, Distributed fault detection for interconnected second-order systems. Automatica **47**(12), 2757–2764 (2011)
182. J. Shang, D. Cheng, J. Zhou, T. Chen, Asymmetric vulnerability of measurement and control channels in closed-loop systems. IEEE Trans. Control Netw. Syst. **9**(4), 1804–1815 (2022)
183. Q. Shen, B. Jiang, P. Shi, Adaptive fault diagnosis for T-S fuzzy systems with sensor faults and system performance analysis. IEEE Trans. Fuzzy Syst. **22**(2), 274–285 (2014)
184. S. Simani, S. Fantuzzi, R.J. Patton, *Model-Based Fault Diagnosis in Dynamic Systems Using Identification Techniques* (Springer, London, 2003)
185. R.S. Smith, Covert misappropriation of networked control systems: presenting a feedback structure. IEEE Control Syst. Mag. **35**, 82–92 (2015)
186. J. Stoustrup, Plug & play control: control technology towards new challenges. Eur. J. Control. **15**(3), 311–330 (2009)
187. S. Tan, J.M. Guerrero, P. Xie, R. Han, J.C. Vasquez, Brief survey on attack detection methods for cyber-physical systems. IEEE Syst. J. **14**, 5329–5339 (2020)
188. D.M.J. Tax, R.P.W. Duin, Support vector data description. Mach. Learn. **54**(1), 45–66 (2004)
189. T.-T. Tay, I. Mareels, J.B. Moore, *High Performance Control* (Springer Science + Business Media, 1998)
190. T.T. Tay, J.B. Moore, R. Horwitz, Indirect adaptive techniques for fixed controller performance enhancement. Int. J. Control **50**(5), 1941–1959 (1989)
191. J.B. Tenenbaum, V. de Silva, J.C. Langford, A global geometric framework for nonlinear dimensionality reduction. Science **290**, 2319–2323 (2000)
192. N. Tishby, N. Zaslavsky, Deep learning and the information bottleneck principle, in *Proceeding of the IEEE Information Theory Workshop* (2015)
193. M. Tsai, L. Chen, A numerical algorithm for inner-outer factorization of real-rational matrices. Syst. Contr. Lett. 209–217 (1993)
194. A. van der Schaft, *L2 - Gain and Passivity Techniques in Nonlinear Control* (Springer, London, 2000)
195. P. Van Overschee, B.D. Moor, *Subspace Identification for Linear Systems* (Kluwer Academic Publishers, USA, 1996)
196. V. Venkatasubramanian, R. Rengaswamy, S. Kavuri, K. Yin, A review of process fault detection and diagnosis part III: process history based methods. Comput. Chem. Eng. **27**, 327–346 (2003)
197. V. Venkatasubramanian, R. Rengaswamy, K. Yin, S. Kavuri, A review of process fault detection and diagnosis part I: quantitative model-based methods. Comput. Chem. Eng. **27**, 293–311 (2003)
198. M. Vidyasagar, *Control System Synthesis: A Factorization Approach* (The MIT press, Cambridge, 1985)
199. G. Vinnicombe, *Uncertainty and Feedback: H_∞ Loop-Shaping and the ν Gap Metric* (World Scientific, 2000)
200. H. Wang, G.-H. Yang, A finite frequency domain approach to fault detection observer design for linear continuous-time systems. Asian J. Control **10**, 1–10 (2008)

201. J. Wang, J. Zhou, X. Chen, *Data-Driven Fault Detection and Reasoning for Industrial Monitoring* (Springer, 2021)
202. J.L. Wang, G.-H. Yang, J. Liu, An LMI approach to h-index and mixed H_-/H_∞ fault detection observer design. Automatica **43**, 1656–1665 (2007)
203. X. Wang, G. Yang, D. Zhang, Optimal stealth attack strategy design for linear cyber-physical systems. IEEE Trans. Cybern. **52**(1), 472–480 (2022)
204. Y. Wang, S.X. Ding, H. Ye, G. Wang, A new fault detection scheme for networked control systems subject to uncertain time-varying delay. IEEE Trans. Signal Proc. **56**, 5258–5268 (2008)
205. Y. Wang, H. Ye, S.X. Ding, Y. Cheng, P. Zhang, G. Wang, Fault detection of networked control systems with limited communication. Int. J. Contr. **82**, 1344–1356 (2009)
206. S. Weerakkody, B. Sinopoli, Detecting integrity attacks on control systems using a moving target approach, in *2015 54th IEEE Conference on Decision and Control (CDC)* (2015), pp. 5820–5826
207. J.C. Willems, P. Rapisarda, I. Markovsky, B.L. De Moor, A note on persistency of excitation. Syst. & Control Lett. **54**(4), 325–329 (2005)
208. J. Willems, Deterministic least squares filtering. J. Econom. **118**, 341–373 (2004)
209. J. Wünnenberg, Observer-based fault detection in dynamic systems. Ph.D. dissertation, University of Duisburg (1990)
210. J. Wünnenberg, P.M. Frank, Sensor fault detection via robust observers, in *System Fault Diagnostics, Reliability and Related Knowledge-Based Approaches*, ed. by Tsafestas, Singh, Schmidt (Reidel Press, 1987), pp. 147–160
211. A. Xu, Q. Zhang, Nonlinear system fault diagnosis based on adaptive estimation. Automatica **40**, 1181–1193 (2004)
212. Y. Xu, S.X. Ding, H. Luo, S. Yin, A real-time performance recovery framework for vision-based control systems. IEEE Trans. Ind. Electr. **68**(2), 1571–1580 (2021)
213. Y. Xu, S.X. Ding, S. Yin, H. Luo, Z. Zhao, Performance degradation monitoring and recovery of vision-based control systems. IEEE Trans. Control Syst. Technol. **29**(6), 2712–2719 (2021)
214. W. Yan, L.K. Mestha, M. Abbaszadeh, Attack detection for securing cyber physical systems. IEEE Internet Things J. **6**(5), 8471–8481 (2019)
215. X.-G. Yan, C. Edwards, Robust sliding mode observer-based actuator fault detection and isolation for a class of nonlinear systems. Int. J. Syst. Sci. **39**, 349–359 (2008)
216. Y. Yang, S.X. Ding, L. Li, On observer-based fault detection for nonlinear systems. Syst. Contr. Lett. **82**, 1399–1410 (2015)
217. Y. Yang, S.X. Ding, L. Li, Parametrization of nonlinear observer-based fault detection systems. IEEE Trans. Autom. Control **61**, 3687–3692 (2016)
218. H. Ye, S.X. Ding, Fault detection of networked control systems with network-induced delay, in *Proceedings Of the 8th International Conference on Control, Automation, Robotics and Vision*, Kunming, China (2004), pp. 294–297
219. S. Yin, S.X. Ding, X. Xie, H. Luo, A review on basic data-driven approaches for industrial process monitoring. IEEE Trans. Ind. Electr. **61**, 6418–6428 (2014)
220. S. Yin, B. Xiao, S.X. Ding, D. Zhou, A review on recent development of spacecraft attitude fault-tolerant control system. IEEE Trans. Ind. Electr. **63**, 3311–3320 (2016)
221. D. Youla, J. Bongiorno, H. Jabr, Modern wiener-hopf design of optimal controllers part i: the single-input-output case. IEEE Trans. Autom. Control **21**(1), 3–13 (1976)
222. D. Youla, H. Jabr, J. Bongiorno, Modern wiener-hopf design of optimal controllers-part ii: the multivariable case. IEEE Trans. Autom. Control **21**(3), 319–338 (1976)
223. D. Zhang, Q.-G. Wang, G. Feng, Y. Shi, A.V. Vasilakos, A survey on attack detection, estimation and control of industrial cyber-physical systems. ISA Trans. **116**, 1–16 (2021)
224. P. Zhang, S.X. Ding, On fault detection in linear discrete-time, periodic, and sampled-data systems (survey). J. Control Sci. Eng. 1–18 (2008)
225. P. Zhang, S.X. Ding, G.Z. Wang, D.H. Zhou, Fault detection for multirate sampled-data systems with time delays. Int. J. Control **75**(18), 1457–1471 (2003)

226. P. Zhang, S.X. Ding, G. Wang, D. Zhou, A freqwuency domain approach to fault detection in sampled-data systems. Automatica **39**, 1303–1307 (2003)
227. P. Zhang, S.X. Ding, G. Wang, D. Zhou, Fault detection of linear discrete-time periodic systems. IEEE Trans. Autom. Control **50**(2), 239–244 (2005)
228. Q. Zhang, K. Liu, Y. Xia, A. Ma, Optimal stealthy deception attack against cyber-physical systems. IEEE Trans. Cybern. **50**(9), 3963–3972 (2020)
229. Q. Zhang, K. Liu, A.M.H. Teixeira, Y. Li, S. Chai, Y. Xia, An online kullback-leibler divergence-based stealthy attack against cyber-physical systems. IEEE Trans. Autom. Control **68**(6), 3672–3679 (2023)
230. X. Zhang, M.M. Polycarpou, T. Parisini, Adaptive fault diagnosis and fault-tolerant control of MIMO nonlinear uncertain systems. Int. J. Contr. **83**, 1054–1080 (2010)
231. X. Zhang, M.M. Polycarpou, T. Parisini, Fault diagnosis of a class of nonlinear uncertain systems with lipschitz nonlinearities using adaptive estimation. Automatica **46**, 290–299 (2010)
232. Y. Zhang, J. Jiang, Bibliographical review on reconfigurable fault-tolerant control systems. Ann. Rev. Control **32**, 229–252 (2008)
233. M. Zhong, S.X. Ding, Q. Han, Q. Ding, Parity space-based fault estimation for linear discrete time-varying systems. IEEE Trans. Autom. Control **55**, 1726–1731 (2010)
234. M. Zhong, S.X. Ding, D. Zhou, X. He, An h_i/h_∞ optimization approach to event-triggered fault detection for linear discrete time systems. IEEE Trans. Autom. Control **65**(10), 4464–4471 (2020)
235. M. Zhong, S. Ding, J. Lam, H. Wang, An LMI approach to design robust fault detection filter for uncertain LTI systems. Automatica **39**, 543–550 (2003)
236. M. Zhong, T. Xue, S.X. Ding, D. Zhou, *Fault Diagnosis for Linear Discrete Time-Varying Systems and Its Applications* (Springer, 2023)
237. M. Zhong, L. Zhang, S.X. Ding, D. Zhou, A probabilistic approach to robust fault detection for a class of nonlinear systems. IEEE Trans. Indus. Elec. **64**, 3930–3939 (2017)
238. M.Y. Zhong, T. Xue, S.X. Ding, A survey on model-based fault diagnosis for linear discrete time varying systems. Neurocomputing **306**, 51–60 (2018)
239. M. Zhong, H. Ye, S.X. Ding, G. Wang, Observer-based fast rate fault detection for a class of multirate sampled-data systems. IEEE Trans. Autom. Control **52**(3), 520–525 (2007)
240. B. Zhou, T. Zhao, On asymptotic stability of discrete-time linear time-varying systems. IEEE Trans. Autom. Control **62**(8), 4274–4281 (2017)
241. C. Zhou, B. Hu, Y. Shi, Y.-C. Tian, X. Li, Y. Zhao, A unified architectural approach for cyberattack-resilient industrial control systems. Proc. IEEE **109**, 517–541 (2021)
242. J. Zhou, J. Shang, T. Chen, Optimal deception attacks on remote state estimators equipped with interval anomaly detectors. Automatica **148**, 110723 (2023)
243. K. Zhou, J. Doyle, K. Glover, *Robust and Optimal Control* (Prentice-Hall, Upper Saddle River, New Jersey, 1996)

Index

S. X. Ding, *Unified Framework of Control and Detection*,
https://doi.org/10.1007/978-3-662-72417-0

MIX
Papier aus verantwortungsvollen Quellen
Paper from responsible sources
FSC® C105338

If you have any concerns about our products,
you can contact us on
ProductSafety@springernature.com

In case Publisher is established outside the EU,
the EU authorized representative is:
Springer Nature Customer Service Center GmbH
Europaplatz 3, 69115 Heidelberg, Germany

Printed by Libri Plureos GmbH
in Hamburg, Germany